AF802606

Ocular Blood Flow

Leopold Schmetterer • Jeffrey W. Kiel
Editors

Ocular Blood Flow

Springer

Editors
Leopold Schmetterer, Ph.D.
Department of Clinical Pharmacology
Center of Medical Physics and
Biomedical Engineering
Medical University of Vienna
Vienna
Austria

Jeffrey W. Kiel, Ph.D., FARVO
Deptartment of Ophthalmology
University of Texas Health Science Center
San Antonio, TX
USA

ISBN 978-3-540-69468-7 ISBN 978-3-540-69469-4 (eBook)
DOI 10.1007/978-3-540-69469-4
Springer Heidelberg New York Dordrecht London

Library of Congress Control Number: 2012934553

© Springer-Verlag Berlin Heidelberg 2012
This work is subject to copyright. All rights are reserved by the Publisher, whether the whole or part of the material is concerned, specifically the rights of translation, reprinting, reuse of illustrations, recitation, broadcasting, reproduction on microfilms or in any other physical way, and transmission or information storage and retrieval, electronic adaptation, computer software, or by similar or dissimilar methodology now known or hereafter developed. Exempted from this legal reservation are brief excerpts in connection with reviews or scholarly analysis or material supplied specifically for the purpose of being entered and executed on a computer system, for exclusive use by the purchaser of the work. Duplication of this publication or parts thereof is permitted only under the provisions of the Copyright Law of the Publisher's location, in its current version, and permission for use must always be obtained from Springer. Permissions for use may be obtained through RightsLink at the Copyright Clearance Center. Violations are liable to prosecution under the respective Copyright Law.
The use of general descriptive names, registered names, trademarks, service marks, etc. in this publication does not imply, even in the absence of a specific statement, that such names are exempt from the relevant protective laws and regulations and therefore free for general use.
While the advice and information in this book are believed to be true and accurate at the date of publication, neither the authors nor the editors nor the publisher can accept any legal responsibility for any errors or omissions that may be made. The publisher makes no warranty, express or implied, with respect to the material contained herein.

Printed on acid-free paper

Springer is part of Springer Science+Business Media (www.springer.com)

Contents

Part I

Anatomy

Anatomy of the Ocular Vasculatures

1

Gerard A. Lutty, Imran Bhutto, and Donald Scott McLeod

Core Messages

- The human eye is an incredible assembly of unique types of tissues, each of which has its own requirements in terms of blood supply.
- The major source of blood for the anterior eye is the anterior ciliary arteries, whereas the central retinal artery and the posterior ciliary arteries, both originating from the ophthalmic artery, is the supply for the posterior eye.
- The iris receives its arterial blood supply from the major arterial circle (MAC). The iris vessels are radially oriented with slightly sinuous courses, which allow them to accommodate to the movements of the pupil. The iris venous channels roughly follow the arterial ones. The collector trunks from these veins enter the ciliary body, where the blood then drains into the vortex system and the ciliary plexus.
- The ciliary body, a part of the uveal tract, is the circumferential tissue posterior to the iris composed of the ciliary muscle and ciliary processes. Its blood supply is the same as iris but the capillaries of the ciliary processes are large and fenestrated like the choriocapillaris.
- There are two sources of blood and nutrient supply for the human retina: the central retinal artery and the choroidal blood vessels. The choroid receives the greatest blood flow (65-85%), which is vital for maintenance of the outer retina. The remaining 20-30% flows to the retina through the central retinal artery in the optic nerve head to nourish the inner retinal layers.
- There are three layers of capillaries in retina: 1) the radial peripapillary capillaries (RPCs) around the optic nerve head only; 2) the inner or superficial plexus; and 3) the outer or deep layer of capillaries. The retinal vasculature is end-arterial: blood moves from artery to arteriole to capillary and precapillary venules drain into venules and then veins. A blood-retinal barrier exists.
- The only avascular area of retina is the foveal avascular zone where the foveolar

G.A. Lutty, Ph.D. (✉) • D.S. McLeod
Wilmer Ophthalmological Institute, Johns Hopkins Hospital, M041 Smith Building, 400 North Broadway, Baltimore, MD 21287-9115, USA
e-mail: glutty@jhmi.edu; dmcleod1@jhmi.edu

I. Bhutto, M.D., Ph.D.
Wilmer Ophthalmological Institute, Johns Hopkins Hospital, M041 Smith Building, 400 North Broadway, Baltimore, MD 21287-9115, USA

Department of Ophthalmology, Wilmer Eye Institute, Johns Hopkins Hospital, M041 Smith Building, 400 North Broadway, Baltimore, MD 21287-9115, USA
e-mail: ibhutto1@jhmi.edu

L. Schmetterer, J.W. Kiel (eds.), *Ocular Blood Flow*,
DOI 10.1007/978-3-540-69469-4_1, © Springer-Verlag Berlin Heidelberg 2012

pit is present in the center of macula, which is surrounded by a ring of capillaries in a single layer.

- The choroidal arteries arise from long and short posterior ciliary arteries and branches of Zinn's circle (around the optic disc). The arteries pierce the sclera around the optic nerve and fan out to form the three vascular layers in the choroid: outer (most scleral), medial and inner capillary system, the choriocapillaris (nearest Bruchs membrane of the pigment epithelium). One or two vortex veins drain each of the 4 quadrants of the eyeball. The vortex veins penetrate the sclera and merge into the ophthalmic vein.
- The choriocapillaris is fenestrated and lobular in the posterior pole, but ladder-like in periphery.

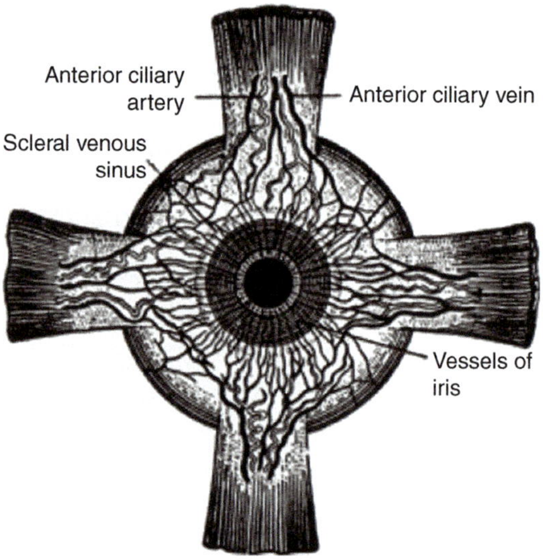

Fig. 1.1 Frontal view schematic of the anterior ciliary arteries and veins (From Wolff [45], p. 68, with permission)

The eye is an incredible assembly of unique types of tissues, each of which has its own requirements in terms of blood supply. The eye contains avascular tissues that must have nutrition like cornea and lens to richly vascularized tissue like iris, retina, and choroid. The sources of blood for the eye are the retinal artery and the posterior and anterior ciliary arteries derived from the ophthalmic artery. The ophthalmic artery is the most often stenosed vessel from the internal carotid artery because of its right-angle branching from the carotid. Eighty percent of the ocular blood goes to the uveal tract and 20% to the retina. The anterior ciliary arteries enter the globe from the rectus muscles to provide blood to the limbal, conjunctival, and the scleral vasculatures (Fig. 1.1). The posterior ciliary arteries supply the choroid.

1.1 Limbus and Conjunctiva

1.1.1 Cornea

Although the cornea is normally avascular in the fetal and adult human [9], there is a ring of vasculature around its border, the limbus. Conjunctival, episcleral, and scleral vessels and terminal branches of the circumcorneal vessels supply the peripheral cornea (Figs. 1.1 and 1.2). The vessels ramify at different levels in the limbus. Most of them supply the sclera in the region of the canal of Schlemm and play a small part in corneal nutrition and metabolism (Fig. 1.2). Drainage of the blood from the limbal, conjunctival, scleral vasculatures and the deep scleral venous plexus adjacent to Schlemm canal is through the episcleral veins. The episcleral veins drain to the anterior ciliary veins and finally into the ophthalmic vein.

Most of the limbal capillaries have a thick wall, but some limbal capillaries have a thin wall and are fenestrated [20]. There is also a lymphatic system at the limbus in which the endothelial cells are thin and their basement membrane is poorly developed, but they are not fenestrated [20]. It has recently been demonstrated using the LYVE-1 antibody (lymphatic vessel endothelial hyaluronic acid receptor) that recognizes lymphatic not vascular endothelium that inflammatory stimuli like a suture can induce both hemangiogenesis and lymphangiogenesis to arise from these plexuses at the limbus and invade the cornea [8].

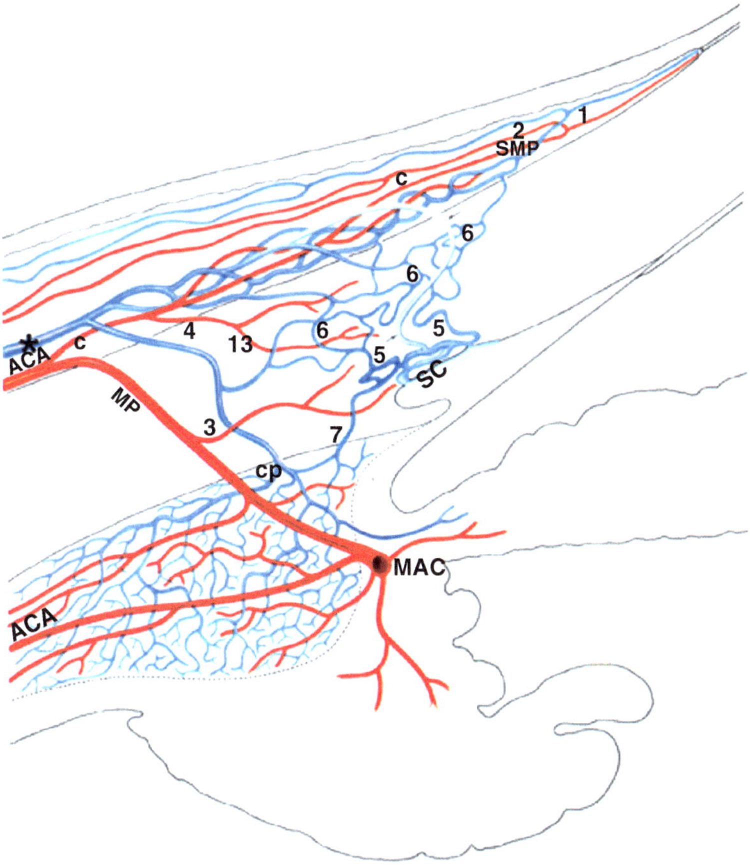

Fig. 1.2 Schematic of the limbal blood supply with arteries represented in *red* and veins in *blue*. The anterior ciliary artery (*ACA*) branches to form episcleral and major perforating (*MP*) artery. The episcleral artery branches to form conjunctival (*C*) and intrascleral blood vessels (*IS*). The conjunctival vessels form the superficial marginal plexus (*SMP*) of the cornea. The SMP is the origin of the peripheral corneal arcades (*1*) and other recurrent vessels which run posteriorly to supply 3–6 mm of perilimbal conjunctiva. The MP goes through the sclera to join the major arterial circle (*MAC*) of the iris. The intrascleral channels of the limbus (*3*) originate as a branch from the MP artery. A branch of the anterior ciliary artery (*4*) often also supplies this area. The episcleral vein (*asterisk*) is the major venous drainage from the limbus. It eventually unites with the ophthalmic veins. Near Schlemm's canal (*SC*) is the deep scleral venous plexus (*5*) from which an aqueous vein (*arrow*) arises and then joins the episcleral veins. In the limbal stroma, an extensive network is formed by the intrascleral venous plexus (*6*). The deep and intrascleral venous plexus drains the ciliary plexus (*CP*) (*7*) (From Hogan et al. [19], p. 120, with permission)

1.1.2 Vasculature Distribution in the Anterior Segment

The blood supply of the anterior segment of the eye has distinctive characteristics. Except for the perforating vessels like the anterior ciliary arteries, the sclera is a relatively avascular structure. It has a low metabolic requirement because of the slow turnover rate of its collagen. The episcleral blood supply is derived mainly from the anterior ciliary arteries anterior to the insertions of the rectus muscles and from the long and short posterior ciliary arteries posterior to these insertions (Figs. 1.1 and 1.2). Scleral stroma contains capillary networks but is supplied by the episcleral and, to a lesser degree, choroidal vascular networks. The arteries, veins, and nerves traverse the sclera through emissary canals. These canals or passageways are separated from the sclera by a thin layer of loose connective tissue. Emissary canals are located more superiorly at 12 o'clock and inferiorly at 6 o'clock than nasally and temporally. Emissary canals provide a passageway for extraocular extensions of intraocular tumors.

On the surface of the sclera are the arteries of the muscles that arise from the ophthalmic artery and run forward as the anterior ciliary arteries. The anterior ciliary arteries pass through the sclera just in front of the insertions of the rectus muscles in a slightly oblique direction from posterior to anterior (Fig. 1.1). Each rectus muscle has two anterior ciliary arteries, except the lateral rectus muscle, which has only one. The seven anterior ciliary arteries meet via their lateral branches 1–5 mm behind the limbus and form the anterior episcleral arterial circle, which feeds the limbal, anterior conjunctival, and anterior episcleral tissues (Fig. 1.2). The anterior episcleral arterial circle broadly resolves into limbal arcades, an anterior conjunctival plexus, a superficial episcleral plexus, and a deep episcleral plexus.

Limbal arcades and anterior conjunctival plexus usually share their origins and form the most superficial layer of vessels (Fig. 1.2). The superficial episcleral plexus lies within the parietal layer of the episclera and anastomoses at the limbus with the conjunctival plexus, branches of the same plexus, and with the deep episcleral plexus. The deep episcleral plexus lies within the visceral layer of episclera and anastomoses with branches of the same plexus. In addition, extensions of the remaining anterior ciliary arterial branches perforate the limbal sclera through emissary canals and meet the long posterior ciliary arteries in the ciliary muscle to form the major arterial circle of the iris (Fig. 1.2). The anterior episcleral arterial circle and the major arterial circle of the iris communicate by scleral perforating anterior ciliary arterial branches, which provide nutrients to the uveal tract.

Intrascleral arteries of the limbus, an incomplete arterial circle, accompany the canal of Schlemm, derived from the superficial and deep terminal branches of the anterior ciliary arteries. The arterioles show a nonfenestrated endothelial lining and a 1–2 layer medial coat.

The limbal venous circle collects blood from the anterior conjunctival veins and limbal arcades and drains into radial episcleral collecting veins. The episcleral collecting veins also receive blood from anterior episcleral veins and perforating scleral veins. Perforating scleral veins emerge from Schlemm's canal, from which they receive aqueous humor. They penetrate the sclera through different emissary canals than do the arteries. These canals, over the ciliary body, often also carry the ciliary nerves. As the episcleral collecting veins run posteriorly across the sclera, they form the anterior ciliary veins, which leave the anterior surface of the globe through the rectus muscles (Fig. 1.1).

1.1.3 The Conjunctiva

The conjunctiva is a thin, translucent mucous membrane that covers the inner surfaces of the upper and lower lids and extends to the limbus on the surface of the globe. Three regions within the conjunctiva are recognized: the palpebral or tarsal region, which lines the inner surface of the lids; the fornical region, which lines the upper and lower surfaces of the recess or cul-de-sac known as the fornix; and the bulbar region, which lines the surface of the sclera between the fornix and the limbus. The conjunctiva has two structural components throughout all regions: the surface epithelium and the substantia propria.

1.1.3.1 The Conjunctival Arterial Supply

Most of the arterial channels are arterioles, which terminate in a complex subconjunctival capillary network. The arterial supply of the conjunctiva is from the peripheral tarsal arcades, the marginal tarsal arcades, the anterior ciliary arteries, and the deep ciliary system. The peripheral perforating branches of the peripheral tarsal arcade pass above the tarsal plate, pierce the palpebral muscle, and divide into ascending and descending conjunctival branches.

The descending branches supply the proximal two-thirds of the tarsal conjunctiva, anastomosing with the shorter branches of the marginal artery, which pierces the tarsal plate at the subtarsal fold. The ascending branches pass up over the fornix to the globe, where they become the posterior conjunctival arteries. These anastomose with the anterior conjunctival arteries from the limbus and together they supply the bulbar conjunctiva (Fig. 1.2).

The marginal arcades send perforating branches through the tarsus to the conjunctiva at the subtarsal fold. These divide into marginal and tarsal secondary branches, which run perpendicularly either to feed the very vascular zone at the lid margin or to meet with corresponding branches of the peripheral arcades. The tarsal conjunctiva is well supplied with blood, hence its red color. The fornix is red and the bulbar conjunctiva is colorless unless congested. The conjunctival capillaries mainly are of the nonfenestrated continuous type that occurs in skeletal muscle and iris. These capillaries are excellent sites to externally view vaso-occlusive processes. In sickle cell disease, vaso-occlusions in conjunctival capillaries are called comma signs and are an excellent diagnostic characteristic for sickle cell disease [32].

1.1.3.2 The Conjunctival Veins

The palpebral veins drain the tarsal conjunctiva, fornix, and posterior bulbar conjunctiva. In the upper lid, a venous plexus between the tendons of the levator drains into the veins of the levator and superior rectus and then into the ophthalmic vein. Immediately behind the limbal arcades and anterior to the episcleral arterial circle lies a perilimbal venous circle, which collects blood from the limbus, marginal corneal arcades, and the anterior conjunctival veins. The circle drains into radial episcleral collecting veins and then into the veins of the rectus muscles.

1.2 Uveal Tract

The ophthalmic artery branches to form the posterior ciliary arteries and the central retinal artery. There are two types of ciliary arteries supplying the uveal tract: 15–20 short posterior ciliary arteries and two long posterior ciliary arteries (Fig. 1.3). Venous outflow of the uveal tract consists of four vortex veins that drain all of the choroid and most of the anterior uveal tract. Some uveal blood from the iris and ciliary body drains through the scleral plexus of anterior scleral veins. The four vortex veins are at the equator and drain to the superior and inferior orbital veins.

1.2.1 The Iris

The iris is the most anterior portion of the uveal tract. It lies in the frontal plane of the eye between the anterior and posterior chamber and is bathed on both surfaces by the aqueous humor.

1.2.1.1 The Major Arterial Circle of the Iris

The iris receives its arterial blood supply from the major arterial circle (MAC), which lies in the stroma of the ciliary body near the iris root (Fig. 1.4). Branches from the major arterial circle enter the periphery of the iris through its root. The blood vessels of the iris are radially oriented with slightly sinuous courses, which allow them to accommodate to the movements of the pupil. The arteries arise mainly from the MAC of the iris. Some authors report that the major circle is not always continuous in man, and the same is true in the macaque monkey. Some arteries also arise directly from the perforating branches of the anterior ciliary arteries after they have pierced the sclera to reach the ciliary body (Figs. 1.1 and 1.2). Anastomosing occasionally, they converge radially from ciliary to pupillary margin. In pupillary miosis, their course is straight, but they become sinuous (or serpentine) as the pupil dilates. The collarette of the iris is a thickened area (0.6 mm), which overlies an incomplete vascular circle (circulus vasculosus iridis minor), and is where the ciliary zone is separated from the papillary zone of the iris (Fig. 1.5) [45]. At the collarette, a few anastomoses occur and, with corresponding venous anastomoses, make an incomplete vascular circle, the so-called circulus

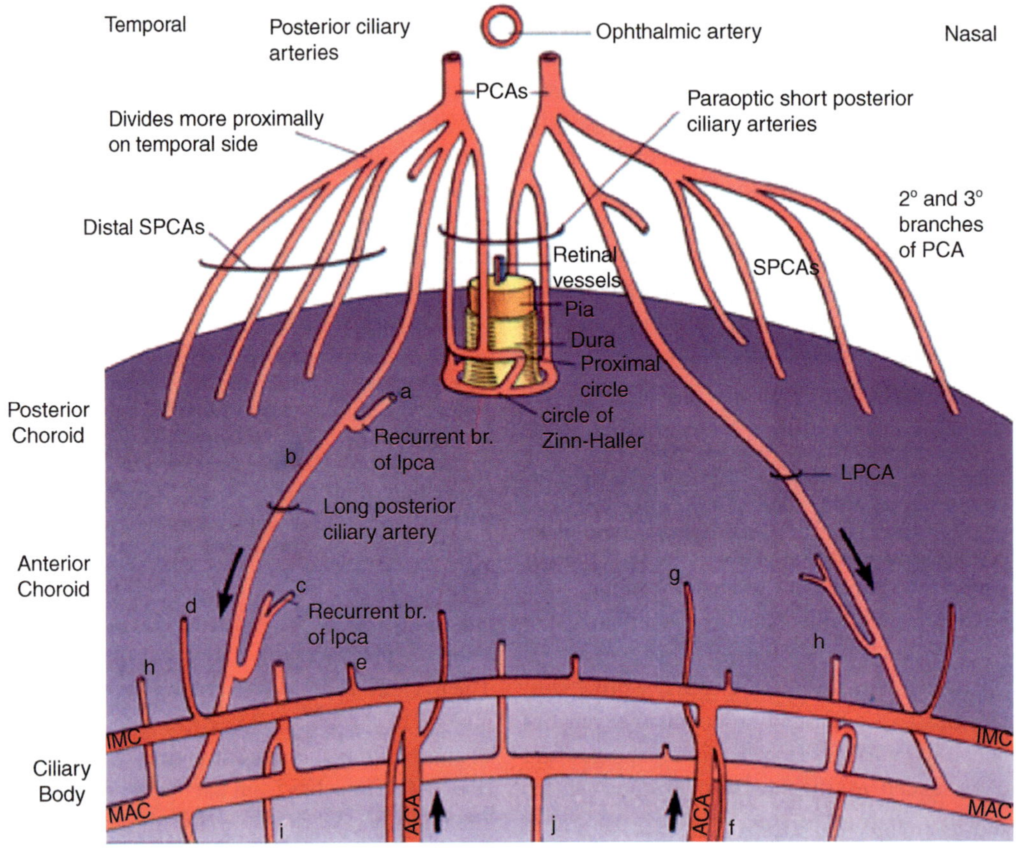

Fig. 1.3 The arterial supply of the uveal tract. The ophthalmic artery provides blood to posterior ciliary arteries (*PCAs*). The paraoptic and distal short posterior ciliary arteries (*SPCAs*) are branches of the PCAs. From the paraoptic SPCAs arise the proximal and distal (Haller and Zinn) arterial circles which supply blood to the capillaries of the optic nerve head and pia directly. The long PCAs (*LPCAs*) arise from the PCAs and supply most of the blood to major arterial circle of the iris (*MAC*) and recurrent arteries to the anterior choroid. The temporal long ciliary arteries (*LPCAs*) also supply the MAC and give off a recurrent branch to the posterior choroids. The major arterial circle in the ciliary body supplies the iris and ciliary processes. The anterior ciliary arteries (*ACA*) branch to form the intramuscular arterial circle (*IMC*), which supplies the outer and more posterior part of the ciliary muscle. Branches of the LPCAs and ACAs anastomose often (From Wolff [45], p. 374, with permission)

arteriousus iridis minor. Most vessels reach the pupillary margin where, after breaking up into capillaries, they bend around into the veins. The iris venous channels roughly follow the arterial ones, the larger veins lying in the anterior stroma, and the smaller ones near the dilator muscle. The collector trunks from these veins enter the ciliary body, where the blood then drains into the vortex system and the ciliary plexus.

The two long posterior ciliary arteries (medial and lateral) arise from the ophthalmic artery and enter the globe. These arteries, together with the nerves, traverse the sclera through emissary canals in an oblique manner, from posterior to anterior, and enter the suprachoroidal space at the equator. They run forward to give arterial supply to the ciliary body and the iris. In addition, they meet the anterior ciliary arteries to form the major arterial circle of the iris (Figs. 1.2 and 1.4). The major arterial circle of the iris is located in the stroma of the ciliary body and gives arterial supply to the iris. Surgery on the vertical, but not the horizontal, rectus muscles may give rise to ischemic defects in the iris. The superior and inferior anterior uvea is at greater risk of ischemia after superior

Fig. 1.4 Vascular cast of the anterior choroid and iris of a rat. A long posterior ciliary artery bifurcates to form the circular iris arteries at the root of the iris. The circumlimbal arterial vessels have been removed from the cast (*P* pupil; scale bar = 1 mm) (From Bhutto [4], Anat Rec, p. 67)

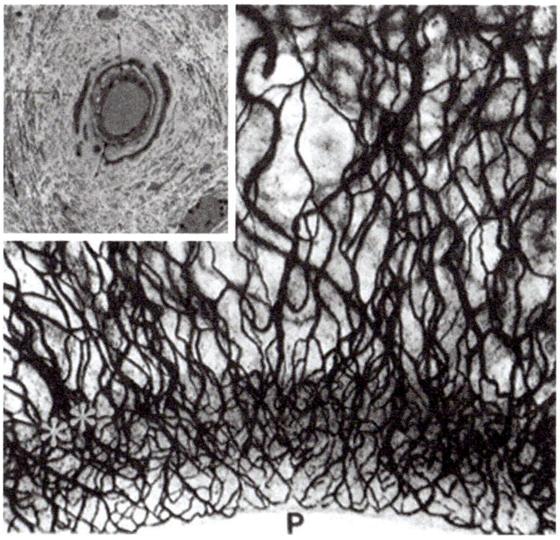

Fig. 1.5 Monkey iris injected with horseradish peroxidase and then prepared as a whole mount. The collarette (**) of the iris lies near the pupillary opening (*P*) (From Wolfe [45], p. 322, with permission). *Inset*: Ultrastructure of a small iris arteriole. The clear perivascular space denoted by the *arrows* contains a dense ground substance that is associated with the endothelial basement membrane. There is a dense circular collagenous zone (*a*) outside the clear zone (From Hogan [19], p. 241, with permission)

and inferior anterior ciliary occlusion following ligation of the respective muscles.

Although the blood vessels of iris are designated artery, vein, and capillary, the assignments of type of blood vessel are based mostly on the size of the blood vessel. Freddo and Raviola found that the iridial blood vessels are homogenous in structure regardless of diameter [11]. They possess no artery/capillary/vein units (ACV). There is a continuous layer of endothelial cells sitting on slender basal lamina and a thick, dense collagenous adventitia (Fig. 1.5, inset). The vessels have been described as a tube within a tube [45]. There is no traditional smooth muscle layer, but instead, there are fibroblasts, melanocytes, and an occasional macrophage arranged in one or two layers (Fig. 1.5, inset). The artery and vein differ only in the lack of muscularis in the vein. This unusual arrangement of the blood vessel wall is thought to adapt to the iris movement [11].

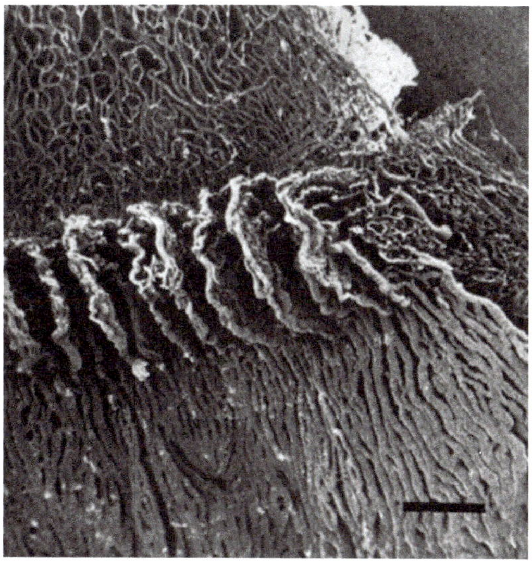

Fig. 1.6 Vascular cast from a monkey viewed from the vitreous perspective showing the posterior aspects of the iris (*top*), ciliary processes (*middle*), and the pars plana (*bottom*). The pars plana has densely packed veins which collect venous blood from the ciliary body (scale = 500 μm) (From Shimizu [37], p. 101, with permission)

1.2.2 Ciliary Body and Processes

The ciliary body is the circumferential tissue posterior to the iris composed of the ciliary muscle and ciliary processes (Fig. 1.6). It is part of the uveal tract—the layer of tissue, which provides most of the nutrients in the eye. It extends from the ora serrata to the root of the iris. It is divided into two main zones: posterior, the smooth pars plana (orbiculus ciliaris), and anterior, the pars plicata (corona ciliaris). It is triangular in horizontal section and is coated by a double layer of the ciliary epithelium. The outer layer of epithelium is amelanotic and is in contact with the vitreous body. The inner layer of epithelium is amelanotic until it reaches the iris and then is highly pigmented, continuous with the retinal pigment epithelium (RPE), and covers the cells of the dilator muscle. The retina ends at the ora serrata portion of the ciliary body, the function of which is to secrete the aqueous humor. There are three sets of ciliary muscles in the eye: the longitudinal, radial, and circular muscles. They are near the front of the eye, above and below the lens. They are attached to the lens by connective tissue called the zonules of Zinn and are responsible for shaping the lens to focus light on the retina. The ciliary body has four functions: accommodation, aqueous humor production, restoration of vitreous mucopolysaccharide, and the production and maintenance of the lens zonules. One of the most essential roles of the ciliary body is the production of the aqueous humor, which is responsible for providing most of the nutrients for the lens and the cornea as well as waste management of these areas.

Nutrients for the ciliary body come from the same blood vessels that supply the iris. Those vessels in the anterior ciliary body also supply blood to the limbal tissues, anterior choroid, and the ciliary body. The main arterial supply to the ciliary body is through the long posterior and the anterior ciliary arteries, which come together to form the major arterial circle of the iris just posterior to the anterior chamber angle recess (Fig. 1.2). From the MAC, branches pass to the anterior part of the ciliary processes, where they form a capillary bed. Branches also go to the stroma and ciliary muscle, and others pass to the iris, anterior limbal region, and anterior choroid. The capillaries of the ciliary processes are large and fenestrated rather like the choriocapillaris. The capillaries in the ciliary muscle are smaller and less fenestrated. The principal route of venous drainage is posteriorly into the choroid and vortex vein system and, to a lesser extent, into the intrascleral venous plexus and episcleral veins in the limbal region.

The ciliary body is difficult to study in vivo because of its location in the eye; a small amount of its anterior surface can be seen indistinctly with the gonioscope, while the pars plana and ora serrata can be seen with the indirect ophthalmoscope and with the three-prism contact lens of Goldmann when the sclera is depressed.

The ciliary body is a main target of drugs against glaucoma, as the ciliary body is responsible for aqueous humor production; lowering aqueous humor production will cause a subsequent drop in the intraocular pressure.

With age, everyone develops a condition known as presbyopia. This occurs as the ciliary body muscle and lens gradually lose elasticity, causing difficulty in reading.

1.2.3 Choroid and Suprachoroid

The choroid's anterior boundary is Bruch's membrane, which separates retina from choroid, and the posterior boundary is the lamina fusca, which is a transition zone composed of layers of long, collagenous, ribbonlike processes that branch and are interconnected and separate choroid from sclera. The choroid is a long, thin, vascular, and pigmented tissue, which forms the posterior portion of the uveal tract (the iris, ciliary body, and choroid). Because of its great vascularity, the choroid has some of the properties of erectile tissue. The choroidal vessels supply blood to and receive blood from the anterior portion of the eye as well as nourish the outer retina and carry waste from the RPE. The choroidal vasculature in primates is composed of the choriocapillaris (an internal or anterior layer), medium-sized vessels (Sattler's layer) measuring 40–90 μm, and large arteries and veins measuring 20–100 μm are posterior (Haller's layer). Its capillaries form a very unusual pattern (Fig. 1.7), being arranged in a single layer restricted to the inner portion of the choroid; this arrangement enables the capillary layer to provide nutrition for the outer retina.

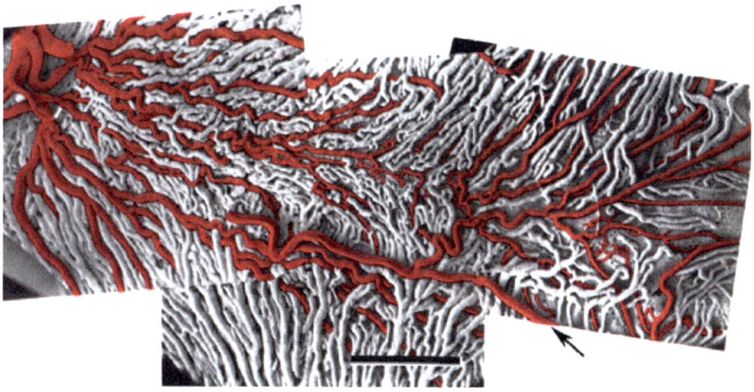

Fig. 1.7 Vascular cast of a monkey viewed from the scleral side. The LPCA (*arrow*), whose intrascleral portion was broken, and the SPCAs supply a large area temporal to the macula. Arteries are colored *red*. The choriocapillaris is visible on the *right side* of the picture (scale bar = 1 mm) (From Shimizu [37], p. 101, with permission)

1.2.3.1 Development of the Choroidal Vasculature

The choroid is derived embryonically from mesoderm, neural crest cells, and neuroectoderm. The condensation of neural crest cells that occurs initially around the anterior region of the optic cup and proceeds posteriorly to the optic stalk differentiates into the cells of the ensheathing choroidal stroma [19]. Endothelium-lined blood spaces appear very early in the mesenchymal tissue and first coalesce anteriorly at the rim of the optic cup to form the embryonic annular vessel. Prior to the fourth week (5.0-mm embryonic stage), the associated mesoderm surrounding the optic cup is undifferentiated.

We have recently demonstrated that the initial human choriocapillaris develops by hemovasculogenesis: differentiation of endothelial, hematopoietic, and erythropoietic cells from a common precursor, the hemangioblast [14]. At 6–7 WG, erythroblasts [nucleated erythrocytes expressing epsilon hemoglobin (Hbε)] were observed within the islands of precursor cells (blood-island-like formations) in the choriocapillaris layer and scattered within the forming choroidal stroma. Often, the same $Hb\varepsilon^{+}$ cells coexpressed endothelial cell (CD31, CD39), hematopoietic (CD34), and angioblast markers (VEGFR-2) suggesting that they had a common progenitor, the hemangioblast. By 8–12 WG, most of the erythroblasts had disappeared, and vascular lumen became apparent in choriocapillaris. Uniquely, the capillaries (choriocapillaris) form first without any association with intermediate or large vessels.

Rudimentary vortex veins appear in the upper and lower nasal and temporal quadrants of the eye during the sixth week [19]. The short posterior ciliary arteries also appear in the mesoderm choroidal condensation. By the 11th WG (50- to 60-mm stage), these arteries have branched extensively throughout the choroid. At 14–23 WG, some endothelial cells were proliferating on the scleral side of choriocapillaris in association with forming deeper vessels, suggesting that angiogenesis is responsible for the apparent anastomosis between the choriocapillaris and intermediate blood vessels.

1.2.3.2 Arteries

There are three main arterial sources of blood to the choroid: long posterior ciliary arteries (LPCAs, temporal, and nasal), short posterior ciliary arteries (SPCAs), and the anterior ciliary artery. The LPCAs follow long, oblique intrascleral courses traveling in the potential suprachoroidal space and send branches from the ora serrata region posteriorly to supply the choroid as far posterior as the equator (Fig. 1.7). A ciliary nerve accompanies each LPCA. There are 15–20 SPCAs in man that supply the choroid from equator to optic nerve (Fig. 1.7) while there are no SPCAs in rodents (Fig. 1.8). The arteries surround the optic nerve (the circle of Zinn) in the posterior pole, penetrating and then branching peripherally in a wheel-shaped arrangement (Fig. 1.3). Triangular watersheds separate the radial areas supplied by the arteries with the apices directed toward the fovea, which is

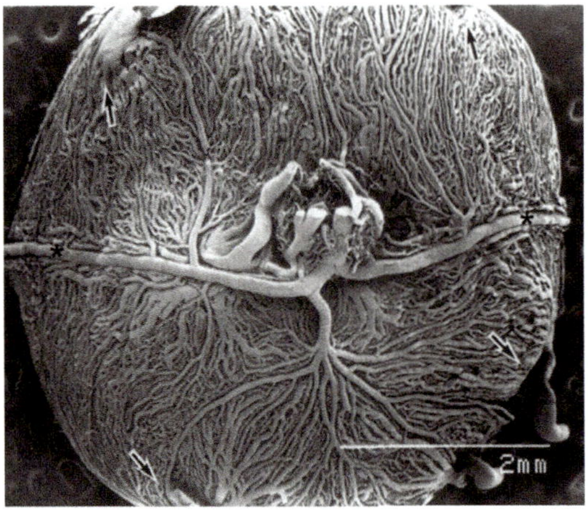

Fig. 1.8 Cast of an entire rat choroidal vasculature viewed from the scleral aspect. Two long ciliary arteries (*asterisks*) originate from the optic nerve and traverse temporally and nasally. There are no SPCAs in rodent, only branches of the two LPCAs. The vortex veins are marked with *arrows* (temporal is *left* and nasal is *right*) (From Bhutto [4], Anat Rec, p. 654)

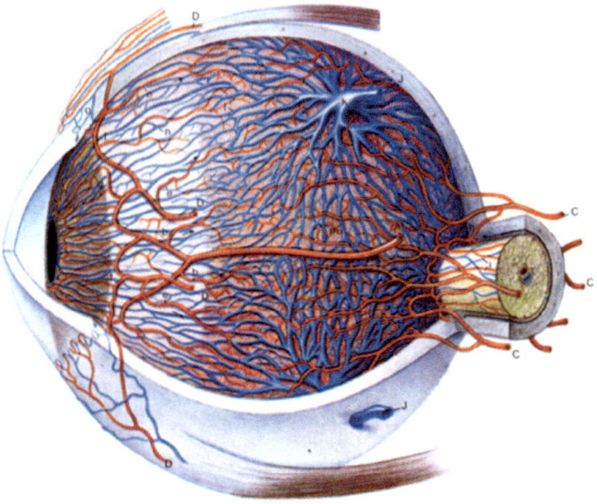

Fig. 1.9 Schematic of the blood vessels of the uveal tract in human. One of the two LPCAs (*A*) is present traveling along the horizontal meridian. It bifurcates at the ora serrata and immediately branches into arteries that traverse back (*arrows*) to supply the anterior choriocapillaris including the equator. The SPCAs enter the choroids near the optic nerve (*c*) and then divide quickly to supply the posterior choriocapillaris (not shown). The ACA (*D*) enters the globe through the rectus muscle and traverse through the sclera into the ciliary body. The ACA yields 8–12 branches (*e*) before joining the major circle of iris (*f*). The major circle of the iris branches anteriorly into the iris (*g*) and posteriorly into the ciliary body. The circle of Zinn (*h*) lies in the sclera and supplies some of the blood to the optic nerve and disk. The vortex veins form an ampulla (*k*) before exiting the globe through the sclera (*J*). The vortex veins not only drain an entire quadrant of the choroid but also drain blood from the iris and ciliary body. Some anterior veins, however, enter the episcleral system of veins (From Hogan [19], p. 326, with permission)

the anatomic basis of the triangular syndrome. A watershed zone in choroid is an area that normally fills and drains slowly with blood that occurs in areas supplied by two or more end arteries. Hayreh, after extensive in vivo experimental studies on choroidal circulation and its watershed zones in monkey and man, argues that the choroidal vascular bed is a strictly segmental and end-arterial system and has watershed zones situated between the various PCAs, the short PCAs, the choroidal arteries, the arterioles, and the vortex veins [17]. There is great variability in humans on the location of watershed zones of choroid, but their significance is that a fall in perfusion pressure in one or more of the arteries in the involved area can result in ischemia due to the poor vascularity of the watershed zone [17]. The nature of the choroidal vasculature and the existence of watershed zones in the choroid are controversial, of great clinical importance, and probably play a significant role in the production of various ischemic lesions in the choroid.

The final arterial source of blood to choroid is the anterior ciliary arteries, which send recurrent branches posteriorly to supply the choroid at 3 o'clock and 9 o'clock soon after they pierce the anterior sclera. At the major iridal circle, anastomoses are found between short, long, and anterior ciliary arteries and the arteries of all three-vessel systems rapidly extend internally via arterioles to supply blood to the choriocapillaris (Fig. 1.3). Mast cells are intimately associated with most choroidal arteries and may provide cytokines, proteolytic enzymes, and potent vasomodulatory substances for these arteries.

1.2.3.3 Choroidal Veins (Vortex Veins)

The main venous drainage of the choroid occurs through four to six vortex veins located at the equator (Fig. 1.9) that drain into superior and inferior ophthalmic veins. In the submacular area, the venous portion predominates over the arterial portion, and venules are closely arranged. The meshwork of the venous plexus becomes less dense with increasing distance from the macula. The vessels are straighter in the extramacular region, losing the tortuosity that is characteristic of the macular region. Vessels of larger lumen form the subcapillaris plexus and eventually flow into the vortex veins (Fig. 1.9). Venous drainage is segmentally organized into quadrants, with watersheds oriented horizontally through the disk and fovea and vertically through the papillomacular region. Arterial and venous choroidal watersheds are under the center of the macula, which may either predispose it to relative ischemia or prevent ischemia through multiple submacular blood supplies.

1.2.3.4 Choriocapillaris

The choriocapillaris, located solely in the internal portion of the choroid, appears as a nonhomogenous network of large (10–38 μm) capillaries [35]. Modern histological and image analysis techniques suggest much smaller diameters for choriocapillaris [34]; for example, using the alkaline phosphatase flat embedded choroid technique [26], we find submacular capillaries to be an average diameter of 14.7 mm (McLeod and Lutty, unpublished results; [26]). This monolayer vascular network, flattened in the anterior-posterior aspect, changes from a dense, honeycomb-like, nonlobular structure in the peripapillary area to a lobule-like pattern in submacular areas and most of the posterior pole and equatorial areas (Fig. 1.10). The choriocapillaris lobules measure 0.6–1.0 mm. In the peripheral area, the pattern of choriocapillaris is a more elongated, palmlike, or fanlike vascular network forming arcades that terminate at the ora serrata [26]. The network of choriocapillaris is supplied by feeding arterioles derived from the short posterior ciliary arteries and drained by the collecting venules (Fig. 1.10). These arterioles and venules form the medium-sized vessels of the choroid occupying the choroidal stroma (Sattler's layer). The majority of these vessels in the peripapillary and submacular areas form a 90° angle with the posterior aspect of the choriocapillaris.

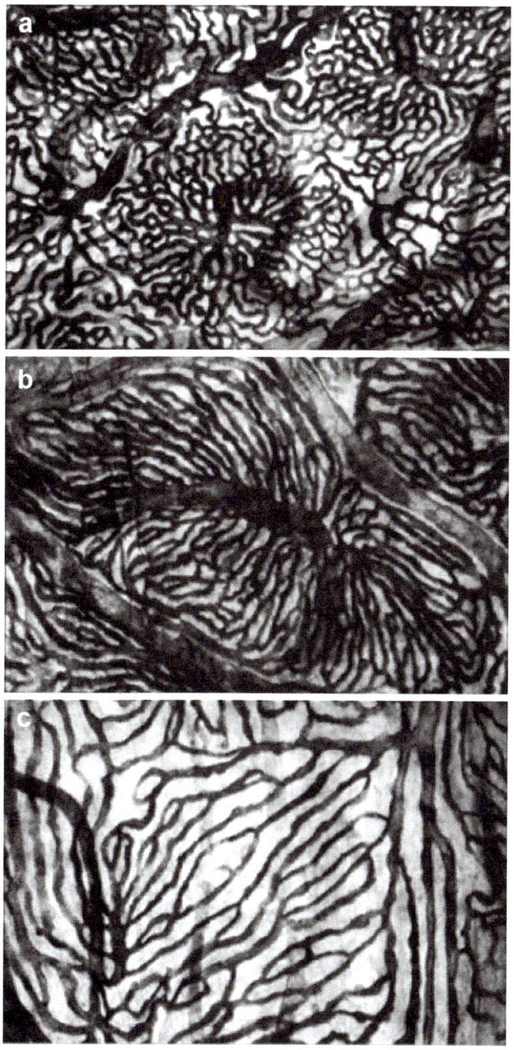

Fig. 1.10 Pattern of choriocapillaris in the posterior pole (**a**), equator (**b**), and periphery (**c**) in a flat mount human choroids incubated for alkaline phosphatase (APase) activity and then bleached. Venous blood vessels and capillaries have the most intense APase activity, while arterial blood vessels have the least. The lobular pattern is apparent in the pole (**a**) and at the equator (**b**), while the capillaries are more ladderlike in the periphery (**c**)

There is controversial evidence that both supports and disproves the idea that "lobules" exist and subdivide the choroid into many functional islands. Wybar [46], in his studies of human eyes, showed that the SPCAs are not terminal because the choriocapillaris is a single, continuous capillary vascular layer. Castro-Correia [6] was the first to describe rose-shaped arteriolar terminations. He thought that these vessels, although anatomically separated, were functionally interconnected. On the other hand, Hayreh [16] has advocated the presence of noncommunicating lobules. On the basis of fluorescein angiography findings, he described a mosaic of lobules, each one containing an arteriole in the middle and a venule at its periphery.

There is disagreement over the location of arterioles and venules in the lobules. Shimizu and Ujiie [36] confirmed the central location of the artery and the peripheral location of the venule. They specified the dimension of the lobules of the choriocapillaris: 200 μm in diameter at the equator, 100 μm at the posterior pole, and as small as 30–50 μm in the submacular region. On the contrary, Krey [22] and McLeod and Lutty [26], who performed histologic studies of the choriocapillaris stained with alkaline phosphatase reaction product, described a lobular organization of the choroid with arterioles and venules located peripherally and centrally, respectively. Uyama et al. [42] described the same lobular structure with a venule in the middle and arterioles in the periphery. Torczinsky and Tso [40, 41], in their histologic and fluorescein angiography studies of the choriocapillaris in albino monkeys, found structural differences between the posterior and peripheral choriocapillaris. Tilton et al. found that rat choriocapillaris had 50% less pericytes than retinal capillaries [39]. Also unlike retinal capillaries, pericyte loss did not occur in rat experimental diabetes [38]. Choriocapillaris is also unique in that the capillaries are fenestrated predominantly on the inner side, i.e., toward retina (Fig. 1.11). This probably is associated with the primary functions of this capillary system: transport of nutrients to RPE and photoreceptors and remove waste from disk shedding and RPE digestion of these disks.

The same sidedness has been observed in the location of VEGF receptors. Blaauwgeers et al. have reported that both VEGF receptor 3 (FLT-4) and VEGF receptor 2 (FLK-1 or KDR) are found on choriocapillaris endothelial cells on the retinal side [5]; however, we have not observed sided expression of VEGFR-2 in fetal or adult human choriocapillaris (unpublished results). Perhaps, sided expression of VEGF receptors is related to the basal production of VEGF by RPE. RPE was actually one of the first cells shown to produce VEGF and upregulate production during hypoxia [1]. Perhaps, the release of VEGF on the retinal side encourages maintenance of fenestrae on the choriocapillaris.

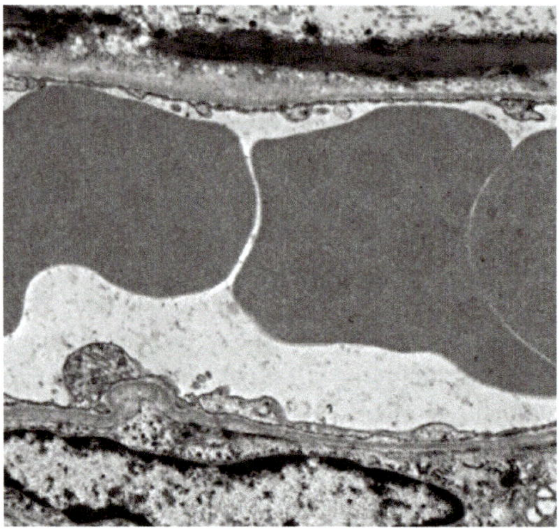

Fig. 1.11 Ultrastructure of a human choriocapillaris lumen. The choriocapillaris is immediately posterior to Bruch's membrane (*top*). Posterior to Bruch's membrane is the basement membrane of the choriocapillaris endothelium, which is richly fenestrated on this side of the lumen. An elongated endothelial cell nucleus is present on the scleral side of the lumen (*bottom*), which contains RBCs. (Original magnification 11,800×) (From Rhonda Grebe, Wilmer Ophthalmic Institute, Johns Hopkins Hospital)

1.3 Optic Nerve Vasculature

The central retinal artery is a direct branch of the ophthalmic artery. The central retinal artery is a small muscular artery with a luminal diameter of 170–245 μm. The ratio of the wall thickness to lumen is 1:4 in the pre–lamina cribrosa area to 1:10 post–lamina cribrosa [3]. There is a

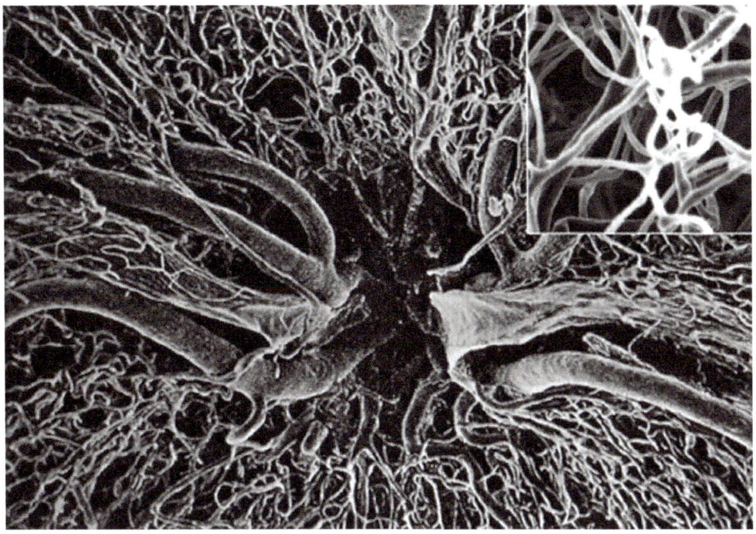

Fig. 1.12 Vascular cast of a monkey's optic nerve head. The retinal arteries branch soon after emerging from the nerve head, and the retinal veins appear flattened as they descend into the optic nerve. The capillaries of the nerve head resemble and appear contiguous with the radial peripapillary capillaries. (Inset) The meshwork of capillaries in the optic nerve head encloses polygonal spaces (From Earl Addicks and Harry Quigley, Wilmer Ophthalmological Institute, Johns Hopkins Hospital)

subendothelial elastic lamina in the central retinal artery, but this is lost after it branches into the major arteries of retina. The central retinal vein has a luminal diameter of 200–245 μm and is classified as a medium caliper vein. It has a few pericytes, and the media has abundant elastic fibers. The papillary capillaries in the human optic nerve head range in size from 7 to 10 μm in lumenal diameter. They form a meshwork of vessels enclosing polygonal spaces in the nerve (Fig. 1.12). They have a typical capillary structure of endothelium, pericytes, and a basement membrane and are invested with astrocytes. However, there are less astrocytes associated with capillaries in the nerve head than at the level of lamina cribrosa.

Much of our knowledge of the optic nerve vasculature comes from the elegant and extensive work of Sohan Hayreh, which is nicely summarized in his Von Sallman Lecture and the source of the following information [18]. There is a great deal of variability in the arrangement of blood vessels in the optic nerve. Blood vessels at the surface of the optic disk are supplied by retinal arterioles. The prelaminar area, the area between the surface and lamina cribrosa, is supplied by the peripapillary choroid (Fig. 1.13). The blood supply is sectoral in this region and is not from the peripapillary choriocapillaris or the central retinal artery. There is a dense capillary plexus (luminal diameters of 10–20 μm) in the lamina cribrosa region, which is supplied by the centripetal branches of the short posterior ciliary arteries or by the circle of Zinn and Haller (Fig. 1.13). From the corrosion casts of Fryczkowski [13] and Olver [31], the circle or perioptic nerve arteriolar anastomoses as Olver called it are supplied by the paraoptic short PCAs. Which PCAs supply it varies by individual, but in the majority of people it is the medial or lateral paraoptic short PCAs. The retrolaminar portion of the nerve is supplied by the pial vessels branches and sometimes branches from the central retinal artery. Therefore, the main source of blood for the optic nerve is from the PCA circulation via the peripapillary choroid and the short PCAs from the circle of Zinn and Haller. This yielded a blood supply of the optic nerve that has a sectoral distribution. The extremely variable pattern of distribution of the PCAs in the choroid and optic nerve may play a role in occurrence of optic neuropathies. Hayreh concludes that derangement of the posterior ciliary circulation in the optic nerve is responsible for most common ischemic optic neuropathies of the optic nerve head [18].

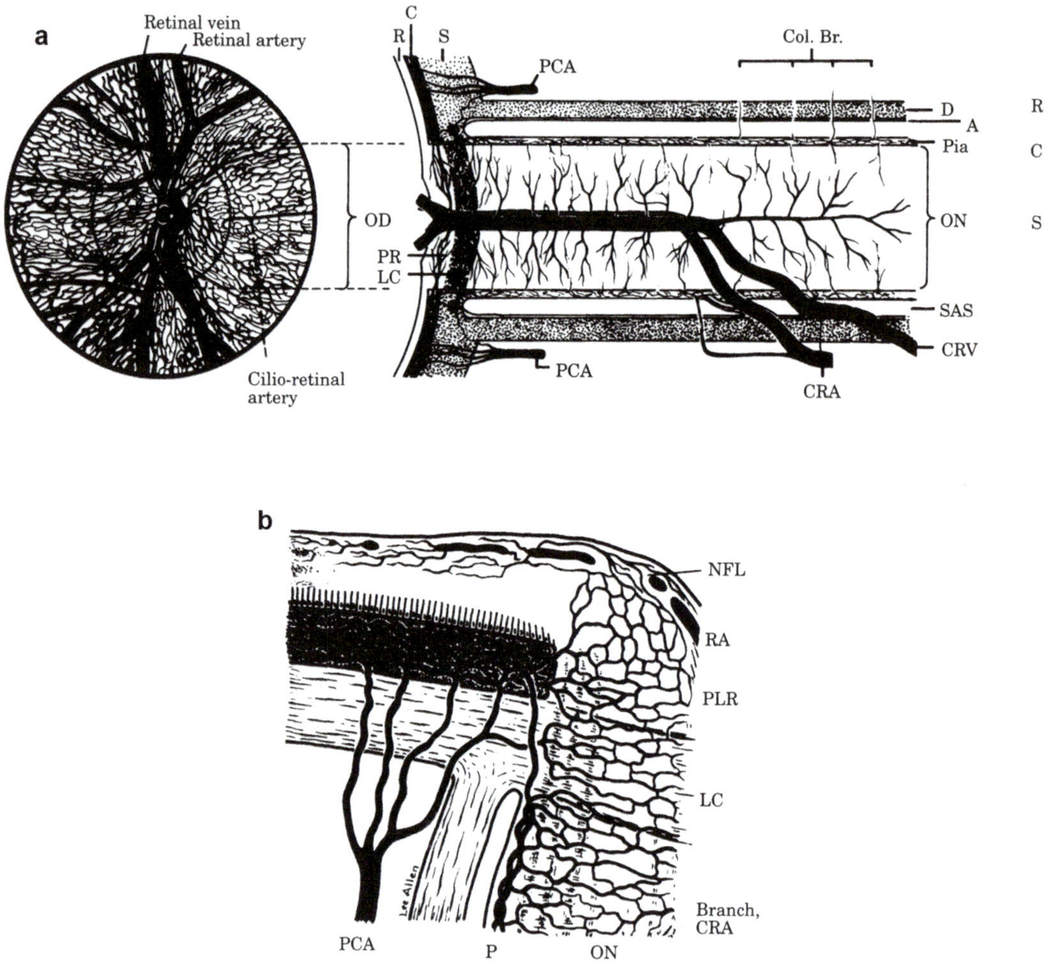

Fig. 1.13 Schematic representations from the work of Sohan Hayreh of the blood supply in the optic nerve and intraorbital sector of the optic nerve (**a**) and optic nerve head and retrolaminar optic nerve (**b**). (*A* arachnoid, *C* choroid, *Col. Br.* collateral branches of the ophthalmic artery or its orbital branches, *CRA* central retinal artery, *CRV* central retinal vein, *D* dura, *LC* lamina cribrosa, *NFL* nerve fiber layer, *OD* optic disk, *ON* optic nerve, *P* pia, *PCA* posterior ciliary artery, *PLR/PR* prelaminar region, *R* retina, *RA* retinal arteriole, *S* sclera, *SAS* subarachnoid space) (From Hayreh [18], Exp Eye Res 61, p. 261, with permission)

1.4 Retina

1.4.1 Development of the Retinal Vasculature

We have demonstrated that angioblasts are present in inner retina of embryonic and fetal human and neonatal dog, and they differentiate and organize to form the primordial retinal vasculature [7, 25, 27]. The angioblasts coalesce and differentiate to form primordial blood vessels or cords. The assembly of these vascular cords occurs in extracellular spaces formed by the innermost Muller cell processes. The astrocyte migration from optic nerve trails formation of the retinal vasculature in dog and man, and blood vessels become invested with astrocytes after they are formed. We have demonstrated that these ecto-ADPase^{+} (CD39^{+}) angioblasts emerge from a large pool of precursors that form the inner neuroblastic layer in man and express CXCR 4 and

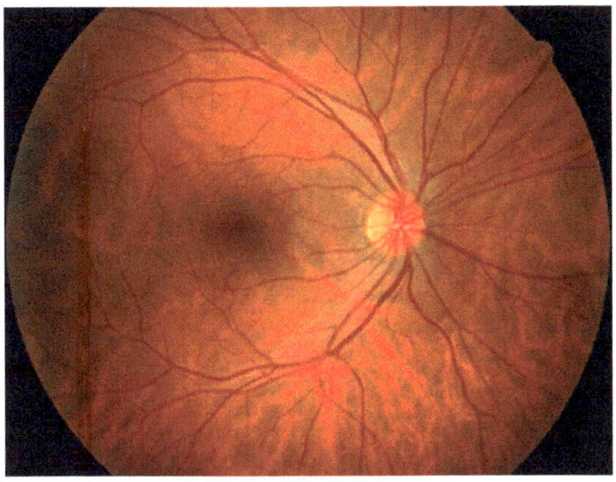

Fig. 1.14 Fundus photograph of a 35-year-old Caucasian male. The major arteries and veins emanate from the optic nerve head. The arteries appear *darker red* and narrower than veins. Each quadrant of the retina is supplied by a major artery and vein. The nasal vessels (*right*) have a straighter and shorter course. The temporal blood vessels run an arcuate path around the macula. The major vessels divide by dichotomous and side arm branches

cKit [15]. It appears that the chemokine stromal-derived factor 1 (SDF-1) and stem cell factor are associated with the differentiation of retinal precursors into angioblasts and their migration to sites of vessel assembly in the nerve fiber layer. The deep capillary plexus forms by angiogenesis in all species, formation of a vasculature from a preexisting vasculature by migration, and proliferation of endothelial cells. The result of vasculogenesis and then angiogenesis in human and dog retina is an arcade-like pattern adult vasculature (Fig. 1.14). The arcades of vasculature form inferior and superior to the macula in man, but the macula is the last area of vasculature to develop [33]. The rodent (rat and mouse) retinal vasculatures form by angiogenesis, budding from the blood vessels in the optic nerve [12]. The new blood vessels use a preexisting astrocyte template to determine their pattern. The resultant adult rodent vasculature is spoke wheel-like.

1.4.2 Adult Retinal Vasculature

Twenty percent of the ocular blood goes to the central retinal artery in primates. The central retinal artery is the source of blood to the retinal vasculature in primates, and blood is drained from retina through the central retinal vein. In general, the central retinal artery lies nasal to the central retinal vein in the optic nerve head. Each quadrant of the primate retina is supplied by a major artery and vein (Figs. 1.12 and 1.13). The major temporal arteries and veins run in an arcuate course inferior and superior to the macula while the nasal major blood vessels have shorter more direct course (Fig. 1.14). Small arterioles and venules may run directly from optic nerve head to macula. The exception to these generalities is a cilioretinal artery, which occurs in 25% of humans and hooks around the temporal margin of the disk and provides blood for a portion of the macula. The retinal vascular system has an end-arterial hierarchy: arteries to arterioles to capillaries then venules and finally veins. Retinal arteries and veins divide by dichotomous and sidearm branching. There are numerous arteriovenous crossings in the retinal vasculature, and the artery is anterior to the vein in most cases. These crossings will be sites of occlusion in branch vein occlusion and sickle cell retinopathy [23, 28].

Retinal arteries have a luminal diameter of 100 μm as they exit the disk and five to seven layers of smooth muscle cells. The number of smooth muscle cell layers decreases from two to three at equator and one to two in periphery. The arteries lack a subendothelial elastic lamina. The retinal veins have a lumenal diameter of 200 μm at the disk and a thin endothelial cell basement membrane (0.1 μm). There is no elastic lamina, but the walls of the veins have widely spaced pericytes. At arteriovenous crossings, the artery and vein share a common coat [2]. Retinal capillaries vary in diameter from 3.5 to 6 μm. They are unique capillaries in that there is a 1:1 ratio of pericytes to endothelial cells. Pericytes and endothelial cells share a basement membrane. Intramural pericytes, formerly known as Rouget's cells or mural cells, communicate with endothelial cells and jointly control vasotonia and mutual cellular quiescence.

The retinal vasculature is present in almost all areas of retina. The noticeable exception is the fovea, which lacks any blood vessels (Fig. 1.15). This is called the foveal avascular zone (FAZ), and obviously, it only exists in primates since only they have foveas. The FAZ is 400–500 μm

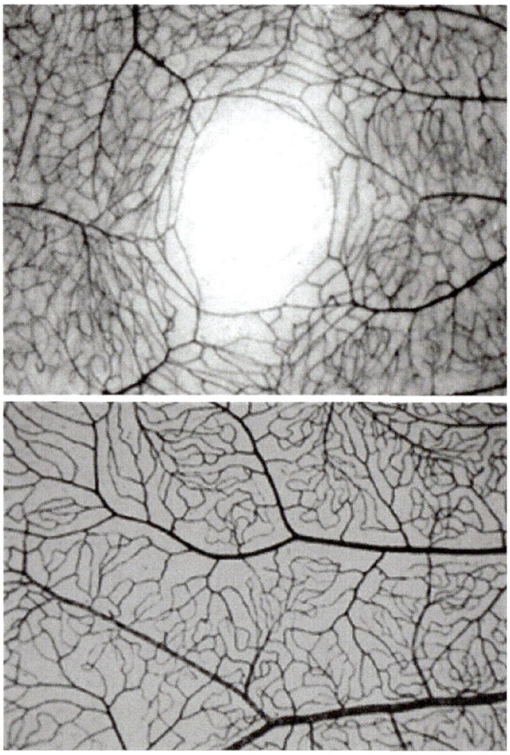

Fig. 1.15 Areas of a human retina from a 19-year-old Caucasian male that were incubated for adenosine diphosphatase activity (ADPase), which is present only in the retinal blood vessels. This is a peripheral region of retina, and the artery has characteristically more intense ADPase reaction product than the vein (*bottom*). The artery has a capillary-free zone around it, and the end-arterial hierarchy is apparent in this area because there is no deep or secondary network in this region of retina. The fovea (*top*) is the only avascular zone in retina, the foveal avascular zone (FAZ)

in diameter. The far peripheral retina at ora serrata also is avascular. The other area that has a sparse vasculature is the capillary-free zones around arteries (Fig. 1.15).

The retinal vasculature occupies only the inner half of the retina and only supplies nutrients for the inner half of retina, contributing little to maintenance of the photoreceptors whose nuclei occupy the outer nuclear layer [43]. There is a superficial or inner vascular plexus and a deep or secondary capillary network. The majority of the arteries and veins lie in the superficial vascular plexus, while the majority of the capillaries lie in the deep plexus. Arterioles and

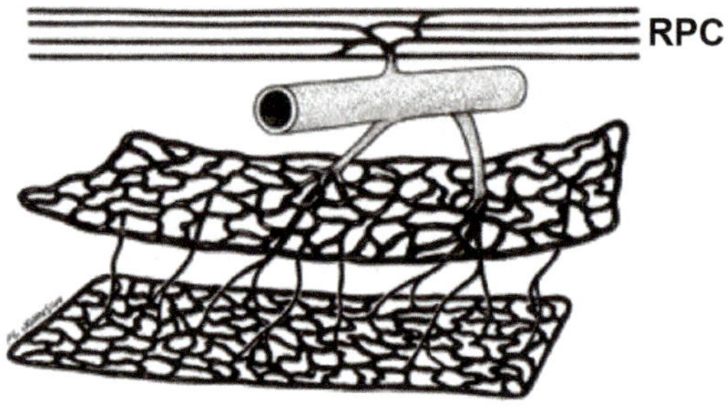

Fig. 1.16 Schematic showing the relationship of radial peripapillary capillaries (*RPCs*) to a large retinal blood vessel and the superficial and deep or secondary capillary systems. The same retinal artery supplies all capillary systems in this area (From Wise [44], p. 28)

venules will communicate between the superficial and deep networks. The exception to this is arteriovenous crossings where the deeper large vessel will lie at the level of the outer plexiform or outer nuclear layer. Lerche has found that volumetrically 2.4% of the inner nuclear layer is capillaries while 0.5% of the inner plexiform layer and 1.3% of the nerve fiber layer are capillaries [24]. The bilayered system exists in most of the retina but not in periphery where the retina is thin and metabolism is lower so there is only a single layer of wide-mesh, broad caliper capillaries (Fig. 1.15). In the foveal slope, there is a single layer of capillaries. Another notable exception to the bilayered system is the peripapillary region where the retina is thickest. In this region, there are three layers plus a discreet capillary network in the superficial portion of the nerve fiber layer (Fig. 1.16). These radial peripapillary capillaries (RPCs) were described by Michaelson and Campbell in 1940 [30]. RPCs are relatively straight and follow the paths of the major superotemporal and inferotemporal vessels and are much shorter and less prominent nasally where they are adjacent to disk. There are no RPCs in the macular region. RPCs differ from other capillaries in retina in that they follow long straight paths instead of being arranged in mesh-like nets, they have few arterioles, and they rarely anastomose with other capillaries, whereas these anastomoses are common in other retinal capillaries [44]. RPCs will cross underlying venules but not arterioles. They presumably nourish the nerve fibers that emanate from the optic nerve in bundles in these areas.

Like the brain microvasculature, the retinal vasculature is a tight vasculature in that macromolecules are not permeable. This occurs because endothelial cells make tight junctions with each other by continuous fusion of the outer leaflets of the cytoplasmic membrane, which results in formation of a blood-retinal barrier (BRB). The actual mechanism is not known, but the lack of permeability depends upon the presence of astrocytes investing the blood vessels [21]. Either loss of astrocytes or damage to the endothelium will cause a breakdown in the BRB.

1.4.3 Nonprimate Adult Retinal Vasculatures

The retinal vasculature, more so than any other ocular vasculature, varies between primates and other mammals. In placental animals, retinal vasculatures can be classified into euangiotic or holangiotic where the entire retina is supplied by either a central retinal artery (primates) or branches of the cilioretinal arteries (rodents, insectivores, and carnivores), angiotic where only part of the retina is vascularized and the vasculature sits anterior to retina proper (rabbits), pseudoangiotic in which blood vessels are small and only around the disk (elephants, horses, manatees, guinea pigs), and anangiotic where the retina is avascular (marsupials, monotremes, hippopotamus, and rhinos) [44]. The ophthalmic artery divides into five to seven retinal arteries in rodents and becomes a central retinal artery in primates.

Euangiotic retinas of primates are similar to man except for the owl monkey (*Aotes trivirgatus*). The owl monkey does not have radial peripapillary capillaries, and the fovea has capillaries. Although the fovea is vascularized, the capillary-free zone around arteries and arterioles is very large compared to other primates [10].

The arterial supply for the dog and cat is from cilioretinal arteries that exit the disk at the margin, not a central retinal artery. The origin of the cilioretinal arteries is the long posterior ciliary artery and not a direct branch of the ophthalmic artery. Both species lack a fovea but have an area centralis, the area with the greatest density of ganglion cells. The area centralis is vascularized by capillaries from the superior and inferior arcades meeting at the horizontal raphe. A major difference between cat and dog is that the dog has far more pericytes than endothelial cells, perhaps as high as a 5:1 ratio.

The rat and mouse are also euangiotic, and the retinal arteries arise from central retinal arteries. The pattern in these rodents is unique being spoke wheel with long arteries and veins running alternately in a pattern from disk to periphery. Michaelson reported that the superficial plexus was mostly arterial and capillaries, not arterioles, and traversed from the superficial to the deep vascular plexus [29]. The deep plexus is predominantly venous, and most blood vessels in the deep plexus are capillaries. There is a central retinal vein that drains rodent retinas.

The rabbit has an angiotic retina, but the single layer of retinal blood vessels are present on the surface of the retina in what is called two medullary rays running nasal and temporal. This system covers only about 40% of the retina, the area in which there are nerve fibers. The rabbit retinal vasculature is not a good choice for studying the retinal vasculature because of its unique anatomy.

1.5 Conclusions

There are unique methods of development in human retinal and choroidal vasculatures, which results in two unique adult vasculatures. Vasculogenesis and then angiogenesis in retina result in an end-arterial vascular system that forms a blood-retinal barrier due to its intimate relationship with astrocytes. The human choroidal vascular develops by hemo-vasculogenesis which yields a lobular vasculature with the choriocapillaris forming a sheet of capillaries adjacent to the RPE. This relationship and cytokines made by the RPE may allow the choroidal capillaries to remain fenestrated. The three vasculatures with fenestrations are found either supplying nutrition for avascular tissues (limbal supplying cornea and ciliary processes supplying epithelial cells adjacent to vitreous and lens) or the source of nutrition for a highly metabolically active tissue like the photoreceptors, which are nourished by the choriocapillaris. The sources of the blood for the various ocular vasculatures and their susceptibility to infarction predict the incidence and severity of the ischemia that occurs in that tissue. All of the unique vasculatures in the eye are supplied by only two sources: the ophthalmic artery and anterior ciliary arteries. Branches from these two sources must become specialized to provide for their unique tissue. The variability in the supply of blood in the eye as in the human variability of PCAs supplying the optic nerve might predict the individual's chances of developing an optic neuropathy if they could be visualized in vivo.

References

1. Adamis AP, Shima DT et al (1993) Synthesis and secretion of vascular permeability factor/vascular endothelial growth factor by human retinal pigment epithelial cells. Biochem Biophys Res Commun 193:631–638
2. Archer DB, Gardiner TA et al (2007) Functional anatomy, fine structure and basic pathology of the retinal vasculature. In: Joussen AM, Gardner TW, Kirchhof B, Ryan SJ (eds) Retinal vascular disease. Springer, Berlin, p 6
3. Becker B, Ley AP (1966) Retinal arteriosclerosis. In: Chowdry EV (ed) Arteriosclerosis. Thomas, Springfield, p 401
4. Bhutto IA, Amemiya T (2001) Microvascular architecture of the rat choroid: corrosion cast study. Anat Rec 264:63–71
5. Blaauwgeers HG, Holtkamp GM et al (1999) Polarized vascular endothelial growth factor secretion by human retinal pigment epithelium and localization of vascular endothelial growth factor receptors on the inner choriocapillaris. Am J Pathol 155:421–428
6. Castro-Correia J (1957) Vascularization of the choroid. Acta Anat 31:238–245
7. Chan-Ling T, McLeod DS et al (2004) Astrocyte-endothelial cell relationships during human retinal vascular development. Invest Ophthalmol Vis Sci 45:2020–2032
8. Cursiefen C, Chen L et al (2004) VEGF-A stimulates lymphangiogenesis and hemangiogenesis in inflammatory neovascularization via macrophage recruitment. J Clin Invest 113(7):1040–1050
9. Cursiefen C, Rummelt C et al (2006) Absence of blood and lymphatic vessels in the developing human cornea. Cornea 25(6):722–726
10. de Oliveira LFN, Ripps H (1968) The "area centralis" of the owl monkey (*Aotes trivirgatus*). Vision Res 8:223
11. Freddo TF, Raviola G (1982) The homogenous structure of blood vessels in the vascular tree of *Macaca mulatta* iris. Invest Ophthalmol Vis Sci 22:279–291
12. Fruttiger M (2002) Development of the mouse retinal vasculature: angiogenesis versus vasculogenesis. Invest Ophthalmol Vis Sci 43:522–527
13. Fryczkowski AW, Grimson BS et al (1984) Scanning electron microscopy of vascular casts of the human scleral lamina cribrosa. Int Ophthalmol 7(2):95–100
14. Hasegawa T, McLeod DS et al (2007) The embryonic human choriocapillaris develops by hemo-vasculogenesis. Dev Dyn 236(8):2089–2100
15. Hasegawa T, McLeod DS et al (2008) Vascular precursors in developing human retina. Invest Ophthalmol Vis Sci 49(5):2178–2192
16. Hayreh SS (1974) Submacular choroidal vascular pattern. Experimental fluorescein fundus angiographic studies. Albrecht Von Graefes Arch Klin Exp Ophthalmol 192:181–196
17. Hayreh SS (1990) In vivo choroidal circulation and its watershed zones. Eye (Lond) 4:273–289
18. Hayreh SS (1995) The optic nerve head circulation in health and disease. Exp Eye Res 61:259–272
19. Hogan MJ, Alvarado JA et al (1971) Histology of the human eye. W.B. Sanders Company, Philadelphia
20. Iwamoto T, Smelser GK (1965) Electron microscope studies on the mast cells and blood and lymphatic capillaries of the human corneal limbus. Invest Ophthalmol 4(5):815–834
21. Janzer RC, Raff MC (1987) Astrocytes induce blood–brain barrier properties in endothelial cells. Nature 325:253–257
22. Krey H (1975) Segmental vascular patterns of the choriocapillaris. Am J Ophthalmol 80:198–202
23. Kumar B, Yu DY et al (1998) The distribution of angioarchitectural changes within the vicinity of the arteriovenous crossing in branch retinal vein occlusion. Ophthalmology 105(3):424–427
24. Lerche W (1965) Die Kapillarisierung der menschlichen Retina. In: Rohen JW (ed) The structure of the eye, vol 11. Schattauer, Stuttgart, p 121
25. McLeod DS, Hasegawa T et al (2006) The initial fetal human retinal vasculature develops by vasculogenesis. Dev Dyn 235(12):3336–3347
26. McLeod DS, Lutty GA (1994) High resolution histologic analysis of the human choroidal vasculature. Invest Ophthalmol Vis Sci 35:3799–3811
27. McLeod DS, Lutty GA et al (1987) Visualization of a developing vasculature. Microvasc Res 33:257–269
28. McLeod DS, Merges C et al (1997) Histopathological features of neovascularization in sickle cell retinopathy. Am J Ophthalmol 124:473–487
29. Michaelson IC (1954) Retinal circulation in man and animals. Chas. C. Thomas, Springfield
30. Michaelson IC, Campbell ACP (1940) The anatomy of the finer retinal vessels. Trans Ophthalmol Soc UK 60:71

31. Olver JM, Spalton DJ et al (1994) Quantitative morphology of human retrolaminar optic nerve vasculature. Invest Ophthalmol Vis Sci 35(11): 3858–3866
32. Paton D (1961) The conjunctival sign of sickle cell disease. Arch Ophthalmol 66:90–94
33. Provis JM (2001) Development of the primate retina. Prog Retin Eye Res 20:799–821
34. Ramrattan RS, van der Schaft TL et al (1994) Morphometric analysis of Bruch's membrane, the choriocapillaris, and the choroid in aging. Invest Ophthalmol Vis Sci 35:2857–2864
35. Sattler H (1876) Ueber den feineren Bau der Choroidea des Menschen nebst Beitragen zur pathologischen and vergleichenden Anatomie der Aderhaut. Albrecht Von Graefes Arch Ophthalmol 22:1–15
36. Shimizu K, Ujiie K (1976) Angioarchitecture of the choroid. Simposio internazionale sull'Uvea, Pravia
37. Shimizu K, Ujiie K (1978) Structure of ocular vessels. Igaku-Shoin, New York
38. Tilton RG, LaRose LS et al (1986) Absence of degenerative changes in retinal and uveal capillary pericytes in diabetic rats. Invest Ophthalmol Vis Sci 27: 716–721
39. Tilton RG, Miller EJ et al (1985) Pericyte form and distribution in rat retinal and uveal capillaries. Invest Ophthalmol Vis Sci 26:68–73
40. Torczynski E, Tso MO (1976) The architecture of the choriocapillaris at the posterior pole. Am J Ophthalmol 81:428–440
41. Tso MO, Torczynski E (1978) The architecture of the choriocapillaris and macularedema. In: Proceedings of the XXIII international congress in ophthalmology, Kyoto, 1978
42. Uyama M, Itotagawa S et al (1978) Angioarchitecture of the choroid studied by plastic casts in the choroidal vasculature. In: Proceedings of the XXIII international congress in ophthalmology, Kyoto, 1978
43. Wangsa-Wirawan ND, Linsenmeier RA (2003) Retinal oxygen. Fundamental and clinical aspects. Arch Ophthalmol 121:547–557
44. Wise GN, Dollery CT et al (1971) The retinal circulation. Harper and Row Publishers, New York
45. Wolff E (1997) Wolff's anatomy of the eye and orbit. Chapman and Hall, London
46. Wybar K (1954) Vascular anatomy of the choroid in relation to selective localization of ocular disease. Br J Ophthalmol 38:513–527

Part II

Measurement of Ocular Blood Flow

Determination of Ocular Blood Flows with the Microsphere Method

2

Siv F.E. Nilsson and Albert Alm

Core Messages

- The advantages with the microsphere method are: (1) it measures blood flow directly, (2) it is suitable for measuring blood flow in small pieces of tissue and in inaccessible tissues, and (3) it does not disturb the normal circulation, if the experiments are properly designed.
- Radioactive, colored, and fluorescent microspheres have been used for determination of ocular blood flow. A promising new development is the use of neutron-activated microspheres.
- For reliable measurements with the microsphere method, the size and number of microspheres should be optimized for the tissue under investigation.
- If few microspheres are trapped in a tissue, due to low blood flow and/or small sample size (e.g., retina and anterior uvea), the precision of the determinations can be increased by more experiments.
- Biological variation contributes more to the error in the measurements than paucity of microspheres.
- Differences in arterial blood pressure, anesthesia, and arterial blood gases contribute to the variation of reported values on ocular blood flows. Other factors such as diseases, age, gender, and hormonal and seasonal variations could contribute to the variation as well.

S.F.E. Nilsson, Ph.D. (✉)
Division of Drug Research/Pharmacology, Department of Medical and Health Sciences – IMH, Faculty of Health Sciences, University of Linköping, Linköping SE-581 85, Sweden
e-mail: siv.nilsson@liu.se

A. Alm, M.D., FARVO
Division of Ophthalmology, Department of Neurosciences, University Hospital, University of Uppsala, Uppsala SE-781 85, Sweden
e-mail: albert.alm@akademiska.se

2.1 Introduction

The basic principle of the microsphere method is simple; microspheres injected into the systemic circulation are distributed and entrapped in tissues in proportion to the blood flow through the tissues. Although various types of solid particles had been used to study the microcirculation previously (see [36]), it was the introduction of plastic radioactive microspheres (RM) that boosted interest in the method. Rudolph and Heymann [77] were the first to use RM to study the blood flow distribution in the fetus of sheep and goat. Makowski [49] later included the collection of a reference blood sample

L. Schmetterer, J.W. Kiel (eds.), *Ocular Blood Flow*,
DOI 10.1007/978-3-540-69469-4_2, © Springer-Verlag Berlin Heidelberg 2012

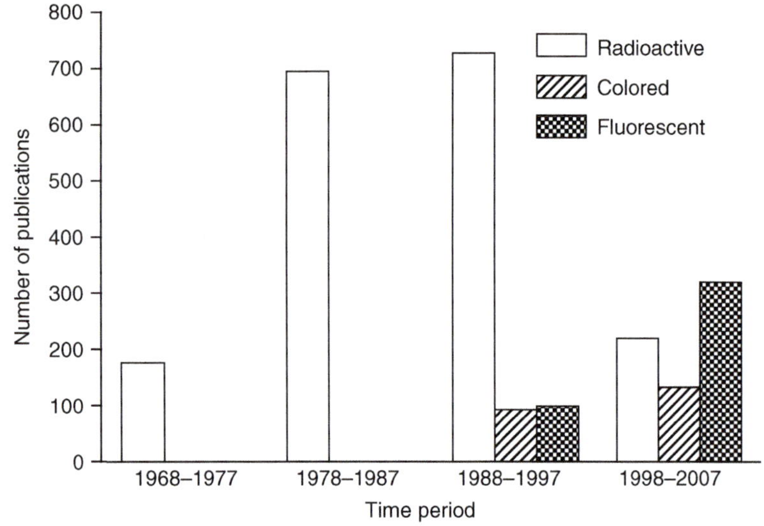

Fig. 2.1 Number of publications with different types of microspheres published between 1968 and 2007. Data is based on a PubMed search on February 11, 2008, using the search terms "radioactive microspheres blood flow," "colored microspheres blood flow," and "fluorescent microspheres blood flow," respectively. Numbers are approximate, as the publications were not checked for their relevance to blood flow measurement with microspheres or original articles versus reviews

from an artery, which made it possible to calculate actual blood flow values as

$$\frac{Q_R}{N_R} = \frac{Q_T}{N_T}$$

or

$$Q_T = \frac{Q_R \times N_T}{N_R}$$

where N_R and N_T are the number of microspheres in the reference sample and tissue, respectively, Q_R is reference flow, and Q_T is blood flow through the tissue. The method soon became a gold standard for measuring regional blood flow in various tissues and was adapted to measure ocular blood flows [4, 67, 93].

During the last decade, the number of studies using radioactive microspheres for blood flow determinations has declined, however (Fig. 2.1). There are several reasons for this; RM are relatively expensive, radiation exposure of personnel, decay during storage, difficulties and high costs associated with the disposal of the radioactive waste, and possible environmental hazards [71]. This has encouraged the search for alternatives, such as colored (CM) [10, 34, 40, 47] and fluorescent microspheres (FM) [1, 24, 32, 35]. Both types of spheres have been used to quantify ocular blood flows [2, 25, 38, 65, 68, 92, 94]. The most recent addition is the use of neutron-activated microspheres (NAM), that is, microspheres containing stable nuclides that emit γ-irradiation when activated by neutron irradiation [41, 73]. In studies of myocardial blood flow, it has been shown that colored [34, 47], fluorescent [1, 24, 32], and neutron-activated [41, 73] microspheres give blood flow values that show a good correlation with those obtained by simultaneous injection of radioactive microspheres.

The potential sources of errors in connection with the use of radioactive and colored/fluorescent microspheres for determination of local blood flows have been extensively discussed elsewhere [11, 36, 71] and are similar regardless of the tissues under investigation. Thus, the present review will focus more on those issues that are of particular interest in relation to the measurement of ocular blood flows and our experience with the method. The terms microspheres and spheres will be used interchangeably in the remainder of the chapter.

2.2 Advantages and Disadvantages with the Method and Different Microspheres

The major advantages with the microsphere method, regardless the type of label, are that it (1) measures blood flow directly, (2) can measure blood flow in small pieces of tissues, and (3) can be used for studies on not easily accessible tissues, without disturbing normal blood flow.

The ability to measure ocular blood flows without any eye surgery is very valuable. This is particularly true in species like the rabbit, which has eyes that are very sensitive to trauma. Measurement of uveal blood flow from a cannulated vortex vein in rabbits gave higher flow values than microspheres and made it difficult to study the effect of vasodilating agents, unless the animals were pretreated with indomethacin [20]. Extracapsular lens extraction and the corresponding sham operation increased blood flow in the posterior as well as the anterior uvea, an effect that could be abolished by aspirin [39].

The disadvantages with the microsphere method are that blood flow cannot be followed continuously and that relatively few measurements can be made in each animal. In addition, there is the risk of disturbing the normal microcirculation by occluding too many vessels with repeated injections, and separation of the different nuclides/colors becomes increasingly difficult with several different spheres (see below).

Colored and fluorescent microspheres can be counted in histological preparations [2, 10, 35, 65, 92] and in aqueous aliquots after digestion of the tissue [25, 34, 38, 40], or the absorbance/fluorescence can be measured after tissue digestion and extraction of dye from the microspheres [1, 24, 32, 47, 89, 94]. Thus, although CM and FM have longer shelf lives and are cheaper than RM, their use is more labor intensive, and hence the total cost may be equal. The waste disposal is simpler with CM and FM, but there are other health hazards for the personnel, as the sample processing requires the use of sodium or potassium hydroxide for digestion of the tissues and organic solvents for extraction of the dye. However, much of the sample processing can be automated [9, 87] to reduce the health risks as well as the costs. The advantages with FM over CM are that the measurement of fluorescence is more sensitive and can be done without spillover correction [32, 71], provided a limited number of differently labeled microspheres are used.

The NAM are cheaper than CM and FM, and no sample processing except drying is necessary. However, the samples have to be collected in special vials and sent to a central laboratory for analysis. To our knowledge, there is presently only one manufacturer of NAM, who also provides the analysis service. The analysis cost per sample is relatively high, but NAM are most likely cost effective for a laboratory that wants do limited studies with few tissues. The major advantages with NAM are long shelf life, little tissue processing, and reduced health and environmental hazards. Furthermore, the samples may be archived and reanalyzed if necessary or used for other analyses [73].

2.3 Stochastic Error in the Entrapment of Microspheres

Provided that the microspheres are properly mixed with the blood, they will distribute according to the blood flow to different tissues. As the entrapments of microspheres are stochastic events, the number of microspheres in the reference and tissue samples will show a Poisson distribution around a mean value, if multiple samples are collected. Early theoretical and experimental data showed that for blood flow calculations to be made with 10% precision at the 95% confidence level, the reference and tissue sample should contain approximately 400 microspheres [23]. However, a later study showed that, providing the number of microspheres in the reference sample is ≥400, the same precision can be achieved with considerably fewer microspheres in the tissue sample [66]. Still, the "400 microspheres/piece rule" has sometimes been considered a "must," although it is only valid when one wants to determine the "true" blood flow through a tissue. If one merely wants to measure heterogeneity in blood flow or correlations, this can be achieved with considerably fewer microspheres [69], that is, if one wants to study differences in blood flow within a tissue, between tissues, or at different time points in the same animal.

For determination of ocular blood flows, we have usually used $1–2 \times 10^6$ spheres per injection in rabbits, cats, and monkeys, which will yield

Table 2.1 Approximate number of microspheres (n)[a] trapped in ocular tissues of rabbit, cat, and monkey after injection of 1×10^6 spheres and theoretical coefficient of variation (CV)[b]

Species	Retina n (CV)	Choroid n (CV)	Iris n (CV)	Ciliary body n (CV)
Rabbit	25 (20%)	2,000 (2%)	130 (9%)	165 (8%)
Cat	75 (12%)	2,000 (2%)	165 (8%)	1,100 (3%)
Monkey	45 (15%)	640 (4%)	15 (26%)	115 (9%)

[a]The number of microspheres has been calculated based on the mean of the blood flow values in the references cited in Table 2.2 and assuming cardiac output to be approximately 450 g/min in rabbits [46, 53, 55, 82], 250 g/min in cats [56, 60], and 600 g/min in monkeys [8]
[b]$CV = (\sqrt{n}/n)*100$

less than 400 microspheres in most ocular tissues, except in the choroid (Table 2.1). Table 2.2 shows values on ocular blood flows obtained in different studies published from the Department of Physiology at Uppsala University during more than 25 years. The experiments were made by several different researchers during a long time period, but the methodology has not changed very much. As expected from the low number of microspheres entrapped in the retina and anterior uvea, the coefficient of variation (CV) for these tissues is high in most studies. Furthermore, the observed CVs in the different studies (Table 2.2) are higher than theoretically expected, based solely on the number of microspheres (Table 2.1) for all tissues, including the choroid. Thus, one can assume that factors other than too few microspheres, most likely biological variability (see below), contribute to the error. Using the rat cochlea, another tissue with low blood flow, as a model, Hillerdal et al. showed that the biological variation contributed more to the error than the paucity of microspheres and that the precision could be improved by conducting more experiments [37].

Figure 2.2 is based on data from two studies in rabbits in which there was no difference in treatment between the left and the right eye during three blood flow determinations [54, 58]. The data shows that there is a good correlation between blood flow in the right and the left eye even in tissues with few microspheres, like the rabbit retina (Table 2.3). Regression analysis (Fig. 2.2) shows that the best fitted line has a slope that is not significantly different from 1.0 for the retina and ciliary body, but for the choroid and iris, the slope is significantly less than 1.0, that is, 0.92 ± 0.04 ($P\leq0.001$) and 0.91 ± 0.08 ($P\leq0.001$), respectively. The reason for the slightly lower blood flow on the right side is not clear. It could be due to anatomical differences or the catheter used for the injection of microspheres could have interfered with normal hemodynamics, since in all experiments it was advanced into the left ventricle from the left brachial artery (see below). The difference between the right and left eye blood flow is too small (~10%) to be detected in the separate experimental series [54, 58], however.

Figure 2.3 is based on data from experiments with intracameral injection of test substances in rabbits [54], and shows the relationship between the first, second, and third blood flow determination, for the retina and choroid. The correlation between the three determinations was very good for the choroid and moderate for the retina (Table 2.4). Regression analysis of the retinal data showed that the slope of the regression line was not significantly different from 1.0, when blood flow was compared between the first and second or third determination, but the confidence interval is large (Fig. 2.3). For the choroid, the slopes of the regression lines were significantly lower than 1.0, when the second and third determinations were compared with the first determination (Fig. 2.3). This is more likely to be due to a decrease in the arterial blood pressure (81 ± 8, 78 ± 8, and 76 ± 8 mm during the first, second, and third determinations, respectively) than to disturbances of local blood flow in the choroid (see below).

Thus, despite the low number of microspheres in some ocular tissues, there is a fairly good correlation between eyes and between repeated determinations. Even with relatively few microspheres in the tissue and a limited number of experiments, it has been possible to detect blood flow changes of about 50% in the retina [55, 83] and of about 20% in anterior uvea [53, 85].

Table 2.2 Data on normal ocular blood flows determined with radioactive microspheres[a] in rabbits, cats, and monkeys in papers published from the Department of Physiology, Uppsala University, from 1973 to 2001

			Blood flow (mg/min)[b]				
Species/anesthesia	MABP (mmHg)	*N*[c]	Retina	Choroid	Iris	Ciliary body	Reference
Rabbit	n.d.	13	8±1	899±77	75±7	50±6	[17]
Conscious	85±2	10	15±2	1,000±63	68±8	72±5	[18]
	71±2	10	10±2	1,063±58	39±4	105±10	[45]
	69±2	10	11±2	1,138±68	44±4	89±13	[45]
	90±3	14	n.d.	1,096±163	84±12	93±15	[19]
	n.d.	5	11±3	779±97	65±14	62±12	[88]
	n.d.	7	n.d.	1,014±101	58±6	100±17	[44]
Range of CV[d]			*42–63*	*17–56*	*27–53*	*22–60*	
Rabbit	71±4	10	18±3	1,020±115	94±24	82±7	[57]
Urethane i.v.	75±4	8	n.d.	1,044±117	62±15	97±16	[46]
	n.d.	8	8±2	678±115	45±8	35±6	[59]
	n.d.	8	n.d.	1,131±284	95±23	116±22	[43]
	n.d.	7	n.d.	842±120	58±10	95±11	[43]
	n.d.	6	n.d.	585±126	39±12	52±10	[44]
	n.d.	6	n.d.	609±113	22±7	47±19	[44]
	n.d.	6	n.d.	578±114	35±10	59±18	[44]
	n.d	5	12±6	567±140	37±13	65±17	[88]
	63±3	10	20±5	700±63	60±10	65±9	[58]
	77±4	12	11±2	805±56	59±10	82±8	[53]
	74±8	6	10±4	811±230	38±16	68±25	[82]
	84±5	11	n.d.	1,101±257	39±10	99±23	[83]
	65±5	8	10±2	729±140	56±14	50±11	[55]
	60±5	8	10±1	633±105	72±20	59±11	[55]
	60±4	9	11±2	552±80	40±10	30±6	[55]
Range of CV			*28–112*	*24–77*	*46–103*	*27–99*	
Cat	113±5	12	15±2	1,070±122	60±11	262±30	[4]
Chloralose i.v.[e]	160±5	11	17±2	1,110±120	23±4	275±13	[59]
	134±9	7	25±3	1,037±104	63±30	318±43	[58]
	67±6	6	28±4	444±94	33±13	242±45	[33]
	111±7	6	17±4	874±123	58±12	306±25	[56]
	97±5	7	19±2	916±144	37±9	306±50	[56]
	98±4	8	23±4	798±92	39±9	263±34	[56]
	115±7	7	18±2	916±170	47±15	272±35	[56]
	104±4	9	17±2	671±56	21±3	201±14	[60]
Range of CV			*28–58*	*25–52*	*43–158*	*13–46*	
Monkey	85±5	5	30±2	360±48	5±1	47±4	[84]
Pentobarbital[f]	100±6	12	32±3	505±75	8±1	73±7	[3]
	91±8	7	16±3	348±70	5±1	28±7	[59]
	85±6	5	32±6	391±36	13±3	68±9	[86]
	73±6	6	25±3	327±51	15±2	132±27	[8]
Range of CV			*15–50*	*21–53*	*33–53*	*19–66*	

[a]The number of microspheres given varied between 0.5 and 3×10^6 but was usually $1–2\times10^6$

[b]Values are the mean ± SEM for groups presented in the cited references. If data on blood pressure or flow values for different groups were not presented in the paper, the data were retrieved from original data whenever possible

[c]Number of animals or eyes

[d]For each study/group, CV was calculated as $(\text{SEM}*\sqrt{n}*100)/\text{mean}$

[e]Induction of anesthesia was achieved by inhalation of chloroform [4, 58, 59], i.m. injection of alphaxalone and alphadolone acetate [33] or ketamine and xylazine [33, 56, 60]

[f]Induction of anesthesia was achieved with i.m. injection of methohexital [3, 59, 84, 86] or ketamine [8]

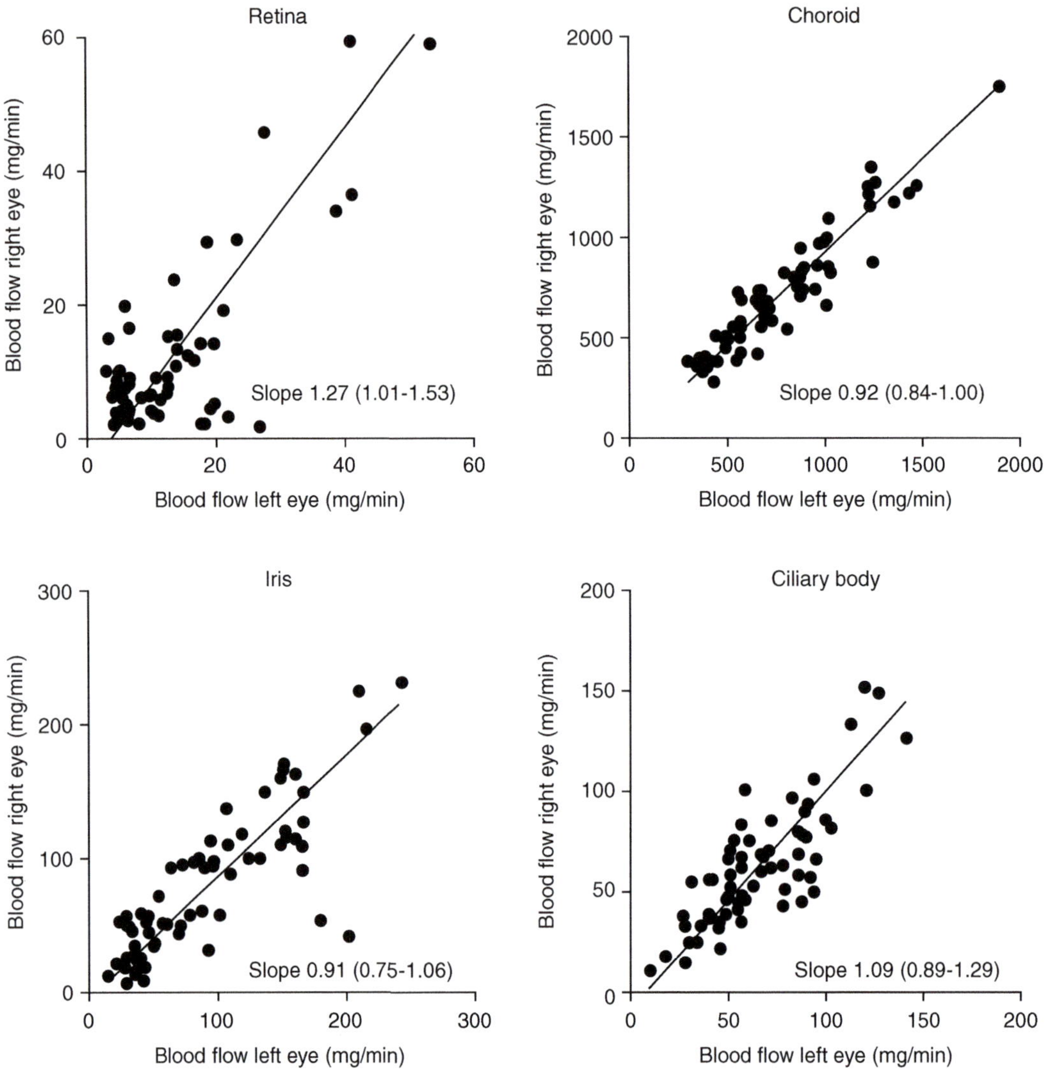

Fig. 2.2 Relationship between blood flow in the right and left eye in rabbits. Data are from 22 animals, with 3 blood flow determinations in each (Plotted from original data [54, 58]). Regression line (Deming regression) with slope and 95% confidence intervals are given

Table 2.3 Correlation coefficients for comparison between ocular blood flows in the right and the left eye in rabbits[a]

Tissue	*r* (CI)[b]	*P* value
Retina	0.77 (0.65–0.86)	$P \leq 0.0001$
Choroid	0.95 (0.91–0.97)	$P \leq 0.0001$
Iris	0.83 (0.73–0.89)	$P \leq 0.0001$
Ciliary body	0.81 (0.71–0.88)	$P \leq 0.0001$

[a]Based on data from experiments in references [54, 58]
[b]Correlation coefficient with the 95% confidence interval

2.4 Methodological Errors and Practical Advice

2.4.1 Size of the Microspheres

Ideally, the number and size of microspheres should be optimized for each tissue and species. As discussed above, a large number of injected microspheres are desirable to get a high level of

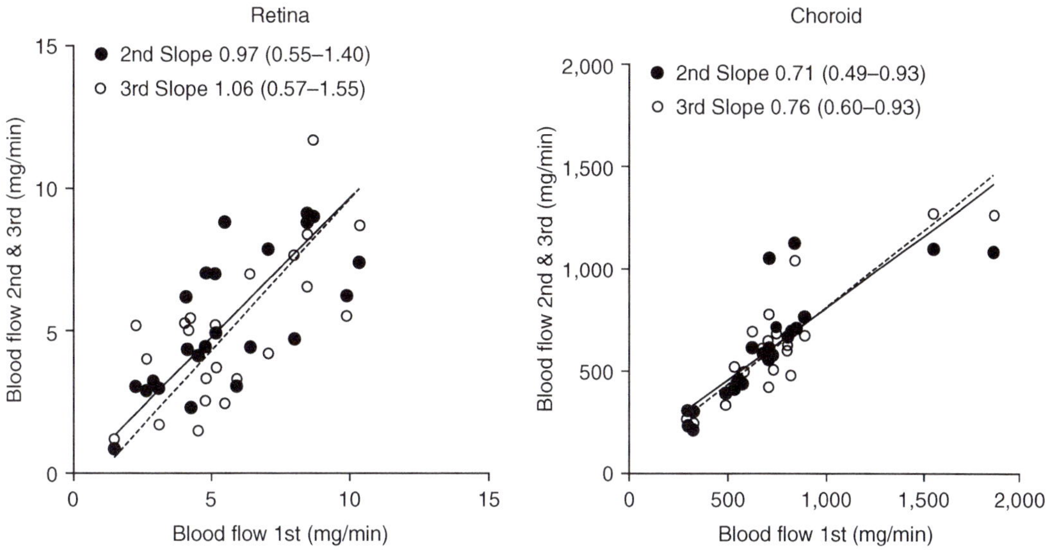

Fig. 2.3 Relationship between first and second blood flow determination (*filled circles, unbroken line*) and between first and third blood flow determination (*open circles, dotted line*) in rabbits. Data are from 12 animals, each contributing 2 eyes each (Plotted from original data [54]). Regression line (Deming regression) with slope and 95% confidence intervals are given

Table 2.4 Correlation coefficients for comparison between the first, second, and third determination of retinal and choroidal blood flow in rabbits[a]

	Retina		Choroid	
Comparison	r (CI)[b]	P value	r (CI)[b]	P value
1st and 2nd	0.72 (0.44–0.87)	$P \leq 0.0001$	0.82 (0.62–0.92)	$P \leq 0.0001$
1st and 3rd	0.70 (0.41–0.86)	$P \leq 0.0002$	0.90 (0.78–0.96)	$P \leq 0.0001$
2nd and 3rd	0.55 (0.16–0.78)	$P \leq 0.0084$	0.92 (0.83–0.97)	$P \leq 0.0001$

[a]Based on data from experiments in references [54, 58]
[b]Correlation coefficient with the 95% confidence interval

precision in the measurements, but the risk of disturbing central and local hemodynamics will increase with the number of microspheres. The likelihood that disturbances of normal hemodynamics will occur is dependent on a combination of the size and number of microspheres and the species under investigation. Smaller animals have a smaller cross-sectional area of their vascular tree, and therefore the risk of disturbing their normal circulation is higher.

Obviously, if the microspheres are too small, they can pass through the tissue and hence cause an underestimation of blood flow. However, it is desirable that the microspheres do not get trapped until they reach the capillary bed of the tissue under investigation. Injection of microspheres that are too large can underestimate flow and disturb local hemodynamics. Ideally, the microspheres should be injected as a bolus injection, but this is not possible as it will increase blood pressure. Thus, a large sphere that gets trapped early in an arteriole could prevent other spheres from entering the capillaries supplied by the arteriole, and hence cause an underestimation of blood flow. Furthermore, as the total cross-sectional area of the arterioles is much smaller than the total cross-sectional area of the capillaries, blood flow to a larger area will become occluded. By occluding many arterioles in many tissues, larger spheres may disturb central hemodynamics by increasing total peripheral resistance and hence blood pressure or the large spheres may impair cardiac blood flow and therefore decrease cardiac performance.

The most commonly used size of microspheres for determination of ocular blood flows is 15 μm, which has been validated in rabbits, cats, dogs, pigs, and monkeys [3, 4, 7, 76, 78, 85]. In cats and monkeys, there is a good correlation between uveal blood flows measured with 15- and 35-μm spheres injected simultaneously [3, 4, 7]. Injection of 8–10-μm and 15-μm spheres in rabbits indicated that the smaller spheres may pass the capillaries in the iris and ciliary body, as the larger spheres gave higher flow values [7, 85]. In dogs, uveal blood flows were not significantly different when measured by 15- and 25-μm spheres, but a small percentage (3%) of the injected 15-μm microspheres was found in the venous effluent from the eye [76]. Total ocular blood flow measured with four different sizes of microspheres (10, 15, 25, and 35 μm) in pigs showed that 10-μm spheres gave significantly lower flow values than 15-μm spheres. The values obtained with the larger spheres were higher but not significantly different from those obtained with the 15-μm spheres [78]. Collection of venous blood from a cannulated vortex vein indicated that more than 99% of 15-μm spheres were trapped in the pig eye [90]. Retinal capillaries are generally smaller in diameter than choroidal capillaries [15, 16, 26, 61–64], which indicate that smaller spheres are to be preferred when the main goal is to study circulation in the retina and optic nerve [31, 42]. Using smaller spheres also makes it possible to inject a larger number of spheres, which will simultaneously improve the reliability of the measurements.

A recent study showed that 10-μm spheres are more suitable than 15-μm spheres for measuring choroidal blood flow in the rat, as many 15-μm spheres were trapped before the choriocapillaries and injection of 10-μm spheres caused a larger number of spheres to be trapped in the choroid [92]. Similarly, as in larger species, smaller spheres (8 μm) were found to be best suited for measurement of blood flow to the retina and optic nerve [92]. Thus, for determination of retinal and choroidal blood flow at the same time in the same animal, one should, if practically possible, simultaneously inject two sizes of microspheres using a higher number of the small-sized spheres [91].

Care regarding the size of the microspheres may be warranted also when one wants to study regional differences within an ocular tissue. In a recent histological study on cat eyes, it was observed that 15-μm spheres did not reach into the tapetum lucidum, whereas in nontapetal choroid, the microspheres were observed in precapillary arterioles just before the choriocapillaries [50]. Furthermore, the diameter of the precapillary arterioles was smaller in the tapetal region (4.7 ± 0.8 μm) than in the nontapetal region (6.2 ± 0.9 μm). Due to these findings, the authors suggested that 15 μm may not be suitable for measuring choroidal blood flow [50]. Although their findings may be of importance when studying regional differences within the choroid, it is most likely not true for determinations of total choroidal blood flow. Larger spheres show axial streaming, that is they move in the center of the vessel. The concentration of microspheres will therefore be lower in the periphery than in the middle of the vessel. This causes disproportionately fewer spheres to enter small branch arteries, which will underestimate flow in the area supplied by the small arteries. Too large spheres may therefore incorrectly show unevenness in the distribution of blood flow within a tissue [36]. Thus, the findings by May and Narfström [50] could be of importance when studying differences in blood flow between tapetal and nontapetal regions of the choroid. However, axial streaming is not likely to contribute to the error when measuring total choroidal blood flow with 15-μm spheres in cats. First, there is a good correlation between choroidal blood flow values obtained with 15- and 35-μm spheres [6, 7], even though axial streaming is more pronounced with larger spheres [36]. Second, it has been shown that flow biasing due to axial streaming is significant only when the feeding vessel to the tissue is just a few times larger than the diameter of microspheres [13]. As the choroidal arteries are much larger than 15 μm, axial streaming ought not to affect the results when total choroidal blood flow is measured.

2.4.2 Physical Characteristics of Microspheres

All microspheres are made of polystyrene, but their physical characteristics differ somewhat, depending on the type of label. RM [36] and NAM [41] are much heavier than blood (density ~1.5 g/ml compared to 1.05 g/ml), whereas the CM and FM have a density close to that of blood [89]. Due to their density and hydrophobic nature, the microspheres will sediment on the bottom of the storage vial and form aggregates. To reduce this problem, the microspheres are suspended in saline with a small amount of surfactant (e.g., Tween) added. Saline with dextran has also been used as a suspending medium to slow the sedimentation of microspheres. Still, the microspheres need to be sonicated and vigorously shaken before being removed from the storage vial, and prevented from sedimenting again before injection. One should be aware that the additives can influence hemodynamics in some species; Tween [51] and dextran [30] have been reported to cause hypotension in dogs and rats, respectively.

In early studies with RM, the variability in the diameter of the microspheres was quite large and not always in agreement with the manufacturers' specifications [36, 52]. Although the variability in size has diminished [71] (an acceptable SD of 0.1 μm for 10- and 15-μm spheres is now common), it is wise to check the size of new batches.

2.4.3 Injection of Microspheres and Collection of Reference Sample

The microspheres should be thoroughly mixed with the blood when they are injected, which is best achieved by injection into the left atrium or ventricle. In earlier studies, we injected the microspheres directly into the left ventricle via a steel cannula connected to a syringe with polyethylene tubing. In larger animals, this procedure demands a thoracotomy and artificial ventilation, whereas in rats, the microspheres can be injected into the left ventricle through the thoracic wall. To be able to measure ocular blood flows in conscious rabbits, we started to do a heart catheterization by introducing a polyethylene tubing from a brachial artery. As this procedure caused less damage to the heart and minimized the risk of injecting the microspheres into the right ventricle, we subsequently used it in anesthetized animals as well. Empirically, we found that a slight bend (approximately 120°) at the tip of the catheter (OD approximately 1 mm) and using the left brachial artery in rabbits and the right brachial artery in monkeys and cats was the easiest way to succeed. To create the bend of the catheter, the tubing was briefly held above the flame of a match or candle (Fig. 2.4). As the catheter was advanced toward the left ventricle, it was gently rotated, and the pulse pressure wave was continuously recorded to register its entry into the ventricle. Before fixing the catheter in place, care was taken to check that an excessive length of catheter had not been advanced into the ventricle, that it did not disturb normal heart rhythm, and that blood could be easily aspirated. After the experiments, the proper location of the catheter was checked by visual inspection. Furthermore, we usually analyzed a piece of lung tissue to reveal possible injection of microspheres into the right ventricle due to damage by the catheter or possible septum defects.

After dispensing from the storage vial, the microspheres were diluted with saline to a volume of approximately 1 ml (for experiments in rabbits, cats, and monkeys) in a test tube. As the microspheres were aspirated into a plastic syringe, a small air bubble was allowed to remain in the syringe. By gently turning the syringe, the air bubble moved up and down and helped in preventing the microspheres from sedimenting in the syringe. After the syringe had been attached to the catheter, blood was withdrawn and mixed with the microspheres by the aid of the air bubble. This procedure is advantageous when one wants to do the blood flow determinations at fixed time points during or after an intervention, e.g., at the start of nerve stimulation or administration of a drug, as it only takes a few seconds before the microspheres start to sediment. Care was taken to hold the syringe in a tip-down position during injection to avoid injecting the air bubble, and after injection,

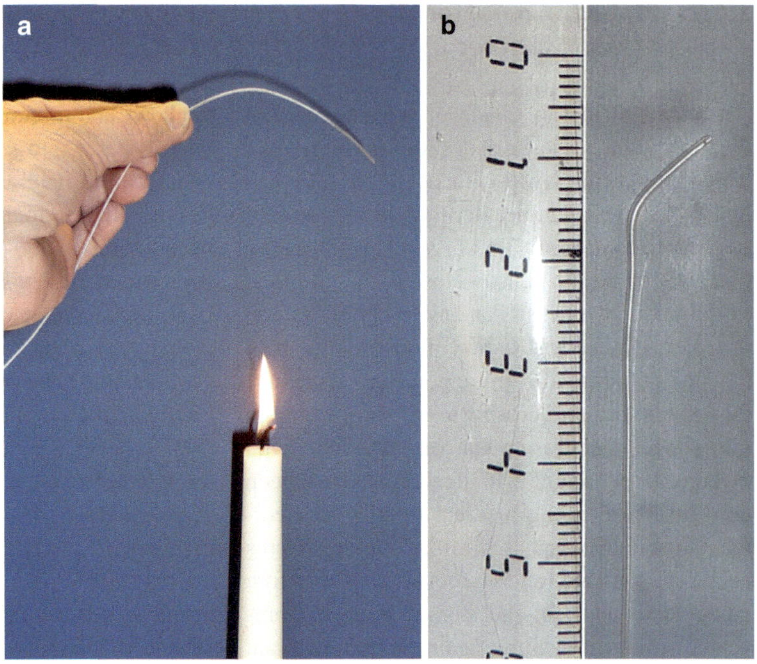

Fig. 2.4 Preparation of a catheter for heart catheterization. By holding the polyethylene tubing above a flame (**a**), a bend of about 120° was created (**b**)

blood was aspirated into the syringe to remove microspheres left in the catheter. If cardiac output was to be measured, triplicate samples (10 μl) were collected from the test tube before aspirating the microspheres into the syringe, and the exact volume in the syringe was noted. After the injection, the content of the syringe was emptied into the same test tube, after the volume had been noted, and triplicate samples were collected. This made it possible to calculate the exact amount of radioactivity injected and hence cardiac output.

Usually, the microspheres were injected during 10–20 s, and the reference sample was collected during 1–2 min. Collection of the reference sample can be done by withdrawal with a syringe pump at a known speed or by free flow from a cannulated artery into preweighed vials. A high reference flow is desired to get a large number of spheres in the reference sample and to make sure that no microspheres remain in the tubing. On the other hand, the reference flow needs to be low enough to avoid disturbing normal hemodynamics. Pump withdrawal has the advantage of making it possible to calculate flows in volume/min, but it involves the transfer of blood from the syringe and rinsing of the syringe. We have been using free flow into preweighed counting vials, collecting blood for 1 min with 10 s for each vial (or 2 min with 20 s for each vial). This procedure has the advantage that in every experiment we could confirm that no microspheres remained in the tubing and we avoided the transfer of the reference sample. Furthermore, with this sampling technique, it is possible to detect recirculation of microspheres. With appropriate-sized tubing, the free flow sample from a femoral artery (rabbits, cats, and monkeys) was usually 1–2 ml, only slightly more than the injected volume and less than 1% of the animal's total blood volume. Usually, the peak radioactivity was seen in tube 1 or 2, with little or no activity in tubes 3–4, and background activity in tubes 5–6.

2.4.4 Dissection

After the experiment, the eyes were enucleated and kept moist until dissection. During dissection, the eyeball was first cleaned of extraocular tissues and was then cut open along the ora serrata when dissecting cat and monkey eyes. The zonular fibers were cut with fine scissors to remove the lens. The iris could then be separated from the ciliary body by gently pulling with a pair of small forceps or by cutting. The ciliary body was scraped off the sclera with a pair of forceps. The posterior part of the eye was divided into four quadrants by cutting from the ora serrata to the optic nerve head. After removal of the vitreous, the retina was gently removed from the choroid, which was subsequently scraped off the sclera. In rabbits, a slightly different technique was used due to anatomical differences. For example, the ciliary muscle is poorly developed, and the ciliary processes extend almost all the way to pupillary margins. Usually, the optic nerve head was removed, and the eyeball was cut open from the posterior pole to the ora serrata by four incisions. The lens is also not as firmly attached in rabbit as in monkeys and cats, which makes it possible to remove it by just a pair of forceps. The vitreous and retina were removed and the iris-ciliary body pulled off the sclera. The ciliary processes were then scraped off the back of the iris with a pair of forceps. In albino rabbits, the iris is almost white, whereas the ciliary processes are blood filled, which makes it relatively easy to see that the processes have been completely removed. The choroid was scraped off the sclera.

As we weighed our blood samples and it is sometimes difficult to completely remove the vitreous from the retina and ciliary body, we usually expressed ocular blood flows as total flow in mg/min, whereas many other investigators use ml/min/g tissue. The disadvantage with total flow is of course that the tissues have to be quantitatively sampled. However, this is not a big problem, and it may even be easier to do this if one lets some of the vitreous remain attached to the retina. After dissection, the tissue samples as well as the blood samples were centrifuged to make sure that the samples were in the bottom of the test tube before counting (see below).

2.4.5 Detection of RM and NAM

Although NAM have several advantages over RM, as concerns occupational and environmental hazards, the mode of their detection is the same, gamma irradiation, and hence the potential errors in their detection are similar. Radioactive disintegrations are stochastic events; that is, measuring the same sample repeatedly will result in a Poisson distribution around a mean value. The contribution of the stochastic nature of radioactivity to the total error is likely to be low, however [11].

Other possible sources of error are dependent on the gamma detector and include errors due to the separation of the energy spectrum of different radionuclides, coincidence, and geometry [11, 27, 36]. The energy spectrum for different radionuclides almost always overlap to some extent, which makes it necessary to correct for the spillover between different counting windows when using more than one nuclide in an experiment. Several different correction methods have been used to correct for the spillover between different radionuclides [12, 36, 80, 95]. With increasing number of radionuclides, the calculations become increasingly extensive. One can imagine that this was a large obstacle when the microsphere method was first introduced, but with modern gamma detectors, including software that automatically corrects for spillover, this ought not to be any major problem for the experimenter. If the radioactivity in a sample is very high, the true radioactivity of the sample may be underestimated due to coincidence; that is, if two disintegrations occur at the same time or very close in time, the detector will only register one event. Considering the low blood flow and/or small tissue pieces that can be obtained from ocular tissues, this is not likely to occur in conjunction with determination of ocular blood flows, unless the radioactivity in the blood sample is very high. Nor is the geometry likely to be of any great concern when determining ocular blood flows. As the radioactive microspheres have a higher density than blood, they will rapidly sediment on the bottom of the test tube that contains the reference blood sample. Thus, during counting, the microspheres will be close to the bottom of the detector. The microspheres in tissues samples will be dispersed through the tissue, however, meaning they will be at different heights in the detector. As most detectors are constructed as wells with a height of a few centimeters and the sensor at the bottom, disintegrations occurring higher up in the well have a lower probability to be detected. A difference in height of 2 cm between blood sample and tissues sample has been shown to cause up to a 13% systematic error in blood flow calculations [27]. Considering the small size of the tissue samples, geometry should normally not be a significant source of error when measuring ocular blood flow, as long as one checks that the samples are at the bottom of the counting vial. Another prerequisite is of course that the automatic movement of samples in and out of the detector functions properly.

2.4.6 Detection of CM and FM

As there are several different ways to quantify the colored and fluorescent microspheres, the potential sources of error vary with the mode of detection. The spheres can be counted in histological preparations manually or by the aid of different imaging systems. After tissue digestion, the spheres can be counted in aqueous aliquots by a hemocytometer or the spheres can be dissolved to release the dye, which then is quantified by spectrophotometry or fluorospectrophotometry.

Ocular blood flows have been studied by counting of microspheres in histological sections of the optic nerve [31, 42, 81, 92] or flat mounts of the retina and choroid [2, 65, 92]. In these types of experiments, different counting errors may contribute to the variability. Spheres that are close together may be counted as one sphere or fail to be detected by the imaging system [65]. In histological sections, spheres may be counted twice if they are cut in half by the sectioning.

If CM or FM are counted in aqueous aliquots or determined by photometry (both methods have been used to quantify ocular blood flows [25, 38, 68]), the tissues must first be digested and the microspheres isolated. For photometry, the dye is subsequently extracted from the microspheres with an organic solvent. During this processing of the tissues, some microspheres may be lost, which could increase the variability in the measured blood flow values. The ease by which different tissues can be digested varies, depending on how dense the tissue is and the fat content of the tissue. In lung tissue, the organic solvent can directly dissolve the dye from the microspheres, without prior digestion of the samples [32]. Other tissues have to be digested in sodium hydroxide [40, 68], potassium hydroxide [1, 9, 24, 32, 35, 47, 87, 94], or ethanolic KOH [70, 89]. Tissues with a high fat content are more difficult to digest, but the digestion can be improved by letting the samples autolyze for 1–2 weeks [70, 89]. Two different methods have been used for the separation of microspheres from the digested tissue: negative pressure filtration [24, 32, 47] and sedimentation [70, 89]. As the former method involves transfer of the sample from one vial to another, it is recommended to add an internal standard to the samples to calculate recovery [24]. The latter method is based on the difference in density between the FM and the ethanolic KOH (1.05 compared to 0.893 g/cm^3). As the FM are heavier than the solvent, they will easily sediment by centrifugation [89]. The supernatant can then be aspirated and the FM dissolved in the same vial [70, 89]. Van Oosterhout et al. [89] achieved almost 100% recovery with the sedimentation method.

In analogy with the overlap of energy spectrum for different radionuclides, there is an overlap in the wavelength of different colors that has to be corrected for when using CM and FM. The spectral overlap is less with FM [32] than with CM [47], and the spillover only occurs between adjacent colors. FM labeled with up to six different colors can be separated without spillover correction [32]; with spillover correction, up to 13 colors have been separated [79]. Colors in the blue and violet regions of the spectrum should be avoided if possible, however [79]. Autofluorescence from the solvent and compounds released from the tissue increase background in the blue region, which increases variability in the calculated blood flow values [79, 89]. Scarlet and violet FM have minor secondary peaks that overlap with the blue spectra [79]. The spillover is also dependent on the intensity of the fluorescence; if two adjacent colors have a big difference in intensity, it can increase spillover [89]. Furthermore, one must establish that there is a linear relationship between the dye concentration (number of microspheres) and the fluorescence, as at higher dye concentrations the relationship becomes curvilinear due to quenching [32]. This must be done for each new batch of fluorescent microspheres, as the dye load may vary from one batch to another, and hence, the relative fluorescence intensity of the different colors and the spillover may vary. The quenching is not merely dependent on the concentration of the dye that is being measured at a given wavelength; increasing the number of colors that are used increases the quenching and decreases the linear range [79]. Other potential sources of errors in conjunction with fluorescence measurements are dilution errors, unclean or unmatched cuvettes, and machine variability.

2.5 Biological Variation

2.5.1 Blood Pressure

Despite the same type of anesthesia, the mean arterial blood pressure (MABP) varies considerably between different studies. Although all data presented in Table 2.2 were determined under control conditions, there are several differences that could have influenced the blood pressure: (1) the extent of surgical intervention before the first blood flow determination varied, which may have influenced the depth of anesthesia needed, (2) in some studies, muscle relaxant drugs were used, and the type of muscle relaxant varied between studies, (3) differences in arterial blood gases, and (4) other factors such as age, gender, and subclinical diseases could have contributed.

The data from urethane-anesthetized rabbits show that there is a significant correlation between MABP and blood flow in the choroid and ciliary body, that is, correlation coefficients of 0.85 (95% confidence interval 0.42–0.97; $P \leq 0.01$) and 0.85 (95% confidence interval 0.43–0.97; $P \leq 0.01$), respectively. No correlation was found between MABP and retinal or iridal blood flow (Fig. 2.5). For the studies done in conscious rabbits, there was no significant correlation between arterial blood pressure and ocular blood flows (data not shown). The reason for this discrepancy between urethane-anesthetized and conscious rabbits is not clear, but it could be due to the relatively few studies in conscious rabbits for which blood pressure data could be retrieved. Another possible explanation could be that in conscious animals, a high blood pressure (possibly due to a higher level of stress) could be a sign of high sympathetic activity, which will prevent choroidal blood flow from rising, whereas retinal blood flow is not expected to correlate with blood pressure within a normal range due to autoregulation [22].

The data from studies with cats show a significant correlation between MABP and choroidal blood flow with a correlation coefficient of 0.84 (95% confidence interval 0.41–0.97; $P \leq 0.01$) (Fig. 2.6). For monkeys, the relationship between MABP and choroidal blood flow was of borderline statistical significance only, with a correlation coefficient of 0.82 (95% confidence interval −0.24–0.99; $P = 0.09$), possibly due to the few studies (Fig. 2.6). There was no significant correlation between blood pressure and blood flow in the retina or in the anterior uvea, neither in data from studies in cats nor in data from studies in monkeys.

Thus, examining the combined data from several microsphere studies reveals the same pattern as had previously been observed in single studies [4, 5]; choroidal blood flow changed passively with the arterial blood pressure in anesthetized animals, whereas blood flow in the anterior uvea and retina was less pressure sensitive, indicating true autoregulation. Hence, some of the variability in choroidal blood flow, within as well as between studies (Table 2.2), may be due to differences in arterial blood pressure. Furthermore, these differences between the ocular tissues have to be considered when one evaluates blood flow data and there are differences in arterial blood pressure.

2.5.2 Influence of Anesthesia

General anesthesia may affect ocular blood flows in several ways: (1) depression of the cardiovascular center in the medulla oblongata or (2) general vasodilatation can decrease the arterial blood pressure and hence decrease blood flow in the choroid and anterior uvea. Some anesthesia may affect ocular blood flows (3) by direct effects on the vascular smooth muscles or (4) by changing

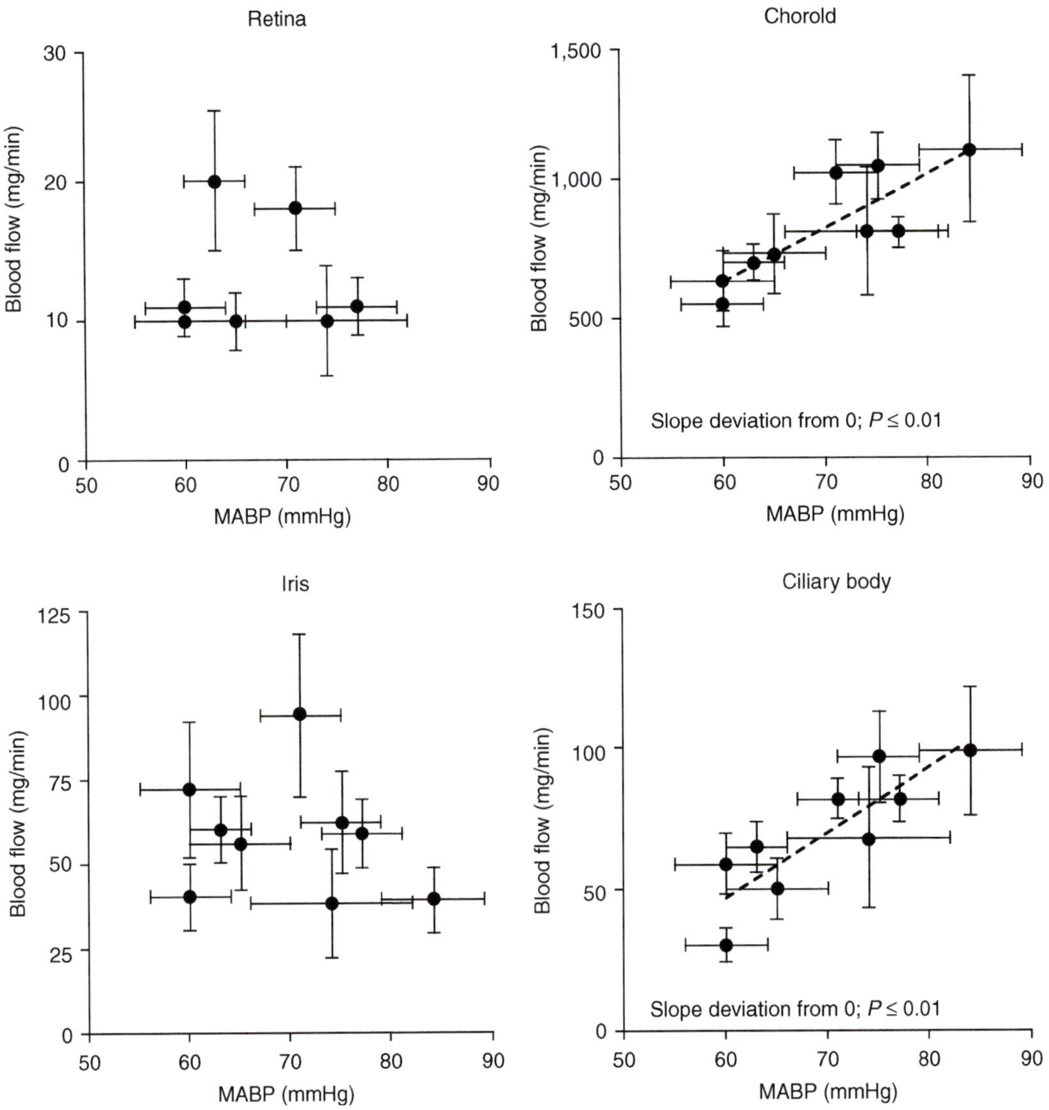

Fig. 2.5 Ocular blood flows in urethane-anesthetized rabbits, as a function of mean arterial blood pressure (MABP) (Plotted from data presented in Table 2.2). *Dotted line* is the linear regression line

neuronal or hormonal control of the vascular tone. For instance, pentobarbital anesthesia causes vasodilatation in the anterior uvea of the rabbit by abolishing a cholinergic vasoconstrictor tone [21]. Much of the rabbit ocular blood flow data that we have obtained are from rabbits anesthetized with urethane. Under this type of anesthesia, there is a slight sympathetic tone to the uveal blood vessels; unilateral sectioning of the cervical sympathetic nerve increases uveal blood flow on the sectioned side [46, 53]. This can be advantageous when one wants to study vasodilatory agents or to determine if drugs or test substances exert their effects via the sympathetic nervous system. However, urethane is classified as carcinogenic, and its use demands special precautions (a special permit is needed in Sweden).

In our search for a suitable replacement for urethane, we tested different inhalation anesthetics. In these experiments, anesthesia was induced by i.v. infusion of a mixture of ketamine (5 mg kg^{-1}) and xylazine (2 mg kg^{-1}). A tracheotomy was

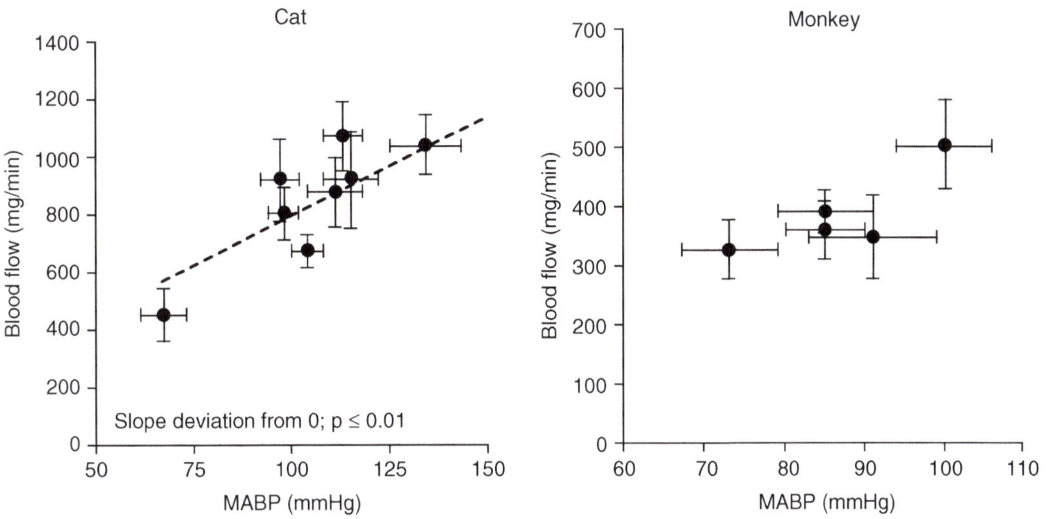

Fig. 2.6 Choroidal blood flows as a function of mean arterial blood pressure (MABP) in cats and monkeys (Plotted from data presented in Table 2.2). *Dotted line* is the linear regression line

made to insert a tracheal tube, and the cervical sympathetic nerve was cut on one side. The rabbit was then artificially ventilated with an Anesthesia WorkStation (AWS, Hallowell EMC, Pittsfield, MA) connected to a vaporizer for administration of the anesthetic. The AWS has a carbon dioxide absorber, as it is constructed as a closed system and intended for using oxygen as the carrier gas. As we wanted to use air as the carrier gas (to keep arterial PO_2 at a normal level), we had to remove the absorber and use a flow of air of 1.5–2.5 $l\,min^{-1}$ to maintain arterial blood gases at normal levels. The respiration frequency was 55–65 breaths min^{-1} and the tidal volume 15–20 ml, adjusted to keep the positive airway pressure at about 10 cmH_2O. The concentration of anesthetic gas was then adjusted to maintain a surgical plane of anesthesia during the remaining part of the preparation: insertion of two arterial catheters for blood pressure registration and reference blood sampling, a venous catheter, and heart catheterization as described above. The depth of anesthesia was judged by checking the reaction to a pinch in the ear, the blinking reflex, and the corneal reflex. During surgery, it was necessary to increase the concentration of anesthetic until the corneal reflex disappeared to achieve surgical anesthesia.

Initial experiments with halothane and isoflurane revealed that when the effect of the initial dose of xylazine wore off, the rabbits tried to breathe spontaneously, making it difficult to adjust the acid–base balance. This could be prevented by a slow i.v. infusion of xylazine (a bolus dose of 0.2 mg/kg followed by continuous infusion of 0.05 mg kg^{-1} min^{-1}), which therefore was used subsequently. Xylazine is an adrenergic α_2-agonist with sedative and muscle relaxant properties. Thus, in addition to causing the necessary muscle relaxant effect to allow artificial ventilation, xylazine caused a slight rise in MABP. Before the blood flow determinations, the concentration of anesthetic was reduced until a weak corneal reflex was observed, which also increased MABP slightly. Unlike halothane and isoflurane, enflurane could be used without the addition of xylazine and without causing difficulties with the ventilation. In the experiments with enflurane, ocular blood flows were therefore determined during enflurane only and at 20 and 50 min after the start of xylazine infusion. Sevoflurane could not be used without the addition of xylazine.

In the experiments with enflurane, the infusion of xylazine increased MABP slightly (47 ± 2, 51 ± 5, and 49 ± 5 mmHg during the three blood flow determinations, respectively), whereas uveal

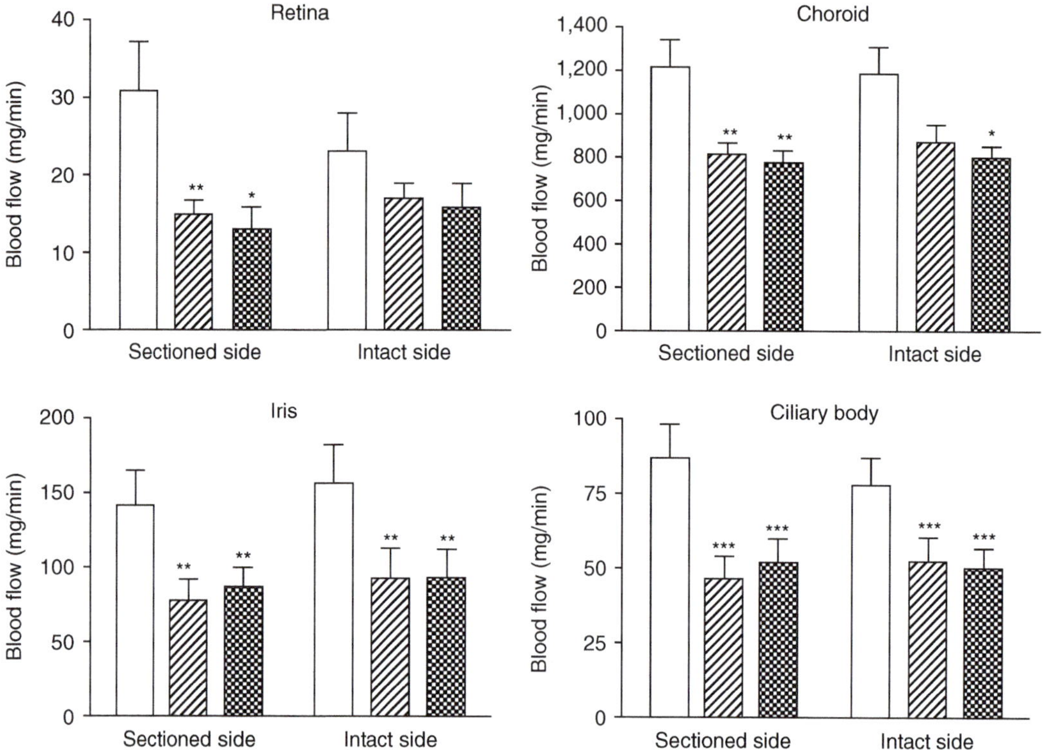

Fig. 2.7 Ocular blood flows during enflurane anesthesia alone and in combination with xylazine (0.2 mg/kg as a bolus dose followed by continuous infusion of 0.05 mg kg^{-1} min^{-1}) in rabbits. Blood flow determinations were made during enflurane alone (*open bars*) and 20 (*hatched bars*) and 50 min (*shaded bars*) after the start of xylazine infusion. Mean values ± SEM are shown ($n = 7$). Statistical comparison was made with one-way ANOVA, with Tukey's test as post-ANOVA test. $^{*}P \leq 0.05$, $^{**}P \leq 0.01$, and $^{***}P \leq 0.001$

blood flows decreased (Fig. 2.7). Retinal blood flow was also decreased by xylazine, but the effect was only statistically significant for the side with the sectioned sympathetic nerve (Fig. 2.7). Thus, xylazine appeared to cause vasoconstriction in the retina as well as in the uvea.

Comparison between the different inhalation anesthetics (with the addition of xylazine) did not reveal any significant differences, except for a significantly higher blood flow in the anterior uvea during halothane anesthesia. There was no significant difference between the eye with the sectioned sympathetic nerve supply and the control eye. Nor was uveal blood flow much different from blood flow during urethane anesthesia when compared to a previous study [53] (Table 2.5). However, during inhalation anesthesia, the MABP was lower (Table 2.5) than in most of the studies with urethane (Table 2.2), and choroidal blood flow was higher (Table 2.5) than expected from the MABP (Table 2.2 and Fig. 2.5). Calculations of uveal vascular resistances showed that the choroidal vascular resistance was significantly lower during halothane, enflurane, and sevoflurane anesthesia than during urethane anesthesia (Fig. 2.8). Vascular resistance in the iris was also lower during inhalation anesthesia, but only halothane was significantly different compared to urethane (Fig. 2.8). Vascular resistance in the ciliary body was not significantly different during inhalation anesthesia compared to urethane (Fig. 2.8). Retinal blood flow tended to be slightly higher (Table 2.5) and retinal vascular resistance lower (data not shown) with halothane, enflurane, and sevoflurane than with isoflurane and urethane.

At least two factors could have contributed to the lower vascular tone in the uvea with the inha-

Table 2.5 Systemic cardiovascular parameters, arterial blood gases, and ocular blood flows during different inhalation anesthesia combined with i.v. infusion of xylazine (0.2 mg/kg as a bolus dose followed by continuous infusion of 0.05 mg kg–1 min–1) in rabbits

	Halothane (1–1.5%)[b] $n=6$	Enflurane (2–3.5%)[b] $n=7$	Isoflurane (1–2%)[b] $n=7$	Sevoflurane (1.6–3.5%)[b] $n=7$	Urethane[a] $n=12$
Cardiovascular parameters					
MABP (mmHg)[c]	56±3**	51±5**	61±4*	64±3	77±4
Heart rate (min^{-1})	189±10**	182±4**	190±10**	161±7**	312±8
Cardiac output (g min^{-1})	457±53	341±26	267±20**	438±36	415±29
TPR (100 mmHg min g^{-1})[d]	13±2*	15±1	24±3	15±1	19±1
Arterial blood gases					
pH	7.42±0.01	7.43±0.01	7.48±0.01	7.48±0.02	7.47±0.01
PCO_2 (kPa)	5.7±0.1**	5.5±0.1**	5.5±0.1**	5.5±0.1**	4.6±0.1
PO_2 (kPa)	10.9±0.7	11.2±0.3	13.2±0.3	11.7±0.5	11.9±0.5
Ocular blood flows[e]					
Retina I	14±1	17±2	11±2	15±3	11±2
Retina S	15±2	15±2	10±2	12±1	13±2
Choroid I	1,129±103	863±92	783±86	1,054±153	805±65
Choroid S	1,127±118	814±56	836±86	1,021±105	951±79[f]
Iris I	136±12*	93±19	52±10	86±10	59±10
Iris S	152±19**	77±15	59±12	86±7	71±12
Ciliary body I	75±12	52±8*	42±2*	51±8*	82±8
Ciliary body S	82±16	47±7**	41±5**	55±9**	105±12[f]

$^*P \le 0.05$, $^{**}P \le 0.01$
[a]Comparison was made with urethane anesthesia (one-way ANOVA, with Dunnett's test as post-ANOVA test); urethane data is from a previously published study [53]
[b]Inhaled concentration that produced a weak corneal reflex
[c]Mean arterial blood pressure
[d]Total peripheral resistance
[e]*I* = intact sympathetic nerve supply, *S* = cervical sympathetic nerve sectioned
[f]Denote a significant difference ($P \le 0.05$) between the two sides (two-tailed students' *t*-test; paired data)

lation anesthetics. A direct vasodilatory effect on the uveal blood vessels is one possibility, but decreased vascular tone due higher PCO_2 in the experiments with inhalation anesthetics (Table 2.5) could also have contributed (see below). Furthermore, the vasoconstrictive effect of xylazine in the enflurane experiments indicates that the vasodilatory effect of the inhalation anesthetics could be even larger.

In the cat, increased inhaled concentrations of enflurane (0.5, 1.0 and 1.5 MAC) caused a concentration-dependent decrease in MABP and choroidal vascular resistance and a significant increase in choroidal blood flow by the highest concentration. Retinal blood flow was not significantly affected, but retinal vascular resistance fell with increasing concentrations [75]. In similar experiments with halothane, MABP and choroidal blood flow decreased with increasing concentration, whereas choroidal vascular resistance was not significantly affected. Retinal blood flow increased and retinal vascular resistance decreased with increased concentration [74]. These data indicate that the two inhalation anesthetics may have different effects on choroidal and retinal vasculature in the cat. Enflurane appears to have a direct vasodilatory effect on choroidal vessels but no or a much smaller effect on retinal vessels. With halothane, the effects seem to be reversed; that is, retinal vessels are dilated, but choroidal vessels are not significantly affected. In our experiments, choroidal vascular resistance was lower than with urethane for both enflurane and halothane, indicating that both anesthetics dilated

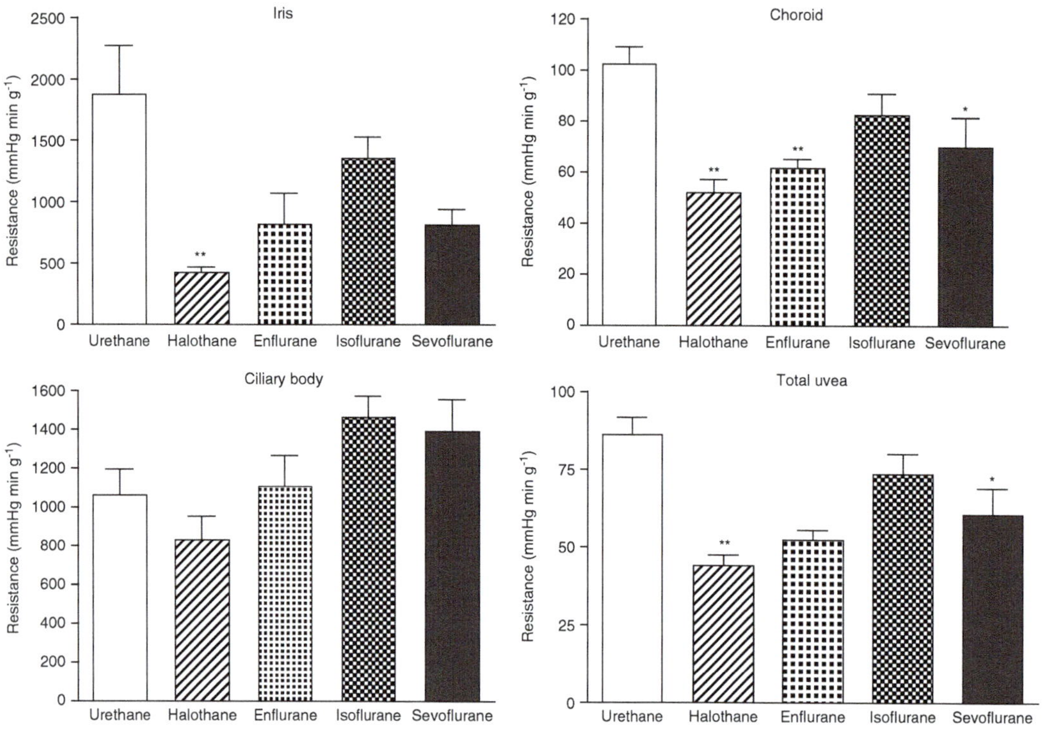

Fig. 2.8 Uveal vascular resistances under different types of inhalation anesthesia in combination with i.v. infusion of xylazine (0.2 mg/kg as a bolus dose followed by continuous infusion of 0.05 mg kg^{-1} min^{-1}) in rabbits. Mean values ± SEM are shown ($n=6$ for halothane; $n=7$ for enflurane, isoflurane, and sevoflurane; and $n=12$ for urethane). Comparison was made with urethane anesthesia (one-way ANOVA with Dunnett's test as post-ANOVA test) (Data from a previously published study [53])

choroidal blood vessels. Thus, it seems the effect of enflurane on choroidal vasculature may differ between rabbits and cats. Furthermore, the high retinal blood flow during enflurane only and the decrease in blood flow caused by xylazine (Fig. 2.7.) indicate that enflurane has a vasodilatory effect on retinal vessels in the rabbit. The inhaled concentrations of enflurane and halothane in our experiments correspond to 0.7–1.2 MAC [28], which is in the same range as used by Roth in cats [74, 75].

2.5.3 Arterial Blood Gases

Differences in arterial oxygen and carbon dioxide tension are other factors that could contribute to the biological variation. In one of the first microsphere studies on ocular blood flow, it was shown that hypercapnia increases retinal as well as uveal blood flows in cats [4]. By the use of the microsphere technique, choroidal vasodilatation during hypercapnia has also been observed in the rabbit [14]. A recent study in the rat, however, indicates that the response is different in this species, as hypercapnia had no effect on choroidal blood flow, but hypocapnia decreased blood flow [91]. Retinal blood flow was decreased by hypocapnia and increased by hypercapnia in the rat [91]. It is well known that the arterial oxygen tension significantly affects retinal vascular tone; hypoxia causing vasodilatation and hyperoxia, vasoconstriction. The effect on uveal vessels is less or absent, however [26]. The number of studies done with microspheres is few, but it has been shown that retinal blood flow is increased by arterial hypoxia in the cat [2] and that the effects of hypoxia on uveal blood flows are minor in the rabbit [48].

It may be worth noting that different anesthesia may affect acid–base balance differently. In our experiments done under urethane anesthesia in rabbits, base excess (BE) was mostly negative, indicating metabolic acidosis. A sodium bicarbonate solution was routinely given to adjust BE and pH to normal values. In the experiments with inhalation anesthetics, BE was mostly positive, indicating metabolic alkalosis. Hence, the arterial PCO_2 was allowed to be slightly higher than normal (Table 2.5), to keep arterial pH closer to normal.

2.5.4 Other Possible Causes for Biological Variability

Age, gender, seasonal and hormonal variations, and diseases are other factors that could contribute to biological variability, but the influence of these factors on the measurements of ocular blood flows are largely unknown. In most of the studies from the Department of Physiology, Uppsala University (Table 2.2), both sexes of animals were used, which could have contributed to the variability. As regards age, the rabbits were mostly about the same age, as they were ordered to be a certain weight/age from special breeders. For cats and monkeys, there could be a much larger variation in age within and between different studies. At the time of the earlier studies, there was no law on approved breeders of laboratory animals in Sweden. As cats therefore could be obtained from several different breeders and even private persons, their age was largely unknown. Similarly, the monkeys were wild-caught animals of unknown age. There is one study for which we know the exact age of every animal though. In this study, ocular blood flows determined in Abyssinian cats with hereditary retinal degeneration were compared with blood flows in normal cats. As the disease progresses slowly, we tried to find age-matched controls. The results showed that uveal blood flows were negatively correlated with age in the Abyssinian cats, most likely as a result of the progression of the disease, but not in the controls. However, the age range was narrower (0.5–4.8 years compared to 0.9–8.1 years) for the controls and not as evenly spread [60]. Thus, a decrease with age in uveal blood flow in normal cats cannot be completely excluded. A decrease in pulsatile ocular blood flow by age has been observed in humans [72], and choroidal blood flow, measured by Laser Doppler flowmetry, is decreased with age in pigeons [29].

Diseases that had not yet given any clinical signs of illness in the animal could also contribute to the biological variation. Over the years, it happened on two or three occasions that we had to stop doing experiments in rabbit, due to respiratory diseases in the animals. The rabbits seemed healthy, when sitting in their cages, but during the experiments, it was impossible to adjust their acid–base balance to normal values. Sampling of blood from an ear artery, in the conscious animals, revealed respiratory insufficiency.

2.6 Summary for the Clinician

The microsphere method is a reliable method for determination of ocular blood flows in experimental animals. The advantages with the method are: (1) it measures blood flow directly, (2) it is suitable for measuring blood flow in small pieces of tissue and in inaccessible tissues, and (3) it does not disturb the normal circulation, if the experiments are properly designed. Radioactive microspheres have been the most commonly used type of microsphere for ocular blood flow determinations as well as for regional blood flow determinations in general. Due to the health and environmental issues connected with the use of radioactive material, the use of radioactive microspheres has declined during the last decade. Colored and fluorescent microspheres have to some extent replaced the radioactive ones, but these are more labor intensive. A promising new development is the use of neutron-activated microspheres, which have the same advantages as radioactive microspheres but lack the disadvantages.

For reliable measurements with the microsphere method, the size and number of microspheres should be optimized for the tissue under investigation. In larger species (e.g., rabbit, cat, monkey), the most commonly used size and injection number are 15 μm and $1–2 \times 10^6$, which is a good compromise if one wants to measure uveal as well as retinal blood flows and cannot use more than one size in an experiment. If retinal blood flow and/or blood flow in the optic nerve head are the primary interest, smaller spheres (10 μm) and a larger number of spheres are preferred. In smaller species (e.g., rat), spheres with a diameter of 10 μm are recommended for measurement of choroidal blood flow and 8-μm spheres for retina and optic nerve. The "400 spheres/piece of tissue rule" is difficult to achieve for the retina and anterior uvea, but doing more experiments can compensate for the paucity of microspheres in the tissues. Furthermore, biological variation contributes more to the error in the measurements than the paucity of microspheres. Differences in arterial blood pressure, anesthesia, and arterial blood gases contribute to the variation of reported values on ocular blood flows. Other factors such as diseases, age, gender, and hormonal and seasonal variations could contribute to the variation as well.

References

1. Abel FL, Cooper RH, Beck RR (1993) Use of fluorescent latex microspheres to measure coronary blood flow distribution. Circ Shock 41:156–161
2. Ahmed J, Pulfer MK, Linsenmeier RA (2001) Measurement of blood flow through the retinal circulation of the cat during normoxia and hypoxemia using fluorescent microspheres. Microvasc Res 62: 143–153
3. Alm A (1980) The effect of topical l-epinephrine on regional ocular blood flow in monkeys. Invest Ophthalmol Vis Sci 19:487–491
4. Alm A, Bill A (1972) The oxygen supply to the retina. II. Effects of high intraocular pressure and of increased arterial carbon dioxide tension on uveal and retinal blood flow in cats. A study with radioactively labelled microspheres including flow determinations in brain and some other tissues. Acta Physiol Scand 84:306–319
5. Alm A, Bill A (1973) Ocular and optic nerve blood flow at normal and increased intraocular pressures in monkeys (*Macaca irus*): a study with radioactively labelled microspheres including flow determinations in brain and some other tissues. Exp Eye Res 15:15–29
6. Alm A, Bill A (1973) The effect of stimulation of the cervical sympathetic chain on retinal oxygen tension and on uveal, retinal and cerebral blood flow in cats. Acta Physiol Scand 88:84–94
7. Alm A, Tornquist P, Stjernschantz J (1977) Radioactively labelled microspheres in regional ocular blood flow determinations. Bibl Anat:24–29
8. Alm A, Lambrou GN, Maepea O et al (1997) Ocular blood flow in experimental glaucoma: a study in cynomolgus monkeys. Ophthalmologica 211:178–182
9. Anetzberger H, Thein E, Walli AK et al (2003) Determination of regional bone blood flow by means of fluorescent microspheres using an automated sample-processing procedure. Eur Surg Res 35:337–345
10. Angelborg C, Slepecky N, Larsen HC et al (1987) Colored microspheres for blood flow determinations twice in the same animal. Hear Res 27:265–269
11. Austin RE Jr, Hauck WW, Aldea GS et al (1989) Quantitating error in blood flow measurements with radioactive microspheres. Am J Physiol 257: H280–H288
12. Baer RW, Payne BD, Verrier ED et al (1984) Increased number of myocardial blood flow measurements with radionuclide-labeled microspheres. Am J Physiol 246:H418–H434
13. Bassingthwaighte JB, Malone MA, Moffett TC et al (1987) Validity of microsphere depositions for regional myocardial flows. Am J Physiol 253:H184–H193
14. Beausang-Linder M (1982) Effects of sympathetic stimulation on cerebral and ocular blood flow. Modification by hypertension, hypercapnia, acetazolamide, PGI2 and papaverine. Acta Physiol Scand 114:217–224
15. Bhutto IA, Amemiya T (1995) Corrosion cast demonstration of retinal vasculature of normal Wistar-Kyoto rats. Acta Anat (Basel) 153:290–300
16. Bhutto IA, Amemiya T (2001) Microvascular architecture of the rat choroid: corrosion cast study. Anat Rec 264:63–71
17. Bill A (1974) Effects of acetazolamide and carotid occlusion on the ocular blood flow in unanesthetized rabbits. Invest Ophthalmol 13:954–958
18. Bill A (1979) Effects of indomethacin on regional blood flow in conscious rabbits – a microsphere study. Acta Physiol Scand 105:437–442
19. Bill A (1984) Effect of acute hemorrhaging in rabbits on blood circulation in the eye and various other tissues. The role of the sympathetic nerves. Klin Monatsbl Augenheilkd 184:305–307
20. Bill A, Nilsson SFE (1985) Control of ocular blood flow. J Cardiovasc Pharmacol 7(Suppl 3):S96–S102
21. Bill A, Stjernschantz J (1980) Cholinergic vasoconstrictor effects in the rabbit eye: vasomotor effects of pentobarbital anesthesia. Acta Physiol Scand 108:419–424
22. Bill A, Linder M, Linder J (1977) The protective role of ocular sympathetic vasomotor nerves in acute arterial hypertension. Bibl Anat:30–35

23. Buckberg GD, Luck JC, Payne DB et al (1971) Some sources of error in measuring regional blood flow with radioactive microspheres. J Appl Physiol 31:598–604
24. Chien GL, Anselone CG, Davis RF et al (1995) Fluorescent vs. radioactive microsphere measurement of regional myocardial blood flow. Cardiovasc Res 30:405–412
25. Chiou GC, Zhao F, Shen ZF et al (1990) Effects of D-timolol and L-timolol on ocular blood flow and intraocular pressure. J Ocul Pharmacol 6:23–30
26. Cioffi GA, Granstam E, Alm A (2003) Ocular circulation. In: Kaufman PL, Alm A (eds) Adler's physiology of the eye, clinical application. Mosby, St. Louis, pp 747–784
27. Dole WP, Jackson DL, Rosenblatt JI et al (1982) Relative error and variability in blood flow measurements with radiolabeled microspheres. Am J Physiol 243:H371–H378
28. Drummond JC (1985) MAC for halothane, enflurane, and isoflurane in the New Zealand white rabbit: and a test for the validity of MAC determinations. Anesthesiology 62:336–338
29. Fitzgerald ME, Tolley E, Jackson B et al (2005) Anatomical and functional evidence for progressive age-related decline in parasympathetic control of choroidal blood flow in pigeons. Exp Eye Res 81:478–491
30. Flaim SF, Morris ZQ, Kennedy TJ (1978) Dextran as a radioactive microsphere suspending agent: severe hypotensive effect in rat. Am J Physiol 235:H587–H591
31. Geijer C, Bill A (1979) Effects of raised intraocular pressure on retinal, prelaminar, laminar, and retrolaminar optic nerve blood flow in monkeys. Invest Ophthalmol Vis Sci 18:1030–1042
32. Glenny RW, Bernard S, Brinkley M (1993) Validation of fluorescent-labeled microspheres for measurement of regional organ perfusion. J Appl Physiol 74: 2585–2597
33. Granstam E, Wang L, Bill A (1993) Vascular effects of endothelin-1 in the cat; modification by indomethacin and L-NAME. Acta Physiol Scand 148:165–176
34. Hale SL, Alker KJ, Kloner RA (1988) Evaluation of nonradioactive, colored microspheres for measurement of regional myocardial blood flow in dogs. Circulation 78:428–434
35. Hale SL, Vivaldi MT, Kloner RA (1986) Fluorescent microspheres: a new tool for visualization of ischemic myocardium in rats. Am J Physiol 251:H863–H868
36. Heymann MA, Payne BD, Hoffman JI et al (1977) Blood flow measurements with radionuclide-labeled particles. Prog Cardiovasc Dis 20:55–79
37. Hillerdal M, Sperber GO, Bill A (1987) The microsphere method for measuring low blood flows: theory and computer simulations applied to findings in the rat cochlea. Acta Physiol Scand 130:229–235
38. Hong SJ, Wu KY, Chen IJ (1998) Ocular hypotensive and vasodilative effects of two beta-adrenergic blockers with intrinsic sympathomimetic activity. Curr Eye Res 17:700–707
39. Jay WM, Aziz MZ, Green K (1985) Alterations in ocular and optic nerve blood flow during intraocular surgery in aspirin pretreated rabbits. Curr Eye Res 4:563–568
40. Kern KB, Lancaster L, Goldman S et al (1990) The effect of coronary artery lesions on the relationship between coronary perfusion pressure and myocardial blood flow during cardiopulmonary resuscitation in pigs. Am Heart J 120:324–333
41. Kingma JG Jr, Simard D, Rouleau JR (2005) Comparison of neutron activated and radiolabeled microsphere methods for measurement of transmural myocardial blood flow in dogs. J Thromb Thrombolysis 19:201–208
42. Kondo M, Wang L, Bill A (1997) The role of nitric oxide in hyperaemic response to flicker in the retina and optic nerve in cats. Acta Ophthalmol Scand 75:232–235
43. Koskinen LO (1985) Effects of raised intracranial pressure on regional cerebral blood flow: a comparison of effects of naloxone and TRH on the microcirculation in partial cerebral ischaemia. Br J Pharmacol 85:489–497
44. Koskinen LO (1986) Effect of low intravenous doses of TRH, acid-TRH and cyclo(His-Pro) on cerebral and peripheral blood flows. Br J Pharmacol 87: 509–519
45. Koskinen LO, Bill A (1983) Regional cerebral, ocular and peripheral vascular effects of naloxone and morphine in unanesthetized rabbits. Acta Physiol Scand 119:235–241
46. Koskinen LO, Bill A (1984) Thyrotropin-releasing hormone (TRH) causes sympathetic activation and cerebral vasodilation in the rabbit. Acta Physiol Scand 122:127–136
47. Kowallik P, Schulz R, Guth BD et al (1991) Measurement of regional myocardial blood flow with multiple colored microspheres. Circulation 83: 974–982
48. Linder J (1982) Effects of cervical sympathetic stimulation on cerebral and ocular blood flows during hemorrhagic hypotension and moderate hypoxia. Acta Physiol Scand 114:379–386
49. Makowski EL, Meschia G, Droegemueller W et al (1968) Measurement of umbilical arterial blood flow to the sheep placenta and fetus in utero. Distribution to cotyledons and the intercotyledonary chorion. Circ Res 23:623–631
50. May CA, Narfström K (2008) Choroidal microcirculation in Abyssinian cats with hereditary rod-cone degeneration. Exp Eye Res 86:537–540
51. Millard RW, Baig H, Vatner SF (1977) Cardiovascular effects of radioactive microsphere suspensions and Tween 80 solutions. Am J Physiol 232:H331–H334
52. Moore CD, Gewertz BL, Wheeler HT et al (1981) An additional source of error in microsphere measurement of regional blood flow. Microvasc Res 21:377–383

53. Nilsson SFE (1991) Neuropeptide Y (NPY): a vasoconstrictor in the eye, brain and other tissues in the rabbit. Acta Physiol Scand 141:455–467
54. Nilsson SFE (1994) PACAP-27 and PACAP-38: vascular effects in the eye and some other tissues in the rabbit. Eur J Pharmacol 253:17–25
55. Nilsson SFE (1996) Nitric oxide as a mediator of parasympathetic vasodilation in ocular and extraocular tissues in the rabbit. Invest Ophthalmol Vis Sci 37:2110–2119
56. Nilsson SFE (2000) The significance of nitric oxide for parasympathetic vasodilation in the eye and other orbital tissues in the cat. Exp Eye Res 70:61–72
57. Nilsson SFE, Bill A (1984) Vasoactive intestinal polypeptide (VIP): effects in the eye and on regional blood flows. Acta Physiol Scand 121:385–392
58. Nilsson SFE, Maepea O (1987) Comparison of the vasodilatory effects of vasoactive intestinal polypeptide (VIP) and peptide-HI (PHI) in the rabbit and the cat. Acta Physiol Scand 129:17–26
59. Nilsson SFE, Linder J, Bill A (1985) Characteristics of uveal vasodilation produced by facial nerve stimulation in monkeys, cats and rabbits. Exp Eye Res 40:841–852
60. Nilsson SFE, Maepea O, Alm A et al (2001) Ocular blood flow and retinal metabolism in abyssinian cats with hereditary retinal degeneration. Invest Ophthalmol Vis Sci 42:1038–1044
61. Ninomiya H, Inomata T (2005) Microvasculature of the hamster eye: scanning electron microscopy of vascular corrosion casts. Vet Ophthalmol 8:7–12
62. Ninomiya H, Inomata T (2006) Microvascular anatomy of the pig eye: scanning electron microscopy of vascular corrosion casts. J Vet Med Sci 68:1149–1154
63. Ninomiya H, Kuno H (2001) Microvasculature of the rat eye: scanning electron microscopy of vascular corrosion casts. Vet Ophthalmol 4:55–59
64. Ninomiya H, Inomata T, Kanemaki N (2005) Microvasculature of the retina, ciliary processes and choroid in the North American raccoon (*Procyon lotor*) eye: a scanning electron microscopic study of corrosion casts. J Vet Med Sci 67:547–554
65. Nork TM, Kim CB, Shanmuganayagam D et al (2006) Measurement of regional choroidal blood flow in rabbits and monkeys using fluorescent microspheres. Arch Ophthalmol 124:860–868
66. Nose Y, Nakamura T, Nakamura M (1985) The microsphere method facilitates statistical assessment of regional blood flow. Basic Res Cardiol 80:417–429
67. O'Day DM, Fish MB, Aronson SB et al (1971) Ocular blood flow measurement by nuclide labeled microspheres. Arch Ophthalmol 86:205–209
68. Orgul S, Cioffi GA, Bacon DR et al (1996) Measurement of optic nerve blood flow with nonradioactive colored microspheres in rabbits. Microvasc Res 51:175–186
69. Polissar NL, Stanford DC, Glenny RW (2000) The 400 microsphere per piece "rule" does not apply to all blood flow studies. Am J Physiol Heart Circ Physiol 278:H16–H25
70. Powers KM, Schimmel C, Glenny RW et al (1999) Cerebral blood flow determinations using fluorescent microspheres: variations on the sedimentation method validated. J Neurosci Methods 87:159–165
71. Prinzen FW, Glenny RW (1994) Developments in non-radioactive microsphere techniques for blood flow measurement. Cardiovasc Res 28:1467–1475
72. Ravalico G, Toffoli G, Pastori G et al (1996) Age-related ocular blood flow changes. Invest Ophthalmol Vis Sci 37:2645–2650
73. Reinhardt CP, Dalhberg S, Tries MA et al (2001) Stable labeled microspheres to measure perfusion: validation of a neutron activation assay technique. Am J Physiol Heart Circ Physiol 280:H108–H116
74. Roth S (1992) The effects of halothane on retinal and choroidal blood flow in cats. Anesthesiology 76: 455–460
75. Roth S, Pietrzyk Z, Crittenden AP (1993) The effects of enflurane on ocular blood flow. J Ocul Pharmacol 9:251–256
76. Roy MS, Harrison KS, Harvey E et al (1989) Ocular blood flow in dogs using radiolabelled microspheres. Int J Rad Appl Instrum B 16:81–84
77. Rudolph AM, Heymann MA (1967) The circulation of the fetus in utero. Methods for studying distribution of blood flow, cardiac output and organ blood flow. Circ Res 21:163–184
78. Saxena PR, Verdouw PD (1985) Tissue blood flow and localization of arteriovenous anastomoses in pigs with microspheres of four different sizes. Pflugers Arch 403:128–135
79. Schimmel C, Frazer D, Glenny RW (2001) Extending fluorescent microsphere methods for regional organ blood flow to 13 simultaneous colors. Am J Physiol Heart Circ Physiol 280:H2496–H2506
80. Schosser R, Arfors KE, Messmer K (1979) MIC-II – a program for the determination of cardiac output, arterio-venous shunt and regional blood flow using the radioactive microsphere method. Comput Programs Biomed 9:19–38
81. Sebag J, Feke GT, Delori FC et al (1985) Anterior optic nerve blood flow in experimental optic atrophy. Invest Ophthalmol Vis Sci 26:1415–1422
82. Seligsohn EE (1992) Adrenergic and non-adrenergic cardiovascular effects of thyrotropin-releasing hormone (TRH) in the anaesthetized rabbit. Acta Physiol Scand 146:107–117
83. Seligsohn EE, Bill A (1993) Effects of NG-nitro-L-arginine methyl ester on the cardiovascular system of the anaesthetized rabbit and on the cardiovascular response to thyrotropin-releasing hormone. Br J Pharmacol 109:1219–1225
84. Stjernschantz J, Bill A (1979) Effect of intracranial stimulation of the oculomotor nerve on ocular blood flow in the monkey, cat, and rabbit. Invest Ophthalmol Vis Sci 18:99–103

85. Stjernschantz J, Alm A, Bill A (1976) Effects of intracranial oculomotor nerve stimulation on ocular blood flow in rabbits: modification by indomethacin. Exp Eye Res 23:461–469
86. Stjernschantz J, Nilsson SF, Astin M (1989) Vasodynamic and angiogenic effects of eicosanoids in the eye. Prog Clin Biol Res 312:155–170
87. Thein E, Raab S, Harris AG et al (2000) Automation of the use of fluorescent microspheres for the determination of blood flow. Comput Methods Programs Biomed 61:11–21
88. Thorig L, Bill A (1986) Effects of B-HT 920 in the eye and on regional blood flows in anaesthetized and conscious rabbits. Curr Eye Res 5:565–573
89. Van Oosterhout MF, Willigers HM, Reneman RS et al (1995) Fluorescent microspheres to measure organ perfusion: validation of a simplified sample processing technique. Am J Physiol 269: H725–H733
90. Wang L, Tornquist P, Bill A (1997) Glucose metabolism in pig outer retina in light and darkness. Acta Physiol Scand 160:75–81
91. Wang L, Grant C, Fortune B et al (2008) Retinal and choroidal vasoreactivity to altered PaCO(2) in rat measured with a modified microsphere technique. Exp Eye Res 86:908–913
92. Wang L, Fortune B, Cull G et al (2007) Microspheres method for ocular blood flow measurement in rats: size and dose optimization. Exp Eye Res 84:108–117
93. Weiter JJ, Schachar RA, Ernest JT (1973) Control of intraocular blood flow. I. Intraocular pressure. Invest Ophthalmol 12:327–331
94. Zhan GL, Lee PY, Ball DC et al (2002) Time dependent effects of sympathetic denervation on aqueous humor dynamics and choroidal blood flow in rabbits. Curr Eye Res 25:99–105
95. Zwissler B, Schosser R, Weiss C et al (1991) Methodological error and spatial variability of organ blood flow measurements using radiolabeled microspheres. Res Exp Med (Berl) 191:47–63

Laser Doppler Flowmetry in Animals

3

Jeffrey W. Kiel and Herbert A. Reitsamer

Core Messages

- Laser Doppler flowmetry (LDF) provides continuous measurements of tissue perfusion in a relatively small sampling volume.
- LDF readings scale linearly with independent measurements of blood flow, but calibration slopes vary from site to site due to LDF's small sampling volume and normal microcirculation heterogeneity, so that LDF readings are expressed in arbitrary perfusion units.
- LDF detects movement and cannot distinguish between moving blood cells and tissue or probe movement; hence, great care must be taken to prevent non-blood flow-related movement during LDF measurements.

3.1 Introduction

This chapter will describe the technique known as laser Doppler flowmetry and its use in animals. Because the technique is based on light, it permits noninvasive measurements of perfusion in a tissue that can be illuminated, but perhaps its greatest strength is that it provides a continuous measurement. However, the technique also has significant weaknesses, and it is technically challenging to use properly. On balance, it has added significantly to our knowledge of ocular blood flow and will likely continue to be a valuable tool to increase our understanding of the physiology and pathophysiology of the eye.

3.2 History

Using a prototype optical device, Riva et al. made the first laser Doppler measurements in an animal eye, starting with blood velocity measurements in rabbit retinal arteries in 1972 [14]. This initial prototype laser Doppler velocimeter was soon refined and utilized to measure blood velocity in human arteries and veins in 1974 [27]. During that study, Doppler spectra were also collected from the microcirculation in the optic nerve head, but the spectra were not analyzed to obtain blood flow information.

The technique for measuring microcirculatory perfusion known as laser Doppler flowmetry arose from a serendipitous series of observations in the early 1970s by a young postdoctoral fellow

J.W. Kiel, Ph.D., FARVO (✉)
Department of Ophthalmology, University of Texas Health Science Center, 7703 Floyd Curl Drive MS 6230, San Antonio, TX 78229-3900, USA
e-mail: kiel@uthscsa.edu

H.A. Reitsamer, M.D.
Department of Ophthalmology and Optometry, Paracelsus Medical University, Müllner Hauptstrasse 48, A-5020 Salzburg, Austria
e-mail: h.reitsamer@salk.at

L. Schmetterer, J.W. Kiel (eds.), *Ocular Blood Flow*, DOI 10.1007/978-3-540-69469-4_3, © Springer-Verlag Berlin Heidelberg 2012

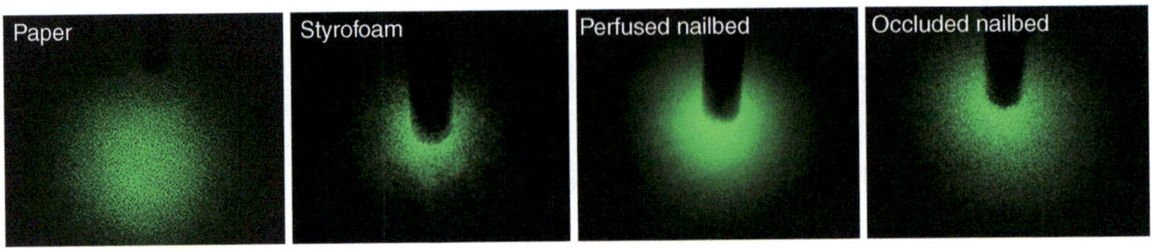

Fig. 3.1 Laser speckle pattern produced by the probe of a fiber optic laser Doppler flowmeter. The speckle pattern visible on and in static materials disappears in a perfused tissue but reappears when the tissue is unperfused. These observations first suggested the possibility that the Doppler effect on moving blood cells could be used to measure tissue perfusion (Green laser is part of a customized Perimed flowmeter system the authors used to measure perfusion at different depths. The "nail bed" locations are the base of the fingernail on the author's index finger without and with a rubber band tourniquet at the first knuckle)

at NIH named Michael D. Stern [19]. While going through some surplus equipment in his mentor's laboratory, Stern came across an old helium-neon laser and wondered if he could possibly get some laser Doppler perfusion measurements from his finger. He constructed a simple system that failed to work, but he noticed that the speckle pattern produced when a stationary laser beam is directed at a static surface failed to occur when the laser beam was directed at his finger. He initially thought the absence of the speckle pattern was because he was not able to hold his finger still but then wondered if light scattering within the tissue eliminated the temporal coherence of laser light. This possibility was disproved by the distinct speckle pattern he obtained when he directed his laser beam at a styrofoam cup or a piece of beefsteak and, ultimately, when he directed the beam at his finger when it was occluded with a rubber band. As Stern later recounted to A.P. Shepherd, "if the naked eye could see the effect of blood flow on the scattered light, it should be easy to measure it with the right hardware" [19]. It was not easy, but Stern persevered and developed prototype systems [24–26] that gave rise to the Bonner-Nossal theory of coherent light scattering in perfused tissue [1, 2] and the commercial laser Doppler flowmeters now available (Fig. 3.1).

Ocular blood flow researchers have two types of laser Doppler flowmeters at their disposal – the systems that use optics to focus the laser beam through the cornea and sample a portion of the light scattered by the illuminated tissue and the systems that use fiber optics to send and retrieve the light. The optical flowmeters were designed specifically for the eye and permit noninvasive measurements of iris and fundus sites in humans and animals. The fiber optic design is utilized by most commercial LDF instruments and is most often used for measurements on exposed tissue surfaces such as skin. With the proper fiber optic probe configuration and light wavelength, measurements can be made through the sclera to assess underlying ocular tissues, or a needle probe can be inserted into the eye with the tip directed at the tissue of interest; eye measurements with the fiber optic instruments are almost exclusively done in animals.

3.3 Theory

A full-fledged treatment of the theoretical underpinnings of laser Doppler flowmetry is beyond the scope of this chapter, and the interested reader is directed to the original Bonner–Nossal paper [1] and the excellent textbook by Shepherd and Öberg for further details [21]. However, some of the more salient aspects of laser light scattering in perfused tissue are important for understanding the advantages and limitations of the technique and will be addressed briefly.

Photons directed from a laser into a nonperfused tissue or static light scattering medium

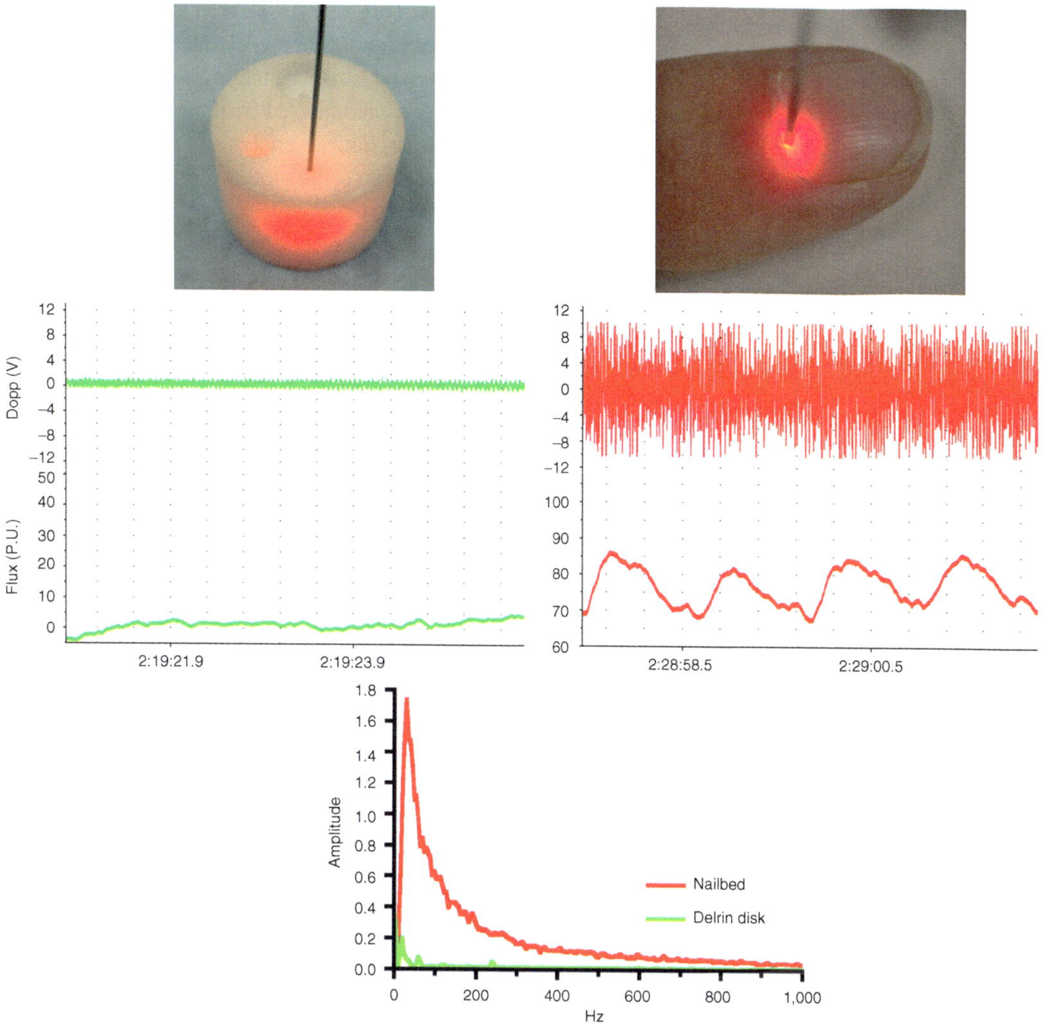

Fig. 3.2 Monochromatic light scattered within a static structure and perfused tissue. Photons scattered by non-moving structures undergo no Doppler shift in frequency, and so those collected by a receiving fiber generate no beat frequencies on a photodetector and no broadening of the frequency spectrum of the light collected. In a perfused tissue, photons are scattered by moving blood cells and static tissue components, producing a mix of frequency-shifted and unshifted photons that generate beat frequencies on a photodetector proportional to the Doppler broadening of the frequency spectrum of the collected light (Measurements done with Perimed PF2b which provides an analog output of the photodetector voltage that was recorded on a PowerLab digital data acquisition system for off-line spectral analysis)

travel in a straight line until they are deflected and randomly scattered by the stationary structural matrix. As shown in Fig. 3.2, the majority of the photons scatter away from their point of entry into the matrix, but a fraction of the photons return back toward their point of entry. Current optical and fiber optic-based laser Doppler flowmeters collect some of the photons scattered back toward their point of entry and convey them to a photodetector. (The distance between the point of photon entry and collection is a determinant of the measurement depth, as will be discussed later.) The different path lengths traveled by the photons do not alter their frequency or

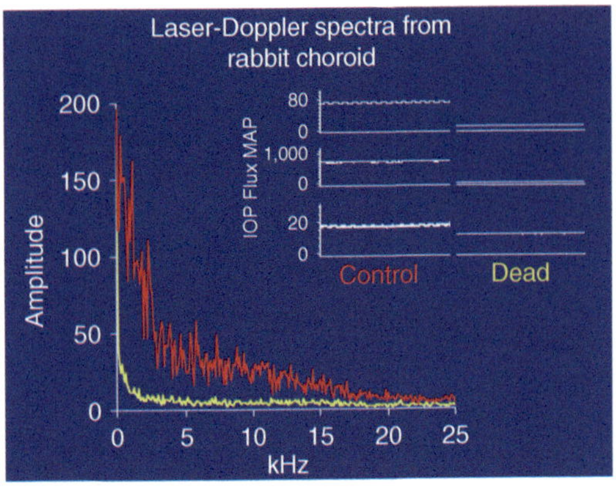

Fig. 3.3 Doppler-broadened frequency spectra collected with a fiber optic flowmeter from one site on the choroid of an anesthetized rabbit before and after euthanasia. In a highly perfused tissue like the choroid, the higher blood cell velocities can extend the frequency spectra to 20 kHz, as opposed to the lower frequency range seen in tissues like the nail bed (see Fig. 3.2) (Measurements made with a customized Perimed PF-2b with an upper-frequency cutoff at 20 kHz and the photodetector voltage analyzed with a digital oscilloscope to obtain the spectra)

wavelength but do produce the positive and negative wave interference that generates the speckle pattern we perceive with our eyes [11]. At a photodetector, a static speckle pattern generates a relatively stable DC signal proportional to the light intensity (Fig. 3.2).

Something different happens if there are moving particles passing through the stationary matrix, such as occurs with red blood cells passing through a perfused tissue like a nail bed. In this situation, the photons encountering moving particles undergo a Doppler shift in frequency proportional to the particle velocities and scattering angles. The Doppler shifting and consequent loss of the original common wavelength diminish the wave interference that gives rise to the speckle pattern. Instead, both visually and as detected with a photodetector, the mixing of photons with different frequencies generates a blurred image and a complex photodector signal composed of beat frequencies proportional to the frequency difference between the unshifted photons and those shifted by interaction with the moving particles (Fig. 3.2). Fourier analysis of the photodetector output reveals a Doppler-broadened frequency spectrum reflecting the range of velocity vectors encountered by the collected photons. By contrast, Fourier analysis of the photodetector signal generated by photons collected from a stationary matrix shows no spectral amplitude beyond the inherent "shot noise" of the photodetector (Fig. 3.2).

It is this spectral broadening of the photodetector signal from laser light collected from a perfused tissue that is the basis of laser Doppler flowmetry. From this spectrum, it is possible to derive the average velocity of the moving particles and the number of moving particles, the product of which is the red blood cell flux within the volume of tissue sampled. With appropriately fast signal processing, laser Doppler flowmeters provide essentially continuous measurements, which is one of the technique's primary strengths. In tissues like the nail bed or the retina and optic nerve head, the vascular density is low and the range of red blood cell velocities relatively small, so the Doppler-broadened spectrum from these tissues is narrow (i.e., there are few Doppler-shifted frequencies greater than 1–2 kHz). However, as Fig. 3.3 shows, in a highly perfused tissue like the choroid, the Doppler-broadened spectrum extends to much higher frequencies, reflecting the higher velocity and greater number of moving red blood cells in the sampled tissue volume (i.e., Doppler-shifted frequencies as high as 20 kHz are not uncommon).

3.4 Validation

The most important requirement of any new technique claiming to measure blood flow is that its measurements agree with those of an established technique. For flow in an accessible artery or vein, the ultimate measurement gold standard is a beaker and stopwatch, which allow timed collection of blood efflux when the vessel is severed. Those volumetric measurements (e.g., ml/min) are used to calibrate the voltage signal generated by extravascular probes that transduce the passage of blood through the probe's sampling volume by detecting the flow-dependent change in energy emitted by the probe (e.g., an electromagnetic field in electromagnetic flowmeter or an ultrasound beam in transit time flowmeters). These relatively direct methods are useful for measuring cardiac output with a probe on the aorta, regional blood flow in a tissue supplied by a major artery (e.g., cerebral blood flow with a probe on the common carotid artery), or blood flow to an organ supplied by a single artery (e.g., renal blood flow with a probe on the renal artery). In the latter case, the postmortem organ weight can be used to calculate the blood flow normalized to a standard tissue weight (e.g., ml/min/g). Less direct but well-established techniques such as venous occlusion plethysmography, the reference sample microsphere entrapment technique, and the clearance of inert gases or tracers are used in more complex situations where a direct technique is inappropriate (e.g., a heterogeneous tissue in which the flow distribution between tissue layers or regions is needed).

These direct and indirect approaches were used to validate the various laser Doppler flowmeters that became commercially available in the mid-1980s. In a wide variety of tissues, a strong linear correlation was found consistently between flux measurements with laser Doppler flowmeters and measurements of tissue or organ blood flow with established techniques. A sampling of those validation results is shown in Fig. 3.4.

3.5 Calibration

Although laser Doppler flowmetry measurements scale linearly with those of established techniques, a universal calibration constant to convert flux readings into volumetric flow units does not exist between tissues (i.e., a 1-V flux signal on a kidney and a liver does not correspond consistently to similar renal and hepatic volumetric blood flows), nor between the same organ in different subjects, or even in the same organ from one site to the next [11]. The reason for the variability between measurement sites is not known with certainty, but it is likely due to the normal tissue variations in vessel density and spatial organization as well as variations in tissue light scattering and light absorption. As discussed below, the volume of tissue sampled with typical fiber optic probes and light wavelengths is approximately 1 mm^3, and so the technique is inherently sensitive to local variations in tissue structure. Strategies to minimize site-dependent variations include averaging readings from multiple sites or using probes with additional collecting fibers around the transmitting fiber; however, the technique is most reliable when used to monitor the acute response to a perturbation (e.g., a change in perfusion pressure or administration of a drug) while continuously monitoring a single site. If laser Doppler flowmetry measurements need to be made during more than one session, great care needs to be taken to insure the same site is measured each time. Caution is also needed when comparing flux readings at the same site before and after tissue remodeling is likely to have occurred due to the passage of time, disease, or treatment (Fig. 3.5).

Although volumetric calibration is not possible, commercial fiber optic-based laser Doppler flowmeters use a two-point calibration procedure to insure that all flowmeters made by a given manufacturer have the same gain and no zero offset. This is accomplished by setting the flowmeter to read zero when the probe is measuring a standardized solid scattering structure (e.g., an opaque plastic cylinder as in Fig. 3.2) and to read a particular flux value (e.g., 250 perfusion units for Perimed flowmeters) when the probe is measuring a standardized suspension of small particles at a fixed concentration and specified temperature such that the Brownian motion of the particles generates a universal reference Doppler-broadened frequency spectrum [11]. The two-point calibration helps to insure that measurement differences between flowmeters are not flowmeter-related.

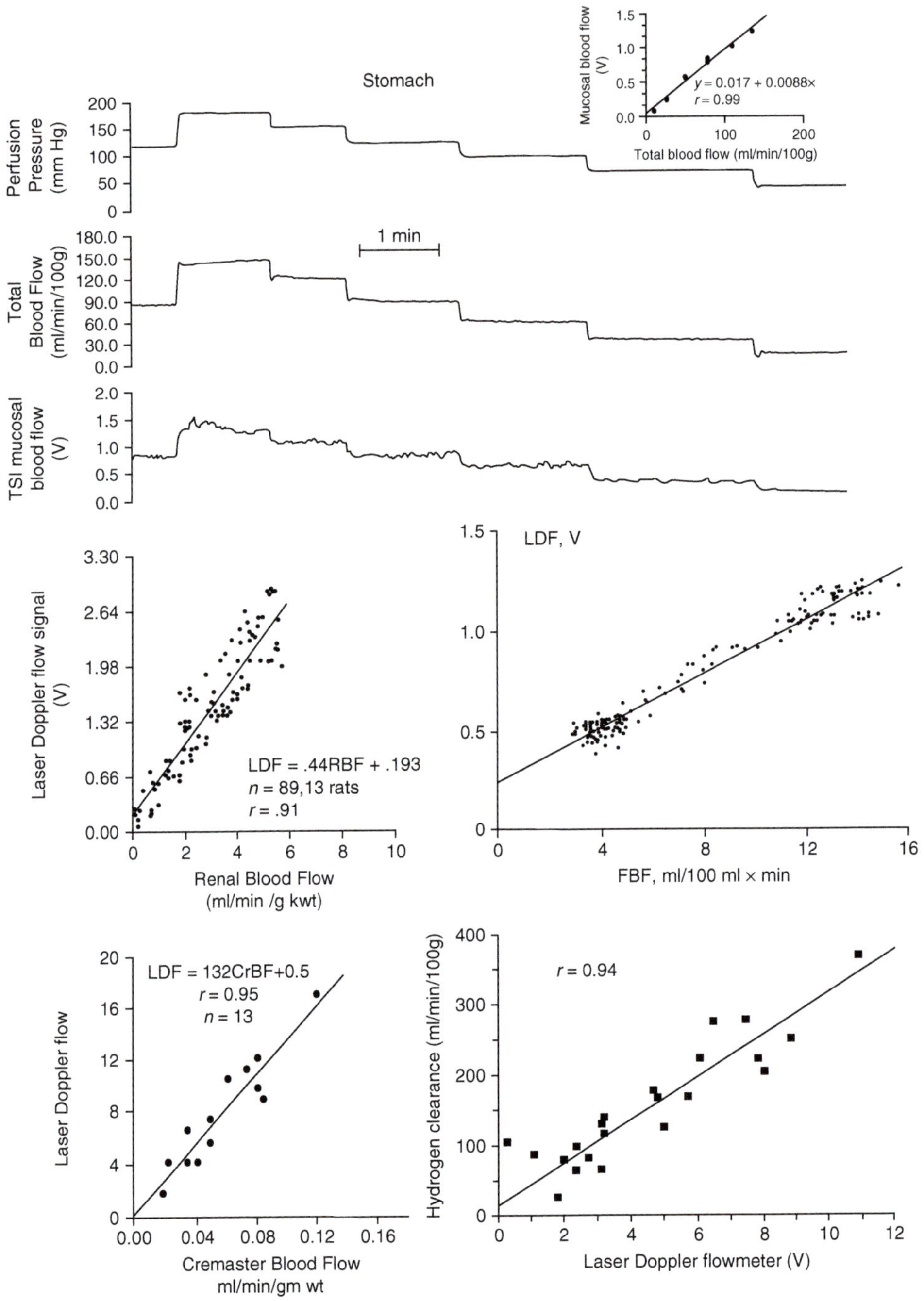

Fig. 3.4 Sampling of results from validation studies in various tissues showing linear correlation between laser Doppler flowmeter measurements and measurements by established techniques (*Top*: isolated canine stomach preparation with mucosal perfusion measured with a TSI flowmeter and total blood flow measured with an electromagnetic flowmeter [6]; *Middle left*: rat renal cortex perfusion measured with a Perimed PF1 and total renal blood flow measured with an electromagnetic flowmeter [18] © Shepherd A, Öberg P (1990) Laser Doppler Blood Flowmetry. Kluwer Academic Publishers, USA; *middle right*: human forearm skin perfusion measured with a Perimed PF1c and total forearm blood flow measured by plethysmography [5]; *bottom left*: rat skeletal muscle perfusion measured with a Perimed PF1d and total muscle blood flow measured with radioactive microspheres [23]; *bottom right*: isolated canine stomach preparation with mucosal blood flow measured with a combined laser Doppler and hydrogen clearance probe [3])

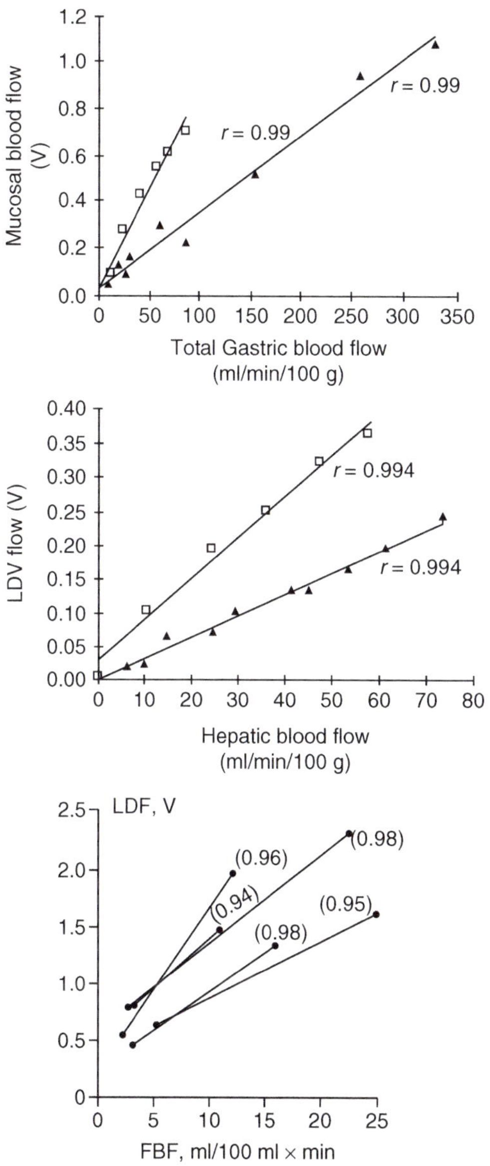

Fig. 3.5 Linear but variable slopes relating simultaneous laser Doppler flowmetry measurements and other techniques precludes calibration. At a given total blood flow, moving the probe from one site to another on many tissues gives different readings likely due to the technique's small sampling volume and normal variations in tissue structure and vascular organization (*Top*: isolated canine stomach preparation with mucosal perfusion measured with a TSI flowmeter and total blood flow measured with an electromagnetic flowmeter [22]; *Middle*: rat liver mucosal perfusion measured with a TSI flowmeter and total blood flow measured with an electromagnetic flowmeter [22]; *Bottom*: human forearm skin perfusion measured with a Perimed PF1c and total forearm blood flow measured by plethysmography [5])

3.6 Zero Offset

The zero reading with laser Doppler flowmeters is more complicated than it might seem. All photodetectors generate a small amount of electrical noise that contributes to the Doppler-broadened frequency spectra. To eliminate this contribution, a low-frequency cutoff filter is applied (e.g., 20 Hz in the Perimed PF-4000 and PF-5000). This low-frequency cutoff does not eliminate Doppler shifts imparted by relative motion of the probe and the zeroing object during the two-point calibration; hence, it is imperative that all motion artifacts be avoided during the zeroing procedure. The elimination of motion artifacts is more difficult when measuring in live tissue under no-flow conditions such as in a nail bed during brachial artery occlusion in humans or in the choroid when IOP is raised above systolic blood pressure in anesthetized animals. Under such conditions, the flux reading approaches but often does not reach zero due to slight movement of the measurement site originating from the subject (e.g., respiratory movement, cardiac pulse, and muscle fasciculations) or within the site (e.g., Brownian motion or settling of trapped red blood cells and vasomotion) or the environment (e.g., building or laboratory vibrations). However, in acute animal experiments, when the animal is euthanized, it is possible to verify that the flux goes to zero at death or, if not, to identify and eliminate any nonbiologic sources of motion or vibration. In some circumstances when it is not possible to eliminate motion artifacts, some flowmeters allow the lower-frequency cutoff to be set higher, which minimizes the motion artifact contribution to the analyzed spectrum but also eliminates the contribution from slower-moving red blood cells, thereby causing a falsely high flux reading [17] (Fig. 3.6).

One device that was marketed as a laser Doppler flowmeter was the Heidelberg retinal flowmeter. The device was a modification of a scanning laser ophthalmoscope, and it should have worked in theory. Unfortunately, as shown in Fig. 3.7, it had a large variable zero offset that was evident in measurements of static objects,

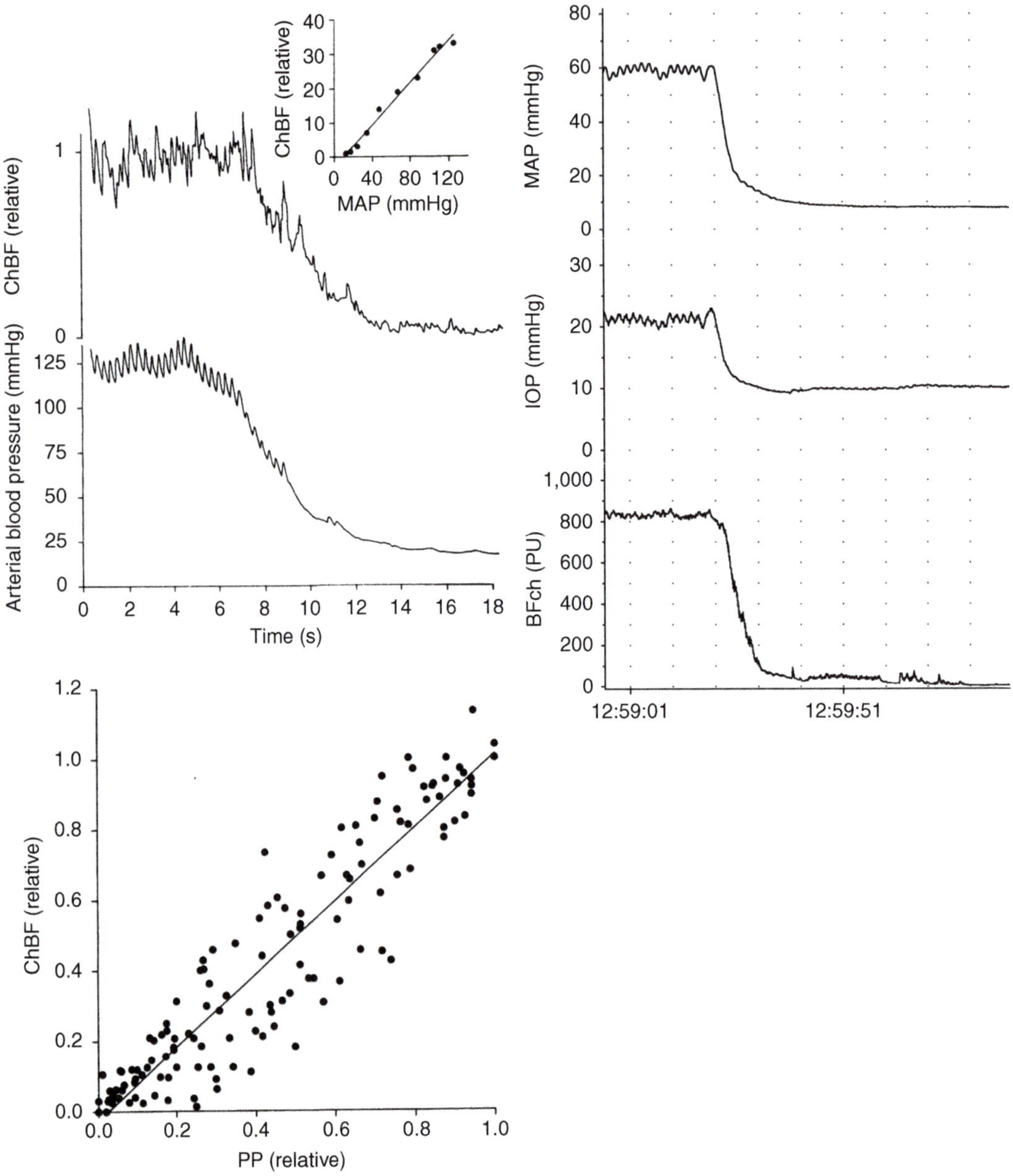

Fig. 3.6 Lack of zero offset in cat and rabbit choroid at euthanasia. *Left panel*: the trace and inset graph show the drop in blood pressure and choroidal flux falling to zero in an anesthetized cat given an overdose of pentobarbital, and the *bottom graph* shows the linear fall in choroidal flux to zero as perfusion pressure falls during euthanasia in 14 cats [16]. *Right panel*: trace shows that a similar parallel fall in blood pressure and choroidal flux occurs in an anesthetized rabbit given an overdose of pentobarbital (authors' unpublished observation). The cat choroidal measurements were made with a modified fundus camera using a Vasomedics BPM 403A for signal processing, while the rabbit choroidal measurements were made with an intravitreal fiber optic probe and a Perimed PF-4000

ham, and dead rabbit eyes [28]. Moreover, as Fig. 3.8 shows, it failed to detect any change in retinal perfusion after complete obstruction to the central retinal artery in rats [29]. The device is no longer marketed.

3.7 Effects of Oxygen

A potential problem with laser Doppler flowmetry that was dispelled early in its development is the sensitivity to changes in blood oxygenation

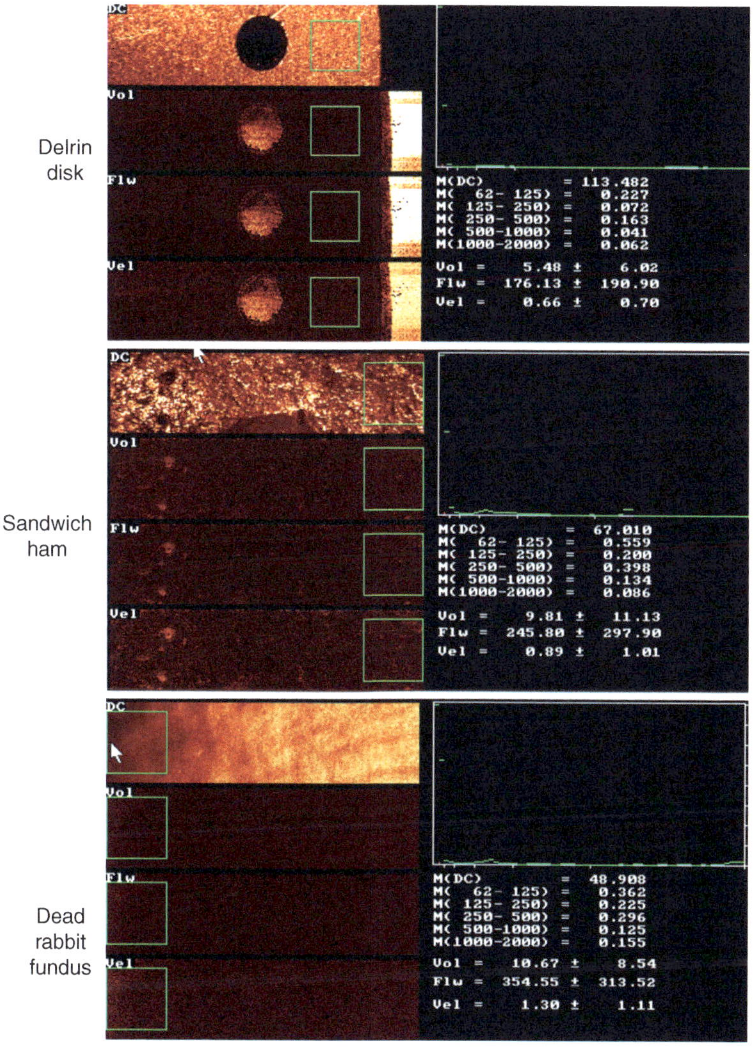

Fig. 3.7 Zero offset in HRF measurements of static objects, lunch meat, and fundus of a dead rabbit. "Flow" readings in live rabbit fundus scans ranged from 200 to 700 perfusion units, but postmortem flow readings in the same rabbits ranged from 150 to 350 perfusion units from one location to another [28]

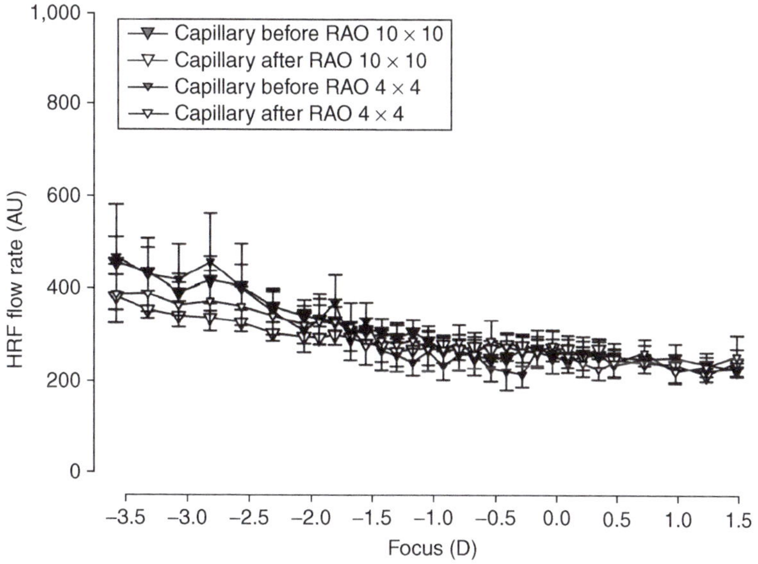

Fig. 3.8 HRF readings from rat retina over a range of focal depth failed to detect a change in retinal perfusion after occlusion of the central retinal artery [29]

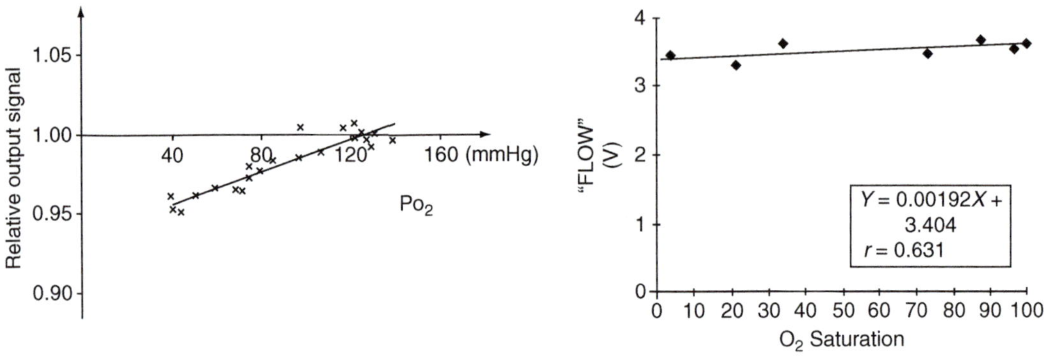

Fig. 3.9 Flux measurements in suspensions of red blood cells with 632.8 nm light (*left*) and 780 nm light (*right*) are relatively insensitive to changes in blood oxygenation [10, 22]

[10, 22]. Assuming the tissue is otherwise motionless, red blood cell movement is the dominant movement in a perfused tissue. The shape of a moving red blood cell influences the deflection angle and consequently the Doppler frequency shift imparted to any photon it encounters. The shape of a red blood cell is, in turn, influenced by its hemoglobin binding with oxygen, and so blood oxygen levels could influence the flux measurement. Moreover, the light wavelengths used in laser Doppler flowmeters are not isobestic for hemoglobin, and so changes in hemoglobin oxygen saturation could alter light absorption and affect the flux measurement as well. However, Fig. 3.9 shows that tests with red blood cell suspensions equilibrated over a wide range of oxygen partial pressure reveal only a slight effect (i.e., <5%) on flowmeters using He-Ne lasers (632.8-nm wavelength) and almost no effect on flowmeters using infrared lasers (780-nm wavelength).

3.8 Effect of Laser Light on Vessel Reactivity

Although low-power lasers (<2 mW) are used in laser Doppler flowmeters, the potential for a direct effect of laser light on vessel reactivity was an early concern. However, this concern was mitigated by experiments with a hamster cremaster muscle preparation in which arterioles exposed to light from a 10-mW He-Ne laser or from a 3-mW infrared laser diode failed to alter vessel diameter [22].

3.9 Measurement Depth and Sampling Volume

The measurement depth for laser Doppler flowmetry is ill-defined, and this has been a source of confusion and controversy since the technique's inception. The confusion was exacerbated in the early years because flowmeters used He-Ne lasers that emit a beautiful red light, and it was easy to see that the light penetrated to a considerable depth in most tissues. When flowmeters switched to infrared laser diodes and the light penetration in tissue was no longer visible, users became more inclined to accept the theoretical arguments for a much more shallow measurement depth. For many users, the misunderstanding arose from assuming that the photon penetration they could see equated to measurement depth, when in fact any photon they could see had escaped capture by the collecting fiber and could not have contributed to the flux measurement. Indeed, as Fig. 3.10 shows, the majority of the photons emitted into a scattering medium or tissue disperse away from the point of entry, and only a small fraction of the photons are scattered such that they travel in the vicinity of the entry point where they can enter the adjacent collecting fiber. Mathematical models indicate that photons with shorter scattering

Fig. 3.10 Explanation for why laser Doppler flowmetry measurement depth is more shallow than the depth of photon penetration. In this example, green photons are delivered via the sending fiber of a typical fiber optic probe to a uniform scattering medium (i.e., a vial of dilute coffee creamer) where the photons diffuse a considerable distance away from the point of entry. In the *middle* photo, a red laser is connected to the receiving fiber to show that it is adjacent to the sending fiber (typical fiber separation is 250 μm). Most of the green photons delivered do not take the short scattering path that leads back to the collecting fiber adjacent to the entry point and so could not contribute to an analyzed frequency spectrum. In other words, any photon that can be seen after leaving the sending fiber has obviously escaped capture by the receiving fiber. The *right* photo shows the nail view of a thumb illuminated from the opposite side with a He-Ne laser via a standard fiber optic probe. Although a large volume of tissue is illuminated, the light that can be seen (i.e., the photons captured by the camera) is not captured by the receiving fiber and so does not contribute to the measurement; it is only the photons that travel less than 500–1,000 μm into the tissue that have a reasonable chance of returning to the receiving fiber

paths have higher probabilities of being caught by the adjacent collecting fiber and that those probabilities diminish significantly at depths beyond 500–1,000 μm for many tissues when measured with typical wavelengths and fiber separations [1, 2, 20].

Figure 3.11 presents some empirical evidence supporting a measurement depth less than approximately 1 mm. An early controversy in laser Doppler flowmetry was whether the technique measured through the entire gastric or intestinal wall or was the measurement depth sufficiently shallow to detect blood flow changes only in the mucosa or muscularis if the probe were placed on that side. Unique flow responses that could only be detected from one side or the other of the gut wall indicted that the measurement depth had to be less than the wall thickness [6, 20], and this was confirmed by placing an unperfused piece of mucosa (approximately 1 mm thick) between the fiber optic probe and the perfused mucosa of an isolated canine stomach preparation [8]. The interposed unperfused tissue eliminated the baseline flux signal and attenuated the reactive hyperemic response to a brief period of ischemia. A similar experiment was done to confirm that the measurement depth is sufficient to measure through the rabbit sclera and detect blood flow changes in underlying tissues like the ciliary body and anterior choroid. In this case, a layer of sclera (approximately 300 μm thick) harvested from the fellow eye was placed between the probe and the anterior sclera over the ciliary body in the intact eye, and although the flux signal was attenuated, it still registered flux changes in response to changing the ocular perfusion pressure by occluding the vena cava [9]. These results are consistent with a measurement depth between 500 and 1,000 μm.

For eye blood flow researchers, the measurement depth estimate of <1 mm is problematic since tissues of interest like the optic nerve head and the retina-choroid are stratified into comparatively thinner layers, and so it is unclear which layers contribute how much to the measurement. For some studies, it is possible to avoid the issue by choosing the right animal model. For example, for choroidal blood flow studies, an animal with an avascular fovea (e.g., primates) or a relatively avascular retina (e.g., rabbit or guinea pig) can be used to eliminate Doppler shifts from blood cells moving in the retinal circulation, which are otherwise a problem for optic systems working through the cornea and for intravitreal or

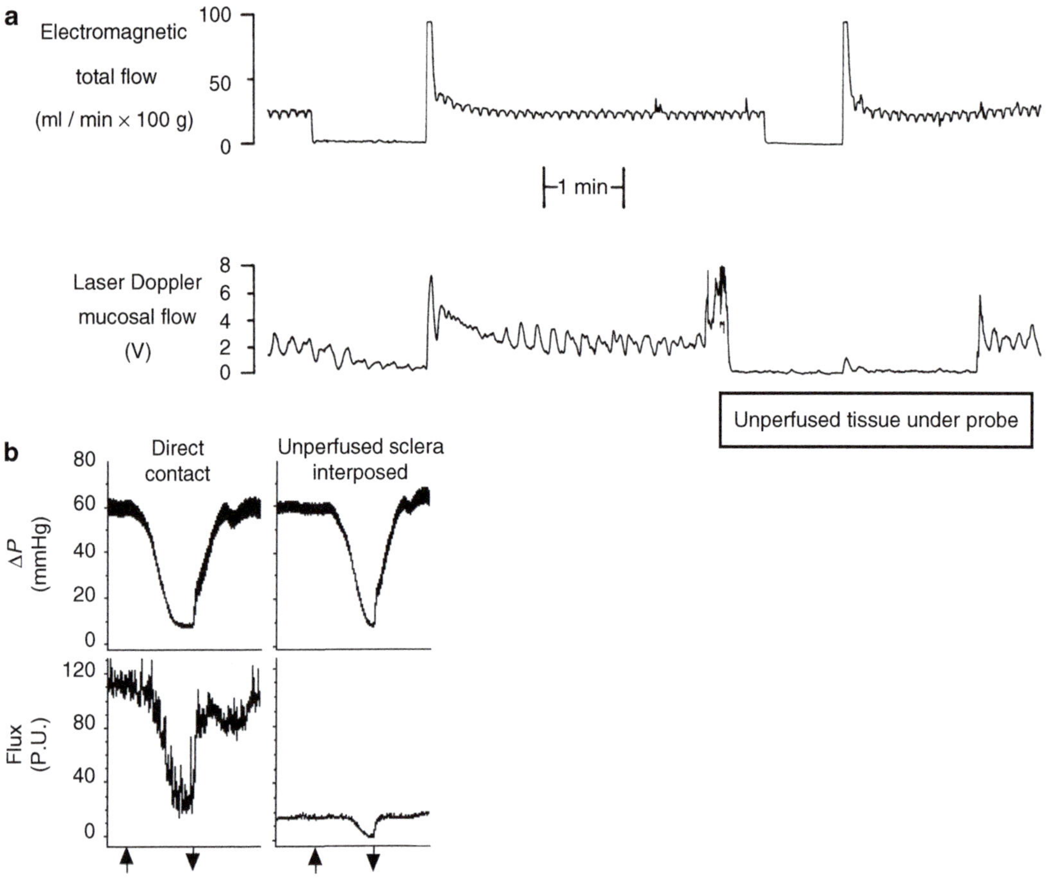

Fig. 3.11 Examples of experiments indicating a measurement depth <1 mm. (**a**) Experimental trace of total blood flow and mucosal flux measured with a prototype laser Doppler flowmeter in an autoperfused, isolated canine stomach preparation. The initial 90-s occlusion of the supply artery elicits the typical reactive hyperemic response in total blood flow and mucosal perfusion. Placing a layer of unperfused mucosa between the fiber optic probe and the perfused mucosa blocked the baseline mucosal signal and attenuated the peak of the second reactive hyperemia. The thickness of the unperfused mucosa was approximately 1 mm [8]. (**b**) Experimental trace of perfusion pressure (ΔP=arterial pressure minus intraocular pressure) and flux measured with a Perimed PF-4000 through the sclera over the ciliary body in an anesthetized rabbit. *Left* and *right panels* show responses to similar occlusions of the vena cava to lower blood pressure (*up* and *down arrows* indicate start and stop of caval occlusions), while the flux was measured without and with a layer of unperfused sclera (approximately 300 μm thick) interposed between the probe and eye. Measuring through two layers of sclera attenuated but did not eliminate the flux signal [9]

transscleral fiber optic probes. In animals with a retinal circulation such as cats, Riva et al. [16] made a cogent argument that "the contribution of the retinal RBCs [red blood cells] should be about one-eighth that of the choroicapillaris RBCs" based on anatomical grounds. Consistent with this proposal, optical measurements through the cornea to the peripheral retina in humans during hyperoxia-induced retinal vasoconstriction were dominated by the choroidal contribution, which is unresponsive to hyperoxia, leading the authors to conclude that the optical flowmeter employed (Oculix 4000) cannot be used to measure retinal capillary blood flow [13].

The ambiguous measurement depth in the optic nerve head makes those measurements difficult to interpret as well since the deeper layers are supplied from the short posterior ciliary arteries and choroid and the more superficial layers are supplied by the retinal circulation. If the measurement depth is 500–1,000 μm, it should be sufficient to measure in both layers, and indirect evidence suggests that is the case [4, 15]. However, in a primate model, the optic nerve head flux failed to decrease when the posterior ciliary arteries were disrupted, although the flux decreased with central retinal artery occlusion and more so when both supply sources were disrupted [12]. These results suggest that most of the Doppler-shifted photons were captured from the superficial layer of the optic nerve head. However, the role of collateral contributions to the flux signal is difficult to assess in this model. Moreover, to avoid inclusion of large vessels, the distance between the incident beam and the collection region was small, which favors a more shallow measurement. Thus, the measurement depth in the optic nerve head remains unclear.

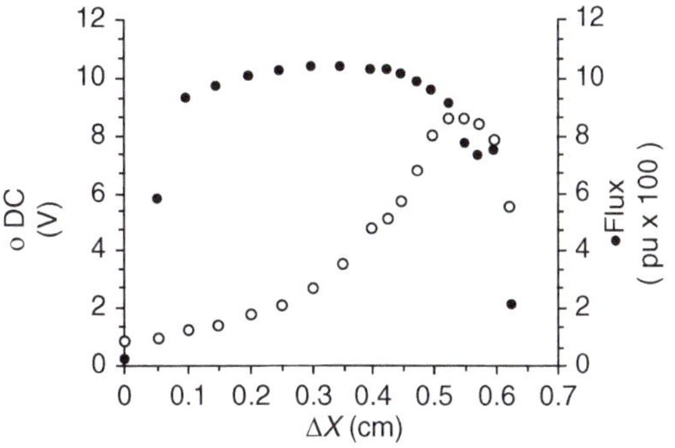

Fig. 3.12 Changes in flux reading and DC voltage from photodetector as an intravitreal probe is moved closer to the retinal surface in an anesthetized rabbit with tissue contact at approximately 0.55 cm [7]

3.10 Caveats

Although the signal processing in laser Doppler flowmeters compensates for small changes in the amount of light returned to the photodetector [21], large changes in collected light alter the DC voltage and can affect the flux measurement. This most often occurs when the distance between the probe and the tissue changes (Fig. 3.12), which in turn changes the volume of tissue sampled as well as the calculation of the number of moving particles, which is derived from the ratio of the AC to DC voltage. Consequently, it is important to maintain the distance from probe to tissue constant, which can be verified by monitoring the DC signal [7]. From a practical standpoint, this means that the probe cannot be held with a fixed manipulator but must be adjusted as needed or held with a device that maintains continual probe-tissue contact with a consistent force. The authors use a modified record player tonearm for ciliary blood flow measurements, while skin blood flow measurements are usually made with the probe held to the tissue with an adhesive plastic disk. A fixed manipulator holding the probe in contact with the tissue is also problematic if the tissue moves, since movement toward the probe will occlude the underlying vessels and movement away from the probe will relieve any baseline pressure on the underlying vessels. Viewed with an operating microscope, it is evident that the eye moves slightly with each cardiac and respiratory cycle, and large changes in systemic blood pressure, as occur with systemic drug administration, move the eye to a greater extent.

Monitoring the DC voltage can also be used to insure the probe remains at the same location. As noted earlier, the flux changes as the probe is moved from one location to another on a tissue, and this can be a problem if measurements are needed for different measuring sessions since the original site must be relocated. However, this can also be useful for selecting a measurement site if the anatomy and flow characteristics under the probe are known, as in the case of transcleral measurements of the anterior structures of the eye which have a distinct flux profile as the probe is moved posteriorly from the limbus (Fig. 3.13) [9]. Once a site is selected though, it is imperative that the probe remain at the same location, since a slight movement of the probe away from the original site may give a very different flux reading derived from a different tissue. Often, but not always, a new site has different light scattering properties revealed by a change in the DC, which can prompt reverification of the probe location. The authors' record player tonearm probe holder helps to maintain a constant probe position.

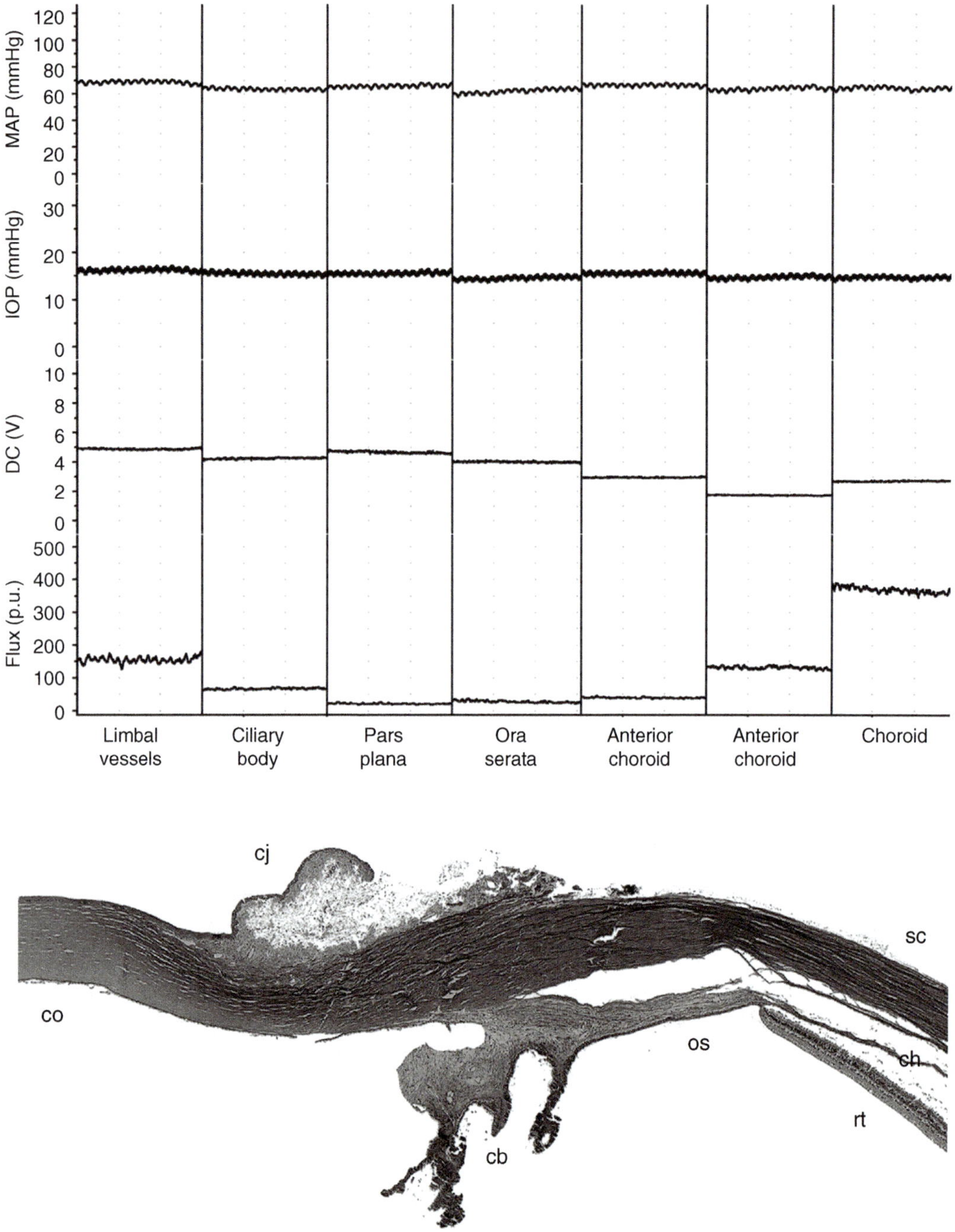

Fig. 3.13 Transcleral flux profile in the anesthetized rabbit corresponds to the underlying tissue structures (*co* cornea, *cj* conjunctiva, *cb* ciliary body, *os* ora serrata, *rt* retina, *ch* choroid, *sc* sclera) (Modified from Kiel et al. [9])

Finally, the greatest virtue and weakness of laser Doppler flowmetry is that it is an exquisitely sensitive motion detector. It cannot distinguish between Doppler shifts caused by moving blood cells and motion artifacts, and so it is the user's responsibility to eliminate all non-blood flow-related movement to insure valid flux measurements. In experiments with anesthetized animals, this can be achieved by pharmacologically paralyzing the animal to eliminate saccadic eye movement and blinks (which requires respiring the animal with a ventilator and regular corneal hydration to prevent desiccation) and placing the head in stereotaxic device to prevent transmission of respiratory movements to the head. The table or platform on which the animal is placed must also be isolated from all sources of vibration such as the respirator or equipment with cooling fans (e.g., the flowmeter) or moving parts (e.g., a capnograph). Euthanizing an animal and verifying that the flux goes to zero is the best way to insure that all extraneous movements have been eliminated in an experimental setup.

References

1. Bonner R, Nossal R (1981) Model for laser Doppler measurements of blood flow in tissue. Appl Opt 20:2097–2107
2. Bonner R, Nossal R (1990) Principles of laser-Doppler flowmetry. In: Shepherd A, Öberg P (eds) Laser Doppler blood flowmetry. Kluwer Academic Publishers, Norwell
3. DiResta G, Riedel G, Kaplan P et al (1987) Hybrid blood flow probe for simultaneous H2 clearance and laser-Doppler velocimetry. Am J Physiol 253:G573–G581
4. Harino S, Riva C, Petrig B (1992) Intravenous nicardipine in cats increases optic nerve head but not retinal blood flow. Invest Ophthalmol Vis Sci 33:2885–2889
5. Johnson J, Taylor W, Shepherd A et al (1984) Laser-Doppler measurement of skin blood flow: comparison with plethysmography. J Appl Physiol 56:798–803
6. Kiel J, Shepherd AP (1990) Gastrointestinal blood flow. In: Shepherd AP, Öberg PA (eds) Laser-Doppler blood flowmetry. Kluwer Academic Publishers, Norwell
7. Kiel JW, Shepherd AP (1992) Autoregulation of choroidal blood flow in the rabbit. Invest Ophthalmol Vis Sci 33:2399–2410
8. Kiel JW, Riedel GL, DiResta GR et al (1985) Gastric mucosal blood flow measured by laser-Doppler velocimetry. Am J Physiol 249:G539–G545
9. Kiel JW, Reitsamer HA, Walker JS et al (2001) Effects of nitric oxide synthase inhibition on ciliary blood flow, aqueous production and intraocular pressure. Exp Eye Res 73:355–364
10. Nilsson G, Tenland T, Oberg P (1980) Evaluation of a laser Doppler flowmeter for measurement of tissue blood flow. IEEE Trans Biomed Eng BME-27: 597–604
11. Öberg P (1990) Innovations and precautions. In: Shepherd A, Öberg P (eds) Laser Doppler blood flowmetry. Kluwer Academic Publishers, Norwell
12. Petrig BL, Riva CE, Hayreh SS (1999) Laser Doppler flowmetry and optic nerve head blood flow. Am J Ophthalmol 127:413–425
13. Polska E, Luksch A, Ehrlich P et al (2002) Measurements in the peripheral retina using LDF and laser interferometry are mainly influenced by the choroidal circulation. Curr Eye Res 24:318–323
14. Riva C, Ross B, Benedek GB (1972) Laser Doppler measurements of blood flow in capillary tubes and retinal arteries. Invest Ophthalmol 11:936–944
15. Riva C, Harino S, Petrig B et al (1992) Laser Doppler flowmetry in the optic nerve. Exp Eye Res 55:499–506
16. Riva CE, Cranstoun SD, Mann RM et al (1994) Local choroidal blood flow in the cat by laser Doppler flowmetry. Invest Ophthalmol Vis Sci 35:608–618
17. Riva CE, Hero M, Titze P et al (1997) Autoregulation of human optic nerve head blood flow in response to acute changes in ocular perfusion pressure. Graefes Arch Clin Exp Ophthalmol 235:618–626
18. Roman R (1990) Renal blood flow. In: Shepherd A, Öberg P (eds) Laser Doppler blood flowmetry. Kluwer Academic Publishers, Norwell
19. Shepherd A (1990) History of laser Doppler flowmetry. In: Shepherd A, Öberg P (eds) Laser Doppler blood flowmetry. Kluwer Academic Publishers, Norwell
20. Shepherd A, Kiel J (1989) Gastrointestinal blood flow-measuring techniques. In: Schultz J, Wood J (eds) Handbook of physiology: the gastrointestinal system. American Physiological Society, Bethesda
21. Shepherd AP, Öberg PA (1990) Laser-Doppler blood flowmetry. Kluwer Academic Publishers, Norwell
22. Shepherd AP, Riedel GL, Kiel JW (1987) Evaluation of an infrared laser-Doppler blood flowmeter. Am J Physiol 252:G832–G839
23. Smits GJ, Roman RJ, Lombard JH (1986) Evaluation of laser-Doppler flowmetry as a measure of tissue blood flow. J Appl Physiol 61:666–672
24. Stern M (1975) In vivo evaluation of microcirculation by coherent light scattering. Nature 254:56–58
25. Stern M, Bowen P, Parma R et al (1979) Measurement of renal cortical and medullary blood flow by laser-Doppler spectroscopy in the rat. Am J Physiol 236:F80–F87
26. Stern M, Lappe D, Bowen P et al (1977) Continuous measurement of tissue blood flow by laser-Doppler spectroscopy. Am J Physiol 232:H441–H448
27. Tanaka T, Riva C, Ben-Sira B (1974) Blood velocity measurements in human retinal vessels. Science 186:830–831
28. van Heuven W, Kiel J, Elliott W et al (1996) Evaluation of the Heidelberg retina flowmeter. Invest Ophthalmol Vis Sci 37:S967
29. Yu DY, Townsend R, Cringle SJ et al (2005) Improved interpretation of flow maps obtained by scanning laser Doppler flowmetry using a rat model of retinal artery occlusion. Invest Ophthalmol Vis Sci 46:166–174

Oxygen Measurements in Animals

4

Robert A. Linsenmeier

Core Messages

- Retinal oxygen measurements in animals and humans, made with several techniques, provide information about the effectiveness of the circulation in meeting the metabolic demands of the retina.
- Under normal conditions, PO_2 in the outer retina of cats and monkeys varies from a high of about 50 mmHg at the choriocapillaris to a low of 5 mmHg or less around the photoreceptor inner segments.
- PO_2 in the inner retina averages about 20 mmHg, but is heterogeneous.
- Hyperoxia dramatically increases PO_2 in the outer retina. Increases in the inner retina are much smaller. Hypercapnia makes all these changes larger.
- Light decreases oxygen utilization in the outer retina by as much as a factor of two, and increases PO_2. Steady light does not change the metabolism of the inner retina, but flickering light can increase it relative to darkness.
- Hypoxemia, retinal vascular occlusions, retinal detachment, and later stages of diabetic retinopathy all lead to decreased retinal PO_2. Changes glaucoma and early diabetic retinopathy are uncertain.

4.1 Introduction

Measurements of oxygen partial pressure (PO_2) and blood oxygen saturation (SO_2) are not equivalent to measurements of blood flow, but, on a moment-to-moment basis, the most critical job of the circulation in most organs is the delivery of oxygen. One of the important reasons for studying blood flow in the first place is the hope that it will be a surrogate for tissue metabolism and the condition of the tissue, but the ability to understand normal physiology or pathophysiology on the basis of blood flow measurements alone is limited. Coupling blood flow measurements with blood oxygen saturation measurements provides more information about retinal metabolism, as discussed further below. Tissue oxygen measurements on their own, without simultaneous blood flow measurements, show the result of what the circulation provides to the cells. Information about oxygen is therefore critical to an understanding of the performance of the retinal and choroidal circulations under physiological and pathological conditions.

For the purposes of oxygen supply and metabolism, the retina is best thought of as two separate

R.A. Linsenmeier, Ph.D.
Departments of Biomedical Engineering, Neurobiology and Ophthalmology, Northwestern University, 2145 Sheridan Road, Evanston, IL 60208-3107, USA
e-mail: r-linsenmeier@northwestern.edu

L. Schmetterer, J.W. Kiel (eds.), *Ocular Blood Flow*,
DOI 10.1007/978-3-540-69469-4_4, © Springer-Verlag Berlin Heidelberg 2012

organs, the inner retina and the outer retina, which are about equal in thickness. The oxygen supply of the inner retina shares many properties with that of the brain, but has unusual aspects because of the proximity of the vitreous, the leakiness of the arterioles to oxygen, and the interaction between the retinal and choroidal circulations in providing oxygen under certain conditions. The oxygen supply to the outer retina (RPE and photoreceptors) is unlike that of any other organ, because its main supply, the choroid, is quite unusual and because the metabolically demanding photoreceptors are in an avascular layer.

Work on the role of altered oxygenation in retinal diseases began in the 1950s by Ashton [24, 26] and Patz [144, 145] for retinopathy of prematurity and arterial occlusion, and the first important measurements of oxygen in the eye of animals were made in the early 1970s by Alm and Bill [14, 16]. Since these publications, there has been a steady stream of work so that good descriptions of many phenomena, both physiological and pathophysiological, can be given, and several reviews have appeared focusing on different aspects of oxygen measurement, oxygen supply, and oxidative metabolism [23, 92, 113, 188, 189, 205, 229]. This section will discuss the several ways in which oxygen measurements are made in animals, explain basic physiological properties of retinal and optic nerve head oxygenation, and relate these to pathophysiology.

4.2 Measurements of PO_2 and Saturation

The importance of oxygen in the eye has stimulated the development of several methods of measurement, each of which has both advantages and disadvantages [92, 205, 229]. The methods discussed here are (1) oxygen electrodes, (2) histology with Hypoxyprobe, (3) magnetic resonance imaging, (4) phosphorescent dyes, and (5) reflection oximetry or spectrophotometry. These are all used in vivo, sometimes in conjunction with measurements of blood flow. For understanding the use of oxygen in the retina, in vitro measurements have also proven to be important and usually involve the use of oxygen electrodes, although not always microelectrodes.

4.2.1 Oxygen Electrodes

Oxygen-sensitive electrodes work on the same principle as those in blood gas machines, that is, polarography. When a noble metal such as gold or platinum is polarized at about –0.7 V with respect to a reference electrode, it becomes a cathode at which oxygen is chemically reduced [45, 209]. Nothing besides oxygen is reduced in tissue, and the electrode is selective for oxygen, unlike the situation when an electrode is polarized at a positive potential and can generate various oxidation products. The reduction of oxygen generates a current proportional to the oxygen tension (PO_2) in the medium, which could be as little as a few picoamps or as much as a few nanoamps, depending on the size of the cathode. While these electrodes consume oxygen, for microelectrodes, it is a small amount that does not substantially change the oxygenation of the tissue. If the stoichiometry at the cathode is $O_2+4e^-+2H_2O \rightarrow 4OH^-$ [45] and the microelectrode produces 6 pA of current, a typical value in the inner retina, then making simple unit conversions allows one to show that the electrode uses the same amount of oxygen as a region 5 μm in radius that has an oxygen consumption of 4 ml O_2-100 g^{-1}-min^{-1}, also a typical value for inner retina. That is, the electrode simply acts like one more cell in the retina. Oxygen electrodes of different types have been used in the vitreous humor (e.g., [12, 16, 64, 158, 224]), the retina (e.g., [6, 37, 116, 156, 227]), and the optic nerve head (e.g., [3, 63, 104, 107, 136, 148, 188]). For intraretinal measurements, there is an advantage in using double-barreled microelectrodes, in which one barrel is an oxygen cathode and the other measures voltage (Fig. 4.1). The voltage barrel records the intraretinal electroretinogram, which indicates the physiological state of the retina and the quality of the electrode penetration. However, to the extent that it can be determined, single-barreled microelectrodes (i.e., without a voltage barrel) give similar intraretinal measurements.

A calibration (i.e., current vs. PO_2) must be obtained for each electrode, and investigators must

Fig. 4.1 Double-barreled oxygen microelectrode. The top barrel has a gold cathode, recessed from the tip. This helps to confine the diffusion field to the recess and allows measurements with high spatial precision. The LMA is a low-melting-point alloy that serves as a substrate onto which gold is plated. The lower barrel is filled with saline and is used to record intraretinal voltages, such as the electroretinogram

assume that the calibration done in vitro applies in vivo. The microenvironment may be slightly different in the eye than in the calibration system in terms of temperature and diffusion properties, however, so PO_2s obtained in this way should be considered good approximations rather than exact values. One can have more confidence in differences between two situations (e.g., light and dark) in the same animal than in those that rely on comparing measurements across animals. However, sometimes, there is no alternative to drawing conclusions from interanimal comparisons. Absolute measurements can be made with Clark-type oxygen electrodes, which have both the anode and cathode in the same environment under a gas-permeable membrane, but these are larger and can only be used in the vitreous humor.

Outweighing the disadvantages of microelectrodes for measurements in animals are their great advantages. These include (1) their spatial resolution in giving intraretinal PO_2, which is unequaled by any other technique; (2) their temporal resolution, with the time required for 90% of the complete response to a PO_2 change of 25–50 ms [169], much faster than any biological changes in PO_2 in the retina [41, 55]; and (3) their ability to measure tissue PO_2 rather than intravascular PO_2. Oxygen electrodes have been used primarily in animals, although a few measurements have been made in the vitreous of humans [94, 126, 168].

4.2.2 Hypoxyprobe

Pimonidazole (Hypoxyprobe) is a chemical that irreversibly forms adducts in tissue when the PO_2 is below some critical level, usually given as 10 mmHg [161]. These pimonidazole-protein adducts can be detected histologically using a monoclonal antibody [52]. Pimonidazole has been used primarily to investigate hypoxia in animal and human tumors, but it has also been used in some normal tissues, including the retina [57, 58, 73, 93].

Pimonidazole histology could be an important compliment to measurements with microelectrodes because there is the possibility of sampling more tissue, of leaving the eye completely undisturbed during measurements, and of making measurements in eyes that are too small or in animals that are too delicate for physiological measurements (e.g., neonatal animals or potentially long-term diabetic rats or mice). The disadvantages are that this technique (1) can give only one set of measurements per animal, (2) only gives a binary readout of PO_2 (i.e., either higher or lower than the cutoff PO_2), and (3) does not have a well-validated cutoff PO_2.

4.2.3 Magnetic Resonance Imaging

Magnetic resonance imaging (MRI) consumes no oxygen and can even be used in very small eyes of neonatal rodents [30]. MRI has been used in three ways. One method employs the property that the T_1-weighted image of hydrogen in the water in the eye is modulated by local PO_2 [29, 96]. The water T_1 is sensitive to factors other than oxygen, so it does not provide an absolute measurement, but changes in this signal when an animal or person breathes oxygen or carbogen (95% O_2 and 5% CO_2) do reflect oxygen changes in the retina. The MRI responses have been called ΔPO_2. Because the retina is thinner than a standard voxel, the voxel containing the retina may include choroid as well as inner and outer retina, so detail cannot be resolved. Consequently, the approach has primarily been to make measurements of the row of voxels in the vitreous in front of the retina [96], on the good assumption that the oxygen there has diffused from the retinal circulation (Fig. 4.2). With longer acquisition times, higher field strengths, and good eye

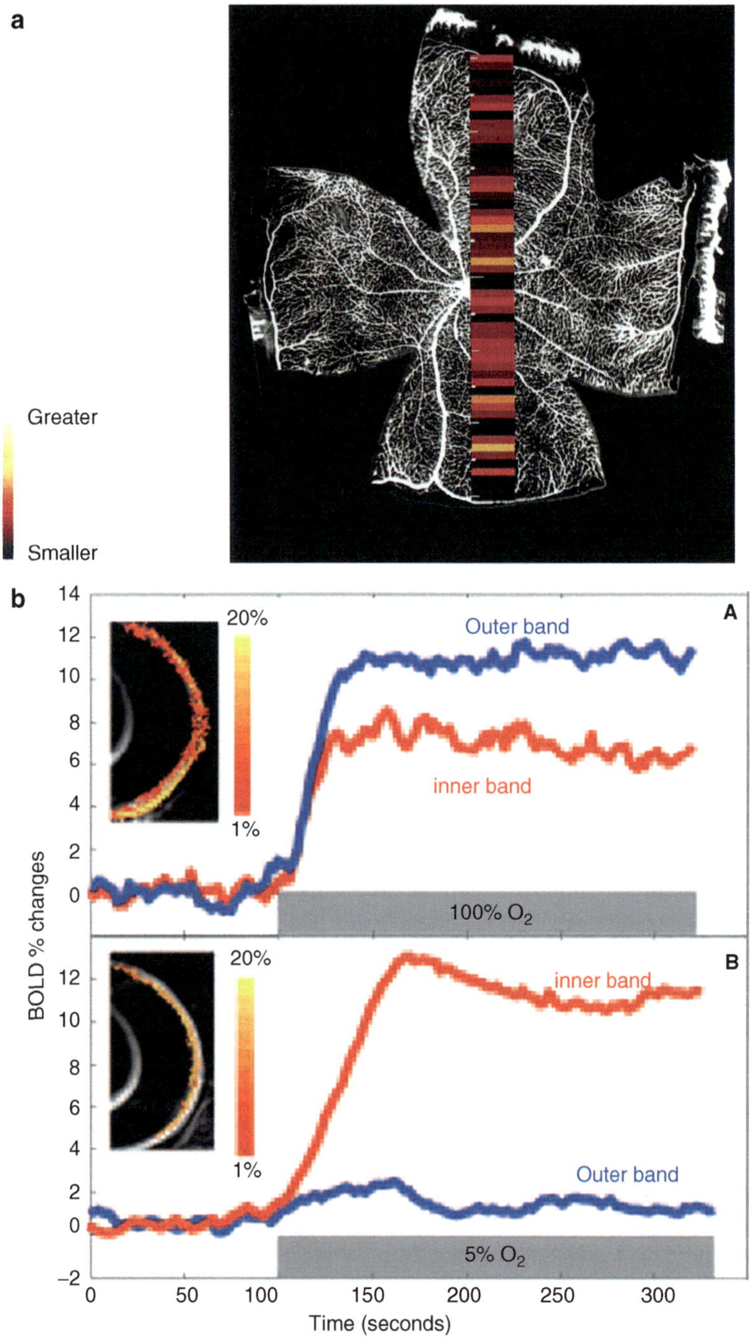

Fig. 4.2 (**a**) Average ΔPO_2 band derived from two mice during a 2-min carbogen challenge was overlaid on a representative FITC-dextran infused flat mount. The median panretinal ΔPO_2 was 102 mmHg. Pixels from the superior portion of the retina are at the *top* of the figure. The *longer white tick mark* in the center of each band represents the posterior pole near the optic nerve; the *shorter tick marks* represent 0.5-mm increments along the retinal surface [96]. Published with kind permission of © Elsevier 2001. (**b**) Differential layer-specific BOLD fMRI of the retina. Lamina-specific BOLD fMRI responses to (*A*) hyperoxia (100% O_2) and (*B*) hypercapnia (5% CO_2 in air) from a normal rat. BOLD percent change maps are overlaid on echoplanar images. The color bar indicates 1–20% BOLD changes [50]. Reprinted with the kind permission of the National Academy of Sciences, U.S.A. © 2006

stabilization, it is possible to detect some structure in the retina itself and even do the second type of measurement, BOLD (Blood-oxygen-level-dependent) imaging [50, 61]. BOLD images are dependent on both blood flow and hemoglobin saturation and again are useful for investigating changes in oxygenation rather than absolute values. In contrast, the third magnetic resonance method can provide absolute values. This technique relies on placing a small droplet of perfluorocarbon, such as perfluorotributylamine (PFTBA), in the vitreous against the retina [96]. One then measures the inverse T_1 of fluorine (^{19}FNMR), which depends only on oxygen and temperature, and can therefore be interpreted directly in terms of PO_2. The fluorine technique yields a measurement of the average vitreal PO_2 close to the retina, which is taken to be an estimate of the average inner retinal PO_2 near the droplet. Measurements made in this way over the human macula give values of 6–9 mmHg [211], and measurements in the vitreous of normal rabbits give values of 22 ± 9 mmHg [71] and 39.4 ± 9.2 mmHg [210] over the vascularized retina near the optic disc, in reasonably good agreement with electrode measurements [168].

4.2.4 Phosphorescence Decay

Oxygen quenches the phosphorescence of certain palladium-porphyrin compounds, so for these molecules, the phosphorescence lifetime following excitation is inversely proportional to PO_2 [121, 196]. Alternatively, in response to time-varying excitation of the dye, one can measure a phase shift in the phosphorescence signal. Unlike the techniques reviewed so far, phosphorescence has been used to give information about intravascular PO_2 rather than tissue PO_2 because the dyes bind to albumin. Compounds with excitation maxima near 500 nm and photon emission in the deep red or near infrared have been bound to albumin and injected intravenously in rodents to image the larger retinal arteries and veins as well as the optic nerve head [175, 176, 212, 213] (Fig. 4.3a). Unfortunately, because of their toxicity, these dyes are not available for use in humans. As expected, results with these dyes show that the PO_2 is higher in retinal arteries than in veins, and it is sometimes possible to see gradients along vessels. This method might also be expected to give PO_2s in retinal capillaries as well, but it is uncertain whether signals in areas between arteries and veins are purely from retinal capillaries or contain a contribution from the choroid vessels. A way around this is with retinal slice imaging combined with a phosphorescent dye [170–172]. In this technique, the excitation beam and viewing angle are obliquely oriented with respect to the retina rather than being along the optic axis. This clearly reveals retinal vessels and choroidal circulation separately (Fig. 4.3b). In one study, a different phosphorescent dye, sodium pyrenebutyrate, was added to the Ringer solution bathing a retina in vitro so that the PO_2 in extravascular tissue could be imaged [237]. Intravitreal injection of the dye may eventually be developed to allow mapping of tissue PO_2 in vivo.

4.2.5 Oximetry

In contrast to all the other techniques, oximetry gives values for hemoglobin saturation, or concentration of oxygen, rather than PO_2. Saturation is related to PO_2 in a nonlinear way that depends on the shape of the hemoglobin saturation curve, but, except in hyperoxia where arterial PO_2 increases without an increase in saturation, PO_2 and saturation can be interconverted if the parameters characterizing the hemoglobin saturation curve are known. Like phosphorescence, oximetry gives an intravascular PO_2 rather than tissue PO_2, but, unlike phosphorescence, it can be used in humans. The principles and results were recently reviewed [83], and this method is considered further elsewhere in this book. A number of investigators have tackled the job of determining oxygen saturation in retinal vessels, which is difficult for a variety of reasons [83], including possible influences of the choroid on the reflected light and differences in background absorption and fundus reflection among subjects. However, the basic information from the various kinds of oximetry is largely consistent with the earliest measurements by Hickam and Frayser [69, 88, 89] in the important result that retinal arterial blood is almost fully saturated and that retinal venous blood has a saturation of about 60%, which is lower than in most organs. While oximetry is sometimes used in animal studies [103], these are often done to validate techniques that are ultimately designed for humans.

4.3 Vitreal, Intraretinal, and Intravascular Oxygen in Holangiotic Retinas

While all vertebrates have a choroidal circulation, not all have a retinal circulation. The species that do have a retinal circulation are all mammals

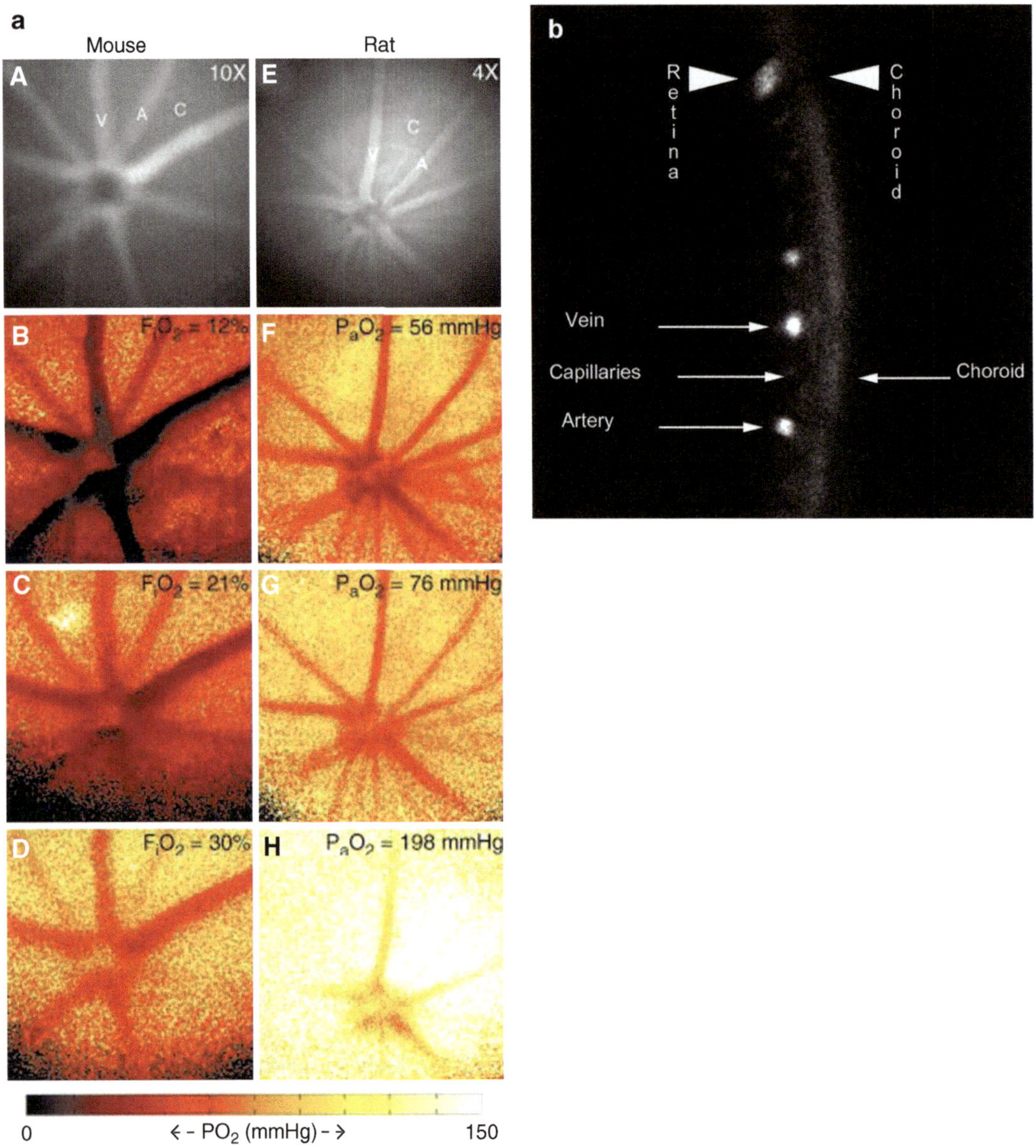

Fig. 4.3 (**a**) Phosphorescence intensity images (A and E) and colorized two-dimensional maps of PO_2 in one mouse retina (B–D) and one rat retina (F–H) at different inspiratory oxygen fractions. Images were taken through a 10× microscope objective (mouse) or 4× microscope objective (rat). F_iO_2 is indicated for each mouse map, while actual arterial blood gas oxygen tensions are indicated for each rat map. The arterial (*A*), venous (*V*), and capillary (*C*) regions are indicated [176]. Reprinted with the kind permission of the Biomedical Engineering Society © 2003. (**b**) Optical section phosphorescence image shows the retinal and choroidal vasculatures displaced in depth. The retina and choroid are on the left and right side of the image, respectively, as indicated by *arrowheads*. Retinal artery, vein, capillaries, and choroid are indicated by the *arrows*. Phosphorescence is quenched more at higher PO_2, so the vein is brighter than the choroid or artery [170]. Reprinted with the kind permission of Informa PLC © 2006

and are said to have holangiotic retinas. These include nonhuman primates, dogs, pigs, cats, rats, mice, cows, and some other species [220]. The following section focuses on results in these species.

4.3.1 Vitreal Oxygen

Measurements of vitreal PO_2 close to the retina or preretinal vitreous PO_2 ($PvrO_2$) date to the 1950s [76]. Vitreal measurements continue to be useful

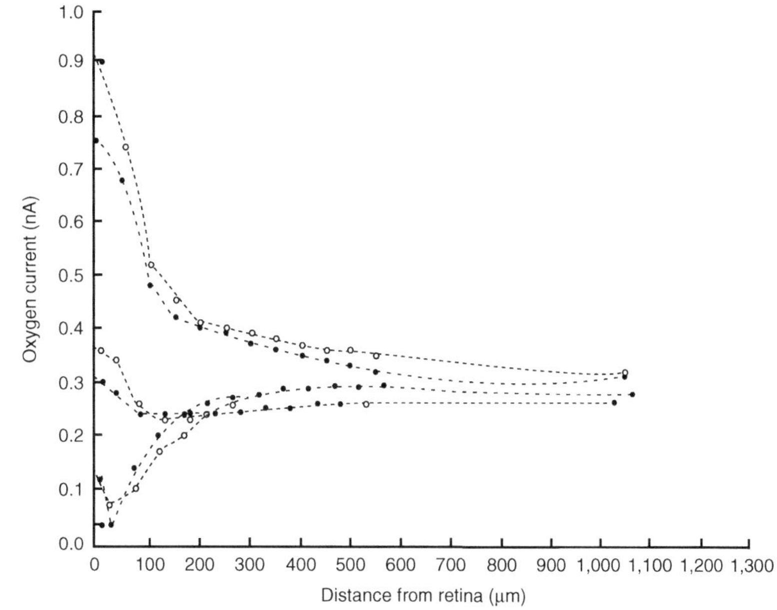

Fig. 4.4 Preretinal vitreal PO_2 profiles during air breathing measured near an artery (*open* and *closed circles*), vein (*open* and *closed squares*), and intermediate position (*open* and *closed diamonds*). The *open symbols* were obtained during the withdrawal of the microelectrode from the retina and the *closed symbols* from a subsequent advance to the retina [7]. Reprinted with the kind permission of Informa PLC © 1985

because they are less invasive and sometimes use Clark-type electrodes that have the advantage of an internal reference [8, 12, 16, 42, 158, 181–183]. In cats, which appear to be representative, the average $PvrO_2$ under baseline conditions (i.e., not hyperoxic, hypoxic, or ischemic) is about 20 mmHg (variously reported as having average values of 18.9 [16], 20–30 [42], 15–20 [114], 20.2 [6], and 19 [182] mmHg). In rat, the average in the midvitreous, which is only about 500 μm from the retina, is 22.6 mmHg [12]. While the oxygen in the vitreous must come largely from the retina [42, 114], the details of the oxygen gradients within a few hundred micrometers of the vitreal-retinal interface are complex [12, 46]. Near arterioles and venules in cat [7] and rat [12], there is a gradient of decreasing PO_2 from the retina into the vitreous (Fig. 4.4). In contrast, in regions away from the ophthalmoscopically visible vessels, the PO_2 is generally lower at the retinal surface than further out in the vitreous in cat [7, 44] and in some (Lau and Linsenmeier, unpublished observations), but not all [12, 225], measurements in rat. These differences in the gradients have at least two interesting implications. First, measurements of $PvrO_2$ tend to overestimate the PO_2 in much of the inner retina, so, when possible, intraretinal measurements are preferable. In a series of measurements in cat that included both the inner retina and vitreous, the vitreal PO_2 was higher than inner retinal PO_2 in normal regions, but frequently lower than inner retinal PO_2 after photocoagulation of the outer retina [44], so, while photocoagulation had a significant effect on intraretinal PO_2, its effect on vitreal PO_2 was not significant. Second, the vitreous must be supplying much of the retina with some oxygen, and this includes at least part of the fovea in primates [2, 233]. Of course, ultimately, this oxygen is derived from the retinal circulation, but from somewhat remote vessels. Ordinarily, the diffusion of oxygen from the vitreous probably provides little of the demand of the inner retina, but the exact amount has not been determined.

4.3.2 Intraretinal Oxygen

Intraretinal microelectrode measurements have been made under many conditions in cat and rat and, to a lesser extent, in primates, pigs, rabbits, and guinea pigs. Examples of oxygen "profiles" recorded in dark adaptation for the central retina of primate, cat, and rat are shown in Fig. 4.5. These profiles were measured during electrode withdrawal from the choroid to the vitreous. There is a relatively high PO_2 at the choroid and

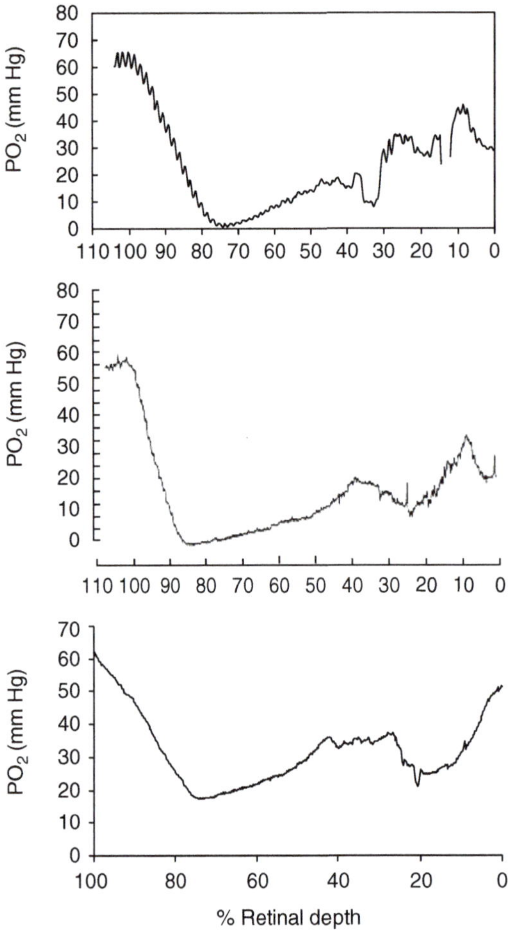

Fig. 4.5 Oxygen profiles, measured during electrode withdrawal from the choroid, from the dark-adapted retinas of monkey (*top*), cat (*middle*), and Long-Evans rat (*bottom*). Zero percent depth is the interface between retina and vitreous, and 100% is the choroid. The outer 50% of the retina is avascular, and the inner 50% contains the retinal circulation, which accounts for the local peaks (*Top*: [37], *middle*: [121], *bottom*: Lau and Linsenmeier, unpublished). *Middle:* Reprinted with kind permission of Springer © 2005

a steep drop across the avascular photoreceptor layers in the outer half of the retina. In the cat and primate retina, the lowest PO_2s average only a few mm Hg around the photoreceptor inner segments [2, 37, 115, 118], and PO_2 rises again through the outer nuclear layer. In rat, the PO_2 is higher in the inner segments [225, 228, 230] (Lau and Linsenmeier, unpublished observations) than it is in cat or primate.

In the inner half of the retina, PO_2 is more variable, consistent with the presence of the capillaries of the retinal circulation. Mean intraretinal PO_2 from many measurements in cat was 18.1 ± 12.6 mmHg (mean and SD) [118]. The distribution was not normal, but had a long tail, so that the data were skewed toward the high end, with some values above 60 mmHg. However, there are also a substantial number of values below 5 mmHg. Evidently, there is little functional consequence of this variation in PO_2. In cat and primate, there are generally two peaks in oxygen in the inner retina [118, 222], but no detectable consistent pattern to the PO_2 gradients. In rat, there is often a distinct minimum between capillary layers that is probably in the inner nuclear layer [55, 225, 229].

As discussed below, illumination has a large effect on retinal metabolism and therefore on outer retinal PO_2, and it may affect inner retinal PO_2 as well.

4.3.3 Oxygen in Retinal and Choroidal Blood

In humans, arterial saturation (S_aO_2) is typically 19 vol.% (i.e., ml O_2/100 ml blood) or greater, but animals often have less hemoglobin and therefore lower maximum saturation. Arterial saturation is about 12.6% in cats [87] and 12.3 [194] to 13.8 vol.% [200] in pigs. The vessels commonly called retinal arteries are actually arterioles, and evidently, enough oxygen leaves the arterioles that capillaries are not needed in their immediate vicinity [86], causing a capillary-free zone. As noted above, average retinal venous saturation ($S_{vR}O_2$) is about 8 vol.% lower than S_aO_2 under normal conditions in humans, and it is about 6.6 vol.% lower than S_aO_2 in pig [200], reflecting a larger arteriovenous difference than in most organs in the body.

In contrast, the average saturation in choroidal veins ($S_{vC}O_2$) is quite high, only about 0.4 vol.% [200] or 0.25 vol.% [194] below S_aO_2 in pigs and 0.94–1 vol.% [14, 198] below S_aO_2 in cat. If the vortex vein saturation were at or below the average choriocapillaris saturation, then the predicted PO_2 of the choriocapillaris would be no lower than 68 mmHg, given the cat hemoglobin saturation curve [87], and choriocapillaris PO_2 would be similar in pig [27]. While choriocapillaris PO_2 is sometimes this high, the average values of choriocapillaris PO_2 from microelectrode measurements are considerably lower than this prediction, 48 ± 13 mmHg (mean and SD) in monkey [37], 54 ± 11.7 [40] and 41.2 ± 16.2 mmHg [141] in cat and 53 mmHg [227] in rat. Thus, there is a discrepancy between choroidal venous saturation and choroidal venous PO_2 that is currently unresolved. The explanation may be in the method of measuring $S_{vC}O_2$, which was not oximetry, as in retinal vessels, but by drawing blood from the vortex veins. This is somewhat problematic because these veins drain not only the choroid but also the anterior uvea. Further, they are global measures that cannot account for any variation of choroidal saturation with eccentricity or any choroidal shunts that bypass the choriocapillaris. If the arteriovenous difference in the anterior uvea is smaller, or if there were substantial shunting around the choriocapillaris, the vortex vein measurements would overestimate $S_{vC}O_2$ and come into better agreement with the PO_2 measurements. Microelectrode measurements also are potentially subject to some error, however, if, for instance, the surgical manipulations of the eye needed to make the measurements impaired choroidal blood flow, or if penetration of the choriocapillaris with the electrode were to compress the circulation locally. In this context, it is worth noting that noninvasive phosphorescence measurements tend to agree with the microelectrode data, giving a choriocapillaris PO_2 of 58 ± 2 mmHg in rat [171]. While the very small oxygen extraction from the choroid is often cited as one of the hallmarks of this circulation, the PO_2 measurements suggest that the arteriovenous difference is somewhat larger and that choriocapillaris saturation is more like 2 or 2.5 vol.% below S_aO_2.

4.4 Oxygen in Avascular Retinas

All animals have a choroidal circulation, but not all homeotherms have a full retinal circulation. Instead, they may be merangiotic, like rabbits, which have retinal circulation only in the medullary wings near the optic disc, paurangiotic, like guinea pigs, which have even shorter retinal vessels, or anangiotic, like bats and birds, which have no retinal circulation [220]. Functionally, most of the retina in all of these animals is avascular. The inner retina of rabbits is well studied and relies heavily on anaerobic glycolysis rather than oxidative metabolism [19, 123, 231], as does the inner retina of guinea pig [226]. The avascular retinas also tend to be thinner than the holangiotic retinas [47], which may be necessary to allow adequate glucose supply or clearance of hydrogen ion and lactate into the choroid. The oxygenation of the outer retina of rabbit and guinea pig is similar to that in cat or rat, but their inner retinas have very low PO_2s [226, 231]. Little is known about PO_2s or oxygen utilization in other animals with minimal retinal circulation. In birds, there has been speculation that the vascularized pecten provides oxygen to the inner retina via diffusion through the vitreous [152], and in fruit bats, there are indentations of the retina by the choroid [149], which may bring more oxygen to the inner retina than in rabbits and guinea pigs.

Very little has been done with respect to retinal oxygenation in poikilotherms, except fish. In the trout retina, and presumably that of other fish species, there is no shortage of oxygen in the inner retina even though there is no retinal circulation. During inspiration of air-saturated water, the PO_2 at the choroid is almost 400 mmHg due to a countercurrent oxygen multiplication system that is believed to be similar to the swim bladder [65] in concentrating oxygen. This allows inner retinal PO_2 to be over 100 mmHg [59].

4.5 Analysis of Retinal Oxygen Utilization

The retina is often said to have the highest oxygen utilization (QO_2), or the highest total metabolic rate, of any tissue. These statements are not easily substantiated, but it is true that the oxygen utilization per gram of tissue is among the highest in the body, similar to cerebral oxygen consumption (3–5 ml O_2-100 g^{-1}-min^{-1}) [110,

125] in monkeys and cats [37]. The inner retina does not appear to have unusual metabolic properties compared to the brain in holangiotic retinas, but the photoreceptors do. Most of the photoreceptor mitochondria, except for a few in the synaptic terminals, are in the inner segments (IS). The IS are no more than a quarter of the total photoreceptor length, so the local QO_2 in that small region is quite high. Furthermore, even the very high QO_2 of the IS does not appear to allow sufficient ATP production, so the photoreceptors require glycolytic energy production as well. Based on the observation that 80% of the glucose is used for lactate production in darkness [18, 200], and that glycolysis yields about 18 times less ATP, glycolysis accounts for about 18% of the total ATP production of photoreceptors, or possibly more in the rat retina in vitro [217].

There are two ways to obtain information about retinal oxygen utilization. The first relies on a mass balance often called the Fick principle: all the oxygen entering a tissue has to either be consumed or leave the tissue, so the difference between the output and input is the utilization. A related technique used in vitro, initially by Warburg [206], but subsequently by others (e.g., [19, 128, 216, 217]), is to put the retina in a closed chamber and measure the rate at which oxygen decreases. The second method is to recognize that under certain conditions, the shape of the oxygen gradients can be interpreted in terms of just a few parameters, one of which is QO_2. Mathematical models of diffusion and consumption can then be developed and fitted to data.

4.5.1 Fick Principle Analyses

For the in vivo application of the mass balance method, one needs to know the amount of oxygen that is entering the tissue in the arteries and the amount leaving in the veins. The entering O_2 is the blood flow rate times S_aO_2, and the exiting O_2 is the blood flow times the venous saturation:

$$QO_2 = F \cdot (S_aO_2 - S_vO_2) \tag{4.1}$$

where QO_2 is oxygen consumption (ml O_2-100 g^{-1}-min^{-1} or comparable units of mass or moles of O_2 per unit weight per time), F is the volumetric blood flow rate (ml-100 g^{-1}-min^{-1}), and S_aO_2 and S_vO_2 are O_2 saturations in arterial and venous blood (ml O_2/ml blood). While this might seem easy enough to apply, and does give reasonable values for whole tissue QO_2, there are at least three complications in using this method. First, part of the O_2 for the retina comes from the retinal circulation, and part from the choriocapillaris, so that values for retinal and choroidal blood flow rates and retinal and choroidal venous saturations are needed. Equation 4.1 therefore needs to be applied separately to choroidal and retinal circulations, which, roughly speaking, supply the outer and inner halves of the retina:

$$Q_{IR} = F_R \cdot (S_aO_2 - S_{vR}O_2) \tag{4.2}$$

$$Q_{OR} = F_C \cdot (S_aO_2 - S_{vC}O_2) \tag{4.3}$$

Here, Q_{IR} and Q_{OR} are the oxygen consumption of the inner and outer retina, respectively, R denotes retinal circulation, and C denotes choroidal circulation. From the published values, one often needs to convert blood flow in milliliters per minute to flow per unit weight of tissue, in ml-100 g^{-1}-min^{-1}. Finally, the unit weight in the denominator of the flow value should be the weight nourished by the circulation, not the total weight of the retina, and not the weight of the choroid, which has a very low metabolic rate itself. For this correction, one needs to know the weight of retina nourished by each circulation, and assuming that each circulation nourishes half is a reasonable approximation. (As discussed below, about 10% of the photoreceptors' oxygen supply in the dark is from the retinal circulation, so Q_{IR} is slightly overestimated, and Q_{OR} is slightly underestimated in this calculation). After this correction, values obtained from microsphere measurements for normal F_R and F_C from cat, pig, and monkey are about 45 ml-100 g^{-1}-min^{-1} [4, 17, 166, 200] and 1,400 ml-100 g^{-1}-min^{-1} [17, 34, 48, 166, 198, 200], respectively. Values for the arteriovenous differences are discussed above. Taking 6.6 vol.% for ($S_aO_2 - S_{vR}O_2$) and 0.5 vol.% (the average of pig and cat values) for ($S_aO_2 - S_{vC}O_2$):

Table 4.1 In vivo values of oxygen consumption of the dark-adapted outer retina (Q_{OR}) and inner retina (Q_{IR}) of species with holangiotic retinas

Species	Dark Q_{OR}	Light Q_{OR}	Dark Q_{IR}	Light Q_{IR}	Light/dark ratio Q_{OR}	Light/dark ratio Q_{IR}	Method	Reference
Rat	3.7	2.4	2.3	2.3	0.53	0.95	Warburg[a]	[128]
Rat	2.2[b]	1.5			0.69		Model fit	[55]
Cat	4.9	2.5			0.51		Model fit	[40]
Cat	4.1[c]	1.4			0.33		Model fit	[118]
Cat	4.4	2.7			0.61		Model fit	[85]
Cat	3.9	1.5	3.5	3.7	0.38	1.06	Model fit	[41]
Cat	6.3	5.3			0.84		Fick principle	[198]
Pig	7.2	4.4			0.61		Fick principle	[200]
Pig			3.8	3.9		1.03	Fick principle	[199]
Macaque	4.9[c]	3.5			0.72		Model fit	[37]
Macaque		2.7[b,d]					Model fit	[233]

Values have been converted to ml O_2-100 g^{-1}-min^{-1} when necessary from units in the original references. The ratio of outer and inner retinal values in strong steady illumination relative to darkness is given when these are available

[a] Q_{OR} was obtained after pharmacological block of the inner retina with 2-amino-phosphonobutyric acid and kynurenic acid, and Q_{IR} was obtained by subtraction from whole retinal value. Medrano and Fox argued that this reduced Q_{IR} to less than 10% of its initial value, so the block of inner retinal metabolism by this technique was relatively complete

[b] Values given "per cm^2" of retina and converted on the assumption that the outer retina is 0.01 cm thick

[c] Values from a regression of Q_{OR} on choroidal PO_2 at a choroidal PO_2 of 50 mmHg

[d] Assumed to be light-adapted

$$Q_{IR} = 2.97 \text{ ml - } 100 \text{ g}^{-1} - \text{min}^{-1}$$

$$Q_{OR} = 7 \text{ ml} - 100 \text{ g}^{-1} - \text{min}^{-1}$$

These values are composites, calculated to illustrate how one arrives at QO_2 by the Fick method, but only Wang et al. [198–200] have obtained all the necessary values in the same retina. Table 4.1 gives these and measurements derived from other methods.

4.5.2 Steady State Diffusion Theory and the Outer Retina

Oxygen is thought to be transported primarily by simple passive diffusion through neural tissue, not by active transport or convection. By creating mathematical models using diffusion theory, and fitting these models to measurements of PO_2 made with microelectrodes, values for Q_{OR} and Q_{IR} can be obtained. Diffusion theory methods can give much more localized values than are possible with Fick principle methods.

Several assumptions are required in these models. First, the tissue is usually considered to be homogeneous because oxygen moves at least as well through membranes as through extracellular space or cytoplasm (which is not true for the diffusion of ions [134, 140]). A further constraint is that diffusion is allowed only in the dimension parallel to the photoreceptor axis (i.e., perpendicular to the retinal surface). This direction is called here the "x" direction. Oxygen enters this region only at the boundaries, which are the choriocapillaris (at $x=0$) and the inner edge of the avascular region (at $x=L$). The PO_2s at these locations, P_C (at $x=0$) and P_L (at $x=L$), are taken to be constant boundary conditions in steady state models. Mathematically, then, the retina is viewed as a slab of tissue with diffusion occurring perpendicular to the slab. This is reasonable because the photoreceptors are relatively homogeneous across at least small sections of retina, and oxygen gradients are not expected perpendicular to the long axis of the photoreceptors (the y and z directions). These considerations lead to the use of Fick's first law for one-dimensional, steady state, O_2 diffusion:

$$d^2PO_2/dx^2 = QO_2/Dk \tag{4.4}$$

where D is the diffusion coefficient for oxygen and k is oxygen solubility.

The first mathematical analysis of retinal O_2 transport solved Eq. 4.4 and made predictions of the O_2 gradients through the outer retina [60], but when this was done, there were no data to validate the model. When data were obtained, it was clear that this analysis would not be sufficient. QO_2 is not uniform across the photoreceptor, so different equations must be used for the outer segments (OS – layer 1 of the model), the inner segments (IS – layer 2 of the model), and the outer nuclear layer (ONL – layer 3 of the model) [85]. Equation 4.4 is used for layer 2, where all the mitochondria are located. There are no mitochondria in layers 1 and 3, so $QO_2=0$, and the equation in those layers reduces to:

$$\mathrm{d}^2PO_2/\mathrm{d}x^2 = 0 \quad (4.5)$$

These equations can be solved to yield PO_2 as a function of x [85]:

$$P_1(x) = \alpha_1 x + \beta_1 \quad 0 \le x \le L_1 \text{ (layer 1)} \quad (4.6)$$

$$P_2(x) = (Q_2O_2/2Dk)x^2 + \alpha_2 x + \beta_2 \quad L_1 \le x \le L_2 \text{ (layer 2)} \quad (4.7)$$

$$P_3(x) = \alpha_3 x + \beta_3 \quad L_2 \le x \le L \text{ (layer 3)} \quad (4.8)$$

In these equations, the subscripts 1, 2, and 3 refer to each layer. The model and its parameters are shown in Fig. 4.6. The constants a_i and b_i are found by applying the boundary conditions and continuity conditions. Continuity requires that PO_2 is the same in both layer 1 and layer 2 at the boundary between these layers, and the flux also has to be the same in the two layers at that boundary. That is:

$$P_1(L_1) = P_2(L_1) \quad (4.9)$$

and

$$D_1k_1\mathrm{d}P_1(x)/\mathrm{d}x = D_2k_2\mathrm{d}P_2(x)/\mathrm{d}x \quad (4.10)$$

Corresponding conditions hold at the layer 2–layer 3 border.

The theory predicts that P should be linear with distance in layers 1 and 3 and a quadratic function of x in layer 2. This is exactly what is observed, as shown in Fig. 4.5. Because Q_2O_2/Dk, L_1, and L_2 are the only unknowns in these equations, fitting these models to data is feasible and allows the extraction of Q_2O_2/Dk under many conditions. Obtaining Q_2O_2 itself from this lumped parameter relies on knowing values for D and k. Measurements of D in the retina showed that it was $1.97 \cdot 10^{-5}$ cm²/s, 71% of the value in water, and that it was independent of location [164], so $D_1=D_2=D_3=D$. The value of k has

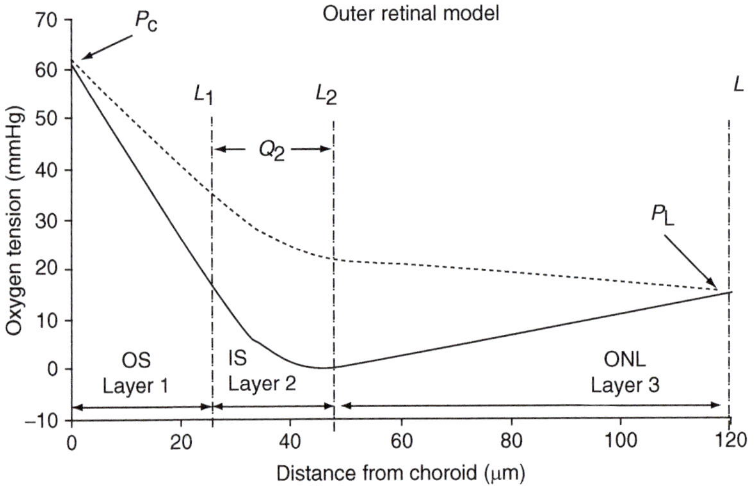

Fig. 4.6 Mathematical model of outer retinal oxygen diffusion and consumption. Parameters of the model are indicated (P_C, P_L, L_1, L_2, L, and Q_2). *Solid curve* is a simulation of a typical oxygen profile in dark-adapted cat retina, in which Q_{OR} is 4.5 ml-100 g^{-1}-min^{-1}. *Dashed curve* is a simulation for light adaptation in which all parameters are the same except Q_{OR}, which is half the value for the dark-adapted case [113]. Copyright © Société Française d'Ophtalmologie. Published by Elsevier Masson SAS. Paris. 2008. All rights reserved

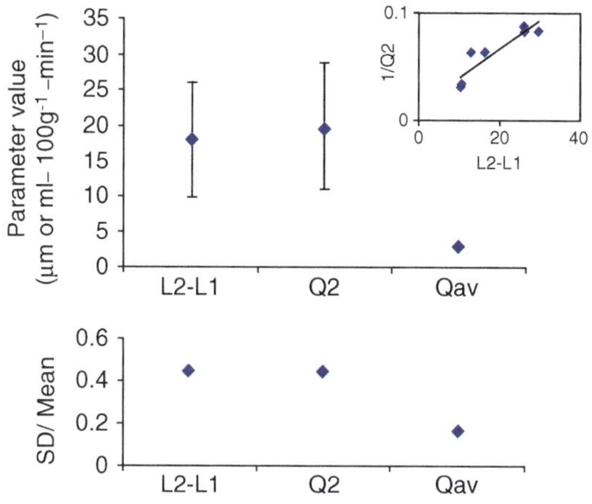

Fig. 4.7 Justification for using Q_{OR}. *Top*, means and SD of important parameters of the oxygen diffusion model obtained from a group of six dark-adapted profiles in one cat retina. Values of the thickness of the consuming region (L_2-L_1) and consumption of the consuming region ($Q_2O_2=Q_2$) vary considerably, but values of their product, Q_{av} (= $Q_2 \cdot (L_2-L_1)/L = Q_{OR}$), have little variation. *Bottom*, the coefficient of variation (SD/mean) of each parameter. The *inset* plots $1/Q_2$ against L_2-L_1 and shows that L_2-L_1 and Q_2 are inversely related, which is why the coefficient of variation of Q_{OR} is small [113]. Copyright © Société Française d'Ophtalmologie. Published by Elsevier Masson SAS. Paris. 2008. All rights reserved

never been measured in the retina, but must be similar to that for blood and brain, 2.4 · 10^{-5} ml O_2-ml tissue^{-1}-mmHg^{-1} [118] at 37°C, and it is not expected to be different in different layers. Errors in *D* and *k* are probably small and do not introduce substantial error into the derived values of QO_2. They would in any case be systematic errors that would not affect comparisons across conditions.

The model can fit data even when the profile is somewhat distorted, presumably because the electrode can stretch or pull on the retina as it is withdrawn, but there is a large coefficient of variation in Q_2O_2. Distortion also creates variability in the thickness of layer 2 (L_2-L_1), whereas the known value from histological measurements is about 25 μm. When the fitting leads to values of L_2-L_1 that are higher, Q_2O_2 for that profile is generally lower (Fig. 4.7). The product of those two parameters, normalized by the total thickness of the outer retina, is the average oxygen consumption across the outer retina:

$$Q_{OR} = Q_2O_2 \cdot (L_2 - L_1) / L$$

Because of the inverse relation between L_2-L_1 and Q_2O_2 across profiles in an animal, Q_{OR} always has a considerably smaller coefficient of variation than Q_2O_2.

4.5.3 Results from Oxygen Profiles and the Outer Retinal Model

Photoreceptors receive O_2 from both circulations, but the choroidal circulation satisfies about 90% of what they use in the dark [118], an estimate that is derived from the relative fluxes across the OS and ONL. By inspection alone (Fig. 4.5), one can see that the gradient through the OS is steeper, partially because the IS are closer to the choroid than to the inner retina and partially because P_C is generally higher than P_L. Because P_C varies across animals, for reasons that are not fully understood, it is difficult to give normal values of P_C and Q_{OR}, but in both cat and monkey, the average value of P_C is about 50 mmHg, and the corresponding average value of Q_{OR} is 4–5 ml O_2-100 g^{-1}-min^{-1} [37, 118]. This agrees reasonably well with Fick principle measurements for Q_{OR} (Table 4.1). The local value of QO_2 in the IS themselves is quite high and must be one of the very highest values in the body, more than 20 ml O_2-100 g^{-1}-min^{-1} in the dark-adapted cat retina [118].

Another finding is that in the dark-adapted cat and monkey retinas, Q_{OR} depends strongly on P_C [37, 118, 141]. The reason for this can be appreciated from the profiles (Fig. 4.5) and from the model (Fig. 4.6). The PO_2 around the inner segments cannot be less than zero. Therefore, when P_C is lower, the PO_2 gradient through the outer retina becomes shallower, so the flux is reduced. When less O_2 is delivered, Q_{OR} must decrease.

4.5.4 Other Diffusion Models

The one-dimensional model of the outer retina (Fig. 4.6) represents many of the important cases, but it is insufficient in some situations. First, if there were lateral variations in QO_2 or the O_2 supply across the retina, it would not give correct values because gradients of PO_2 would be expected in the y and z directions, violating one of the assumptions of the model. For instance, the edges of a region of photoreceptor loss following photocoagulation would not be handled by the standard model because the photocoagulated region has no oxygen consumption, whereas adjacent normal retina has the usual QO_2. A region which has a large druse under the retina also distorts the outer retinal layers and is not handled by the standard model. In these cases, a more appropriate geometry is a central cylinder containing the lesion or the druse, surrounded by an annulus of normal tissue. There may be gradients of O_2 in or out of this cylinder. The three-dimensional model cannot be solved analytically and requires finite element methods [43, 146]. A second situation in which the three layer model has proven to be somewhat insufficient is retinal detachment. In this case, the fluid layer under the retina is a fourth layer in terms of O_2 transport. The fluid layer was originally modeled as an extension of layer 1 [120] because, like layer 1, the subretinal fluid does not consume O_2. However, there may be convection in this layer, and the diffusion coefficient for oxygen is higher than in the retina. The slope of the O_2 gradient is shallower in this layer because of the higher diffusion coefficient, so a new model incorporates these properties [201].

The model described above is also inadequate to deal with the inner retina, where the supply of O_2 comes from retinal capillaries that are embedded within the tissue in a complex three-dimensional geometry. Again, this violates the assumptions of a homogenous medium and of one-dimensional diffusion. One way to extend the analysis to the inner retina is to eliminate the retinal circulation and then use an additional layer or layers for the inner retina. Experimental data for such a model can be provided by occluding a retinal artery and making the animal hyperoxic so that inner retinal oxygen is provided by the choroid [11, 41, 156]. There are at least two assumptions in this approach. First, it assumes that Q_{IR} is still normal during vascular occlusion, although in reality, it may decrease somewhat because the amount of oxygen available may be just barely adequate [11, 41, 108, 156] and the inner retina acidifies during arterial occlusion [36]. Second, mitochondria are not uniformly distributed in the inner retina [100], so a single inner retinal layer is a simplification. Nevertheless, a model with one additional layer for the inner retina still gives good fits [41], and relative inner and outer retinal oxygen consumption values are similar to those obtained by other methods (Table 4.1).

Another approach to the inner retina, used in rat [55] and monkey [233], has been to divide it into five additional layers, each with its own value of QO_2, some of which are zero. Deriving the appropriate equations for these layers is not difficult, but fitting this model to selected pieces of the O_2 profile assumes that the capillary beds lie in discrete locations and that their position can be identified from the O_2 profiles. In addition, there are likely to be three-dimensional gradients, and these are not considered in the model. Thus, results of this model should be interpreted with caution. Roos [165] modeled the inner retina by including a negative consumption, an O_2 supply, to represent retinal blood flow. This model is reasonable, but gives a uniform value of PO_2 across the inner retina. It has been used only for simulations and not to fit data.

4.6 Physiological Variations in Retinal Oxygen

Studies that have used one or more of the methods outlined above have led to a reasonably complete understanding of the influences of many variables on retinal O_2 supply and consumption.

4.6.1 Light

In all species, light increases the PO_2 of the outer retina, as shown for monkey in Fig. 4.8. Because the outer retina is avascular and the boundary values do not change much with light, the increase in PO_2 must mean that photoreceptor oxygen utilization (QO_2) decreases with light. This conclusion is supported by diffusion modeling [37, 40, 118, 153, 229, 236] and in vivo and in vitro mass balances [19, 84, 128, 177, 200].

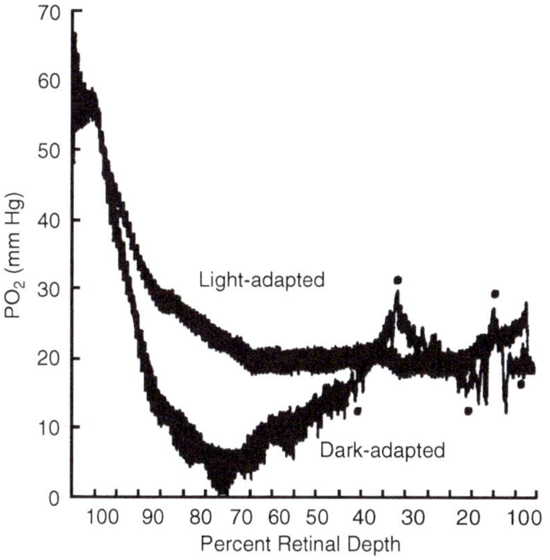

Fig. 4.8 Oxygen profiles from the monkey (*M. fascicularis*) perifoveal retina during dark and light adaptation. *Dots* indicate the dark-adapted profile [2]. Reprinted by kind permission of the Association for Research in Vision and Ophthalmology © 1993

The ATP produced in the dark fuels many cellular processes [18, 138], but the largest component is the operation of the Na/K ATPase in the IS, which extrudes the Na^+ that enters through the light-dependent channels in the OS, as well as the Na^+ that enters through the $Ca^{+2}/Na^+/K^+$ exchanger of the outer segment (e.g., [106]). This uses at least half of the total energy [19, 84, 105, 236]. The second most demanding process is probably the turnover of cGMP that holds these channels open [19, 84]. The activity of the Na/K ATPase decreases in light, but the turnover of cGMP increases [19, 20, 75, 84], so the decrease in Q_{OR} is not as great as the decrease in the pump rate. The maximum size of the light vs. dark difference in Q_{OR} appears to be species dependent, possibly because these processes are balanced differently (Table 4.1).

The time required for the metabolism to adjust from the dark-adapted to the light-adapted state is relatively fast. The time constant for the metabolic adjustment is 19 s in rat [55], 10–13 s in cat [41], and 26 s in monkey [205]. The time course of recovery from the light-adapted Q_{OR} to the dark-adapted Q_{OR} is also relatively rapid [55, 115], unless the illumination is above rod saturation, when recovery can take several minutes [115, 205].

The time constants of the two light-dependent processes (cGMP turnover and Na^+ pumping) are evidently similar in mammals, but the two metabolic events can be seen distinctly in poikilotherms [84, 105, 153]. In isolated toad retina, the increase in metabolic rate due to increased cGMP turnover is faster, having a time constant of 60 s, compared to 180 s for the slowing of the Na^+ pump [84]. In individual salamander rods, there is also a clear increase in QO_2 that is faster than the larger decrease [153]. The signal that tells the photoreceptor to change its metabolic rate is probably not feedback from ATP or ADP levels, because ATP levels in salamander photoreceptors were unchanged when QO_2 decreased and phosphocreatine was decreasing at that time, which is inconsistent with decreased ADP [153]. Changes in intracellular sodium or calcium may be involved in metabolic regulation [153, 174], but there are no conclusive data.

Cones contain more mitochondria than rods at all eccentricities [90], but this does not necessarily mean that cones have higher rates of oxidative metabolism than rods. The excess mitochondria change the refractive index of the cones and may contribute to waveguide properties more than to increased metabolism (e.g., [90]). Per unit area of retina, the photoreceptors in the primate fovea, which are of course cones, appear to use slightly less O_2 than the parafoveal photoreceptors [37, 233]. Despite the large amount of mitochondria, it was shown recently that ground squirrel cones still produce lactate at high rates [219].

In the *inner* retina of cat, there is a tendency for PO_2 to be lower during illumination [115, 118], but at present this is a weak conclusion because the largest data set used different animals for light and dark measurements [118] rather than paired comparisons. Unlike the situation in the outer retina, differences in inner retinal PO_2 cannot be interpreted in terms of QO_2, because blood flow may also change during illumination. All studies indicate, in fact, that the inner retinal utilization of oxygen (Q_{IR}) is approximately the same in steady light as it is in darkness (Table 4.1) [41, 127, 199]. The relative values of Q_{IR} and Q_{OR}, however, are not the same in all studies. In cat, Q_{IR} was about the same as the Q_{OR} in darkness, while in rat and pig, Q_{IR} was lower, similar to Q_{OR} in light adaptation. These studies were done with three quite different methods: diffusion modeling in cat, Fick principle measurements in pig, and in vitro Warburg measurements in rat, so whether the different value in cat is a species difference or a methodological difference is not clear.

Time-varying illumination should lead to a greater excitation of inner retinal neurons and cause an increase in Q_{IR}, but there is no direct information on this point. It is known that retinal blood flow increases when light is flickered rather than kept steady [163] and that inner retinal deoxyglucose uptake is greater during flicker than during darkness or steady illumination [34]. These findings are consistent with increased metabolism.

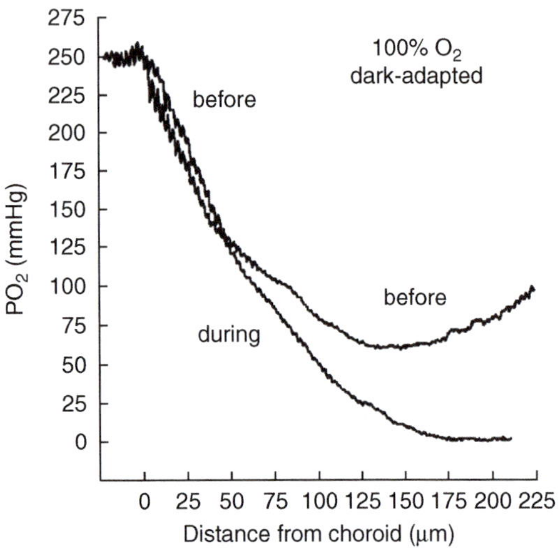

Fig. 4.9 Oxygen profiles from dark-adapted cat retina during inspiration of 100% O_2, before and during occlusion of the retinal arteriole supplying the region. The choriocapillaris PO_2 during these experiments was 203 ± 45 mmHg (mean and SD; $n = 8$ cats) during hyperoxia before occlusion. The mean inner PO_2 in the innermost 25% of the retina before occlusion during hyperoxia was 50 ± 28 mmHg. In these experiments, the corresponding values during air breathing before occlusion were 54 ± 12 mmHg for choriocapillaris and 17 ± 8 mmHg for inner retina [40]. Reprinted by kind permission of the Association for Research in Vision and Ophthalmology © 1995

4.6.2 Hypoxia

The IS are in a region of low PO_2 under normal conditions, and even a small decrease in P_aO_2 (hypoxemia) leads to a decrease in choroidal PO_2, which reduces the flux of oxygen to the photoreceptors and decreases Q_{OR} [118]. In some tissues, increased blood flow during hypoxemia allows the arteriovenous difference to decrease, with the purpose of keeping venous and tissue PO_2 relatively normal. This is the situation that applies in the inner retina, where tissue PO_2 is protected against hypoxemia until P_aO_2 falls below about 35 mmHg [62, 118]. However, with the very small arteriovenous oxygen difference in the choroid, increased blood flow during hypoxemia would be of no benefit and does not occur [33]. Photoreceptor metabolism is also compromised when intraocular pressure (IOP) is elevated. In this situation, choroidal blood flow decreases, the choroidal arteriovenous saturation difference increases, and choroidal PO_2 decreases [10, 222], as in hypoxemia.

The strong Pasteur effect (i.e., increased glycolysis at low PO_2) in the photoreceptors (e.g., [19, 53, 142, 217]) compensates for much of the loss of oxidatively derived ATP, and evidently, this is what prevents hypoxia from causing too much damage. The ERG a-wave, which originates in photoreceptors, is remarkably resistant to hypoxemia, even though Q_{OR} falls [101]. However, it is now clear that chronic hypoxemia can kill adult rat photoreceptors [190, 208], and there is evidence that elevated IOP is damaging to photoreceptors [137]. In birds, there is evidence that decreased choroidal blood flow leads to photoreceptor damage [91, 173].

4.6.3 Hyperoxia

During hyperoxia (defined to be any inspired percentage, F_iO_2, greater than 21% O_2), choroidal and outer retinal PO_2s are elevated dramatically [40, 51, 117, 156, 227], as shown in Figs. 4.9 and 4.10, primarily because the choroidal circulation lacks metabolic regulation. Q_{OR} is not increased in hyperoxia [40, 117, 156, 227] which means that the O_2 demand of the photoreceptors must be satisfied during normoxia. In animals breathing 100% O_2, choroidal PO_2 is high enough that some O_2 from the choroid spills over into the inner retina [11, 40, 157, 227], adding to the increased oxygen available from the retinal circulation. The average increase in inner retinal and vitreal PO_2 is

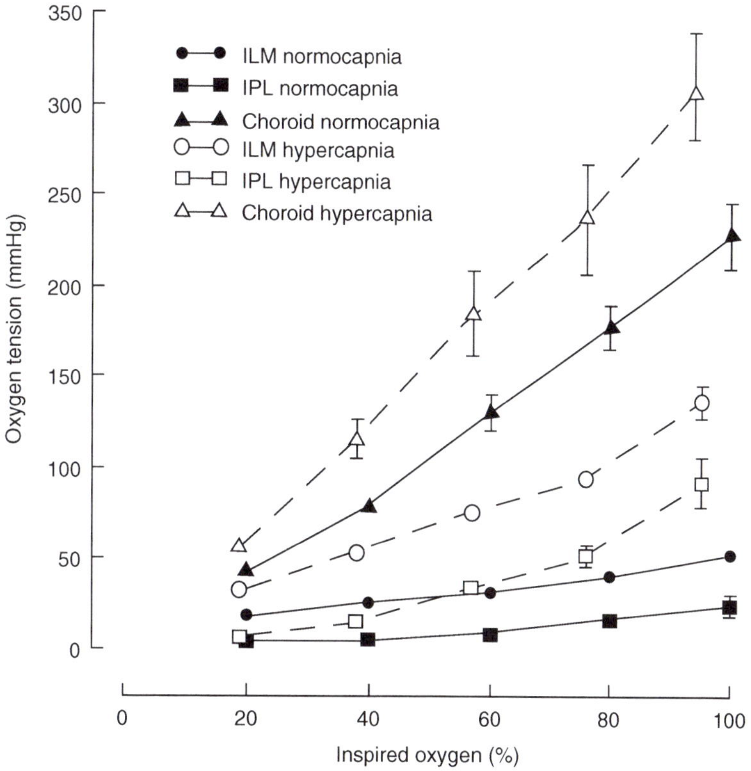

Fig. 4.10 Mean values (±SEM) of PO_2 at three locations in the retina of light-adapted Sprague Dawley rats during normoxia (21%) and hyperoxia, with or without hypercapnia (5% CO_2). The *symbols* indicate the inner limiting membrane (*ILM*), the site of the minimum PO_2 in the inner plexiform layer (*IPL*), and the choroid. *Filled symbols* represent normocapnic conditions ($n=10$), and *open symbols* represent hypercapnic conditions ($n=6$) [229]. Reprinted by kind permission of the Association for Research in Vision and Ophthalmology © 1999

30–40 mmHg in cat [40, 117], 30–88 mmHg in rat [30, 227], about 26 mmHg in monkey [233], but only about 2 mmHg in pig [160], possibly because of very strong regulation of the retinal circulation.

Hyperoxia is known to be toxic in some tissues, but strong evidence of this in the retina was lacking until recently. It now appears that 100% O_2, like hypoxemia, can cause apoptosis of photoreceptors in the adult rat retina [190, 208]. Whether this will limit the therapeutic value of hyperoxia is not clear, because lower percentages of inspired O_2, such as 70%, also increase outer retinal PO_2 [227, 233] and may not be toxic. In their study of the detached retina in cats given 70% O_2 to breathe for several days, Mervin et al. [130] noted no photoreceptor apoptosis in attached regions of the retina.

4.6.4 Hypercapnia

Adding CO_2 to the inspired gas increases blood flow through both circulations, but seems to have a more pronounced effect in dilating the retinal circulation than the choroidal circulation (Fig. 4.2). As shown in Fig. 4.10, when 5% CO_2 is combined with elevated O_2, it increases PO_2 in both the inner and outer retina more than occurs with 100% O_2 [51, 160, 227]. The mechanism for this in the inner retina is that carbogen limits or abolishes the vasoconstriction of the retinal circulation that is usually caused by hyperoxia. Elevation of local CO_2 without systemic hyperoxia can be produced either by inspiration of air plus 5% CO_2 or by an intravenous carbonic anhydrase inhibitor such as dorzolamide [147]. Both cause increased retinal PO_2, as shown for CO_2 inspiration in Fig. 4.10.

4.7 Pathophysiology and Retinal Oxygen

4.7.1 Vascular Occlusion

Under normal conditions, O_2, rather than any other substrate, is the limiting factor in vision. Vision is lost in human subjects after only 5–10 s when all the ocular circulation is occluded by raising intraocular pressure (IOP) above blood pressure. It can be prolonged to only about 35 s if the subject breathes 100% O_2 at 3 atm before the occlusion [22, 49]. Clinically and experimentally, occluding just the retinal circulation also rapidly causes loss of retinal function and vision [70], mainly because of anoxia [11, 40, 108, 144].

The inner retina is subject to arterial, venous, and capillary occlusions either as the primary event due to thrombosis or atherosclerosis or an accompaniment to other disease processes. As one would expect, complete ischemia caused by a central or branch *artery occlusion* reduces PO_2 through the entire inner retina to zero in darkness [11, 40, 108]. There is so little choroidal O_2 left over after the photoreceptor demand is satisfied that the choroidal supply has little impact during arterial occlusion if the inspired gas is air [40]. Light increases the outer retinal PO_2, but even then, there is not enough O_2 from the choroid to make a substantial difference to the inner retina. *Venous occlusion* also causes inner retinal hypoxia [158, 186], which may be less severe than in arterial occlusion, but venous occlusion is more likely to cause retinal and iris neovascularization.

Ordinarily, little O_2 from the choroid makes its way to the inner retina, and little O_2 from the retinal circulation supplies the photoreceptors. However, during hyperoxia, choroidal O_2 could be important in supplying the inner retina during acute retinal vascular occlusions, allowing survival of the inner retina until the occlusion resolves. Work on this dates back to Patz in the 1950s [144], but the diffusion of O_2 from the choroid to the vitreous under hyperoxic conditions was first shown with vitreal PO_2 measurements by Landers [108], and intraretinal PO_2 distributions in this situation were reported later [11, 40]. These measurements showed that the average PO_2 of the inner retina could be restored to almost normal in animals during occlusion (Fig. 4.9). The ERG could not be fully normalized by hyperoxia, but 70% or 100% O_2 was able to markedly increase the ERG amplitude and allowed better survival of the ERG after the occlusion [35, 40], even if the hyperoxia was delayed relative to the onset of the occlusion.

Hyperoxia as a clinical treatment has not gained a substantial following. There have been negative results of hyperoxic interventions [22, 81], but in many cases, the duration of oxygen therapy has been too brief to afford a full test of this therapy. There have also been successes with both hyperoxia [144] and hyperbaric O_2 [5, 28, 207], even 8–12 h after the occlusion. A possible advantage of hyperbaric therapy over hyperoxia at 1 atm is that the vitreous would become a reservoir that could provide some O_2 to the retina longer than the hyperbaric episode itself. In this respect, it would be similar to the technique of vitreoperfusion with a solution containing elevated O_2 [38], which has proven effective in animals. However, hyperbaric therapy may simply appear more successful than hyperoxia at this time because the hyperbaric treatments have tended to be somewhat longer than treatments with simple hyperoxia. The PO_2 profiles recorded in animals suggest that O_2 at 1 atm fully oxygenates the retina, and it should therefore provide as much benefit as hyperbaric O_2 if it were maintained continuously.

4.7.2 Diabetes

Capillary leakage and occlusion followed by neovascularization are hallmarks of diabetic retinopathy in both type I and type II diabetes. Neovascularization also occurs after capillary occlusion in sickle cell retinopathy. There are no direct measurements of O_2 in sickle cell disease and few in diabetes. Most of the PO_2 measurements in diabetic animals have been made in the vitreous relatively early in the disease [13, 64, 182, 184] and have not revealed tissue hypoxia, although some disturbance in the usual gradients was noted [13]. These measurements may not have had the necessary spatial resolution, however, and may have been made before significant circulatory changes. In contrast, regions of hypoxia, correlated with histological evidence of leukostasis, were found in the inner retina in cats that had been diabetic more than 7 years (Fig. 4.11), and the average PO_2 in the inner retina of diabetics was lower than that in control animals. These animals had no ophthalmoscopically visible regions of capillary dropout on fluorescein angiography [119]. In this condition of background retinopathy, vascular endothelial

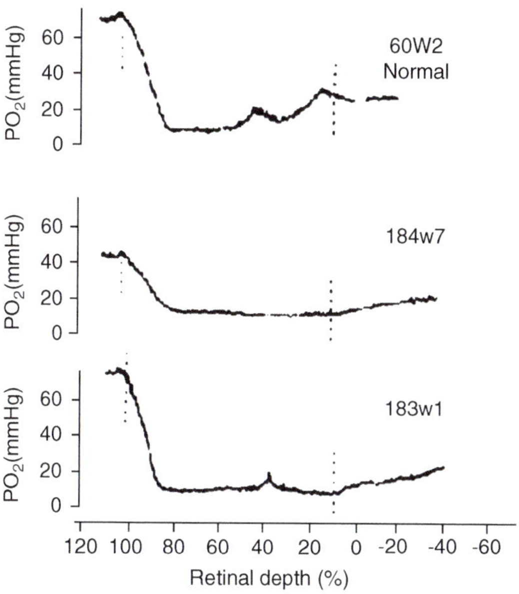

Fig. 4.11 Oxygen profiles from the dark-adapted retina of one normal cat (*top*) and two cats with long-standing diabetes, showing the low inner retinal PO_2 observed in many profiles (Modified from Linsenmeier et al. [119]). Reprinted by kind permission of the Association for Research in Vision and Ophthalmology © 1998

growth factor (VEGF) is already elevated in humans [21, 127]. Recently, de Gooyer et al. [58] used Hypoxyprobe to demonstrate modest hypoxia in the retina of diabetic mice. Thus, it appears that hypoxia can be a component of diabetic retinopathy relatively early in its progression. There is indirect evidence that the retina is hypoxic as well, involving increased hypoxia-inducible factor 1 (HIF-1) in animals [154], increased contrast sensitivity when diabetics are given oxygen to breathe [82], and decreased retinal blood flow after photocoagulation [78]. Diabetic rats studied with MR imaging have a subnormal ΔPO_2 response to a hyperoxic episode, which also supports the presence of hypoxia [31, 32], although this result may have other explanations, and in humans, a supernormal ΔPO_2 response has been reported [195], which complicates this finding. Consequently, the onset and severity of tissue hypoxia through the early stages of diabetic retinopathy is not clear.

There is little question that when capillary dropout is substantial, the retina becomes severely hypoxic. There have been no demonstrations of this in animals, which rarely develop retinopathy of this severity. Vitreal measurements (from midvitreous) in humans confirm that the retina is hypoxic in proliferative retinopathy [94]. It should be possible to make measurements of inner retinal QO_2 with a Fick principle approach, by measuring retinal venous saturation and blood flow during diabetes, but the time course of retinal blood flow changes in diabetes is still controversial, and a complete study of this type is lacking. Retinal venous saturation decreased when diabetic humans without retinopathy were made hyperglycemic [193]. This was interpreted as an increase in QO_2. The metabolic rate of the inner retina in diabetics relative to normal subjects, either during normoglycemia or hyperglycemia, is not known.

Diffusion of choroidal O_2 to the inner retina has been recognized as an explanation for the beneficial effects of panretinal photocoagulation (PRP) in reducing neovascularization. A great deal of evidence from measurements in the vitreous or from changes in blood flow [9, 71, 78, 80, 129, 132, 159, 180, 182] supported the hypothesis [189, 221] that PRP worked by removing the very demanding photoreceptors and allowing O_2 to diffuse into the inner retina. Only recently has it been shown directly that intraretinal PO_2 increases following PRP, immediately after lesions in experiments on rabbits [232] and after lesions have healed in cats [44]. This does not rule out additional benefits of PRP, such as removing cells that produce VEGF, but does strongly support the O_2 hypothesis. Intraretinal measurements in cats also suggested that PRP may sometimes be ineffective [68] because PRP damages not only the photoreceptors but also the choroid itself [44, 109, 214, 215] and lowers choroidal PO_2. Vitrectomy can also be beneficial in preventing vasoproliferation, and the hypothesis is that it works by allowing convection in the vitreous, which improves the distribution of oxygen from relatively normoxic to relatively hypoxic regions [126, 180, 185, 189].

4.7.3 Retinal Detachment

Any condition that reduces the gradient of O_2 from the choroid to the IS should reduce Q_{OR} and have a similar effect as hypoxemia. For example, retinal detachment decreases the O_2 gradient by increasing the distance between the choroid and the IS. This effect of retinal detachment was first predicted by using the diffusion model to simulate O_2 environments in the retina with different degrees of detachment [120]. Subsequently, in cats, detachment was shown to decrease photoreceptor metabolism [202] and to cause photoreceptor apoptosis and Muller cell reactions [111]. Hyperoxia during detachment increases both inner retinal and choroidal PO_2 [202] and reduces apoptosis and Muller cell responses [111, 112, 130, 167], even if hyperoxia is delayed by a day relative to the onset of detachment.

Drusen in age-related macular degeneration (AMD) may have a similar effect as detachment by moving the IS further from the choroid. The situation for AMD is not as well documented, but a druse that is 50 μm high is predicted to reduce Q_{OR} significantly [146].

4.7.4 Retinal Degenerative Diseases

An effect of choroidal O_2 on the retinal circulation has been demonstrated in animal models of retinitis pigmentosa. As in humans [79], loss of photoreceptors eventually leads to the almost complete obliteration of the retinal circulation in animals [39, 74, 133, 143]. Penn et al. [151] showed that the loss of retinal capillaries could be prevented in a rat model by making the animal somewhat hypoxic, so the choroidal PO_2 would not be so high. Direct measurements of retinal PO_2 during the progression of photoreceptor degeneration in RCS rats [228] and Abyssinian cats [143] showed that with a greater loss of photoreceptors, the PO_2 gradients from the choroid reach into the inner retina. It seems likely that the earliest change is simply vasoconstriction of retinal vessels caused by the relative hyperoxia, but this clearly progresses to vasoobliteration.

4.7.5 Retinopathy of Prematurity

In contrast to the beneficial effect of choroidal O_2 in PRP, choroidal O_2 has been implicated negatively in retinopathy of prematurity (ROP). Elevated choroidal PO_2 was hypothesized to be responsible for the failure of the retinal circulation to develop in premature infants [25, 145]. Reducing inspired PO_2 in neonatal care units decreased the incidence of ROP [97], presumably by allowing the retinal circulation to grow normally. Consistent with this idea, making neonatal mice hyperoxic and then withdrawing the excess O_2 first prevented vessel growth and then produced neovascularization [178]. However, simply exposing neonatal rats to continuous hyperoxia proved to be less damaging than exposing them to fluctuating levels of inspired O_2 [150], which was not expected from the original hypothesis. In addition, normalizing O_2 in the neonatal unit did not prevent all ROP [124]. Finally, hypoxia can be demonstrated in the vitreous of rats when vessels are developing, but hypoxia was no worse in regions of neovascularization than in normally developing retinas, suggesting that hypoxia is important in vessel growth, but not an explanation of abnormal vasculogenesis [235]. Consequently, aspects of ROP are still not understood.

4.8 Retinal Molecular Changes Related to Oxygen

The molecular effects of both decreased and increased retinal oxygen are being worked out and cannot be covered in depth here, but understanding the connection between physiology and molecular biology is an important topic for the future. Hypoxia is known to increase hypoxia-inducible factor 1 (HIF-1) in the retina [56, 77, 139], which leads to many changes, including elevation of vascular endothelial growth factor. HIF-1 also increases in retinal diseases, including diabetes [23]. Some retinal gene and protein changes can be directly ascribed to hypoxia [56, 102, 192], but a comprehensive comparison remains to be done between the molecular changes in hypoxia and those in pathophysiological conditions that may include hypoxia, such as diabetic retinopathy [99, 155], retinal detachment [66, 95, 162, 234], and glaucoma [1, 98, 131, 223].

Oxygen radicals and the damage they cause are another connection between oxygen levels and molecular effects of oxygen. Because of the

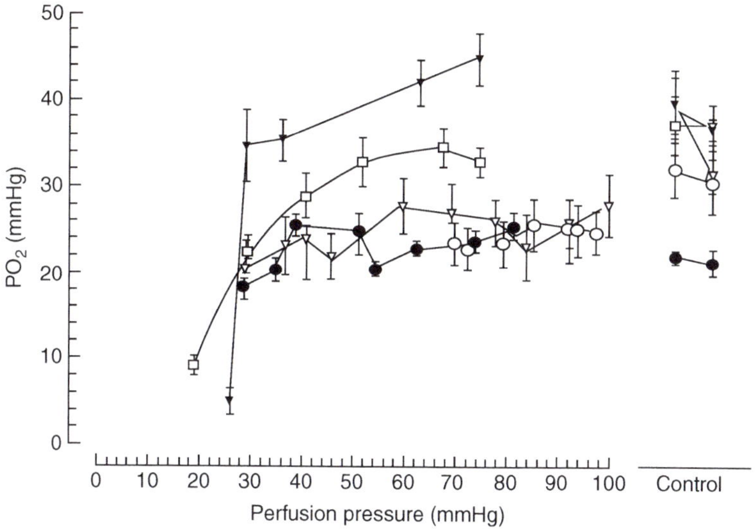

Fig. 4.12 Regulation of PO_2 in the optic nerve head of cats, measured by phosphorescence lifetime decay. Each set of connected points is one animal, in which IOP was raised in order to lower perfusion pressure (PP) in five cats. Each data point was found by averaging the integrated phosphorescence intensity in the optic nerve head region from the intensity map. For each cat, control PO_2 distribution was mapped before the pressure-controlling needles were inserted. The two PO_2 values below 10 mmHg at low PP in two cats were taken shortly after the blood flow stopped [175]. Reprinted by kind permission of the Association for Research in Vision and Ophthalmology © 1992

high PO_2 in the outer retina, there are multiple antioxidant systems [72, 191] to deal with the large number of radicals that are expected to be present in the RPE and outer retina even under normal conditions. Recently, Winkler [218] proposed that the lack of glutathione in photoreceptors leads to oxidative damage and is the reason that outer segments need to be continuously renewed. Further oxidative damage may occur at both increased and decreased levels of oxygen and is an active area of research.

4.9 Oxygen in the Optic Nerve Head

The local oxygen environment of the optic nerve head is of interest largely because of the possible role of ischemia in causing axonal damage and ganglion cell loss in glaucoma [67] and the possibility that some of the medications for glaucoma may improve aspects of the microenvironment in this region. The nerve head is generally divided into a prelaminar region, the laminar region, and the retrolaminar region. While it has been possible to assess blood flow in each one by using microspheres in monkeys [179], measurements of oxygen tension have been confined to the prelaminar region adjacent to the retina and inside the globe or to the vitreous in front of that region. These measurements show that the PO_2s are normally similar to those in the inner retina [3, 175] and that the regulation is also similar, with modest [3] or negligible [63] increases in PO_2 during hyperoxia. Hypercapnia increases optic nerve head PO_2 [187]. There is no difference in optic nerve head PO_2 between darkness and steady illumination [3, 54], but it decreases during flicker [3], presumably because the increase in blood flow, which is well established in the optic nerve head [197], does not quite keep pace with the increased metabolism required when ganglion cells are firing more rapidly.

In the normal cat eye, phosphorescence quenching measurements showed that the PO_2 in the optic nerve head is well maintained during acute increases of intraocular pressure (IOP) that lower perfusion pressure (mean arterial blood pressure minus IOP) to 30–40 mmHg (Fig. 4.12) [175]. The actual perfusion pressure at which regulation was impaired must have been even lower because ophthalmic artery pressure is lower than mean arterial pressure. Similarly, good regulation over a wide range of perfusion pressures has been observed with microelectrodes in cat and pig during IOP elevation [3, 107] and in monkey during reduction of blood pressure [63]. The efficiency of this pressure regulation is similar to that observed in the neural retina [10, 175, 222]. Unfortunately, no oxygen measurements have been made in an animal with any form of chronic pressure elevation, so there are no data that address the question of whether optic nerve head regulation is impaired by glaucoma as one might predict.

In the optic nerve head, there has been considerable interest in the carbonic anhydrase (CA) inhibitors dorzolamide and acetazolamide, which lower IOP [188], for their potential additional effects on blood flow and PO_2. These agents increase optic nerve head PO_2 (measured in the vitreous near the nerve head) between 40% and twofold in pigs [107, 148, 187]. Their action is probably due to increased local PCO_2 [148] because metabolic acidosis has no effect. The effectiveness of dorzolamide on the optic nerve stimulated experiments that showed that it could also be effective in increasing retinal PO_2 after a branch vein occlusion [135]. Few other intravenous drugs have any effect on inner retinal PO_2 [15], so the CA inhibitors may have some value, although CA inhibition can also markedly reduce retinal pH [204].

References

1. Ahmed F, Brown KM, Stephan DA, Morrison JC, Johnson EC, Tomarev SI (2004) Microarray analysis of changes in mRNA levels in the rat retina after experimental elevation of intraocular pressure. Invest Ophthalmol Vis Sci 45:1247–1258
2. Ahmed J, Braun RD, Dunn R Jr, Linsenmeier RA (1993) Oxygen distribution in the macaque retina. Invest Ophthalmol Vis Sci 34:516–521
3. Ahmed J, Linsenmeier RA, Dunn R Jr (1994) The oxygen distribution in the prelaminar optic nerve head of the cat. Exp Eye Res 59:457–465
4. Ahmed J, Pulfer MK, Linsenmeier RA (2001) Measurement of blood flow through the retinal circulation of the cat during normoxia and hypoxemia using fluorescent microspheres. Microvasc Res 62: 143–153
5. Aisenbrey S, Krott R, Heller R, Krauss D, Rossler G, Heimann K (2000) Hyperbaric oxygen therapy in retinal artery occlusion. Ophthalmologe 97:461–467
6. Alder VA, Cringle SJ, Constable IJ (1983) The retinal oxygen profile in cats. Invest Ophthalmol Vis Sci 24:30–36
7. Alder VA, Cringle SJ (1985) The effect of the retinal circulation on vitreal oxygen tension. Curr Eye Res 4:121–129
8. Alder VA, Niemeyer G, Cringle SJ, Brown MJ (1986) Vitreal oxygen tension gradients in the isolated perfused cat eye. Curr Eye Res 5:249–256
9. Alder VA, Cringle SJ, Brown M (1987) The effect of regional retinal photocoagulation on vitreal oxygen tension. Invest Ophthalmol Vis Sci 28:1078–1085
10. Alder VA, Cringle SJ (1989) Intraretinal and preretinal PO_2 response to acutely raised intraocular pressure in cats. Am J Physiol 256:H1627–H1634
11. Alder VA, Ben-Nun J, Cringle SJ (1990) PO_2 profiles and oxygen consumption in cat retina with an occluded retinal circulation. Invest Ophthalmol Vis Sci 31:1029–1034
12. Alder VA, Yu DY, Cringle SJ (1991) Vitreal oxygen tension measurements in the rat eye. Exp Eye Res 52:293–299
13. Alder VA, Yu DY, Cringle SJ, Su EN (1991) Changes in vitreal oxygen tension distribution in the streptozotocin diabetic rat. Diabetologia 34:469–476
14. Alm A, Bill A (1970) Blood flow and oxygen extraction in the cat uvea at normal and high intraocular pressures. Acta Physiol Scand 80:19–28
15. Alm A (1972) Effects of norepinephrine, angiotensin, dihydroergotamine, papaverine, isoproterenol, histamine, nicotinic acid, and xanthinol nicotinate on retinal oxygen tension in cats. Acta Ophthalmol (Copenh) 50:707–719
16. Alm A, Bill A (1972) The oxygen supply to the retina. I. Effects of changes in intraocular and arterial blood pressures, and in arterial PO_2 and PCO_2 on the oxygen tension in the vitreous body of the cat. Acta Physiol Scand 84:261–274
17. Alm A, Bill A (1972) The oxygen supply to the retina. II. Effects of high intraocular pressure and of increased arterial carbon dioxide tension on uveal and retinal blood flow in cats. A study with radioactively labelled microspheres including flow determinations in brain and some other tissues. Acta Physiol Scand 84:306–319
18. Ames A (1992) Energy requirements of CNS cells as related to their function and to their vulnerability to ischemia: a commentary based on studies on retina. Can J Physiol Pharmacol 70(Suppl):S158–S164
19. Ames A, Li YY, Heher EG, Kimble CR (1992) Energy metabolism of rabbit retina as related to function: high cost of Na transport. J Neurosci 12:840–853
20. Ames A 3rd, Walseth TF, Heyman RA, Barad M, Graeff RM, Goldberg ND (1986) Light-induced increases in cGMP metabolic flux correspond with electrical responses of photoreceptors. J Biol Chem 261:13034–13042

21. Amin RH, Frank RN, Kennedy A, Eliott D, Puklin JE, Abrams GW (1997) Vascular endothelial growth factor is present in glial cells of the retina and optic nerve of human subjects with nonproliferative diabetic retinopathy. Invest Ophthalmol Vis Sci 38: 36–47
22. Anderson B Jr, Saltzman HA, Heyman A (1965) The effects of hyperbaric oxygenation on retinal arterial occlusion. Arch Ophthalmol 73:315–319
23. Arjamaa O, Nikinmaa M (2006) Oxygen-dependent diseases in the retina: role of hypoxia-inducible factors. Exp Eye Res 83:473–483
24. Ashton N, Ward B, Serpell G (1953) Role of oxygen in the genesis of retrolental fibroplasia; a preliminary report. Br J Ophthalmol 37:513–520
25. Ashton N (1954) Pathological basis of retrolental fibroplasia. Br J Ophthalmol 38:385–396
26. Ashton N (1954) Animal experiments in retrolental fibroplasia. Trans Am Acad Ophthalmol Otolaryngol 58:51–53; discussion 53–54
27. Bartels H, Harms H (1959) Oxygen dissociation curves of the blood of mammals (human, rabbit, guinea pig, dog, cat, pig, cow and sheep). Pflugers Arch 268:334–365
28. Beiran I, Goldenberg I, Adir Y, Tamir A, Shupak A, Miller B (2001) Early hyperbaric oxygen therapy for retinal artery occlusion. Eur J Ophthalmol 11: 345–350
29. Berkowitz BA, Wilson CA (1995) Quantitative mapping of ocular oxygenation using magnetic resonance imaging. Magn Reson Med 33:579–581
30. Berkowitz BA (1996) Adult and newborn rat inner retinal oxygenation during carbogen and 100% oxygen breathing. Comparison using magnetic resonance imaging delta PO_2 mapping. Invest Ophthalmol Vis Sci 37:2089–2098
31. Berkowitz BA, Kowluru RA, Frank RN, Kern TS, Hohman TC, Prakash M (1999) Subnormal retinal oxygenation response precedes diabetic-like retinopathy. Invest Ophthalmol Vis Sci 40:2100–2105
32. Berkowitz BA, Roberts R, Luan H, Peysakhov J, Knoerzer DL, Connor JR, Hohman TC (2005) Drug intervention can correct subnormal retinal oxygenation response in experimental diabetic retinopathy. Invest Ophthalmol Vis Sci 46:2954–2960
33. Bill A (1962) Aspects of physiological and pharmacological regulation of uveal blood flow. Acta Soc Med Ups 67:122–132
34. Bill A, Sperber GO (1990) Control of retinal and choroidal blood flow. Eye (Lond) 4(Pt 2):319–325
35. Birol G, Budzynski E, Wangsa-Wirawan ND, Linsenmeier RA (2004) Hyperoxia promotes electroretinogram recovery after retinal artery occlusion in cats. Invest Ophthalmol Vis Sci 45:3690–3696
36. Birol G, Budzynski E, Wangsa-Wirawan ND, Linsenmeier RA (2005) Retinal arterial occlusion leads to acidosis in the cat. Exp Eye Res 80: 527–533
37. Birol G, Wang S, Budzynski E, Wangsa-Wirawan ND, Linsenmeier RA (2007) Oxygen distribution and consumption in the macaque retina. Am J Physiol Heart Circ Physiol 293:H1696–H1704
38. Blair NP, Shaw WE, Dunn R Jr, Tsukarhara Y, Floro C, Rusin MM (1991) Limitation of retinal injury by vitreoperfusion initiated after onset of ischemia. Arch Ophthalmol 109:113–118
39. Blanks JC, Johnson LV (1986) Vascular atrophy in the retinal degenerative rd mouse. J Comp Neurol 254:543–553
40. Braun RD, Linsenmeier RA (1995) Retinal oxygen tension and the electroretinogram during arterial occlusion in the cat. Invest Ophthalmol Vis Sci 36: 523–541
41. Braun RD, Linsenmeier RA, Goldstick TK (1995) Oxygen consumption in the inner and outer retina of the cat. Invest Ophthalmol Vis Sci 36:542–554
42. Briggs D, Rodenhauser J (1973) Distribution and consumption of oxygen in the vitreous body of cats. In: Kessler M, Bruley D, Clark LJ, Lubbers D, Silver I, Strauss J (eds) Oxygen supply: theoretical and practical aspects of oxygen supply and microcirculation of tissue. Urban and Schwarzenberg, Munich, pp 265–269
43. Budzynski E (2005) Effects of photocoagulation on PO_2 in the cat retina. Ph.D. thesis, Biomedical Engineering, Northwestern University, Evanston
44. Budzynski E, Smith JH, Bryar P, Birol G, Linsenmeier RA (2008) Effects of photocoagulation on intraretinal PO_2 in cat. Invest Ophthalmol Vis Sci 49:380–389
45. Buerk DG (1993) Biosensors. Technomic, Lancaster
46. Buerk DG, Shonat RD, Riva CE, Cranstoun SD (1993) O_2 gradients and countercurrent exchange in the cat vitreous humor near retinal arterioles and venules. Microvasc Res 45:134–148
47. Buttery RG, Hinrichsen CF, Weller WL, Haight JR (1991) How thick should a retina be? A comparative study of mammalian species with and without intraretinal vasculature. Vision Res 31:169–187
48. Caprioli J, Sears M, Mead A (1984) Ocular blood flow in phakic and aphakic monkey eyes. Exp Eye Res 39:1–7
49. Carlisle R, Lanphier EH, Rahn H (1964) Hyperbaric oxygen and persistence of vision in retinal ischemia. J Appl Physiol 19:914–918
50. Cheng H, Nair G, Walker TA, Kim MK, Pardue MT, Thule PM, Olson DE, Duong TQ (2006) Structural and functional MRI reveals multiple retinal layers. Proc Natl Acad Sci USA 103:17525–17530
51. Chung CK, Linsenmeier RA (2007) Effect of carbogen (95% O_2/5% CO_2) on retinal oxygenation in dark-adapted anesthetized cats. Curr Eye Res 32:699–707
52. Cline JM, Thrall DE, Page RL, Franko AJ, Raleigh JA (1990) Immunohistochemical detection of a hypoxia marker in spontaneous canine tumours. Br J Cancer 62:925–931
53. Cohen LH, Noell WK (1965) Relationships between visual function and metabolism. In: Graymore CN (ed) Biochemistry of the retina. Academic, New York
54. Cranstoun SD, Riva CE, Munoz JL, Pournaras CJ (1997) Continuous measurements of intra-vascular

PO_2 in the pig optic nerve head. Klin Monatsbl Augenheilkd 210:313–315
55. Cringle SJ, Yu DY, Yu PK, Su EN (2002) Intraretinal oxygen consumption in the rat in vivo. Invest Ophthalmol Vis Sci 43:1922–1927
56. Crosson L, Kroes R, Moskal J, Linsenmeier R (2009) Gene expression patterns in normoxic, hypoxic and post-hypoxic adult rat retina with special reference to the NMDA receptor and its interactome. Mol Vis 15:296–311
57. Danylkova NO, Pomeranz HD, Alcala SR, McLoon LK (2006) Histological and morphometric evaluation of transient retinal and optic nerve ischemia in rat. Brain Res 1096:20–29
58. de Gooyer TE, Stevenson KA, Humphries P, Simpson DA, Gardiner TA, Stitt AW (2006) Retinopathy is reduced during experimental diabetes in a mouse model of outer retinal degeneration. Invest Ophthalmol Vis Sci 47:5561–5568
59. Desrochers PE, Pratt KA, Fromm PO, Hoffert JR (1985) Oxygen diffusion in the trout retina. Exp Eye Res 41:607–618
60. Dollery CT, Bulpitt CJ, Kohner EM (1969) Oxygen supply to the retina from the retinal and choroidal circulations at normal and increased arterial oxygen tensions. Invest Ophthalmol 8:588–594
61. Duong TQ, Pardue MT, Thule PM, Olson DE, Cheng H, Nair G, Li Y, Kim M, Zhang X, Shen Q (2008) Layer-specific anatomical, physiological and functional MRI of the retina. NMR Biomed 21:978–996
62. Enroth-Cugell C, Goldstick TK, Linsenmeier RA (1980) The contrast sensitivity of cat retinal ganglion cells at reduced oxygen tensions. J Physiol 304:59–81
63. Ernest JT (1977) Optic disk oxygen tension. Exp Eye Res 24:271–278
64. Ernest JT, Goldstick TK, Engerman RL (1983) Hyperglycemia impairs retinal oxygen autoregulation in normal and diabetic dogs. Invest Ophthalmol Vis Sci 24:985–989
65. Fairbanks MB, Hoffert JR, Fromm PO (1974) Short circuiting the ocular oxygen concentrating mechanism in the teleost salmo gairdneri using carbonic anhydrase inhibitors. J Gen Physiol 64:263–273
66. Farjo R, Peterson WM, Naash MI (2008) Expression profiling after retinal detachment and reattachment: a possible role for aquaporin-0. Invest Ophthalmol Vis Sci 49:511–521
67. Flammer J, Orgul S, Costa VP, Orzalesi N, Krieglstein GK, Serra LM, Renard JP, Stefansson E (2002) The impact of ocular blood flow in glaucoma. Prog Retin Eye Res 21:359–393
68. Frank RN (2004) Diabetic retinopathy. N Engl J Med 350:48–58
69. Frayser R, Hickam JB (1964) Retinal vascular response to breathing increased carbon dioxide and oxygen concentrations. Invest Ophthalmol 3:427–431
70. Fujino T, Hamasaki DI (1965) The effect of occluding the retinal and choroidal circulations on the electroretinogram of monkeys. J Physiol 180:837–845
71. Funatsu H, Wilson CA, Berkowitz BA, Sonkin PL (1997) A comparative study of the effects of argon and diode laser photocoagulation on retinal oxygenation. Graefes Arch Clin Exp Ophthalmol 235:168–175
72. Ganea E, Harding JJ (2006) Glutathione-related enzymes and the eye. Curr Eye Res 31:1–11
73. Gardiner TA, Gibson DS, de Gooyer TE, de la Cruz VF, McDonald DM, Stitt AW (2005) Inhibition of tumor necrosis factor-alpha improves physiological angiogenesis and reduces pathological neovascularization in ischemic retinopathy. Am J Pathol 166:637–644
74. Gerstein DD, Dantzker DR (1969) Retinal vascular changes in hereditary visual cell degeneration. Arch Ophthalmol 81:99–105
75. Goldberg ND, Ames AA 3rd, Gander JE, Walseth TF (1983) Magnitude of increase in retinal cGMP metabolic flux determined by 18o incorporation into nucleotide alpha-phosphoryls corresponds with intensity of photic stimulation. J Biol Chem 258:9213–9219
76. Goren SB, Krause AC (1956) The effects of hypoxia and hyperoxia upon the oxygen tension in the vitreous humor of the cat. Am J Ophthalmol 42:764–769
77. Grimm C, Wenzel A, Groszer M, Mayser H, Seeliger M, Samardzija M, Bauer C, Gassmann M, Reme CE (2002) HIF-1-induced erythropoietin in the hypoxic retina protects against light-induced retinal degeneration. Nat Med 8:718–724
78. Grunwald JE, Riva CE, Brucker AJ, Sinclair SH, Petrig BL (1986) Effect of panretinal photocoagulation on retinal blood flow in proliferative diabetic retinopathy. Ophthalmology 93:590–595
79. Grunwald JE, Maguire AM, Dupont J (1996) Retinal hemodynamics in retinitis pigmentosa. Am J Ophthalmol 122:502–508
80. Guven D, Ozdemir H, Hasanreisoglu B (1996) Hemodynamic alterations in diabetic retinopathy. Ophthalmology 103:1245–1249
81. Haddad HM, Leopold IH (1965) Effect of hyperbaric oxygenation on microcirculation: use in therapy of retinal vascular disorders. Invest Ophthalmol 4:1141–1151
82. Harris A, Arend O, Danis RP, Evans D, Wolf S, Martin BJ (1996) Hyperoxia improves contrast sensitivity in early diabetic retinopathy. Br J Ophthalmol 80:209–213
83. Harris A, Dinn RB, Kagemann L, Rechtman E (2003) A review of methods for human retinal oximetry. Ophthalmic Surg Lasers Imaging 34:152–164
84. Haugh-Scheidt LM, Griff ER, Linsenmeier RA (1995) Light-evoked oxygen responses in the isolated toad retina. Exp Eye Res 61:73–81
85. Haugh LM, Linsenmeier RA, Goldstick TK (1990) Mathematical models of the spatial distribution of retinal oxygen tension and consumption, including changes upon illumination. Ann Biomed Eng 18:19–36
86. Henkind P (1966) The retinal vascular system of the domestic cat. Exp Eye Res 5:10–20
87. Herbert DA, Mitchell RA (1971) Blood gas tensions and acid–base balance in awake cats. J Appl Physiol 30:434–436

88. Hickam JB, Sieker HO, Frayser R (1959) Studies of retinal circulation and a-V oxygen difference in man. Trans Am Clin Climatol Assoc 71:34–44
89. Hickam JB, Frayser R (1966) Studies of the retinal circulation in man: observations on vessel diameter, arteriovenous oxygen, and mean circulation time. Circulation 33:302–316
90. Hoang QV, Linsenmeier RA, Chung CK, Curcio CA (2002) Photoreceptor inner segments in monkey and human retina: mitochondrial density, optics, and regional variation. Vis Neurosci 19:395–407
91. Hodos W, Miller RF, Ghim MM, Fitzgerald ME, Toledo C, Reiner A (1998) Visual acuity losses in pigeons with lesions of the nucleus of Edinger-Westphal that disrupt the adaptive regulation of choroidal blood flow. Vis Neurosci 15:273–287
92. Hogeboom van Buggenum IM, van der Heijde GL, Tangelder GJ, Reichert-Thoen JW (1996) Ocular oxygen measurement. Br J Ophthalmol 80:567–573
93. Holcombe DJ, Lengefeld N, Gole GA, Barnett NL (2008) The effects of acute intraocular pressure elevation on rat retinal glutamate transport. Acta Ophthalmol 86:408–414
94. Holekamp NM, Shui YB, Beebe D (2006) Lower intraocular oxygen tension in diabetic patients: possible contribution to decreased incidence of nuclear sclerotic cataract. Am J Ophthalmol 141:1027–1032
95. Hollborn M, Francke M, Iandiev I, Buhner E, Foja C, Kohen L, Reichenbach A, Wiedemann P, Bringmann A, Uhlmann S (2008) Early activation of inflammation- and immune response-related genes after experimental detachment of the porcine retina. Invest Ophthalmol Vis Sci 49:1262–1273
96. Ito Y, Berkowitz BA (2001) MR studies of retinal oxygenation. Vision Res 41:1307–1311
97. James S, Lanman JT (1976) History of oxygen therapy and retrolental fibroplasia. Prepared by the American Academy of Pediatrics, Committee on Fetus and Newborn with the Collaboration of Special Consultants. Pediatrics 57:591–642
98. Johnson EC, Jia L, Cepurna WO, Doser TA, Morrison JC (2007) Global changes in optic nerve head gene expression after exposure to elevated intraocular pressure in a rat glaucoma model. Invest Ophthalmol Vis Sci 48:3161–3177
99. Joussen AM, Huang S, Poulaki V, Camphausen K, Beecken WD, Kirchhof B, Adamis AP (2001) In vivo retinal gene expression in early diabetes. Invest Ophthalmol Vis Sci 42:3047–3057
100. Kageyama GH, Wong-Riley MT (1984) The histochemical localization of cytochrome oxidase in the retina and lateral geniculate nucleus of the ferret, cat, and monkey, with particular reference to retinal mosaics and on/off-center visual channels. J Neurosci 4:2445–2459
101. Kang Derwent J, Linsenmeier RA (2000) Effects of hypoxemia on the a- and b-waves of the electroretinogram in the cat retina. Invest Ophthalmol Vis Sci 41:3634–3642
102. Kaur C, Sivakumar V, Foulds WS (2006) Early response of neurons and glial cells to hypoxia in the retina. Invest Ophthalmol Vis Sci 47:1126–1141
103. Khoobehi B, Beach JM, Kawano H (2004) Hyperspectral imaging for measurement of oxygen saturation in the optic nerve head. Invest Ophthalmol Vis Sci 45:1464–1472
104. Kiilgaard JF, Pedersen DB, Eysteinsson T, la Cour M, Bang K, Jensen PK, Stefansson E (2004) Optic nerve oxygen tension: the effects of timolol and dorzolamide. Br J Ophthalmol 88:276–279
105. Kimble EA, Svoboda RA, Ostroy SE (1980) Oxygen consumption and ATP changes of the vertebrate photoreceptor. Exp Eye Res 31:271–288
106. Krizaj D, Copenhagen DR (1998) Compartmentalization of calcium extrusion mechanisms in the outer and inner segments of photoreceptors. Neuron 21:249–256
107. la Cour M, Kiilgaard JF, Eysteinsson T, Wiencke AK, Bang K, Dollerup J, Jensen PK, Stefansson E (2000) Optic nerve oxygen tension: effects of intraocular pressure and dorzolamide. Br J Ophthalmol 84:1045–1049
108. Landers MB 3rd (1978) Retinal oxygenation via the choroidal circulation. Trans Am Ophthalmol Soc 76:528–556
109. Lee C, Smith JH, Kang Derwent JJ, Budzynski E, Linsenmeier RA (2011) Decreased circulation in the feline choriocapillaris underlying retinal photocoagulation lesions. Invest Ophthalmol Vis Sci 52(6):3398–3403, Northwestern
110. Lenzi GL, Frackowiak RS, Jones T, Heather JD, Lammertsma AA, Rhodes CG, Pozzilli C (1981) $CMRO_2$ and CBF by the oxygen-15 inhalation technique. Results in normal volunteers and cerebrovascular patients. Eur Neurol 20:285–290
111. Lewis G, Mervin K, Valter K, Maslim J, Kappel PJ, Stone J, Fisher S (1999) Limiting the proliferation and reactivity of retinal Muller cells during experimental retinal detachment: the value of oxygen supplementation. Am J Ophthalmol 128:165–172
112. Lewis GP, Talaga KC, Linberg KA, Avery RL, Fisher SK (2004) The efficacy of delayed oxygen therapy in the treatment of experimental retinal detachment. Am J Ophthalmol 137:1085–1095
113. Linsenmeier R, Pournaras C (2008) Consommation Et Diffusion De L'oxygene Retinien. In: Pournaras C (ed) Pathologies vasculaires oculaires. Masson, Paris, pp 99–107
114. Linsenmeier RA, Goldstick TK, Blum RS, Enroth-Cugell C (1981) Estimation of retinal oxygen transients from measurements made in the vitreous humor. Exp Eye Res 32:369–379
115. Linsenmeier RA (1986) Effects of light and darkness on oxygen distribution and consumption in the cat retina. J Gen Physiol 88:521–542
116. Linsenmeier RA, Yancey CM (1987) Improved fabrication of double-barreled recessed cathode O_2 microelectrodes. J Appl Physiol 63:2554–2557

117. Linsenmeier RA, Yancey CM (1989) Effects of hyperoxia on the oxygen distribution in the intact cat retina. Invest Ophthalmol Vis Sci 30:612–618
118. Linsenmeier RA, Braun RD (1992) Oxygen distribution and consumption in the cat retina during normoxia and hypoxemia. J Gen Physiol 99:177–197
119. Linsenmeier RA, Braun RD, McRipley MA, Padnick LB, Ahmed J, Hatchell DL, McLeod DS, Lutty GA (1998) Retinal hypoxia in long-term diabetic cats. Invest Ophthalmol Vis Sci 39:1647–1657
120. Linsenmeier RA, Padnick-Silver L (2000) Metabolic dependence of photoreceptors on the choroid in the normal and detached retina. Invest Ophthalmol Vis Sci 41:3117–3123
121. Linsenmeier RA (2005) Retinal Bioengineering, Chap. 13 of Neural Engineering, ed. B. He. New York: Kluwer Academic, pp. 421–484
122. Lo LW, Koch CJ, Wilson DF (1996) Calibration of oxygen-dependent quenching of the phosphorescence of Pd-meso-tetra (4-carboxyphenyl) porphine: a phosphor with general application for measuring oxygen concentration in biological systems. Anal Biochem 236:153–160
123. Lowry OH, Roberts NR, Lewis C (1956) The quantitative histochemistry of the retina. J Biol Chem 220:879–892
124. Lucey JF, Dangman B (1984) A reexamination of the role of oxygen in retrolental fibroplasia. Pediatrics 73:82–96
125. Madsen PL, Holm S, Herning M, Lassen NA (1993) Average blood flow and oxygen uptake in the human brain during resting wakefulness: a critical appraisal of the kety-schmidt technique. J Cereb Blood Flow Metab 13:646–655
126. Maeda N, Tano Y (1996) Intraocular oxygen tension in eyes with proliferative diabetic retinopathy with and without vitreous. Graefes Arch Clin Exp Ophthalmol 234(Suppl 1):S66–S69
127. Mathews MK, Merges C, McLeod DS, Lutty GA (1997) Vascular endothelial growth factor and vascular permeability changes in human diabetic retinopathy. Invest Ophthalmol Vis Sci 38:2729–2741
128. Medrano CJ, Fox DA (1995) Oxygen consumption in the rat outer and inner retina: light- and pharmacologically-induced inhibition. Exp Eye Res 61: 273–284
129. Mendivil A (1997) Ocular blood flow velocities in patients with proliferative diabetic retinopathy after panretinal photocoagulation. Surv Ophthalmol 42(Suppl 1):S89–S95
130. Mervin K, Valter K, Maslim J, Lewis G, Fisher S, Stone J (1999) Limiting photoreceptor death and deconstruction during experimental retinal detachment: the value of oxygen supplementation. Am J Ophthalmol 128:155–164
131. Miyahara T, Kikuchi T, Akimoto M, Kurokawa T, Shibuki H, Yoshimura N (2003) Gene microarray analysis of experimental glaucomatous retina from cynomolgus monkey. Invest Ophthalmol Vis Sci 44:4347–4356
132. Molnar I, Poitry S, Tsacopoulos M, Gilodi N, Leuenberger PM (1985) Effect of laser photocoagulation on oxygenation of the retina in miniature pigs. Invest Ophthalmol Vis Sci 26:1410–1414
133. Narfstrom K (1985) Progressive retinal atrophy in the Abyssinian cat. Clinical characteristics. Invest Ophthalmol Vis Sci 26:193–200
134. Nicholson C, Phillips JM (1981) Ion diffusion modified by tortuosity and volume fraction in the extracellular microenvironment of the rat cerebellum. J Physiol 321:225–257
135. Noergaard MH, Bach-Holm D, Scherfig E, Bang K, Jensen PK, Kiilgaard JF, Stefansson E, la Cour M (2008) Dorzolamide increases retinal oxygen tension after branch retinal vein occlusion. Invest Ophthalmol Vis Sci 49:1136–1141
136. Noergaard MH, Pedersen DB, Bang K, Jensen PK, Kiilgaard JF, Stefansson E, la Cour M (2008) Indomethacin decreases optic nerve oxygen tension by a mechanism other than cyclo-oxygenase inhibition. Br J Ophthalmol 92:126–130
137. Nork TM, Ver Hoeve JN, Poulsen GL, Nickells RW, Davis MD, Weber AJ, Vaegan SSH, Lemley HL, Millecchia LL (2000) Swelling and loss of photoreceptors in chronic human and experimental glaucomas. Arch Ophthalmol 118:235–245
138. Okawa H, Sampath AP, Laughlin SB, Fain GL (2008) ATP consumption by mammalian rod photoreceptors in darkness and in light. Curr Biol 18:1917–1921
139. Ozaki H, Yu AY, Della N, Ozaki K, Luna JD, Yamada H, Hackett SF, Okamoto N, Zack DJ, Semenza GL, Campochiaro PA (1999) Hypoxia inducible factor-1alpha is increased in ischemic retina: temporal and spatial correlation with VEGF expression. Invest Ophthalmol Vis Sci 40:182–189
140. Padnick-Silver L, Linsenmeier RA (2002) Quantification of in vivo anaerobic metabolism in the normal cat retina through intraretinal pH measurements. Vis Neurosci 19:793–806
141. Padnick-Silver L, Linsenmeier RA (2003) Effect of acute hyperglycemia on oxygen and oxidative metabolism in the intact cat retina. Invest Ophthalmol Vis Sci 44:745–750
142. Padnick-Silver L, Linsenmeier RA (2005) Effect of hypoxemia and hyperglycemia on pH in the intact cat retina. Arch Ophthalmol 123:1684–1690
143. Padnick-Silver L, Derwent JJ, Giuliano E, Narfstrom K, Linsenmeier RA (2006) Retinal oxygenation and oxygen metabolism in Abyssinian cats with a hereditary retinal degeneration. Invest Ophthalmol Vis Sci 47:3683–3689
144. Patz A (1955) Oxygen inhalation in retinal arterial occlusion; a preliminary report. Am J Ophthalmol 40:789–795
145. Patz A (1955) Retrolental fibroplasia; experimental studies. Trans Am Acad Ophthalmol Otolaryngol 59:25–34; discussion 40–21
146. Peddada R, Linsenmeier RA (2005) Diminished oxygen diffusion to outer retina in the presence of

Drusen in age-related macular degeneration (AMD): computational model and histopathologic correlation. Association for Research in Vision and Ophthalmology, Ft. Lauderdale
147. Pedersen DB, Koch Jensen P, la Cour M, Kiilgaard JF, Eysteinsson T, Bang K, Wiencke AK, Stefansson E (2005) Carbonic anhydrase inhibition increases retinal oxygen tension and dilates retinal vessels. Graefes Arch Clin Exp Ophthalmol 243:163–168
148. Pedersen DB, Stefansson E, Kiilgaard JF, Jensen PK, Eysteinsson T, Bang K, la Cour M (2006) Optic nerve pH and PO_2: the effects of carbonic anhydrase inhibition, and metabolic and respiratory acidosis. Acta Ophthalmol Scand 84:475–480
149. Pedler C, Tilley R (1969) The retina of a fruit bat (*Pteropus giganteus Brunnich*). Vision Res 9:909–922
150. Penn JS, Henry MM, Wall PT, Tolman BL (1995) The range of PaO_2 variation determines the severity of oxygen-induced retinopathy in newborn rats. Invest Ophthalmol Vis Sci 36:2063–2070
151. Penn JS, Li S, Naash MI (2000) Ambient hypoxia reverses retinal vascular attenuation in a transgenic mouse model of autosomal dominant retinitis pigmentosa. Invest Ophthalmol Vis Sci 41:4007–4013
152. Pettigrew JD, Wallman J, Wildsoet CF (1990) Saccadic oscillations facilitate ocular perfusion from the avian pecten. Nature 343:362–363
153. Poitry S, Tsacopoulos M, Fein A, Cornwall MC (1996) Kinetics of oxygen consumption and light-induced changes of nucleotides in solitary rod photoreceptors. J Gen Physiol 108:75–87
154. Poulaki V, Qin W, Joussen AM, Hurlbut P, Wiegand SJ, Rudge J, Yancopoulos GD, Adamis AP (2002) Acute intensive insulin therapy exacerbates diabetic blood-retinal barrier breakdown via hypoxia-inducible factor-1alpha and VEGF. J Clin Invest 109:805–815
155. Poulaki V, Joussen AM, Mitsiades N, Mitsiades CS, Iliaki EF, Adamis AP (2004) Insulin-like growth factor-I plays a pathogenetic role in diabetic retinopathy. Am J Pathol 165:457–469
156. Pournaras CJ, Riva CE, Tsacopoulos M, Strommer K (1989) Diffusion of O_2 in the retina of anesthetized miniature pigs in normoxia and hyperoxia. Exp Eye Res 49:347–360
157. Pournaras CJ, Tsacopoulos M, Riva CE, Roth A (1990) Diffusion of O_2 in normal and ischemic retinas of anesthetized miniature pigs in normoxia and hyperoxia. Graefes Arch Clin Exp Ophthalmol 228:138–142
158. Pournaras CJ, Tsacopoulos M, Strommer K, Gilodi N, Leuenberger PM (1990) Experimental retinal branch vein occlusion in miniature pigs induces local tissue hypoxia and vasoproliferative microangiopathy. Ophthalmology 97:1321–1328
159. Pournaras CJ, Tsacopoulos M, Strommer K, Gilodi N, Leuenberger PM (1990) Scatter photocoagulation restores tissue hypoxia in experimental vasoproliferative microangiopathy in miniature pigs. Ophthalmology 97:1329–1333
160. Pournaras JA, Petropoulos IK, Munoz JL, Pournaras CJ (2004) Experimental retinal vein occlusion: effect of acetazolamide and carbogen (95% O_2/5% CO_2) on preretinal PO_2. Invest Ophthalmol Vis Sci 45:3669–3677
161. Raleigh JA, Dewhirst MW, Thrall DE (1996) Measuring tumor hypoxia. Semin Radiat Oncol 6:37–45
162. Rattner A, Toulabi L, Williams J, Yu H, Nathans J (2008) The genomic response of the retinal pigment epithelium to light damage and retinal detachment. J Neurosci 28:9880–9889
163. Riva CE, Logean E, Falsini B (2005) Visually evoked hemodynamical response and assessment of neurovascular coupling in the optic nerve and retina. Prog Retin Eye Res 24:183–215
164. Roh HD, Goldstick TK, Linsenmeier RA (1990) Spatial variation of the local tissue oxygen diffusion coefficient measured in situ in the cat retina and cornea. Adv Exp Med Biol 277:127–136
165. Roos MW (2007) Theoretical estimation of retinal oxygenation during retinal detachment. Comput Biol Med 37(6):890–896
166. Roth S, Pietrzyk Z (1994) Blood flow after retinal ischemia in cats. Invest Ophthalmol Vis Sci 35:3209–3217
167. Sakai T, Lewis GP, Linberg KA, Fisher SK (2001) The ability of hyperoxia to limit the effects of experimental detachment in cone-dominated retina. Invest Ophthalmol Vis Sci 42:3264–3273
168. Sakaue H, Negi A, Honda Y (1989) Comparative study of vitreous oxygen tension in human and rabbit eyes. Invest Ophthalmol Vis Sci 30:1933–1937
169. Schneiderman G, Goldstick TK (1978) Oxygen electrode design criteria and performance characteristics: recessed cathode. J Appl Physiol 45:145–154
170. Shahidi M, Shakoor A, Blair NP, Mori M, Shonat RD (2006) A method for chorioretinal oxygen tension measurement. Curr Eye Res 31:357–366
171. Shahidi M, Wanek J, Blair NP, Mori M (2009) Three-dimensional mapping of chorioretinal vascular oxygen tension in the rat. Invest Ophthalmol Vis Sci 50(2):820–825
172. Shakoor A, Blair NP, Mori M, Shahidi M (2006) Chorioretinal vascular oxygen tension changes in response to light flicker. Invest Ophthalmol Vis Sci 47:4962–4965
173. Shih YF, Fitzgerald ME, Reiner A (1993) Effect of choroidal and ciliary nerve transection on choroidal blood flow, retinal health, and ocular enlargement. Vis Neurosci 10:969–979
174. Shimazaki H, Oakley B 2nd (1986) Decline of electrogenic Na^+/K^+ pump activity in rod photoreceptors during maintained illumination. J Gen Physiol 87:633–647
175. Shonat RD, Wilson DF, Riva CE, Cranstoun SD (1992) Effect of acute increases in intraocular pressure on intravascular optic nerve head oxygen tension in cats. Invest Ophthalmol Vis Sci 33: 3174–3180

176. Shonat RD, Kight AC (2003) Oxygen tension imaging in the mouse retina. Ann Biomed Eng 31:1084–1096
177. Sickel W (1972) Retinal metabolism in dark and light. In: Fuortes MGF (ed) Physiology of photoreceptor organs handbook of sensory physiology. Springer, Berlin, pp 227–727
178. Smith LE, Wesolowski E, McLellan A, Kostyk SK, D'Amato R, Sullivan R, D'Amore PA (1994) Oxygen-induced retinopathy in the mouse. Invest Ophthalmol Vis Sci 35:101–111
179. Sperber GO, Bill A (1985) Blood flow and glucose consumption in the optic nerve, retina and brain: effects of high intraocular pressure. Exp Eye Res 41:639–653
180. Stefansson E, Landers MB 3rd, Wolbarsht ML (1981) Increased retinal oxygen supply following pan-retinal photocoagulation and vitrectomy and lensectomy. Trans Am Ophthalmol Soc 79:307–334
181. Stefansson E, Wolbarsht ML, Landers MB 3rd (1983) In vivo O_2 consumption in rhesus monkeys in light and dark. Exp Eye Res 37:251–256
182. Stefansson E, Hatchell DL, Fisher BL, Sutherland FS, Machemer R (1986) Panretinal photocoagulation and retinal oxygenation in normal and diabetic cats. Am J Ophthalmol 101:657–664
183. Stefansson E (1988) Retinal oxygen tension is higher in light than dark. Pediatr Res 23:5–8
184. Stefansson E, Peterson JI, Wang YH (1989) Intraocular oxygen tension measured with a fiber-optic sensor in normal and diabetic dogs. Am J Physiol 256:H1127–H1133
185. Stefansson E (1990) Oxygen and diabetic eye disease. Graefes Arch Clin Exp Ophthalmol 228:120–123
186. Stefansson E, Novack RL, Hatchell DL (1990) Vitrectomy prevents retinal hypoxia in branch retinal vein occlusion. Invest Ophthalmol Vis Sci 31:284–289
187. Stefansson E, Jensen PK, Eysteinsson T, Bang K, Kiilgaard JF, Dollerup J, Scherfig E, la Cour M (1999) Optic nerve oxygen tension in pigs and the effect of carbonic anhydrase inhibitors. Invest Ophthalmol Vis Sci 40:2756–2761
188. Stefansson E, Pedersen DB, Jensen PK, la Cour M, Kiilgaard JF, Bang K, Eysteinsson T (2005) Optic nerve oxygenation. Prog Retin Eye Res 24:307–332
189. Stefansson E (2006) Ocular oxygenation and the treatment of diabetic retinopathy. Surv Ophthalmol 51:364–380
190. Stone J, Maslim J, Valter-Kocsi K, Mervin K, Bowers F, Chu Y, Barnett N, Provis J, Lewis G, Fisher SK, Bisti S, Gargini C, Cervetto L, Merin S, Peer J (1999) Mechanisms of photoreceptor death and survival in mammalian retina. Prog Retin Eye Res 18:689–735
191. Taylor A, Jacques PF, Dorey CK (1993) Oxidation and aging: impact on vision. Toxicol Ind Health 9:349–371
192. Thiersch M, Raffelsberger W, Frigg R, Samardzija M, Wenzel A, Poch O, Grimm C (2008) Analysis of the retinal gene expression profile after hypoxic preconditioning identifies candidate genes for neuroprotection. BMC Genomics 9:73
193. Tiedeman JS, Kirk SE, Srinivas S, Beach JM (1998) Retinal oxygen consumption during hyperglycemia in patients with diabetes without retinopathy. Ophthalmology 105:31–36
194. Tornquist P, Alm A (1979) Retinal and choroidal contribution to retinal metabolism in vivo. A study in pigs. Acta Physiol Scand 106:351–357
195. Trick GL, Edwards P, Desai U, Berkowitz BA (2006) Early supernormal retinal oxygenation response in patients with diabetes. Invest Ophthalmol Vis Sci 47:1612–1619
196. Vinogradov SA, Lo LW, Jenkins WT, Evans SM, Koch C, Wilson DF (1996) Noninvasive imaging of the distribution in oxygen in tissue in vivo using near-infrared phosphors. Biophys J 70:1609–1617
197. Van Vo T, Riva CE (1995) Variations of blood flow at optic nerve head induced by sinusoidal flicker stimulation in cats. J Physiol 482(Pt 1):189–202
198. Wang L, Kondo M, Bill A (1997) Glucose metabolism in cat outer retina. Effects of light and hyperoxia. Invest Ophthalmol Vis Sci 38:48–55
199. Wang L, Tornquist P, Bill A (1997) Glucose metabolism of the inner retina in pigs in darkness and light. Acta Physiol Scand 160:71–74
200. Wang L, Tornquist P, Bill A (1997) Glucose metabolism in pig outer retina in light and darkness. Acta Physiol Scand 160:75–81
201. Wang S (2006) Effects of hyperoxia on retinal metabolism and function in the detached cat retina. Biomedical Engineering, Northwestern University, Evanston
202. Wang S, Linsenmeier RA (2007) Hyperoxia improves oxygen consumption in the detached feline retina. Invest Ophthalmol Vis Sci 48:1335–1341
203. Wang S, Birol G, Budzynski E, Flynn R, Linsenmeier RA (2010) Metabolic responses to light in monkey photoreceptors. Current Eye Research 35(6):510-518
204. Wangsa-Wirawan N, Padnick-Silver L, Budzynski E, Linsenmeier R (2001) pH regulation in the intact cat outer retina. Invest Ophthalmol Vis Sci 42:S367
205. Wangsa-Wirawan ND, Linsenmeier RA (2003) Retinal oxygen: fundamental and clinical aspects. Arch Ophthalmol 121:547–557
206. Warburg O, Posener K, Negelein E (1924) Uber Den stoffwechsel Der carcinomzelle. Biochem Z 152: 309–343
207. Weinberger AW, Siekmann UP, Wolf S, Rossaint R, Kirchhof B, Schrage NF (2002) Treatment of acute central retinal artery occlusion (CRAO) by hyperbaric oxygenation therapy (HBO) – pilot study with 21 patients. Klin Monatsbl Augenheilkd 219:728–734
208. Wellard J, Lee D, Valter K, Stone J (2005) Photoreceptors in the rat retina are specifically vulnerable to both hypoxia and hyperoxia. Vis Neurosci 22:501–507
209. Whalen WJ, Riley J, Nair P (1967) A microelectrode for measuring intracellular PO_2. J Appl Physiol 23:798–801
210. Wilson CA, Berkowitz BA, Hatchell DL (1992) Oxygen kinetics in preretinal perfluorotributylamine. Exp Eye Res 55:119–126

211. Wilson CA, Berkowitz BA, McCuen BW 2nd, Charles HC (1992) Measurement of preretinal oxygen tension in the vitrectomized human eye using fluorine-19 magnetic resonance spectroscopy. Arch Ophthalmol 110:1098–1100
212. Wilson DF, Vinogradov SA, Grosul P, Kuroki A, Bennett J (2005) Imaging oxygen pressure in the retina of the mouse eye. Adv Exp Med Biol 566:159–165
213. Wilson DF, Vinogradov SA, Grosul P, Vaccarezza MN, Kuroki A, Bennett J (2005) Oxygen distribution and vascular injury in the mouse eye measured by phosphorescence-lifetime imaging. Appl Opt 44:5239–5248
214. Wilson DJ, Green WR (1987) Argon laser panretinal photocoagulation for diabetic retinopathy. Scanning electron microscopy of human choroidal vascular casts. Arch Ophthalmol 105:239–242
215. Wilson DJ, Finkelstein D, Quigley HA, Green WR (1988) Macular grid photocoagulation. An experimental study on the primate retina. Arch Ophthalmol 106:100–105
216. Winkler BS (1981) Glycolytic and oxidative metabolism in relation to retinal function. J Gen Physiol 77:667–692
217. Winkler BS et al (1995) A quantitative assessment of glucose metabolism in the isolated rat retina. In: Christen Y, Doly M, Droy-Lefaix M (eds) Les Seminaires Ophthalmologiques D'ipsen, Vision Et Adaptation. Elsevier, Paris, pp 78–96
218. Winkler BS (2008) An hypothesis to account for the renewal of outer segments in rod and cone photoreceptor cells: renewal as a surrogate antioxidant. Invest Ophthalmol Vis Sci 49:3259–3261
219. Winkler BS, Starnes CA, Twardy BS, Brault D, Taylor RC (2008) Nuclear magnetic resonance and biochemical measurements of glucose utilization in the cone-dominant ground squirrel retina. Invest Ophthalmol Vis Sci 49:4613–4619
220. Wise G, Dollery CT, Henkind P (1971) The retinal circulation. Harper and Row, New York
221. Wolbarsht ML, Landers MB 3rd (1980) The rationale of photocoagulation therapy for proliferative diabetic retinopathy: a review and a model. Ophthalmic Surg 11:235–245
222. Yancey CM, Linsenmeier RA (1989) Oxygen distribution and consumption in the cat retina at increased intraocular pressure. Invest Ophthalmol Vis Sci 30:600–611
223. Yang Z, Quigley HA, Pease ME, Yang Y, Qian J, Valenta D, Zack DJ (2007) Changes in gene expression in experimental glaucoma and optic nerve transection: the equilibrium between protective and detrimental mechanisms. Invest Ophthalmol Vis Sci 48:5539–5548
224. Yu DY, Cringle SJ, Alder VA (1990) The response of rat vitreal oxygen tension to stepwise increases in inspired percentage oxygen. Invest Ophthalmol Vis Sci 31:2493–2499
225. Yu DY, Cringle SJ, Alder VA, Su EN (1994) Intraretinal oxygen distribution in rats as a function of systemic blood pressure. Am J Physiol 267:H2498–H2507
226. Yu DY, Cringle SJ, Alder VA, Su EN, Yu PK (1996) Intraretinal oxygen distribution and choroidal regulation in the avascular retina of guinea pigs. Am J Physiol 270:H965–H973
227. Yu DY, Cringle SJ, Alder V, Su EN (1999) Intraretinal oxygen distribution in the rat with graded systemic hyperoxia and hypercapnia. Invest Ophthalmol Vis Sci 40:2082–2087
228. Yu DY, Cringle SJ, Su EN, Yu PK (2000) Intraretinal oxygen levels before and after photoreceptor loss in the RCS rat. Invest Ophthalmol Vis Sci 41:3999–4006
229. Yu DY, Cringle SJ (2001) Oxygen distribution and consumption within the retina in vascularised and avascular retinas and in animal models of retinal disease. Prog Retin Eye Res 20:175–208
230. Yu DY, Cringle SJ (2002) Outer retinal anoxia during dark adaptation is not a general property of mammalian retinas. Comp Biochem Physiol A Mol Integr Physiol 132:47–52
231. Yu DY, Cringle SJ (2004) Low oxygen consumption in the inner retina of the visual streak of the rabbit. Am J Physiol Heart Circ Physiol 286:H419–H423
232. Yu DY, Cringle SJ, Su E, Yu PK, Humayun MS, Dorin G (2005) Laser-induced changes in intraretinal oxygen distribution in pigmented rabbits. Invest Ophthalmol Vis Sci 46:988–999
233. Yu DY, Cringle SJ, Su EN (2005) Intraretinal oxygen distribution in the monkey retina and the response to systemic hyperoxia. Invest Ophthalmol Vis Sci 46:4728–4733
234. Zacks DN, Han Y, Zeng Y, Swaroop A (2006) Activation of signaling pathways and stress-response genes in an experimental model of retinal detachment. Invest Ophthalmol Vis Sci 47:1691–1695
235. Zhang W, Ito Y, Berlin E, Roberts R, Berkowitz BA (2003) Role of hypoxia during normal retinal vessel development and in experimental retinopathy of prematurity. Invest Ophthalmol Vis Sci 44:3119–3123
236. Zuckerman R, Weiter JJ (1980) Oxygen transport in the bullfrog retina. Exp Eye Res 30:117–127
237. Zuckerman R, Cheasty JE, Wang Y (1993) Optical mapping of inner retinal tissue PO_2. Curr Eye Res 12:809–825

Measurement of Ocular Blood Flow: Angiography

5

Sebastian Wolf

Core Messages

- High-resolution video fluorescein angiography allows the assessment of retinal microcirculation.
- Data for global retinal microcirculation can be derived from estimations of the arm-retina time, the arteriovenous passage time, and mean arterial dye velocity. Perifoveal capillary microcirculation can be accessed from the transit of hypofluorescent segments in the capillary macular network.
- In retinal pathologies such as diabetic retinopathy, retinal vein occlusion assessment of retinal blood flow by angiographic techniques resulted in significant differences between patients and healthy subjects as well as between the groups.
- The interindividual variability of retinal blood flow data is quite large.
- Prognostic information for the individual patient from retinal blood flow measurement using angiographic techniques is very limited.

Fluorescein angiography has been used for the past 30 years to quantitative retinal hemodynamics [9, 10, 15, 17, 22, 28, 33]. The introduction of the video technique has improved the temporal resolution of fluorescein angiograms. The combination of video fluorescein angiography and digital image analysis allowed for precise assessment of retinal circulation [17, 23, 30]. Furthermore, the introduction of the scanning laser technique for recording of high-speed fluorescein angiograms has improved the techniques for quantification of retinal circulation [13, 14, 29, 32]. Several studies have attempted to quantify retinal hemodynamics from angiograms by dye dilution techniques [10, 16, 18, 24]. However, these studies used standard photographic techniques to record fluorescein angiograms with a time resolution of 1–2 images per second. The analysis of the arterial inflow into the retinal vascular bed was not possible but offered the first insights into retinal circulation. Data derived from these measurements included the arm-retina time and the arteriovenous passage time. The arm-retina time gives a rough estimate of the vascular system supplying the eye. Especially in patients with carotid artery obstructions, a significant increase of the arm-retina time has been demonstrated [27]. The retinal hemodynamics in greater arterioles and venules can be quantified by measurement of arteriovenous passage time and mean dye velocity from the early phase of the fluorescein angiograms [17, 19, 23, 30]. Several studies have demonstrated the correlation between clinical

S. Wolf, M.D., Ph.D.
Universitätsklinik für Augenheilkunde, Inselspital, University of Bern, Freiburgstrasse, CH-3010 Bern, Switzerland
e-mail: sebastian.wolf@insel.ch

L. Schmetterer, J.W. Kiel (eds.), *Ocular Blood Flow*,
DOI 10.1007/978-3-540-69469-4_5, © Springer-Verlag Berlin Heidelberg 2012

findings and alterations of arteriovenous passage time and mean dye velocity [11, 31].

Beside dye dilution techniques, direct visualization of retinal capillary microcirculation in perifoveal capillaries is possible using high-speed scanning laser techniques [8, 12, 20, 21, 25, 26, 32]. Perifoveal capillary flow velocities give data on the macular microcirculation. By means of the scanning laser technique, flow velocities of segments of low fluorescence were quantified in perifoveal capillaries. Biomicroscopic recordings of conjunctival and periungual capillaries [4] clearly show segmentation corresponding to erythrocytes (rouleaux formations) versus plasma. From these findings, it is clear that segmentation in the fluorescence intensity corresponds to segments of erythrocytes and cell-free plasma [3]. Measurements of capillary flow velocities with fluorescein techniques correspond to the velocity of segments of cell-free plasma in the bloodstream.

In the following, the measuring techniques as well as clinical data from blood flow measurements using angiographic techniques will be described.

5.1 Measuring Technique

High-speed digital fluorescein angiography is performed using a scanning laser ophthalmoscope after intravenous injection of 5 ml sodium fluorescein (10%) with a 10 ml saline flush. Digital picture analysis of high-speed digital fluorescein angiograms allows for quantitative assessment of retinal circulation. Retinal macrocirculation can be characterized by the arm-retina time, the arteriovenous passage time, arterial mean dye velocity, and the arterial vessel diameters. In contrast, the measurement of capillary flow velocity provides data on retinal microcirculation. Macrocirculatory measurements are carried out in the 40° observation field. This mode allows for imaging the entire posterior pole. For the assessment of capillary flow velocities, high-definition angiograms of the perifoveal capillary network are necessary. Therefore, the 15–20° field of the scanning laser ophthalmoscope is used. For all measurements, the video signal generated by the scanning laser ophthalmoscope is converted into digital information and recorded digitally.

For the measurement of retinal macrocirculation, density variations in the fluorescein angiograms are analyzed by means of digital image processing system. After correction for eye movements, the digital image processing system measures the entire angiogram sequence, recording the intensity of fluorescein at various locations. Six points are interactively selected for measurement. The computer then analyzes the entire angiogram frame by frame. Fifty frames per second are evaluated. For each image, the program recorded the mean intensity levels at each of the six selected locations. Intensity curves are obtained by plotting the collected data against the time axis. The time of the first appearance of fluorescein is evaluated from the intensity curves. According to the location of the measuring points, several parameters of retinal circulation are assessed.

Two points are selected for measurement on the superiotemporal and inferiotemporal arteries (0.5 disc diameter) from the disc margin. Two more distally located points, two disc diameters from the disc margin on each artery are similarly monitored. The differences in appearance time and the actual distance between the proximal and distal marking point at each artery are used for calculation of the mean arterial dye velocity. The time elapsed between the appearance of dye at the proximal reference point at the temporal arteries and an adjacent point at its corresponding vein is used to determine the arteriovenous passage times.

Measurement of retinal microcirculation of the 15–20° field of the scanning laser ophthalmoscope is used. In these digital high-definition angiograms, segments of low and high fluorescence can be observed moving through the perifoveal network. The sequences are processed off-line to evaluate the mean capillary flow velocity and coefficient of variation of mean capillary flow velocity. The measurement of flow velocity is based on the determination of transit time "Δt" between two measuring points, separated by a known distance "Δs". The actual distance "Δs" were measured by a digital image processing system counting all pixels on the capillary between

Table 5.1 Arm-retina time, arteriovenous passage time, and mean arterial dye velocity of healthy subjects and the interindividual variation

	No.	Arm-retina time (s)	Arteriovenous passage time (s)	Mean arterial dye velocity (mm/s)
Healthy subjects	221	10.9±2.6	1.58±0.4	6.67±1.59
Interindividual variation		24%	27%	24%

the measuring points. The velocity of the moving hypofluorescent segments was calculated as $v=\Delta s/\Delta t$. The assessment of the capillary flow velocities was performed in each patient in 15 different vessels. All measurements were performed in the monolayer capillary network preventing confounding errors from oblique vessels. The velocity of ten different segments of low and high fluorescence in each capillary is quantified. Every value of the mean blood flow velocity (v) is based on 150 single measurements. All of these are performed within a time period of less than 5 s in the early transit phase of the angiograms. Measurements of actual distances were corrected for the refractive error.

5.2 Normal Values

Fluorescein angiograms of 221 healthy volunteers (116 male; 105 female, age: 30.9 ± 12.8 years) were performed to establish reference values for the arm-retina time, arteriovenous passage time, and mean arterial dye velocity (Table 5.1).

The interindividual variation was calculated and showed a 24% variation for the arm-retina time, a 27% variation for the arteriovenous passage time, and a 24% variation for the mean arterial dye velocity. The correlation between the arm-retina time and the arteriovenous passage time was weak ($r=0.17$; $p<0.05$). With increasing mean arterial dye velocity, a decrease of the arteriovenous passage time was noted ($r=-0.31$; $p<0.01$).

In a group of 90 healthy subjects, fluorescein angiography for assessment of capillary flow parameters was performed. The capillary flow velocity was 2.89±0.41 mm/s. The interindividual variation of the capillary flow velocity was 14.2%. With increasing arteriovenous passage time, the capillary flow velocity decreased slightly ($r=-0.25$; $p<0.05$).

The influence of age, blood pressure, heart rate, and intraocular pressure was evaluated by multiple stepwise regression analysis. A significant relation between age and arm-retina time ($r=0.331$; $p<0.01$) and arteriovenous passage time ($r=0.164$; $p<0.01$) was found. Both arm-retina time and arteriovenous passage time were observed to increase slightly with age. Blood pressure, pulse rate, and intraocular pressure showed no additional significant influence on the dynamic data. Comparing the 10% youngest and oldest subjects (age: 21 ± 1 years vs. 63 ± 8 years), no significant differences could be found between the two groups (arteriovenous passage time: 1.6±0.4 s vs. 1.7±0.4 s, $p>0.05$).

The mean capillary flow velocity showed a dependence on age ($r=-0.296$; $p<0.01$). The mean capillary flow velocity decreased with increasing age. No additional significant influence of the blood pressure, heart rate, and intraocular pressure was found.

5.3 Retinal Pathologies

5.3.1 Diabetes Mellitus

Bertram et al. [11] report retinal blood flow data from a study in patients with diabetic retinopathy. For this study, the system described above was used. In a group of 124 patients, with type I diabetes mellitus (71 male, 53 female age: mean 35±12 years), retinal circulation times were assessed. The severity of diabetic retinopathy was evaluated by ophthalmoscopy, fundus photography, and angiography. The diabetic patients were distributed into four groups: (1) no retinopathy (no DR); (2) mild to moderate nonproliferative

Table 5.2 Arm-retina time, arteriovenous passage time, and mean arterial dye velocity of diabetic patients and reference values

	No.	Arm-retina time (s)	Arteriovenous passage time (s)	Mean arterial dye velocity (mm/s)
Reference values	221	10.9 ± 2.6	1.58 ± 0.4	6.67 ± 1.59
Diabetic patients	124	11.5 ± 3.4	2.35 ± 0.9	4.33 ± 2.20
		n.s.	$p < 0.01$	$p < 0.05$

Table 5.3 Arm-retina time, arteriovenous passage time, and mean arterial dye velocity of patients with CRVO and reference values

	No.	Arm-retina time (s)	Arteriovenous passage time (s)	Mean arterial dye velocity (mm/s)
Reference values	221	10.9 ± 2.6	1.58 ± 0.4	6.67 ± 1.59
CRVO	173	13.3 ± 3.9	6.47 ± 3.25	1.92 ± 0.72
Ischemic CRVO	57	13.2 ± 3.1	6.72 ± 2.79	1.71 ± 0.75
Nonischemic CRVO	116	13.3 ± 4.3	6.35 ± 3.45	2.04 ± 0.69

retinopathy (mild NPDR); (3) severe nonproliferative retinopathy (severe NPDR); and (4) proliferative retinopathy (PDR). Table 5.2 shows the values for the arm-retina time, arteriovenous passage time, and mean arterial dye velocity for the diabetic patients and the reference values.

No significant difference of arm-retina time was found between the groups with different stage of diabetic retinopathy. The arteriovenous passage time was significantly prolonged in all four groups as compared with the normal subjects. There was an increase of arteriovenous passage time with progressing diabetic retinopathy. The differences in arteriovenous passage time between no retinopathy and mild NPDR ($p < 0.05$), no retinopathy and severe NPDR ($p < 0.05$), no retinopathy and PDR ($p < 0.01$), and mild NPDR and PDR ($p < 0.01$) were significant.

Various studies [1, 2, 5–7, 12, 21, 25, 26] have demonstrated significant differences of perifoveal microcirculation in patients with diabetic retinopathy as compared with healthy subject. It has been demonstrated that capillary flow velocity is significantly reduced in patients with diabetes mellitus.

5.3.2 Central Retinal Vein Occlusion

A group of 173 patients (97 male, 76 female age: mean 64 ± 13 years) with acute central retinal vein occlusion (CRVO) retinal circulation times were assessed with the above described system. A total of 116 patients suffered from a nonischemic CRVO. Table 5.3 shows the values for the arm-retina time, arteriovenous passage time, and mean arterial dye velocity for the patients with CRVO.

The statistical analysis confirms significant differences between patients and the reference values but no differences between ischemic and non-ischemic CRVO.

5.4 Summary

Fluorescein angiography with a scanning laser ophthalmoscope allows the assessment of retinal hemodynamics. The evaluation of the high-speed digital fluorescein angiograms by means of digital picture analysis provides reproducible data on retinal hemodynamics. The arm-retina time gives a rough estimate of the vascular system supplying the eye. Especially in patients with carotid artery obstructions, a significant increase of the arm-retina time has been demonstrated. The retinal hemodynamics in greater arterioles and venules can be quantified by measurement of arteriovenous passage time and mean dye velocity from the early phase of the fluorescein angiograms. Several studies have demonstrated the correlation between clinical findings and alterations of arteriovenous passage time and mean dye velocity. Perifoveal capillary flow velocities give data on the macular microcirculation. By means of the scanning laser technique, flow velocities of segments of low fluorescence were quantified in perifoveal capillaries.

In retinal pathologies such as diabetic retinopathy, retinal vein occlusion assessment of retinal blood flow by angiographic techniques resulted in significant differences between the groups. However, the interindividual variability of retinal blood flow data is quite large. Therefore, prognostic information for the individual patient from retinal blood flow measurement using angiographic techniques is very limited.

Proprietary Interests None.

References

1. Arend O, Harris A, Wolf S (1994) Capillary blood flow velocity measurements in cystoid macular edema with the scanning laser ophthalmoscope. Am J Ophthalmol 117(6):819–820
2. Arend O, Remky A, Harris A, Bertram B, Reim M, Wolf S (1995) Macular microcirculation in cystoid maculopathy of diabetic patients. Br J Ophthalmol 79:628–632
3. Arend O, Wolf S (1992) Segmentation of fluorescence in the retinal microcirculation – is it a valid indicator of blood cell flow? – reply. Br J Ophthalmol 76: 510–511
4. Arend O, Wolf S, Bertram B, Schulte K, Hoberg A, Jung F, Reim M (1991) Konjunktivale Mikrozirkulation und Hämorheologie bei Patienten mit venösen Verschlüssen der Retina. Fortschr Ophthalmol 88:243–247
5. Arend O, Wolf S, Harris A, Reim M (1995) The relationship of macular microcirculation to visual acuity in diabetic patients. Arch Ophthalmol 113:610–614
6. Arend O, Wolf S, Krantz M, Bertram B, Schulte K, Reim M (1992) Influence of the perifoveal microcirculation on visual acuity in patients with diabetes mellitus. Invest Ophthalmol Vis Sci 33:1366
7. Arend O, Wolf S, Remky A, Sponsel WE, Harris A, Bertram B, Reim M (1994) Perifoveal microcirculation with non-insulin dependent diabetes mellitus. Graefes Arch Clin Exp Ophthalmol 232:225–231
8. Arend O, Wolf S, Schulte K, Bertram B, Reim M (1993) The macular microcirculation with progression of diabetic retinopathy. Eur J Ophthalmol 3:162
9. Bateman WA, Kruger RA (1984) Blood flow measurement using digital angiography and parametric imaging. Med Phys 11(2):153–157
10. Ben-Sira I, Riva CE (1973) Fluorophotometric recording of fluorescein dilution curves in human retinal vessels. Invest Ophthalmol Vis Sci 12:310–312
11. Bertram B, Wolf S, Fiehöfer S, Schulte K, Arend O, Reim M (1991) Retinal circulation times in diabetes mellitus type I. Br J Ophthalmol 75:462–465
12. Funatsu H, Sakata K, Harino S, Okuzawa Y, Noma H, Hori S (2006) Tracing method in the assessment of retinal capillary blood flow velocity by fluorescein angiography with scanning laser ophthalmoscope. Jpn J Ophthalmol 50(1):25–32
13. Gabel VP, Birngruber R, Nasemann J (1988) Fluorescein angiography with the scanning laser ophthalmoscope (SLO). Lasers Light Ophthalmol 2(1):35–40
14. Harris A, Arend O, Kopecky K, Caldemeyer K, Wolf S, Sponsel W, Martin B (1994) Physiological perturbation of ocular and cerebral blood flow as measured by scanning laser ophthalmoscopy and color Doppler imaging. Surv Ophthalmol 38:81–86
15. Harris A, Chung HS, Ciulla TA, Kagemann L (1999) Progress in measurement of ocular blood flow and relevance to our understanding in glaucoma and age-related macular degeneration. Prog Retin Eye Res 18(5):670–687
16. Hitchings RA, Spaeth GL (1977) Fluorescein angiography in chronic simple and low-tension glaucoma. Br J Ophthalmol 61:126–132
17. Jung F, Kiesewetter H, Körber N, Wolf S, Reim M, Müller G (1983) Quantification of characteristic blood-flow parameters in the vessels of the retina with a picture analysis system for video-fluorescence angiograms: initial findings. Graefes Arch Clin Exp Ophthalmol 211:133–136
18. Kohner EM, Hamilton AM, Saunders SJ, Sutcliffe BA, Bulpitt CJ (1975) The retinal blood flow in diabetes. Diabetologia 11:27–33
19. Koyama T, Matsuo N, Shimizu K, Iishi A, Shiraga F, Okayama H, Tsuji T, Mihara M, Tsuchida Y (1986) A measuring system of retinal mean circulation time by means of videofluorescence angiography and image analyser. Ther Res 5(4):685–690
20. Nasemann JE (1989) Observation du débit érythrocytaire dans les vaisseaux rétiniens au cours de l´angiographie en fluorescence. J Fr Ophthalmol 12(10):629–634
21. Noma H, Funatsu H, Sakata K, Harino S, Mimura T, Hori S (2008) Macular microcirculation in hypertensive patients with and without branch retinal vein occlusion. Acta Ophthalmol 87(6): 638–642
22. Novotny HR, Alvis DL (1960) A method of photographing fluorescence in circulating blood of the human eye. Am J Ophthalmol 50:176
23. Preußner PR, Richard G, Darrelmann O, Weber J, Kreissig I (1983) Quantitative measurement of retinal blood flow in human beings by application of digital image-processing methods to television fluorescein angiograms. Graefes Arch Clin Exp Ophthalmol 221:110–112
24. Riva CE, Feke GT, Ben-Sira I (1978) Fluorescein dye-dilution technique and retinal circulation. Am J Physiol 234:H315–H322
25. Sakata K, Funatsu H, Harino S, Noma H, Hori S (2006) Relationship between macular microcirculation and progression of diabetic macular edema. Ophthalmology 113(8):1385–1391

26. Sakata K, Funatsu H, Harino S, Noma H, Hori S (2007) Relationship of macular microcirculation and retinal thickness with visual acuity in diabetic macular edema. Ophthalmology 114(11):2061–2069
27. Wolf S, Arend O, Bertram B, Ringelstein EB, Reim M (1992) Retinal hemodynamics and hemorheological findings in patients with internal carotid occlusion. Clin Hemorheol 12(3):488
28. Wolf S, Arend O, Reim M (1994) Measurement of retinal hemodynamics with scanning laser ophthalmoscopy: reference values and variation. Surv Ophthalmol 38:95–100
29. Wolf S, Arend O, Toonen H, Bertram B, Reim M (1992) Measurement of retinal micro- and macrocirculation in patients with diabetes mellitus with scanning laser ophthalmoscopy. Clin Vis Sci 7(6):461–469
30. Wolf S, Jung F, Kiesewetter H, Körber N, Reim M (1989) Video fluorescein angiography: method and clinical application. Graefes Arch Clin Exp Ophthalmol 227:145–151
31. Wolf S, Schulte K, Arend O, Glöckner WM, Handt S, Jung F, Reim M (1992) Correlation between retinal microcirculation, plasma viscosity and visual function in patients with macroglobulinemia. Clin Hemorheol 12(5):725–731
32. Wolf S, Toonen H, Arend O, Jung F, Kaupp A, Kiesewetter H, Meyer-Ebrecht D, Reim M (1990) Zur Quantifizierung der retinalen Kapillardurchblutung mit Hilfe des Scanning-Laser-Ophthalmoskops. Biomed Tech 35:131–134
33. Wolf S, Toonen H, Koyama T, Meyer-Ebrecht D, Reim M (1990) Scanning laser ophthalmoscopy for the quantification of retinal blood-flow parameters: a new imaging technique. In: Naseman JE, Burk ROW (eds) Scanning laser ophthalmoscopy and tomography. Quintessenz, München, pp 91–96

Measurement of Retinal Vessel Diameters

6

Gerhard Garhöfer and Walthard Vilser

Core Messages

- Given that local vascular tone is an important regulator of blood flow, exact determination of vessel diameter is crucial. The development of new and sophisticated instruments allow now for the exact and non-invasive determination of vascular tone in vivo.

6.1 Introduction

Since the development of the first method for funduscopy by Hermann von Helmholtz in 1851, both clinicians and researchers were interested in a method for the assessment of retinal vessel diameters. It has very early been recognized that morphological and functional changes of retinal vessels do not only reflect ocular vascular pathologies. Moreover, changes of the retinal vascular system may also serve as an indicator for a couple of systemic vascular-related disorders and their associated risks factors. In this context, alterations of the vascular system may be seen in either morphological changes of the retinal vasculature or in alterations of the vascular function. Hence, given that, in the eye, noninvasive investigation of the microcirculation is possible, it provides a unique possibility for interdisciplinary vessel diagnostic in microcirculation. Using new optical technologies with high resolution down to the microscopic range, physiological vascular regulation mechanisms and alterations of the microcirculation can be investigated.

Beside the clinical use of methods to visualize the ocular fundus, the exact determination of retinal vessel size is in particular crucial for the measurement of retinal blood flow. Given that there is increasing evidence that blood flow alteration may contribute to the pathogenesis of several ocular diseases, the assessment of ocular blood flow has received more and more importance. Blood flow in the eye, as in every vascular bed, is mainly dependent on perfusion pressure and vascular resistance. The latter is regulated by the local vascular tone, in particularly, in small resistance vessels. Thus, the regulation of local blood flow is necessarily dependent on an intact regulation of vascular tone in the tissue. This underlines the importance of measuring precisely the retinal vessel size when attempting to assess blood flow in a specific vascular bed.

The regulation of vascular tone is based on a complex interaction between local, systemic, and

G. Garhöfer, M.D. (✉)
Department of Clinical Pharmacology,
Medical University of Vienna,
Waehringer Guertel 18-20,
Vienna A-1090, Austria
e-mail: gerhard.garhoefer@meduniwien.ac.at

W. Vilser, Ph.D.
IMEDOS Systems UG (haftungsbeschränkt),
Am Nasstal 4, Jena D-07751, Germany
e-mail: info@imedos.de

L. Schmetterer, J.W. Kiel (eds.), *Ocular Blood Flow*,
DOI 10.1007/978-3-540-69469-4_6, © Springer-Verlag Berlin Heidelberg 2012

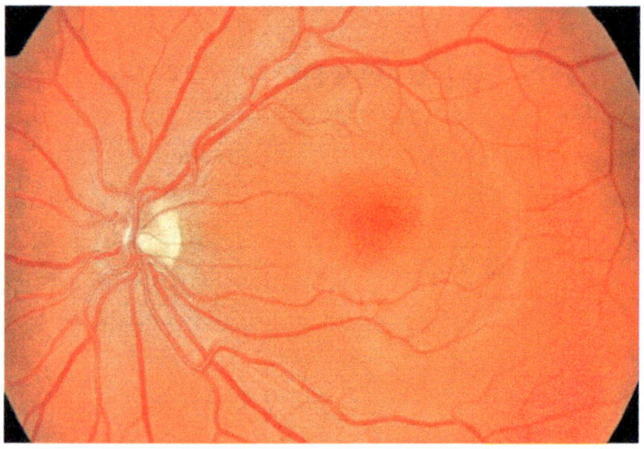

Fig. 6.1 Normal human fundus

neural components compensating for differing metabolic demands and changes in perfusion pressure. As a part of these regulation systems, it has been shown that the retina and the optic nerve head have the ability to adapt their blood flow to increasing metabolic demands caused by augmented neural activity. There is increasing evidence now that not only basal vascular tone is important to preserve physiological function but also the ability of the vascular system to adapt to changes in metabolic demands and that these mechanisms may be corrupted in several systemic and ocular diseases. The developments of sophisticated stimulation techniques allow now not only for the precise determination of vessel size but also for vascular and endothelial function. This may in future help for the early diagnosis and treatment of ocular and systemic vascular diseases.

6.1.1 Anatomy

The retinal vasculature is a classical end-artery system and supplies the inner layers of the neural retina. Observed ophthalmologically, it consists of a typically arranged network of vessel branches, entering the eye with the optic nerve head (Fig. 6.1). The central retinal artery is derived from the ophthalmic artery, which, in turn, is a branch of the internal carotid. Based on blood flow measurements, it has been calculated that the diameter of the central retinal artery is in the range from 134 to 208 μm in healthy subjects [8]. The central retinal artery enters the optic nerve approximately 10 mm behind the globe and divides near the lamina cribrosa into upper and lower main branches. Further divisions lead to the typical appearance of the retinal vasculature divided into four major branches, each supplying one quadrant of the neural retina. The retinal arteries are approximately 200 μm in diameter and, in an anatomical point of view, arterioles. The general structure of the retinal vessels is comparable to that of muscular arteries. Near the optic disc, the vessel wall consists of five to seven layers of smooth muscle cells, decreasing to two or three layers at the equator [23]. As a peculiarity, the retinal vessels lack any autonomic nerve supply.

Retinal blood flow is drained almost exclusively by the central retinal vein, which subsequently empties into the superior ophthalmic vein. Retinal veins are usually larger in diameter (up to 300 μm) and lack a well-developed smooth muscle covering. Calculations have revealed that total retinal blood flow in healthy subjects is in the range of 38–80 μl/min [8, 11] and may be considerably impaired under pathological conditions.

6.2 Vessel Diameter Measurements Based on Optical Images of Indirect Ophthalmoscopes

Already 1 year after the development of the ophthalmoscope by Hermann von Helmholtz, a first approach to measure retinal vessel diameters in vivo has been introduced by Ruete and Landolt. The measurement principle utilizes an optical image plane of the indirect ophthalmoscopy placing reference markers of well-defined size into it. The markers can be viewed in focus together with the vessels to gauge the vessel diameter by direct comparison. Subsequent modicifactions apply measurement oculars with measurement scales or moveable measurement markers like the Mikuni ocular [42].

Later on, the "Lobeck ocular" represented the first serious optical measurement device, based on an optical image splitting method and

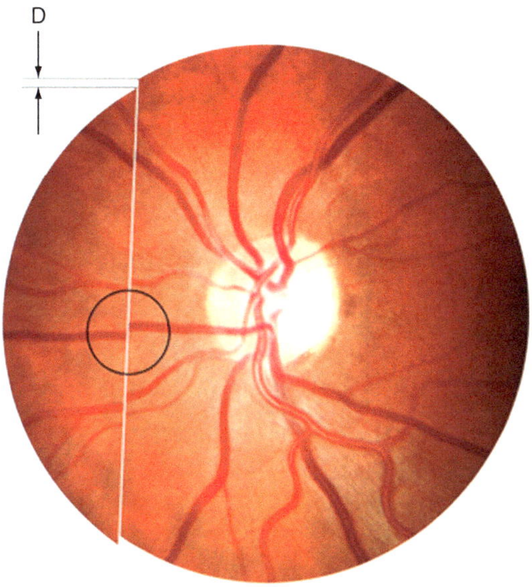

Fig. 6.2 Image-splitting principle (Modified from Vilser et al. [61])

developed by Lobeck [37] and manufactured at Carl Zeiss Jena during the 1930s. This principle was enhanced later to the use of a parallel plate micrometer [59] (Fig. 6.2).

The first electrooptical system for measuring retinal vessel diameters was presented by Delori [7]. In the image plane of an indirect ophthalmoscope, a light scanning system picks up brightness profiles perpendicular to the vessel course. The electronic signals representing these vessel cross sections are then analyzed in order to identify the vessel borders by the means of the half-height algorithm described in detail below. The specific purpose of this device was the measurement of oxygen saturation by use of different absorption profiles of oxy/desoxyhemoglobin including the diameter measurements. Another similar development was based on a modification of a commercially available photographic retinal camera (Carl Zeiss Jena, Germany) presented on the Leipzig fair in 1987 [58]. The device was equipped with a linear CCD sensor to gain brightness information necessary for diameter assessment.

However, diameter measurements in the live image of the fundus as visualized by indirect ophthalmoscopic devices are difficult, time consuming, and allow usually only for assessment one single-measurement location. An additional problem of these approaches is that the systems are very sensitive to eye movements. Thus, the reproducibility of the technique is strongly dependent on exact fixation abilities of the subject measured. The considerable numbers of disadvantages in combination with considerable error sources have limited the application of these early devices strictly to research purposes.

6.3 Vessel Diameter Measurements Based on Photographic and Digitally Stored Images

The development of retinal cameras and the advantage of the photographic or digital storage of fundus images with high temporal and spatial resolutions lead to improved measurement systems for vessel diameters and finally to the introduction of modern retinal vessel analysis. In particular, fundus photography for the first time allowed for the assessment of vessel diameters without the problems of a live examination. Acquired fundus images can be analyzed offline by trained persons, which allows for the anonymized analysis in double-masked clinical studies.

6.3.1 Basics for Measurements on Stored Images

Basically, all methods for measurement based on photographic and digital images follow a similar pattern: In a first step, the fundus image acquisition takes place, which is then followed by vessel diameter measurement based on the analysis of the recorded images.

6.3.1.1 Measuring Principle

Figure 6.3 demonstrates the basic principle of measurement, on which most of the techniques currently available are based by use of retinal cameras. Imaging of the column of the red blood in the fundus image is dominated by light absorption. The brightness profile from the vessel cross section (vessel profile) is analyzed to detect the

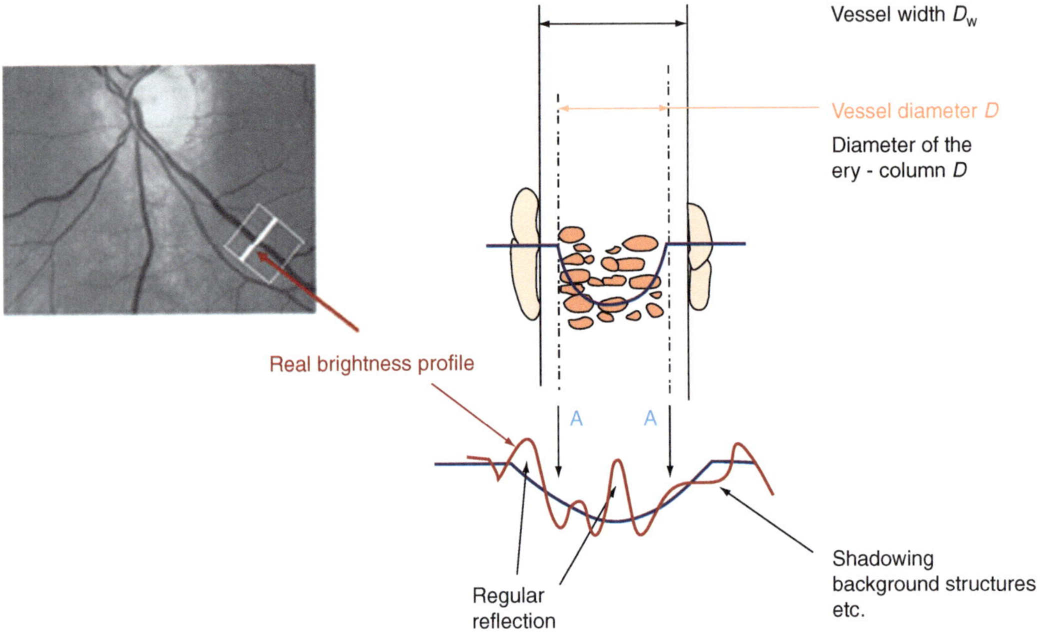

Fig. 6.3 Measuring principle

vessel edges and to define the measuring points inside these edges and therefore in the edges of the red blood cell column. The diameter of a vessel segment is estimated as distance between the measuring points on a perpendicular line to the flow direction of the vessel (Fig. 6.3).

The column of red blood cells is absorbing light backscattered from the retinal layers and results in an ideal brightness profile. However, in reality, that ideal profile is interfered with disturbances like local and global regular reflections and background structures. The optical properties of the eye, imaging system, and electronic image sensor influence the vessel profile by applying low-pass filtering and noise to the fundus image. The result is the real vessel profile, more or less disturbed like shown in the figure and changed by the optical magnification of the eye and optical device.

Imaging systems different from regular fundus cameras (e.g., laser scanning systems) usually introduce different imaging conditions, which make modifications of the measuring principle necessary. An exceptional position among the approaches holds the methods based on fluorescein angiography, which requires the application of a fluorescent dye. Given that fluorescein interfuses the whole vessel including the otherwise invisible plasma edge stream, the fluorescence pattern light represents the whole lumen, not only the width of the red blood cell column. This results in a different brightness profile, and a different diameter definition must be considered. Thus, the data assessed with fluorescein angiography are not entirely comparable with the data gained with other methods.

In the past, depending on the image acquisition and analysis technique, a variety of methods were developed and applied. The methodological properties of the different methods are determined both by the image acquisition system (retinal camera, laser scanner) and the measurement methods itself.

6.3.1.2 Definition of Retinal Vessel Diameter and Vessel Width

Although arbitrary, the terminology for defining vessel size is important. The vessel diameter is defined as the diameter of the red blood cell column and the vessel width as the diameter of the column of a contrast agent at a particular point in time for a given local vessel segment.

Because the vessel diameter and width differ, it is necessary to clearly distinguish measuring results using contrast agents (i.e., width) from results obtained without the use of contrast agents (i.e., diameter). It is assumed that the thickness of the plasma zone is constant under a wide range of hemodynamic conditions, but the validity of this assumption needs to be verified. Another important assumption is that the measured vessel has a circular cross section. This seems a reasonable assumption for the pressure conditions near the optic nerve head, but whether this assumption holds true for all parts of the ocular vascular system has yet to be shown.

6.3.1.3 Vessel Edge Definition and Methods for Estimation of Measuring Marks in the Vessel Edges

For the correct determination of vessel diameters, a correct identification of vessel borders is crucial. As illustrated in Fig. 6.3, the vessel edges are not sharp brightness transitions from the vessel to the surrounding tissue, but rather they appear as gradual transitions from bright to dark. This edge brightness transition can be as much as 1/3 of the vessel diameter. This edge can be further obscured by the brightness profile of the whole image.

Defining the vessel edge requires proper placement of measuring marks at the edges of brightness transition. The ideal vessel edge definition (A-A) for the chosen definition of vessel diameter is shown in Fig. 6.2 and marks the beginning and the end of the red blood cell column. However, the position of the measuring marks determined by measuring algorithms automatically or manually by an examiner are often misplaced. This is a common cause of systematic and random errors in retinal vessel diameter measurements and the impetuous for the development of automatic algorithms immune to brightness disturbances and with high reproducibility. One main aim for the development of automatic algorithms is to be sufficiently robust to the brightness disturbances to get a high reproducibility [60].

In principle, two different methods for vessel border determination are available: a subjective and an (semi)automatic, objective one.

Subjective Edge Definition of Manual Measurement Directly in the Acquired Fundus Image

In the simplest approach, the grader has to adjust a measuring mark until it fits the vessel border according to his visual impression [22, 24]. Obviously, it is crucial to ensure a perpendicular cross sectional line between the marks. Alternatively, a circle can be drawn with the center on one vessel edge, and the radius has to be increased until the circle is tangent to the opposite vessel edge [24]. Manual measurements are time consuming and suffer from subjective systematic measurement errors, dependent on the individual perception of the edge position by the grader. As one consequence, there are different measurement results between graders (intraindividual variability of graders). A more subtle error is that graders tend to place the measuring marks towards the darker part of the profile. A manual assessment of the same vessel shown bright on a dark background will result in a wider diameter estimate than if the vessel is shown dark on a bright background. Such misperceptions introduce systematic errors between measurements in positive and negative images. Furthermore, vessels in fluorescein angiographic images are measured wider than in images without dye independent from difference caused by plasma zone and by steeper edge brightness transitions. Finally, it should be noted that examination conditions like different image brightness and contrast, ambient light, and different styles of measuring marks may introduce errors and difference between measuring results [59, 60].

Objective Edge Definition

To improve reproducibility by reducing a number of errors due to perception of the grader, some methods apply objective criterions for diameter assessment. Those methods can be classified into the following three groups:

1. *Half-height maximum*
   The measuring points for both edges are defined as the half height of the brightness of the surrounding tissue and the vessel center on each side of the vessel. This definition was used by densitometric and scanning techniques [5].
2. *Filter algorithms for automatic measurements*
   One group of algorithms for automatic measurements of vessel diameter are special filter kernels [34] to mark the measuring points in the vessel edges of the brightness profiles crossing the vessel.
3. *Model-based algorithms for automatic measurements*
   A modeled vessel profile is approximated to the acquired real brightness profile. Extracted model parameters from the best fitting represent the vessel diameter [39]. Models can also account for regular reflections and other systematic error sources. A principal drawback is the time need of such methods.

6.3.1.4 Problems and Measuring Errors

Errors can arise from the individual properties of the eye, the imaging system, and image processing. Depending on the measuring algorithms and examination conditions, there are different sources for systematic and random errors. Different examination or measuring conditions can turn systematic errors into random errors and vice versa.

Magnification Error

The magnification scale of the fundus image is strongly affected by the anatomical properties of the eye. As a nominal value, the magnification scale should be given for the Gullstrand normal eye adapted to the far which is well defined. Deviations from that model can be quantified as axial ametropia and refractive ametropia, changing the optical magnification [1, 36]. A process of biological self-correction may counterbalance the effects of the two forms of ametropia, resulting in an emmetropic eye. Both the ametropic and the emmetropic eye with an axial length different to the Gullstrand eye will induce magnification errors due to the optical system of the imaging device. The amount of error is influenced by the optical system of the fundus camera [60]. It is not sufficient to estimate merely total and axial ametropia of the system (eye) in order to correct magnification errors. Furthermore, deviations in corneal shape like astigmatism or keratocone may induce additional errors. The influence of various magnification-related errors can be minimized by the use of local or temporal relative values like the arteriovenous ratio.

Errors of Measurements in Angiographic Pictures

Typically, measurements are performed in photographic images (colored or black and white). Alternatively, it has been proposed to use fluorescein angiography to enhance the contrast of the vessel edge against its retinal background [22, 55]. However, the use of fluorescein results in the measurement of vessel width as discussed above. In general, measurements of vessel width in angiographic images face a lot of error sources which are difficult to control. The dye filling in arteries expands from the vessel center, whereas in the veins it remains heterogenous. Only over a short period of time the acquisition of images for vessel diameter measurement is possible. Most of the angiographic images lead to considerable measurement errors.

Additionally, fluorescein angiographic images often are overexposed in order to get good contrast and to display the capillary bed. This leads to problems in the identification of the vessel edges in the brightness profiles which feigns better reproducibility but induces additional errors. These fundamental drawbacks of fluorescein angiographic imaging methods remain independent of the methods for vessel diameter measurement applied.

6.3.1.5 Physiological Variability of Vessel Diameter

Variance of the measurement of retinal vessel diameters is not only caused by errors of the assessment technique itself. It has been shown that retinal vessels exhibit changes in diameter during the cardiac cycle. Pulsation of the central retinal vein is visible on direct ophthalmoscopy in a large number of persons, and pulse-related changes in retinal vessel width can be measured by assessing diameter information at different phases of the cardiac cycle. Therefore, assessing retinal vessel diameters – regardless which technique was used – at random points in the pulse cycle may result in an unrecognized source of variation in the measurements of retinal vessel diameters between subjects and over time in the same individual. It has been shown that the arterial diameter can vary up to 3.5% and up to 4.8% in retinal veins of healthy subjects [6]. The same study described that the venular diameter was smallest in early systole, reaching a maximum in early diastole, whereas the arteriolar diameter peaked slightly earlier. Considering that the calculation of blood flow from vessel diameter and blood speed uses the square of the vessel diameter, the potential error introduced in such a calculation can be up to 9%, if this factor is not taken into account.

Several techniques have been proposed to overcome the pulse-depending changes. Dumsky and colleagues have introduced a technique which uses electrocardiographically synchronized fundus photography [9]. For this purpose, a dedicated electrocardiographic synchronization unit was built to trigger the camera at a preset time interval after the R wave of the ECG. Using six or more synchronized fundus photographs, the authors could detect diameter changes induced by exercise as small as 1.4% in a group of six volunteers. Knutson and colleagues showed that pulse synchronous digital images triggered by the means of a ear clip as triggering device can significantly reduce the variation of measurements [30].

Another cause of biological diameter variability is vasomotions and the blood pressure waves (Meyer waves) with periods near 10 and 20 s. In young people, the magnitudes are considerably higher than the pulsation magnitudes. These waves are different in arteries and veins and therefore cause changes in arteriovenous ratio. Newer instruments (DVA and RVA Imedos, Jena), which use the real time assessment of retinal vessel diameter, allow for the correlation of vessel diameter course to a wide number of biological signals such as the ECG. Alternatively, given that these systems are measuring retinal vessel size up to 25 times per second, retinal vessel diameters can simply be determined by averaging the measuring results over several seconds.

The brief overview should give an introduction into the methodology of vessel diameter measurements. It is by far not capable of reflecting the multitude of approaches and results worldwide regarding this topic.

6.3.2 Methods

Different imaging systems and different technical solutions to measure in the photographic or digital images with different edge definitions result in different methods, which are summarized in groups.

6.3.2.1 Optical Micrometric Measurements Based on Photographic Negatives

The optical micrometric determination of retinal vessel diameters was used in clinical studies for many years. Basically, the micrometric determination of the vessel diameter or width uses photographic image negatives of the fundus, viewed under a microscope. By the means of a micrometer ocular attached to the microscope and a known magnification of the instrument, the observer visually estimates the measuring points in the vessel edges (manual measurement).

Several improvements for the micrometric assessment have been proposed. In a technique also known as projection micrometry, a photographic fundus picture was taken with a fundus camera and then projected onto a screen with known magnification (up to 35 times). Vessel diameter measurement was performed manually using a caliper [4]. More sophisticated approaches utilized dedicated computer systems to translate the position of the caliper in measurable units.

As discussed already, common drawbacks of those manual methods are a variety of subjective errors and their extensive time need.

6.3.2.2 Microdensitometry Based on Photographic Negatives

In the so-called microdensitometric techniques, the vessel edges are defined based on a densitometric trace of the vessel images crossing the vessel. The result is similar to the brightness profile. The crucial point in the microdensitometric technique is the objective determination of the position of the measuring points in the vessel edges by means of the half-height diameter method enabling an objective measurement. The disadvantage of densitometry was again the considerable time need. So today, also this method is replaced by modern methods of digital image processing [5].

6.3.2.3 Measurements Based on Digital Images

During the changeover from photographic to electronic imaging, high-resolution image scanners were used to enable digital image processing on photographic fundus images. Nowadays, this technology has kept its relevance for the postprocessing of major studies from the past. Time need and errors from photographic and scanning processes are the drawbacks of this method.

The development and application of sensitive high-resolution image sensors like CCDs turned conventional photographic retinal cameras into digital imaging systems. This enables a direct computer processing of digital fundus images to assess vessel diameter in normal or vessel width in fluorescein angiographic images.

But even for the evaluation of digital fundus images, visual-measuring methods similar to the methods of optical micrometry were a common approach. The image is displayed on a monitor, and the user has to place the measuring marks (lines or circles) using a mouse [24]. The respective display magnification properties of the display monitor and the various subjective influences may have adverse impact on the measurement results.

Nowadays, vessel measurement is performed using more and more automated methods. Usually, the examiner has to select the measurement location by mouse click. There, the brightness profiles are acquired from the vessel cross sections and evaluated automatically using the methods discussed already. Sources of errors for these methods are the imaging conditions and effects of the optical devices and imaging sensors. Further developments of the digital systems for vessel diameter measurement lead to the introduction of vessel analysis.

6.4 Diameter Assessment for Blood Flow

Until now, volumetric blood flow in major retinal vessels cannot be measured directly. One of the most widely used approaches is to calculate blood flow from measured velocity and vessel diameter in the same vessel segment. One has, however, to note that this reflects blood flow in one single vessel. If one wants to determine total retinal blood flow, this procedure has to be carried out in each single vessel separately.

6.4.1 Assessment of Flow by Use of Doppler Technique (CLBF)

Another possibility to measure blood flow is the combination of Doppler techniques with measurements of vessel diameter in photographic fundus images [19]. The Canon laser Doppler blood flowmeter (CLDF, Canon, Tokyo, Japan) combines retinal blood velocity measurement according to the laser Doppler principle with a system for the measurement of retinal vessel diameters [10]. It is the first commercial device that is capable for measuring simultaneously vessel size and blood speed. This allows for the calculation of retinal blood flow in µl/min based on the Poiseuille principle in the selected vessel segments with high reproducibility [20] and can therefore provide an immediate measure of retinal blood flow in actual units of µl/min. Basically, the CLDF consists of a modified fundus camera, which is equipped with two lasers. The red blood cell speed is determined by bidirectional laser Doppler velocimetry, which is described in detail elsewhere in this book. Briefly, Doppler-shifted light scattered from the flowing blood cells in the target vessel is detected simultaneously in two directions separated by a fixed angle. The Doppler shift in the backscattered laser light is a function of red blood cell velocity. A red 675-nm-diode laser is used for velocity measurement.

Retinal vessel diameters are determined automatically by computer analysis of a vessel cross section recorded by a CCD line scan sensor connected to the fundus camera system. For the determination of vessel diameter, the half height of the vessel brightness profile is used. Furthermore, the diameter measurements can be corrected for the refractive error of the eye. The instrument is equipped with an automatic vessel tracking system that maintains alignment of the laser beam on the target vessel during the measurement.

6.5 Retinal Vessel Analysis

Given that several ocular and systemic vascular-related diseases are associated with concomitant changes in the microvasculature, the exact quantification of the retinal vessel size has gained more and more attention. Several lines of evidence indicate that changes in retinal vessel diameters reflect not only ocular diseases but also may serve as an early predictor for systemic diseases such as systemic hypertension or stroke. Along this line of thought, a lot of effort has been put into the investigation of structural alterations of retinal blood vessels and their association to ocular and systemic diseases.

6.5.1 Basics of Retinal Vessel Analysis

In principle, two different approaches for automatic vessel analysis are available (Fig. 6.4). The analysis of one singe image of the ocular fundus is usually referred to as static vessel analysis. Based on this technology, several parameters describing structural alterations of the retinal vascular system, such as generalized arterial narrowing, can be assessed. However, the limitation of this approach is that no information can be gained about vascular function. The further development of this technique led to the approach of dynamic vessel analysis. In principle, dynamic vessel analysis is based on the analysis of retinal vessel

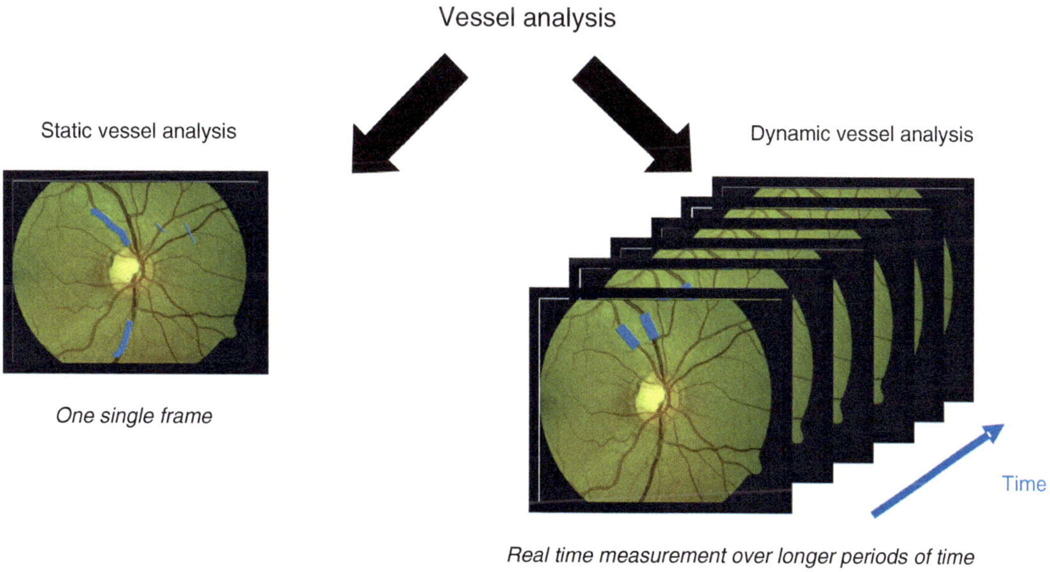

Fig. 6.4 Two approaches for retinal vessel analysis are available: static and dynamic vessel analysis

diameters in relation to time and location on the vessel. For this purpose, image sequences from seconds up to several minutes are analyzed, allowing for the investigation of time and location dependent vessel behavior. Additionally, as one of the advantages of this technique, the latter approach also allows for the use of different provocation methods such as flicker stimulation, squatting, or breathing tests, to assess vascular function in vivo.

The basic principles of diameter measurements have been described before in this section. Whereas for single measurements it may be sufficient to manually mark the chosen vessel segment, this approach is no longer feasible for the high number of measurements necessary for vessel analysis. To overcome this problem, sophisticated methods and algorithms of image processing with a high degree of automation have been introduced and are currently subject to research. Pursuing various approaches and different objectives, they share the following essential steps:

1. Identification of vessel segments (recognition of vessel segments inside of images)
2. Identification of connected segments or building vessel trees
3. Vessel classification (differentiation between arteries and veins)
4. Definition of reference points and recognition of eye movements
5. Correction of eye movements between images
6. Plausibility tests
7. Analysis of measured vessel diameter values and estimate of characteristic vessel parameters to describe vascular properties of vessel state and function such as the AVR

6.5.2 Static Vessel Analysis

Given that the eye is the unique site in the human body where a direct sight on the vessels is possible, special emphasis has been directed toward the early identification of morphological changes in several vascular-related diseases. In particular, it is known for a long time that changes in blood pressure, diabetes, or other diseases are reflected in morphologic changes of the ocular fundus. These changes include arteriolar narrowing, arteriovenous nicking, cotton wools spots, microaneurysm, blot hemorrhages, and others. Whereas morphologic changes such as microaneurysms and hemorrhages need to be analyzed by a trained specialist, the assessment of retinal vessel diameters can be done automatically by analyzing fundus images.

One of the first approaches to assess a disease risk factor by measuring the retinal vascular system was introduced by Quigley and colleagues [50]. For this purpose, the so-called pressure attenuation index (PAI) was defined. The PAI reflects the pressure loss along the arterial vessels, based on diameter measurements of retinal vessels. It was stated that low-end arterial pressures (high PAI values) may be a protective factor in diabetic retinopathy and that PAI may be use as risk indicator for the development and progression of the diabetic retinopathy.

Today, the most widely used approach to describe retinal vessels state is based on the observation that a decrease in arterial vessel caliber may reflect generalized arterial narrowing in other vascular beds and predict systemic and ocular diseases. Unfortunately, representative measurements of retinal vessel diameters are difficult to perform and complicated by several different factors [24]. First and most importantly, the angioarchitecture and the branching pattern of the retinal vessels differ considerably between subjects. This is of special importance because the total cross section of the arteriolar system increases with each bifurcation of the vessels leaving the optic nerve head. Thus, in order to achieve comparable and reliable results, the measurement procedures should also account for the specific branching pattern of the subject under study. Secondly, measurements of vessel size by the means of a fundus camera or fundus photography may be influenced by an improper focus or the individual refractive error of the subject under measurement.

To overcome these limitations, the arteriovenous ratio (AVR), a relative measure to assess arteriolar narrowing has been introduced. The AVR is currently the most widely used parameter for static vessel analysis and allows individual vessel

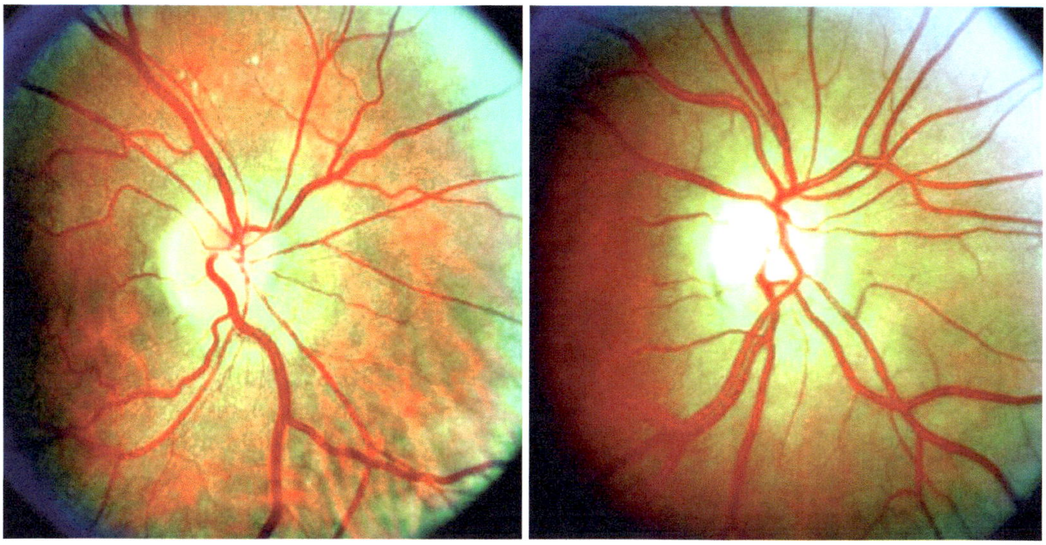

Fig. 6.5 Retinal photographs showing a fundus with generalized arteriolar narrowing with an AVR of 0.64 (*left*) and a fundus with normal retinal arterioles with an AVR of 1.08 (*right*) (Wong et al. [68])

diameters to be combined into summary indices reflecting the average arteriolar and venular diameters of the eye [24]. To assess the AVR, the quotient of the so-called central retinal arterial equivalent (CRAE) and central retinal venous equivalent (CREV) is calculated. The first approach to calculate the CRAE has been introduced by Parr and Spears [46, 47]. The formula used for CRAE calculation is based on theoretical models and includes factors such as the vessel width and the number of times the arteries have branched for the calculation of the CRAE. However, the studies by Parr and Spears are limited by the fact that only the CRAE has been used to assess arterial narrowing and the retinal veins have not been included in the analysis.

Later, extending the approach from Parr and Spears, a corresponding formulas for the calculation of CREV have been developed by Hubbard and colleagues [24]. As a further improvement, the AVR has been included calculated as a quotient of the CRAE and CREV. The authors suggest that the AVR is a more appropriate measure to reflect arterial narrowing because it also includes information of retinal veins [24]. The main advantage of this approach is that the AVR, as a relative factor, is presumed to be largely unaffected by the arteriolar branching pattern in a certain individual, magnification of the fundus photo due to refractive errors, or the broadening of the vessel diameter due to the opacity of the ocular medias [24]. An AVR of 1 indicates that, on average, arteriolar diameters are the same as venular diameter in that eye, whereas a smaller AVR indicates arteriolar narrowing (Fig. 6.5). However, this approach is based on the assumption that the diameter of retinal veins remains relatively constant despite other influencing factors such as blood pressure. Whether this assumption holds true in all cases will be discussed later in this chapter.

Further modifications leading to revised formulas for calculation of CRAE and CREV have been proposed by other authors. For example, Knudtson and colleagues have pointed out that the above mentioned formulas for the calculation of the CRAE and the CREV are not completely independent from number of vessel measured [31]. Consequently, the authors have developed new formulas which also account for differences in branching of the vessels included [31].

To guarantee comparable results, several prerequisites for the calculation of the AVR have to be fulfilled. Most importantly, the image used for the analysis has to be taken centered on the optic disc (Fig. 6.6). Given that, as stated above, the

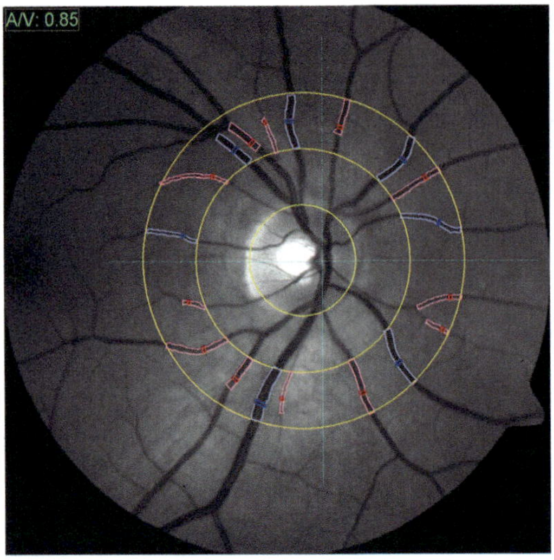

Fig. 6.6 Fundus image with marked measuring circular area and measured vessel diameters (Software VM2/ Imedos GmbH Jena/Germany)

individual branching pattern plays a crucial role for the assessment of the AVR, only vessels coursing through a specified area surrounding the optic disc are taken into consideration. At their first descriptions of the static vessel analysis [24], digitized photographic images and a manual diameter measurement had been used. The technique was then improved by the development of automatic algorithms [67].

Newer technologies are based on the analysis of digitally taken fundus pictures and specially developed software programs for the automatic detection and classification of the vessels and the concomitant diameter measurement (*X1-IVAN-software of the Wisconsin University*). However, up to now, the software is still dependent on the input of an experienced user to confirm the measurements or to correct false or missing vessel detections, wrong classifications, and obvious measurement errors. While the range of use of the "IVAN" software (University of Wisconsin) is strictly limited to public research, Imedos GmbH (Jena, Germany) introduced a commercially available software "Vesselmap" and an integrated device system for static vessel analysis "static vessel analyzer" (SVA) (Fig. 6.6). Other software solutions are provided by Thalia and Medivision. The most recent development in static vessel analysis is a follow-up tool (Imedos) capable of performing automatic individual progression observations. Relative changes of arterial and venous vessel diameters are determined related to a reference examination. In this way, a consideration of central equivalents for individual diagnosis becomes possible.

Today, the standardized assessment of the AVR based on arterial and venous central equivalents got widely accepted and became a gold standard in epidemiologic studies. However, examination protocols, measuring restrictions, procedures, conditions as well as used formulas, and retina cameras must be considered when attempting to compare results of different studies.

6.5.3 Results and Limits of Static Vessel Analysis

The importance of retinal vessel diameter in the diagnosis of systemic vascular-related diseases was recognized already in the 1930s, especially the relevance of the ratio between corresponding artery and vein [37]. Unfortunately, a number of interesting scientific study results from Lobeck and others related to diabetes, arterial hypertension, and nephrological diseases and other fell into oblivion [38]. However, the observation that patients suffering from systemic hypertension shows decreased retinal arterial diameters, which has been used as a diagnostic criterion and a grading scale for hypertensive retinopathy for many years [56]. For a long time, this decrease in retinal arterial diameter was explained as a counter regulatory response of the retinal arteries to increased perfusion pressure caused by the elevated systemic blood pressure. Since it is now possible to exactly quantify the changes in vessel size, these observations have been further investigated by large epidemiological studies.

One of the fist studies investigating the association between retinal vessel size and systemic blood pressure in a large epidemiological design was the Arteriosclerosis Risk in Communities Study (ARIC). Including more than 11,000

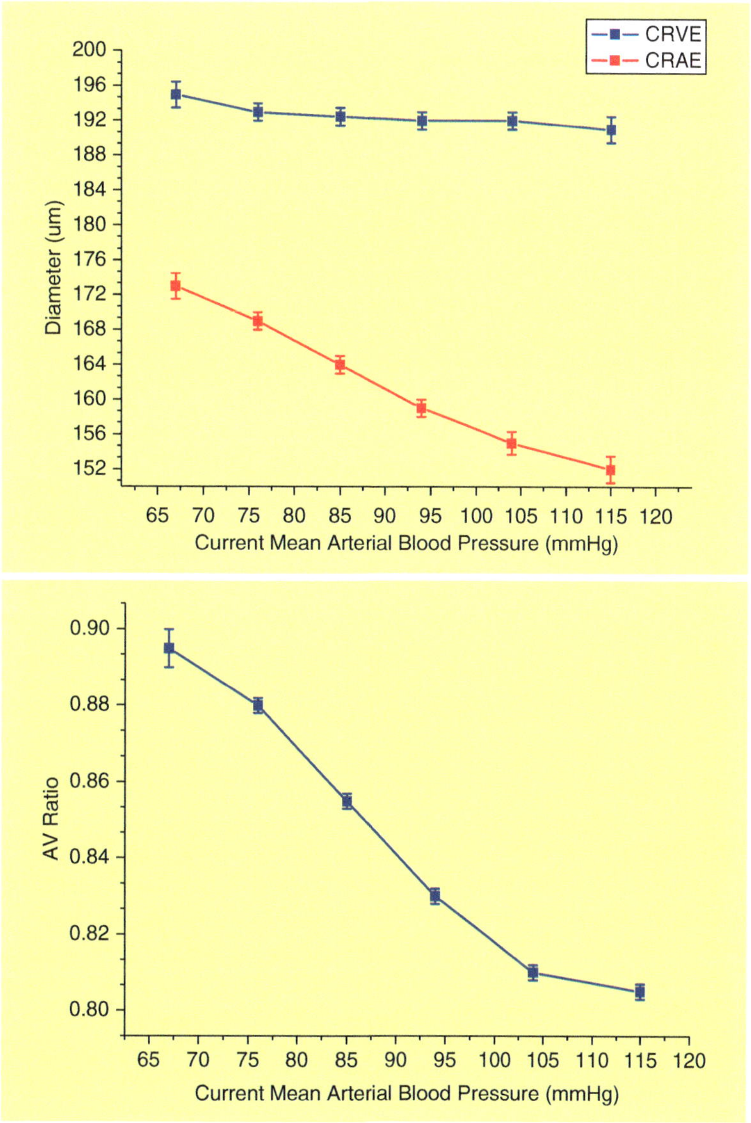

Fig. 6.7 Average central retinal artery equivalent (CRAE) and venous equivalent (CRVE) by mean arterial blood pressure and AV ratio versus mean arterial pressure (Data are derived from the Atherosclerosis Risk in Communities Study (n=9,040). Both figures are modified from Hubbard et al. [24])

participants in four examination centers, retinal vascular abnormities were documented using fundus photography. The AVR was then calculated based on digitized pictures of the fundus photographs. The data of the ARIC study indicate that the AVR was strongly associated with current blood pressure in both women and men (Fig. 6.7) [57].

Further evidence from the Beaver Dam Eye Study, a population-based prospective cohort study including 2,450 subjects, reported that narrowed retinal arterioles are associated with long-term risk of systemic hypertension. In particular, subjects with smaller retinal arteriolar diameters were more likely to develop hypertension than people with larger arteriolar diameters, independent of other known risk factors for hypertension [68]. For the analysis, the AVR was categorized into quarters, the first quarter representing the most pronounced arterial narrowing and the fourth quarter representing the reverence value. It was observed that, after adjustment for age and sex, subjects with a low AVR (first quarter) have a threefold higher risk of developing systemic

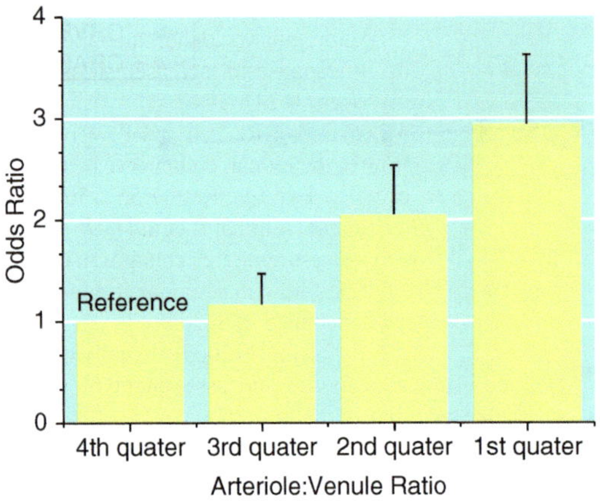

Fig. 6.8 Odds ratio for incident hypertension in relation to retinal arteriole-venule ratio (Wong et al. [68])

hypertension (odds ratio 2.95) compared to those with the highest AVR (Fig. 6.8). Based on these results, the authors suggested that structural alterations of the microvasculature may be linked to the development of hypertension and that arterial narrowing may precede the development of systemic hypertension.

The Blue Mountains Eye Study, a large population-based cohort study including more than 3,600 subjects, revealed that a low AVR is associated with increasing age, which was interpreted as a generalized arterial narrowing in the elderly [33]. These results are in keeping with the data from other large epidemiologic studies, such as data from the ARIC study or the Cardiovascular Health Study [64, 66] (CHS) also indicating for decreasing retinal arterial diameters with increasing age. Furthermore, a low AVR was found to be associated with other cardiovascular risk factors, such as cigarette smoking or systemic blood pressure.

Several lines of evidence indicate that retinal vessel diameters may also reflect systemic pathological changes in the body. In particular, data of the ARIC study show that nonspecific inflammatory markers, such as increased white blood cell and fibrinogen levels, are associated with a smaller AVR, indicating a for generalized arterial narrowing, independent from other known risk factors [29]. In contrast, the Rotterdam study showed increased white blood cell count lead to a particularly vasodilatation in retinal veins, whereas retinal arteries dilate only in a lesser extent [25]. These findings are consistent with the data of the Beaver Dam Eye Study, also indicating a relationship between increased inflammation markers and venular arterial diameters [28]. However, the results of the latter studies need to be interpreted with caution, given that no information is available whether the subjects under study had acute infections or some subgroups took any anti-inflammatory medication. Furthermore, given that the above mentioned studies were cross-sectional, temporal sequence of inflammatory-induced endothelial dysfunction and venular diameter cannot be determined. Data of an interventional study in healthy subjects showed that after experimental induction of inflammation by the means of administration of low dose of *Escherichia coli* endotoxin, retinal veins significantly dilated, paralleled by an increase of white blood cell count [32].

Interestingly, data from recent experiments indicate that retinal vessel diameters may carry also information about other vascular beds. It has been shown that narrowing of retinal vessels is associated with lower myocardial blood flow and perfusion reserve in asymptomatic subjects [63]. The authors from this study conclude that retinal arteriolar narrowing may serve as a marker of coronary microvascular disease. Along these line of thought, a couple of studies have shown that changes in retinal vessel diameters can predict the risk of coronary heart disease, stroke, and stroke mortality [64–66]. Pooled data from the Beaver Dam Eye Study and the Blue Mountains Eye Study showed that smaller arterial diameters and larger retinal venous diameters are associated with an increased risk of stroke mortality [62]. These data clearly support the idea that retinal vessel diameters may serve as a predictor for event in other vascular beds such as the heart or the brain.

With regard to ocular diseases, it has been hypothesized that generalized arterial narrowing can predict the development of open-angle glaucoma. Given that – as outlined in detail in another chapter of this book – reduced blood flow may be involved in the pathogenesis of glaucoma, much emphasis has been put into the investigation of vessel diameters and glaucoma. Using manual diameter measurements of projected retinal images, it was reported that eyes with open-angle glaucoma have significantly reduced peripapillary retinal vessel diameters compared with a healthy, age-matched control group [27]. Further studies revealed a general narrowing of retinal vessel diameters in more than 50% of patients with early stage glaucoma, whereas these alterations were only visible in 15% of normal eyes [51]. However, the interpretation of those studies is hampered by the fact that the studies were neither population based nor adequately adjusted for blood pressure. Furthermore, subjective methods for the assessment of retinal vessel size have been used. Using an automatic technique based on the analysis of digitized fundus images, the Blue Mountains Eye Study examined the relationship between retinal vessel diameter and open-angle glaucoma. The result of this study indicate that generalized arterial narrowing is significantly associated with optic nerve head damage in patients with open-angle glaucoma, independent of age, gender, smoking, or blood pressure [43]. Whether these changes in diameter reflect an ischemic process leading to the optic nerve head damage, or occur secondary to the neuron loss caused by the disease, is however unclear. The Rotterdam Study failed to show a predictive value of retinal baseline diameter to changes of the optic disc [26]. The authors of the study conclude that the Rotterdam Study does not provide evidence for a retinal vascular role in the pathogenesis of open-angle glaucoma. One needs, however, to consider that retinal vessel diameter may not be an adequate parameter of retinal perfusion status in glaucoma.

Recent analysis, however, have changed the view of the AVR as the optimal measure for generalized arteriolar narrowing. The calculation of the AVR is mainly based on the assumption that the venous diameter remains relatively constant in response to blood pressure, age, or other factors. The above mentioned results indicate that this assumption may not hold true for every cases [35]. In particular, the fact that retinal veins dilate in response to increase inflammatory markers indicate that venules may carry different information than arteries and should consequently be analyzed separately. Consequently, as an alternative, it has been proposed to analyze retinal arterial and venous diameters separately.

Furthermore, static vessel analysis allows for the stratification of microvascular risk factors based on measurement of stationary vessel diameters. However, a further limitation of static vessel analysis is that these measured parameters allow only for a very limited conclusion about function alterations of the vasculature. Given that several physiological variations such as vasomotor changes or alterations in ocular perfusion pressure will interfere with the measured values, an exact determination of vessel function is dependent on the measurement of additional parameters as it is done in dynamic vessel analysis.

6.5.4 Results and Limits of Dynamic Vessel Analysis

In contrast to this static vessel analysis, reflecting a snapshot of retinal vessel diameter, carrying information on vascular tone, the so-called dynamic vessel analysis has been introduced to assess alterations of vascular function. For the dynamic vessel analysis, images of the fundus are recorded over a longer period of time ranging from a few seconds up to several minutes. During the measurement, period provocation tests, i.e., stimulation of the eye with flickering light, exercise, or breathing gases with variable mixtures of O_2, CO_2, and others, are used to assess vascular function. As compared to static vessel analysis, dynamic vessel analysis can provide several advantages. First and most importantly, changes of retinal vessel diameter can be measured as a function of time and position along the vessel.

6.5.4.1 Stimulation with Flicker Light

It is known for a long time that retinal vessels have the ability to adapt to changes in ocular

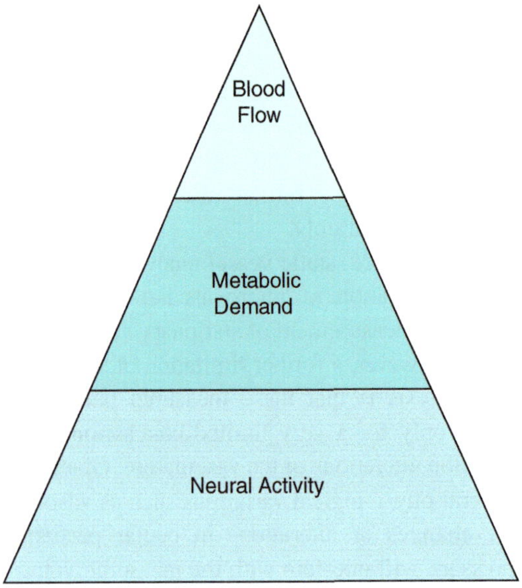

Fig. 6.9 Concept of neurovascular coupling. Increased neural activity leads to increased metabolic demand and in turn to augmented blood flow

perfusion pressure, which is commonly referred to as autoregulation. Whereas the phenomenon of blood flow autoregulation has been investigated in several vascular beds including the eye, it has recently become clear that retinal vessels can also adapt to changes in metabolic demands. Experimental evidence for a coupling between increased metabolism and increased blood flow in the eye was first suggested by the findings that glucose metabolism was enhanced in the retinal ganglion cells by flickering light. According to the current concept of neurovascular coupling, increased neural activity during stimulation with flicker light leads to an increased ganglion cell activity and to augmented metabolic demand in the retina (Fig. 6.9). This subsequently leads to retinal vasodilatation and increased blood flow. Whereas a detailed review of the current view of neurovascular coupling and its possible mediators has been published recently [53], this article will focus on the effect of flicker stimulation on retinal vessel diameters.

Initially, the effect of flicker light on the retinal circulation was investigated using photographs of the human eye fundus taken in red-free light [12]. These photos were synchronized with the cardiac pulse and taken after 1 min of flicker stimulation. After digitalization of the pictures, the diameters of straight segments of retinal vessels were analyzed before and during illumination with flickering light. The authors report a significant increase in retinal vessel diameters due to the stimulation with flickering light.

Subsequent investigations of the flicker-induced retinal vessel diameter response were conducted with an automatic tool for the retinal vessel analysis (RVA, Imedos, Jena, Germany) [44, 49]. Initially, flicker stimulation was achieved using light from a xenon arc lamp that was chopped with a rotating sector disc [49] or by light flashes [15] introduced into the illumination pathway of the fundus camera by a fiber optics. Using an optical filter system to spectrally differentiate the flicker light and the fundus illumination, the flicker was superposed on the continuous fundus illumination needed to measure the vessel diameter. Later, to increase the stimulus contrast, the flicker light was generated by electronic chopping of the fundus illumination at a frequency of 12.5 Hz [44].

Although the exact mechanism of flicker-light-induced vasodilatation is still a matter of controversy, it has been used as a provocation test for the ability of retinal vessels to adapt to different metabolic situations in health and disease. A variety of studies have shown that flicker-induced vasodilatation is altered in ocular and systemic diseases.

It has been shown that flicker light responses are diminished in patients with early stage glaucoma. In this study, 31 patients with early stage glaucoma and 31 age- and sex-matched healthy volunteers were included [16]. To avoid any vascular effects of glaucoma medication, a wash-out period was scheduled for all patients. Compared with age-matched control group, the study provides evidence that flicker responses are diminished in patients with early stage glaucoma, independently of the glaucoma medication administered. Interestingly, the reduced flicker response was only observed in retinal veins, not in retinal arteries. A further remarkable observation of this study is that flicker responses are already diminished already in

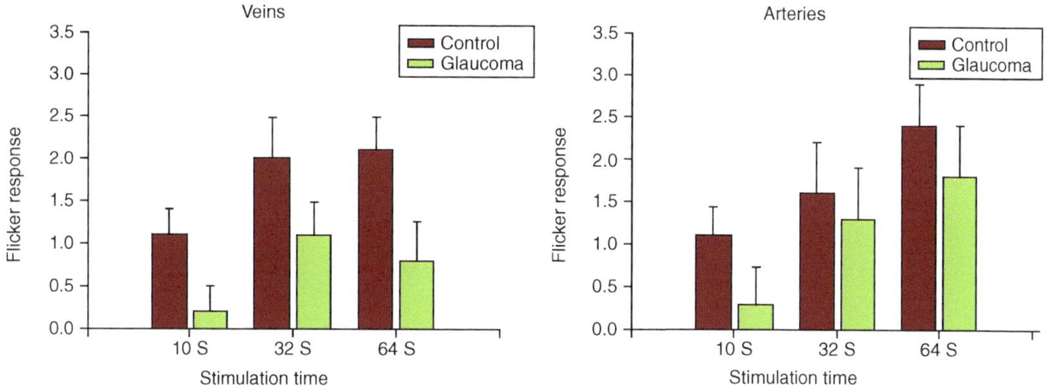

Fig. 6.10 Flicker-light-induced vasodilatation of retinal arteries and retinal veins in percent change from baseline in patients with open-angle glaucoma and a healthy control group (Modified from Garhofer et al. [16])

patients with early stage glaucoma, having only moderate glaucomatous changes. Whether the decreased flicker response in patients with glaucoma can be attributed to a vascular dysregulation or to a reduction of neural activity caused by ganglion cell loss, as it appears in glaucoma, has yet to be clarified.

The latter results were later confirmed by the work of Riva and colleagues. The authors report that flicker-evoked response measured at the optic-disc rim is reduced in ocular hypertension and early glaucoma [54]. These studies are of particular interest because it has been shown that a short-time increase of IOP by the means of an episcleral suction cup does not alter flicker-induced vasodilatation in young, healthy volunteers [14]. These results indicate that factors other than increased intraocular pressure must be responsible for the missing flicker-induced vasodilatation.

Given that vasospasm, defined as inappropriate constriction of insufficient vasodilatation, is a major risk factor for developing glaucoma, Gugleta and colleague have investigated flicker-induced vasodilatation in young, healthy women with vasospastic syndrome [21]. The authors show that in subjects with vasospasm as identified by nail-fold capillaroscopy, the maximum dilatory response to flicker stimulation was significantly reduced compared to the control group. Whether this reduced response is related to a higher risk for developing glaucoma or other vascular-related diseases has to be investigated.

Decreased flicker responses were also observed in patients with early stage diabetic retinopathy. Patients suffering from diabetes show decreased flicker-induced vasodilatation in retinal arteries compared to a healthy control group [17] (Fig. 6.10). However, no significant difference between the diabetes group and the control group was detected in retinal veins. Whether this is simply related to the smaller percentage change in retinal veins during flicker stimulation or to another mechanism has yet to be investigated. The results of this study are keeping with evidence from a recent study that showed also diminished flicker response in retinal vessel from diabetic patients in a larger group of diabetic patients [41]. Furthermore, the present study demonstrates that the vasodilatation of retinal arteries and veins under the flickering light decreases continuously with increasing stages of diabetic retinopathy. This has led to the suggestion that the effect of flicker-evoked vasodilatation could be used as an early screening tool to detect vascular dysregulation in patients with glaucoma [41].

Along this line of thought, Nagel and colleagues investigated the effect of age, systemic blood pressure, and baseline retinal vessel diameters on flicker-induced vasodilatation [44]. This is of special interest because it is known that the retinal vessels constrict with increasing blood pressure. The finding of the study indicates that flicker-induced vasodilatation is significantly

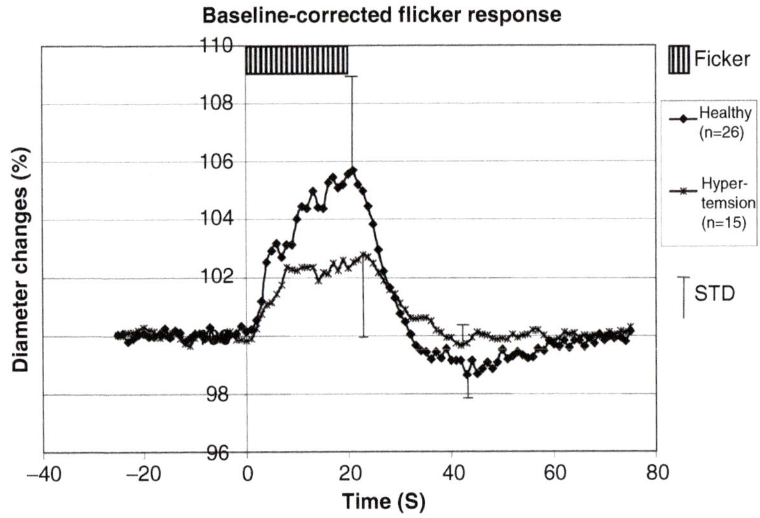

Fig. 6.11 Mean flicker response curve in healthy subjects and untreated hypertensive subjects (Nagel et al. [44])

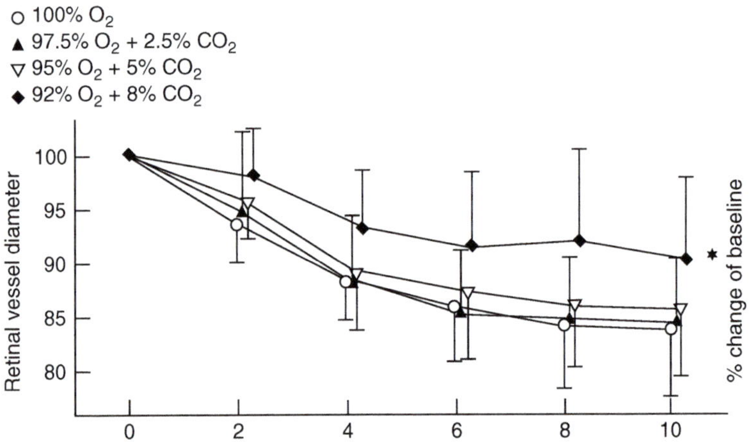

Fig. 6.12 Retinal vessel diameter during breathing of different mixture of O_2 and CO_2. Modified from Luksch et al. [40]

diminished with increasing blood pressure, whereas baseline retinal diameter had no influence on flicker responses (Fig. 6.11). Surprisingly, no significant correlation was observed between flicker response and age. This was attributed by the authors to the small sample size of the study.

6.5.4.2 Other Provocation Tests

Other provocation tests than flicker stimulation have also been used to assess vascular function in vivo. Although a detailed description of all methods and results are beyond the scope of this chapter, some examples for other provocation methods will be given.

It is known that increasing the tissue pO_2 by means of inhalation of 100% oxygen induces a pronounced vasoconstrictor effect in retinal vessels [52] (Fig. 6.12). Given that oxygen is nontoxic to adults and widely available in the clinical setting, a couple of studies have used oxygen to test vascular function. It has been reported that in diabetic patients, the vasoconstrictor response decreases with increasing stage of the disease and improves after panretinal photocoagulation [18], indicating for an impaired vascular regulation in the diabetic eye.

Another common approach to test vascular function is to change ocular perfusion pressure. Given that ocular perfusion pressure is determined by intraocular pressure and blood pressure, this can either be done by changing blood pressure – pharmacologically or by the means of

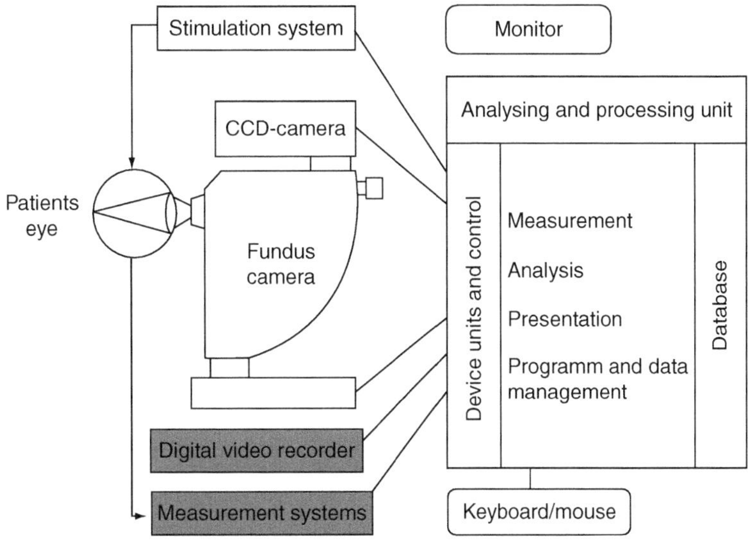

Fig. 6.13 Device unit of dynamic vessel analyzer (DVA, Imedos GmbH Jena, Germany)

isometric exercise – or by changing intraocular pressure. It is known that arteries constrict with increasing systemic blood pressure, reflecting an autoregulatory response to maintain perfusion pressure [2]. It has, however, been shown that this vascular answer is impaired under high blood glucose levels, indicating for a vascular dysregulation [3]. Additionally, evidence has been provided that the diameter response of retinal vessels to increased blood pressure is reduced in patients with type 2 diabetes [13].

In addition, regulation of vessel diameter can be tested by modifying intraocular pressure (IOP). Based on the observation that regulation of vascular tone is altered in patients with glaucoma, several experiments have focused on investigating diameter changes of retinal vessel in response to altered IOP. Nagel and colleagues have used the episcleral suction cup technique to induce an increase in IOP. Subsequently, retinal arterial and venous diameters were measured in healthy subjects, patients with open-angle glaucoma, and patients with ocular hypertension. It was observed that the change in retinal vessel diameter induced by short-time increase of IOP was significantly different among the three investigated groups [45]. Interestingly, short-time increase of intraocular pressure again by the means of a suction cup does not modify the response of retinal vessel diameter to flicker stimulation [14]. The latter study reveals that the response of retinal vessel diameters to flicker stimulation is maintained up to an IOP of 43 mmHg. This indicates that even at high IOPs, blood flow is responding to neural stimulation caused by flickering light. Accordingly, based on this data, it appears that a reduced flicker response as seen in patients with glaucoma in previous studies [16, 17] is not obligatory a direct consequence of increased IOP. However, one needs to be careful to directly apply the data of the present study to the results observed in glaucoma patients, because long-term changes in IOP were not mimicked in the present study. Because changes in IOP as induced by the suction cup technique is uncomfortable for the subject and time consuming, this approach is limited to research purposes and is currently not used in clinical practice.

6.5.5 Systems Available for Dynamic Vessel Analysis

There are currently 3 commercial systems available for dynamic vessel analysis: RVA (Retinal Vessel Analyzer), DVA (Dynamic Vessel Analyzer plus), and DVA-light (all Imedos GmbH in Jena, Germany). The schematic setup of the DVA is shown in Fig. 6.13.

The DVA is a modified fundus imaging system with the capability for dynamic and static vessel analysis. The main component of the vessel analysis system is an optical device for illumination and imaging of the fundus. For that purpose the DVA currently utilizes a fundus camera. The optical image from the fundus camera is received by a special CCD-camera attached to the system. The resulting digitized image (or image sequence) is further analyzed by the processing unit. This unit integrates all aspects of the measurement and analysis and controls the fundus camera as well as additional hardware.

For diameter measurements in the static and dynamic vessel analysis, the DVA applies dedicated adaptive filter algorithms, able to support both the actual measurement and vessel detection. The algorithms used are largely independent of changes in contrast and brightness. The spatial resolution in terms of the size of segments along the vessel is up to 12 μm, spatial resolution in the direction of measurement (perpendicular to the vessel) is better than 1 μm, and temporal resolution is 40ms.

By default the DVA is equipped with a device for flicker stimulation, with an option to connect further measurement systems (i.e. ECG, blood pressure). Flicker stimulation is achieved by an electro-optical shutter module inserted into the optical path of the fundus camera. The resulting flicker stimulus has a frequency of 12.5 Hz and a contrast ratio of about 25:1. Flicker stimulation can be used with a standardized protocol to examine vessel function.

Based on a standard or high performance imaging system (Visualis), standalone versions of dynamic vessel analysis (Retinal Vessel Analyzer, RVA) and static vessel analysis (SVA) are possible. DVA, RVA and DVAlight combine high reproducibility with a high temporal and spatial resolution [48].

6.6 Further Perspectives

The fast technical development of new instruments and software for the automatic and highly reproducible assessment of retinal vessel diameters has brought a new and interesting perspective for the early diagnosis of vascular related diseases. In particular, the observation that changes in retinal vessel size can reflect ocular pathologies and predict future disease progression underlines the importance of this new field of research. Dynamic vessel analysis further broadens the potential applications These techniques are non-invasive and easy to perform with high reproducibility and accuracy. Thus, retinal vessel analysis may be of interest for both ophthalmology and internal medicine and may help in the early identification of high risk patients and the early diagnosis of vascular related diseases.

References

1. Bengtsson B, Krakau CE (1992) Correction of optic disc measurements on fundus photographs. Graefes Arch Clin Exp Ophthalmol 230:24–28
2. Blum M, Bachmann K, Wintzer D, Riemer T, Vilser W, Strobel J (1999) Noninvasive measurement of the Bayliss effect in retinal autoregulation. Graefes Arch Clin Exp Ophthalmol 237:296–300
3. Blum M, Brandel C, Muller UA (2005) Myogenic response reduction by high blood glucose levels in human retinal arterioles. Eur J Ophthalmol 15:56–61
4. Bracher D, Dozzi M, Lotmar W (1979) Measurement of vessel width on fundus photographs. Albrecht Von Graefes Arch Klin Exp Ophthalmol 211:35–48
5. Brinchmann-Hansen O, Engvold O (1986) Microphotometry of the blood column and the light streak on retinal vessels in fundus photographs. Acta Ophthalmol Scand 179(suppl):9–19
6. Chen H, Patel V, Wiek J, Rassam S, Kohner E (1994) Vessel diameter changes during the cardiac cycle. Eye (Lond) 8:97–103
7. Delori FC, Fitch KA, Feke GT, Deupree DM, Weiter JJ (1988) Evaluation of micrometric and microdensitometric methods for measuring the width of retinal vessel images on fundus photographs. Graefes Arch Clin Exp Ophthalmol 226:393–399
8. Dorner GT, Polska E, Garhofer G, Zawinka C, Frank B, Schmetterer L (2002) Calculation of the diameter of the central retinal artery from noninvasive measurements in humans. Curr Eye Res 25:341–345
9. Dumskyj M, Aldington S, Doré C, Kohner E (1996) The accurate assessment of changes in retinal vessel diameter using multiple frame electrocardiograph synchronised fundus photography. Curr Eye Res 15:625–632
10. Feke GT, Goger DG, Tagawa H, Delori FC (1987) Laser Doppler technique for absolute measurement of blood speed in retinal vessels. IEEE Trans Biomed Eng 34:673–680

11. Feke GT, Tagawa H, Deupree DM, Goger DG, Sebag J, Weiter JJ (1989) Blood flow in the normal human retina. Invest Ophthalmol Vis Sci 30:58–65
12. Formaz F, Riva C, Geiser M (1997) Diffuse luminance flicker increases retinal vessel diameter in humans. Curr Eye Res 16:1252–1257
13. Frederiksen CA, Jeppesen P, Knudsen ST, Poulsen PL, Mogensen CE, Bek T (2006) The blood pressure-induced diameter response of retinal arterioles decreases with increasing diabetic maculopathy. Graefes Arch Clin Exp Ophthalmol 244:1255–1261
14. Garhofer G, Resch H, Weigert G, Lung S, Simader C, Schmetterer L (2005) Short-term increase of intraocular pressure does not alter the response of retinal and optic nerve head blood flow to flicker stimulation. Invest Ophthalmol Vis Sci 46:1721–1725
15. Garhofer G, Zawinka C, Huemer KH, Schmetterer L, Dorner GT (2003) Flicker light induced vasodilatation in the human retina – Effect of lactate and changes in mean arterial pressure. Invest Ophthalmol Vis Sci 44:5309–5314
16. Garhofer G, Zawinka C, Resch H, Huemer KH, Schmetterer L, Dorner GT (2004) Response of retinal vessel diameters to flicker stimulation in patients with early open angle glaucoma. J Glaucoma 13:340–344
17. Garhofer G, Zawinka C, Resch H, Kothy P, Schmetterer L, Dorner GT (2004) Reduced response of retinal vessel diameters to flicker stimulation in patients with diabetes. Br J Ophthalmol 88:887–891
18. Grunwald JE, Riva CE, Brucker AJ, Sinclair SH, Petrig BL (1984) Altered retinal vascular response to 100% oxygen breathing in diabetes mellitus. Ophthalmology 91:1447–1452
19. Grunwald JE, Riva CE, Sinclair SH, Brucker AJ, Petrig BL (1986) Laser Doppler velocimetry study of retinal circulation in diabetes mellitus. Arch Ophthalmol 104:991–996
20. Guan K, Hudson C, Flanagan JG (2003) Variability and repeatability of retinal blood flow measurements using the Canon Laser Blood Flowmeter. Microvasc Res 65:145–151
21. Gugleta K, Zawinka C, Rickenbacher I, Kochkorov A, Katamay R, Flammer J, Orgul S (2006) Analysis of retinal vasodilation after flicker light stimulation in relation to vasospastic propensity. Invest Ophthalmol Vis Sci 47:4034–4041
22. Hodge JV, Parr JC, Spears GF (1969) Comparison of methods of measuring vessel widths on retinal photographs and the effect of fluorescein injection on apparent retinal vessel calibers. Am J Ophthalmol 68:1060–1068
23. Hogan MJ, Feeney L (1963) The ultrastructure of the retinal blood vessels. I. The large vessels. J Ultrastruct Res 39:10–28
24. Hubbard LD, Brothers RJ, King WN, Clegg LX, Klein R, Cooper LS, Sharrett AR, Davis MD, Cai J (1999) Methods for evaluation of retinal microvascular abnormalities associated with hypertension/sclerosis in the Atherosclerosis Risk in Communities Study. Ophthalmology 106:2269–2280
25. Ikram MK, de Jong FJ, Vingerling JR, Witteman JC, Hofman A, Breteler MM, de Jong PT (2004) Are retinal arteriolar or venular diameters associated with markers for cardiovascular disorders? The Rotterdam study. Invest Ophthalmol Vis Sci 45:2129–2134
26. Ikram MK, de Voogd S, Wolfs RC, Hofman A, Breteler MM, Hubbard LD, de Jong PT (2005) Retinal vessel diameters and incident open-angle glaucoma and optic disc changes: the Rotterdam study. Invest Ophthalmol Vis Sci 46:1182–1187
27. Jonas JB, Nguyen XN, Naumann GO (1989) Parapapillary retinal vessel diameter in normal and glaucoma eyes. I. Morphometric data. Invest Ophthalmol Vis Sci 30:1599–1603
28. Klein R, Klein BE, Knudtson MD, Wong TY, Tsai MY (2006) Are inflammatory factors related to retinal vessel caliber? The Beaver Dam Eye Study. Arch Ophthalmol 124:87–94
29. Klein R, Sharrett AR, Klein BE, Chambless LE, Cooper LS, Hubbard LD, Evans G (2000) Are retinal arteriolar abnormalities related to atherosclerosis?: The Atherosclerosis Risk in Communities Study. Arterioscler Thromb Vasc Biol 20:1644–1650
30. Knudtson MD, Klein BE, Klein R, Wong TY, Hubbard LD, Lee KE, Meuer SM, Bulla CP (2004) Variation associated with measurement of retinal vessel diameters at different points in the pulse cycle. Br J Ophthalmol 88:57–61
31. Knudtson MD, Lee KE, Hubbard LD, Wong TY, Klein R, Klein BE (2003) Revised formulas for summarizing retinal vessel diameters. Curr Eye Res 27:143–149
32. Kolodjaschna J, Berisha F, Lung S, Schaller G, Polska E, Jilma B, Wolzt M, Schmetterer L (2004) LPS-induced microvascular leukocytosis can be assessed by blue-field entoptic phenomenon. Am J Physiol Heart Circ Physiol 287:H691–H694
33. Leung H, Wang JJ, Rochtchina E, Wong TY, Klein R, Mitchell P (2004) Impact of current and past blood pressure on retinal arteriolar diameter in an older population. J Hypertens 22:1543–1549
34. Li H, Hsu W, Lee ML & Wong TY (2005) Automatic grading of retinal vessel caliber. IEEE Trans Biomed Eng 52:1352–1355.
35. Liew G, Sharrett AR, Kronmal R, Klein R, Wong TY, Mitchell P, Kifley A, Wang JJ (2007) Measurement of retinal vascular caliber: issues and alternatives to using the arteriole to venule ratio. Invest Ophthalmol Vis Sci 48:52–57
36. Littmann H (1982) Zur Bestimmung der wahren Größe eines Objektes auf dem Hintergrund des lebenden Auges. Klin Monatsbl Augenheilkd 180:286–289
37. Lobeck E (1935) Über Messungen am Augenhintergrund. Graefes Arch Clin Exp Ophthalmol 133:152–156
38. Lobeck E (1938) Die Breite der Netzhautgefäße als Differentialdiagnosticum bei Hochdruck- und Nierenkrankheiten. Klin Monatsbl Augenheilkd 1938:765
39. Lowell J, Hunter A, Steel D, Basu A, Ryder R, Kennedy RL (2004) Measurement of retinal vessel

widths from fundus images based on 2-D modeling. IEEE Trans Med Imaging 23:1196–1204
40. Luksch A, Garhofer G, Imhof A, Polak K, Polska E, Dorner GT, Anzenhofer S, Wolzt M, Schmetterer L (2002) Effect of inhalation of different mixtures of O(2) and CO(2) on retinal blood flow. Br J Ophthalmol 86:1143–1147
41. Mandecka A, Dawczynski J, Blum M, Muller N, Kloos C, Wolf G, Vilser W, Hoyer H, Muller UA (2007) Influence of flickering light on the retinal vessels in diabetic patients. Diabetes Care 30:3048–3052
42. Mikuni M (1959) Eine Methode zur Messung der Netzhautgefäßweite. Klin Monatsbl Augenheilkd 135:205–211
43. Mitchell P, Leung H, Wang JJ, Rochtchina E, Lee AJ, Wong TY, Klein R (2005) Retinal vessel diameter and open-angle glaucoma: the Blue Mountains Eye Study. Ophthalmology 112:245–250
44. Nagel E, Vilser W, Lanzl I (2004) Age, blood pressure, and vessel diameter as factors influencing the arterial retinal flicker response. Invest Ophthalmol Vis Sci 45:1486–1492
45. Nagel E, Vilser W, Lanzl IM, Lanzi IM (2001) Retinal vessel reaction to short-term IOP elevation in ocular hypertensive and glaucoma patients. Eur J Ophthalmol 11:338–344
46. Parr JC, Spears GF (1974) General caliber of the retinal arteries expressed as the equivalent width of the central retinal artery. Am J Ophthalmol 77:472–477
47. Parr JC, Spears GF (1974) Mathematic relationships between the width of a retinal artery and the widths of its branches. Am J Ophthalmol 77:478–483
48. Polak K, Dorner GT, Kiss B, Polska E, Findl O, Rainer G, Eichler HG, Schmetterer L (2000) Evaluation of the Zeiss retinal vessel analyser. Br J Ophthalmol 84:1285–1290
49. Polak K, Schmetterer L, Riva CE (2002) Influence of flicker frequency on flicker induced changes of retinal vessel diameters. Invest Ophthalmol Vis Sci 43:2721–2726
50. Quigley M, Cohen S (1999) A new pressure attenuation index to evaluate retinal circulation. A link to protective factors in diabetic retinopathy. Arch Ophthalmol 117:84–89
51. Rader J, Feuer WJ, Anderson DR (1994) Peripapillary vasoconstriction in the glaucomas and the anterior ischemic optic neuropathies. Am J Ophthalmol 117:72–80
52. Riva CE, Grunwald JE, Sinclair SH (1983) Laser Doppler Velocimetry study of the effect of pure oxygen breathing on retinal blood flow. Invest Ophthalmol Vis Sci 24:47–51
53. Riva CE, Logean E, Falsini B (2005) Visually evoked hemodynamical response and assessment of neurovascular coupling in the optic nerve and retina. Prog Retin Eye Res 24:183–215
54. Riva CE, Salgarello T, Logean E, Colotto A, Galan EM, Falsini B (2004) Flicker-evoked response measured at the optic disc rim is reduced in ocular hypertension and early glaucoma. Invest Ophthalmol Vis Sci 45:3662–3668
55. Sandor T, Rhie FH, Soeldner JS, Gleason RE, Rand LI (1981) Reproducibility of the densitometric analysis of fluorescein angiograms. Int J Biomed Comput 12:401–418
56. Scheie HG (1953) Evaluation of ophthalmoscopic changes of hypertension and arteriolar sclerosis. AMA Arch Ophthalmol 49:117–138
57. Sharrett AR, Hubbard LD, Cooper LS, Sorlie PD, Brothers RJ, Nieto FJ, Pinsky JL, Klein R (1999) Retinal arteriolar diameters and elevated blood pressure: the Atherosclerosis Risk in Communities Study. Am J Epidemiol 150:263–270
58. Vilser W (1987) New diagnostic possibilities with a retinal measuring system. Jena Rev 32:76–78
59. Vilser W, Königsdörffer E, Brandt H (1979) Längenmessungen mittels Planplattenmikrometrie am menschlichen Augenhintergrund. Graefes Arch Clin Exp Ophthalmol 212:109–115
60. Vilser W, Tirsch P, Münch K, Kleen W, Klein S (1990) Automatische Gefäßweitenmessung. Folia Ophthalmol 15:297–303
61. Vilser W, Münch K, Saleh K, Kassner Ch, Seifert BU, Henning G (2004) Vessel model for the validation of retinal vessel analysis. Biomed Tech, 49: (Erg.-Bd.2,T.2)–816
62. Wang JJ, Liew G, Klein R, Rochtchina E, Knudtson MD, Klein BE, Wong TY, Burlutsky G, Mitchell P (2007) Retinal vessel diameter and cardiovascular mortality: pooled data analysis from two older populations. Eur Heart J 28:1984–1992
63. Wang L, Wong TY, Sharrett AR, Klein R, Folsom AR, Jerosch-Herold M (2008) Relationship between retinal arteriolar narrowing and myocardial perfusion: multi-ethnic study of atherosclerosis. Hypertension 51: 119–126
64. Wong TY, Hubbard LD, Klein R, Marino EK, Kronmal R, Sharrett AR, Siscovick DS, Burke G, Tielsch JM (2002) Retinal microvascular abnormalities and blood pressure in older people: the Cardiovascular Health Study. Br J Ophthalmol 86:1007–1013
65. Wong TY, Klein R, Couper DJ, Cooper LS, Shahar E, Hubbard LD, Wofford MR, Sharrett AR (2001) Retinal microvascular abnormalities and incident stroke: the Atherosclerosis Risk in Communities Study. Lancet 358:1134–1140
66. Wong TY, Klein R, Sharrett AR, Duncan BB, Couper DJ, Tielsch JM, Klein BE, Hubbard LD (2002) Retinal arteriolar narrowing and risk of coronary heart disease in men and women. The Atherosclerosis Risk in Communities Study. JAMA 287:1153–1159
67. Wong TY, Knudtson MD, Klein R, Klein BE, Meuer SM, Hubbard LD (2004) Computer-assisted measurement of retinal vessel diameters in the Beaver Dam Eye Study: methodology, correlation between eyes, and effect of refractive errors. Ophthalmology 111:1183–1190
68. Wong TY, Shankar A, Klein R, Klein BE, Hubbard LD (2004) Prospective cohort study of retinal vessel diameters and risk of hypertension. BMJ 329:79

Laser Doppler Techniques for Ocular Blood Velocity and Flow

7

Charles E. Riva

Core Messages

This chapter describes the noninvasive techniques of laser Doppler velocimetry (LDV) and flowmetry (LDF) for the measurement of blood flow and the regulation of this flow in the vascular systems of the eye fundus. After providing the theoretical background underlying LDV and LDF, the implementation of the techniques for blood flow measurements in the retina, optic nerve, and subfoveal choroid of the human eye is presented. The conditions that maximize the range of applications, the limitations inherent to LDV and LDF, and the variability of the data obtained in various groups of subjects and patients are presented and discussed. Further developments of LDV and LDF are needed before these techniques become routine clinical tools for the early assessment of ocular pathologies of vascular origin and the effect of various treatments on the fundus circulation.

C.E. Riva, D.Sc.
Clinica Oculistica, Dipartimento di Discipline Chirurgiche, Rianomatorie e dei Trapianti "Antonio Valsalva", Università degli Studi di Bologna, It, Les Combes 71, Grimisuat 1971, Switzerland

Department of Medicine, University of Lausanne, Les Combes 71, Grimisuat 1971, Switzerland
e-mail: charles.riva@netplus.ch

7.1 Introduction

The measurement of blood flow in the ocular fundus is of scientific as well as clinical interest. Its scientific value lies in the possibility of gaining insight into the physiology of deep vascular beds that are under local and central nervous control. Its clinical potential is in the early assessment of alterations of blood flow, whether associated with specific ocular diseases or resulting from systemic disorders and in the evaluation of treatments acting on the disturbed blood flow.

Ideally, to be of clinical value, the measurement of blood flow in the eye should be reproducible, accurate, and sensitive enough to reveal early pathologic changes. Furthermore, the spatial resolution of such measurements should permit the assessment of blood flow at localized sites of the retinal, optic nerve, and choroidal vascular systems. The temporal response should be fast enough to provide a detailed assessment of the regulatory responses of blood flow evoked by various physiological stimuli.

This chapter we will review the current status of the laser Doppler velocimetry (LDV) and laser Doppler flowmetry (LDF) techniques that are based uniquely on lasers of long coherence length. The more recent developments, which make use of lasers of short coherence length, are discussed elsewhere in this book.

The laser has made it possible to detect with high resolution the Doppler shift that light undergoes when scattered by moving objects. With the

L. Schmetterer, J.W. Kiel (eds.), *Ocular Blood Flow*,
DOI 10.1007/978-3-540-69469-4_7, © Springer-Verlag Berlin Heidelberg 2012

fast development of this device and optical techniques associated with it, this principle has led over the last 35 years to the measurement of blood flow in a number of tissues of the body [1]. In 1972, the retina provided the first measurements of blood velocity obtained by means of the Doppler shift [2]. LDV measurement of blood velocity in human retinal vessels was published 2 years later [3]. Subsequent to the pioneering work of Stern [4], who assessed blood flow in the tissue of the skin, Riva et al. [5] described a method to measure blood velocity in the human optic nerve microcirculation. This method was then extended to the measurement of blood flow in the vascular bed of the cat optic nerve head (ONH) [6], the human ONH [7, 8], and the subfoveal choroid [9]. In 1995, Michelson and Schmauss first reported mapping of blood flow in the microcirculation of the human retina by means of scanning LDF [10].

The present chapter is by no means an exhaustive description of LDV and LDF. Regrettably, due to limitation of space, a choice of papers and reviews among the extensive laser Doppler literature related to the eye had to be made. Part of the material of this chapter has been reproduced with permission from previous publications [11–14].

7.2 Retinal Laser Doppler Velocimetry

7.2.1 The Doppler Effect

The basis of LDV and LDF is the Doppler effect, first described in 1842 by the Austrian physicist Christian Doppler in an article entitled "On the Colored Light of Double Stars and Some Other Heavenly Bodies." Doppler describes the frequency shift that a sound or a light wave undergoes when emitted from an object that is moving away or towards an observer. It manifests itself, for example, in the increase in the pitch of the siren of an approaching ambulance.

Consider a particle, such as a red blood cell (RBC), moving at velocity $\vec{V}$ (Fig. 7.1). A laser beam of single frequency, f_0, is incident on this particle. This beam, which is defined by the wave vector $\vec{K}_i$, will be scattered by the particle in various directions. The beam scattered in the direction of the detector, defined by $\vec{K}_s$, is shifted in frequency by an amount:

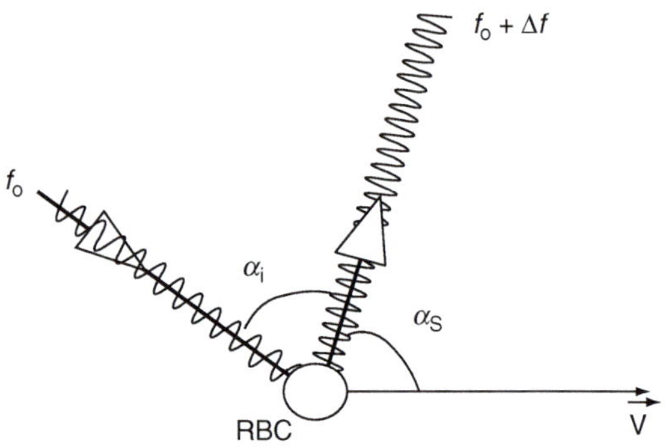

Fig. 7.1 The Doppler effect: The frequency of the light scattered by the particle moving at the velocity $\vec{V}$ in the direction defined by a_s is shifted in frequency by an amount Δf compared to that of the incident light (frequency f_0 and direction defined by a_i with V)

$$\Delta f = \frac{1}{2\pi}(\vec{K}_s - \vec{K}_i) \bullet \vec{V} = \frac{1}{2\pi}(\vec{K}) \bullet \vec{V} \qquad (7.1)$$

where the symbol (•) represents the scalar product and $\vec{K} = \vec{K}_s - \vec{K}_i \bullet \; |\vec{K}_s| \cong |\vec{K}_i| \cong 2\pi n/\lambda \bullet \; n$ is the refractive index of the medium and λ is the wavelength in *vacuo* of the incident laser light.

In order to determine Δf, the scattered light must be detected by a sensor, which transforms the electric field of the scattered light incident upon it into a current. This current is in turn processed to determine its frequency content. The technique by which Δf is determined from the scattered field is known as "optical mixing spectroscopy (OMS)" or "light beating spectroscopy" [15, 16].

Understanding the application of OMS to the measurement of RBC velocity in retinal vessels requires a description of the electric field of the scattered light, followed by that of the current at the output of the detector. This description will be first carried out in the simplified case of singly scattering particles moving in a glass capillary tube and then extended to multiply scattering RBCs.

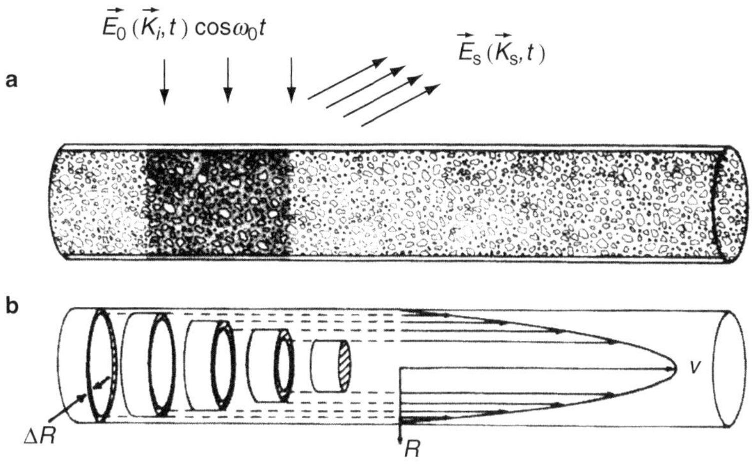

Fig. 7.2 (**a**) Schematic representation of a volume of a capillary tube illuminated by an incident wave defined by $\vec{E}_0 \cos\omega_0 t$ and with the wave vector $\vec{K}_i$. The light scattered by the particles moving in the tube and the vessel wall is defined by $\vec{K}_s$. (**b**) Assumed parabolic velocity profile of the particles in the capillary tube. The cylindrical shell ΔR contains particles moving at velocity V, the value of which is given by Poiseuille's law (Adapted from Riva and Feke [13] with permission from the Publisher)

7.2.2 Electric Field Scattered by Singly Scattering Particles Moving in a Capillary Tube

The electric field, $\vec{E}_s(\vec{K}_s,t)$, of the light scattered in direction $\vec{K}_s$ at time t by a dilute solution of singly scattering particles moving at constant velocity through a glass capillary tube (Fig. 7.2) consists of light scattered by both the particles and the glass wall. The incident monochromatic beam (single frequency, f_0), $\vec{E}_0 \cos\omega_0 t$ ($\omega_0 = 2\pi / f_0$), is assumed to illuminate uniformly the entire cross section of the tube.

If the particles are much smaller than the capillary radius, R_0, their velocity profile obeys Poiseuille' law, i.e., $V(R)$ (for simplification, $V(R) = |\vec{V}(R)|$) is a parabolic function of the radial distance, R, from the axis of the tube:

$$V(R) = V_{\max}\left\{1 - \left(\frac{R}{R_0}\right)^2\right\} \tag{7.2}$$

$V_{\max}$ is the centerline (maximum) velocity of the particles. Due to this distribution of velocities, the scattered light consists of a range of Doppler-shifted frequencies.

With Poiseuille flow, if one divides the range of velocities (0 to $V_{\max}$) into N equal increments, ΔV, (Fig. 7.2) the particles with speeds between V and $V+\Delta V$ are located in a shell of radial thickness ΔR, where from Eq. 7.2, $\Delta R = R_0^2 \Delta V / 2RV_{\max}$. Furthermore, if the concentration of particles across the tube is uniform, there will be equal numbers of particles flowing at each velocity range [2].

The field $\vec{E}_s(\vec{K},t)$ is the sum of the field scattered by the particles and the field scattered by the vessel wall:

$$\vec{E}_S(\vec{K},t) = \sum_{i=0}^{N} \vec{E}_i(\vec{K},t) + \vec{E}_{LO}(\vec{K},t) \tag{7.3}$$

$\vec{E}_i(\vec{K},t)$, the field scattered by the particles with velocity V_i has the frequency $2\pi(f_0 + \Delta f_i)$. $\vec{E}_{LO}(\vec{K},t)$ is the field of the light scattered from the tube wall. It has the same frequency as the incident light since the tube is not moving and acts as a reference beam, conventionally called "local oscillator (LO)."

Under the conditions of both uniform concentration of the particles and laser illumination of the tube, each $\vec{E}_i(\vec{K},t)$ has the same amplitude, $A(\vec{K})$ and thus:

$$\vec{E}_i(\vec{K},t) = A(\vec{K})\cos 2\pi(f_0 + \Delta f_i)t \tag{7.4}$$

Similarly, $\vec{E}_{LO}(\vec{K},t)$ can be written as:

$$\vec{E}_{LO}(\vec{K},t) = A_{LO}(\vec{K})\cos 2\pi f_0 t \tag{7.5}$$

Therefore:

$$\begin{aligned}\vec{E}_s(\vec{K},t) &= A_{LO}(\vec{K})\cos 2\pi f_0 t \\ &+ A(\vec{K})\sum_{i=0}^{N}\cos 2\pi(f_0 + \Delta f_i)t\end{aligned} \tag{7.6}$$

The power spectral density of the scattered light $P_{Es}(f)$ detected for the scattering geometry defined by $\vec{K}$ is obtained by applying Fourier analysis:

$$P_{Es}(f) = \frac{A_{\mathrm{LO}}^{2}}{2}\delta(f - f_0) + \frac{A_{\mathrm{LO}}A}{2}\sum_{i=0}^{N}\delta\left[f - (f_0 + \frac{i\Delta f_{\max}}{N})\right] \quad (7.7)$$

with $f_0 \le f \le f_0 + \Delta f_{\max}$

$\Delta f_{\max}$ corresponds to the maximum velocity $V_{\max}$, and $\delta(f - f_0)$ is the delta function at f_0. Thus, for our model, the optical spectrum of $\vec{E}_s(K,t)$ consists of a d function at f_0, a flat part of magnitude $A_{\mathrm{LO}}\,A/2$ from f_0 to $f_0+\Delta f_{\max}$ and zero beyond $f_0+\Delta f_{\max}$. Since f_0 is approximately 5×10^{14} Hz, and Δf_i for retinal vessels (see later) ranges between 0 and $\cong 5\times10^{4}$ kHz, the required resolution ($\Delta f/f_0$) for its detection, namely <<10^{-10}, was far beyond the capability of spectroscopic techniques available until $\cong 1965$. However, with OMS, the information contained in the optical spectrum could be translated from the high frequencies of light waves down to sufficiently low frequencies, where instruments operate with adequate resolution to resolve the spectrum of Doppler shifts.

7.2.3 Doppler Shift Power Spectrum (DSPS) of the Photocurrent

In the application of OMS, $\vec{E}_s(K,t)$ is directed onto the surface (S) of a square wave photodetector. This field generates a photocurrent, $i(\vec{r},t)$, where $\vec{r}$ is a point on S. The time average $\langle\ \rangle$ of this current is:

$$\langle i(t)\rangle = \int_S \beta\vec{E}_s(\vec{K},t)dS \quad (7.8)$$

β is the responsivity of S. All the information on the particles velocities can be extracted from $\langle i(t)\rangle$. This extraction, which is complex and beyond the scope of this chapter, leads to the following expression for the Doppler shift power spectral density DSPS $= P_i(\Delta f)$ of the photocurrent [13]:

$$P_i(\Delta f) = \frac{1}{4}\beta^2 S^2\left[A_{\mathrm{LO}}^{2} + A_s^{2}\right]^2 \delta(\Delta f) + \frac{1}{2}e\beta S\left[A_{\mathrm{LO}}^{2} + A_s^{2}\right] + \beta S S_{coh\mathrm{LO}} A_{\mathrm{LO}}^{2} A^2 \sum_{i=1}^{N}\delta\left(\Delta f - \frac{i\Delta f_{\max}}{N}\right)$$

in the range $0 \le \Delta f \le \Delta f_{\max}$ and

$$P_i(\Delta f) = \frac{1}{2}e\beta S\left[A_{\mathrm{LO}}^{2} + A_s^{2}\right] \quad \text{for} \quad \Delta f > \Delta f_{\max}. \quad (7.9)$$

$\Delta f_{\max}$ is defined as the cutoff frequency of $P_i(\Delta f)$. e is the charge of the electron, and S_{cohLO} is the coherence area of the local oscillator at the detector surface.

The third term of $P_i(\Delta f)$ in Eq. 7.9 is a squared replica of the second term of $P_{Es}(f)$ in Eq. 7.7, but shifted down towards low frequencies, so that the particle velocities can be determined from measurements of $P_i(\Delta f)$. Thus, the process of mixing at the detector surface of the light scattered by the particles with the light from the glass wall (local oscillator) results in the downshifting of the spectrum from optical frequencies to frequencies in the kHz range (Fig. 7.3), where high-resolution spectral analysis can be performed with conventional devices. The term $\frac{1}{2}e\beta S\left[A_{\mathrm{LO}}^{2} + A^2\right]$ is the photodetector shot noise, which is due to the stochastic nature of the photon emission.

In general, the local oscillator is much more intense than the field scattered by the particles so that $A_{\mathrm{LO}} \gg A$. As a result, the shot noise is $\frac{1}{2}e\beta S A_{\mathrm{LO}}^{2}$. In practice, the ability to measure $P_i(\Delta f)$ with high resolution requires that the power spectral density of the signal term be larger than that of the shot noise term. A measure of this ability is the intrinsic signal-to-noise ratio (SNR), which is:

$$\mathrm{SNR}(\Delta f) = \frac{2\beta}{e} S S_{coh\mathrm{LO}} A^2 \sum_{i=1}^{N}\delta\left(\Delta f - \frac{i\Delta f_{\max}}{N}\right) \quad (7.10)$$

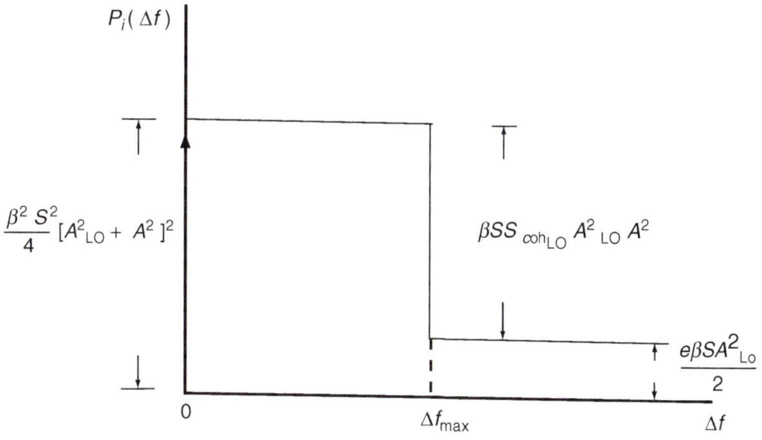

Fig. 7.3 Theoretically predicted photocurrent power spectral density, $P_i(\Delta f)$, explicitly showing DC component $\frac{\beta^2 S^2}{4}[A^2_{LO}+A^2]^2$ at $\Delta f = 0$, signal component, $\beta SS_{cohLO}A^2_{LO}A^2$, extending to f_{max} and shot noise component, $\frac{e\beta SA^2_{LO}}{2}$ (Adapted from Riva and Feke [13] with permission from the Publisher)

For given β, S, and S_{cohLO}, the only way to increase the SNR is to increase A^2, i.e., the intensity of the incident laser light.

7.2.4 Absolute V_{max} Measurements

To obtain $V_{max} = |\vec{V}_{max}|$ from Δf_{max} using Eq. 7.1, the angle between $\vec{K}_i$ and $\vec{V}_{max}$ and the one between $\vec{K}_s$ and $\vec{V}_{max}$ must be known. This is difficult in the living eye due mainly to the difficulty of determining accurately the direction of blood flow in a retinal vessel.

This problem is solved by means of a bidirectional detection scheme [17, 18], in which Δf_{max} is measured for two directions of the scattered light. V_{max} is then obtained from the formula:

$$V_{max} = \frac{\lambda \Delta^* f_{max}}{n \Delta\alpha \cos\beta} \qquad (7.11)$$

In this equation, $\Delta^* f_{max} = \Delta f_{max,2} - \Delta f_{max,1}$. $\Delta\alpha$ is the intraocular angle between the two scattering directions, $\vec{K}_1$, $\vec{K}_2$, which can be obtained from the corresponding extraocular angle [17–19]. The calculation of $\Delta\alpha$ requires the measurement of the axial length of the eye, a routine procedure in ophthalmology [18, 20]. β is the angle between $\vec{V}_{max}$ and the plane defined by the vectors $(\vec{K}_1, \vec{K}_2)$. n is the index of refraction of the medium. Implementation of the bidirectional scheme to a fundus camera is shown in Fig. 7.4, top.

7.2.5 Experimental Test of the Bidirectional LDV Technique

With the aim of testing the validity of the bidirectional scheme [11], polystyrene latex spheres (diameter=0.3 μm) suspended in water were passed at known $|\vec{V}_{max}|$ through a glass capillary tube (diameter=200 μm) placed in the retinal plane of a Topcon model eye. The incident laser beam (helium-neon, 632.8 nm) was focused on the tube which was placed at different locations of the fundus. The direction of the flow was also varied stepwise from 0° (tube horizontal) to 180°. Measurements were performed for an emmetropic and an axially ametropic (± 4 diopters) eye. DSPS were recorded for two scattering directions by means of a two-channel digital signal spectrum analyzer (Fig. 7.4, bottom). These DSPS are characterized by sharp cutoffs from which $\Delta f_{max,1}$ and $\Delta f_{max,2}$ can be accurately determined. The value of V_{max} calculated using Eq. 7.11 was found to be in good agreement (<3% difference) with the value obtained from the known flow through the tube and the cross section of the tube. The difference was attributed to errors in $\Delta\alpha$ and $\Delta^* f_{max}$. Feke et al. demonstrated similar results and also confirmed the linear relationship between $\Delta^* f_{max}$ and $\cos\beta$ for a given V_{max} and scattering geometry ($\Delta\alpha$) [19].

The increase in spectral power in the region of the cutoff frequencies for the DSPS shown in Fig. 7.4 bottom is probably due to a nonuniform illumination of the tube by the laser beam, which was smaller than the tube internal diameter. As a

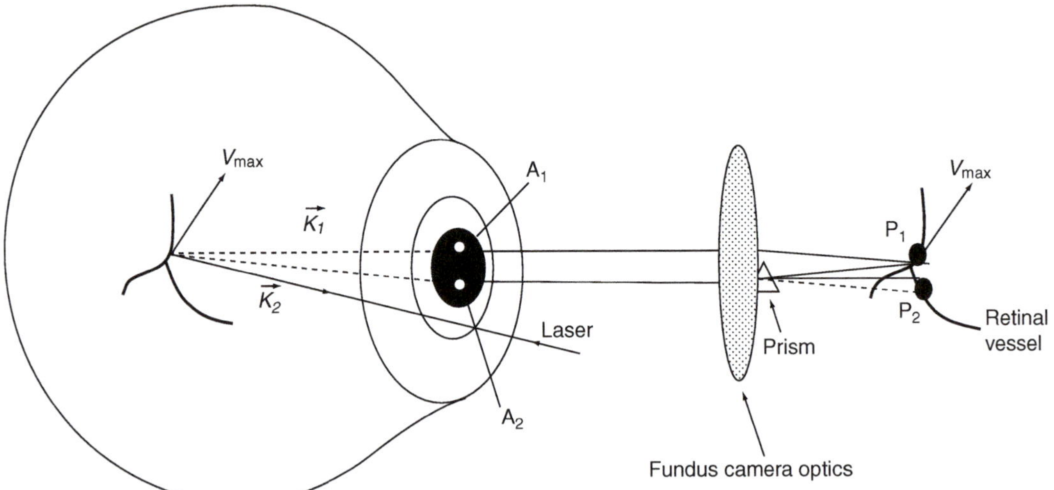

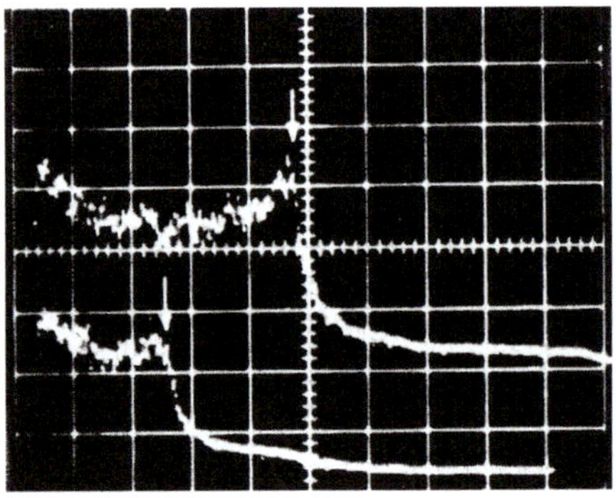

Fig. 7.4 *Top*: Fundus-camera-based scheme for the detection of light scattered from blood moving with velocity V_{max} along two directions $\vec{K}_1$ and $\vec{K}_2$ defined by apertures A_1 and A_2 in the plane of the eye pupil. The beam through A_2 is focused at P_2 after deflection by a prism in the collecting optics. The direction P_1-P_2 is aligned with that of A_1-A_2. The direction A_1-A_2 can be rotated in the plane of the pupil for alignment with the direction of the vessel (Adapted from Riva et al. [25] with permission from the Publisher). *Bottom*: DSPS obtained simultaneously along two directions of the scattered light from a suspension of polystyrene spheres in water through a 200-µm-internal-diameter glass capillary tube mounted in the "retinal" plane of a model eye (Adapted from Riva et al. [18] with permission from the Publisher)

result, the faster particles received more light relative to the slower ones located close to the vessel wall. The effects of laser beam size, eccentricity, as well as absorption of light by the flowing medium, have been reported by Petrig and Follonier [21].

7.2.6 The Output Signal-to-Noise Ratio of the DSPS and Its Dependence on the Recording Time

Due to the stochastic nature of the photodetector emission process, the photocurrent is a fluctuating quantity so that there is an uncertainty in the measurement of the DSPS when the measurement is performed in a finite time. This phenomenon is illustrated by four DSPS recorded in different times from a 0.1 suspension of 1-µm-diameter polystyrene spheres flowing with a $V_{max} = 3.61\ \mathrm{cm/s}$ through a 200-µm-internal-diameter glass capillary tube (Fig. 7.5). As expected, the fluctuations of $P_i(\Delta f)$ decrease as the measurement time increase. It can be shown that, based on a previous analysis [13, 16], if $\Delta \nu \times T \gg 1$,

$$\frac{1}{\sigma(P_i(\Delta f))} \propto \sqrt{T} \qquad (7.12)$$

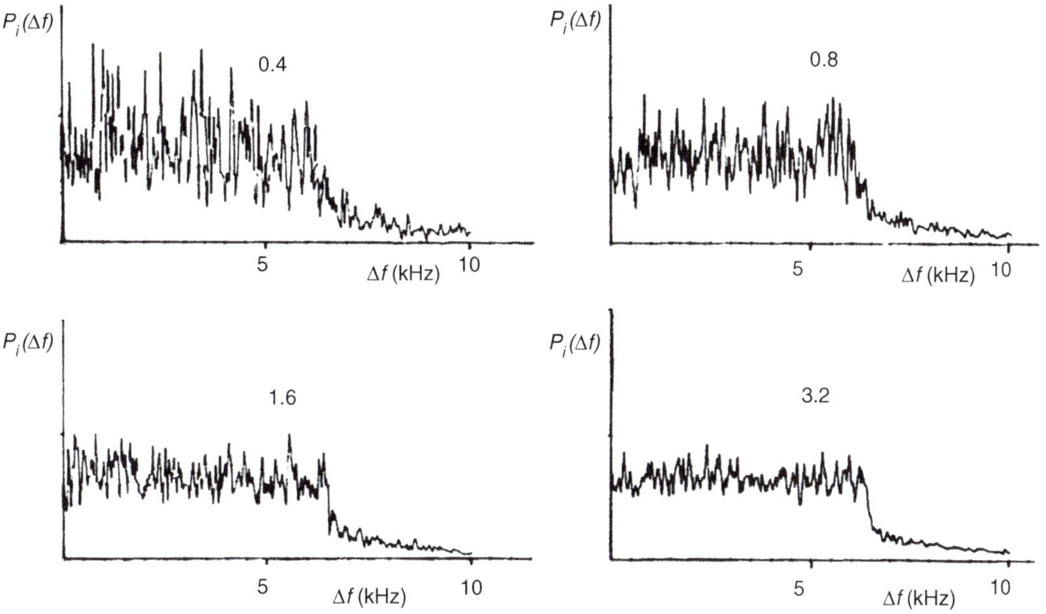

Fig. 7.5 DSPS ($P_i(\Delta f)$) obtained in measurement times of 0.4, 0.8, 1.6, and 3.2 s from 0.1% suspension of polystyrene spheres in water flowing through a 200-μm-internal-diameter glass capillary tube. The precision of $P_i(\Delta f)$ at each Δf increases with the measurement time (Adapted from Riva et al. [13] with permission from the Publisher)

s is the standard deviation, T the measurement time, and Δv the resolution bandwidth of the spectrum analyzer. The linear relationship expressed by Eq. 7.12 has been verified with excellent accuracy for the recordings shown in Fig. 7.5 [13].

7.2.7 The DSPS for RBCs Moving in a Retinal Vessel

7.2.7.1 Multiple Scattering of Blood

Equation 7.10 is based on a model that assumes single scattering of the incident laser light by the RBCs. The validity of this assumption, however, may be questionable in view of previous studies that demonstrate a predominance of multiple scattering when visible light interacts with whole blood. Thus, in a blood layer of 100 μm, the photon mean free path length at a wavelength $\rightarrow = 0.6328$ μm is only 7 μm [22], and multiple scattering of light by the RBCs is expected to be important [23]. Therefore, for blood vessels with diameters typically between 50 and 200 μm, photons would be expected to have undergone a great number of scattering events and, consequently, of Doppler shifts before being detected.

7.2.7.2 DSPS from RBCs Flowing in a Glass Capillary Tube

Four DSPS from whole blood (hematocrit, 41%) flowing at a maximum speed of 1.44 cm/s through a glass capillary tube with a 200-μm internal diameter are shown in Fig. 7.6. The expected Δf_{max} was 5.9 kHz. Clearly, the DSPS obtained in 51.2 s does not exhibit the rectangular shape of the DSPS predicted for a dilute suspension of latex spheres (Fig. 7.5). There is no discernible cutoff at the expected Δf_{max} but a rather a monotonic decrease in the amplitude at the higher frequencies. The spectral power beyond 5.9 kHz is most likely the result of multiple scattering as the laser light penetrates into and exits from the flowing medium [2, 24]. The DSPS obtained in 3.2 s has essentially the same characteristics. However, since the measurement time is much shorter, the statistical fluctuations in $P_i(\Delta f)$ are increased, as expected. One also observes that the DSPS obtained in 0.4 s exhibits a greater increase in the

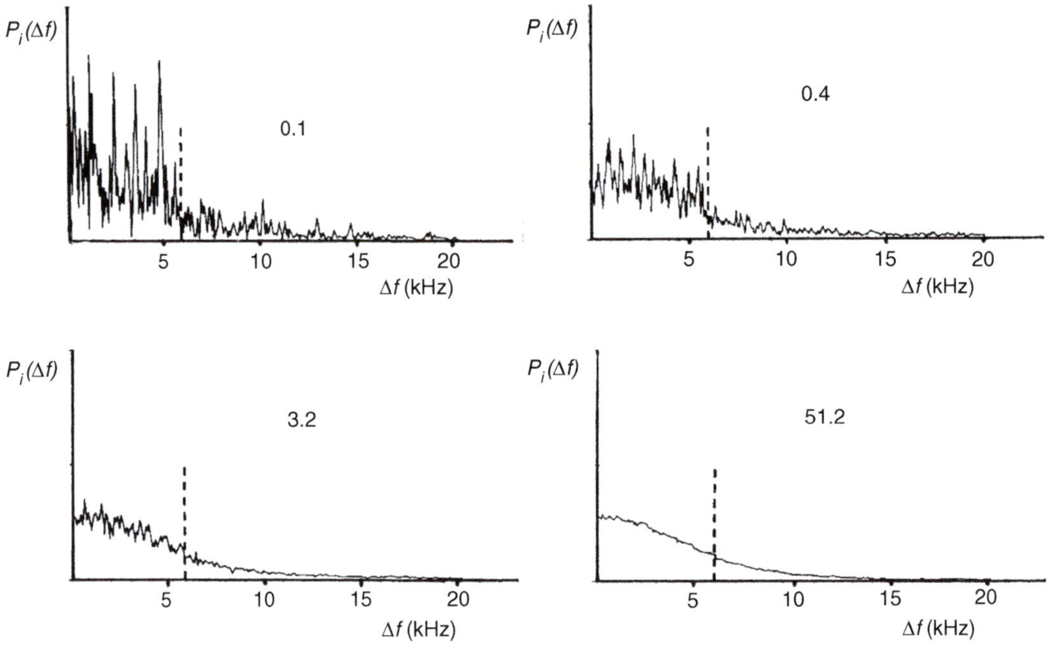

Fig. 7.6 DSPS obtained in 0.1, 0.4, 3.2, and 51.2 s from whole blood flowing through a 200-μm internal diameter glass capillary tube. *Dashed lines* are the expected Δf_{max} (Adapted from Riva et al. [13] with permission from the Publisher)

fluctuations in $P_i(\Delta f)$ in the region $\Delta f<\Delta f_{max}$ than in the region $\Delta f>\Delta f_{max}$. Consequently, the transition in the fluctuations, which becomes discernible at 5.9 kHz, is now clear in the DSPS obtained in 0.1 s. This phenomenon has been explained based on the hypothesis that in the region of the DSPS corresponding to $\Delta f<\Delta f_{max}$, $P_i(\Delta f)$ arises primarily from single scattering, whereas in the region $\Delta f>\Delta f_{max}$, $P_i(\Delta f)$ is only due to the contribution of multiple scattering. A mathematical description of this experimentally observed phenomenon, which allows accurate determination of Δf_{max} of the RBCs by using short measurement times, is beyond the scope of this chapter but can be found in Appendix A of Riva and Feke [13].

7.2.7.3 DSPS from Human Retinal Vessels

DSPS obtained from a human retinal vein using recording times of 0.1, 0.2, and 0.8 s display similar characteristics as those recorded from whole blood in a glass capillary tube (Fig. 7.7). The fluctuations in $P_i(\Delta f)$ at frequencies up to approximately 6.5 kHz increase dramatically as the analysis time is shortened. A recording time <0.2 s displays a clear break in the magnitude of the fluctuations, allowing adequate determination of Δf_{max}.

7.2.7.4 Exploring the Scattering Process

DSPS with nearly ideal rectangular shape can be obtained from cat retinal vessels (diameter <120 μm) if detection of the laser light that has been doubly transmitted through the vessels can be prevented [22], allowing single backscattering to be the predominant process [25]. For vessels with diameter ≈ 120 μm or bigger, single backscattering becomes increasingly less predominant than multiple backscattering, with ensuing degradation of the sharpness of the cutoffs.

The cat retina offers the opportunity to record DSPS generated either through single scattering or through multiple scattering. Multiple scattering occurs when the incident laser is focused on a retinal vessel coursing in front of the tapetum, a highly light reflecting layer. The light transmitted through such a vessel is retransmitted through it after reflection at the tapetum. This doubly forward transmitted light involves predominantly

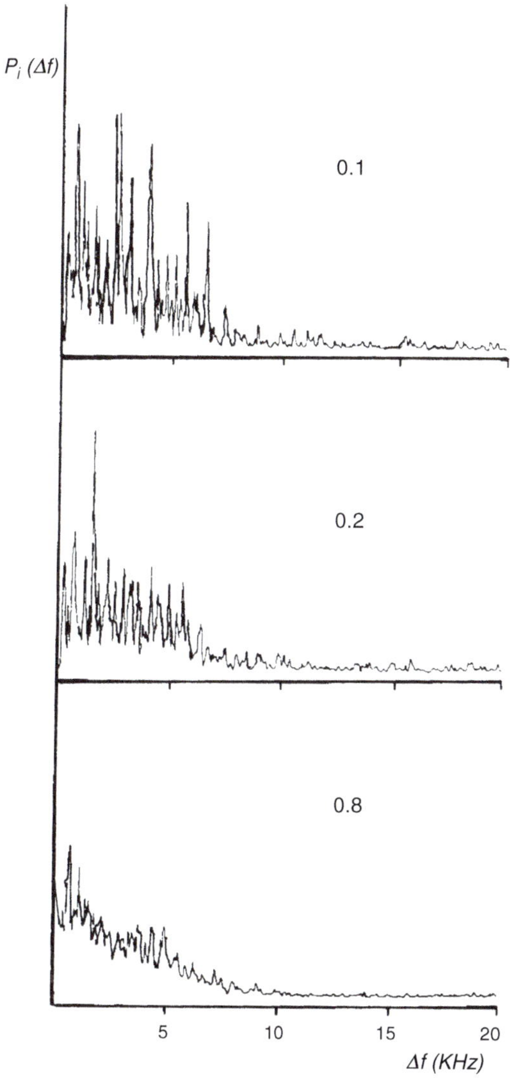

Fig. 7.7 DSPS from a human retinal vein obtained in 0.1, 0.2, and 0.8 s, demonstrating the effect of decreasing the measurement time. With a short measurement time, the cutoff frequency Δf_{max} can be more precisely detected (Adapted from Riva et al. [13] with permission from the Publisher)

multiple scattering of light by the RBCs. The detected light is depolarized, and the corresponding DSPS does not present a sharp cutoff but rather has a smoothly declining shape (Fig. 7.8), which does not vary with the scattering angle. Such DSPS conform to the model of Bonner and Nossal when multiple scattering is the dominant process [26].

In the lower half of the cat's retina, the retinal vessels course in front of a heavily pigmented layer that absorbs nearly all the incident laser light transmitted through the vessel. Practically, only light backscattered by the RBCs and vessel wall is detected. In this case, this light maintains most of the polarization of the incident light. The DSPS have a rectangular shape, and the sharp cutoff varies with the scattering angle, as expected from the Doppler formula (Fig. 7.8). The detected light represents probably pseudo-singly backscattered light by the RBCs, a scattering process occurring in vessels with diameters up to at least 120 µm [27].

7.2.8 Computer Modeling of the DSPS for Automatic Determination of Äf$_{max}$

In the early days of retinal blood velocity recordings by LDV, the photocurrent was fed to a tape recorder, and a loudspeaker and the DSPS were recorded during playback. Typically, for veins and arteries during diastole and often during systole, the frequencies of the DSPS are in the audio range. Therefore, during playback, only those portions of the tape were analyzed for which a clearly identifiable pulsatile pitch (for arteries) or a monotonous, high-frequency pitch (for veins) could be heard [28]. The DSPS were obtained with a hardware spectrum analyzer, one pair at a time, and successively displayed on an oscilloscope screen. An examiner visually determined the Δf_{max}, one channel at a time, by moving a cursor along the frequency axis to the frequency value where a sharp decline in the power spectral density and variance was identifiable. Each estimate of Δf_{max} (mean and standard deviation) was based on 10–20 pairs of DSPS. Such a procedure was time-consuming, especially for retinal arteries, for which several Δf_{max} were measured at different phases of the heart cycle to obtain average V_{max} during the heart cycle. In addition, masking of the examiner with respect to the type of patient and experimental protocol to eliminate possible bias was an additional time-consuming procedure.

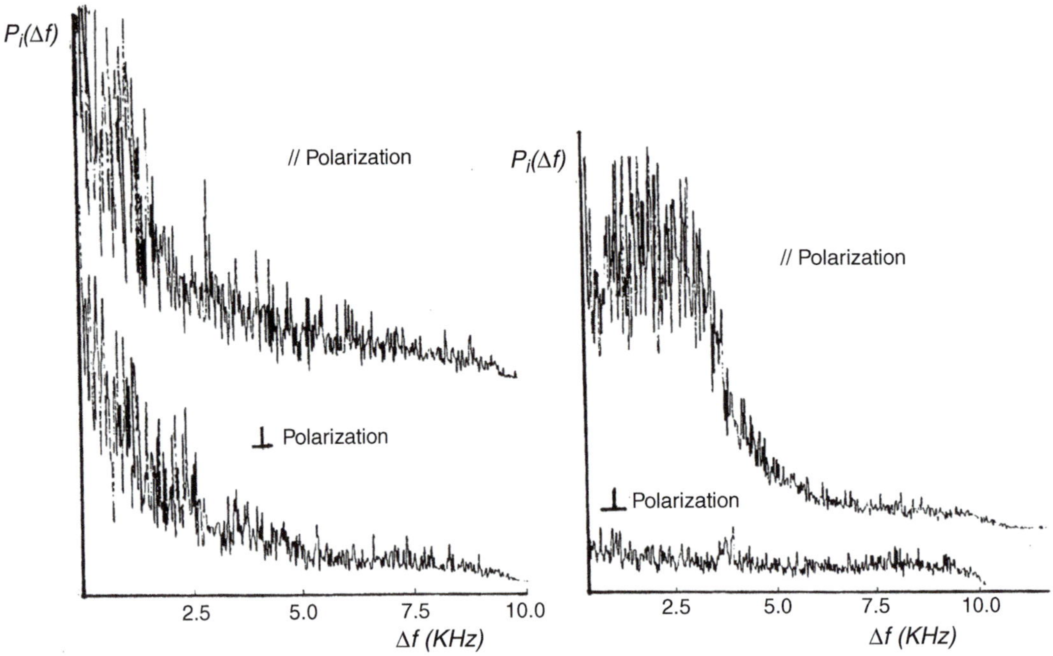

Fig. 7.8 *Left*: DSPS obtained from a vein in the tapetal region of the cat retina. The tapetum is a highly light reflecting layer. For the upper DSPS, the scattered light was detected in the same plane of polarization (//) as the incident light. For the lower DSPS, these polarizations were perpendicular (//). Light transmitted through the blood is highly reflected and some of it retransmitted through the blood. This double transmission represents predominantly multiply scattered light. *Right*: DSPS from a vein in the pigmented (highly absorbing) region of the fundus. Notice the disappearance of the spectral power when the polarization of the scattered light was perpendicular (⊥) to that of the incident light. For vessels above the pigment, only the light backscattered by the blood column reaches the detector. The light transmitted through this column is totally absorbed by the pigment behind the vessel (Adapted from Riva et al. [22] with permission from the Publisher)

Later, to automate the analysis, a computer algorithm to calculate V_{max} based on the aforementioned rectangular shape model of the DSPS was implemented on a NeXT computer [29]. It includes data acquisition, eye blink rejection, power spectrum analysis, and display of the V_{max} data and of its change during the heart cycle.

More recently, a digital signal processor (DSP)-based approach, which addresses the need for higher temporal resolution of V_{max}, combined with blink rejection, was reported [30]. In brief, blink rejection is based on the fact that because the laser light is out of focus at the pupil plane, closing the lid scatters much less laser light back to the detector than does the fundus. This results in a markedly reduced DC value of the photocurrent. Whenever the DC value of the Doppler signal falls below a user-adjustable threshold, the data is automatically excluded from further analysis and presentation.

7.2.9 Instrumentation

A bidirectional retinal laser Doppler velocimeter (RLDV) consists basically of distinct optical systems with the functions of: (1) aiming a laser beam (red or near-infrared) at a main retinal vessel; (2) collecting and detecting some of the light scattered by the blood in the vessel along two directions of scattering; (3) observing the eye fundus and the laser beam; and (4) providing a pinpoint fixation target to be observed by the eye being tested in order to precisely aim the laser at the desired site [13].

Bidirectional LDV systems have been incorporated into commercially available slit lamps [17, 19], standard fundus cameras [18, 31], as well as a portable fundus camera [32]. LDV measurements have been performed using a 632.8-nm HeNe laser [18], a 675-nm laser diode [31], as well as a near-infrared diode (wavelength around 800 nm) [33]. The latter allows measurements to be performed in darkness [33].

With the Oculix LDV(Arbaz, Switzerland), a research instrument based on a Topcon TRC+fundus camera, blood flow, Q (µl/min), is obtained from V_{max} and the diameter, D, of the blood column of the target vessel, the latter measured from fundus photographs (see below). The Canon laser Doppler flowmeter (CLBF 100, Canon, USA) is an instrument able to measure simultaneously V_{max} and D of large retinal arteries, providing at once Q-values in µl/min [20, 34]. D is measured by scanning the vessel perpendicularly with a 543-nm HeNe laser. The CLBF is equipped with an automatic vessel tracking system that employs a linear sensor to monitor the target vessel and maintain the probing laser beam at all times on the vessel during measurement. Results are acquired at 50 measurements per second for 2 s, after which the measurement is repeated to verify data consistency. The CLBF automatically corrects refractive error using the subject's axial length value and the machine's astigmatic correction dial.

7.2.10 Blood Flow in Retinal Vessels

Blood flow (Q) in a retinal vessel is calculated as:

$$Q = \frac{\pi D^2}{4} V_{mean} \qquad (7.13)$$

V_{mean} is the mean velocity of the RBCs in the vessel. For a parabolic profile of the RBCs velocity (Poiseuille flow), $V_{mean} = \frac{V_{max}}{2}$.

The presence of Poiseuille flow depends on the flow rate, the diameter of the vessel, and the concentration (hematocrit) of the RBCs. Various techniques have been recently developed to determine the velocity profile of the RBCs in the human eye, not only with the aim to determine Q accurately but also because a measurement of this profile could be of help in the early diagnosis and treatment of various retinal circulatory impairments. In particular, determination of the velocity gradient at the vessel wall could provide valuable information on the wall shear rate, a quantity that plays an important role in the control of blood flow [35, 36].

A number of techniques have recently demonstrated that the velocity profile of the RBCs in a retinal artery can be satisfactorily fitted with a parabolic function [37–39]. Measured with a confocal scanning LDV system, velocity profiles of RBCs in retinal arteries and veins were fitted with a more general function of the type [40]:

$$V(r) = V_{max}\left[1 - \left|\frac{r}{R_i}\right|^K\right] \qquad (7.14)$$

In this equation, $V(r)$ is the velocity at a distance r from the inner wall, R_i the tube inner radius, and K is a parameter related to the shape of the velocity profile. $K=2$ corresponds to a parabola and $K>2$ to a blunted profile. Average values of K for a vein with a 150-µm diameter were 2.13±0.42 (conf. limits), 2.38±0.5, and 1.94±.0.8 for a 100-µm artery during systole and diastole, respectively (Fig. 7.9) [41].

7.2.11 Precision and Reproducibility of V_{max}

The precision of V_{max} (i.e., how close several values of V_{max} are to each other) was assessed by the coefficient of variation (CV). CV depends on the steadiness of target fixation, the diameter of the vessel, the duration of measurement, and the retinal laser irradiance. With the early method of analysis [28], the measurement of V_{max} was based on at least ten pairs of DSPS for each vessel, which required between 1 min for a subject with good target fixation and several minutes for a subject with poor target fixation. Smaller vessels are

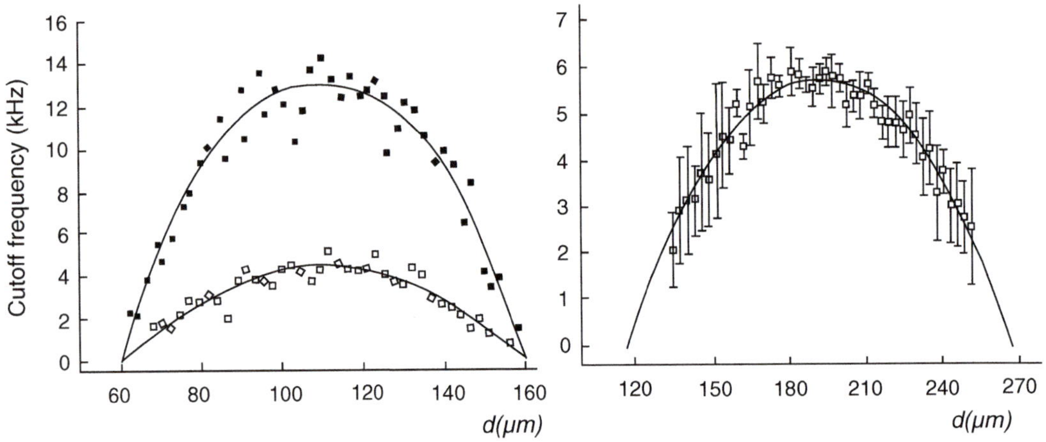

Fig. 7.9 *Left*: Average cutoff frequency (proportional $V_{highest}$) versus scanning distance *d* obtained from 17 scans across a human retinal artery (diameter 100 μm). *Black dots*: systolic phase of the cardiac cycle; *open circles*: diastolic phase. *Continuous lines* are the fit based on Eq. 7.14, which provided a value of $K=2.38\pm0.5$ in systolic and 1.94 ± 0.8 (95% CL) in diastolic phase, respectively. *Right*: Average cutoff frequency (proportional $V_{highest}$) versus scanning distance *d* obtained from five scans across a human retinal vein (diameter 150 μm). *Error bars* are the 95% CL. *K* was found to be 2.13 ± 0.42 (95% CL) (Adapted from Logean et al. [41] with permission from the Publisher)

more difficult to measure, as they require more precise target fixation. Increasing laser retinal irradiance increases proportionally the signal-to-noise ratio of the DSPS [13] and consequently the sharpness of the cutoff frequencies. Intraobserver and interobserver CV(V_{max}) were found to be about 10% and 15%, respectively [28].

Variability of V_{max} due to lateral head motion was estimated from measurements in a capillary tube placed in the "retinal" plane of the model eye. Head motion may move the entrance of the laser beam and the exits of the scattered beams across the pupil. The CV of the mean V_{max} obtained for five horizontal positions of the LDV camera relative to the eye was 5%. The reproducibility of V_{max} and *Q*, evaluated at different times, depends on the constancy with time of retinal blood flow. Measurements of venous V_{max} in a normal subject during 50 min and over a period of 70 days showed retinal blood flow to be remarkably constant over both periods of time [28]. Two measurements of retinal blood flow performed at 1 month interval in five subjects confirmed this finding by reporting an average difference of 5.3% (range: 0.7–9.4%) [19]. An extensive study with the CLBF on 20 normal subjects demonstrated a repeatability for blood flow in major temporal retinal arterioles of 18% [42] and 23% [43].

The full potential of the LDV technique in terms of precision and short-term reproducibility is best demonstrated by a unidirectional LDV recording of Δf_{max} in a retinal arteriole of an anesthetized minipig breathing 100% O_2 (Fig. 7.10) [44]. The rapid (<1 s) variations of Δf_{max} are due to the systolic/diastolic changes in blood pressure. The slower change starting at time 0 min is the response to 100% O_2 breathing.

7.2.12 Limitations, Safety, and Future Directions of the LDV Technique

Three factors currently prevent the LDV technique from being applicable in routine clinical work: ocular media opacities, pupillary dilatation of less than approximately 5 mm, and poor target fixation. The effect of media opacities is to decrease the intensity of the detected scattered light. The decrease of retinal irradiance results from absorption of light and from the scattering by the ocular media. From Eq. 7.10, a decrease in signal-to-noise ratio (SNR) and consequently of the sharpness of the cutoffs of the DSPS can be expected. It is therefore

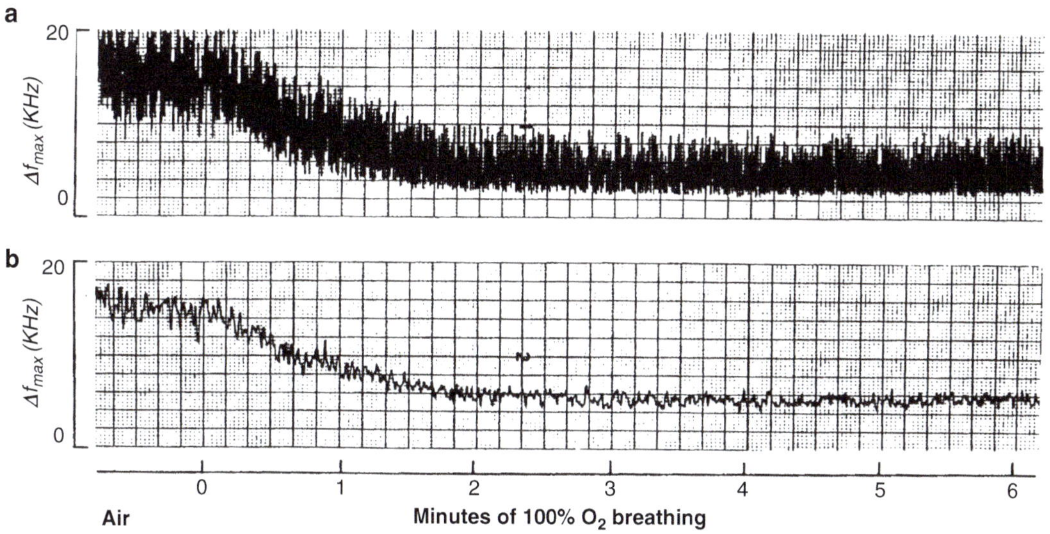

Fig. 7.10 Effect of oxygen breathing on Δf_{max} = constant × V_{max} in a retinal artery of the miniature pig. At time 0 min, the anesthetized pig is given 100% oxygen to breathe. The diastolic and systolic limits (**a**) and the time average (**b**) of Δf_{max} are presented (Adapted from Petrig and Riva [44] with permission from the Publisher)

advantageous to use as much laser power as possible within the limit imposed by the maximum permissible level of retinal irradiance. Assuming that photodetectors of similar sensitivity are used, there is an advantage to use lasers in the near-infrared region of the spectrum because the maximum permissible exposure for continuous illumination is at least 20 times higher there than that at the wavelength of the helium-neon laser [45, 46].

Head motion and poor target fixation lengthen considerably the duration of measurements. For routine clinical applicability of LDV, improvements in speed and automation of the LDV data recording are required. Rapidity of the LDV procedure minimizes cost and patient fatigue and maximizes the number of vessels that can be measured within a given time. Automated data collection and analysis procedures reduce the number of people needed to obtain the desired information and ensure standardization of the techniques among laboratories. Clearly with the eye-tracking mechanism of the CLBF, the majority of eye motion artifacts can be compensated [43], thus considerably decreasing the measurement time. Another approach that does not require stabilization of the laser beam is to use a computer algorithm that not only determines the cutoff frequencies but also identifies those times during which the laser beam is appropriately centered on the vessel and extracts only the Δf_{max} values recorded during this time, along the line described elsewhere [30].

Resolution and accuracy of Δf_{max} measurements could be improved by increasing the effective scattering angle. Both the slit lamp and the fundus camera could be modified for this purpose, but it would require increasing the frequency range of the detection system within the limits set by the pupil size. The quality of the DSPS can be markedly increased by using lasers in the near-infrared region of the spectrum at increased power [33]. These lasers are also much smaller, allowing a more compact laser delivery system.

7.2.13 Physiologic and Clinical Applications (Brief Overview)

Some of the findings obtained in normal volunteers have provided new insights into the physiology of the retinal circulation. For example, they demonstrated high temporal resolution recordings of the pulsatile time course of RBC velocity in retinal vessels [13, 17, 19, 24, 43, 44, 47] and established the relationship between V_{mean} and D. The relationship between Q and D [28] was found to be in good agreement with Murray's law [48], which stipulates that Q varies as D^3 for a vascular system that minimizes its resistance for a given blood volume.

The LDV technique is particularly suitable to investigate retinal blood flow regulation in response to various physiological stimuli, as revealed by Fig. 7.10. Thus, investigations were devoted to the effect of acute increases in mean ocular perfusion pressure (PP_m, the mean pressure driving blood through the eye) achieved using isometric exercises[49], decreases in PP_m induced by increasing the intraocular pressure [50], increases in arterial oxygen and carbon dioxide tension [51], as well as and light/dark transitions [33, 47, 52–54].

A number of LDV studies have led to a better understanding of the effect of diabetes on the retina. These have examined (1) the retinal circulatory changes during the natural history of diabetes [55, 56], (2) the effect of poor glycemic control on retinal hemodynamics [57], (3) the response of retinal blood flow to hyperoxia in patients with various degrees of retinopathy [58], and (4) the effect of various treatment modalities, such as panretinal laser photocoagulation [59, 60]. The most important LDV findings obtained in diabetic patients have been reviewed by Grunwald and Riva [61]. The effect of various antihypertensive medications on the retinal circulation and its autoregulation in normal volunteers and patients with ocular hypertension and primary open-angle glaucoma has been investigated [62–64].

7.3 Real-Time Laser Doppler Flowmetry in the Optic Nerve and Subfoveal Choroid

7.3.1 The DSPS for RBCs Moving in the Microvascular Bed of a Tissue

When a laser beam illuminates a tissue containing a network of microvessels having RBCs moving at various velocities in various directions, the light scattered by the RBCs consists of a summation of waves at different frequencies due to the numerous Doppler shifts undergone by the incident light. To understand the Doppler shifts of light scattered from the surface of a tissue, knowledge of the scattering of light by the tissue is necessary. Critical questions as discussed by Bonner and Nossal [65] are: (1) What are the characteristic scattering angles that determine the relation of Doppler shift to RBC speed? (2) What is the optical path length within the tissue? (3) What tissue volume is sampled by the detected light? (4) For a given tissue, do these parameters significantly vary from one location or individual to another?

Describing the relationship between the laser Doppler signals and the mean RBC speed, blood flow and concentration of RBCs in a tissue is beyond the scope of this chapter. The main features derived from the work of Bonner and Nossal [26, 65, 66] will be now summarized.

In a schematic diagram of light diffusing through a vascularized tissue (Fig. 7.11), photons penetrating into a tissue undergo many collisions with static tissue elements. Collisions with moving RBCs are considered to be only occasional. Furthermore, the network of microvessels is assumed to be random on a length scale defined by the mean distance between RBC scattering events. Consequently, it is assumed that the RBC velocities are also randomly distributed in direction. The specific assumptions of Bonner and Nossal's model are: (1) light is randomized in direction, independently of blood flow and blood volume; (2) the Doppler shifts due to blood flow are due to the scattering by RBCs; (3) the scattering angle by the RBCs is highly anisotropic (mean value ~6° in the forward direction); (4) the mean number of RBCs by which a typical detected photon is Doppler shifted increases linearly with the number density of RBCs; and (5) multiple scattering of light from more than one moving RBC broadens the DSPS.

Most of the light emerging from the tissue has been scattered predominantly by static structural components of the tissue. Thus, this light is nonshifted in frequency and acts as a reference signal that is mixed at the surface of the photodetector

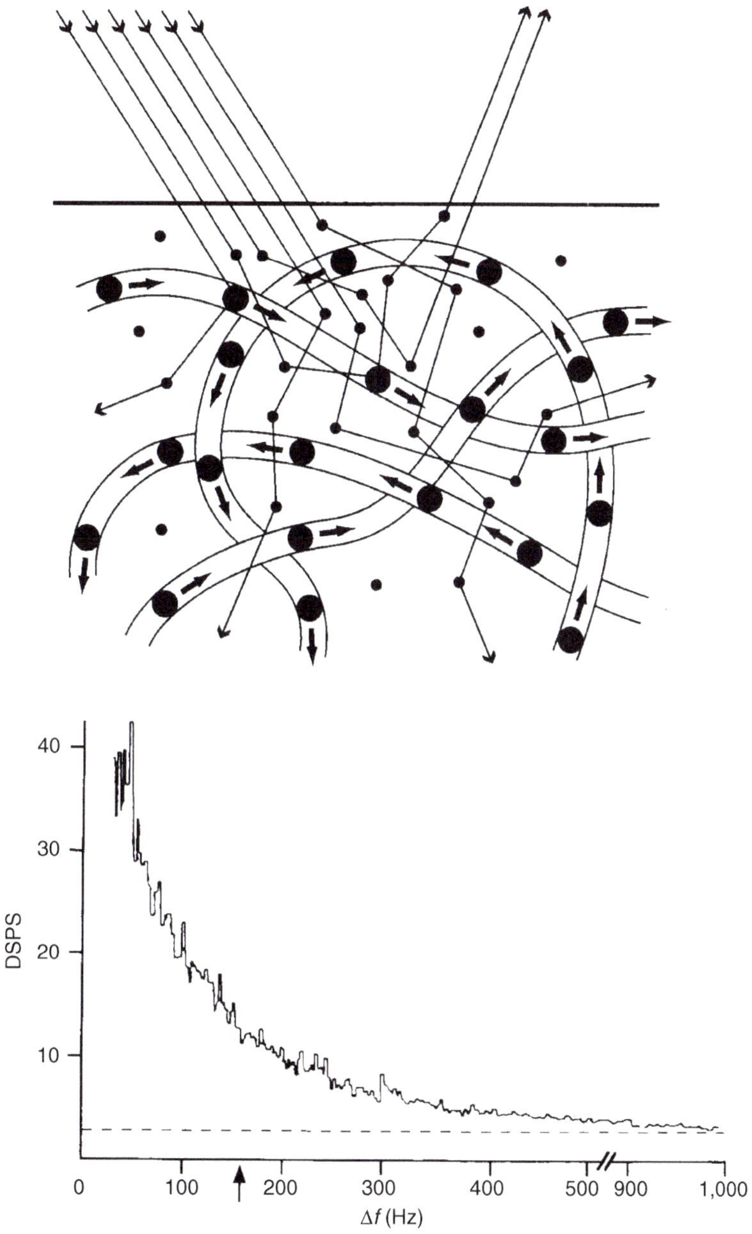

Fig. 7.11 *Top*: Schematic diagram according to Bonner and Nossal [65] shows laser light impinging on a tissue and leaving it in the direction of the detector, after having been scattered by static tissue structures (*small dots*) and RBCs moving in blood vessels (*large dots*); *Bottom*: DSPS from ONH tissue [6]. *Arrow* indicates the mean frequency, which is proportional to the mean velocity of the RBCs

with the scattered light shifted by the RBCs. Applying OMS (see Sect. 1.3), the detector output photocurrent contains only components oscillating at the Doppler shift frequencies. Since the detector does not discriminate between positive and negative frequency shifts, the DSPS only spans the positive frequencies (Fig. 7.11).

7.3.2 Hemodynamic Parameters Derived from the DSPS

In the application of LDF to the measurement of blood flow in the tissue of the eye fundus, such as the optic nerve head (ONH) or the subfoveal choroid, a probing laser beam is focused at the desired site, namely the rim of the optic disk or the fovea, respectively. The light scattered from the illuminated volume of tissue is collected and guided to a photodetector [5, 67].

Various means have been used to process the output photocurrent. Early studies in cats used the electronic systems of a commercial laser Doppler flowmeter (PeriFlux PF3, Perimed, Inc., Stockholm, Sweden or a TSI LaserFlo, Vasamedics, Minneapolis, USA) [6, 68]. More recent methods process the photocurrent using an algorithm developed for the NeXT computer [29]. Since this computer is no longer available, this algorithm is now implemented on a PC-based system [30].

The LDF parameters are calculated according to the procedure described by Bonner and Nossal [65]. These parameters include (1) the "velocity" (*Vel*) of the RBCs, which is a measure of the broadening of the DSPS. For a tissue where the blood volume represents only a small percentage of the tissue volume (this is the case for the ONH), *Vel* is proportional to the root mean square of the RBC velocities; (2) "volume" (*Vol*), the number of moving RBCs in the sampled volume, which is proportional to the area under the DSPS; and (3) *Flow*, which is proportional to the product *Vel* × *Vol*. *Flow* is usually referred to as "blood flow." However, what *Flow* actually measures is the *flux* of the RBCs. Blood flow is directly proportional to RBC flux only if the hematocrit remains constant during an experiment. Before the calculation of these parameters, the DSPS is corrected for the shot noise power present in the photocurrent.

The LDF parameter *Vel* is expressed in Hz. The parameters *Vol* and *Flow* are in arbitrary units. The inability of LDF to provide absolute flow measurements is due to the fact that laser radiation incident on a tissue undergoes scattering and absorption, both processes influencing the penetration pattern of the light. Penetration may differ from one region of a tissue to another, depending upon the optical properties of the tissue. Thus, spatial or temporal variations in tissue structure and vascularization, as is the case for example in the ONH in glaucoma, will affect the LDF parameters. Furthermore, direct comparison between the LDF parameters obtained from different tissues may not be valid due to variations in optical properties resulting from differences in tissue structure and composition. In addition, light scattering by the ocular media will also influence the LDF parameters, as demonstrated by scanning laser Doppler flowmetry performed in the presence of simulated light scatter induced by polystyrene microspheres placed in a cell in front of the eye of volunteers. This scatter induced an artifactual increase in the *Flow* in the retinal capillaries [69].

In addition to the LDF parameters, the mean of the photocurrent (so-called direct current, DC) is also recorded. The DC is proportional to the intensity of the light incident on the detector (shifted + non-shifted scattered light). For repeated measurements from the same site of the fundus, it is important to maintain this parameter constant (arbitrarily within 10% of the mean) to insure that the variations of the LDF parameters are not due to variations in the intensity of the probing beam at the fundus or changes of the measurement site. Constancy of the DC can be further improved by maintaining constant the location of the entrance of the probing beam at the pupil of the tested subject, particularly in the measurements of ONH blood flow [70].

7.3.3 Detection Scheme for Optic Nerve and Subfoveal Choroidal Blood Flow

Two LDF measurement modes have been implemented. The first is the continuous mode for on-line, continuous recording of the LDF parameters at a discrete site of the vascular beds of the ONH, subfoveal region of the choroid, or the iris. The second mode was based on a scanning technique, which provides a two-dimensional image of the RBC flux in the capillaries of the ONH and retina. This technique is no longer available. This chapter describes only the former one.

7.3.3.1 Real-Time Recording of the LDF Signal in the ONH and Subfoveal Choroid (SFCH)

With the fundus-camera-based LDF system, originally described by Riva et al., to record the LDF signal from the ONH and the SFCH [6, 9], a laser beam from a HeNe (623 nm) or diode laser

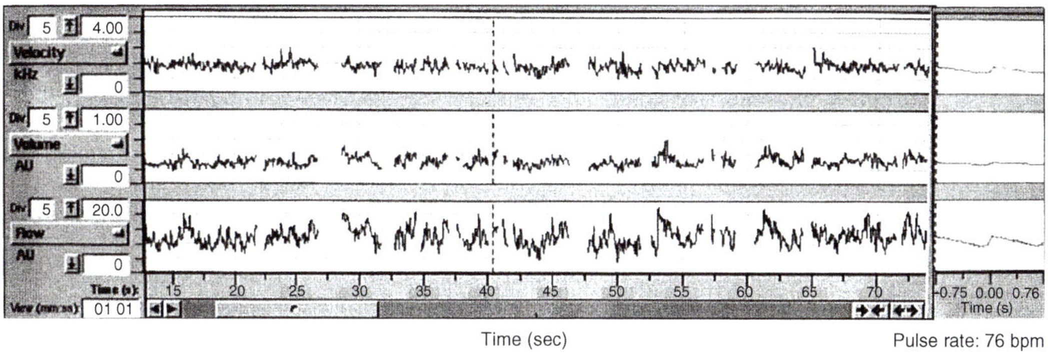

Fig. 7.12 NeXT computer output of LDF measurements in human optic nerve head (ONH) tissue. *Vel* (kHz), *Vol* (AU = arbitrary units), and *Flow* (AU) are shown as a function of time (21 samples/s). *Gaps* in data are caused by *blinks* during which recording is automatically disabled. Waveforms on the *right side* are the data (within a user-defined time period), averaged according to the heart pulse (i.e., each point is the average of all data recorded at the same delay in the heart cycle), showing the systolic and diastolic values of the flow parameters (Adapted from Petrig and Riva [71] with permission from the Publisher)

(670 nm) is delivered to a discrete site (about 150 μm at the disk or at the fovea) [8]. The scattered light is collected by an optical fiber with the image of its aperture (approximately 150 μm in diameter) focused onto the illuminated site. This fiber guides the scattered light to a photodetector. An area of the fundus (30° in diameter) is illuminated in red-free light for observation of the fundus and positioning of the laser beam. For some applications requiring measurements in darkness, a laser probing beam in near-infrared (between 750 and 805 nm) is used, and the fundus is also illuminated in near-infrared light (826 nm). Observation is achieved by means of a CCD camera and video monitor presentation [33, 70]. In addition, the position of entrance of the laser beam at the pupil can be monitored by means of a CCD finger camera [70]. For precise placement of the laser beam at the desired site in ONH blood flow measurements, a point-like target consisting of the aperture of a 50-μm-diameter optical fiber illuminated by a red or green diode can be focused in the retinal plane of the ophthalmoscopic lens and moved in this plane. For the investigation of the effect of increased retinal activity on ONH blood flow, a system for delivering visual stimuli (flicker and contrast reversal pattern) has been incorporated to the fundus camera-based instrument [70].

A NeXT computer system with dedicated software is used for LDF in the human eye [71]. This software allows averaging of the Doppler signal in phase with the heart cycle so that precise measurements of RBC flux variations during the systolic and diastolic phases can be obtained. Recordings of *Vel*, *Vol*, and *Flow* from the ONH of a normal volunteer are shown in Fig. 7.12.

More recently, a compact confocal laser Doppler flowmeter has been developed for the measurement of SFCH blood flow (Fig. 7.13). This instrument detects the light scattered by the RBCs within an annulus centered on the illumination site of the laser at the fovea [72]. A miniaturized version of this instrument has been mounted on a helmet, thus facilitating SFCH blood flow measurements during various types of dynamical physiological maneuvers, such as biking [73]. In both instruments, the illuminated laser spot and the detection annulus are self-aligned. The pupil of the optical system is small enough to allow SFCH blood flow measurement without pupil dilatation.

7.3.4 Critical Questions Regarding the Application of LDF to Ocular Blood Flow

7.3.4.1 LDF Sample Volume

A central question in the application of LDF to the ONH is the depth of the sampling volume. This depth determines the relative contribution to the Doppler signal from the superficial layers,

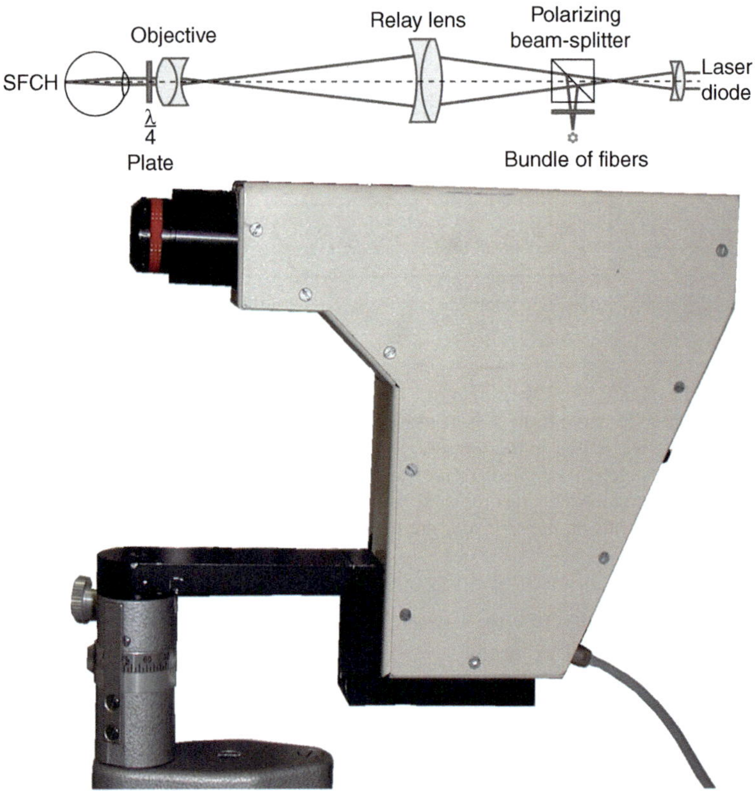

Fig. 7.13 Confocal laser Doppler flowmeter for subfoveal choroidal (*SFCH*) blood flow. *Top*: Diagram of the optics [72]. *Bottom*: Device mounted on a slit lamp support (Courtesy of Mr M. Geiser)

those supplied by the central retinal artery, and the deeper layers supplied by the posterior ciliary arteries. These two vascular beds may have different blood flow regulation properties. Furthermore, the deep layers of the ONH appear to be particularly susceptible to ischemic disorders, including glaucoma.

Investigations on a model system suggests that when the light collecting aperture coincides with the tissue volume illuminated by the probing laser, layers of the ONH tissue as deep as 300 μm contribute to the LDF signal [74]. In man, however, although the depth of tissue sampling in the ONH remains to be experimentally assessed, it appears that the LDF technique detects predominantly the motion of RBCs within the intraocular region of the ONH [75]. A study on monkey eyes concluded that LDF is predominantly sensitive to blood flow changes in the superficial layers of the ONH and less to those layers of the prelaminar and deeper regions of the ONH, and their relative proportions are still unknown [76]. The weaker signal from the deep layers cannot be separated from the dominant signal from the superficial layers to exclusively study the circulation in the deep layers.

7.3.4.2 Linearity of LDF

LDF does not provide absolute measurement of blood flow. Nevertheless, valid measurements of the changes in blood flow are obtained if *Flow* varies linearly with the actual blood flow. Linearity between *Flow* and actual blood flow has been documented for various tissues, such as the skin, skeletal muscle, cerebral cortex, nerves, and others [1].

To test this linearity for the ONH and SFCH, *Flow* was measured in both tissues of the cat eye while the mean ocular perfusion pressure was rapidly decreased (within a few seconds) by means of a lethal injection of pentobarbital. It was assumed that during this rapid decrease, the vascular resistance would remain constant so that the change in *Flow* would be only due to the change in this pressure. The measurements demonstrated for both tissues a highly significant linear ($p<0.01$) decrease in *Flow* as this pressure decreased to zero [6, 77].

7.3.4.3 Scattering Scheme

In conventional LDF in nonocular tissues, where the tissue is directly exposed, the laser beam is delivered through an optical fiber (input fiber), and the scattered light is collected by a second optical fiber (output fiber). Both fibers, which are in direct contact with the tissue, are separated by a distance r of 250 or 500 μm [6]. Separating the illumination and collection areas provides an optimal compromise in terms of signal-to-noise ratio and depth of the sampled volume [78]. Smaller r yields too shallow depth, and greater r decreases the signal-to-noise ratio by reducing the total scattered light impinging on the photodetector.

In ONH LDF, because of the presence of numerous retinal arterioles or venules close to each other in the disk tissue, such a laser delivery-scattered light detection scheme cannot be implemented, and the detecting fiber is placed on top of the directly illuminated volume [5, 79]. This scheme has the advantage of providing greater scattered light intensity, at the expense of the presence of specular reflection of the incident light and a reduction of the volume sampled by the laser light [6].

7.3.5 Reproducibility of LDF

The potential of LDF is mainly realized in the assessment of the changes in blood flow induced by physiological stimuli or by pathologies which do not alter the optical properties of the tissue. In this type of application, the minimum statistically significant change that can be determined in response to a given stimulus for a given population of subjects (sensitivity) depends mainly upon the quality of target fixation, the stimulus, the site of measurements, the technical experience, and other factors. The reproducibility of LDF for ONH blood flow has been appraised in humans by Joos et al. [80]. By averaging five measurements for each session, an intrasession variation of 18% for *Vel* and 24% for *Flow*, and an intersession variation of 12% for *Vel* and 32% for *Flow* were reported. Interexaminer variance was small. The measured variability includes components from both technical/measurement error and physiologic variation. Sample size estimates were computed for experimentally induced changes in *Flow* in single and multiple sessions: to detect a 15% difference in *Flow* with 80% power by means of a paired test, seven subjects would be needed to evaluate changes within one session, whereas 43 subjects would be needed to detect a change between two sessions. Therefore, LDF was found to be useful in evaluating ONH circulation in humans, particularly when acute perturbation experiments within a single session were assessed. Grunwald et al. [81] found that LDF is practical for assessing differences between patient populations. Comparing measurements in glaucoma subjects and controls and using power calculations based on an independent test with a pooled variance estimate, the statistical power to detect a 20% decrease in the glaucoma patients was 80% or greater for a sample size of 24 glaucoma subjects and 14 control subjects. Variability in terms of coefficient of variation (CV) of the ONH blood flow parameters measured at 3 measurement sites in normal controls (n =13) and glaucoma patients with (10) and without systemic hypertension (12) provided the following values: CV(*Vel*) = 16%, 17%, and 12%; CV(*Vol*) = 20%, 15%, and 22%; and CV(*Flow*) = 21%, 20% and 13%, respectively.

The reproducibility of ONH *Flow* responses to flicker was determined based on consecutive trials performed during sessions of less than 19 min duration [82]. The laser beam was aimed at the temporal rim of the ONH in the right eye.

Using two different stimuli, a pure red and a pure green illuminance flicker, the coefficient of variation of the *Flow* responses was 26% ($n=5$ trials) and 12% ($n=4$).

The CVs of SFCH *Vel*, *Vol*, and *Flow* were reported to be 14%, 25%, and 18%, respectively, in subjects with AMD, and 8%, 18%, and 13%, respectively, in normal controls. No statistically significant differences were observed between these CVs for subjects with AMD and those for normal controls [83]. Another study in eyes with AMD of increasing severity provided a CV(*Flow*) of 10.3% ± 7.2% ($n=3$ trials) [84].

With the confocal choroidal flowmeter (Fig. 7.13), the sensitivity of SFCH *Flow*, calculated based on measurements in 10 subjects, was between 7.5% and 17.5% for *Vel*, 14.5–37.8% for *Vol*, and 19.2–31.6% for the *Flow* [85]. Intrasubject variability was less than 8% for *Vel* and *Vol* and less than 18% for *Flow*, and the sensitivity of the measurements (21 subjects) was 3.3% for *Vel*, 7.1% for *Vol*, and 7.4% for *Flow* [86].

7.3.6 Applications of LDF

Continuous, real-time LDF because of its high spatial and temporal resolution appears most suitable for the investigation of the regulatory processes of blood flow in response to various physiological stimuli. The following investigations illustrate some of the applications. Regrettably, because of space limitation, it is not possible to mention all the important studies performed until now.

LDF investigations of ONH blood flow in humans include the responses to: (1) changes in ocular perfusion pressure (increase and decrease) [75, 87, 88]; (2) hyperoxia, breathing carbogen, and mixtures of O_2 and CO_2 [89]; and (3) increased neuronal activity [90].

They also include a number of clinical studies, such as ONH blood flow and its regulation in response to various stimuli in normal, ocular hypertensive, and glaucomatous eyes [91–97] and in children with cerebral malaria [98, 99].

LDF measurements of subfoveal choroidal blood flow in humans are recent and include studies of the effect of (1) increases and decreases of ocular perfusion pressure [100–104]; (2) Valsalva maneuvers [105]; (3) breathing of various gas mixtures (pure O_2, various mixtures of O_2 and CO_2) [86]; and (4) the effect of light [106–108]. Investigations of the effect of aging, age-related macular degeneration, and choroidal neovascularization have been reported [83, 84, 109–111].

7.4 Summary for the Clinician

The LDV/LDF techniques have been applied over a number of years in normal subjects and patients with various ocular and systemic diseases. They provide a noninvasive means of investigating changes in blood flow in the retina, ONH, and subfoveal choroid. Both techniques have a high sensitivity and a temporal response fast enough to reveal the changes in blood flow during the cardiac cycle and in response to acute changes in various physiological stimuli. Their limitations include dilatation of the pupil, particularly for bidirectional LDV, good subject collaboration in terms of target fixation, and head stability as well as clear media. These techniques open new avenues in the investigation of the regulation of blood flow in the vascular tissues of the eye fundus in response to various physiological stimuli and the effect of a number of treatments on the ocular circulation.

Acknowledgement The author thanks the "Fondazione Cassa di Risparmio in Bologna, Italy" for its generous financial support.

References

1. Shepherd AP (1990) History of laser-Doppler blood flowmetry. In: Shepherd AP, Öberg PA (eds) Laser-Doppler blood flowmetry. Kluwer Academic Publishers, Boston, pp 1–16
2. Riva CE, Ross B, Benedek GB (1972) Laser Doppler measurements of blood flow in capillary tubes and retinal arteries. Invest Ophthalmol Vis Sci 11:936–944

3. Tanaka T, Riva CE, Ben-Sira I (1974) Blood velocity measurements in human retinal vessels. Science 186:830–831
4. Stern MD (1975) In vivo evaluation of microcirculation by coherent light scattering. Nature 254:56–58
5. Riva CE, Grunwald JE, Sinclair SH (1982) Laser Doppler measurement of relative blood velocity in the human optic nerve head. Invest Ophthalmol Vis Sci 22:241–248
6. Riva CE, Harino S, Petrig BL, Shonat RD (1992) Laser Doppler flowmetry in the optic nerve. Exp Eye Res 55:499–506
7. Riva CE, Falsini B, Geiser MH, Petrig BL (1999) Optice nerve head blood flow response to flicker: characteristics and possible mechanisms. In: Pillunat LE, Harris A, Anderson DR, Greve EL (eds) Current concepts on ocular blood flow in glaucoma. Kugler Publications, The Hague, pp 191–196
8. Petrig BL, Riva CE (1999) Laser Doppler flowmetry in the optic nerve head. Principle and technique. In: Pillunat LE, Harris A, Anderson DR, Greve EL (eds) Current concepts on ocular blood flow in glaucoma. Kugler Publications, The Hague, pp 171–182
9. Riva CE, Cranstoun SD, Grunwald JE, Petrig BL (1994) Choroidal blood flow in the foveal region of the human ocular fundus. Invest Ophthalmol Vis Sci 35:4273–4281
10. Michelson G, Schmauss B, Langhans MJ, Harazny J, Groh MJM (1996) Principle, validity, and reliability of scanning laser Doppler flowmetry. J Glaucoma 5:99–105
11. Riva CE, Petrig BL (1990) Retinal blood flow: laser Doppler velocimetry and blue field simulation technique. In: Masters BR (ed) Noninvasive diagnostic techniques in ophthalmology. Springer, New York, pp 390–409
12. Riva CE, Petrig BL (2003) Laser Doppler techniques in ophthalmology – Principles and applications. In: Fankhauser F, Kwasniewska S (eds) Lasers in ophthalmology – Basic, diagnostic and surgical aspects. Kugler Publications, The Hague, pp 51–59
13. Riva CE, Feke GT (1981) Laser Doppler velocimetry in the measurement of retinal blood flow. In: Goldman L (ed) The biomedical laser technology and clinical applications. Springer, New York, pp 135–161
14. Riva CE, Petrig BL, Grunwald JE (1989) Retinal blood flow. In: Shepherd AP, Öberg PA (eds) Laser-Doppler blood flowmetry. Kluwer Academic Publishers, Boston, pp 349–383
15. Benedek GB (1969) Optical mixing spectroscopy with applications to problems in physics, chemistry, biology and engineering Polarization, Matter and Radiation (The Jubilee Volume in Honor of Alfred Kastler). Presses Universitaires de France, Paris, pp 49–84
16. Cummins HZ, Swinney HL (1970) Light beating spectroscopy. In: Wolf E (ed) Progress in optics. North-Holland Publishing Company, Amsterdam, pp 133–200
17. Riva CE, Feke GT, Eberli B, Benary V (1979) Bidirectional LDV system for absolute measurement of blood speed in retinal vessels. Appl Opt 18:2301–2306
18. Riva CE, Grunwald JE, Sinclair SH, O'Keefe K (1981) Fundus camera based retinal LDV. Appl Opt 20:117–120
19. Feke GT, Goger DG, Tagawa H, Delori FC (1987) Laser Doppler technique for absolute measurement of blood speed in retinal vessels. IEEE Trans Biomed Eng BME-34:673–680
20. Garcia JPS Jr, Garcia PT, Rosen RB (2002) Retinal blood flow in the normal human eye using the canon laser blood flowmeter. Ophthalmic Res 34:295–299
21. Petrig BL, Follonier L (2005) Ray tracing model for the estimation of the power spectral properties in laser Doppler velocimetry of retinal vessels and its potential application to retinal vessel oximetry. Opt Express 13:10642–10651
22. Riva CE, Shonat RD, Petrig BL, Pournaras CJ (1989) Scattering process in LDV from retinal vessels. Appl Opt 28:1078–1083
23. Van de Hulst HC (1957) Light scattering by small particles. Wiley, New York
24. Feke GT, Riva CE (1978) Laser Doppler measurements of blood velocity in human retinal vessels. J Opt Soc Am 68:526–531
25. Riva CE, Grunwald JE, Petrig BL (1985) Laser Doppler measurement of retinal blood velocity: validity of the single scattering model. Appl Opt 24:605–607
26. Bonner R, Nossal R (1981) Model for laser Doppler measurements of blood flow in tissue. Appl Opt 20:2097–2107
27. Stern MD (1985) Laser Doppler velocimetry in blood and multiply scattering fluids: theory. Appl Opt 24: 1968–1983
28. Riva CE, Grunwald JE, Sinclair SH, Petrig BL (1985) Blood velocity and volumetric flow rate in human retinal vessels. Invest Ophthalmol Vis Sci 26: 1124–1132
29. Petrig BL, Riva CE (1994) New continuous real-time analysis system for laser Doppler flowmetry and velocimetry in the ocular fundus using a digital processor. In: Vision science and its application technical digest series. Optical Society of America, Washington, DC, pp 238–241
30. Petrig BL, Gehrig JP, Pompili P (2000) New multichannel DSP-based laser Doppler flowmetry analysis system for quantification of ocular blood flow. Proc SPIE 4156:318–327
31. Yoshida A, Feke GT, Mori F et al (2003) Reproducibility and clinical application of a newly developed stabilized retinal laser Doppler instrument. Am J Ophthalmol 135:356–361
32. Logean E, Geiser MH, Petrig BL, Riva CE (1997) Portable ocular laser Doppler red blood cell velocimeter. Rev Sci Instrum 68:2878–2882
33. Riva CE, Petrig BL, Grunwald JE (1987) Near-infrared retinal laser Doppler velocimetry. Lasers Ophthalmol 1:211–215

34. Yoshida A, Feke GT, Feke GD, McMeel JW (1998) A new laser Doppler system for examining optic nerve head circulation. J Biomed Opt 3:396–400
35. Koller A, Kaley G (1991) Endothelial regulation of wall shear stress and blood flow in skeletal muscle microcirculation. Am J Physiol 260:H862–H868
36. Nerem RM, Alexander RW, Chappell DC, Medford RM, Varner SE, Taylor WR (1998) The study of the influence of flow on vascular endothelial biology. Am J Med Sci 316:169–175
37. Yazdanfar S, Rollins AM, Izatt JA (2000) Imaging and velocimetry of the human retinal circulation with color Doppler optical coherence tomography. Opt Lett 25:1448–1450
38. Leitgeb RA, Schmetterer L, Drexler W, Fercher AF (2003) Real-time assessment of retinal blood flow with ultrafast acquisition by color Doppler Fourier domain optical coherence tomography. Opt Express 11:3116–3121
39. Ferguson RD, Hammer DX, Elsner AE, Webb RH, Burns SA, Weiter JJ (2004) Wide-field retinal hemodynamic imaging with the tracking scanning laser ophthalmoscope. Opt Express 12:5198–5208
40. Bishop JJ, Nance PR, Popel AS, Intaglietta M, Johnson PC (2001) Effect of erythrocyte aggregation on velocity profiles in venules. Am J Physiol Heart Circ Physiol 280:H222–H236
41. Logean E, Schmetterer L, Riva CE (2003) Velocity profile of red blood cells in human retinal vessels using confocal scanning laser Doppler velocimetry. Laser Phys 13:45–51
42. Feke GT, Tagawa H, Deupree DM, Goger DG, Sebag J, Weiter JJ (1989) Blood flow in the normal human retina. Invest Ophthalmol Vis Sci 30:58–65
43. Guan K, Hudson C, Flanagan JG (2003) Variability and repeatability of retinal blood flow measurements using the canon Laser blood flowmeter. Microvasc Res 65:145–151
44. Petrig BL, Riva CE (1988) Retinal laser-Doppler velocimetry: towards its computer assisted clinical use. Appl Opt 27:1126–1134
45. Delori FC, Pomerantzeff O, Mainster MA (1980) Light levels in ophthalmic diagnostic instruments. Proc SPIE 229:154–160
46. Delori FC, Webb RH, Sliney DH (2007) Maximum permissible exposures for ocular safety (ANSI 2000), with emphasis on ophthalmic devices. J Opt Soc Am 24:1250–1265
47. Garhöfer G, Zawinka C, Resch H, Huemer KH, Dorner GT, Schmetterer L (2004) Diffuse luminance flicker increase blood flow in major retinal arteries and veins. Vision Res 44:833–838
48. Murray CD (1926) The physiological principle of minimum work. I. The vascular system and the cost of blood volume. Proc Natl Acad Sci U S A 12: 207–214
49. Robinson F, Riva CE, Grunwald JE, Petrig BL, Sinclair SH (1986) Retinal blood flow autoregulation in response to an acute increase in blood pressure. Invest Ophthalmol Vis Sci 27:722–726
50. Riva CE, Sinclair SH, Grunwald JE (1981) Autoregulation of retinal circulation in response to decrease of perfusion pressure. Invest Ophthalmol Vis Sci 21:34–38
51. Pakola SJ, Grunwald JE (1993) Effects of oxygen and carbon dioxide on human retinal circulation. Invest Ophthalmol Vis Sci 34:2866–2870
52. Riva CE, Grunwald JE, Petrig BL (1983) Reactivity of the human retinal circulation to darkness: a laser Doppler velocimetry study. Invest Ophthalmol Vis Sci 24:737–740
53. Feke GT, Zuckerman R, Green GJ, Weiter JJ (1983) Responses of human retinal blood flow to light and dark. Invest Ophthalmol Vis Sci 24:136–141
54. Riva CE, Logean E, Petrig BL, Falsini B (2000) Effet de l'adaptation à l'obscurité sur le flux rétinien. Klin Monatsbl Augenheilkd 216:309–310
55. Feke GT, Tagawa H, Yoshida A et al (1985) Retinal circulatory changes related to retinopathy progression in insulin-dependent diabetes mellitus. Ophthalmology 92:1517–1522
56. Grunwald JE, Riva CE, Sinclair SH, Brucker AJ, Petrig BL (1986) Laser Doppler velocimetry study of retinal circulation in diabetes mellitus. Arch Ophthalmol 104:991–996
57. Grunwald JE, Riva CE, Baine J, Brucker AJ (1992) Total retinal volumetric blood flow rate in diabetic patients with poor glycemic control. Invest Ophthalmol Vis Sci 33:356–363
58. Grunwald JE, Riva CE, Brucker AJ, Sinclair SH, Petrig BL (1984) Altered retinal vascular response to 100% oxygen breathing in diabetes mellitus. Ophthalmology 91:1447–1452
59. Feke GT, Green GL, Goger DG, McMeel JW (1982) Laser Doppler measurements of the effect of panretinal photocoagulation on retinal blood flow. Ophthalmology 89:757–762
60. Grunwald JE, Riva CE, Brucker AJ, Sinclair SH, Petrig BL (1986) Effect of panretinal photocoagulation on retinal blood flow in proliferative diabetic retinopathy. Ophthalmology 93:590–595
61. Grunwald JE, Riva CE, Belcaro GV, Hoffmann U, Bollinger A, Nicolaides AN (1994) Retinal blood flow in diabetes Laser Doppler. Med-Orion Publishing Company, London, pp 223–247
62. Grunwald JE (1991) Effect of two weeks of timolol maleate treatment on the normal retinal circulation. Invest Ophthalmol Vis Sci 32:39–45
63. Grunwald JE, Delehanty J (1992) Effect of topical carteolol on the normal human retinal circulation. Invest Ophthalmol Vis Sci 33:1853–1856
64. Grunwald JE, Mathur S, Dupont J (1997) Effects of dorzolamide hydrochloride 2% on the retinal circulation. Acta Ophthalmol Scand 75:236–238
65. Bonner RF, Nossal R (1990) Principles of laser Doppler flowmetry. In: Shepherd AP, Öberg PA (eds) Laser-Doppler blood flowmetry. Kluwer Academic Publishers, Boston, pp 57–72
66. Bonner RF, Nossal R, Havlin S, Weiss GH (1987) Model for photon migration in turbid biological media. J Opt Soc Am A 4:423–432

67. Petrig BL, Riva CE (1991) Near-infrared retinal laser Doppler velocimetry and flowmetry: new delivery and detection techniques. Appl Opt 30:2073–2078
68. Riva CE, Harino S, Shonat RD, Petrig BL (1991) Flicker evoked increase in optic nerve head blood flow in anesthetized cats. Neurosci Lett 128:291–296
69. Venkataraman ST, Hudson C, Harvey E, Flanagan JG (2005) Impact of simulated light scatter on scanning laser Doppler flowmetry. Br J Ophthalmol 89: 1192–1195
70. Logean E, Geiser MH, Riva CE (2005) Laser Doppler instrument to investigate retinal neural activity-induced changes in optic nerve head blood flow. Opt Lasers Eng 43:591–602
71. Petrig BL, Riva CE (1996) Optic nerve head laser Doppler flowmetry: principles and computer analysis. In: Kaiser HJ, Flammer J, Hendrickson P (eds) Ocular blood flow. Karger, Basel, pp 120–127
72. Geiser MH, Diermann U, Riva CE (1999) Compact laser Doppler choroidal flowmeter. J Biomed Opt 4:459–464
73. Geiser MH, Moret F, Riva CE (2001) Helmet-mounted Choroidal Laser Doppler flowmeter. Proc SPIE 4263:91–97
74. Koelle JS, Riva CE, Petrig BL, Cranstoun SD (1993) Depth of tissue sampling in the optic nerve head using laser Doppler flowmetry. Lasers Med Sci 8:49–54
75. Riva CE, Hero M, Titzé P, Petrig BL (1997) Autoregulation of human optic nerve head blood flow in response to acute changes in ocular perfusion pressure. Graefes Arch Clin Exp Ophthalmol 235: 618–626
76. Petrig BL, Riva CE, Hayreh SS (1999) Laser Doppler flowmetry and optic nerve head blood flow. Am J Ophthalmol 127:413–425
77. Riva CE, Cranstoun SD, Mann RM, Barnes GE (1994) Local choroidal blood flow in the cat by laser Doppler flowmetry. Invest Ophthalmol Vis Sci 35:608–618
78. Nilsson GE (1990) Perimed's LDV Flowmeter. In: Shepherd AP, Öberg PÅ (eds) Laser-Doppler blood flowmetry. Kluwer Academic Publishers, Boston, pp 57–72
79. Sebag J, Delori FC, Feke GT et al (1986) Anterior optic nerve blood flow decreases in clinical neurogenic optic atrophy. Ophthalmology 93(6):858–865
80. Joos KM, Pillunat LE, Knighton KW, Anderson DR, Feuer WJ (1997) Reproducibility of laser Doppler flowmetry in the human optic nerve head. J Glaucoma 6:212–216
81. Grunwald JE, Piltz J, Hariprasad SM, Dupont J, Maguire MG (1999) Optic nerve blood flow in glaucoma: effect of systemic hypertension. Am J Ophthalmol 127:516–522
82. Riva CE, Falsini B, Logean E (2001) Flicker-evoked responses of human optic nerve head blood flow: luminance versus chromatic modulation. Invest Ophthalmol Vis Sci 42:756–762
83. Grunwald JE, Hariprasad SM, Dupont J et al (1998) Foveolar choroidal blood flow in age-related macular degeneration. Invest Ophthalmol Vis Sci 39:385–390
84. Grunwald JE, Metelitsina TI, DuPont JC, Ying GS, Maguire MG (2005) Reduced foveolar choroidal blood flow in eyes with increasing AMD severity. Invest Ophthalmol Vis Sci 46:1033–1038
85. Gugleta K, Orgül S, Flammer I, Gherghel D, Flammer J (2002) Reliability of confocal choroidal laser Doppler flowmetry. Invest Ophthalmol Vis Sci 43:723–728
86. Geiser MH, Riva CE, Dorner GT, Diermann U, Luksch A, Schmetterer L (2000) Response of choroidal blood flow in the foveal region to hyperoxia and hyperoxia-hypercapnia. Curr Eye Res 21:669–676
87. Pillunat LE, Anderson DR, Knighton RW, Joos KM, Feuer WJ (1997) Autoregulation of human optic nerve head circulation in response to increased intraocular pressure. Exp Eye Res 64:737–744
88. Movaffaghy A, Chamot SR, Petrig BL, Riva CE (1998) Blood flow in the human optic nerve head during isometric exercise. Exp Eye Res 67:561–568
89. Harris A, Anderson DR, Pillunat L et al (1996) Laser Doppler flowmetry measurement of changes in human optic nerve head blood flow in response to blood gas perturbations. J Glaucoma 5:258–265
90. Riva CE, Logean E, Falsini B (2005) Visually evoked hemodynamical response and assessment of neurovascular coupling in the optic nerve and retina. Prog Retin Eye Res 24:183–215
91. Schmetterer L, Wolzt M, Lexer F et al (1995) The Effect of hyperoxia and hypercapnia on fundus pulsations in the macular and optic disc region in healthy young men. Exp Eye Res 61:685–690
92. Grunwald JE, Piltz J, Hariprasad SM, Dupont J (1998) Optic nerve and choroidal circulation in glaucoma. Invest Ophthalmol Vis Sci 39:2329–2336
93. Piltz-Seymour JR (1999) Laser Doppler flowmetry of the optic nerve head in glaucoma. Surv Ophthalmol 43:S191–S198
94. Piltz-Seymour JR, Grunwald JE, Hariprasad SM, Dupont J (2001) Optic nerve blood flow is diminished in eyes of primary open-angle glaucoma suspects. Am J Ophthalmol 132:63–69
95. Pournaras CJ, Riva CE (2001) Studies of the hemodynamics of the optic head nerve using laser Doppler flowmetry. J Fr Ophtalmol 24:199–205
96. Riva CE, Salgarello T, Logean E, Colotto A, Galan E, Falsini B (2004) Flicker-evoked response measured at the optic disk rim is reduced in ocular hypertension and early glaucoma. Invest Ophthalmol Vis Sci 45:3662–3668
97. Pournaras CJ, Riva CE, Bresson-Dumont H, de Gottrau P, Bechetoille A (2004) Regulation of optic nerve head blood flow in normal tension glaucoma patients. Eur J Ophthalmol 14:226–235
98. Movaffaghy A, Lochhead J, Riva CE et al (2002) Feasibility of LDF measurements of optic nerve head blood flow in children with cerebral malaria. Microvasc Res 64:247–253
99. Beare NAV, Riva CE, Taylor TE et al (2006) Changes in optic nerve head blood flow in children with cerebral malaria and acute papilloedema. J Neurol Neurosurg Psychiatry 77:1288–1290

100. Riva CE, Titzé P, Hero M, Movaffaghy A (1997) Choroidal blood flow during isometric exercises. Invest Ophthalmol Vis Sci 38:2338–2343
101. Riva CE, Titzé P, Petrig BL (1997) Effect of acute decrease of perfusion pressure on choroidal blood flow in humans. Invest Ophthalmol Vis Sci 38:1752–1760
102. Fuchsjäger-Mayrl G, Luksch A, Malec M, Polska E, Wolzt M, Schmetterer L (2003) Role of endothelin-1 in choroidal blood flow regulation during isometric exercice in healthy humans. Invest Ophthalmol Vis Sci 44:728–733
103. Fuchsjäger-Mayrl G, Kally B, Georgopoulos M et al (2004) Ocular blood flow and systemic blood pressure in patients with primary open-angle glaucoma and ocular hypertension. Invest Ophthalmol Vis Sci 45:834–839
104. Longo A, Geiser MH, Riva CE (2004) Posture changes and subfoveal choroidal blood flow. Invest Ophthalmol Vis Sci 45:546–551
105. Riva CE, Petrig BL (1995) Choroidal blood flow by laser Doppler flowmetry. Opt Eng 34(3):746–752
106. Longo A, Geiser MH, Riva CE (2000) Subfoveal choroidal blood flow in response to light-dark exposure. Invest Ophthalmol Vis Sci 41:2678–2683
107. Fuchsjäger-Mayrl G, Polska E, Malec M, Schmetterer L (2001) Unilateral light-dark transitions affect choroidal blood flow in both eyes. Vision Res 41:2919–2924
108. Fuchsjäger-Mayrl G, Malec M, Amoako-Mensah T, Kolodjaschna J, Schmetterer L (2003) Changes in choroidal blood flow during light/dark transitions are not altered by atropine or propranolol in healthy subjects. Vision Res 43:2185–2190
109. Grunwald JE, Hariprasad SM, Dupont J (1998) Effect of aging on foveolar choroidal circulation. Arch Ophthalmol 116:150–154
110. Pournaras CJ, Logean E, Riva CE (2004) Choroidal circulatory failure in AMD. Ophthalmic Res 36:201
111. Pournaras CJ, Logean E, Riva CE et al (2006) Regulation of subfoveal choroidal blood flow in age-related macular degeneration. Invest Ophthalmol Vis Sci 47:1581–1586

Color Doppler Imaging

8

Ingeborg Stalmans, Selim Orgül,
and Leopold Schmetterer

Core Messages

- The CDI technique is used to measure blood velocities in the retrobulbar arteries: Ophthalmic artery, central retinal artery, nasal and temporal posterior ciliary arteries.
- Widely used variables are peak-systolic, end-diastolic and mean flow velocities, as well as resistance index.
- The advantages of the CDI technique are: (1) not dependent on optic media, (2) non-invasive, (3) good reproducibility with an experienced user.
- It can be used in a number of vascular-related ocular pathologies, from glaucoma to central vein occlusion.
- Caution should be used when comparing results obtained with different ultrasound machines/probes, as data may not be interchangeable.
- Velocity should not be read as flow, as this correlation implies vessel diameter. Such variable has not been reproducible using the current technology.

8.1 Principles

Color Doppler imaging (CDI) is a technique that combines B-scan ultrasonography for the conventional imaging of tissue with velocity extraction based on the acoustic Doppler effect. In the eye, this technique can be employed for the visualization of blood velocities in the retrobulbar vessels including the ophthalmic artery (OA), the posterior ciliary arteries (PCAs), and the central retinal artery (CRA). With CDI, ultrasound waves with frequencies of several MHz are used. Analysing the time elapsed between emission and return can lead to a quantification of the reflecting structure depth. B-scan ultrasound can then be used to produce gray-scale images of the structures of the human eye. If a structure within the scattering volume of the ultrasound is moving, a Doppler shift is induced (Fig. 8.1). If the reflecting object is moving toward the transducer, the frequency of the returning sound wave is increased as compared to the emitted sound wave. If the reflecting object is, however, moving away from the ultrasound probe, the frequency of

I. Stalmans, M.D. Ph.D. (✉)
Department of Ophthalmology, University Hospitals Leuven, Kapucijnenvoer 33, Leuven 3000, Belgium
e-mail: ingeborg.stalmans@uz.kuleuven.ac.be

S. Orgül, M.D.
Department of ophthalmology,
University of Basel Eye Clinic,
Mittlere-Strasse 91, Basel CH-4031, Switzerland
e-mail: sorguel@uhbs.ch

L. Schmetterer, Ph.D.
Department of Clinical Pharmacology, Center of Medical Physics and Biomedical Engineering, Medical University of Vienna, Weahringer Guertel 18-20, Vienna A-1090, Austria
e-mail: leopold.schmetterer@meduniwien.ac.at

L. Schmetterer, J.W. Kiel (eds.), *Ocular Blood Flow*,
DOI 10.1007/978-3-540-69469-4_8, © Springer-Verlag Berlin Heidelberg 2012

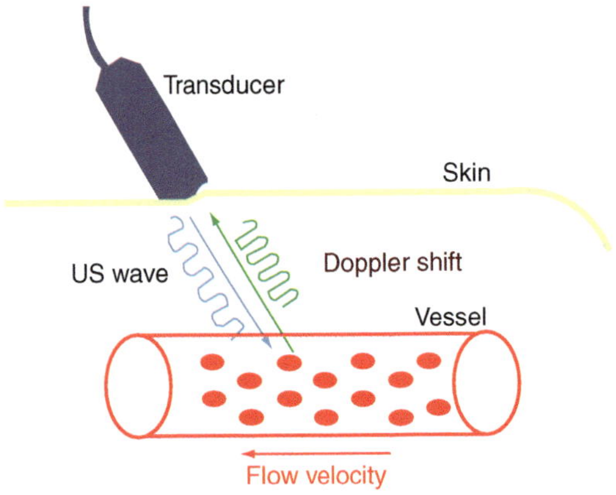

Fig. 8.1 Doppler principles

the returning sound wave is smaller than that of the emitted sound wave. In tissue, the main source of Doppler shifting is related to the movement of red blood cells within the tissue. Hence, blood flow in retrobulbar vessels can be visualized, and blood velocity can be extracted. The Doppler shift Δf depends on the velocity of the moving erythrocytes (v), the frequency of the incoming ultrasound waves f_0, the velocity of the ultrasound wave within the tissue (C), and the Doppler angle. The velocity of the blood can be calculated as

$$v = \Delta f.C / (2 \times f_{0.}\cos\theta)$$

From this equation, it becomes clear that the highest Doppler shift is induced when the transducer is parallel to the vessel and the angle $q=0$ because in this case cos $q=1$. When the vessel is, however, perpendicular to the incoming sound wave, no Doppler shift is detected because cos $q=0$. In practice, measurement of Doppler frequencies are usually done at angles between 30° and 60° [36]. When the angle q is close to 0°, problems arise from total reflection of the sound wave at the vessel wall. When the angle q is, however, close to 90°, the frequency shift becomes largely dependent on the Doppler angle. In this situation, only small errors in measuring the Doppler angle may induce large errors in the frequency.

In ultrasound, there is an inverse relation between penetration depth and resolution. The higher the frequency of the ultrasound, the higher the resolution within the tissue. Since the amount of attenuation per unit distance is also increasing with tissue, higher frequencies are, however, associated with lower penetration depths. Given that the retrobulbar vessels are located several centimeters behind the front surface of the eye, relatively high penetration depth is required, associated with limited resolution.

8.2 Instrumentation

Nowadays, commercial CDI machines use pulsed-wave Doppler. Pulsed-wave ultrasound probes emit series of pulses. These sound pulses are transmitted in the tissue, and the time until the reflected pulse wave is detected by the same ultrasound probe is measured. If a moving structure is present, there is not only a frequency shift in the returning ultrasound wave according to the Doppler effect but also a relative phase shift. Normally, the latter effect is used to extract the velocity data. The maximum velocity that can be detected with a pulsed Doppler probe depends on the pulse repetition rate, the vessel depth, the transmitter frequency, and the Doppler angle. If the velocity is too high, aliasing may occur, resulting in erroneous velocity extraction, but given that retrobulbar vessel is relatively small, this is not a limitation in ophthalmic use of CDI.

Since the delay of the incoming and the reflected wave is related to the depth of the reflecting interface, velocities can be measured at different depths within tissues. Combing pulsed-wave Doppler with B-scan imaging is called Duplex imaging. This technique of overlapping the Doppler signal with the gray-scale reflectance image allows for anatomical allocation of the velocity information. The direction of flow is normally translated into a color scale on the ultrasound image (CDI). With this technique, the colored flow information is visible in parallel to the gray-scale reflection image. Usually, flow toward the probe is depicted in red and is arterial, while flow away from the probe is depicted in blue and is venous. This means that blood flowing from the heart shows up in red, whereas blood flowing toward the heart shows up in blue.

Since CDI allows for relatively high time resolution [50], the blood velocity can be displayed as a function of time. Given the pulsatile nature of blood velocity in retrobulbar vessels, the systolic and the diastolic parts of the velocity signal can easily be identified.

CDI uses a linear array transducer consisting of linearly arranged, sequentially excited piezoelectric elements. As mentioned above, the frequency of the Doppler probe is chosen as a compromise between resolution and penetration depth. A typical transducer for retrobulbar CDI has a frequency of 7.5 MHz, but some investigators have used up to 12.5 MHz, thereby providing better resolution but also weaker Doppler signals.

8.3 Procedure

When the patient is examined in a lying position, legs should be uncrossed to avoid influences on venous return. The patient is instructed to look straight while the eyelids are closed. The examiner is seated behind the head of the patient while the base of the examiner's hand rests on the patient's forehead, with a finger is placed on the patient's cheek (Fig. 8.2). The tip of the probe is covered with a sufficient amount of acoustic coupling gel to provide adequate contact between the probe and the skin. The probe needs to be gently positioned on the closed upper eyelid in order to avoid mechanical force on the eyeball. This may increase intraocular pressure and thereby perfusion pressure, leading to a change in perfusion pressure as discussed later in this chapter.

The anatomy of the eye and the optic nerve head are identified using the gray scale images in the B-scan mode (Fig 8.3a). Color Doppler is used to visualize the flow within the vessels and allows for identification of the appropriate vessels (Fig 8.3b). The sample volume is placed in the center of the vessel, and the angle is set parallel to the vessel to account for the Doppler angle (Fig 8.3c and d).

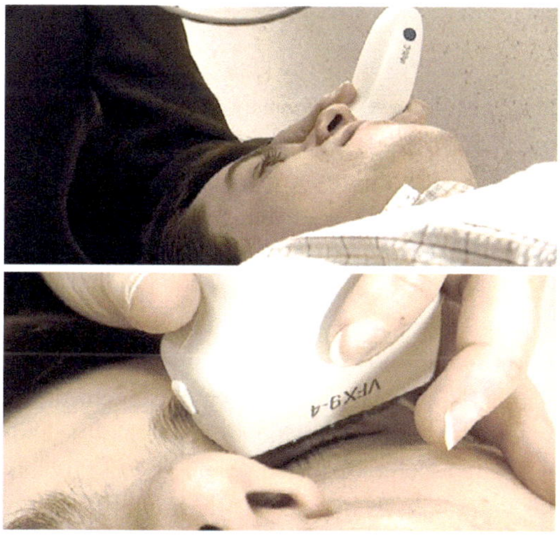

Fig. 8.2 Positioning

In order to obtain reliable and reproducible measurements using CDI, it is important to have a thorough knowledge of the retrobulbar vascular anatomy, as well as the characteristic waveforms of the different vessels, and the specific locations that are conventionally chosen for measurement [51] (Fig. 8.4). The central retinal artery (CRA) and its corresponding vein lie close together in the middle of the optic nerve and cannot be measured separately by CDI. Therefore, a double waveform is obtained with a distinct pulsatile arterial waveform above the zero line and a gentle sinusoidal venous variation below the zero line (Fig 8.5a). The nasal and temporal short posterior ciliary arteries (NPCA and TPCA) are located on both sides of the optic nerve and should be measured at a position that is close to the optic nerve and as anterior as possible without receiving interference from the choroid. It is important to realize that individual short posterior ciliary vessels cannot be distinguished by CDI. Therefore, the obtained waveform represents the mass effect produced by a bundle of vessels rather than from individual ciliary vessels. These arteries produce a more uniform arterial pulse without a venous wave (Fig 8.5b). The ophthalmic artery (OA) is situated deeper in the orbit, and by convention should be measured on the nasal side of the optic nerve, immediately after it crosses the optic nerve (Fig 8.5c).

8.4 Outcome Variables

Usually peak systolic and end diastolic velocities (PSV, EDV) are extracted from the time slope of the Doppler shift. Since the time course of the velocity within a vessel may also carry information on the vascular resistance distal to these vascular beds, a resistance index ($RI = (PSV - EDV)/PSV$) is calculated to quantify vascular resistance (Fig. 8.6). In addition, mean flow velocity (MFV) is often calculated as the time mean of the spectral outline over a heart cycle. Data are measured separately for the CRA, the nasal PCA, the temporal PCA, and the OA. As mentioned above, individual short posterior ciliary vessels cannot

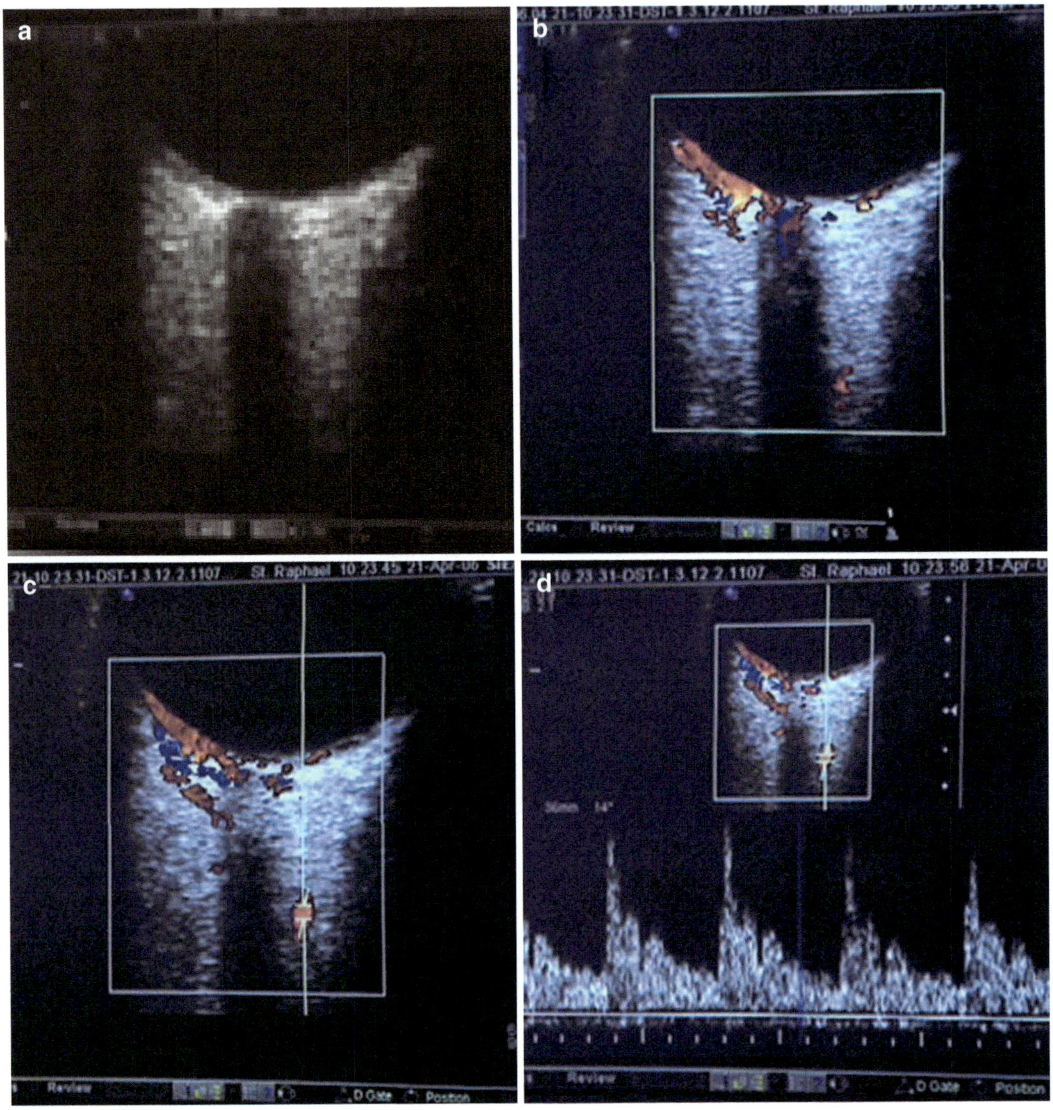

Fig. 8.3 Imaging of retrobulbar structures and flow

be distinguished by CDI because of their small size. Hence, bundles of vessels are examined, and the number of arteries contributing to the signal remains uncertain. Some authors also do not distinguish between nasal and temporal PCAs.

When measuring blood velocities using CDI, it needs to be considered that the Doppler angle influences the results. The direction of the sampling gate therefore has to be in good alignment with the angle of the measured vessel. If this is not the case, a measurement error is introduced, which is the more severe the larger the Doppler angle. In the CRA and the OA, the angle can usually be determined easily. In the PCA, however, the problem is more severe because of their smaller size, the more tortuous course, and the uncertain number of vessels within the probing volume.

The RI offers the advantage that it is independent of the Doppler angle. It is a dimensionless parameter that can take values between 0 and 1. The relation between RI and vascular resistance is, however, not entirely clear. Particularly, it seems that the value is largely dependent of the vascular compliance, which is defined as the ability of a vessel to distend and increase intravascular volume with increasing transmural pressure. In an in vitro model, where liquid was moved using a pulsatile pump, RI was independent of

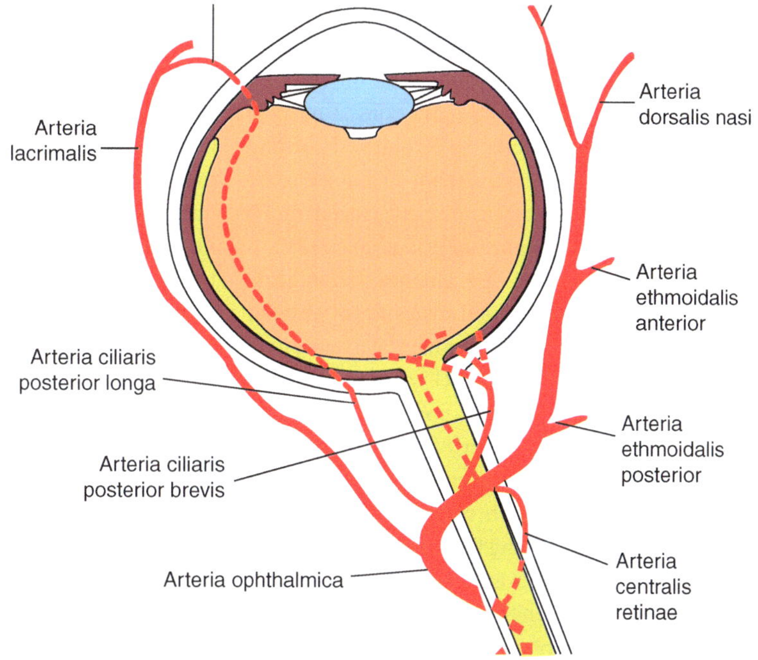

Fig. 8.4 Anatomy retrobulbar vessels

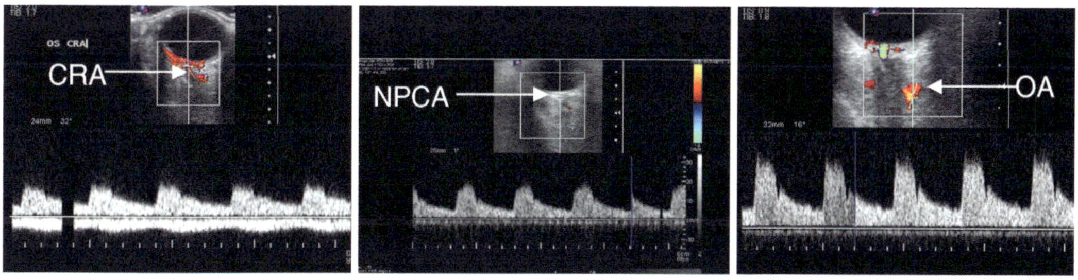

Fig. 8.5 Identification of retrobulbar vessels

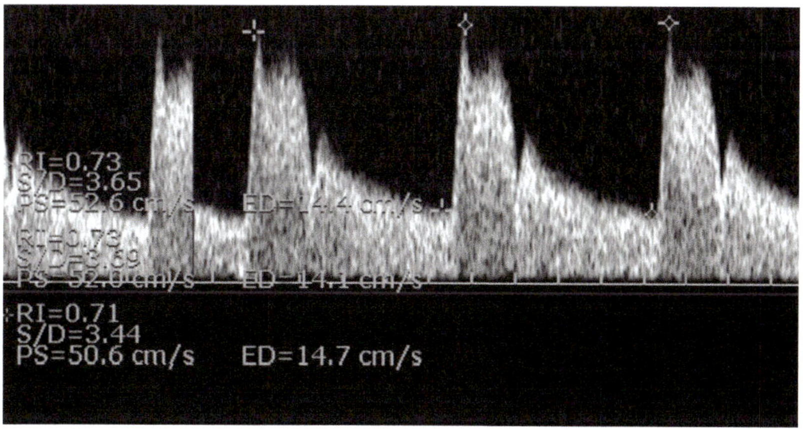

Fig. 8.6 Measured parameters

vascular resistance in the absence of vascular compliance. The higher the compliance, however, the more dependent RI became on vascular resistance [7]. In another model consisting of a branching tubing network, the RI showed a strong dependence on the downstream cross-sectional diameter. It was, however, not influenced much by alterations in resistance or compliance [8].

A number of animal experiments also indicate that RI is not always a good measure of vascular resistance. In isolated rabbit kidneys perfused with a pulsatile perfusion system, RI was dependent on vascular resistance but only under conditions of pronounced vasoconstriction [80]. The RI changed, however, strongly when the pulse pressure amplitude was changed, which may be expected to a certain degree considering that systolic blood pressure drives PSV and diastolic blood pressure drives EDV. Obviously, when flow is nonpulsatile and the pulse pressure amplitude is zero, the RI also becomes zero with any given vascular resistance.

Most of the published studies focused on the reliability of RI as a measure of vascular resistance in the kidney [79]. For the eye, only few data on the reliability of RI have been published. The dependence of RI in retrobulbar vessels on the pulse pressure amplitude has been confirmed in a variety of studies. A reduction in pulse pressure amplitude induced by physiological stimuli was associated with a decrease in the RI in the OA and the PCAs reflecting the change in blood pressure profile [72]. The same phenomenon was observed in studies where angiotensin II and the angiotensin1 receptor antagonist losartan were intravenously administered in healthy subjects [49, 56]. When angiotensin II was administered, an increase in MFV in the OA, together with a dose-dependent decrease in RI, was observed. Considering the very potent vasoconstrictor effects of this drug, this unlikely reflects peripheral vasodilatation. More likely, the increase in MFV reflects local vasoconstriction of the OA at the site of measurement, and the increase in RI is very well compatible with the increase in pulse pressure index observed after systemic angiotensin II infusion.

One study compared the retinal vascular resistance in response to 100% O_2 breathing to the change of RI induced by the same stimulus [66]. The former was calculated based on measurements of retinal vessel diameters using the Retinal Vessel Analyzer and measurements of blood velocities using bidirectional laser Doppler velocimetry as described in other chapters of this book. Calculating total retinal blood flow by measuring all retinal veins entering the optic nerve head revealed an increase in retinal vascular resistance of approximately 100%. By contrast, RI only increased in the order of 10%, making it doubtful whether RI is indeed a good measure of retinal vascular resistance.

8.5 Reproducibility

Short-term reproducibility of the technique is acceptable with data in the OA and CRA less variable than in the PCAs [3, 6, 31, 61, 68]. Some studies also looked at diurnal variability [54, 67], 2–7 day variability [62, 69], and 1 month variability [75], with results being comparable to the short-terms studies. Interobserver variability is, however, considerably higher, highlighting the subjective nature of the technique [3]. Training is major issue in CDI measurements, and long-term experience of the operator significantly reduces variability [11].

8.6 Physiological and Pharmacological Stimuli

Breathing of CO_2 induces pronounced vasodilatation in the ocular vasculature. Accordingly, most studies reported an increase in flow velocities in retrobulbar vessels with systemic hypercapnia, paralleled by a decrease in RI [35, 38, 70]. Hyperoxia on the other hand results in a decrease in flow velocities in most studies, although the results are not unequivocal [19, 73] (Hosking et al. 2004). Some studies also used carbogen with variable percentages of CO_2 ranging from 5% to 8%, with either no effect on blood velocities or a small increase associated with a decrease in RI [32, 73].

Changes in posture have also been used to study retrobulbar flow velocities during changes in ocular perfusion pressure, thereby getting insight into autoregulatory phenomena in the eye. When subjects change from a seated to a supine position, the hydrostatic effect induced by the change in gravity on the cranial circulation is not fully compensated by the baroreflex. Hence, cranial arterial pressure rises, as does cranial venous pressure and IOP. The IOP increase is, however, smaller than the arterial pressure increase, resulting in an increase in OPP. In the OA, an increase in EDV was observed, together with a decrease in RI. In the CRA, neither EDV nor PSV was changed, but a decrease in RI was seen [20]. Given the complex changes in pressures at both the arterial and the venous side during posture, it is, however, difficult to interpret these results in terms of autoregulation.

An experimental increase in IOP, as obtained by a scleral suction cup, results in a decrease of PSV, EDV, and MFV and an increase in RI in the CRA and SPCA but not in the OA of healthy individuals [21, 33, 39]. Hence, changes in IOP affect perfusion in the supplying artery only to a small degree, indicating that OA perfusion is not a good indicator of blood flow in the eye in general. Isometric exercise increases EDV in retrobulbar vessels with little effect on PSV, thereby reducing RI [5, 72]. As mentioned above, the latter effect is most likely related to the change in blood pressure profile and does not reflect a change in vascular resistance [72]. The same holds true for changes in retrobulbar blood velocities during the Valsalva maneuver. The pronounced increase in diastolic blood pressure is reflected in an increase in EDV in the OA and the PCAs as well as a reduction in RIs.

Dynamic autoregulation requires that both changes in blood pressure and changes in flow velocities are continuously measured before and after a step change in blood pressure. In one study, bilateral thigh cuffs were inflated to a pressure approximately 20 mmHg above peak systolic blood pressure for 3 min and subsequently deflated rapidly. In the middle cerebral artery, flow velocities return to baseline earlier than systemic blood pressure, indicating peripheral vasodilatation due to an autoregulatory response [64]. In the OA, flow velocities, however, return to baseline later than blood pressure, suggesting peripheral vasoconstriction due to an unknown mechanism [47]. Whether this, however, is related to the reactivity of the ocular blood vessels is doubtful given that the OA supplies mostly extraocular tissues.

It has already been shown decades ago that carbonic anhydrase inhibitors are potent vasodilators in the brain and systemic administration of acetazolamide induces a pronounced increase in blood velocities in cerebral vessels [15]. In the ocular vessels, the vasodilator effect of carbonic anhydrase inhibitors seems to be less pronounced. In one study, acetazolamide given orally at a dose of 1,000 mg did not change PSV and EDV in the OA or the CRA [32], whereas an increase in MFV in the OA and a significant decrease in RI [15] was found after intravenous administration of the carbonic anhydrase inhibitor in another study.

8.7 Results in Patients with Disease

A large variety of studies has been published reporting retrobulbar flow velocities in patients with glaucoma. Most of these studies found that blood flow velocities are decreased, and RIs are increased in patients with primary open-angle glaucoma [9, 12, 28, 44, 57, 65, 76]. Only few data are, however, available in patients with ocular hypertension. One small-scale study found slightly reduced flow velocities and increased RIs associated with a significant thinning in the RNFL in the inferior quadrant [10]. A number of smaller CDI studies in glaucoma indicate that reduced flow velocities in retrobulbar vessels are associated with the progression of the disease. In one study, baseline end diastolic blood flow velocity in the central retinal artery correlated with the progression rate of visual field damage independent of the extent of glaucomatous damage and intraocular pressure [71]. Galassi and coworkers [29] studied the relationship between retrobulbar CDI, performed shortly after the diagnosis of primary open-angle glaucoma, and the progression of visual field loss for 7 years. Patients with a stable visual field over this observation period had higher EDVs and lower resistivity indices in the OA compared with those with a deteriorating visual field. This result is in good agreement with a study of Zeitz and coworkers indicating that progressive glaucoma patients have reduced EDV and PSV in the PCAs and also reduced PSV in the CRA as compared to nonprogressive patients [86]. In progressive primary open-angle glaucoma patients with controlled IOP, a correlation between EDV in the OA and the CRA and systemic blood pressure was found, which was absent in stable patients and healthy control subjects (Gherghel et al. 2000). This result is indicative of altered ocular blood flow regulation in progressive open-angle glaucoma [25, 26].

A number of studies also investigated retrobulbar flow velocities in patients with age-related macular degeneration (AMD). An early study revealed reduced blood velocities and increased RIs in the PCAs of AMD patients [22]. Based on these measurements and other observations, a hemodynamic model of the pathogenesis of AMD has been formulated [23, 24]. In patients with nonexudative AMD in one eye and a CNV secondary to AMD in the other eye, RI of the nasal and temporal PCAs was higher in the more affected eyes [81].

In patients with diabetes, most authors also reported reduced retrobulbar flow velocities [16, 17, 43, 55, 60], although increased velocities have been reported as well [85]. In interpreting these results, one needs, however, to consider that diabetes is associated with increased retinal branch arterial and venous diameters [47]. Whether the CRA, PCA, and OA diameter is also increased at the site of CDI measurements is, however, unknown. As such, it is unclear whether reduced flow velocities in diabetes truly represent reduced blood flows through the retrobulbar arteries.

Many studies have focused on flow velocities in patients with vascular occlusive disease. As expected, carotid artery stenosis results in reduced anterograde OA flow and in severe cases also reversal of the flow in the OA [13, 27, 34, 37]. After carotid endarterectomy, blood flow velocities in the CRA, PCAs, and OA increase [14, 58, 82]. In patients with OA stenosis, the narrowing of the vessel may be noted on the CDI image if it is severe. As expected, acceleration of blood flow velocities can be seen across the stenotic segment of the vessel associated with poststenotic turbulence [14, 74].

In patients with central retinal artery occlusion, flow velocities in CRA and CRV are strongly reduced with an increase in RI [52, 53]. In patients with central retinal vein occlusion, a decrease in venous velocity in the CRV was reported, which was more pronounced in ischemic than in nonischemic cases [2, 4, 45, 83]. An EDV lower than 3 cm/s in the CRV had a sensitivity of 75% and a specificity of 86% in predicting iris neovascularization in ischemic central retinal vein occlusion [83]. In patients with branch vein occlusions, venous flow velocity was, however, not significantly different compared with healthy control subjects [2, 77, 78]. Patients with central retinal vein occlusion also show reduced flow velocities in the CRA. The increase in RI in the CRA was more marked in ischemic central retinal vein occlusion than in nonischemic [2]. One year after the first presentation, however, the RI in the CRA had normalized in all types of central retinal vein occlusion.

Only few studies are available for patients with nonarteritic anterior ischemic optic neuropathy (NAION). In one study, the PSV and EVD of CRA and PSV of the nasal PCA were significantly decreased in NAION patients when compared to healthy control, whereas no differences were detectable for the temporal PCAs or the OA [42]. In the acute phase of optic neuritis, increased RI was reported in the CRA [1, 18] and the OA [40], whereas during the chronic phase alterations, retrobulbar blood flow velocities were observed [1, 84].

8.8 Advantages and Limitations

CDI has an important advantage in that it is noninvasive and not affected by poor ocular media. Hence, measurements can be done in most clinical cases even with high degrees of cataract.

Generally, CDI of retrobulbar vessels measures velocity only. Hence, CDI data cannot be interpreted as blood flow values because the diameter of the vessels is not known. Attempts have been made to measure absolute blood flow in the OA [41, 59, 63, 87] (Fig. 8.3), but limited reproducibility may limit the use of this technique in clinical studies [87]. Hence, an increase in flow velocities may indicate an increase in blood flow through this vessel but may also arise from local vasoconstriction at the site of measurement as it has been reported for local stenosis of the OA [14]. One also has to consider that the velocities measured with CDI are based on theoretical parabolic distribution of flow rates across the lumen. Whereas this may well be fulfilled in healthy subjects given the size of the measured vessels, it is less clear whether this is also true in diseased patients. Additionally, the validity of RI as a measure of vascular resistance is uncertain as discussed in more detail above.

CDI measurements are subjective, and adequate training is required for obtaining reliable and reproducible results. An important source of error is incorrect angle of incidence, which is the more severe the lower the angle. Another potential source of error is increased intraocular pressure caused by excessive pressure applied to the eyelid. Currently, no data are available quantifying the effect of a routine CDI examination on IOP. In addition, no data exist comparing results with different CDI machines, and the comparability between different instruments remains doubtful. One study indicates that even with different probes, different results are obtained [6].

Finally, one needs to consider that information is only gained from the supplying arteries. For instance, the OA supplies many extraocular tissues, and only approximately 25% of flow through this vessel goes to the eye [46]. The same limitation applies to measurements in the PCAs when deducing information on optic nerve head blood flow. It is clear that the major part of the flow goes to the uveal tract, thereby also supplying the choroid. Given that blood flow in the optic nerve and the choroid underlies very different regulation mechanisms, it is therefore unlikely that any change in PCA blood velocity reflects what is going in the optic nerve head.

References

1. Akarsu C, Tan FU, Kendi T (2004) Color Doppler imaging in optic neuritis with multiple sclerosis. Graefes Arch Clin Exp Ophthalmol 242:990–994
2. Arséne S, Giraudeau B, Le Lez ML, Pisella PJ, Pourcelot L, Tranquart F (2002) Follow up by colour Doppler imaging of 102 patients with retinal vein occlusion over 1 year. Br J Ophthalmol 86:1243–1247
3. Baxter GM, Williamson TH (1995) Color Doppler imaging of the eye: normal ranges, reproducibility, and observer variation. J Ultrasound Med 14:91–96
4. Baxter GM, Williamson TH (1996) The value of serial Doppler imaging in central retinal vein occlusion: correlation with visual recovery. Clin Radiol 51:411–414
5. Beck D, Harris A, Evans D, Martin B (1995) Ophthalmic arterial hemodynamics during isometric exercise. J Glaucoma 4:317–321
6. Boehm AG, Helmke K, Berry CC, Weinreb RN (2002) Comparison of two transducers for color Doppler imaging of the retrobulbar vessels. J Glaucoma 11:148–153
7. Bude RO, Rubin JM (1999) Relationship between the resistive index and vascular compliance and resistance. Radiology 211:411–417
8. Bude RO, Rubin JM (1999) Effect of downstream cross-sectional area of an arterial bed on the resistive index and the early systolic acceleration. Radiology 212:732–738
9. Cellini M, Possati GL, Profazio V, Sbrocca M, Caramazza N, Caramazza R (1997) Color Doppler imaging and plasma levels of endothelin-1 in low-tension glaucoma. Acta Ophthalmol Scand Suppl 224:11–13
10. Cellini M, Bernabini B, Carbonelli M, Zamparini E, Campos EC (2007) Optical coherence tomography, frequency-doubling technology, and colour Doppler imaging in ocular hypertension. Eye (Lond) 21:1071–1077
11. Ciulla TA, Kagemann L, Harris A (2003) Imaging Fundamentals. Retina and Optic Nerve Imaging. In: Ciulla TA, Regillo CD, Harris A (eds) Lippincott Williams and Wilkins Philadelphia, pp 1–11
12. Costa VP, Sergott RC, Smith M, Spaeth GL, Wilson RP, Moster MR, Katz LJ, Schmidt CM (1994) Color Doppler imaging in glaucoma patients with asymmetric optic cups. J Glaucoma 3(1):S91
13. Costa VP, Kuzniec S, Molnar LJ, Cerri GG, Puech-Leão P, Carvalho CA (1997) Clinical findings and hemodynamic changes associated with severe occlusive carotid artery disease. Ophthalmology 104:1994–2002
14. Costa VP, Kuzniec S, Molnar LJ, Cerri GG, Puech-Leão P, Carvalho CA (1999) The effects of carotid endarterectomy on the retrobulbar circulation of patients with severe occlusive carotid artery disease. An investigation by color Doppler imaging. Ophthalmology 106:306–310

15. Dallinger S, Bobr B, Findl O, Eichler HG, Schmetterer L (1998) Effects of acetazolamide on choroidal blood flow. Stroke 29:997–1001
16. Dimitrova G, Kato S, Tamaki Y, Yamashita H, Nagahara M, Sakurai M, Kitano S, Fukushima H (2001) Choroidal circulation in diabetic patients. Eye (Lond) 15:602–607
17. Dimitrova G, Kato S, Yamashita H, Tamaki Y, Nagahara M, Fukushima H, Kitano S (2003) Relation between retrobulbar circulation and progression of diabetic retinopathy. Br J Ophthalmol 87:622–625
18. Elvin A, Andersson T, Söderström M (1998) Optic neuritis. Doppler ultrasonography compared with MR and correlated with visual evoked potential assessments. Acta Radiol 39:243–248
19. Evans DW, Harris A, Danis RP, Arend O, Martin BJ (1997) Altered retrobulbar vascular reactivity in early diabetic retinopathy. Br J Ophthalmol 81:279–282
20. Evans DW, Harris A, Garrett M, Chung HS, Kagemann L (1999) Glaucoma patients demonstrate faulty autoregulation of ocular blood flow during posture change. Br J Ophthalmol 83:809–813
21. Findl O, Strenn K, Wolzt M, Menapace R, Vass C, Eichler HG, Schmetterer L (1997) Effects of changes in intraocular pressure on human ocular haemodynamics. Curr Eye Res 16:1024–1029
22. Friedman E, Krupsky S, Lane AM, Oak SS, Friedman ES, Egan K, Gragoudas ES (1995) Ocular blood flow velocity in age-related macular degeneration. Ophthalmology 102:640–646
23. Friedman E (1997) A hemodynamic model of the pathogenesis of age-related macular degeneration. Am J Ophthalmol 124:677–682
24. Friedman E (2008) The pathogenesis of age-related macular degeneration. Am J Ophthalmol 146:348–349
25. Fuchsjäger-Mayrl G, Wally B, Georgopoulos M, Rainer G, Kircher K, Buehl W, Amoako-Mensah T, Eichler HG, Vass C, Schmetterer L (2004) Ocular blood flow and systemic blood pressure in patients with primary open-angle glaucoma and ocular hypertension. Invest Ophthalmol Vis Sci 45:834–839
26. Fuchsjäger-Mayrl G, Georgopoulos M, Hommer A, Weigert G, Pemp B, Vass C, Garhöfer G, Schmetterer L (2010) Effect of dorzolamide and timolol on ocular pressure: blood flow relationship in patients with primary open-angle glaucoma and ocular hypertension. Invest Ophthalmol Vis Sci 51:1289–1296
27. Fujioka S (2003) Use of orbital color Doppler imaging for detecting internal carotid artery stenosis in patients with amaurosis fugax. Jpn J Ophthalmol 47:276–280
28. Galassi F, Sodi A, Ucci F, Renieri G, Pieri B, Masini E (2000) Ocular haemodynamics and nitric oxide in normal pressure glaucoma. Acta Ophthalmol Scand Suppl 232:37–38
29. Galassi F, Sodi A, Ucci F, Renieri G, Pieri B, Baccini M (2003) Ocular hemodynamics and glaucoma prognosis: a color Doppler imaging study. Arch Ophthalmol 121:1711–1715
30. Gherghel D, Orgül S, Gugleta K, Gekkieva M, Flammer J (2000) Relationship between ocular perfusion pressure and retrobulbar blood flow in patients with glaucoma with progressive damage J. Am J Ophthalmol 130(5): 597-605.
31. Harris A, Williamson TH, Martin B, Shoemaker JA, Sergott RC, Spaeth GL, Katz JL (1995) Test/retest reproducibility of color Doppler imaging assessment of blood flow velocity in orbital vessels. J Glaucoma 4:281–286
32. Harris A, Tippke S, Sievers C, Picht G, Lieb W, Martin B (1996) Acetazolamide and CO2: acute effects on cerebral and retrobulbar hemodynamics. J Glaucoma 5:39–45
33. Harris A, Joos K, Kay M, Evans D, Shetty R, Sponsel WE, Martin B (1996) Acute IOP elevation with scleral suction: effects on retrobulbar haemodynamics. Br J Ophthalmol 80:1055–1059
34. Ho AC, Lieb WE, Flaharty PM, Sergott RC, Brown GC, Bosley TM, Savino PJ (1992) Color Doppler imaging of the ocular ischemic syndrome. Ophthalmology 99:1453–1462
35. Hosking SL, Harris A, Chung HS, Jonescu-Cuypers CP, Kagemann L, Roff Hilton EJ, Garzozi H (2004) Ocular haemodynamic responses to induced hypercapnia and hyperoxia in glaucoma. Br J Ophthalmol 88:406–411
36. Hradílek P, Stourac P, Bar M, Zapletalová O, Skoloudík D (2009) Colour Doppler imaging evaluation of blood flow parameters in the ophthalmic artery in acute and chronic phases of optic neuritis in multiple sclerosis. Acta Ophthalmol 87:65–70
37. Hu HH, Sheng WY, Yen MY, Lai ST, Teng MM (1993) Color Doppler imaging of orbital arteries for detection of carotid occlusive disease. Stroke 24: 1196–1203
38. Huber KK, Adams H, Remky A, Arend KO (2006) Retrobulbar haemodynamics and contrast sensitivity improvements after CO2 breathing. Acta Ophthalmol Scand 84:481–487
39. Joos KM, Kay MD, Pillunat LE, Harris A, Gendron EK, Feuer WJ, Steinwand BE (1999) Effect of acute intraocular pressure changes on short posterior ciliary artery haemodynamics. Br J Ophthalmol 83:33–38
40. Karaali K, Senol U, Aydin H, Cevikol C, Apaydin A, Lüleci E (2003) Optic neuritis: evaluation with orbital Doppler sonography. Radiology 226:355–358
41. Katamay R, Fleischlin C, Gugleta K, Flammer J, Orgül S (2009) Volumetric blood flow measurement in the ophthalmic artery using colour Doppler. Klin Monbl Augenheilkd 226:249–253
42. Kaup M, Plange N, Arend KO, Remky A (2006) Retrobulbar haemodynamics in non-arteritic anterior ischaemic optic neuropathy. Br J Ophthalmol 90: 1350–1353
43. Kawagishi T, Nishizawa Y, Emoto M, Konishi T, Maekawa K, Hagiwara S, Okuno Y, Inada H, Isshiki G, Morii H (1995) Impaired retinal artery blood flow in IDDM patients before clinical manifestations of diabetic retinopathy. Diabetes Care 18:1544–1549
44. Kerr J, Nelson P, O'Brien C (1998) A comparison of ocular blood flow in untreated primary open-angle

glaucoma and ocular hypertension. Am J Ophthalmol 126:42–51
45. Keyser BJ, Flaharty PM, Sergott RC, Brown GC, Lieb WE, Annesley WH Jr (1994) Color Doppler imaging of arterial blood flow in central retinal vein occlusion. Ophthalmology 101:1357–1361
46. Kiss B, Dallinger S, Polak K, Findl O, Eichler HG, Schmetterer L (2001) Ocular hemodynamics during isometric exercise. Microvasc Res 61:1–13
47. Klein R, Klein BE, Moss SE, Wong TY, Hubbard L, Cruickshanks KJ, Palta M (2004) The relation of retinal vessel caliber to the incidence and progression of diabetic retinopathy: XIX: the Wisconsin epidemiologic study of diabetic retinopathy. Arch Ophthalmol 122:76–83
48. Kolodjaschna J, Berisha F, Lung S, Schima H, Polska E, Schmetterer L (2005) Comparison of the autoregulatory mechanisms between middle cerebral artery and ophthalmic artery after thigh cuff deflation in healthy subjects. Invest Ophthalmol Vis Sci 46:636–640
49. Krejcy K, Wolzt M, Kreuzer C, Breiteneder H, Schütz W, Eichler HG, Schmetterer L (1997) Characterization of angiotensin-II effects on cerebral and ocular circulation by noninvasive methods. Br J Clin Pharmacol 43:501–508
50. Lawrence JP (2007) Physics and instrumentation of ultrasound. Crit Care Med 35:S314–S322
51. Lieb WE, Cohen SM, Merton DA, Shields JA, Mitchell DG, Goldberg BB (1991) Color Doppler imaging of the eye and orbit. Technique and normal vascular anatomy. Arch Ophthalmol 109:527–531
52. Lieb WE, Flaharty PM, Ho A, Sergott RC (1992) Color Doppler imaging of the eye and orbit. A synopsis of a 400 case experience. Acta Ophthalmol Suppl 204:50–54
53. Lieb WE (1998) Color Doppler imaging of the eye and orbit. Radiol Clin North Am 36(6):1059–1071
54. Luksch A, Lasta M, Polak K, Fuchsjäger-Mayrl G, Polska E, Garhöfer G, Schmetterer L (2009) Twelve-hour reproducibility of retinal and optic nerve blood flow parameters in healthy individuals. Acta Ophthalmol 87:875–880
55. MacKinnon JR, McKillop G, O'Brien C, Swa K, Butt Z, Nelson P (2000) Colour Doppler imaging of the ocular circulation in diabetic retinopathy. Acta Ophthalmol Scand 78:386–389
56. Matulla B, Streit G, Pieh S, Findl O, Entlicher J, Graselli U, Eichler HG, Wolzt M, Schmetterer L (1997) Effects of losartan on cerebral and ocular circulation in healthy subjects. Br J Clin Pharmacol 44:369–375
57. Martínez A, Sánchez M (2005) Predictive value of colour Doppler imaging in a prospective study of visual field progression in primary open-angle glaucoma. Acta Ophthalmol Scand 83:716–722
58. Mawn LA, Hedges TR 3rd, Rand W, Heggerick PA (1997) Orbital color Doppler imaging in carotid occlusive disease. Arch Ophthalmol 115:492–496
59. Michelson G, Schuierer G (1991) Absolute blood flow in the ophthalmic artery. Fortschr Ophthalmol 88:687–689
60. Modrzejewska M, Pieńkowska-Machoy E, Grzesiak W, Karczewicz D, Wilk G (2008) Predictive value of color Doppler imaging in an evaluation of retrobulbar blood flow perturbation in young type-1 diabetic patients with regard to dyslipidemia. Med Sci Monit 14:MT47–MT52
61. Németh J, Kovács R, Harkányi Z, Knézy K, Sényi K, Marsovszky I (2002) Observer experience improves reproducibility of color Doppler sonography of orbital blood vessels. J Clin Ultrasound 30:332–335
62. Niwa Y, Yamamoto T, Kawakami H, Kitazawa Y (1998) Reproducibility of color Doppler imaging for orbital arteries in Japanese patients with normal-tension glaucoma. Jpn J Ophthalmol 42:389–392
63. Orge F, Harris A, Kagemann L, Kopecky K, Sheets CW, Rechtman E, Zalish M (2002) The first technique for non-invasive measurements of volumetric ophthalmic artery blood flow in humans. Br J Ophthalmol 86:1216–1219
64. Panerai RB (1998) Assessment of cerebral pressure autoregulation in humans–a review of measurement methods. Physiol Meas 19:305–338
65. Plange N, Remky A, Arend O (2003) Colour Doppler imaging and fluorescein filling defects of the optic disc in normal tension glaucoma. Br J Ophthalmol 87:731–736
66. Polska E, Kircher K, Ehrlich P, Vecsei PV, Schmetterer L (2001) RI in central retinal artery as assessed by CDI does not correspond to retinal vascular resistance. Am J Physiol Heart Circ Physiol 280:H1442–H1447
67. Polska E, Polak K, Luksch A, Fuchsjager-Mayrl G, Petternel V, Findl O, Schmetterer L (2004) Twelve hour reproducibility of choroidal blood flow parameters in healthy subjects. Br J Ophthalmol 88:533–537
68. Quaranta L, Harris A, Donato F, Cassamali M, Semeraro F, Nascimbeni G, Gandolfo E, Quaranta CA (1997) Color Doppler imaging of ophthalmic artery blood flow velocity: a study of repeatability and agreement. Ophthalmology 104:653–658
69. Rankin SJ, Walman BE, Buckley AR, Drance SM (1995) Color Doppler imaging and spectral analysis of the optic nerve vasculature in glaucoma. Am J Ophthalmol 119:685–693
70. Roff EJ, Harris A, Chung HS, Hosking SL, Morrison AM, Halter PJ, Kagemann L (1999) Comprehensive assessment of retinal, choroidal and retrobulbar haemodynamics during blood gas perturbation. Graefes Arch Clin Exp Ophthalmol 237:984–990
71. Satilmis M, Orgül S, Doubler B, Flammer J (2003) Rate of progression of glaucoma correlates with retrobulbar circulation and intraocular pressure. Am J Ophthalmol 135:664–669
72. Schmetterer L, Dallinger S, Findl O, Strenn K, Graselli U, Eichler HG, Wolzt M (1998) Noninvasive investigations of the normal ocular circulation in humans. Invest Ophthalmol Vis Sci 39:1210–1220
73. Schüttauf F, Cobet U, Klemenz A, Krause A (1998) Duplex ultrasound examinations of retinal circulation after inhalation of various mixed respiratory gases. Ophthalmologe 95:225–228

74. Sergott RC, Flaharty PM, Lieb WE Jr, Ho AC, Kay MD, Mittra RA, Savino PJ, Bosley TM (1992) Color Doppler imaging identifies four syndromes of the retrobulbar circulation in patients with amaurosis fugax and central retinal artery occlusions. Trans Am Ophthalmol Soc 90:383–398
75. Stalmans I, Siesky B, Zeyen T, Fieuws S, Harris A (2009) Reproducibility of color Doppler imaging. Graefes Arch Clin Exp Ophthalmol 247:1531–1538
76. Stalmans I, Harris A, Fieuws S, Zeyen T, Vanbellinghen V, McCranor L, Siesky B (2009) Color Doppler imaging and ocular pulse amplitude in glaucomatous and healthy eyes. Eur J Ophthalmol 19(4):580–587
77. Tranquart F, Arsene S, Giraudeau B, Piquemal R, Eder V, Le Lez ML, Rossazza C, Pourcelot L (2000) Initial color Doppler findings in retinal vein occlusion. J Clin Ultrasound 28:28–33
78. Tranquart F, Bergès O, Koskas P, Arsene S, Rossazza C, Pisella PJ, Pourcelot L (2003) Color Doppler imaging of orbital vessels: personal experience and literature review. J Clin Ultrasound 31:258–273
79. Tublin ME, Bude RO, Platt JF (2003) Review. The resistive index in renal Doppler sonography: where do we stand? AJR Am J Roentgenol 180:885–892
80. Tublin ME, Tessler FN, Murphy ME (1999) Correlation between renal vascular resistance, pulse pressure, and the resistive index in isolated perfused rabbit kidneys. Radiology 213:258–264
81. Uretmen O, Akkin C, Erakgün T, Killi R (2003) Color Doppler imaging of choroidal circulation in patients with asymmetric age-related macular degeneration. Ophthalmologica 217:137–142
82. Ward JB, Hedges TR 3rd, Heggerick PA (1995) Reversible abnormalities in the ophthalmic arteries detected by color Doppler imaging. Ophthalmology 102:1606–1610
83. Williamson TH, Baxter GM (1994) Central retinal vein occlusion, an investigation by color Doppler imaging. Blood velocity characteristics and prediction of iris neovascularization. Ophthalmology 101:1362–1372
84. Williamson TH, Harris A (1996) Color Doppler ultrasound imaging of the eye and orbit. Surv Ophthalmol 40:255–267
85. Yilmaz Ovali G, Ersoy B, Tuncyurek O, Urk V, Ozkol M, Ozhan B, Baser E, Pabuscu Y (2008) Doppler ultrasonography imaging of hemodynamic alteration of retrobulbar circulation in type 1 diabetic children and adolescents without retinopathy. Diabetes Res Clin Pract 79:243–248
86. Zeitz O, Galambos P, Wagenfeld L, Wiermann A, Wlodarsch P, Praga R, Matthiessen ET, Richard G, Klemm M (2006) Glaucoma progression is associated with decreased blood flow velocities in the short posterior ciliary artery. Br J Ophthalmol 90: 1245–1248
87. Zeitz O, Vilchez SE, Matthiessen ET, Richard G, Klemm M (2006) Volumetric colour Doppler imaging: a useful tool for the determination of ocular blood flow in glaucoma patients? Eye (Lond) 20: 668–673

Other Approaches

9

Gerhard Garhöfer and Leopold Schmetterer

Core Messages

- Beside the classical techniques to assess ocular blood flow, several methods exists which aim to measure specific components of ocular blood flow. In this chapter, the blue field entoptic technique, the laser specle technique and methods that assess the pulsatile ocular blood flow will be covered.

9.1 Blue Field Entoptic Technique

The blue field entoptic technique is a semiquantitative, subjective method that uses the optical blue field entoptic phenomenon to estimate retinal capillary blood flow in retinal perifoveal vessels [27]. Basically, entoptic phenomena are defined as visual effects, observed under certain illumination conditions, whose source lies within the eye itself. The blue field effect is the most well known among these entoptic phenomena.

G. Garhöfer, M.D. (✉)
Department of Clinical Pharmacology, Medical University of Vienna, Waehringer Guertel 18-20, Vienna A-1090, Austria
e-mail: gerhard.garhoefer@meduniwien.ac.at

L. Schmetterer, Ph.D.
Department of Clinical Pharmacology, Center of Medical Physics and Biomedical Engineering, Medical University of Vienna, Waehringer Guertel 18-20, Vienna A-1090, Austria
e-mail: leopold.schmetterer@meduniwien.ac.at

The blue field entoptic phenomenon can easily be seen when looking into a blue light with a narrow optical spectrum at a wavelength of approximately 430 nm. In daily life, the blue field effect can be produced when looking into the blue sky on a bright, sunny day. Under these illumination conditions, many tiny corpuscles that move in a flowing manner with synchronous acceleration corresponding to the cardiac rhythm can be observed around an area of the center of the fovea. The particles often show a tiny dark tail, seem to appear suddenly, and follow fixed, often curving paths before disappearing.

Given that entoptic images are generated within the observer's own eye, it has been hypothesized that the observed effects may carry information about the anatomical and physiological properties of the observer's eye. In particular, it was suggested that the speed and the density of the moving particles observed could reflect quantitative information about blood flow in the retina. However, given that entoptic phenomena are produced within the subject's eye, the subject cannot share the specific view of the phenomenon with others. Thus, the quantification of the blue field effect remained impossible for a long time. Furthermore, the cellular origin of the entoptically produced bright particles remained a matter of controversy for a long time. Although it has been hypothesized since the early interpretations of the blue field effect by Helmholtz and others that the blue field effect is caused by circulating leukocytes, the assumptions concerning the source of the entoptic phenomenon were mainly

L. Schmetterer, J.W. Kiel (eds.), *Ocular Blood Flow*,
DOI 10.1007/978-3-540-69469-4_9, © Springer-Verlag Berlin Heidelberg 2012

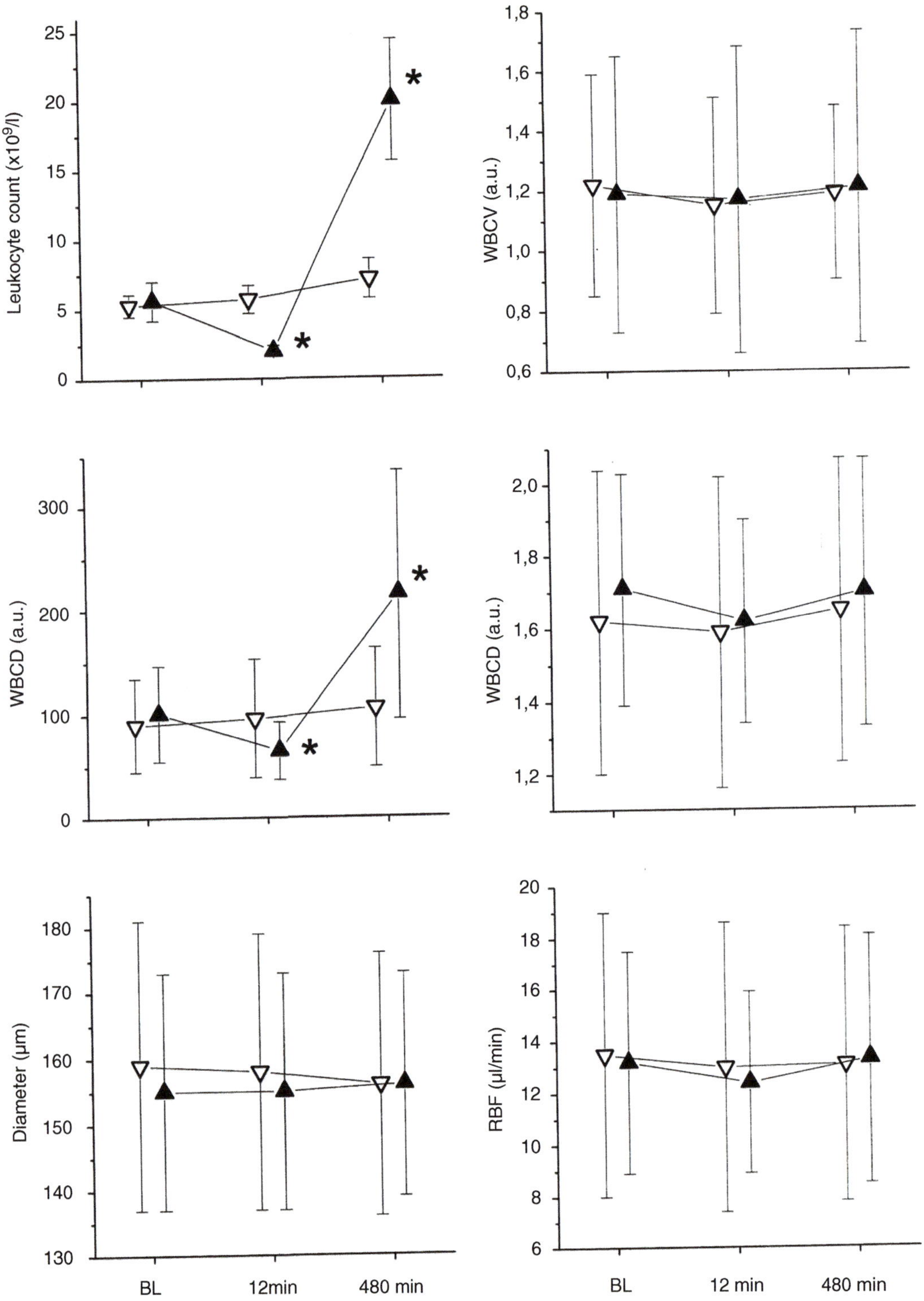

Fig. 9.1 Effects of granulocyte colony stimulating factor (G-CSF) or placebo on retinal hemodnymic parameters in healthy subjects. The time course of leukocyte counts, white blood cell velocity (WBCV), white blood cell density (WBCD), mean retinal red blood cell velocity (Vel), retinal venous diameter (Diameter) and retinal blood flow through a major retinal vein (Flow) after administration of G-CSF (solid up triangles) or placebo (open down triangles) is shown. Data are presented as means ± SD (n=15 per group). Asterisks indicate significant effects of G-CSF versus placebo

based on physiological and optical considerations without experimental evidence.

In particular, the movement patterns of the observed particles share similarities to histologically identifiable capillary loops. Furthermore, the observation that the corpuscles are not visible in the innermost of the fovea, which may represent the foveal avascular zone, strengthens the hypothesis that moving particles in the blood stream cause the entopic effect. Evidence that leukocytes cause the blue field effect is even nowadays mainly derived from indirect evidence. Bauermann observed that in patients with leukemia, the number of particles observed under blue field conditions is paralleled by the leukocyte count in the peripheral blood [2]. Furthermore, it has been shown in animal experiments that the blue field entoptic phenomenon can be reproduced in microvascular preparations [35]. More precisely, the authors used a video microscopic setup with lighting conditions similar to those under which the entopic phenomenon is visualized within the human eye. Under these conditions, the cellular flow within small blood vessels in a wing of a hibernating bat and a rat cremaster muscle was studied. In both preparations, effects of bright moving particles, which could microscopically be identified as leukocytes, were observed [35].

Evidence from an interventional study in healthy humans confirmed these results. In this experiment, granulocyte colony-stimulating factor was infused intravenously in a randomized, placebo-controlled double-masked study in healthy young volunteers in order to increase the leukocyte count in the peripheral blood [9]. Leukocyte movement was then assessed by the blue field technique. The authors of the study found a strong correlation between granulocyte colony-stimulating factor-induced changes in leukocyte count in the peripheral blood and leukocyte density as assessed with the blue field entoptic technique (Fig. 9.1). This was observed in the absence of effects on retinal vessel diameters or red blood cell velocities as assessed with bidirectional laser Doppler velocimetry. This is a clear indication that the blue filed phenomenon reflects leukocyte movement in the perifoveal retinal vasculature [9].

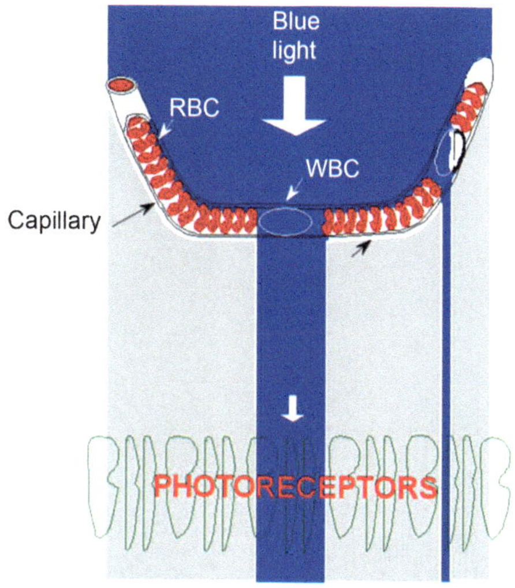

Fig. 9.2 Origin of the blue field entoptic phenomenon (for details see text, courtesy of Martial Geiser)

Today, there is general agreement that the blue field entoptic phenomenon is produced by the different absorption properties of red and white blood cells when the retina is illuminated with blue light. Passing white blood cells do not absorb the short wavelength light, whereas the red blood cells do. Accordingly, white blood cells transversing retinal capillaries that are in front of the photoreceptors are perceived as a flying corpuscle (Fig. 9.2). Hence, the technique is capable of gaining insight into white blood cell movement in retinal perifoveal capillaries. Whether this is proportional to retinal red blood cell movement or retinal blood flow under all clinical conditions remains, however, doubtful.

Several methodological approaches have been proposed to quantify leukocyte movement based on the blue field entoptic effect. Given that the blue field phenomenon is produced by the leukocytes moving in the vessel of the observer, it cannot be made visible by others. Thus, all approaches that have been introduced for the quantification of the blue field phenomenon are strictly subjective in their nature. In an early approach, Riva and Loebl used the blue field phenomenon to investigate autoregulation of the retinal circulation.

For this purpose, subjects were asked to compare the white blood cell velocity observed in one eye with that seen in the other eye. Then the subjects were asked to raise their intraocular pressure (IOP) by pressing a tonometer probe against

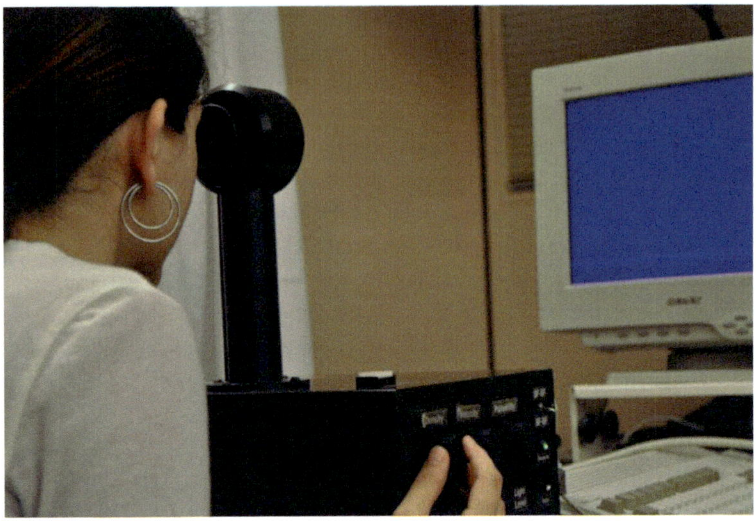

Fig. 9.3 Photograph of the blue field entoptic system

the sclera until a reduction in speed of the moving particles was observed [26]. The IOP at which the white blood cell speed started to decrease was defined by the authors as the point where the autoregulation of retinal blood flow was not sufficient any more to maintain normal blood flow.

Later, Riva and Petrig developed a more refined technique that has been made commercially available (Blue Field Simulator, Oculix Sarl, Arbaz, Switzerland, Fig. 9.3). This approach uses a computer system to simulate a field with corpuscles, similar to the field that is observed by the subjects under blue field conditions to extract quantitative data. For this purpose, the eye is illuminated with light of a center wavelength of 430 nm. By the means of a connected computer system and a video monitor, a simulated particle field is shown to the subject under study. Then, either the simulated field by the computer system or the subjects' own perception of the blue field phenomenon is alternately shown. The subject is asked to adapt the computer image by adjusting speed and number of the moving particles till the computer image reflects his own perception. By comparing and adjusting the white blood cell (WBC) density and the WBC velocity of the particles in the observed simulated particle field with their own perception, an estimate of perimacular WBC flux can be obtained. WBC flux is calculated as

$$\text{WBC flux} = \text{WBC density} \times \text{WBC velocity}$$

These outcome parameters characterize WBC dynamics in perimacular retinal capillaries in arbitrary units.

One of the most important advantages of the blue field technique is that, in contrast to other techniques available, the blue field technique is largely independent from opaque media. Thus, it has been hypothesized that the blue field technique may be used to predict postoperative macular function in patients undergoing future cataract operation [36]. Several experiments indicate that when the blue field phenomenon is observed by patients, retinal function is intact to a large degree [1, 36]. Thus, the blue field technique was proposed to be applicable for the decision whether surgical intervention may be successful for the patients in terms of the postoperative visual outcome, even if opaque ocular media do not allow funduscopy. It has, however, also been reported that the blue field often fails to detect vision loss caused by macular holes and moderate macular dysfunction [21]. Because of these limitations, the blue field entoptic technique has not become generally accepted as a clinical routine test, although it might provide useful information about retinal function in some patient subgroups.

The blue field entoptic system was one of the first methods that allowed for semiquantitative, noninvasive estimation of retinal blood flow blood flow in humans. Thus, the system has been widely used for the investigation of retinal blood flow and its regulation in health and disease. In particular, studies with the blue field instrument were among the first to indicate that the parafoveal circulation is autoregulated in response to changes of perfusion pressure [26]. The technique was subsequently employed to gain insight into retinal autoregulation during both an increase or decrease in IOP [28, 10]. In addition, correlations of white blood cell flux data with mean arterial pressure data have been published [24].

In recent years, the blue field technique was employed in studies in which an uncoupling between erythrocyte and leukocyte flux was induced experimentally. As mentioned above, one of these conditions is obtained when granulocyte colony-stimulating factor is administered to healthy subjects [9]. An increase in retinal leukocyte density is also observed in a human model of systemic inflammation, where endotoxin is administered to healthy subjects [16, 17, 22].

When interpreting the results obtained with the blue field entoptic technique, several limitations have to be considered. The first and most important limitation of the blue field technique is related to the method itself. Given that only the subject under investigation is able to see the blue field effect, the method is subjective in nature and requires sufficient cooperation of the subject. Thus, it is clear that the reproducibility is largely dependent on the ability of the subject to perform the test correctly and may vary widely among subjects. Consequently, in most studies, subjects are asked to perform the test at baseline conditions four to five times. Based on these preliminary tests, which are usually performed during screening, the reproducibility of the measurements and the subjects' ability to perform the test correctly can be ensured. In case the reproducibility is poor (i.e., less than 15%), the results of the testing are normally rejected.

An average variability of 13% with white blood cell velocity measurements has been reported in healthy young volunteers with good visual acuity. Measurements of white blood cell density and white blood cell flux are less reproducible [20, 24]. Given that the variability of the method is strongly dependent on the ability of the subject to perform the measurement tasks correctly, it can be speculated that in elderly patients, or in patients with poor visual acuity, the variability of the method may be significantly larger.

Another important limitation relates to the point that leukocyte flux may not be proportional to erythrocyte flow and to retinal blood flow under all clinical conditions as mentioned above. This is particularly true in diseases in which subclinical inflammation of the retina may play a role such as diabetic retinopathy, HIV-related retinopathy, or age-related macular degeneration. As such, a change in leukocyte flux may well represent processes other than a change in blood flow, including inflammation or altered leukocyte-endothelium interaction.

9.1.1 Laser Speckle Technique

The introduction of laser Doppler velocimetry allowed for the first time the noninvasive measurement of red blood cell velocity in vivo and can – when combined with diameter measurements – provide insight in the volumetric blood flow of individual vessels [29]. However, the latter approach is limited to measurement of major retinal vessels down to a size of approximately 70 μm in diameter. Vessels with a diameter smaller than this cannot be measured. Thus, blood flow measurements in the microcirculation are not possible with laser Doppler velocimetry.

Laser speckle flowgraphy is a noninvasive technique that allows for the real-time, two-dimensional measurement of relative blood flow velocity of ocular microcirculation using the laser speckle phenomenon. The laser speckle phenomenon is one of several interference phenomena observed when coherent light, such as emitted from a laser, is scattered by a diffusing surface. When the eye is illuminated with laser light, a speckle pattern appears, caused by the backscattered light from the rough surface of the ocular fundus.

Basically, the laser speckle phenomenon is based on the principle that when living tissue is illuminated by coherent laser light, reflections off tissue irregularities interfere constructively and destructively, which results in a random

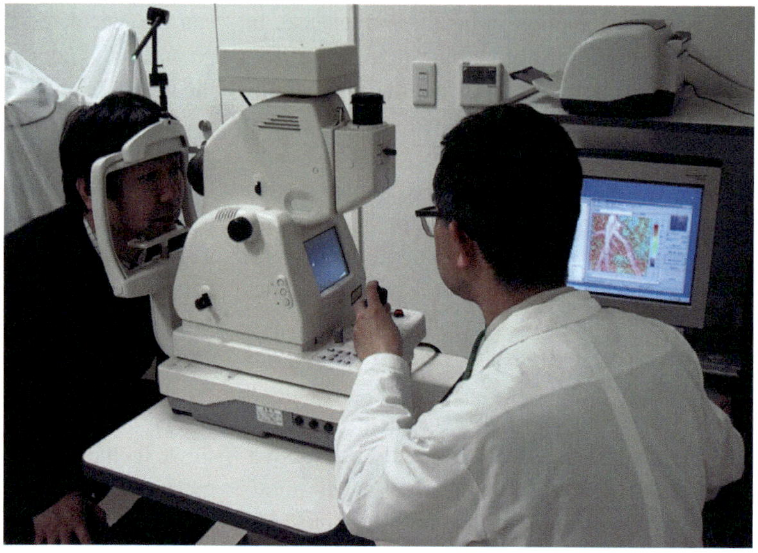

Fig. 9.4 Photograph of the laser speckle system

interference pattern. When a stationary object is illuminated, the pattern remains static and has a sharp contrast between areas of constructive and destructive interference, producing a typical random speckle pattern. When the illuminated object is moving or contains moving particles such as erythrocytes, the structure of the speckle pattern will vary rapidly. These variations of the speckle pattern depend on the velocity of the moving parts in the scattering surface. Thus, the rate of variation in the speckle pattern gives an estimate of blood cell velocity and can therefore be used for the quantification of retinal blood flow [4].

Based on these finding, Briers and Fercher were the first to introduce an instrument for the mapping of retinal blood flow [4] by using laser speckle flowgraphy. The main advantage of this approach is that, in contrast to laser Doppler velocimetry, not only blood velocity in major retinal vessels but also in the retinal microcirculation can be measured. Furthermore, the technique is capable of giving an overall map of the velocity distribution of the retina, whereas other techniques can only determine velocities in one single point of measurement (laser Doppler flowmetry) or require scanning procedures (scanning laser Doppler flowmetry). The method of Briers and Fercher, however, only allowed for a semiquantitative evaluation of the retinal circulation and did not enable analysis of changes over time.

Based on this technology, Tamaki and colleagues have developed an instrument for noncontact, two-dimensional, and quantitative analysis of ocular blood velocity in vivo [39, 41]. This system is based on a fundus camera (TRC-NW5S, Topcon, Tokyo) equipped with a diode laser with a wavelength 808 nm, an image sensor (100×100 pixels, BASIS, Canon, Tokyo, Japan), an infrared charge-coupled device (CCD) camera, and a high-resolution digital CCD (LMD-1000, Victor, Tokyo) camera. For the determination of the speckle effect, a blue component argon laser (wavelength 488 nm, maximum power 3 mW, 2002–3SLL; Uniphase, San Jose, CA) is integrated into the illumination pathway of the fundus camera (Figs. 9.4 and 9.5). Given that a blue-component argon laser is mostly scattered in the superficial retina and absorbed in the retinal pigment epithelium, the signal is mainly derived from the superficial layers of the retina.

Experiments in Dutch rabbit eyes have confirmed that measurements obtained using a blue-component argon laser represent the retinal circulation, being little affected by the underlying choroid [38]. The fundus, which is illuminated by the laser, is observed by an infrared CCD camera. A high-resolution digital CCD camera is used for the measurement of retinal vessel diameters and recording of fundus photographs. The backscattered laser light is imaged on the image

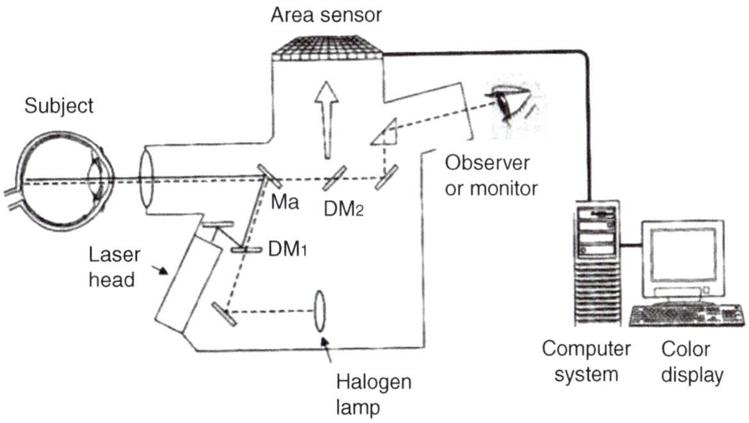

Fig. 9.5 Schematic view of the laser speckle system

sensor, on which the varying speckle pattern appears. The system is capable of measuring both retinal and optic nerve head blood flow velocity.

The blood flow velocity is calculated using first order statistics as square blur ratio (SBR), which is proportional to the reciprocal of the squared speckle contrast. This can also be approximated by using the so-called normalized blur (NB), which is a quantitative index of the blurring of a speckle pattern. The NB values are then translated into a two dimensional color-coded image, in which red areas represent areas with high blood velocities. Only recently, a system became commercially available allowing for a better resolution of 400×400 pixels by employing a usual CCD array (Softcare Ltd., Iizuka, Japan).

Whereas measurements of retinal blood flow have been realized in animal models, the extensive glare prevents measurements using blue lasers in humans. Alternatively, a diode laser can be used to illuminate the fundus. If this laser beam is directed toward a fundus location containing both retinal and choroidal blood vessels, the two vascular beds cannot be separated. If the beam is, however, directed toward the center of the macula lacking retinal microvessels, the subfoveal choroid can be measured [13]. Alternatively, the laser beam can be directed toward the optic nerve head, thereby assessing optic nerve head blood velocity ([37], Fig. 9.6). With some modification, blood flow in the iris can also be measured by focusing the beam toward the iris vasculature and increasing the power of the laser beam [42].

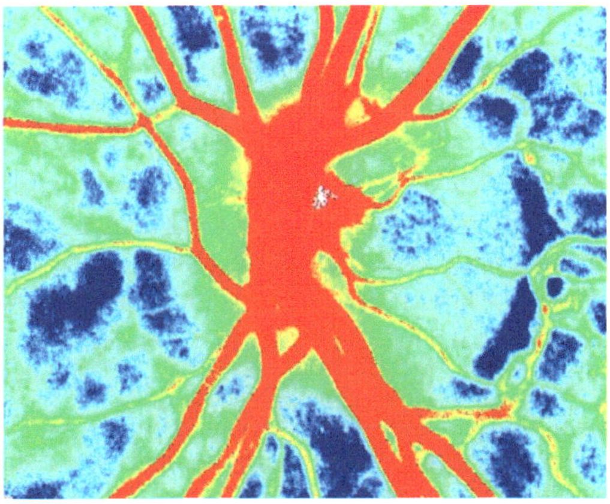

Fig. 9.6 A color code map showing the SBR values of ocular fundus around ONH in a healthy subject on the display of the laser speckle system

The reproducibility of the system has been published in several reports [38, 40]. For measurements in the rabbit retina, the coefficients of reproducibility of the main outcome parameter NB have been shown to be 8.2±1.8% for 5-min intervals and 13.8±2.0% for 24-h intervals, respectively [38]. In humans, the coefficients of reproducibility of 1-min and 24-h interval measurements of NB in the ONH rim tissue were 11.7±3.3% and 13.0±3.0%, respectively. For the choroid-retina, the corresponding values were 8.7±1.5% and 9.7±2.5% (n=12 eyes of six healthy subjects, [41]). In another group of 18 eyes of nine healthy volunteers, the coefficients of reproducibility of 21-day interval measurements of NB in the optic nerve head rim tissue were 8–18% for each time of measurement [41].

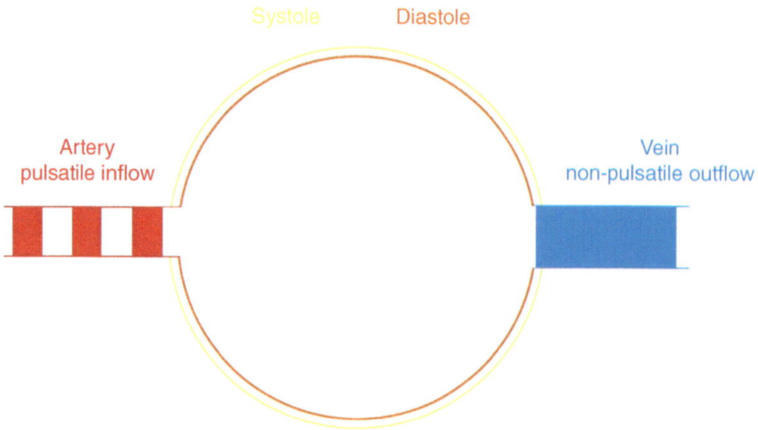

Fig. 9.7 Pulsatile inflow through the arteries and non-pulsatile outflow during the cardiac cycle. During systole ocular volume increases associated with an increase in intraocular pressure

As with most other techniques for noninvasive blood flow assessment in humans, the quality of the readings is strongly dependent on clear ocular media. Furthermore, although the software can adjust for small eye movements, good fixation abilities for the subject under study are needed. Thus, measurements in patients with poor central vision will result in greater variability. As with laser Doppler flowmetry, the outcome variables are given in arbitrary units. Thus, it is difficult to directly compare the data obtained from different sites of measurements or from different subject eyes. Moreover, the technique provides velocity measurements only, although studies have shown that there is a high degree of correlation between measurements of NB and those obtained with hydrogen clearance [37] or the microsphere technique [38]. For measurements in the peripheral human retina, the relative contributions of the choroid and the retina cannot be separated. The use of a confocal system may potentially overcome this problem, but such a system has not yet been realized.

9.1.2 Pulsatile Ocular Blood Flow

Another method for the assessment of ocular blood flow is to measure the changes in ocular volume and pressure during the cardiac cycle. The widespread use of the Goldmann applanation tonometer allows the ophthalmologist to easily observe the pulsatility of the IOP in daily practice and its variation depending on different physiological and pathological states. It was recognized early that differences in these pressure variations may occur in several ocular pathologies. The first approach to quantify these changes in pulsatility of the IOP was introduced by Perkins [23]. The ocular pulse amplitude was measured using a Goldmann applanation tonometer. In particular, the tonometer tip was modified and connected to a pressure transducer to allow for a continuous transmission of the ocular pulse, which in turn generated a pulse trace. Using these data, pulsatility was compared between the investigated eye and the fellow eye in order to identify carotid insufficiency [23].

This technique was further developed by Langham and colleagues to assess the pulsatile choroidal blood flow based on pneumotonometry. The basic principle of this technique is based on the observation that if a bolus of fluid (as it happens during systole) enters the eye, the eye will expand (Fig. 9.7). Consequently, given the assumption that the eye is elastic and the fluid is incompressible, the eye will increase by the volume of the bolus associated with an increase in IOP. When the bolus of fluid leaves the eye again, this will in turn lead to a decrease in eye volume. Thus, if the relation between IOP change and volume change is known (as it is described by the Friedenwald equation), then pressure measurements can be transformed to blood flow measurements [34].

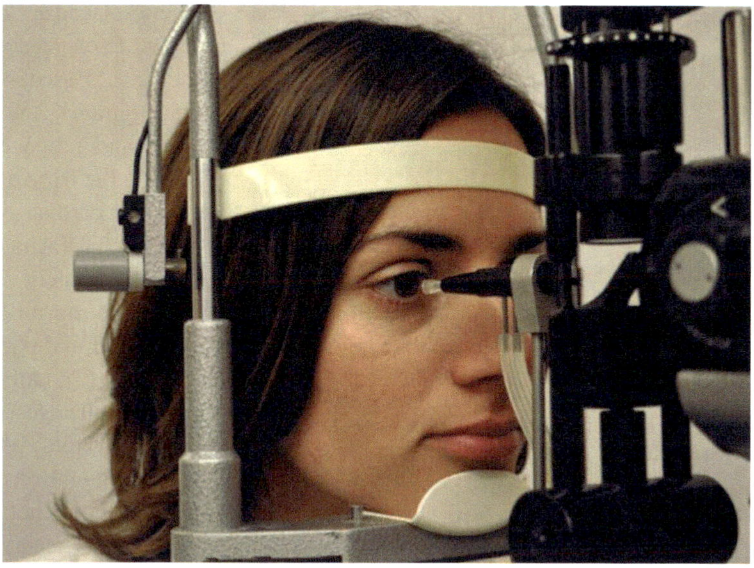

Fig. 9.8 Photograph of the pneumotonometer used for the measurement of pulsatile ocular blood flow

Based on these considerations, the pneumotonometric OBF system has been introduced for the measurement of the pulsatile component of choroidal blood flow. This instrument records the waveform of the ocular pulse amplitude over time. This commercially available blood flow system (Paradigm Medical Industries, Inc., Salt Lake City, UT, Fig. 9.8) assesses changes in intraocular pressure, which are caused by the rhythmic filling of the intraocular vessels, with a pneumatic applanation tonometer as described above. Hereby, the maximum IOP change during the cardiac cycle is called pulse amplitude. Based on the pulse amplitude, the wave form, and the heart rate, the pulsatile component of choroidal blood flow can be calculated.

These calculations are based on a theoretical model eye, which is dependent on several assumptions [34]. First, the measurement of the IOP and its variation over time need to be valid. Second, the relationship between volume and pressure in the eye is taken to be known so that changes in pressure can be converted into changes of volume. This means that the ocular rigidity, which is used to derive ocular volume changes from changes in IOP, is assumed to be equal for all subjects. Third, the hydrodynamic model is based upon the assumption that venous outflow from the eye is nonpulsatile.

An advantage of the pneumotonometric system is that pulsatile ocular blood flow and IOP can be measured at the same time. It has been shown that the amplitude of the ocular pulse is influenced to a major extent by the axial length, and hence, the refraction of the eye [14]. This strong dependence of the pulse amplitude on the eye length can be explained by the differences in intraocular volume between smaller and larger eyes and is not necessarily related to reduced choroidal blood flow or altered ocular rigidity in myopic subjects [3]. A major limitation of all pulsatile ocular blood flow techniques is that only the pulsatile component of the choroidal circulation is assessed and no information on the steady component of ocular blood flow is obtained. Estimates of the pulsatile component of flow in comparison to total blood flow in the choroid vary between approximately 80% and 50% [18, 19]. Another major limitation derives from the conversion of the IOP change over the cardiac cycle to volume changes over the cardiac cycle. As mentioned above, the ocular rigidity is assumed to be equal in all measured subjects. However, experimental evidence indicates that ocular rigidity may considerably vary among subjects [6]. In diseases like age-related macular degeneration [8] and glaucoma [12], evidence has accumulated that sclera stiffness is increased.

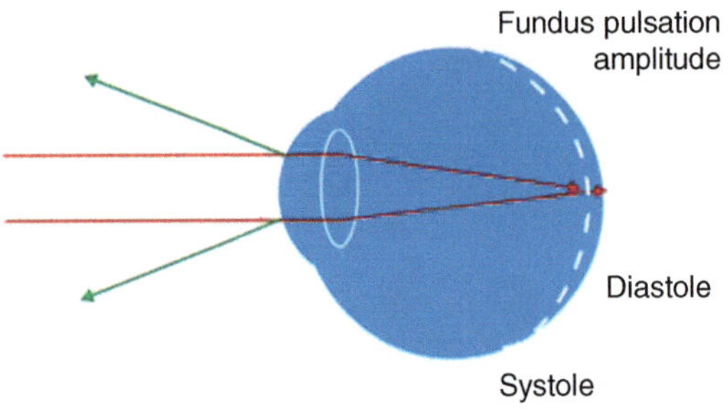

Fig. 9.9 Schematic drawing showing the pathways of light for the measurement of ocular fundus pulsations. The eye is illuminated by a parallel laser beam. The light is reflected at both the front side of the cornea (spherical wave) and the retina (plane wave). The waves form interference fringes

9.1.2.1 Laser Interferometry

Information regarding pulsatile choroidal blood flow can also be obtained by measuring distance changes between cornea and retina during the cardiac cycle [32, 33]. Given that the corneal-retinal distance changes with the periodic filling of the intraocular vessels, the distance change between cornea and retina contains information of the pulsatile component of choroidal blood flow. The ocular volume increases when the arterial inflow exceeds venous outflow during systole. This increase can be mainly attributed to choroidal swelling during systole and leads to a reduction of the distance between cornea and retina. This phenomenon is called ocular fundus pulsation, and the maximum distance change during the cardiac cycle is named fundus pulsation amplitude (FPA).

Technically, the eye is illuminated by the parallel beam of a single mode laser diode with a wavelength of 783 nm. This laser light with a high temporal and spatial coherence is then reflected by both the front side of the cornea and the ocular fundus. The wave reflected at the front side of the cornea is spherical, whereas the wave reflected from the retina is plane (Fig. 9.9). Hence, the two waves produce nonlocalized circular interference fringes, which is then subject of further analysis. The path difference between the two waves is twice the optical length of the eye. The interferences produced from the two reemitted waves are detected via a CCD camera, placed in the center of the interference fringes. Each readout is then plotted along the time axis. Based on these so-called synthetic interferograms, the distance changes between cornea and retina can be evaluated by simply counting the fringes moving inward and outward during the cardiac cycle.

The beam focus on the retina has a diameter of approximately 50–100 μm. The reflection from the posterior segment most likely occurs from the retinal pigment epithelium or Bruch's membrane [7]. Only the interferogram resulting from the strongest reflection of the posterior pole is visible. Whereas other interferogram systems may also arise from other retinal interfaces, they are not visible because of the low intensities. The main reflection from the cornea occurs at the front site of the tear film [11].

The system comprises a fundus camera, allowing for the real-time inspection of the ocular fundus (Fig. 9.10). Thus, measurements can be performed in predefined areas of the ocular fundus. However, given that the two reflected laser beams have to be superimposed, measurements are restricted to an area approximately 25° around the macula. In practice, measurements of FPA have been performed in the macular region and in the region of the optic nerve head. FPA in the macula is in the order of 4 μm in healthy young subjects, whereas in the optic nerve head region, FPA reaches up to 10 μm [32]. The higher FPA in the optic nerve head has been explained by the increased elastic properties in the optic nerve head region of the eye [30]. In the peripheral retina, FPA is slightly lower than in the macula [32].

Several studies indicate that FPA measurements provide high reproducibility and sensitivity in healthy subjects as well as in patients [25, 31, 33]. However, given that currently no gold standard exists regarding the measurement of choroidal blood flow, estimation of validity of the method is more difficult. Studies comparing laser interferometric measurements with pneumotonic measurement of pulsatile ocular blood flow reveal a strong correlation between FPA and POBF, indicating that FPA is a valid index of pulsatile choroidal blood flow in humans [30].

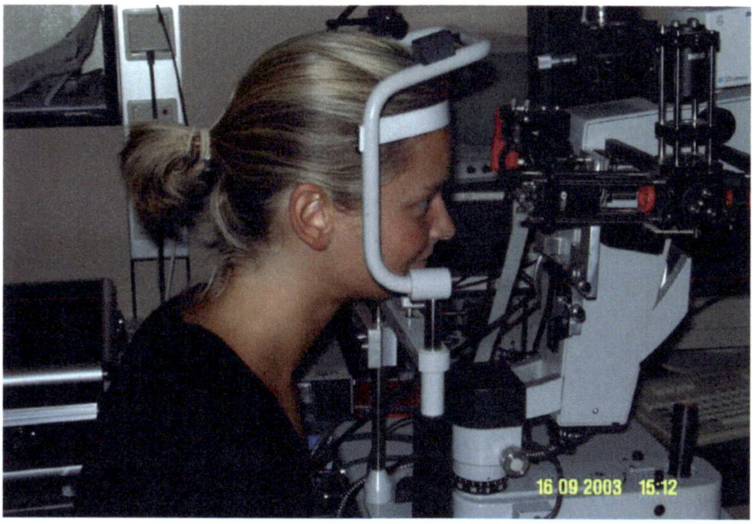

Fig. 9.10 Photograph of the laser interferometer used for the measurement of ocular fundus pulsation

Given that the measurement principle is based on a laser interferometric assessment of the eye length changes, pulsatile choroidal blood flow can be determined in a very high resolution compared to other techniques for the assessment of pulsatile blood flow. Additionally, longitudinal movements of the eye as caused, for example, by head movements do not influence the measurements. Obviously, the main limitation of the technique is also related to the lack of information on the nonpulsatile component of ocular blood flow. Furthermore, it is assumed that the inward movement of the retina at the posterior during the cardiac cycle is a good indicator of scleral outward movement. Whether this holds true for the human eye remains unclear. Employing a hemodynamic model of the human eye [3, 15], however, provides realistic estimates for pulsatile ocular blood flow. The problem may be overcome by using a technique called low-coherence tissue interferometry (LCTI), which uses laser light of short temporal coherence and allows for the measurement of movements of selected retinal layers during the cardiac cycle [5].

References

1. Adams DF, Shock JP (1986) Prediction of postoperative visual acuity in cataract patients using the flying corpuscle viewer. Ophthalmic Surg 17:509–512
2. Bauermann V (1980) Über das Phänomen der entoptisch sichtbaren Blutbewegung. Klin Monatsbl Augenheilkd 137:621
3. Berisha F, Findl O, Lasta M, Kiss B, Schmetterer L (2010) A study comparing ocular pressure pulse and ocular fundus pulse in dependence of axial eye length and ocular volume. Acta Ophthalmol 88(7):766–772
4. Briers JD, Fercher AF (1982) Retinal blood-flow visualization by means of laser speckle photography. Invest Ophthalmol Vis Sci 22:255–259
5. Dragostinoff N, Werkmeister RM, Groschl M, Schmetterer L (2009) Depth-resolved measurement of ocular fundus pulsations by low-coherence tissue interferometry. J Biomed Opt 14:054047
6. Edmund C (1988) Corneal elasticity and ocular rigidity in normal and keratoconic eyes. Acta Ophthalmol (Copenh) 66:134–140
7. Fercher AF, Hitzenberger CK, Drexler W, Kamp G, Sattmann H (1993) In vivo optical coherence tomography. Am J Ophthalmol 116:113–114
8. Friedman E (1997) A hemodynamic model of the pathogenesis of age-related macular degeneration. Am J Ophthalmol 124:677–682
9. Fuchsjager-Mayrl G, Malec M, Polska E, Jilma B, Wolzt M, Schmetterer L (2002) Effects of granulocyte colony stimulating factor on retinal leukocyte and erythrocyte flux in the human retina. Invest Ophthalmol Vis Sci 43:1520–1524
10. Grunwald JE, Sinclair SH & Riva CE (1982) Autoregulation of the retinal circulation in response to decrease of intraocular pressure below normal. Invest Ophthalmol Vis Sci 23: 124–127
11. Hitzenberger CK, Baumgartner A, Drexler W, Fercher AF (1994) Interferometric measurement of corneal thickness with micrometer precision. Am J Ophthalmol 118:468–476
12. Hommer A, Fuchsjager-Mayrl G, Resch H, Vass C, Garhofer G, Schmetterer L (2008) Estimation of ocular rigidity based on measurement of pulse amplitude

using pneumotonometry and fundus pulse using laser interferometry in glaucoma. Invest Ophthalmol Vis Sci 49:4046–4050
13. Isono H, Kimura Y, Aoyagi K, Fujii H, Konishi N (1997) [Analysis of choroidal blood flow by laser speckle flowgraphy]. Nihon Ganka Gakkai Zasshi 101:684–691
14. James CB, Trew DR, Clark K, Smith SE (1991) Factors influencing the ocular pulse–axial length. Graefes Arch Clin Exp Ophthalmol 229:341–344
15. Kiss B, Dallinger S, Polak K, Findl O, Eichler HG, Schmetterer L (2001) Ocular hemodynamics during isometric exercise. Microvasc Res 61:1–13
16. Kolodjaschna J, Berisha F, Lasta M, Polska E, Fuchsjager-Mayrl G, Schmetterer L (2008) Reactivity of retinal blood flow to 100% oxygen breathing after lipopolysaccharide administration in healthy subjects. Exp Eye Res 87:131–136
17. Kolodjaschna J, Berisha F, Lung S, Schaller G, Polska E, Jilma B, Wolzt M, Schmetterer L (2004) LPS-induced microvascular leukocytosis can be assessed by blue-field entoptic phenomenon. Am J Physiol Heart Circ Physiol 287:H691–H694
18. Krakau CE (1995) A model for pulsatile and steady ocular blood flow. Graefes Arch Clin Exp Ophthalmol 233:112–118
19. Langham ME, Farrell RA, O'Brien V, Silver DM, Schilder P (1989) Blood flow in the human eye. Acta Ophthalmol Suppl 191:9–13
20. Luksch A, Lasta M, Polak K, Fuchsjager-Mayrl G, Polska E, Garhofer G, Schmetterer L (2009) Twelve-hour reproducibility of retinal and optic nerve blood flow parameters in healthy individuals. Acta Ophthalmol 87:875–880.
21. Murphy GE (1983) Limitations of blue field entoptoscopy in evaluating macular function. Ophthalmic Surg 14:1033–1036
22. Pemp B, Polska E, Karl K, Lasta M, Minichmayr A, Garhofer G, Wolzt M, Schmetterer L (2010) Effects of antioxidants (AREDS medication) on ocular blood flow and endothelial function in an endotoxin-induced model of oxidative stress in humans. Invest Ophthalmol Vis Sci 51:2–6
23. Perkins ES (1985) The ocular pulse and intraocular pressure as a screening test for carotid artery stenosis. Br J Ophthalmol 69:676–680
24. Polak K, Dorner GT, Kiss B, Polska E, Findl O, Rainer G, Eichler HG, Schmetterer L (2000) Evaluation of the zeiss retinal vessel analyser. Br J Ophthalmol 84: 1285–1290
25. Polska E, Polak K, Luksch A, Fuchsjager-Mayrl G, Petternel V, Findl O, Schmetterer L (2004) Twelve hour reproducibility of choroidal blood flow parameters in healthy subjects. Br J Ophthalmol 88:533–537
26. Riva CE, Loebl M (1977) Autoregulation of blood flow in the capillaries of the human macula. Invest Ophthalmol Vis Sci 16:568–571
27. Riva CE, Petrig B (1980) Blue field entoptic phenomenon and blood velocity in the retinal capillaries. J Opt Soc Am 70:1234–1238
28. Riva CE, Sinclair SH, Grunwald JE (1981) Autoregulation of retinal circulation in response to decrease of perfusion pressure. Invest Ophthalmol Vis Sci 21:34–38
29. Riva CE, Grunwald JE, Sinclair SH, Petrig BL (1985) Blood velocity and volumetric flow rate in human retinal vessels. Invest Ophthalmol Vis Sci 26: 1124–1132
30. Schmetterer L, Dallinger S, Findl O, Eichler HG, Wolzt M (2000) A comparison between laser interferometric measurement of fundus pulsation and pneumotonometric measurement of pulsatile ocular blood flow. 1. Baseline considerations. Eye 14(Pt 1):39–45
31. Schmetterer L, Dallinger S, Findl O, Strenn K, Graselli U, Eichler HG, Wolzt M (1998) Noninvasive investigations of the normal ocular circulation in humans. Invest Ophthalmol Vis Sci 39:1210–1220
32. Schmetterer L, Lexer F, Unfried C, Sattmann H, Fercher AF (1995) Topical measurement of fundus pulsations. Opt Eng 34:711–716
33. Schmetterer L, Wolzt M, Salomon A, Rheinberger A, Unfried C, Zanaschka G, Fercher AF (1996) Effect of isoproterenol, phenylephrine, and sodium nitroprusside on fundus pulsations in healthy volunteers. Br J Ophthalmol 80:217–223
34. Silver DM, Farrell RA (1994) Validity of pulsatile ocular blood flow measurements. Surv Ophthalmol 38(Suppl):S72–S80
35. Sinclair SH, Azar-Cavanagh M, Soper KA, Tuma RF, Mayrovitz HN (1989) Investigation of the source of the blue field entoptic phenomenon. Invest Ophthalmol Vis Sci 30:668–673
36. Sinclair SH, Loebl M, Riva CE (1979) Blue field entoptic phenomenon in cataract patients. Arch Ophthalmol 97:1092–1095
37. Sugiyama T, Utsumi T, Azuma I, Fujii H (1996) Measurement of optic nerve head circulation: comparison of laser speckle and hydrogen clearance methods. Jpn J Ophthalmol 40:339–343
38. Tamaki Y, Araie M, Kawamoto E, Eguchi S, Fujii H (1994) Noncontact, two-dimensional measurement of retinal microcirculation using laser speckle phenomenon. Invest Ophthalmol Vis Sci 35:3825–3834
39. Tamaki Y, Araie M, Kawamoto E, Eguchi S, Fujii H (1995) Non-contact, two-dimensional measurement of tissue circulation in choroid and optic nerve head using laser speckle phenomenon. Exp Eye Res 60:373–383
40. Tamaki Y, Araie M, Tomita K, Nagahara M, Tomidokoro A (1997) Effect of topical beta-blockers on tissue blood flow in the human optic nerve head. Curr Eye Res 16:1102–1110
41. Tamaki Y, Araie M, Tomita K, Nagahara M, Tomidokoro A, Fujii H (1997) Real-time measurement of human optic nerve head and choroid circulation, using the laser speckle phenomenon. Jpn J Ophthalmol 41:49–54
42. Tomidokoro A, Araie M, Tamaki Y, Tomita K (1998) In vivo measurement of iridial circulation using laser speckle phenomenon. Invest Ophthalmol Vis Sci 39:364–371

Part III

Physiology

10 Systemic Determinants

John V. Lovasik and Hélène Kergoat

Core Messages

The choroid is the sole vasculature that provides nourishment to the subfoveal cones essential for normal central vision.

To sustain normal photoreceptor function in all areas of the fundus, choroidal blood flow may be distributed differentially across the fundus to meet regional differences in metabolic demand.

Regular physical exercise over several years promotes retinal blood flow through a measurable dilation of retinal arteries and veins.

Studies of vessel caliber must be interpreted in light of an individual's exercise habits. Exercise "conditions" retinal vessels to respond more efficiently to the metabolic requirements of the retina resulting from photic stimulation.

Rods are more vulnerable than cones to alterations in the ocular perfusion pressure.

1. Rod dysfunction may be the initial sign/symptom of a systemic vascular disorder.
2. Patients with vascular disorders should have regular visual field assessments or scotopic electroretinograms, including oscillatory potentials to assess rod-dominated peripheral vision.

The subfoveal choroidal blood flow decreases significantly when the retina changes from light to full dark adaptation.

1. Patients complaining of an inability to see "normally" at nighttime may simply be reporting this physiological redistribution of blood from the subfoveal choroid to the retinal periphery to support rod-dominated vision in darkness and not a rod-cone dystrophy.
2. Because preclinical ARMD has been linked to decreased subfoveal choroidal blood flow, any treatment for manifest ARMD may be helped by having the patient sleep with a small light adjusted to a luminance that prevents full dark adaptation and therefore prevents the physiological decrease in subfoveal choroidal blood flow that occurs in dark adaptation.

J.V. Lovasik, O.D., Ph.D. (✉) • H. Kergoat, O.D., Ph.D.
School of Optometry, University of Montreal,
C.P. 6128, Succursale Centre-Ville, Montréal,
QC H3C 3J7, Canada
e-mail: john.vincent.lovasik@umontreal.ca;
helene.kergoat@umontreal.ca

L. Schmetterer, J.W. Kiel (eds.), *Ocular Blood Flow*,
DOI 10.1007/978-3-540-69469-4_10, © Springer-Verlag Berlin Heidelberg 2012

A normal range of systemic blood pressure is critical to normal visual function. The combination of a chronically low systemic BP with a high normal IOP places the retina at risk for hypoxic damage through subnormal retinal perfusion.

1. Clinically, the optic nerve head may develop cupping pathognomonic of glaucoma in spite of a normal IOP level. This should trigger an examination of the cardiovascular system to ensure adequate perfusion to the eye and brain.
2. Similarly, when treating systemic hypertension, care must be taken not to change a favorable balance between the OPP and the IOP so as to place the eye at risk for glaucoma; a moderate IOP level combined with a large reduction in the OPP favors the development of low tension glaucoma.

Healthy individuals between 20 and 80 years of age show a progressive narrowing of retinal vessels.

This age-related vasoconstriction must be considered when determining the presence or absence of vascular disease in the elderly.

Smoking tobacco causes a transient decrease in the subfoveal blood flow after cigarette smoking.

Chronic use of cigarettes may increase the risk for ischemic maculopathy.

10.1 Introduction

Oxygenated blood enters the retina through the central retinal artery (CRA) and is dispersed throughout the retinal arterial vasculature. Deoxygenated blood is evacuated from the retina by way of the central retinal vein (CRV) and the orbital venous plexus. The principal route for arterial blood flow into the retina starts with the large diameter internal carotid artery that gives rise to the smaller ophthalmic artery together with its finer offshoot, the CRA. This artery pierces the optic nerve behind the eyeball, travels a short distance, and then enters the eye by way of the optic nerve head (ONH). The CRV also courses through the ONH adjacent to the CRA to move devitalized blood from the eye back into the cardiopulmonary stream for reoxygenation. Once inside the eye, the CRA and CRV subdivide into four arteries and veins coursing in pairs, one for each of the retinal quadrants. The retinal vasculature perfuses the inner two-thirds of the retina with blood containing oxygen and other metabolites essential for supporting the very high metabolic rate that uniquely characterizes the human retina (Fig. 10.1b). The multiple short posterior ciliary arteries that also have their origin in the ophthalmic artery pierce the sclera around the optic nerve to link with the choroid that distributes oxygen and other metabolites to the outer third of the retina via the choriocapillaris (Fig. 10.1d). Choroidal venous effluent leaves the eye primarily via the 4–5 vortex veins at the equator.

With respect to the control of blood flow, it is generally thought that the retinal vasculature has autoregulatory ability, i.e., the intrinsic ability of the retina to maintain a constant level of blood flow despite changes in the ocular perfusion pressure (OPP), principally by modulation of the caliber of retinal vessels. The retinal circulation does not have autonomic innervation. Consequently, it is believed that retinal blood flow is controlled by autoregulatory mechanisms linked to metabolism (metabolic regulation), transmural pressure (myogenic autoregulation), and shear stress. In contrast, the choroid is well endowed with autonomic nerves and is under both sympathetic and parasympathetic control. It also responds to vasoactive hormones and drugs indicating that it is under humoral control. Lastly, the choroid can also respond to changes in perfusion pressure in a manner consistent with autoregulatory control. Clear evidence of choroidal myogenic autoregulation has been reported in animals under neurohumoral blockade [1]; however, unequivocal evidence of choroidal autoregulation in humans is lacking because it is difficult to alter the OPP without concomitant changes in neurohumoral

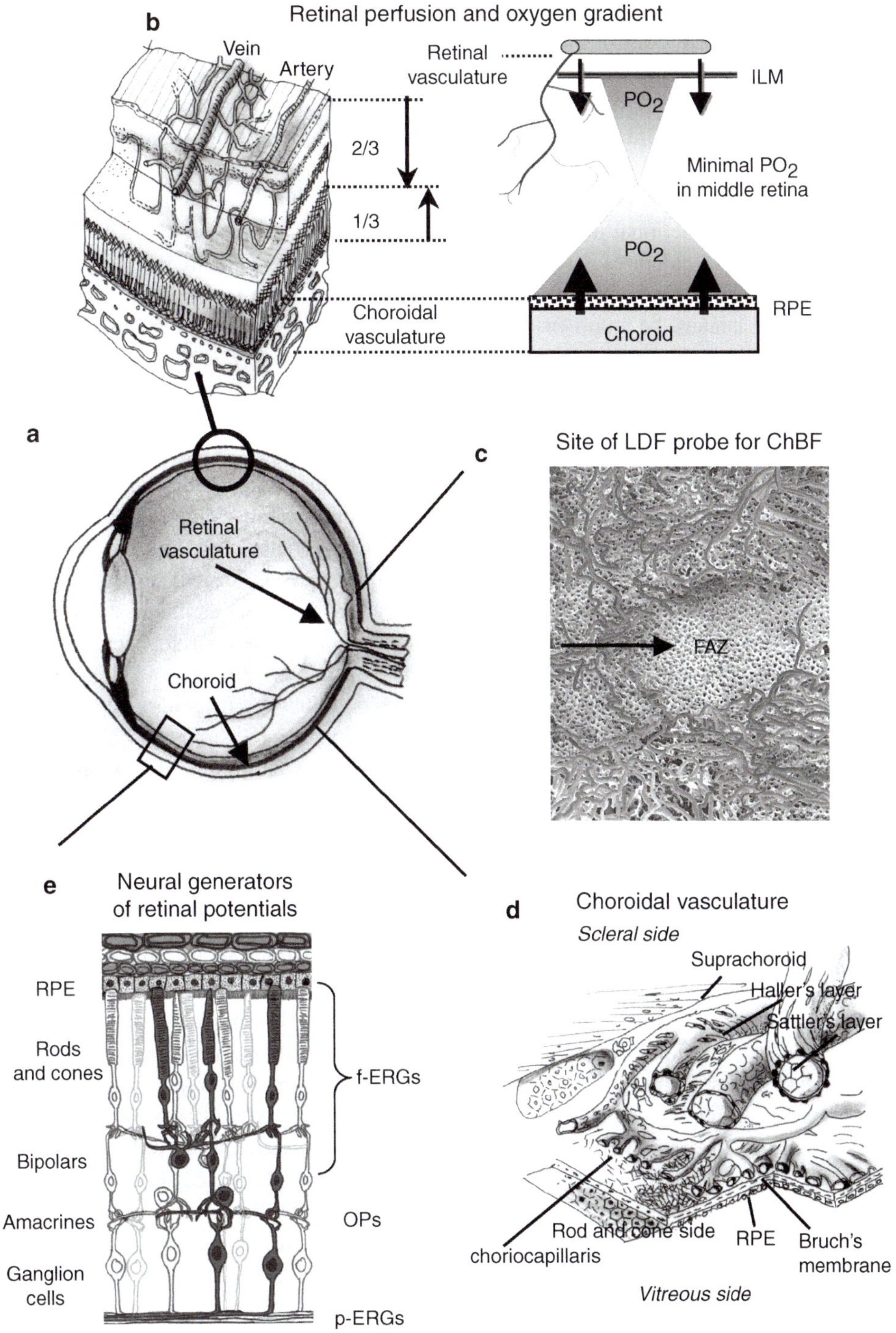

Fig. 10.1 Diagrammatic illustrations of: (**a**) the spatial relationship of the retinal and choroidal vasculatures in the human eye. (**b**) The relative depths to which the choroidal and retinal vasculatures perfuse the retina (*Left diagram*), and the sources and relative dissipation of blood PO_2 across the retina (*Right diagram*). (**c**) An exploded view of the foveal pit revealing the absence of retinal capillaries in the FAZ; recordings of the subfoveal ChBF were made here by NIR-LDF. (**d**) The large retrobulbar choroidal vessels that coalesce into the monolayer of capillaries forming the choriocapillaris. (**e**) An exploded view of a cross section through the retina showing the neural organization, and the cells contributing to the visually evoked retinal potentials that were used to quantify neural responses

input. Therefore, to reflect this unknown mix of extrinsic and intrinsic control mechanisms, we will use the term "choroidal regulation" when discussing the human studies in which choroidal blood flow (ChBF) was either unchanged or altered minimally during physiologically relevant changes in the OPP.

The structural and functional integrity of the retinal and choroidal vasculatures is essential for normal retinal function. Blood flow deficits in the choroidal system cannot be compensated by adjustments in the retinal vasculature, and similarly, deficits in retinal blood flow cannot be offset by modulation of the ChBF. Because of the location of the choroidal and retinal vasculatures with respect to the retina, deficits in either system result in unique abnormalities in the neural function of the cells in the outer third or the inner two-thirds of the retina. Clinically, neural dysfunction of a specific subpopulation of retinal neurons can be identified by noninvasive electrophysiological measurements referred to as "visually evoked retinal potentials." The neural generators of the most frequently employed visually evoked retinal potentials and their subcomponents are illustrated schematically in Fig. 10.1e.

Because of the high metabolic rate of the retina, both the choroid and retinal vasculatures must maintain adequate blood flow to preserve normal structure and visual function. In fact, the pathophysiology of leading causes of blindness in the human eye is believed to involve abnormal vascular perfusion of various retinal layers and the optic nerve *per se*. Consequently, much of the research into ocular blood flow has been dedicated to exploring physiological factors that influence the degree of blood flow in the human eye. To a lesser degree, but perhaps much more relevant to vascular diseases of the eye, is research that is focused on the interaction between neural function and ocular blood flow, the so-called "neurovascular coupling". Experimentally, this direction of research has involved examining the effects of transient changes in the OPP on blood flow in the choroidal and retinal vascular beds that perfuse opposite sides of the retina and the consequent effect on neuroretinal function indexed by visually evoked retinal potentials. With the advances in noninvasive imaging techniques, current research into ocular hemodynamics has also involved direct real-time measurements of vasomotor reactivity (dilation and constriction) to photic stimulation at rest and during transient experimental physiological stress. For general reference, the gross structure of the eye and sources of blood into the retina are illustrated schematically in Fig. 10.1a, b. The recording site for ChBF with a commonly used noninvasive technique, near infrared (NIR) laser Doppler flowmetry (LDF), is illustrated schematically in Fig. 10.1c. The anatomical relationship between the choriocapillaris, the innermost single layer of choroidal capillaries that lies adjacent to and provides metabolites for the retinal pigment epithelium (RPE)-photoreceptor complex, is shown in Fig. 10.1d. The larger choroidal vessels located in the suprachoroidal space are suitable candidates for controlling blood flow in the choroid because even small changes in their caliber could significantly modify blood flow in the choriocapillaris. Such a regulatory role for suprachoroidal vessels does not preclude fine-tuning of blood flow to the photoreceptors globally or regionally by the unique anatomical and functional attributes of the choriocapillaris [2].

10.1.1 Ocular and Systemic Blood Flow

Ocular blood flow must be adjusted continuously in order to sustain normal retinal function in the face of moment-to-moment changes in systemic blood pressure (BP). Practically speaking, OPP defines the pressure head available to drive blood through the ocular vasculature to nourish retinal neurons. The OPP is a positive pressure measured in mmHg that results from the interaction between the force driving blood into the eye and the intraocular pressure (IOP) opposing it. There are several ways to calculate the OPP for the human eye [3]. Currently, the most widely accepted formulation is $OPP=2/3(BP_{mean})-IOP$, where BP_{mean} represents the mean systemic BP measured in the brachial artery by sphygmomanometry, and the IOP is the intraocular pressure measured by

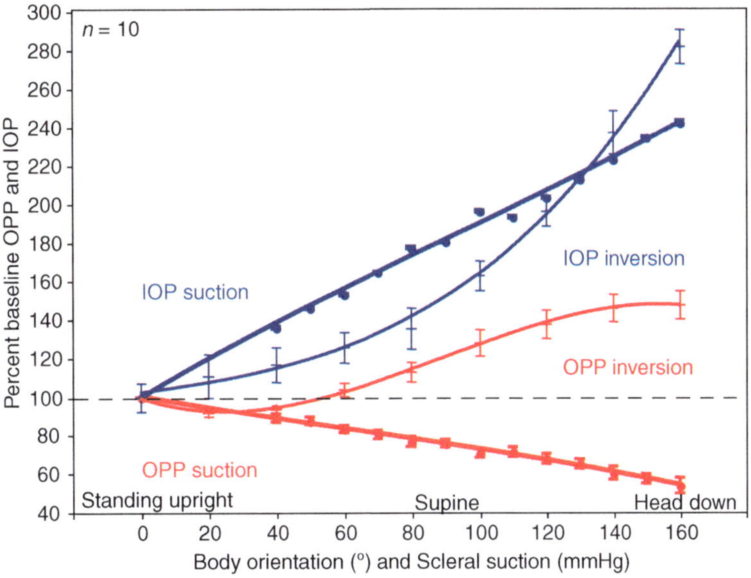

Fig. 10.2 Changes in the IOP and OPP as a function of scleral suction and backward tilt of upright subjects. Increasing levels of scleral suction increased the IOP and progressively *decreased* the OPP due to the constancy of the systemic BP. While body inversion also increased the IOP, the OPP was *increased* due to the larger increase in the systemic BP

tonometry. The fraction 2/3 is a correction factor needed to equate the brachial BP measured with the arm alongside the body to the pressure value at the eye level, i.e., in the central retinal artery. This reduction in the calculated pressure is solely due to the hydrostatic effect of the column of blood within the arteriovenous network within the arm. The BP_{mean} is derived according to: $BP_{mean} = BP_{diast} + 1/3(BP_{syst} - BP_{diast})$, where BP_{diast} and BP_{syst} are the diastolic and systolic BP, respectively. As such, a careful manipulation of either the systemic BP or the IOP can be used to alter the OPP systematically. The fraction 1/3 is a time-averaging constant for the different duration of the systolic *vs.* diastolic phases in the cardiac cycle. Figure 10.2 illustrates that the OPP can be reduced by scleral suction since that increases the IOP and partially offsets the pressure head driving blood into the eye. Conversely, the OPP and IOP are increased through a progressive declination of a subject from an upright standing position. The effect of body tilt is such that the net increase in BP exceeds the increase in IOP. The result is an increase in the OPP that is linked to the degree of body tilt. Figure 10.2 illustrates the curvilinear relationship between body orientation and the OPP.

Blood flow in the systemic or ocular vessels is normally modified by a change in vascular resistance that is manifest anatomically as a change in vessel caliber. Blood flow (Q) through a tubular vessel is calculated as: $Q = \Delta P \pi r^4 / 8nl$, where P = pressure difference between the ends of the vessel, r = vessel radius, n = viscosity of blood, and l = length of the vessel [4]. This formula can be simplified to $Q = \Delta P/R$, where $R = 8nl/\pi r^4$.

Contemporary LDF techniques for quantifying ocular hemodynamics provide only *relative* measurements of blood volume, velocity, and flow. Consequently, a search for, or an evaluation of, the functional state of vascular autoregulation in an eye necessarily involves safe levels of physiological stress where the OPP is changed over a predetermined range, and ocular blood flow is measured simultaneously during the stress interval. If blood flow measurements normalized to pretest values remain constant over time, a normally functioning autoregulatory mechanism can be *confirmed*. However, if the percent change in OPP is matched by the percent change in blood flow, the existence of vascular autoregulation can be *rejected*. With this in mind, attempts to compare blood flow in control and test subjects in any study must invoke some physiological challenge to determine the degree of response in each subject and thereby compare the integrity of blood flow responsiveness across the subjects.

Experimentally, changes in the OPP have been induced by a variety of noninvasive and minimally invasive procedures to assess their interaction with blood flow and hemodynamic parameters that come into play to control blood flow. The OPP in a normal healthy individual can be modulated up or down from its resting value by a variety of dynamic factors most notably aerobic exercise (e.g., stationary biking, running) that increases the OPP by the increase in systemic BP, temperature stimulation of a part of the body to elicit autonomic reflexes that influence vasomotor responses, the Valsalva maneuver (VM) where forced expiration against a closed glottis increases the OPP through a concomitant increase in the mean ophthalmic artery pressure, and systemic absorption of a substance like nicotine through tobacco smoking. Changes in the OPP can also be induced more *passively* through changes in body orientation and scleral suction. The degree to which blood flow in the test vasculature is influenced by changes in the OPP depends on the functional integrity of the inherent autoregulatory system, if one exists, and the percent change in OPP. In the absence of autoregulation, or beyond the range of autoregulation, experimentally induced transient changes in the ocular blood flow may affect neural retinal function. Such effects can be measured objectively using electrophysiological responses from distinct groups of retinal neurons. Changes in the amplitude and conduction time of evoked potentials that emanate from discrete subpopulations of retinal neurons signal the origin and degree of physiological dysfunction linked to the changes in the resting OPP. The results of research initiatives that have examined the interaction between ocular blood flow and neurophysiological function will be presented in later sections of this chapter.

A second line of research into the neurovascular coupling at the ocular level involves an examination of the interdependence of blood flow and neuroretinal function in senescence. For obvious reasons, it is not practical to perform a series of predetermined tests at regular intervals over many years. However, the effects of normal aging on both neural structure and function can be gleaned from what amounts to cross-sectional measurements in a cohort of healthy volunteers that represents the time span over which measurements of neural function would be made. The results from recent studies on the effects of senescence on ocular blood flow, neural structure, and function of the human eye will be presented in a later section of this chapter.

10.2 Local Skin Cooling Effect

The systemic BP is controlled by a group of neurons in the medulla oblongata that collectively form the "vasomotor center." Descending tracks from the cerebral cortex connect to this vasomotor center and relay in the hypothalamus. Brain lesions causing autonomic dysfunction may also cause myocardial damage, cardiac arrhythmia, and disturbances in the mechanisms regulating arterial BP.

10.2.1 Choroidal Blood Flow

The cold pressor test (CPT) is a powerful noninvasive clinical procedure used to detect autonomic abnormalities in cardiovascular function [5], subclinical hypertension, or even predict subsequent neurogenic cardiovascular disease [6]. Typically, it involves immersing a subject's hand into ice water and then quantifying the changes in arterial BP and heart rate (HR). Such localized cooling of skin triggers sympathetic vasoconstrictor nerve impulses to systemic blood vessels to control regional blood flow and increase the systemic BP through the release of catecholamines [7]. Because the CPT abruptly increases the systemic BP noninvasively, it has also been used as a research tool to provoke changes in the OPP in order to study the hemodynamic response parameters in both the retinal and choroidal vasculatures.

The human choroid has been reported to have regulatory behavior during aerobic [8] and isometric [9, 10] exercise that acutely raised the OPP. In these studies, it was hypothesized that blood flow was kept close to resting value through sympathetic vasoconstriction. As a follow-up to this hypothesis, Lovasik et al. [11] examined the degree to which blood flow in the choroid was regulated by stimulation of the sympathetic nervous system through localized cooling of the skin. Specifically, they measured the subfoveal ChBF during physiological provocation involving three consecutive trials of acute opposing changes in the OPP each for 5 min. Because the resting diameter of vessels were unknown prior to testing, the degree of cold-induced vasoconstriction, and hence the increase in OPP, was likely to differ across subjects and could be blunted if the test vessels were partially constricted due to higher sympathetic tone. To mitigate this possibility, warm compresses (40°C) were placed atop and under the forearm to maximize vasodilation so as to obtain maximal increments in the OPP when cold (4°C) compresses were used to increase the OPP. In their experiment, Lovasik et al. [11] used an initial 5-min warm phase that was intended to reduce the BP and OPP, ostensibly by maximizing the vasodilation across subjects. A 5-min cold phase was started immediately after the initial vasodilation to assess ChBF responses during sympathetic stimulation. Finally, the second 5-min warm phase was done to quantify ocular hemodynamic recovery from sympathetic stimulation. Changes in the subfoveal choroidal hemodynamics were recorded continuously with a NI-LDF system during the experimental perturbations in the OPP. The IOP was measured prior to and at the end of experimentation, and the BP was recorded electronically at 1-min intervals to derive the variables for calculating the OPP. These measurements allowed precise comparisons between provoked changes in OPP and the compensatory changes in ChBF, volume, and velocity. The changes in OPP and the subfoveal ChBF across the three test conditions are illustrated in Fig. 10.3.

The initial room air-to-warm provocation reduced the OPP to ~9.3% below baseline, but the ChBF remained close to baseline (Fig. 10.3a) through an increase in volume of ~6.6% (Fig. 10.3b) and a decrease in velocity of ~3.9% (Fig. 10.3c) with respect to their baseline values. During the warm-to-cold phase, the OPP and ChBF both exceeded baseline within 2 min by ~9.7% and ~7.7%, respectively. Thereafter, the OPP gradually returned to baseline just prior to the end of the cold phase, but the ChBF remained above baseline, indicating poor choroidal regulation. In the subsequent cold-to-warm phase, even though the OPP decreased by ~9.3% below baseline within 3 min, the ChBF was regulated and remained at baseline through a combined reduction in velocity of ~4.3% below baseline and an increase in volume of ~9% above baseline.

These results indicated that sympathetic activation by way of cold stimulation of the forearm prevented an effective and rapid regulation of the subfoveal ChBF. Follow-up studies are needed to better understand the amplitude and timing of regulation during such sympathetic stimulation. The trend toward a slight reduction in ChBF during the cold phase suggests that regulation might have occurred if cold stimulation had been continued for a few more minutes. Furthermore, it is not clear whether the behavior of ChBF during cold stimulation involved sympathetic neural activity alone or concomitant release of vasoactive substances that facilitated or impeded vasoconstriction.

In an earlier study by Marinier et al. [12], the effects of sympathetic stimulation on ChBF were examined without the preliminary warm phase to dilate vessels. Specifically, Marinier et al. [12] examined the effect of a *cold-to-warm* water immersion of the hand to induce changes in the OPP. This sequence of stimulation produced an increase in the OPP of about ~17.4% and ~12.5% in the subfoveal ChBF. This showed that the ChBF was only partly regulated, but to a level that was greater than the results reported in the preceding paragraph where provocation had been carried out in reverse order (i.e., warm to cold). These results suggested that sympathetic activation induced changes in vascular resistance that were about the same whether they were initiated during transition from normal room temperature to cold, or from warm to cold. Taken together, the results of these two studies led to the conclusion that the ChBF is poorly or only partially regulated when the increase in the OPP is mediated by cold-induced changes in vascular resistance. Further elaboration of the vascular physiology in the eye during the CPT needs to take into account factors such as body mass, experimental stress, and consumption of caffeinated drinks before experimentation to minimize intersubject response variability to a common stimulus.

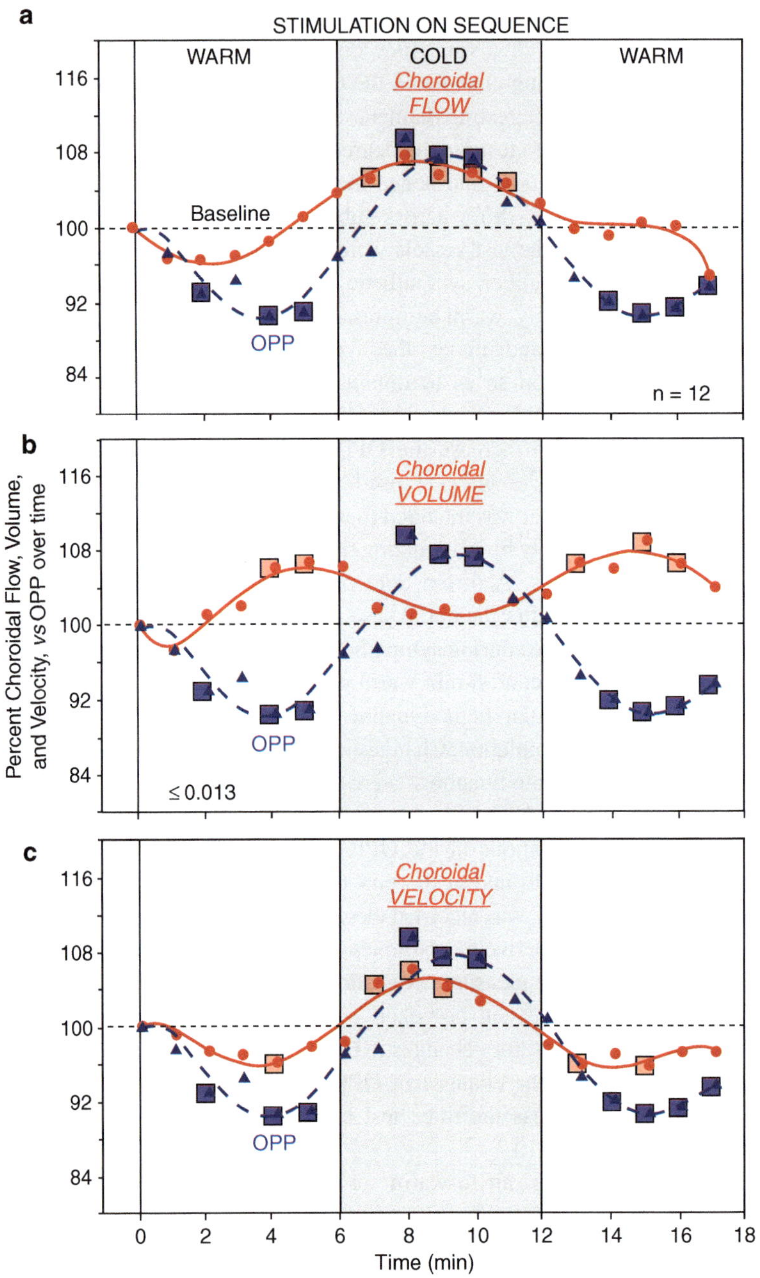

Fig. 10.3 Test results on vascular regulation in the human choroid through physiological provocation involving transient changes in the OPP. Systematic perturbations in the OPP were induced by changes in the systemic BP through the application of warm compresses on the forearm (40°C), then cold compresses (4°C), and a return to warm compresses each for 5 min. Simultaneously measured ChBF remained unchanged during the transient decrease in OPP (**Frame a,** warm compresses) mainly through an increase in blood volume (**Frame b**) but also a decrease in blood velocity (**Frame c**). The cold-induced increase in ChBF with increased OPP was not adequately offset by the reduction in blood volume (**Frame b**) and increase in blood velocity (**Frame c**). Data points within *squares* identify data that achieved statistical difference from baseline

10.2.2 Retinal Blood Flow

Inasmuch as blood flow at a location in a vessel is directly related to the flow velocity (V) and the vessel diameter (D), $Q = V \cdot \pi D^2/4$, changes in the diameter of retinal vessels reflect changes in blood flow if blood velocity is constant. With a Retinal Vessel Analyzer (RVA; Imedos Inc.) system, it is now technologically possible to make real-time measurements of changes in retinal vessel caliber during cold provocation. Figure 10.4a displays a sample real-time measurement of the vasomotor activity during cold provocation as recorded with an RVA system.

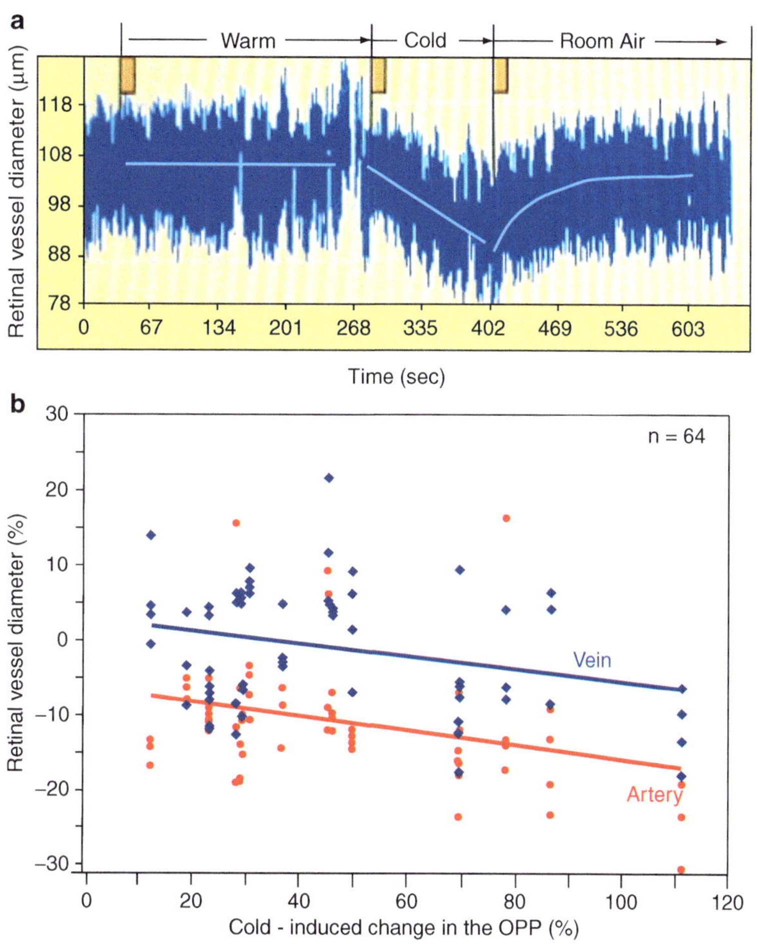

Fig. 10.4 (**a**) Dynamic changes in vessel caliber in response to the CPT. The instantaneous changes in a retinal artery were recorded with an RVA system during an increase in the OPP induced by applying ice-cold compresses to the forearm after an initial warm-up period of several minutes. The cold-induced increase in OPP was offset by a vasoconstriction of ~16% for this subject. The artery subsequently dilated slowly in room air to reach a diameter slightly smaller than measured during the warm-up phase. (**b**) Changes in the caliber of retinal arteries and veins for a cohort of subjects in response to changes in the resting OPP elicited by the CPT. The linear regression lines through the respective data sets revealed a vasoconstriction in both arteries and veins in response to a transient increase in the OPP

Figure 10.4b shows the spread of maximal change in OPP induced through cold provocation, as well as the corresponding maximal changes in retinal and venous diameters for a cohort of 64. These data reveal an inverse relationship between cold-induced changes in OPP and the resulting vessel diameter. Cold provocation elicited vasoconstriction that averaged to be ~10% in both arteries and veins [13]. The constriction of retinal vessels during a transient increase in the OPP, and the inverse relationship, were consistent with a retinal *regulatory* response aimed at keeping blood flow close to pretest levels. Evidence for actual changes in blood flow would require measurements of blood velocity in the test vessel by laser Doppler velocimetry as well as the instantaneous vessel diameter.

10.3 Aerobic Exercise

10.3.1 Choroidal Blood Flow

Stationary biking can be used to increase the systemic BP with a view to measuring ChBF responses during the real-time increase in OPP. The results of one study using this method are shown in Fig. 10.5. The subfoveal ChBF, as measured by a portable LDF device, revealed that the ChBF increased by only ~5%, while the simultaneously measured OPP increased progressively up to a maximum of 42% (Fig. 10.5a) [8]. The dissociation between OPP and ChBF pointed to vascular regulation that kept ChBF close to its pretest levels. The principal site of ChBF regulation was likely in the larger vessels leading to the choroid (Fig. 10.1d), while

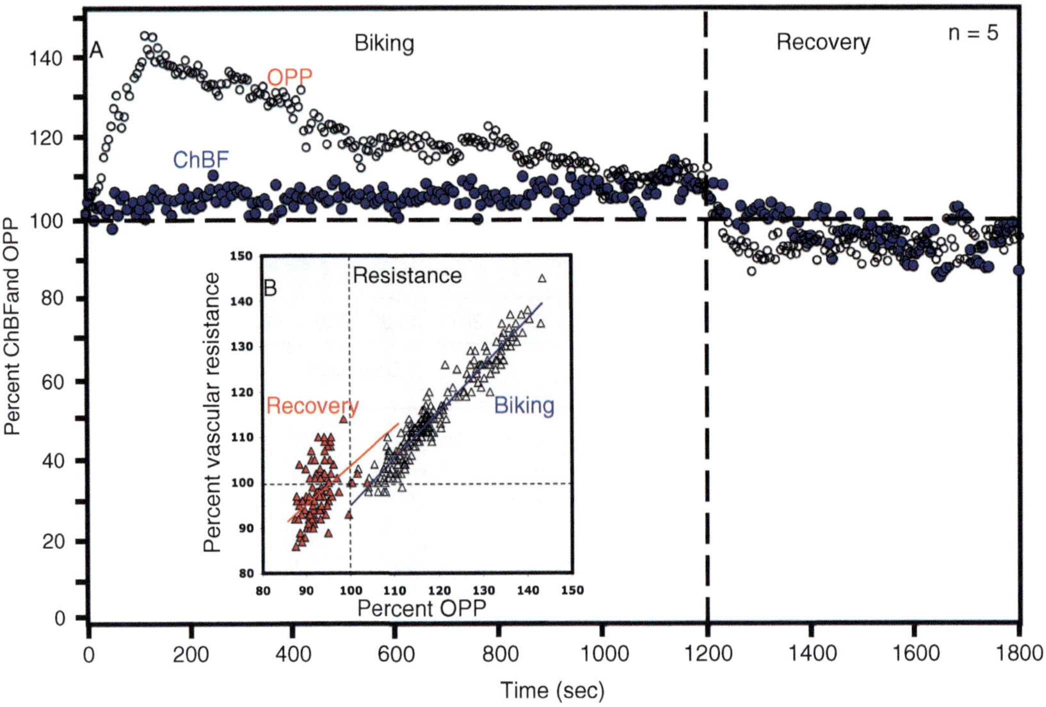

Fig. 10.5 Vascular regulation in the human choroid. (**a**) While stationary biking increased the OPP by ~42%, the simultaneously recorded ChBF remained within ~5% of baseline throughout testing. (**b**) The linear increase in the vascular resistance throughout biking was likely due to a sympathetically driven vasoconstriction. The resistance was much lower during recovery

"fine tuning" of blood flow may have occurred within the monolayer of capillaries comprising the choriocapillaris at the foveal avascular zone (FAZ) (Fig. 10.1c). Inasmuch as a one-to-one correlation was found between the OPP and vascular resistance during biking (Fig. 10.5b), it seemed likely that a sympathetically mediated vasoconstriction was the likely mechanism that regulated blood flow during stationary biking.

10.3.2 Macular Blood Flow

Control of blood flow in the macular capillary network during a transient increase in the OPP can be quantified psychophysically with a blue field entoptoscope (BFE). To quantify a subject's leukocyte density and speed of their movement, a subject first quantifies his entoptically viewed leukocytes at rest and matches them to a computer-generated pattern of leukocyte-like spots by adjusting velocity and density controls in the BFE. Subsequently, the same adjustments are made during experimental alterations in the OPP to determine if there is a change in blood flow in the perifoveolar capillary network. This approach was used in a study evaluating changes in capillary blood flow throughout a large increase in the OPP [14]. In that study, stationary biking increased the resting HR up to a target level of 140 bpm. Inasmuch as the level of physical fitness varied across subjects, the increase in HR was associated with a range of changes in the systemic BP and hence the OPP levels. This variation in physical exertion made it possible to derive a spread of OPP values needed to evaluate blood flow regulation over a range of augmented OPP levels. In order to evaluate the degree of autoregulation across subjects using the hemodynamic parameters provided by this device, blood

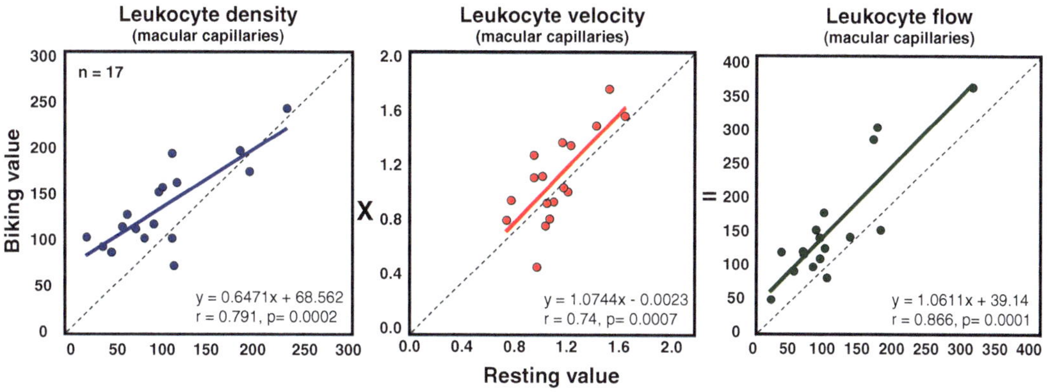

Fig. 10.6 Blood flow regulation in the perifoveal capillary network. The increase in OPP induced by stationary biking translated into an increase in the density of white blood cells in the macular capillary network. The velocity of white blood cells was virtually unchanged from resting values. The product of these two parameters showed that blood flow remained close to pretest values. Vasoconstriction of upstream arteries likely accounted for this regulated blood flow

flow was defined as the product of leukocyte density and velocity. When vascular autoregulation compensates fully for transient changes in OPP, graphs of leukocyte density at rest *vs.* leukocyte density during biking should fit a linear regression model with a slope of 1.0. The same linear interaction should be found for leukocyte velocity and flow at rest *vs.* biking.

The results of this study are depicted in the three plates comprising Fig. 10.6. The regression line through data sets for leukocyte density and velocity revealed slopes that were slightly below and above 1.0, respectively. The product of leukocyte density and velocity revealed a linear relation between flow during biking and flow at rest with a slope close to 1.0 and displaced vertically above the theoretical ideal response. This offset indicated that blood flow during biking was higher than blood flow at rest. Given that the group-averaged OPP increased by ~30% above resting value, the linear correlation between the degree of blood flow while biking and at rest with only a small displacement beyond the resting value pointed to the existence of an autoregulatory mechanism that kept blood flow close to resting value. Therefore, constriction of the larger upstream retinal arterioles was the presumed mechanism by which blood flow was maintained close to resting value during the transient increments in the OPP [15]. Alternatively, it is possible that increased vascular resistance may have been mediated by the action of pericytes on the capillaries [16].

10.3.3 Retinal Blood Flow

Since vessel diameter is the principal factor that modifies blood flow, real-time quantification of vessel caliber is a useful indirect index of retinal blood flow. Because retinal vessels can be visualized directly by standard fundus photography techniques and simultaneously digitized, special software to measure frame-to-frame changes in vessel diameter provides a valuable index of retinal blood flow. The objective of a recent study was to compare cardiovascular function *at the vessel level* in a cohort of ten healthy volunteers and ten long-term runners. The two groups of subjects were matched for age, gender, and OPP level. Because it is virtually impossible to record vasodynamic responses in retinal vessels during running exercise, an alternate way to differentiate these two groups by vessel reactivity was needed. Earlier, it had been reported that runners had higher concentrations of nitric oxide (NO) in their bloodstream compared to normal healthy individuals with a more sedentary lifestyle [17]. Based on this information, it was thought that the cardiovascular status in these two groups could be differentiated by their retinal vessel responses to flicker stimulation of the retina. Consequently, an RVA system was used to compare retinal vasomotor responses to 12.5-Hz flicker presented over 60 s and a 60-s recovery interval. In addition to these measurements of dynamic vessel reactivity, the caliber of retinal vessels was also measured over a ~2,700-μm distance with a VesselMap system (Imedos) in these two distinct cohorts of subjects.

At the *functional* level, the dynamics of *venous* dilation were the same in runners and controls, but the amplitude of *arterial* dilation was smaller in runners [18]. The group-averaged results in Fig. 10.7 show the amplitude and timing of retinal vasomotor responses to flicker in controls and runners. The retinal vessel response profile was characterized as a biphasic increase in diameter during flicker followed by a largely curvilinear constriction after flicker in the recovery phase. The degree of flicker-induced vasodilation was not correlated with the baseline caliber of arteries or veins. Inasmuch as the flicker-induced arterial dilation was lower in runners than control subjects, it was initially proposed that the vascular system in runners was less efficient for meeting the metabolic demands of the retina during sustained photic stimulation. An alternate possibility was that retinal vessels were larger in the runner group and therefore required less dilation to meet the metabolic demands of flicker-induced retinal activity. The insets in Fig. 10.7a, b show the group-averaged cross-sectional diameters of arteries and veins in the superior and inferior quadrants combined over a ~2,700-μm length of each vessel. The cross-sectional diameter of both arteries and veins was significantly larger in runners than in controls. This observation was consistent with the finding that runners had an elevated blood level of NO [17] and thereby required less dilation than the controls to increase blood flow by the same volume. Whatever the physiological mechanisms underlying these differences, these findings suggested facilitated neurovascular coupling in the retina of runners.

10.4 Neural Activation

10.4.1 Valsalva Maneuver

The standard VM involves forced expiration against a closed glottis to an expiration pressure level of ~40 mmHg. It is used clinically to test the *parasympathetic* control of HR, while the CPT is used to assess *sympathetic* control of the systemic BP [19]. From an experimental point of view, the increase in systemic BP during both of these procedures can be used to induce the changes in OPP needed to test for the existence of vascular regulation in the choroid. The four sequential changes in the systemic BP that occur during a VM have been described by Porth et al. [19] as follows: "Phase 1 is a positive phase in BP arising from an increase in the systolic and diastolic BP, likely due to the transmission of increased intrathoracic and intra-abdominal pressure to the left heart and arterial tree. Phase 2 is the longest, negative phase in BP arising from a rapid reversal of the increased systolic and diastolic BP via a decrease in the venous return. The decreased BP initiates a sympathetically mediated reflex to increase HR and peripheral vasoconstriction some 7 s into expiration. Phase 3 occurs immediately at the end of forced expiration and represents a very steep and brief decrease in arterial pressure due to a release of the intrathoracic pressure on the heart and aorta. Phase 4 is the last phase and represents an overshoot in the mean arterial BP due to the continuation of increased sympathetic tone and systemic vascular resistance. A reflex bradycardia follows due to stimulation of the arterial baroreceptors such that the mean BP and HR return to basal value."

In the absence of choroidal regulation, ChBF would parallel the changes in the OPP. However, a recent study [20] on the effect of increasing levels of forced expiration on ChBF reported that blood flow in the choroid exhibited a very different response pattern than that predicted if the choroid was a passive vascular bed directly reflecting the changes in systemic BP and thus the OPP. Subjects in that study performed a Valsalva-like maneuver wherein the force of expiration was increased from 20 to 120 mmHg in ~20 mmHg steps each held for ~10 s. During each level of forced expiration, the subfoveal ChBF and systemic BP were measured in real time with a confocal LDF and electronic sphygmomanometer, respectively.

The effects of these forced expiration levels on retinal vessel diameter were also measured with an RVA system that allows quantification of real-time changes in a retinal vessel diameter

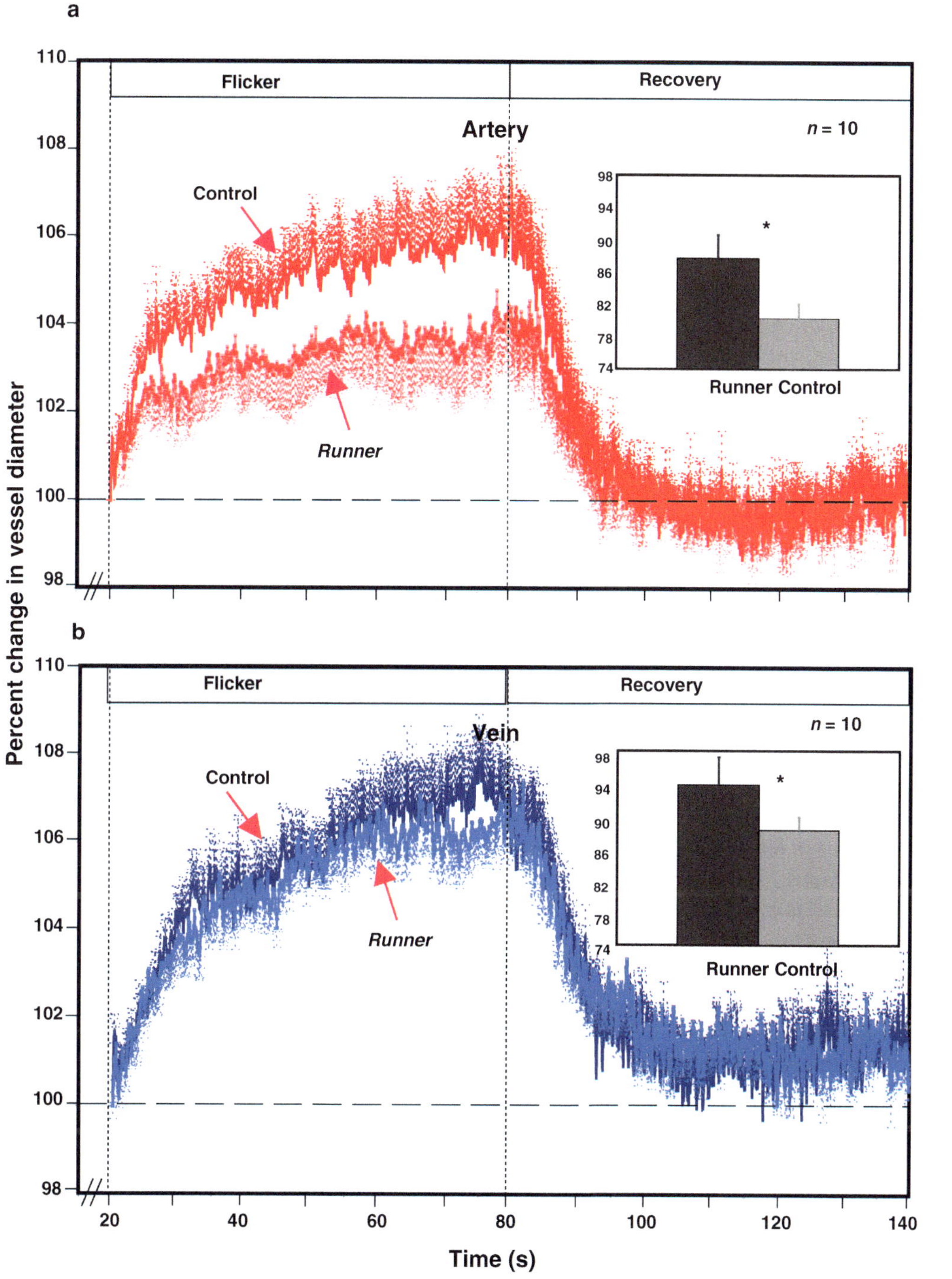

Fig. 10.7 Group-averaged data showing changes in the diameter of retinal arteries (**Frame a**) and veins (**Frame b**) during 60 s of flicker and a recovery interval of 60 s in normal healthy adults ($n = 10$) and a cohort of healthy runners ($n = 10$) matched for age, gender, and OPP level. The overall response profile was the same in both groups for retinal veins. However, the peak amplitude of the arterial dilation was smaller in runners than controls. To determine whether this difference in arterial response had a physiological basis, the cross-sectional diameter of retinal arteries and veins was then measured in these two groups with a VesselMap system. The results shown in the *insets* of frames **a** and **b** revealed that the arterial diameters in the control group were smaller than those in runners. This was interpreted as a higher efficiency in the vasculature of runners because it was capable of responding to higher metabolic demands by a smaller change in vessel diameter

through digital imaging techniques. This was done to compare the responses of the choroidal and retinal vasculatures to the same levels of a vascular stress for the purpose of identifying any response profiles that could be interpreted as signs of vascular regulation.

The results of the different levels of forced expiration on a retinal artery and vein and the HR are illustrated for one subject in Fig. 10.8a, b, respectively. Overall, increased levels of forced expiration transiently increased the diameter of retinal vessels, with a much greater effect occurring in retinal veins than in arteries. For VMs of 60 mmHg and greater, the peak of venous dilation above baseline was followed by a rapid constriction of about the same amplitude below baseline at the end of expiration. Because veins are devoid of constrictor muscle fibers, the rapid venous constriction illustrated in Fig. 10.8a was likely the result of a passive elastic constriction secondary to the rapid reduction in BP that occurs in Phase III of the VM. A comparison of the venous and arterial profiles in Fig. 10.8a showed that arterial and venous dilation started simultaneously. In contrast, the vertical dashed line drawn from the starting point for arterial constriction into the venous record revealed that arterial constriction preceded venous constriction by ~3.9 s and took ~2.6 s longer to develop maximal constriction. This arterial constriction was interpreted as an autoregulatory response to offset the increased perfusion pressure during the VM. The rapid arterial constriction at the end of each VM was also interpreted as an elastic response of the artery due to the rapid drop in BP in phase III of the VM. The increase in the arterial diameter between 20 and 80 mmHg with a progressive decrease in diameter during successive VMs may have been caused by a displacement of the head within the RVA apparatus during the difficult test procedures. Alternatively, a progressive accumulation of vasoconstrictor substance during the higher levels of forced expiration could account for the trend toward a smaller arterial caliber over time. Our experimental design could not confirm or reject this possibility.

The functional integrity of the baroreflex was illustrated by the effects of increased intrathoracic pressure and forced expiration against a closed glottis on the HR. Expiration between 20 and 60 mmHg caused a steep linear increase in the peak HR. Between 60 and 120 mmHg, the increased force of expiration in 20-mmHg steps also caused a linear increase in the peak HR but the increase in HR proceeded at a lower rate. These changes in peak HR during VMs in 20-mmHg steps are shown in Fig. 10.8b. The dashed lines atop HR recordings show the linear changes in peak HR across the levels of expiration.

The effects of forced expiration against a closed glottis (i.e., the VM) on the subfoveal ChBF and the systemic BP are shown in Fig. 10.9a, while a comparison of the effects of increasing forces of expiration on the systolic and diastolic BP are illustrated in Fig. 10.9b. The response profiles of the subfoveal ChBF (upper black recordings) *vs.* changes in systemic BP (corresponding lower red recordings) for VMs between 20 and 100 mmHg were consistent with regulatory responses in the ChBF inasmuch as the ChBF moved in directions opposite to those in the BP and the associated OPP levels. For example, at 40 mmHg, phase I of the VM was characterized by a rapid and brief increase in BP followed by a monotonic drop in the BP together with a progressive decrease in the pulse amplitude. During phase III of the VM, there was an abrupt decrease in the BP followed in phase IV by a slow increase in BP and pulse amplitude. The corresponding trace for changes in the ChBF, recorded in arbitrary units, was almost a mirror image of the changes in BP recorded in mmHg over time. Before the end of the VM, the ChBF had returned to baseline, a finding interpreted as evidence for blood flow regulation. (The green highlighted areas in each ChBF recording indicate sections where ChBF was regulated back to baseline; the red highlighted section of the ChBF for 120 mmHg shows failed blood flow regulation.) This unique interaction between the systemic BP (with associated changes in OPP) and ChBF, together with a return of the ChBF to baseline before the end of each VM, was seen for expiration forces up to and including 100 mmHg. At this level, regulated ChBF was seen for only a brief interval toward the end of the VM. At 120 mmHg, the largest increase in BP recorded during phase I and the subsequent

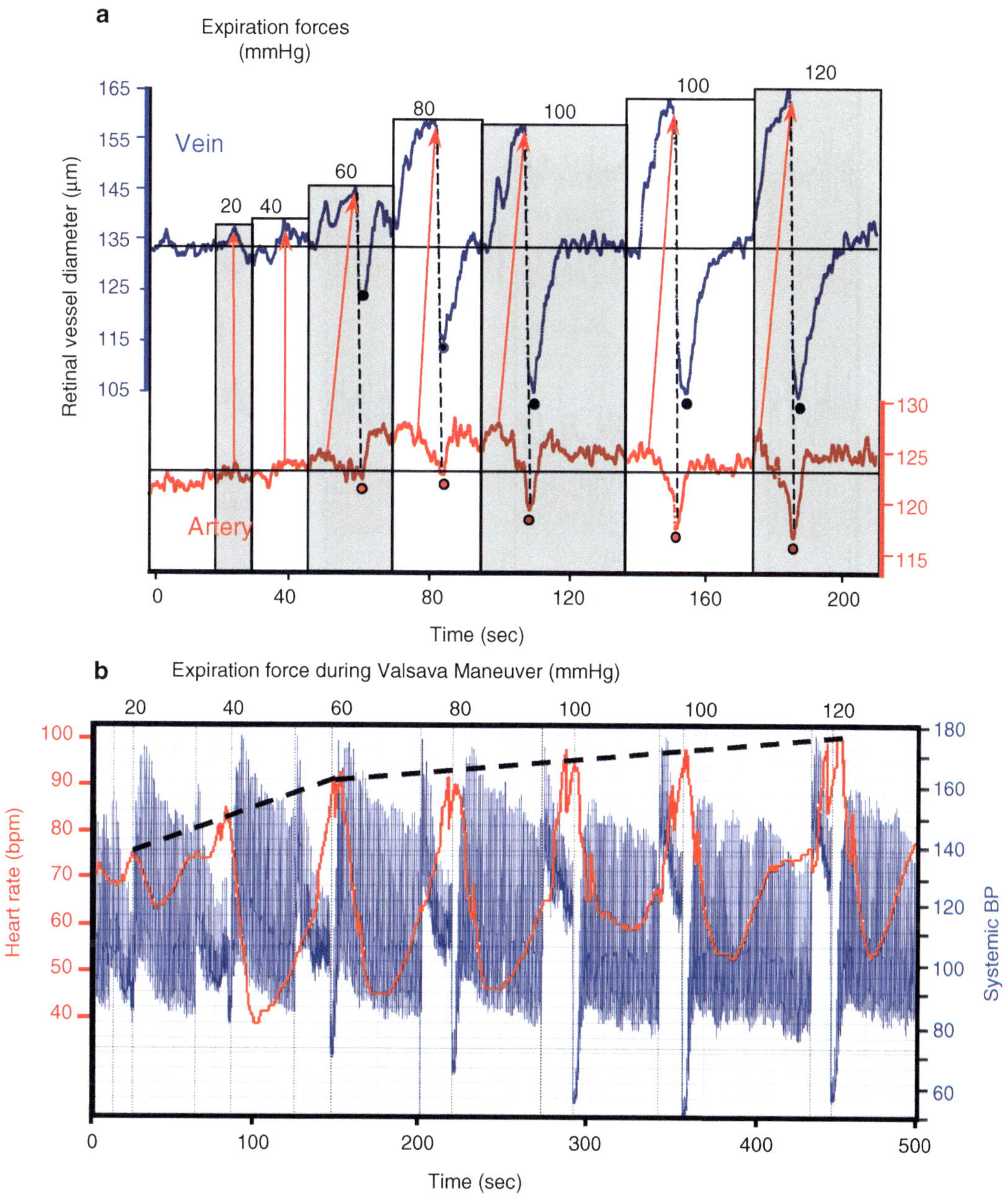

Fig. 10.8 **Frame a** displays sample records showing that VMs with increased force of expiration caused progressively larger dilations in veins at the onset of the VM, followed by a very rapid constriction below baseline of about the same amplitude. The absolute increase in vessel diameter was much larger in veins than arteries. For the retinal artery, the largest dilation occurred between the 60 and 80 mmHg VMs, and thereafter the diameter slowly decreased as the force of expiration increased. The *red oblique arrows* between the arterial and venous recordings highlight the finding that arterial constriction started an average of ~3.9 s before the very rapid venous constriction at the end of each VM. The onset of venous constriction corresponded in time with the greatly increased speed of arterial constriction seen readily at the end of VMs above 40 mmHg. **Frame b** shows the biphasic linear increase in the peak HR with increased force of expiration. The rhythmical sine wave changes in HR correspond to discrete phases in the VM. When the BP first increases in phase I of the VM, the HR decreases, and when the BP decreases in phase II and III, the HR increases. The twin peaks seen at the HR maximum likely correspond to the rapid BP changes in phase III and IV. During the following rebound increase of BP in phase IV, the HR drops again and then increases slowly once again as the BP drops in the late part of phase IV. Over the six levels of VM, the HR increased from a resting value of ~75 to ~100 bpm for the 120 mmHg VM. Over the entire set of VMs, the HR was modulated over a group averaged range of ~40 bpm

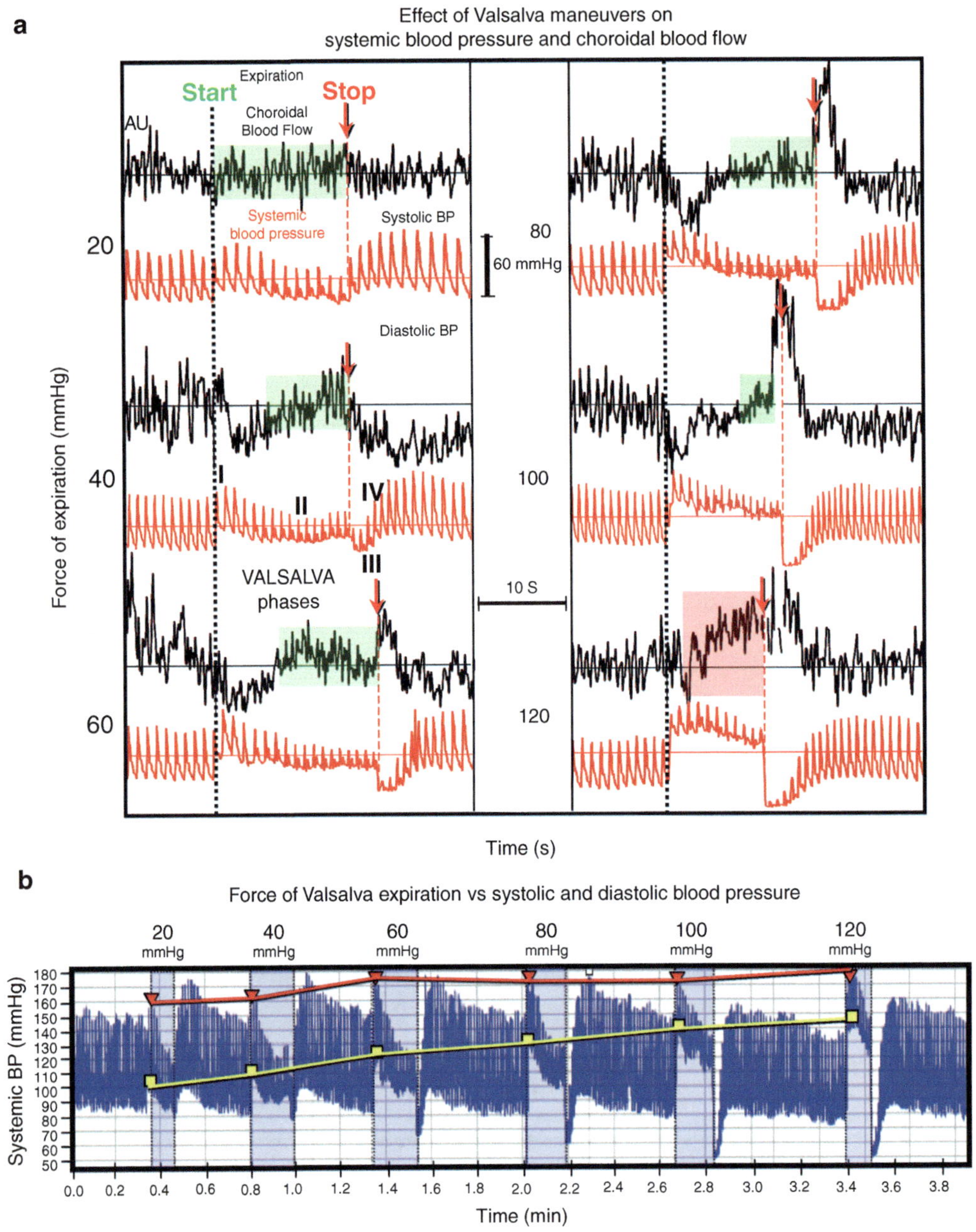

Fig. 10.9 **Frame a** shows the changes in the subfoveal ChBF (in arbitrary units, AU) and the systemic BP (in mmHg) recorded simultaneously during six incremented levels of forced expiration against a closed glottis (i.e., VM). Each VM caused changes in the BP that corresponded to four distinct phases as illustrated in the BP record for the 40 mmHg VM. The changes in the ChBF during each VM was consistent with blood flow regulation inasmuch as the ChBF moved in directions opposite to those seen in the BP, and the ChBF regained baseline well before the end of each VM until regulation was no longer possible. Sections where the ChBF regained baseline during each VM are highlighted in *green*. The absence of ChBF changes opposite to those in the BP, and failure to regain baseline ChBF before the end of a VM together were interpreted as failed blood flow regulation in the choroid. By these criteria, regulation failed during a VM of 120 mmHg (section highlighted in *red*). **Frame b** shows the continuous recording of BP during the six forced levels of expiration. The *lines* through the peak systolic (*inverted triangles*) and peak diastolic (*squares*) values in each VM revealed that the systolic BP reached a constant maximum pressure of ~170 mmHg after the 60 mmHg VM, but the diastolic BP continued to increase as the force of expiration increased. This indicated that the increase in OPP via the VM was driven principally by the increase in diastolic BP

decrease in BP in phase II of the VM were associated with a steep linear increase in the ChBF above baseline, without a return to baseline during the VM, a finding interpreted as the upper end of blood flow regulation in the choroid. It is noteworthy that at 120 mmHg of expiration, the BP throughout most of phase II was greater than the systolic peak of each cardiac pulse measured at baseline. This was the only test condition where the BP remained this high throughout most of the VM, and the likely reason why the OPP would have exceeded the range for blood flow regulation in the choroid. However, without simultaneous measurements of blood flow in the principal feeder arteries to the eye, the involvement of retrobulbar vessels such as the ophthalmic artery in contributing to blood flow regulation in the human choroid cannot be excluded.

Figure 10.9b presents the BP segments from Fig. 10.9a in their original continuous recording format. The line passing through the inverted triangles that identify the systolic peaks during each VM shows that the maximal increase in the systolic BP was constant across all forces of expiration except for the 20 and 40 mmHg recordings. In contrast, the line through the squares identifying the diastolic BP level at each VM reveals that there was a trend for a linear increase in the diastolic BP with each increase in force of expiration. This indicated that the increase in OPP across VMs was principally determined by the increase in diastolic BP. This increase in diastolic BP also suggested that choroidal regulation also had to compensate for a backup of blood volume during each VM.

10.4.2 Nicotine

Nicotine is known to affect both systemic and ocular blood flow. The absorption of nicotine that occurs during smoking of tobacco causes cardiovascular dysfunction, peripheral vasoconstriction, and cold hands [21]. At the ocular level, this psychoactive ingredient has been reported to increase blood flow velocity in the ophthalmic artery [22], lateral short posterior ciliary artery, and the central retinal vein [23]. However, it has also been reported to decrease blood flow velocity in the ophthalmic and posterior ciliary arteries [24]. These differences in effects may be physiological or due to differences in the sensitivity of methods used to measure blood flow velocity or differences between subject populations.

Because LDF measurements of blood flow are *relative* measurements, the effects of nicotine on choroidal hemodynamics in a preliminary study were assessed by comparing the effects of CPT on the ChBF before and after consumption of nicotine. The CPT was selected as the method to change the OPP because it induces a predictable sympathetically mediated systemic vasoconstriction and, as described in a preceding section, elicits blood flow changes in the choroid. In order to minimize any carryover effects of habitual smoking during provocation, a comparison of choroidal responses was set up for both smokers and nonsmokers. In the nonsmokers, the blood level of nicotine was increased by chewing two nicotine gum tablets (~6 mg of nicotine) that equated to smoking two cigarettes for the smokers [25]. Changes in the subfoveal ChBF were expressed as a percentage with respect to baseline to allow a comparison of results on a common scale. For both test groups, the subfoveal ChBF was measured prior to, during, and after immersion of the subject's hand into 4°C water followed by a 2-min recovery period in 40°C water. This procedure was carried out prior to and after consumption of nicotine in two separate test sessions. The results of the CPT in nonsmokers and smokers are presented in Fig. 10.10a, b, respectively. In both groups, the thicker lines represent the changes in the resting ChBF induced by the CPT before ingestion of nicotine. Inasmuch as the CPT is known to increase the systemic BP, the OPP would also be increased. Perfect choroidal regulation would promote vasoconstriction to keep blood flow close to resting values. However, the ChBF first increased by an average of ~17% for the nonsmokers and smokers combined and then decreased slightly below basal value within 60 s into the warm recovery phase. After consumption of nicotine, the cold-induced increase in ChBF was less in both groups of subjects.

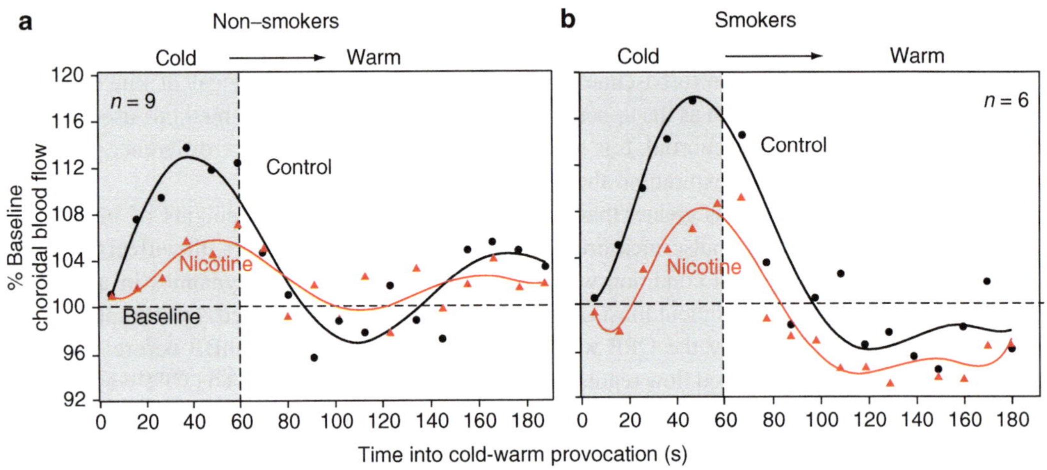

Fig. 10.10 Nicotine affects the subfoveal ChBF. Systemic absorption of a fixed dose of nicotine attenuated the ChBF response to the CPT in both nonsmokers (**Frame a**) and smokers (**Frame b**). Nicotine likely induced vasoconstriction throughout the systemic and ocular vasculature. These data suggest that habitual smoking of nicotine-containing tobacco compromises blood flow in the macula and places central visual function at risk for ischemic damage

Nicotine induces its effects through activation of the nicotinic cholinergic receptor sites. Its physiological effect with respect to blood flow is complicated since it first acts as an agonist-stimulating receptor in the peripheral sympathetic and parasympathetic ganglia, the adrenal gland, and the brain [26]. Secondly, it blocks synaptic transmission at these same sites. Therefore, its effects represent the sum of positive and negative effects on physiologically competing sites. Generally, nicotine increases the HR, enhances alertness, and induces a feeling of well-being likely through the release of dopamine [27]. Since the increase in ChBF induced by cold water was reduced by nicotine, this suggests that it serves as a vasoconstrictor in the choroid.

The observation that subfoveal ChBF was reduced by nicotine is yet another warning against chronic tobacco smoking since it may place the retina at risk for ischemic retinopathy. Tobacco smoking may also be related to decreased central visual function [28]. Hopefully, these preliminary findings of a reduction in ChBF at the fovea with smoking will incite further research into the effect of nicotine consumption on ocular blood flow and neural retinal function.

10.5 Blood Pressure Versus Ocular Perfusion Pressure

10.5.1 Increased Ocular Perfusion Pressure

It has been reported that head-down tilt is an effective noninvasive procedure for increasing the OPP in order to study vascular autoregulation [29]. The changes in OPP result from an increase in the IOP and the systemic BP. While body inversion can triple the resting IOP in some individuals, the greater increase in systemic BP results in an overall increase in OPP [30]. Figure 10.2 shows the effect of body declination on the IOP and OPP. The OPP can also be decreased noninvasively by scleral suction to study the effects of *reduced* perfusion on vascular autoregulation. The interaction between scleral suction, the IOP, and the OPP is also shown in Fig. 10.2. Together, body declination and scleral suction provide a simple experimental approach to increase or decrease the OPP for studies of vascular autoregulation or the consequences of altered perfusion on neuroretinal function.

10.5.1.1 Choroidal Blood Flow

Body declination has been used as physiological provocation to study regulation of the pulsatile and nonpulsatile components of blood flow in the choroid. During the orientation-induced increase in the OPP, the pulsatile component of choroidal flow, i.e., the pulsatile choroidal blood flow (POBF), was measured with a UK-OBF system while the nonpulsatile component was derived from LDF measurements of the ChBF [31]. When the OPP was increased by ~34%, the POBF was attenuated by ~47%. This reduction in the POBF remains constant for as long as the increase in OPP is maintained [32]. The pulsatile component of ChBF as measured by LDF also decreased by ~39% with the nonpulsatile component remaining unchanged for the same increase in the OPP. Inasmuch as the nonpulsatile component of ChBF remained unchanged during a large increase in the OPP, regulation of blood flow in the choroid was indicated. During full-body inversion of this study, the resting IOP was more than doubled. This large increase in IOP may have had a damping effect on the pulsatile component of ChBF measured by either technique. For a similar increase in OPP, the retinal vasculature at the level of the macular capillaries also demonstrated blood flow regulation. Blood flow in the macular network was assessed by the BFE where leukocytes become visible when the retina is illuminated with blue light. For an increase in the OPP of some ~39%, the macular blood flow remained constant when expressed as leukocyte density × velocity [33].

10.5.2 Decreased Ocular Perfusion Pressure

Experimentally, applying scleral suction to the eye increases the IOP because the physiological fluid within the eye, like all other fluids, is not compressible. This is accompanied by a decrease in the OPP since the BP is not changed by this procedure. Step changes in scleral suction are accomplished with a funnel-shaped cup approximately 1 cm in diameter with a cylindrical stub that is connected to a narrow flexible hose and a calibrated vacuum device. The following section will describe hemodynamic reactions of the choroidal and ONH vasculatures in response to transient reductions in the OPP as induced by scleral suction.

10.5.2.1 Choroidal Blood Flow

The basic scleral suction device typically used in ophthalmodynamometry or ocular pneumoplethysmography [34] can be modified by the addition of a stepper motor and logic board to change the OPP rapidly and in precisely controlled levels. Using such a programmable scleral suction system, Lovasik [35] recently reported the effects of the rate and degree of change in the OPP on choroidal hemodynamics. When the OPP was attenuated to a maximum of ~50% through moderate *ramped* increments of the IOP, the overall change in ChBF was only ~25%. In contrast, when the resting OPP was reduced by as much as ~90% using short and small decrements in the OPP, the ChBF remained within ~18% of its resting value (Fig. 10.11). From these observations, it was concluded that the effect of a reduction in the OPP on ChBF differed significantly when the reduction was induced by large rapid decrements in the OPP (ramped suction levels) *vs.* small square-wave decrements in the OPP. These observations suggested that the effectiveness of choroidal regulation is related to the amplitude and temporal attributes of changes in the transmural pressure in choroidal vessels [35]. *Clinically, these observations may have significant implications for the management of patients who are suspect for abnormalities in ocular or systemic blood flow. Specifically, retinal function may be abnormal even if the calculated OPP level is normal because blood flow to the eye is variable over time (turbulent or irregular) due to an obstruction in major vessels such as the internal carotid and ophthalmic arteries.*

10.5.2.2 Optic Nerve Head Blood Flow

A useful index of blood flow in the ONH is the so-called "chromatic pulse (CP)" [36]. The CP refers to the variation in the total area of a target tissue that reflects light with the same hue, saturation, and brightness. For the case of the ONH, it has been shown that there is a spontaneous variation in an area with the same chromaticity at a frequency that is linked in time to an individual's HR. This variation in the CP is measured from real-time color digital images (30 fps) of the ONH with a dedicated image analysis routine to define the

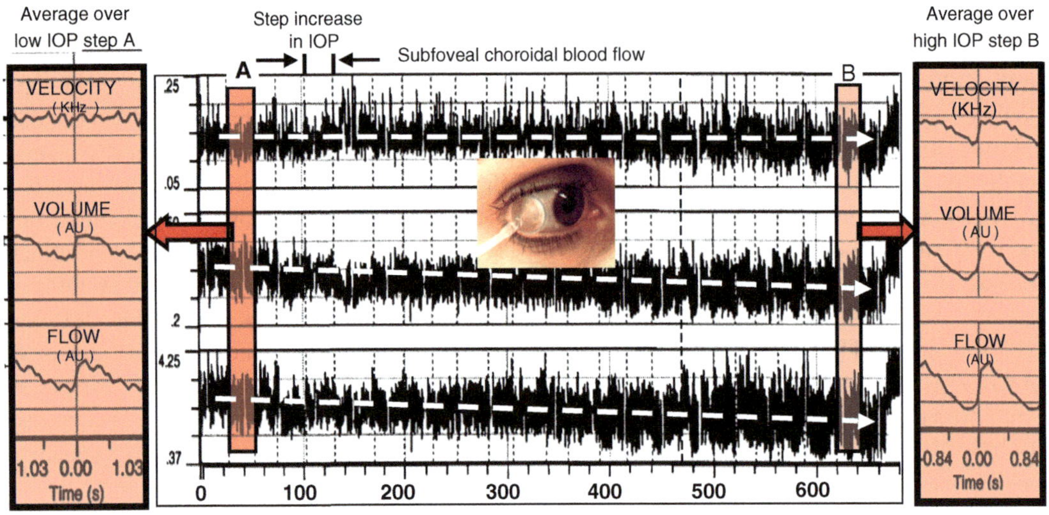

Fig. 10.11 Scleral suction decreased the OPP by a maximum of ~90% in ~3–5% steps over a 680-s interval. The stepped increase in IOP caused only a small reduction in blood velocity but larger changes in blood volume and blood flow. The increasing hypertonia also caused a monotonic increase in the pulse amplitude for blood velocity, volume, and flow. The change in pulse amplitude is best illustrated by the time-averaged data within the 20-s sample records taken near the start (A) and end (B) of blood flow recordings. An exploded view of these sections is shown in the panels on the *left-* and *right-hand sides* of the central record. These data sets show that the pulsatility almost doubled by the end of testing. Removal of the scleral suction cup resulted in large transient overshoots of each blood flow parameter

moment-to-moment variation in the total area with the same chromatic attributes. It has been reported that a physiological provocation to induce transient changes in the OPP elicits specific blood flow changes in the ONH that are readily measured by LDF [37]. It has also been reported that the LDF procedure measures blood flow to a depth of up to ~1 mm in the ONH within an area about ~150 μm in diameter [38]. Measurements with the CP show changes in pulse amplitude and timing that are nearly identical to those obtained by LDF during a transient reduction in the OPP induced by scleral suction. Scleral suction increases the IOP and reduces the OPP in proportion to the degree of suction, assuming no changes in the BP. Sample changes in the CP during transient reduction in the resting OPP are shown for one subject in Fig. 10.12. In this figure, areas of pallor within the ONH that have the same chromatic properties are shown as small islands demarcated by dotted lines. Because the fundus images were acquired at 30 fps, the spontaneous change in the size of these areas within 1 s is defined by 30 points. When these values are plotted as a function of time at rest, the 30 points produce a well-defined triangular profile that repeats at a frequency equal to the subject's HR. The rhythmical variation in the area with the same chromaticity has been given the name "chromatic pulse." The first record at the bottom left side of Fig. 10.12 shows a rhythmical series of triangles forming a "sawtooth" pattern. When the resting OPP was reduced by an increased suction level, the total area of pallor increased as indicated by the increasing size and number of islands in the ONH. This was seen as an increase in the CP amplitude and a vertical displacement of the record at each level of provocation (OPP decreased in 15% steps). When the OPP was allowed to return to its resting level, i.e., "scleral suction cup removed," the degree and extent of pallor decreased abruptly and the record dropped close to the resting level. From this, it was concluded that the CP reflects the temporal variation in the blood perfusing the ONH since the changes in pulse amplitude paralleled those measured by LDF for similar test conditions. A practical advantage of the CP is that it reveals blood flow changes over a much larger area within the ONH that is possible by LDF. Inasmuch as LDF measures blood flow in depth, and provocation induced the same pattern of

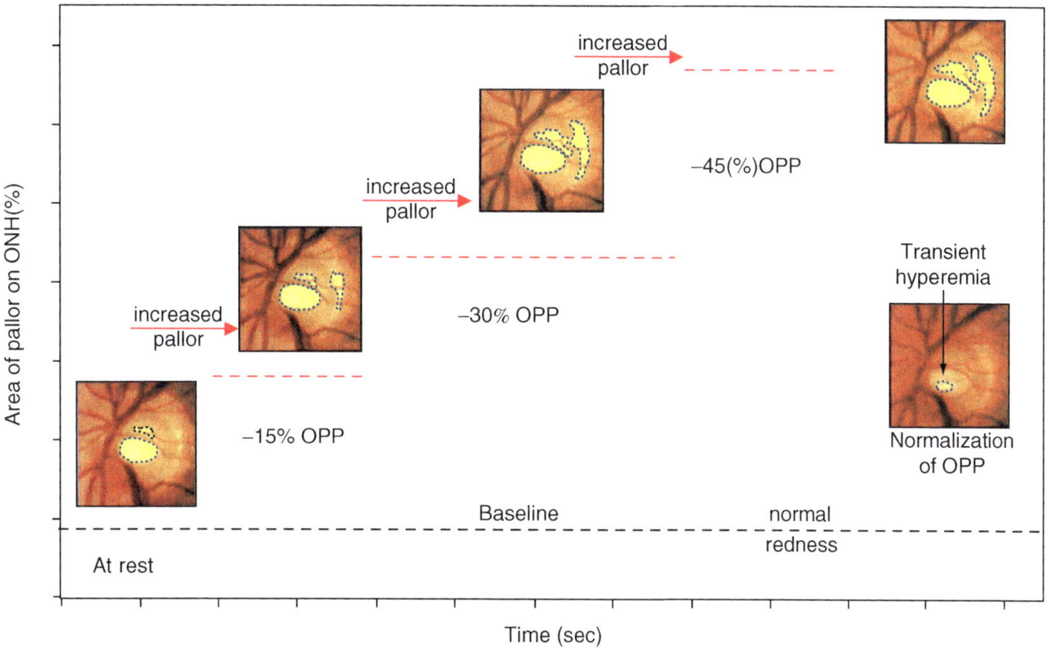

Fig. 10.12 Quantification of relative changes in ONH perfusion by the CP. The CP defines real-time changes in a region of interest (ROI) on the ONH with the same hue, saturation, and brightness. For example, plotting the area of pallor for the ONH as a function of time results in a sawtooth pattern with a periodicity that equals the HR. A stepwise reduction in the OPP increased the CP and caused a vertical displacement of the record that indicated an overall increase in the ROI. An abrupt normalization of the OPP resulted in a transient hyperemic response shown in this figure as an area that was smaller than that seen at rest and slowly returned to resting value. The sawtooth pattern for this phase was that recorded after the short-lived hyperemic response

change in the CP amplitude, the latter likely measures changes in blood flow not only on the surface but also within the ONH.

It is possible to get some idea of the depth of the vessels that contribute to the pinkish appearance of the ONH from clinical ophthalmodynamometry. When the IOP is increased to determine the diastolic and systolic pressures in the CRA, there are perceptible changes in the chromaticity of the ONH with each increment in the IOP. The observation that the progressive attenuation of the OPP was correlated with an increase in the ONH pallor supports the hypothesis that the CP represents changes in blood flow in the anterior portion of the ONH. The persistence of the pink appearance of the ONH even when the superficial retinal capillaries are compressed implies that blood is still present in the pial capillaries in the anterior portion of the optic nerve. Further elevation of the IOP results in an intense pallor of the ONH suggesting that light is no longer reflected through perfused capillaries. When the IOP reaches systolic value, the patient usually reports a loss of peripheral vision because only the central fovea remains perfused by the choroid. Thus, changes in ONH coloration are determined by the degree of tissue perfusion.

10.5.3 Neural Retinal Function

Normal function of the various neurons in the human retina relies on an adequate blood supply to all ten functionally interrelated layers of the retina. Neurons in the outer layer of the retina including the rod and cone photoreceptors rely on oxygen and metabolites derived principally from the choriocapillaris of the choroid. Ganglion cells, bipolar cells, and amacrine cells in the innermost retinal layers rely on blood flow from the retinal vasculature (Fig. 10.1). The choroid supplies the outer third of the retina while the retinal vessels with their capillary network perfuse the inner two-thirds. Physiologically, it is noteworthy that impaired blood flow in one system cannot be compensated by increased blood flow in the other. As such, choroidal ischemia results in unique neural dysfunction with classical clinically detectable alterations in retinal responsiveness to light. Similarly, subnormal retinal blood flow results in unique alterations in the function of the inner retina that can be detected by evoked potentials.

The dependence of visual neurons on blood flow for normal functioning is referred to as neurovascular coupling. Clinically, it is noted that visual functions attributed to different neural processing pathways may be compromised long before others when blood flow to the eye is reduced by subclinical pathology. For example, motion detection of coherent patterns that involves the retinal ganglion cells is compromised early in glaucoma [39]. This highlights a differential vulnerability of neuronal subgroups in the retina to ischemic processes. The integrity of neurovascular coupling in the human eye can be assessed electrophysiologically through stimuli and retinal adaptation that *functionally isolate* specific subpopulations of retinal neurons.

Psychophysical studies have shown the existence of phenomena-like binocular rivalry that likely has its origin at the cortical level. Neural pathways for the control of ChBF to each eye have been reported in subprimate species [40]. To date, there is no evidence for a binocular interaction arising from compromised blood flow to one eye. Consequently, this issue was addressed in a study where blood flow to one eye was reduced to identify any central modulation to both eyes. To do so, the responsivity of a cell population that is exclusively perfused by the retinal vasculature was examined as blood flow to one eye was reduced. The innermost retinal layer is populated by the ganglion cell bodies and their axons that converge from peripheral zones to make up the neural components of the optic nerve. The function of these cell populations can be measured electrophysiologically by the pattern reversal electroretinogram (pERG) in light-adapted conditions. In an earlier study, it was found that a decrease in blood flow to one eye influenced the reactivity of amacrine cells in the contralateral eye also [41]. This study was extended to include an evaluation of the vulnerability of the ganglion cell population to a transient decrease of the OPP in one eye while the pERG was recorded simultaneously from both eyes. To elicit the pERG, subjects fixated a red dot in the center of a high-contrast black-white checkerboard presented on a high-resolution monitor. Following ISCEV guidelines, the checkerboard was adjusted to subtend 29° and the checks were phase reversed at 5 Hz. A clinical signal averager was used to extract the 5–10 μV pERG signal from the ambient and biological noise. The effect of a transient decrement in the OPP on the neurovascular coupling was evaluated during a maximal reduction in the OPP of 60% in four steps of 15% as subjects fixated the checkerboard with both eyes through corrective lenses. The OPP was decreased in the test eye only with a scleral suction cup [42].

The results of this testing are shown in Fig. 10.13. For the test eye, the 60% reduction in the OPP progressively attenuated the pERG amplitude by a maximum of ~45%. Removing the suction cup caused a rapid return to pretest conditions. Remarkably, the amplitude of the contralateral pERG decreased by ~27% at the first suction level and remained near that level throughout further reduction in the OPP. Removal of the suction cup did not change the amplitude or profile of the contralateral pERG throughout the 8-min recovery phase. The persistence of the contralateral pERG at a level below baseline even after removal of the suction cup from the test eye raises several possibilities for the observed phenomenon. A plausible explanation for the contralateral attenuation of the pERG is a centrally mediated attenuation of blood flow to the contralateral eye. It could also be argued that the suction cup elicited an oculocardiac-like reflex, reduced the perfusion pressure for both eyes, and thereby caused the contralateral effect. However, the fact that ganglion cell function in the test eye normalized within 2 min while the contralateral eye remained attenuated throughout the 8-min recovery period suggests that changes in perfusion pressure alone could not explain the results. Additionally, the contralateral attenuation may have been the result of centrifugal fibers

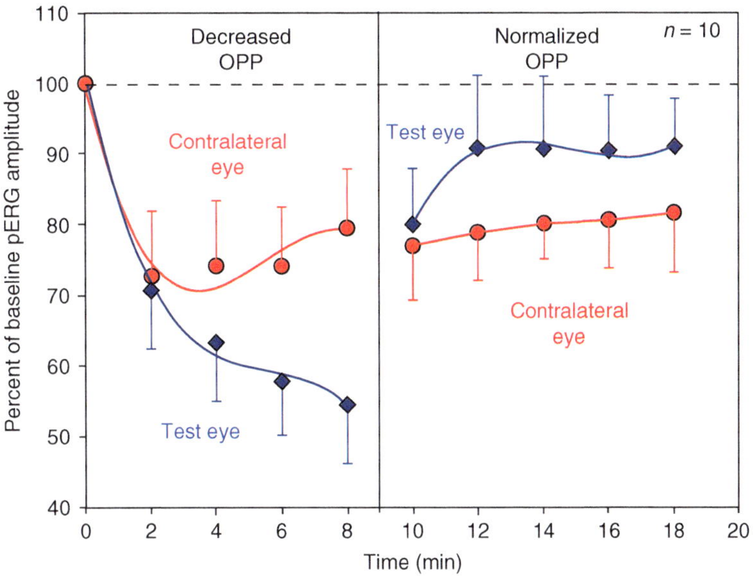

Fig. 10.13 A progressive reduction in the OPP caused by scleral suction, reduced ganglion cell function (indexed by the pERG) in both the test and contralateral eyes, with the attenuation being almost twice as large in the test eye. These changes in neural function implied that the retinal blood flow to the inner retinal layers could not compensate for the effects of decreased OPP. Because ganglion cell function was also affected in the untouched contralateral eye, the possibility of a central neural modulation of blood flow in the contralateral eye was raised. The cause of the prolonged attenuation of the contralateral pERGs after normalization of OPP in the test eye remains unexplained

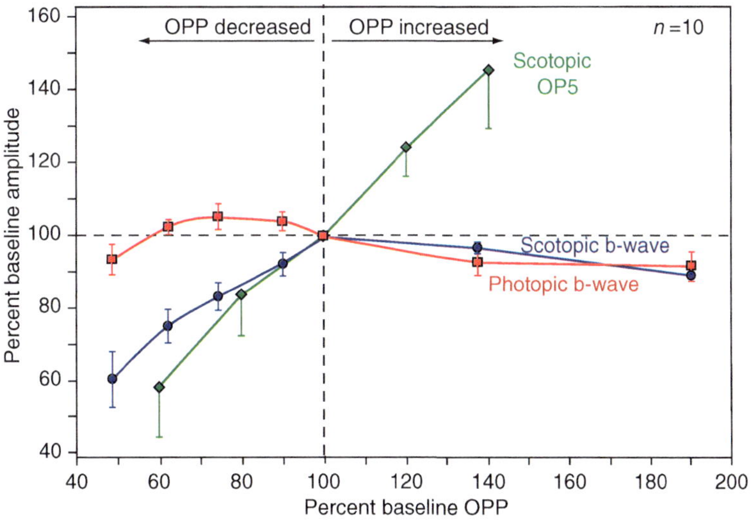

Fig. 10.14 Graphical illustration of the differential effects of an increase or decrease in the OPP on photopic *vs.* scotopic fERGs and the OP5 component that was the most sensitive to alterations in the OPP. Whether the OPP is decreased or increased, retinal function is attenuated to a greater degree when the retina is dark adapted

attempting to balance neural retinal activity in each eye. Some avian species are known to have efferent fibers to each eye for regulating ChBF in response to changing light conditions [43].

The effects of body inversion-induced hyperperfusion on neural retinal function have also been studied. A graphical summary of the principal effects of increased *vs.* decreased OPP on neural retinal function is displayed in Fig. 10.14. An increased OPP as large as ~90% caused minimal changes in the photoreceptor-bipolar reaction to light when indexed by the flash ERG (fERG) for both light- and dark-adapted states [44]. Inasmuch as some components of the fERG originate in the photoreceptor layer, and others reflect the activity of neurons populating the inner retinal layer, any changes in this potential are likely associated with changes in choroidal and retinal blood flow. In contrast to these minimal changes in neural function secondary to hyperperfusion, there were more dramatic changes in retinal reactivity to hypoperfusion. Furthermore, it is noteworthy that the neural responsiveness of the retina to transient decreases in the OPP differs dramatically when the retina is switched from a light- to dark-adapted state. Perhaps the most significant and interesting finding concerning the relationship between neural function and blood flow was the heightened vulnerability of rod photoreceptors to transient decrements in the OPP. When the retina was light adapted, the photopic fERG was not affected by either a transient increase or decrease in the OPP. However, when the retina was dark adapted and the neural response of the retina to light was driven exclusively by rod photoreceptors, both the a-wave and b-wave components of the scotopic fERG were significantly attenuated by a progressive decrease in the OPP [44]. This observation may reflect a change in the topographic distribution of blood, blood volume, and oxygen available to rods [45] in dark adaptation.

The changes in scotopically matched red and blue fERGs during 10% step decrements in the OPP, and the recovery pattern at 1-min intervals are shown in Fig. 10.15. The biphasic shape of the scotopic red fERG reveals the early cone and slower rod contributions to the fERG b-wave, while the monophasic blue fERGs show the rod-isolated responses. The rod components of the scotopic fERGs were seen to decrease when the OPP was reduced by as little as 10%. Further reductions in the OPP caused a progressive reduction in the rod contribution while the cone contribution remained unchanged until the OPP was reduced by ~40%. Furthermore, the cone contribution to the scotopic red fERG was still visible when the rod contribution was extinguished by a 50% reduction in the resting OPP. In addition, when the resting OPP was returned, cone recovery to resting values was more rapid than the rod recovery. Because increased scleral suction was used to reduce the OPP, the cause of the increased vulnerability of scotopic fERGs to decrements in the OPP may also have involved the concomitant increase in pressure of the vitreous against the retina as the IOP was elevated through scleral suction [44].

The increased susceptibility to ischemia of the inner retinal layers during retinal dark *vs.* light adaptation was also observed through measurements of scotopic *vs.* photopic oscillatory potentials (OPs), which reflect the functional status of the amacrine cells. During transient experimental increased OPP as large as 70%, amacrine cell function indexed by the amplitude of photopic OPs remained largely unchanged [46]. In darkness, however, a transient reduction in the OPP attenuated all components of the OP complex, while a transient increase in the OPP induced by body declination caused an increase in the amplitude of OP5 [30] (Fig. 10.14). While the physiological basis for this component-specific vulnerability of the scotopic OPs to variations in the resting OPP remains to be determined for the human eye, these findings nonetheless highlight a unique diagnostic capability of the scotopic OPs for increased or decreased blood flow in the inner retinal layers.

10.6 Blood Gases

The inspired gas content can easily be modified to alter the concentration of naturally occurring gases in arterial blood. Several studies have used such provocations to alter the concentrations of oxygen (O_2), carbon dioxide (CO_2), or nitrogen content in blood to study the effect on blood flow regulation [47, 48]. The ability of a vascular bed to adjust its blood flow parameters during transient changes in O_2 saturation (SaO_2) and/or other metabolites to preserve normal physiological function is referred to as "metabolic regulation." Metabolic regulation in other organs or tissues during a period of hyper/

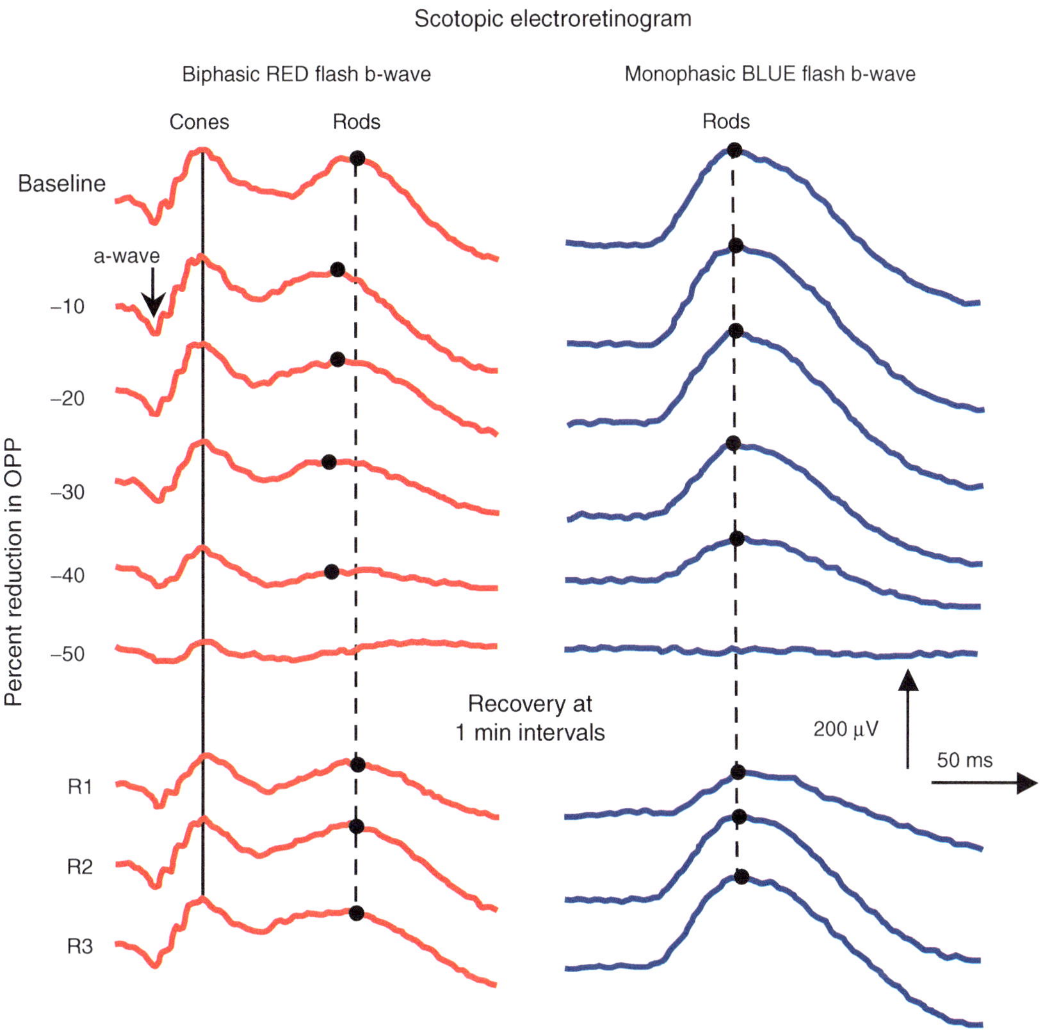

Fig. 10.15 Differential attenuation of cone and rod function in dark adaptation. A progressive reduction of the OPP preferentially attenuated rod photoreceptor function indexed by the second peak in scotopic *red* fERGs and the rod isolated *blue* fERGs

hypoxia or/and hyper/hypocapnia may be reflected by the vasomotor changes in the retinal vasculature. Because retinal vessels comprise the only vasculature in the human body that can be seen directly using noninvasive procedures, it is worthwhile for future research to determine whether changes in retinal vasodynamics and/or structure can be used clinically to diagnose subclinical changes in the systemic vasculature pathognomonic of life-threatening cardiovascular disease. This idea is realistic since it is well known that narrowing of retinal arterioles or abnormal arteriovenous crossings are clinical indices of systemic hypertension.

This section will present studies on the retinal vessel dynamics [49], ONH perfusion [50], and POBF [51] during inhalation of 100% O_2, as well as the changes in POBF following inhalation of carbogen (5% CO_2 in 95% O_2) [51] and a hypoxic gas (12% O_2 in 88% nitrogen) [52].

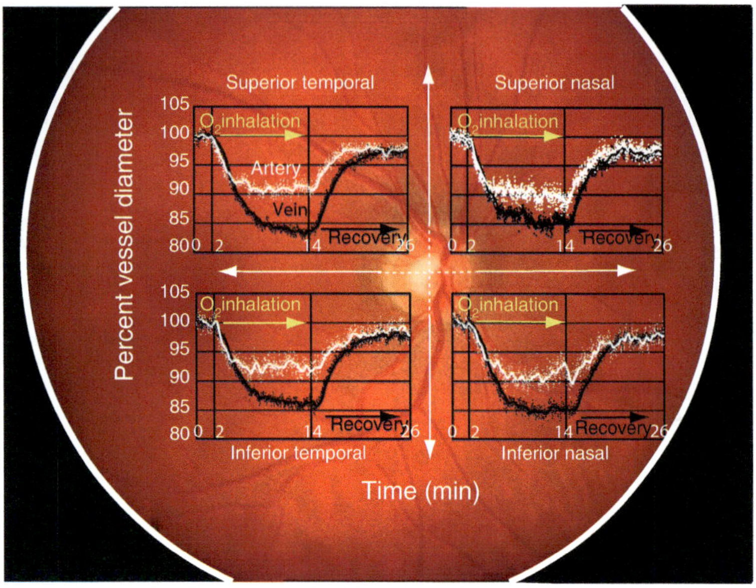

Fig. 10.16 Inhalation of 100% O_2 induced vasoconstriction in both retinal arteries and veins. The reduction in vessel diameter was ~5% greater in veins than arteries, and recovery from O_2 occurred at the same rate in both vessel types. This profile of change in vessel diameter during inhalation of O_2 and the recovery phase was the same in the four retinal quadrants centered on the ONH. The vasoconstriction response to inhalation of 100% O_2 is interpreted as a regulatory response of blood flow to maintain constant levels of oxygen tension

10.6.1 Hyperoxia and Blood Flow

Inhalation of 100% O_2 caused a progressive monotonic constriction of retinal vessels that reached a plateau within 4 min. The data in Fig. 10.16 show that veins constricted about 5% more than arteries, but their overall response profiles were similar. The plateau of constriction was sustained throughout the remaining time of O_2 inhalation. When room air was reintroduced, both arteries and veins redilated monotonically to within ~2% of baseline caliber during the recovery interval. This profile of retinal vessel reactivity to O_2 inhalation was the same for vessels populating each of the principal fundus quadrants relative to the ONH. This O_2-induced vasoconstriction was interpreted as evidence for a mechanism regulating blood flow presumably to maintain the level of O_2 required for normal metabolism.

The reactivity of the blood vessels perfusing the ONH was also examined in a separate study involving inhalation of 100% O_2 with perfusion monitored by the CP. Transient systemic hyperoxia increased the area of pallor in the ONH, suggesting vasoconstriction of superficial and deeper capillaries and decreased perfusion of the anterior portion of the optic nerve [50]. However, the choroidal flow as indexed by the POBF [51] or the ChBF [53] was not altered during systemic hyperoxia.

10.6.2 Hypercapnia-Hyperoxia and Choroidal Blood Flow

Inhalation of carbogen (O_2 + 5 to 7% CO_2), a gas mixture with increased concentration of CO_2, increases the POBF [51], and the subfoveal ChBF [54], as well as the fundus pulsation amplitude (FPA), an index of choroidal pulsatile flow, in the macula [55] in man. Inhalation of CO_2 in air has also been shown to increase blood flow in the macula, as measured by FPA [56, 57]. Similar results indicating an increase in ChBF with increased CO_2 (in air, or in O_2) were reported in animal studies [58–60], although others have reported that CO_2 did not increase ChBF [61].

10.6.3 Hypoxia and Pulsatile Choroidal Blood Flow

Early studies into vascular regulation reported that systemic hypoxia caused an increase in both the perifoveal capillary blood flow [62] and retinal blood flow [63]. However, inhalation of 12% O_2 in nitrogen [mean SaO_2 = 89.0%] did not affect ChBF as indexed by measurements of the POBF [52]. The absence of changes in the POBF could be interpreted to indicate that the pulsatile component of ChBF is unaffected by transient mild systemic hypoxia because of the large volume of oxygenated blood in the choroid. However, it remains possible that mild systemic hypoxia could influence the nonpulsatile component of ChBF. This latter possibility awaits confirmation by measurements of the ChBF with LDF after inhalation of 12% O_2 in nitrogen. Overall, the present state of knowledge suggests that the retinal circulation is likely more sensitive than the choroid to transient mild systemic hypoxia.

10.6.4 Hyperoxia, Hypercapnia, and Retinal Function

In addition to studies on the effects of altered blood gases on ocular blood flow, the effects of hyperoxia and hypercapnia on the neural retinal function have also been assessed. Systemic hyperoxia had minimal effects on the function of the photoreceptors and bipolar cells as indexed by photopic fERG a- and b-waves [64]. However, when the retina was dark adapted, hypercapnia significantly reduced the amplitude of OP5, one index of amacrine cell function [65]. These results provide additional support to the growing body of evidence indicating that neural function in the retina is more vulnerable to changes in blood gases and vascular perfusion levels when the retina is dark adapted.

10.6.5 Hypoxia, Hyperoxia, and Retinal Function

While transient hypoxia attenuates the function of the neural generators of the photopic fERG b-wave and specific OPs, it does not affect the amplitude or the implicit time of the photopic ERG a-wave [66]. This is consistent with the fact that the choroidal blood oxygen tension is highest at the RPE-photoreceptor junction and rapidly decreases toward zero by the middle of the retina [45, 67] (Fig. 10.1b). Consequently, bipolar cells that populate the retinal zone where the choroidal and retinal vasculatures provide the lowest oxygen tension would be particularly vulnerable to the effects of transient systemic hypoxia as was indicated by the reduction in the photopic fERG b-wave. While transient systemic hypoxia does not alter the function of photoreceptors in the outer retinal layer, it does reduce the amplitude and delays the N95 component of the pERG that is principally generated by the ganglion cells in the innermost retinal layer [68]. This greater vulnerability of the ganglion cells to transient hypoxia may be explained by the much smaller oxygen tension in the retinal blood perfusing the inner two-thirds of the retina. The consequence of this reduced level of oxygen tension in retinal blood is that a mild transient systemic hypoxia is likely enough physiological stress to compromise ganglion cell function and thus results in compromised pERGs. The additional observation that transient systemic *hyperoxia* did not affect the function of ganglion cells supported the interpretation that the much smaller degree of oxygen tension in retinal blood flow may be just adequate to support normal neural function in the innermost retinal layer. Thus, transient systemic hyperoxia providing more oxygen to the retinal vasculature would not be expected to compromise the function of ganglion cells found in the inner retinal layer, and this is what was found experimentally [68].

10.7 Regional Choroidal Perfusion

While vascular autoregulation in the retinal vasculature has been demonstrated in many studies over the last 20 years or so, evidence for vascular regulation in the human choroid remains limited but the notion for choroidal regulation is gaining broader acceptance in the ocular blood flow section of the vision science community. For many years, it was thought that there was no need for regulation of ChBF because of the large choroidal volume and flow rate relative to the retinal vasculature. However, seminal work by Linsenmeier et al. [45] revealed a partial pressure of O_2 (PO_2) that decreased rapidly from the choroid and approached a zero value near the middle of the retina (Fig. 10.1b). This implied that the retina consumed virtually all of the O_2 that came from the choroid, and any disruption of blood flow to the RPE-photoreceptor complex could have serious consequences on neural retinal function and vision.

Fortunately, a global reduction in ChBF is not a common clinical occurrence. However, there is increasing evidence that a regional reduction in ChBF occurs in patients with age-related macular degeneration (ARMD) even before there is any clinical evidence of a significant disruption in the foveomacular anatomy [69]. Thus, a subnormal choroidal perfusion of the macula may be a major risk factor for this age-related macular disease [70]. It may be that a reduction in subfoveal ChBF prior to the typical clinical signs and symptoms of ARMD is the result of defective regulation. If this hypothesis is correct, traditional fluorescein and choroidal angiographies may not be adequate for detecting blood flow that is not capable of sustaining normal neural function during transient vascular stress. *Consequently, a more productive method for detecting weakened ChBF would be to quantify choroidal hemodynamic responses during an increased metabolic demand as induced by flicker or safe levels of altered OPP to determine the integrity of choroidal regulation. Absence of regulatory responses could identify individuals at high risk for macular disease and the need to initiate procedures to preserve macular structure and function.*

The proposal that choroidal regulation may also involve a differential perfusion across the ocular fundus is based on several previous observations in blood flow. In subhuman primates, flicker stimulation elicited differential perfusion by the *retinal* vasculature according to regional metabolic demands [71]. As such, only differential ChBF across the retina may be able to satisfy large differences in local metabolic demand. At the *systemic* level, it is well established that blood flow is shunted to regions with increased metabolic demand such as preferential blood flow to the large leg muscles while running, even to the detriment of blood flow to internal organs [72]. Finally, data in Sects. 10.7.3 and 10.7.4 present findings that support the hypothesis of differential choroidal perfusion in favor of retinal areas with increased metabolic requirements.

10.7.1 Cones Versus Rods: Structure and Function

The degree to which a flash elicits an electrophysiological response from the retina depends on two variables: (1) the physical properties of the flash and (2) the anatomical location and density distribution of the cone and rod photoreceptors across the retina. Variables in the flash stimulus include wavelength composition, luminance level, and presentation frequency (Hz). With respect to retinal location, the short, medium, and long wavelength sensitive cones ("Blue," "Green," and "Red" cones) totaling ~6 million [73], are distributed within the central ~20° around the foveola. The G and R cones have the highest density in the central 5° (~200 K cones/μm^2) while the blue cones for the same central area have a density of ~2 K cones/μm^2 [74]. Rhodopsin-containing rods estimated at ~120 million [73] are found in all regions of the retina except the foveola and occur in greatest density within a concentric annular zone that extends between ~5° and ~50° from the foveola, and thereafter decrease in density toward the periphery.

This differential distribution of cones and rods across the retina is the basis for the rationale presented in a later section of this chapter that repeated stimulation of extra foveal blue-sensitive cones and especially the blue-sensitive rods greatly increased the metabolic need of more eccentric retinal sites, and consequently, through some unknown mechanism, the ChBF was directed away from the fovea in favor of the mid-periphery of the retina. As such, it was concluded that the *choroid* that nourishes the RPE-photoreceptor complex can preferentially divert blood flow in the macula from one site to another to support the metabolic demands of retinal sites with heightened neuronal activity, as was reported to be the case for the retinal vasculature [71].

10.7.2 Choroidal Angioarchitecture

The angioarchitecture of the choroid at the posterior pole is described as a honeycomb, nonlobular structure that gradually changes into a mosaic of clearly defined lobules with anastomosing capillaries from the peripapillary zone to the retinal periphery. While the distinct lobular pattern of the choriocapillaris in the periphery determines how blood will flow from one point to another, the direction of blood flow in the foveal zone is less obvious. The choriocapillaris in the foveomacular zone has a homogenous structure and blood flow in that area does not match what is seen by fluorescein and indocyanine angiographies [2]. It is likely that foveomacular blood flow is determined by pressure gradients and differences in metabolic requirements within the macular retina [2]. Thus, the choroidal lobuli comprising the choriocapillaris layer of the choroid determine the flow of blood by both structure (lobuli) and function (neuroretinal activity).

The technological challenge for this new level of clinical diagnostics based on blood flow measurements requires the development of a system that can penetrate the RPE (IR probe) and make measurements of blood flow in the retina and/or choroid in arbitrary sites of the retina, as well as a series of sites that correspond to visual field thresholds for vision. Perhaps most importantly, such a system should provide variable luminance focal flicker to determine threshold changes in blood flow, as well as stimulus–response profiles for suspect areas of the fundus.

10.7.3 Dark Adaptation

Photopic vision is mediated by the cone photoreceptors that provide high spatial resolution and color perception. Scotopic vision is a function of rod photoreceptors and is characterized by achromatic vision and low spatial resolution. The inverse distribution of cone *vs.* rod density across the retina, and the drop in choriocapillaris vessel density from the fovea to the periphery, may be an evolutionary adaptation that attempts to balance metabolic needs with energy supplies. Inasmuch as the retina consumes most of the O_2 that the choroid can provide [75], topographic change in choriocapillaris vessel density may reflect the metabolic needs of the photoreceptors in the overlying retina. *To date, a correlation between ChBF in the foveal and the perifoveal area and the transition from cone to rod function has not been demonstrated in man.* However, a recent study did investigate such a coupling between central photoreceptor function and ChBF in the human retina [76]. A continuously recording NI-LDF was used to record foveomacular ChBF changes throughout a 26-min dark adaptation interval. A 10-s recording of the ChBF was made in the foveal and perifoveal zones every 2 min for 26 min of dark adaptation. The group averaged ChBF decreased by ~17% (Fig. 10.17). This indicated that the gain in light sensitivity by the retina in the dark was accompanied by a decrease in the resting level of ChBF in the cone-rich foveomacular zone and that volumetric change in the flow presumably went to the rod-rich area because rod sensitivity to light was maximized after dark adaptation. Since rods outnumber cones by ~20:1 in the human retina, and the highest rod density is found ~15–18° peripheral to the fovea, it was hypothesized that the reduction in foveal ChBF was a manifestation of ChBF regulation wherein blood was moved toward a retinal site with higher physiological activity and metabolic demand. Since retinal blood flow increases in darkness [77, 78], presumably to support the metabolic needs of rod photoreceptors, and the photoreceptor "dark current," choroidal blood that nourishes photoreceptors from the scleral side may also move in the same direction, i.e., away from the fovea, toward the peripherally located rods in proportion to their activity. The reduction in the subfoveal ChBF during dark adaptation raised the possibility that choroidal blood was shunted radially outward toward the paramacular annular zone that is densely populated by rod photoreceptors. Measurements of ChBF in perimacular regions during dark adaptation would be needed to confirm empirically the hypothesis of a redistribution of choroidal blood from cone-dominated to rod-dominated areas in the retina. Figure 10.17 presents the graphical illustration of the changes in subfoveal ChBF as the retina transited from light to dark adaptation over a 26-min recording interval. During dark adaptation, the subfoveal ChBF decreased by about 6% in the first 10 min of dark

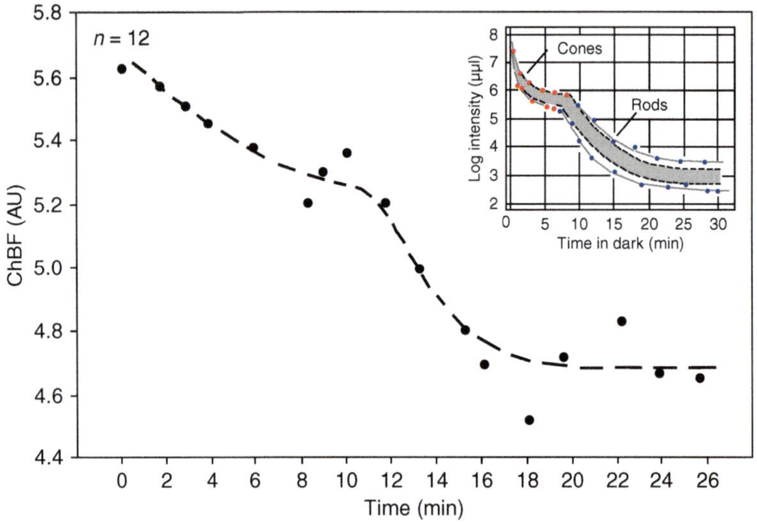

Fig. 10.17 Group-averaged subfoveal ChBF as a function of time into dark adaptation. ChBF was measured by NI-LDF. The dim speckle appearance of the probing laser beam does not affect dark adaptation of the retina. In the first 6 min, the ChBF decreased by about 6%, and then by another ~11% during the remaining time. The *dashed biphasic line* has been drawn to show the resemblance between the change in the subfoveal ChBF in dark adaptation and the psychophysically measured dark adaptation curve (see *Inset*). A linear regression model through the data points confirmed the reduction in ChBF throughout dark adaptation. The reduction in subfoveal choroidal blood suggests a routing of this blood outward toward retinal zones with increasing metabolic needs. Rod photoreceptors populating the perifoveal retinal zones are physiologically logical recipients of the subfoveal blood shunted to eccentric retinal sites because of their increasing metabolic requirements

adaptation and then decreased rapidly by about 11% within the next 8.5 min. Thereafter, the ChBF remained at a constant level of ~17% below the initial ChBF measurement in normal ambient lighting. The biphasic attenuation of the subfoveal ChBF during dark adaptation resembled the psychophysical dark adaptation curve for the human retina shown in the inset of Fig. 10.17. *This likely was the first study that raised the possibility of a correlation between the changes in the subfoveal ChBF during the switch from cone to rod vision in the human retina with the psychophysically measured changes in light detection thresholds across the retina as it changed from a light to dark-adapted state.* Because the amplitude and timing of the rod-cone break in dark adaptation is strongly influenced by the level of light adaptation of the retina, and because blood flow changes likely precede perceptual changes in light thresholds by yet unknown amounts, the degree and timing of the subfoveal ChBF changes in dark adaptation likely do not correlate perfectly with the subjective measurements of retinal dark adaptation. *Nonetheless, the objective quantification of blood flow changes that may precede functional changes in vision could provide significant advantages in the differential diagnoses for photoreceptor dysfunction, neuronal atrophy, and ultimately permanent vision loss. Further research is required to evaluate the clinical utility of this objective procedure for subfoveal changes in the ChBF during dark adaptation.*

10.7.4 Protracted Blue Flicker

Recent studies reported that flicker stimulation of the retina increased blood flow in the retinal vasculature and the ONH [79–81], but not in the choroid [82, 83]. It was rationalized that the absence of blood flow change in the choroid meant that it was independent of changes in retinal metabolism.

Of particular importance in evaluating these data is the fact that these studies used flicker composed of wavelengths that included all or large parts of the photopic spectral sensitivity curve. Such flicker would have activated the blue-, green-, and red-sensitive cones (or short-, medium-, and long-wave-sensitive cones) in proportion to their respective spectral response profiles [74]. Anatomically, the "green" and "red" cones are mainly located in the foveal zone, while the "blue" cones are found in highest density outside the fovea in a ~5° perifoveal ring, and in half the density in a second ring that extends a further 10° outward. By population, ~64% of cones are red, ~32% green, and ~2% blue. The spectral property of the flicker used in those studies allows a prediction that the foveal retina would likely have been stimulated most strongly, followed by the perifovea, and then the more peripheral retina to the lowest degree. This differential stimulation across the retina has particularly important implications for predicting the retinal zones where the most significant changes in blood flow could be expected to occur.

In a recent study, blue flicker was presented over a broad luminance range to selectively stimulate rods and short-wave cones [84]. The hypothesis in that study was that narrowband blue flashes were more likely to activate rods populating the greater macular periphery, and therefore the subfoveal ChBF would be redistributed toward the retinal sites with greater metabolic activity. Subfoveal choroidal hemodynamics were quantified by LDF while the retina was stimulated by dim blue flashes delivered at low-to-high frequencies and then step increments in luminance using the same frequency sequence. In the last stage of the experiment, the order of step changes in flicker luminance was reversed, keeping the same order of increasing frequency. Flash ERGs elicited by the same wavelength (= 473 ± 11 nm) were subsequently recorded in response to a series of bright-to-dim and dim-to-bright blue flashes (over a 4.0-log range) to help explain the different ChBF responses.

The resulting data shown in the top graph in Fig. 10.18 revealed that blue flicker with increasing luminance steps caused a ~32% reduction in the subfoveal ChBF that was caused by a reduction in volume but no change in velocity. Surprisingly, the same flicker frequencies with luminance steps presented in reverse order (high-to-low) had no measurable effect on the ChBF, volume, or velocity. The absence of blood flow

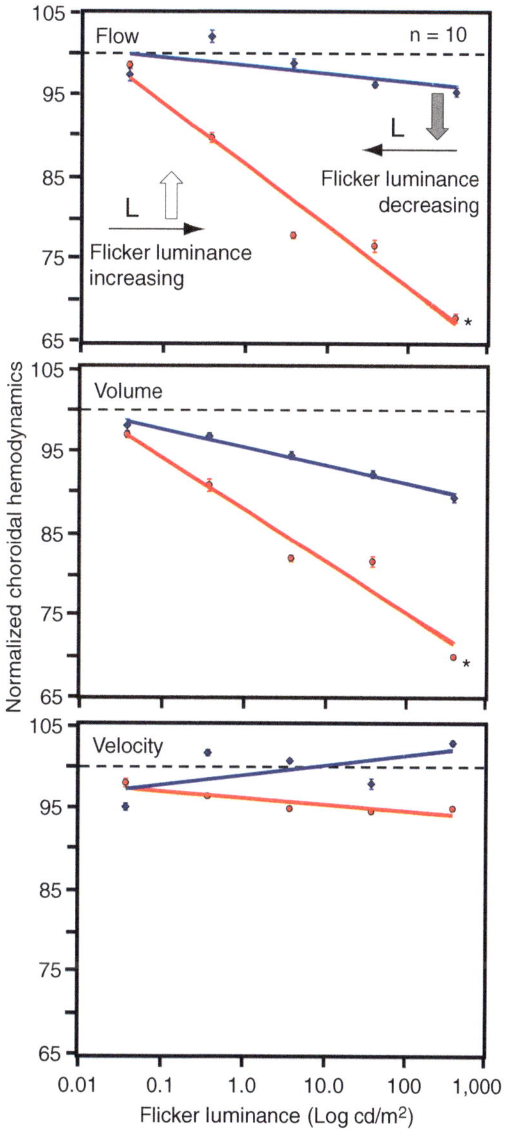

Fig. 10.18 The subfoveal ChBF as changed by shortwave narrow-bandpass flicker. *Blue flicker* with luminance increased from 0.0375 to 375 cd/m^2 in equal steps caused a linear reduction in the subfoveal ChBF due to a decrease in blood volume. In contrast, blue flicker with changes in luminance delivered in reverse order stimulus, i.e., from 375 to 0.0375 cd/m^2, had no effect on the subfoveal ChBF

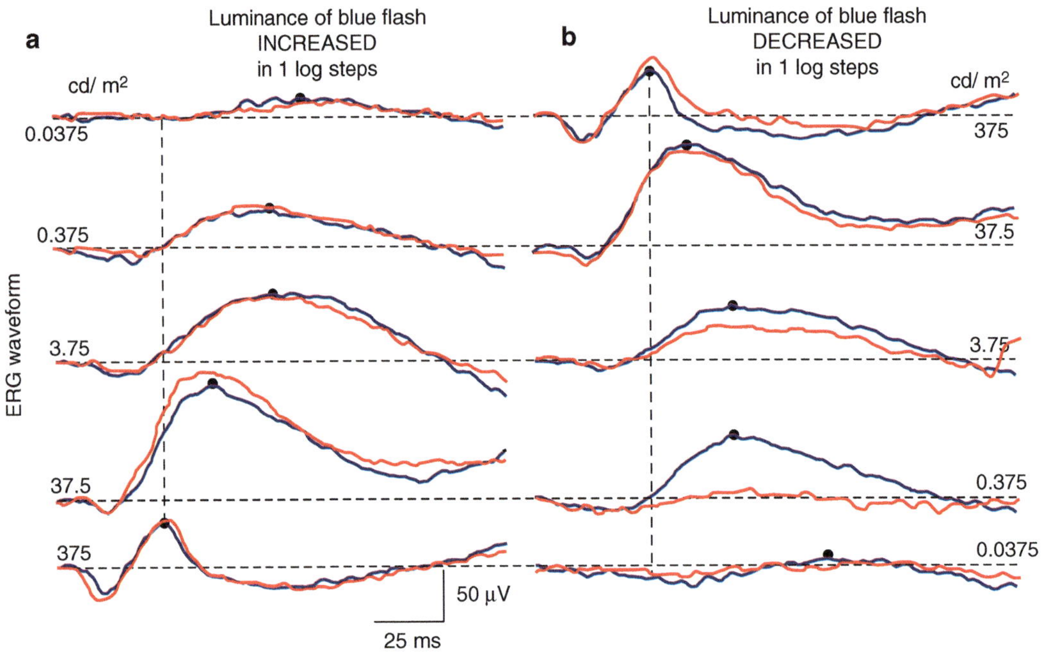

Fig. 10.19 Objective recordings of retinal responses to diffuse *blue flashes* of increasing luminance (**Frame a**) *vs*. decreasing luminance (**Frame b**). *Blue traces* indicate ERGs recorded before the frequency flicker series, while the *red traces* show ERGs taken after the flicker series. These ERG waveforms confirmed that the same stimuli could elicit different retinal responses (waveform, amplitude, and timing) if they were preceded by flashes presented in reverse order of luminance. These findings supported the conclusion that blood flow measurements could also show opposite trends when the flash intensity series was presented in reverse order

changes during the reversal in luminance steps appeared erroneous but was subsequently validated by fERGs. These electrophysiological indices of neural function provided objective measurements of changes in photoreceptor function that also explained the profiles of change in blood flow. Stated succinctly, blue flashes of increasing luminance elicited progressively larger rod-dominated retinal responses from the blue-sensitive rhodopsin containing rods (Fig. 10.19a; blue recordings). Such flashes elicited ERGs with a progressive increase in the b-wave component for each 1.0-log increase in flash luminance. At the highest flash luminance, rod activity was suppressed and only blue-sensitive cones contributed to the fERG. As such, these flashes progressively increased metabolic activity in rods that populate the far macular zone and thereby drew blood away from the subfoveal choroid. In contrast, flashes of decreasing luminance first strongly activated foveally located blue-sensitive cones with the brightest flash and subsequently had smaller effects on rods with decreasing flash luminance as is shown by the rapid attenuation of the fERG b-wave (Fig. 10.19b; blue recordings). As the effect on rods was reduced, the flashes progressively increased their effect on centrally located photoreceptors and thereby required the subfoveal ChBF to remain unchanged. The red recordings in Frames a and b show the ERGs elicited at each flash intensity at the end of each flicker frequency series. A comparison of the red ERG recordings for identical flash intensities in Frame a vs. b reveals that the ERGs differed widely for the 3.75 and 0.375 flash intensities. For example, in Frame a, a 0.375-cd/m^2 flash elicited virtually equal amplitude ERGs before and after the flicker frequency series. In contrast, in Frame b, the same flash intensity presented after the flicker frequency series elicited an ERG that was nearly extinguished.

Overall, the blood flow findings combined with the blue fERGs provided compelling evidence that ChBF must have been shifted away from the subfovea, likely toward retinal regions with the highest metabolic activity, in this case toward eccentric rods because the stimulus wavelength was selected to match the spectral sensitivity of rhodopsin that is found exclusively in rods. Confirmation that blood moves from the subfoveal choroid to the peripheral choroid during a similar series of blue flashes would require simultaneous ChBF measurements in the periphery and the subfovea, which is technically very challenging. The decrease in subfoveal ChBF during the blue flash stimuli that increased activity in rod photoreceptors parallels the progressive reduction in subfoveal ChBF measured during physiological dark adaptation of the retina where the site of increasing metabolic activity was increasingly far from the fovea, in the rod-dominated zone of the far macula.

10.8 Aging

10.8.1 Structure

Normal aging is accompanied by a reduction in the number of arterioles and venules and an associated neural loss such as a dropout in retinal ganglion cells [85]. Along with this reduction in retinal vessels, there is a modification in the geometry of arterial bifurcations that likely alters blood flow and leads to blood rarefaction of the microvasculature [86]. In addition to these changes in retinal vessels, the density and caliber of the vessels forming the choriocapillaris in the human macula also decrease, thereby causing a thinning of the choroid [87]. There is also a significant reduction in the sympathetic innervation to the choroid [88] and a loss of endothelial cells in the choriocapillaris [89]. With normal aging, a thickening of Bruch's membrane is known to occur [87, 89] and is subject to accumulation of cholesterol, particularly in the macular region [90], thereby impairing normal diffusion of blood between the RPE and the choroid [91, 92]. All of these histological findings are compatible with an alteration in ocular hemodynamics in senescence.

10.8.2 Blood Flow

Recent technological advances such as the RVA have led to noninvasive measurements of real-time changes in retinal vessel diameter during flicker. Using this technology, Polak et al. [93] recently reported that retinal vessels dilate in response to flicker between 1 and 60 Hz. Shortly afterward, Nagel and Vilser [94] reported on a clinically usable procedure for quantifying the degree of vasodilation in retinal arteries and veins during a 12.5-Hz flicker. These latter studies formed the basis of a study that compared the timing and amplitude of change in retinal vessel caliber in healthy volunteers 20–80 years of age. In order to compare the degree of dilation between arteries and veins within and across subjects in different age groups, the data sorted by decade of life were normalized such that the prestimulus vessel diameter was designated as 100%, and subsequent changes in caliber were expressed as a percentage of that value.

The group-averaged maximal dilation for an artery and vein in response to 60 s of flicker presented at 12.5 Hz was ~6% and ~8%, respectively. This difference in the degree of vasodilation between arteries and veins was maintained across all age groups from 20 to 80 years of age. However, the overall dilation measured in arteries and veins was less for the subjects in the 60- and 70-year age groups. A sample comparison of the vasodilation/vasoconstriction response profiles to flicker for a young and an elderly subject are presented in Fig. 10.20. The outstanding difference in these response profiles was the delayed and slower vasoconstriction of both artery and vein during the recovery phase for the elderly subject. Consequently, neither vessel returned to baseline within the 60-s recovery phase.

The vasodilation to flicker has been interpreted as a response to increased metabolic demand of the retina and hence a need for increased blood flow for more oxygen and metabolites needed to sustain neural activity. The failure of retinal arterioles to dilate as much as the younger subjects may be attributed to a reduction in the amount of circulating substances such as NO with age or a dropout of muscle fibers, thereby affecting the reactivity of the

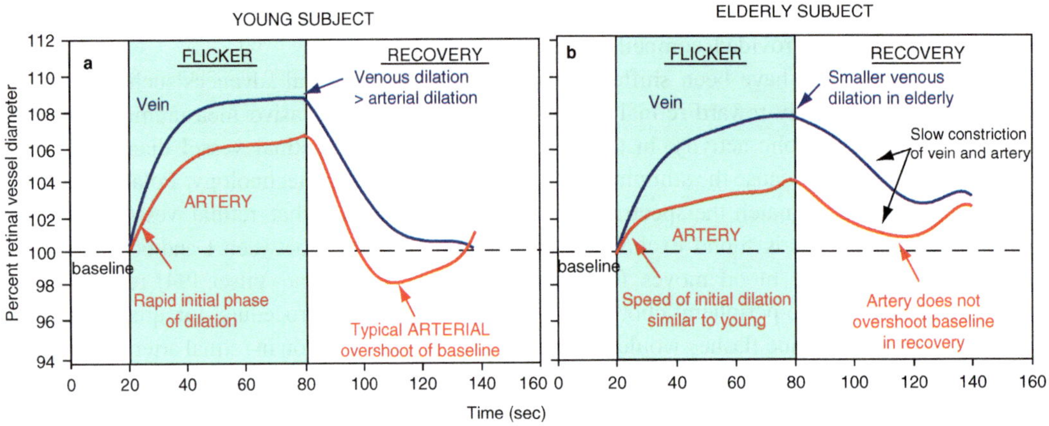

Fig. 10.20 Representative retinal vessel dilation and recovery profiles in response to a 12.5-Hz flicker in young (**Frame a**) *vs.* elderly subjects (**Frame b**). Results from an ongoing study have revealed that retinal veins dilate biphasically as do arteries during flicker, but veins dilate to a greater degree than arteries in both young and elderly subjects. Furthermore, both arteries and veins tend to dilate more in younger subjects than in the elderly. Interestingly, retinal arteries in younger subjects overshoot baseline diameter during the vasoconstriction recorded in the recovery phase, but arteries in elderly subjects do not show such an overshoot in recovery. In fact, neither arteries nor veins in elderly subjects regain baseline vessel diameter in the recovery interval

vessels and its ability to vasodilate. Alternatively, because there is an attrition of photoreceptors with age, it is possible that the neural retinal response to standardized flicker luminance was less in subjects over 60 years of age compared to younger subjects, thereby requiring a smaller increase in blood flow. Further studies are needed to determine the precise cause of these new findings on retinal hemodynamics in senescence. Whatever the reason for reduced arterial dilation, this vasomotor deficit may be projected to blood flow in the brain and may also indicate subtle subclinical cardiovascular deficits in senescence.

At the structural level, retinal vessel diameters were also measured in the same group of subjects in an effort to determine structural changes with age and to determine the correlation between retinal hemodynamics and structure. The retinal vessel diameter was quantified over a ~2,700-μm length to provide a more precise measure of age-related changes in vessel caliber because previous studies of vessel size were based on very narrow cross-sectional values. For the data presented in Fig. 10.21, high-resolution digital fundus images were first taken with the Imedos Visualis system and then vessel diameters were quantified with their VesselMap

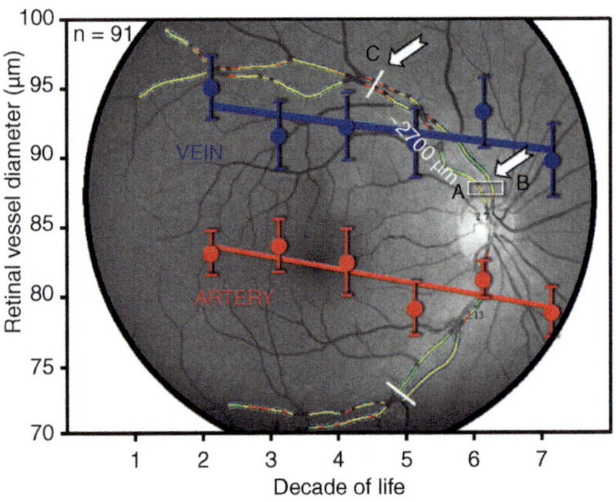

Fig. 10.21 A: optic nerve head, B: white rectangle identifies the location where the first measurement of the vessel caliber was made in the course of determining the taper of the target vessel over a 2700 μm distance from the optic nerve head toward the retinal periphery. C: white line through the retinal vessel identifies the most peripheral site where the measurement of vessel caliber was made. The two solid white arrows highlight the 2700 μm length of vessel along which changes in vessel caliber were determined

software. The group-averaged vessel diameter for paired arteries and veins for each age group in this study are presented in Fig. 10.21. Overall, there is a trend for a reduction in diameter in both arteries and veins for subjects between 20 and 80 years of age. This narrowing of retinal vessels may be an adaptation for the concomitant increase in the systemic BP and hence an increase in the OPP, although still within normal range. Inasmuch as all subjects were free of systemic and ocular diseases, it was presumed that the decrease in vessel diameter was a response to increased OPP [95].

10.8.3 Retinal Function

The normal aging process is accompanied by physiological decline in many organs, thereby decreasing the functional reserves and increasing their susceptibility to disease. However, the boundary between the functional decline linked with aging and the development of pathology is not well described in a variety of ocular diseases.

Recent studies have concluded that aging *per se* is associated with a variety of significant nonpathological changes in neural structure and function in the eye of healthy subjects. Specifically, the function of all major retinal neurons, including rod and cone photoreceptors, bipolar/Mueller cells, amacrine cells [96], ganglion cells [97], the retinal nerve fiber layer [98], and the retinal ganglion cell axons within the ONH [99], is compromised in senescence. These findings allow a distinction between the effects of normal aging and the onset of ocular pathology that occurs more frequently in the elderly.

While these findings have defined the anatomical and physiological changes in the aging eye, changes in the ocular blood flow associated with aging *per se* are yet to be quantified. This identifies an area of fundamental and clinical research that is essential for correctly diagnosing subclinical ocular pathology related to vascular dysfunction or impaired blood flow that occurs more frequently in the aging population. ARMD is a prime example of a prominent ocular disease thought to have a vascular origin. However, a correct diagnosis of the onset of any vascular disease of the eye must rule out any effects of normal aging on retinal or choroidal perfusion and abnormalities in the systemic vasculature that directly affect ocular blood flow. The work described above represents some of the first initiatives to define changes in vascular perfusion at different levels within the retina and their effect on neural function in the aging eye. Additional studies are required for a complete description of changes in neurovascular coupling as a function of age.

Abbreviations

ARMD	Age-related macular degeneration
BFE	Blue Field Entoptoscope
BP	Blood pressure
BPdiast	Diastolic blood pressure
BPmean	Mean systemic blood pressure
BPsyst	Systolic blood pressure
ChBF	Choroidal blood flow
CO_2	Carbon dioxide
CP	Chromatic pulse
CPT	Cold pressor test
CRA	Central retinal artery
CRV	Central retinal vein
D	Diameter
FAZ	Foveal avascular zone
fERG	Flash electroretinogram
FPA	Fundus pulsation amplitude
HR	Heart rate
IOP	Intraocular pressure
ISCEV	International Society for Clinical Electrophysiology of Vision
LDF	Laser Doppler flowmetry
NIR	Near infrared
NO	Nitric oxide
O_2	Oxygen
ONH	Optic nerve head
OPP	Ocular perfusion pressure
OPs	Oscillatory potentials
pERG	Pattern reversal electroretinogram
PO_2	Partial pressure for oxygen
POBF	Pulsatile ocular blood flow
Q	Blood flow
RPE	Retinal pigment epithelium
RVA	Retinal Vessel Analyzer
SaO_2	Saturation of oxygen
V	Velocity
VM	Valsalva maneuver

Acknowledgements We are grateful to the NSERC, CIHR, FRSQ, and CFI for providing research funds needed to carry out the various projects reported in this chapter. We are particularly grateful for the expert research assistance provided by Mireille Parent throughout our studies in aging. We also acknowledge our research collaborators, the many research assistants and student research assistants that have worked in our labs and the graduate students who have made significant contributions to our research programs. We also thank Dr. Alain Savoie for his beautiful illustrations in Fig. 10.1. Finally, we extend our appreciation to Denis Latendresse, Micheline Gloin, Marc Melillo, Normand Lalonde, and François Vaillancourt for their expert help in graphic design, electronics, computing, and technical design throughout our research at the ÉOUM.

References

1. Kiel JW (1999) Modulation of choroidal autoregulation in the rabbit. Exp Eye Res 69:413–429
2. Fryczkowski AW (1994) Anatomical and functional choroidal lobuli. Int Ophthalmol 18:131–141
3. Lovasik JV, Kothe AC, Kergoat H (1993) A comparison of non-invasive methods to derive the mean central retinal artery pressure in man. Optom Vis Sci 70:1005–1011
4. Guyton AC, Hall JE (1996) Overview of the circulation: medical physics of pressure, flow, and resistance. In: Textbook of medical physiology. WB Saunders, Philadelphia
5. Al Suwaidi J, Higano ST, Holmes DR Jr et al (2001) Pathophysiology, diagnosis, and current management strategies for chest pain in patients with normal findings on angiography. Mayo Clin Proc 76:813–822
6. Menkes MS, Matthews KA, Krantz DS et al (1989) Cardiovascular reactivity to the cold pressor test as a predictor of hypertension. Hypertension 14:524–530
7. Mizushima T, Tajima F, Nakamura T et al (1998) Muscle sympathetic nerve activity during cold pressor test in patients with cerebral vascular accident. Stroke 29:607–612
8. Lovasik JV, Kergoat H, Riva CE et al (2003) Choroidal blood flow during exercise-induced changes in the ocular perfusion pressure. Invest Ophthalmol Vis Sci 44:2126–2132
9. Riva CE, Titze P, Hero M et al (1997) Choroidal blood flow during isometric exercises. Invest Ophthalmol Vis Sci 38:2338–2343
10. Polska E, Simader C, Weigert G et al (2007) Regulation of choroidal blood flow during combined changes in intraocular pressure and arterial blood pressure. Invest Ophthalmol Vis Sci 48:3768–3774
11. Lovasik JV, Kergoat H, Gauthier C, Dion I (2009) Regulation of choroidal blood flow fails during cold stimulation. Optom Vis Sci 86:E-abstract 85985
12. Marinier JA, Kergoat H, Lovasik JV (2000) Increased foveal-choroidal blood flow in man during bilateral cold pressor testing. Optom Vis Sci Suppl 77:231
13. Lovasik JV, Kergoat H, Greendale C et al (2003) Changes in retinal vessel diameter during a cold pressor test. Optom Vis Sci Suppl 80
14. Forcier P, Kergoat H, Lovasik JV (1998) Macular hemodynamic responses to short-term acute exercise in humans. Vision Res 38:181–186
15. Kergoat H, Lovasik JV (1995) Response of parapapillary retinal vessels to exercise. Optom Vis Sci 72:249–257
16. Pournaras CJ, Rungger-Brändle E, Riva CE et al (2008) Regulation of retinal blood flow in health and disease. Prog Retin Eye Res 27:284–330
17. Maeda S, Miyauchi T, Kakiyama T et al (2001) Effects of exercise training of 8 weeks and detraining on plasma levels of endothelium-derived factors, endothelin-1 and nitric oxide, in healthy young humans. Life Sci 69:1005–1016
18. Lovasik JV, Kergoat H, Racine N, et al (2007) Neurovascular coupling in long-term joggers versus healthy non-joggers. Invest Ophthalmol Vis Sci 49:E-Abstract 2273
19. Porth CJ, Bamrah VS, Tristani FE et al (1984) The Valsalva maneuver: mechanisms and clinical implications. Heart Lung 13:507–518
20. Lovasik JV, Kergoat H, Riva CE, et al (2002) Correlation between the intra-thoracic pressure and choroidal blood flow. Invest Ophthalmol Vis Sci 49:E-Abstract 3315
21. Benowitz NL (2003) Cigarette smoking and cardiovascular disease: pathophysiology and implications for treatment. Prog Cardiovasc Dis 46:91–111
22. Rojanapongpun P, Drance SM (1993) The effects of nicotine on the blood flow of the ophthalmic artery and the finger circulation. Graefe's Arch Ophthalmol 231:371–374
23. Kaiser HJ, Schoetzau A, Flammer J (1997) Blood flow velocity in the extraocular vessels in chronic smokers. Br J Ophthalmol 81:133–135
24. Steigerwalt RD Jr, Laurora G, Incandela L et al (2000) Ocular and orbital blood flow in cigarette smokers. Retina 20:394–397
25. Lafleur J, Lovasik JV (2000) Nicotine modifies reactivity of foveal choroidal blood flow. Optom Vis Sci Suppl 77:155
26. Balbini APS, Montovani JC (2005) Methods for smoking cessation and treatment of nicotine dependence. Rev Bras Otorrinolaringol 6:820–826
27. Benowitz NL (2008) Clinical pharmacology of nicotine: implications for understanding, preventing, and treating tobacco addiction. Clin Pharmacol Ther 83:531–541
28. Delcourt C, Diaz JL, Ponton-Sanchez A et al (1998) Smoking and age-related macular degeneration. The POLA Study. Pathologies Oculaires Liées à l'Age. Arch Ophthalmol 116:1031–1035
29. Lovasik JV, Kothe AC, Spafford MM (1987) Vascular and neural changes during body inversion: preliminary findings. Can J Optom 49:133–140

30. Kergoat H, Lovasik JV (1990) The effects of altered retinal vascular perfusion pressure on the white flash scotopic ERG and oscillatory potentials in man. Electroencephalogr Clin Neurophysiol 75: 306–322
31. Lovasik JV (1999) Assessment of the vascular autoregulatory properties in the human choroid with two measurement systems. Optom Vis Sci Suppl 76:246
32. Kergoat H, Lovasik JV (2005) Seven-degree head-down tilt reduces choroidal pulsatile ocular blood flow. Aviat Space Environ Med 76:930–934
33. Lovasik JV, Kergoat H (1994) Gravity induced homeostatic reactions in the macular and choroidal vasculature of the human eye. Aviat Space Environ Med 65:1010–1014
34. Gee W (1985) Ocular pneumoplethysmography. Surv Ophthalmol 29:276–292
35. Lovasik JV (2004) Choroidal dynamics during rapid vs slow changes in ocular perfusion. Optom Vis Sci Suppl 81:167
36. Lovasik JV, Gagnon M, Kergoat H (1994) Correlation of the chromaticity of the human ocular fundus with changes in the intraocular pressure, choroidal blood flow, and visual neural function. Surv Ophthalmol 38:S35–S51
37. Lovasik JV, Kergoat H (1996) Laser Doppler measurements of blood flow in the optic nerve head during cardiac shunting of blood. Optom Vis Sci Suppl 73:77
38. Koelle JS, Riva CE, Petrig BL et al (1993) Depth of tissue sampling in the optic nerve head using laser Doppler flowmetry. Laser Med Sci 8:49–54
39. Silverman SE, Trick GL, Hart WM Jr (1990) Motion perception is abnormal in primary open-angle glaucoma and ocular hypertension. Invest Ophthalmol Vis Sci 31:722–729
40. Fitzgerald ME, Vana BA, Reiner A (1990) Control of choroidal blood flow by the nucleus of Edinger-Westphal in pigeons: a laser Doppler study. Invest Ophthalmol Vis Sci 31:2483–2492
41. Kergoat H, Lovasik JV (1994) Unilateral ocular vascular stress in man and retinal responsivity in the contralateral eye. Ophthalmic Physiol Opt 14:401–407
42. Lovasik JV, Kergoat H, Gagnon M (2005) Experimentally reduced perfusion of one eye impairs retinal function in both eyes. Optom Vis Sci 82:850–857
43. Fitzgerald ME, Gamlin PD, Zagvazdin Y et al (1996) Central neural circuits for the light-mediated reflexive control of choroidal blood flow in the pigeon eye: a laser Doppler study. Vis Neurosci 13:655–669
44. Lovasik JV, Kothe AC (1989) Neural effects of transiently raised intraocular pressure: the scotopic and photopic flash electroretinogram. Clin Vis Sci 4:313–321
45. Linsenmeier RA (1986) Effects of light and darkness on oxygen distribution and consumption in the cat retina. J Gen Physiol 88:521–542
46. Kothe AC, Lovasik JV (1988) Neural effects of body inversion: photopic oscillatory potentials. Curr Eye Res 7:1221–1229
47. Kiss B, Polska E, Dorner G et al (2002) Retinal blood flow during hyperoxia in human revisited: concerted results using different measurement techniques. Microvasc Res 64:75–85
48. Gilmore ED, Hudson C, Nrusimhadevara RK et al (2007) Retinal arteriolar diameter, blood velocity, and blood flow response to an isocapnic hyperoxic provocation in early sight-threatening diabetic retinopathy. Invest Ophthalmol Vis Sci 48:1744–1750
49. Jean-Louis S, Lovasik JV, Kergoat H (2005) Systemic hyperoxia and retinal vasomotor responses. Invest Ophthalmol Vis Sci 46:1714–1720
50. Kergoat H, Lovasik JV, Justino L (2001) Effects of hyperoxia on the blood flow and neural function of the inner retina. Optom Vis Sci Suppl 78:99
51. Kergoat H, Faucher C (1999) Effects of oxygen and carbogen breathing on choroidal hemodynamics in humans. Invest Ophthalmol Vis Sci 40:2906–2911
52. Kergoat H, Marinier JA, Lovasik JV (2005) Effects of transient mild systemic hypoxia on the pulsatile choroidal blood flow in healthy young human adults. Curr Eye Res 30:465–470
53. Geiser MH, Riva CE, Dorner GT et al (2000) Response of choroidal blood flow in the foveal region to hyperoxia and hyperoxia-hypercapnia. Curr Eye Res 21:669–676
54. Riva CE, Cranstoun SD, Grunwald JE et al (1994) Choroidal blood flow in the foveal region of the human ocular fundus. Invest Ophthalmol Vis Sci 35:4273–4281
55. Schmetterer L, Lexer F, Findl O et al (1996) The effect of inhalation of different mixtures of O_2 and CO_2 on ocular fundus pulsations. Exp Eye Res 63:351–355
56. Schmetterer L, Wolzt M, Lexer F et al (1995) The effect of hyperoxia and hypercapnia on fundus pulsations in the macular and optic disc region in healthy young men. Exp Eye Res 61:685–690
57. Schmetterer L, Findl O, Strenn K et al (1997) Role of NO in the O_2 and CO_2 responsiveness of cerebral and ocular circulation in humans. Am J Physiol 273: R2005–R2012
58. Trokel S (1965) Effect of respirator gases upon choroidal hemodynamics. Arch Ophthalmol 73:838–842
59. Friedman E, Chandra SR (1972) Choroidal blood flow III. Effects of oxygen and carbon dioxide. Arch Ophthalmol 87:70–71
60. Alm A, Bill A (1972) The oxygen supply to the retina. II. Effects of high intraocular pressure and of increased arterial carbon dioxide tension on uveal and retinal blood flow in cats. A study with radioactively labelled microspheres including flow determinations in brain and some other tissues. Acta Physiol Scand 84:306–319
61. Wang L, Grant C, Fortune B et al (2008) Retinal and choroidal vasoreactivity to altered $PaCO_2$ in rat measured with a modified microsphere technique. Exp Eye Res 86:908–913
62. Fallon TJ, Maxwell D, Kohner M (1985) Retinal vascular autoregulation in conditions of hyperoxia and hypoxia using the blue field entoptic phenomenon. Ophthalmology 92:701–705
63. Strenn K, Menapace R, Rainer G et al (1997) Reproducibility and sensitivity of scanning laser Doppler flowmetry during graded changes in PO_2. Br J Ophthalmol 81:360–364

64. Kergoat H, Tinjust D (2004) Neuroretinal function during systemic hyperoxia and hypercapnia in humans. Optom Vis Sci 81:214–220
65. Faucher C, Kergoat H (2002) Modulation of the scotopic electroretinogram and oscillatory potentials with systemic hyperoxia and hypercapnia in humans. Curr Eye Res 24:376–386
66. Tinjust D, Kergoat H, Lovasik JV (2002) Investigation of neuroretinal function during mild systemic hypoxia in man. Aviat Space Environ Med 73:1189–1194
67. Pournaras CJ, Riva CE, Tsacopoulos M et al (1989) Diffusion of O_2 in the retina of anesthetized miniature pigs in normoxia and hyperoxia. Exp Eye Res 49:347–360
68. Kergoat H, Hérard MÈ, Lemay M (2006) RGC sensitivity to mild systemic hypoxia. Invest Ophthalmol Vis Sci 47:5423–5427
69. Grunwald JE, Hariprasad SM, DuPont J et al (1998) Foveolar choroidal blood flow in age-related macular degeneration. Invest Ophthalmol Vis Sci 39: 385–390
70. Metelitsina TI, Grunwald JE, DuPont JC et al (2008) Foveolar choroidal circulation and choroidal neovascularization in age-related macular degeneration. Invest Ophthalmol Vis Sci 49:358–363
71. Kiryu J, Asrani S, Shahidi M et al (1995) Local response of the primate retinal microcirculation to increased metabolic demand induced by flicker. Invest Ophthalmol Vis Sci 36:1240–1246
72. McArdle WD, Katch FI, Katch VL (1996) Functional capacity of the cardiovascular system. In: Exercise physiology: energy, nutrition and human performance. Williams & Wilkins, Baltimore
73. Østerberg G (1935) Topography of the layer of rods and cones in the human retina. Acta Ophthalmol Suppl 6:1–103
74. Ahnelt PK (1998) The photoreceptor mosaic. Eye 12:531–540
75. Linsenmeier RA, Padnick-Silver L (2000) Metabolic dependence of photoreceptors on the choroid in the normal and detached retina. Invest Ophthalmol Vis Sci 41:3117–3123
76. Kergoat H, Lovasik JV, Bitton E (2002) Reduction in choroidal blood flow in the foveal and perifoveal area during dark adaptation. Invest Ophthalmol Vis Sci 49:E-Abstract 3300
77. Feke GT, Zuckerman R, Green GJ et al (1983) Response of human retinal blood flow to light and dark. Invest Ophthalmol Vis Sci 24:136–141
78. Riva CE, Grunwald JE, Petrig BL (1983) Reactivity of the human retinal circulation to darkness: a laser Doppler velocimetry study. Invest Ophthalmol Vis Sci 24:737–740
79. Formaz F, Riva CE, Geiser M (1997) Diffuse luminance flicker increases retinal vessel diameter in humans. Curr Eye Res 16:1252–1257
80. Falsini B, Riva CE, Logean E (2002) Flicker-evoked changes in human optic nerve blood flow: relationship with retinal neural activity. Invest Ophthalmol Vis Sci 43:2309–2316
81. Michelson G, Patzelt A, Harazny J (2002) Flickering light increases retinal blood flow. Retina 22: 336–343
82. Longo A, Geiser M, Riva CE (2000) Effect of light on choroidal blood flow in the fovea centralis. Klin Monatsbl Augenheilkd 216:311–312
83. Garhofer G, Huemer KH, Zawinka C et al (2002) Influence of diffuse luminance flicker on choroidal and optic nerve head blood flow. Curr Eye Res 24:109–113
84. Lovasik JV, Kergoat H, Wajszilber MA (2005) Blue flicker modifies the subfoveal choroidal blood flow in the human eye. Am J Physiol Heart Circ Physiol 289:H683–H691, Epub 2005 Apr 1
85. Ibrahim YW, Bots ML, Mulder PG et al (1998) Number of perifoveal vessels in aging, hypertension, and atherosclerosis: the Rotterdam study. Invest Ophthalmol Vis Sci 39:1049–1053
86. Stanton AV, Wasan B, Cerutti A et al (1995) Vascular network changes in the retina with age and hypertension. J Hypertens 13:1724–1728
87. Ramrattan RS, van der Schaft TL, Mooy CM et al (1994) Morphometric analysis of Bruch's membrane, the choriocapillaris, and the choroid in aging. Invest Ophthalmol Vis Sci 35:2857–2864
88. Nuzzi R, Finazzo C, Grignolo FM (1996) Changes in adrenergic innervation of the choroid during aging. J Fr Ophtalmol 19:89–96
89. Killingsworth MC (1987) Age-related components of Bruch's membrane. Graefes Arch Clin Exp Ophthalmol 225:406–412
90. Curcio CA, Leigh Millican C, Bailey T et al (2001) Accumulation of cholesterol with age in human Bruch's membrane. Invest Ophthalmol Vis Sci 42:265–274
91. Moore DJ, Hussain AA, Marshall J (1995) Age-related variation in the hydraulic conductivity of Bruch's membrane. Invest Ophthalmol Vis Sci 36:1290–1297
92. Starita C, Hussain AA, Patmore A et al (1997) Localization of the site of major resistance to fluid transport in Bruch's membrane. Invest Ophthalmol Vis Sci 38:762–767
93. Polak K, Schmetterer L, Riva CE (2002) Influence of flicker frequency on flicker-induced changes of retinal vessel diameter. Invest Ophthalmol Vis Sci 43: 2721–2726
94. Nagel E, Vilser W (2004) Flicker observation light induces diameter response in retinal arterioles: a clinical methodological study. Br J Ophthalmol 88:54–56
95. Lovasik JV, Kergoat H, Boutin T, et al (2008) Retinal arterial constriction with aging may modulate vascular perfusion of the eye. Invest Ophthalmol Vis Sci 49:E-Abstract 2090
96. Kergoat H, Kergoat MJ, Justino L (2001) Age related changes in the flash electroretinogram and oscillatory potentials in individuals 75 years of age and older. J Am Geriatr Soc 49:1–6
97. Justino L, Kergoat H, Kergoat MJ (2001) Changes in the retinocortical evoked potentials in subjects 75 years of age and older. Clin Neurophysiol 112:1343–1348
98. Lovasik JV, Kergoat MJ, Justino L et al (2003) Neuroretinal basis of visual impairment in the very elderly. Graefe's Arch Clin Exp Ophthalmol 241:48–55
99. Kergoat H, Kergoat MJ, Justino L et al (2001) Age-related topographical changes in the normal human optic nerve head measured by scanning laser tomography. Optom Vis Sci 78:431–435

11 Local Determinants

Jeffrey W. Kiel

Core Messages

Local determinants of ocular blood flow include physical factors like arterial pressure and intraocular pressure, which determine the perfusion pressure across the ocular circulations, and active mechanisms inherent to ocular blood vessels and their interaction with their immediate environment. Based on evidence from more accessible circulations, the possible active mechanisms include:

- Myogenic local control
- Metabolic local control
- Flow-mediated vasodilation
- Flow control by intercellular conduction

There is evidence of local control in the ocular circulations, but which local mechanisms are responsible and how they interact is still under investigation.

11.1 Introduction

The local determinants of blood flow through the ocular circulations include both passive and active mechanisms. The most obvious passive mechanism is the compressing force exerted by the intraocular pressure (IOP) on the intraocular blood vessels, particularly the veins. By contrast, the active "local control" mechanisms in the ocular circulations are more difficult to define. Based on other tissues with more accessible circulations, the possible local control mechanisms include vascular responses linked to nearby tissue metabolism (e.g., reactive hyperemia, functional hyperemia, and autoregulation), transmural pressure (e.g., myogenic response, reactive hyperemia, and autoregulation), shear stress (flow-dependent vasodilation), and intercellular conduction. The relative contributions of these active mechanisms in the ocular circulations are hard to define given the complex vascular organization and relative inaccessibility. Ocular blood flow measurement is technically challenging, and discrete perturbations that elicit unambiguous responses characteristic of a particular mechanism in a single vascular bed are difficult to achieve. Eliminating confounding neurohumoral inputs is a further, though not insurmountable, challenge. For these reasons, we can often infer that a blood flow response in an ocular circulation is locally mediated, but the relative contributions of the underlying local control mechanisms are ill-defined. Nonetheless, there is evidence of local control behavior in the prelaminar optic nerve, choroid, retina, ciliary body, and iris.

J.W. Kiel, Ph.D., FARVO
Department of Ophthalmology, University of Texas Health Science Center, 7703 Floyd Curl Drive MS 6230, San Antonio, TX 78229-3900, USA
e-mail: kiel@uthscsa.edu

L. Schmetterer, J.W. Kiel (eds.), *Ocular Blood Flow*,
DOI 10.1007/978-3-540-69469-4_11, © Springer-Verlag Berlin Heidelberg 2012

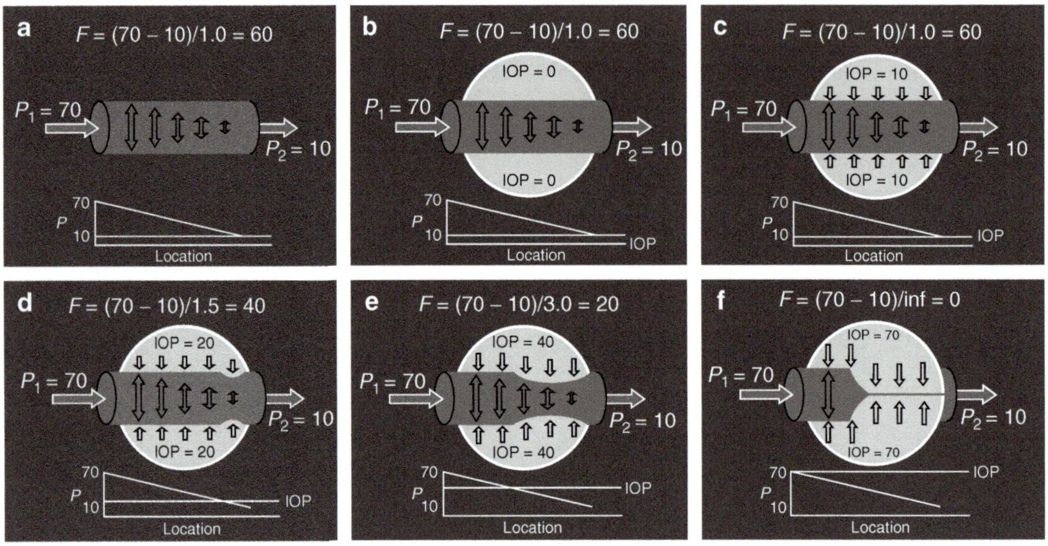

Fig. 11.1 Ocular Starling resistor. (**a**) Flow through a vessel is a function of the pressure gradient along the vessel divided by the resistance. (**b**, **c**) If the vessel passes through an organ (e.g., an eye) with a low tissue pressure (e.g., IOP), the pressure inside the vessel exceeds the pressure outside the vessel (i.e., the transmural pressure gradient) and so the vessel remains distended. (**d**, **e**) If the tissue pressure is somewhat higher and begins to exceed the pressure at the lowest point inside the vessel (i.e., at the "venous" end), that region of the vessel will begin to collapse, which will increase the resistance to flow in that segment thereby raising the intravessel pressure until the transmural pressure is again slightly positive. (**f**) If the tissue pressure exceeds the arterial input pressure, the vessel inside the organ will collapse completely, the resistance will be infinite, and flow through the vessel ceases [5]

11.2 Ocular Perfusion Pressure, IOP, and the Ocular Starling Resistor Effect

The arteriovenous pressure gradient and the vascular resistance determine blood flow through a vessel or tissue circulation. For the ocular circulations, the pressure gradients begin with the arterial pressure in the supply arteries (i.e., the central retinal artery, the short posterior ciliary arteries, the long posterior ciliary arteries, and the anterior ciliary arteries) and end with the venous pressures in the exiting veins (i.e., primarily the central retinal vein and the vortex veins). The vascular resistance is a function of the viscosity of the blood, and the net length, branching pattern, and cross-sectional area of the vasculature. In most organs, including the eye, the pressure drop from the systemic circulation to the capillary level indicates the small arteries and arterioles are the primary site of resistance. However, unlike most organs, the veins within the eye are exposed to a significant compressing force, the intraocular pressure (IOP), and so they behave as Starling resistors, i.e., the pressure in the veins just before they exit the eye must exceed the IOP or they collapse (Fig. 11.1) [1–5]. Consequently, the effective venous pressure for the ocular circulations is the IOP. It is for this reason that the ocular perfusion pressure (OPP) is defined as the mean arterial pressure (MAP, at eye level) minus the IOP. From this definition, it follows that raising the IOP while holding MAP constant at different levels should generate a family of pressure-flow curves, each going to zero when the IOP equals MAP, and that the family of curves should resolve into a single curve when flow is plotted as a function of OPP. This behavior has been demonstrated in the rabbit choroid (Fig. 11.2) and is assumed to be qualitatively similar in the other ocular circulations across species [6] (Fig. 11.3).

11.3 Types of Local Control

By definition, local control of blood flow refers to mechanisms intrinsic to the blood vessels and the adjacent parenchymal cells. These

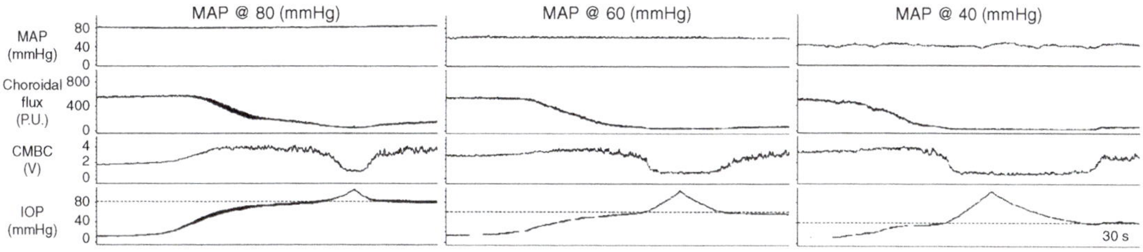

Fig. 11.2 IOP effect on an ocular circulation. Traces show rabbit choroidal (primarily choriocapillaris) blood flow (choroid flux) and blood volume index (CMBC, concentration of moving blood cells) responses to raising IOP by intravitreal saline infusion (30 μl/min) while holding mean arterial pressure (MAP) constant at ≈80, 60, and 40 mmHg. Blood flow ceases and blood is expelled (i.e., the vessels collapse) when the IOP reaches the MAP and the ocular perfusion pressure goes to zero. When the infusion is stopped, the IOP spontaneously declines until the OPP is again greater than zero, then blood flow resumes and begins refilling the vessels (Author's observations using laser Doppler flowmetry to measure choroidal blood flow)

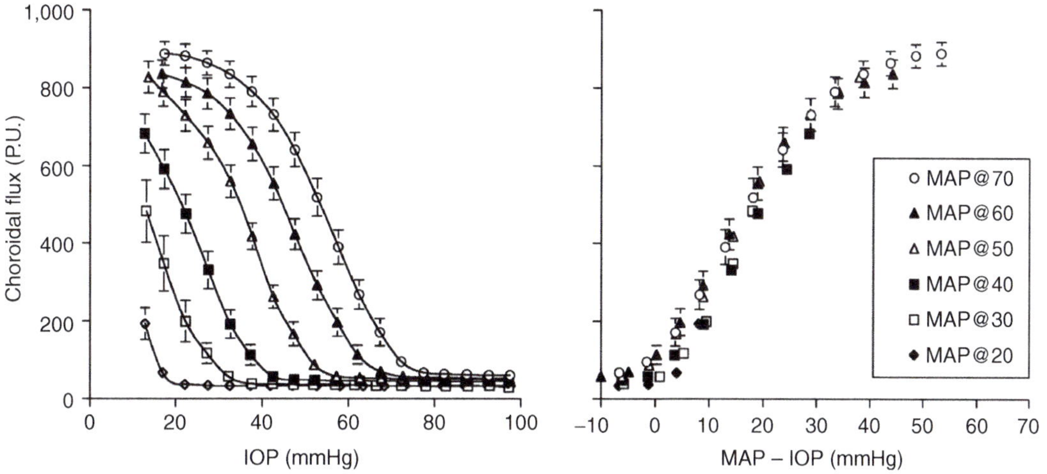

Fig. 11.3 Choroidal blood flow (choroidal flux) changes with IOP and ocular perfusion pressure (MAP–IOP) in the rabbit. (*Left*) Choroidal blood flow response to increasing IOP at different MAP levels. Choroidal blood flow falls as IOP approaches the MAP and stops when IOP exceeds the MAP. (*Right*) The same data with choroidal blood flow plotted against MAP–IOP. The curves superimpose showing that MAP–IOP is the effective ocular perfusion pressure [6]

mechanisms include the myogenic local control, metabolic local control, flow-mediated vasodilation, and flow control by intercellular conduction.

11.3.1 Myogenic Local Control

> "The muscular coat of the arteries reacts, like smooth muscle in other situations, to a stretching force by contraction. It also reacts to a diminution of tension by relaxation, shown, of course, only when in a state of tone. These reactions are independent of the central nervous system, and are of a myogenic nature." "The peripheral powers of reaction possessed by the arteries is of such a nature as to provide as far as possible for the maintenance of a constant flow of blood through the tissues supplied by them, whatever may be the height of the general blood-pressure, except in so far as they are directly overruled by impulses from the central nervous system." (Baylis [7])

As noted by Baylis over a century ago, vascular smooth muscle contracts in response to stretch and relaxes when imposed stretch is released, and this inherent property tends to

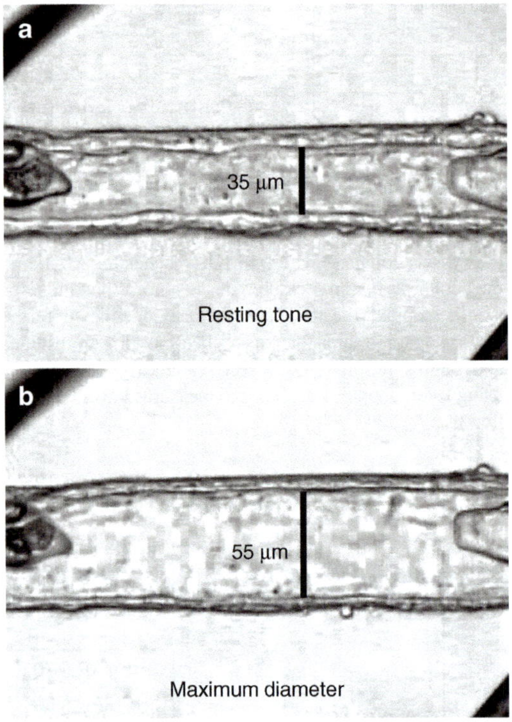

Fig. 11.4 Isolated human retinal arteriole cannulated with glass micropipettes. (**a**) The vessel was allowed to develop resting basal tone (35 μm internal diameter) at 55 cmH_2O intraluminal pressure. (**b**) Maximum diameter (55 μm internal diameter) of the vessel was established in Ca^{2+}-free solution with 0.1 mM sodium nitroprusside [8]

promote constant blood flow when the arterial segment of a vascular bed is exposed to changing arterial pressure. The myogenic mechanism, also called the myogenic response, has been demonstrated in vessels from a wide variety of tissues and organs, including the eye (Fig. 11.4). In vitro experiments indicate that the myogenic response requires extracellular calcium but does not require an intact endothelium (Fig. 11.5). The myogenic response is also relatively quick, occurring within seconds to minutes (Fig. 11.6).

In many tissues, blood flow remains relatively constant over a range of arterial pressure in the absence of any neurohumoral input – a phenomenon known as autoregulation. As noted by Baylis, the myogenic mechanism has the potential to help maintain blood flow if arterial pressure changes, and there is ample evidence indicating a myogenic role in autoregulation [13]. However, the myogenic mechanism responds to stretch, or more specifically vascular wall tension, and consequently, it can also exacerbate the flow response to pressure change under some circumstances. For example, if venous pressure rises, upstream pressures also rise and trigger an arterial myogenic response such that the fall in blood flow due to the

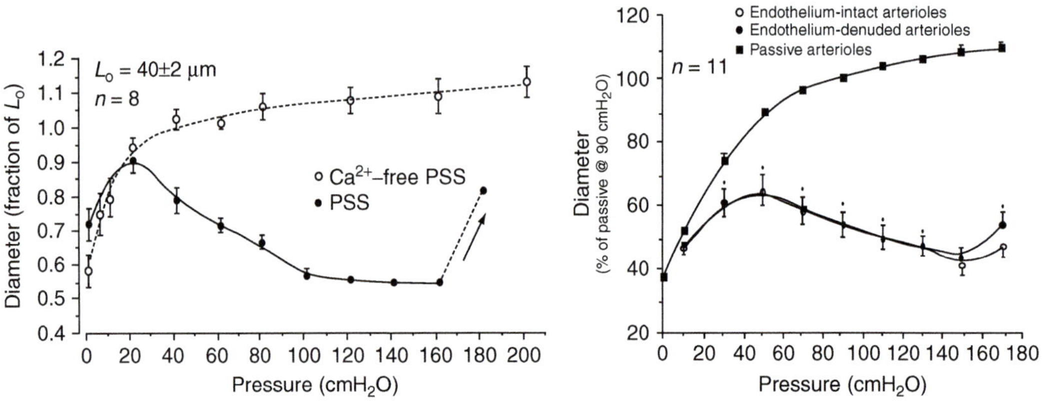

Fig. 11.5 Effect of increasing intraluminal pressure on vessel diameter. (*Left*) Arterioles isolated from hamster cheek pouch fail to vasoconstrict in response to increasing intraluminal pressure in the absence of extracellular calcium. (*Right*) Arterioles isolated from rat cremaster muscle vasoconstrict in response to increasing intraluminal pressure in the presence and absence of endothelium [9, 10]

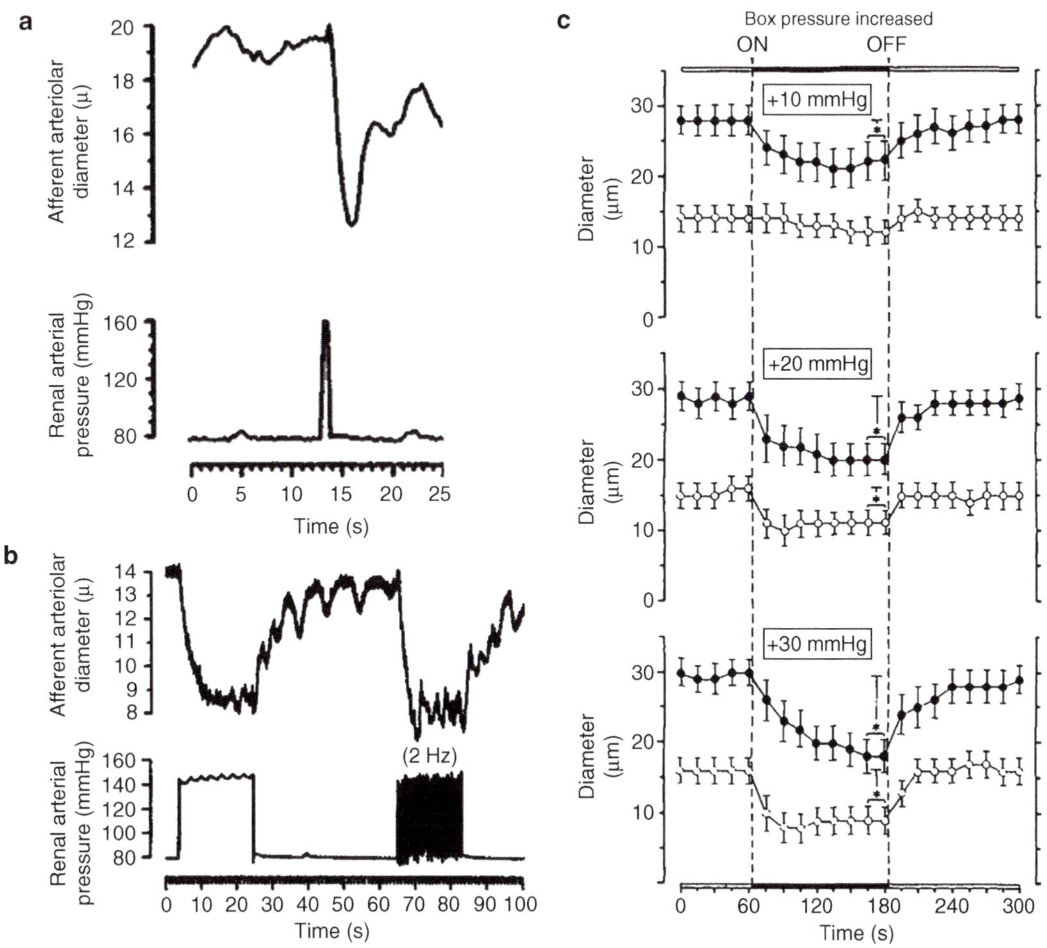

Fig. 11.6 Rapidity of myogenic contraction. (**a**) Afferent arteriole constriction in rat kidney preparation to 1 s arterial pressure spike and (**b**) to 20 s sustained and pulsatile step-increases in arterial pressure. (**c**) Rat cremaster third-order (*filled circles*) and fourth-order (*open circles*) arteriole constrictions to step increases in transmural pressure of 10, 20, and 30 mmHg. Myogenic response rate and magnitude of vasoconstriction vary with location in vascular tree and size of pressure step [11, 12]

decreased perfusion pressure is worsened by the myogenic vasoconstriction [14]. Figure 11.7 shows another example of myogenic nonautoregulatory behavior in a rat cremaster arteriole using the pressurized box preparation [15]. In this preparation, the animal is placed inside an airtight box and the cremaster muscle is exteriorized for viewing on an inverted microscope stage. When the box pressure is raised, there is an equivalent increase in arterial and venous pressures, thereby raising the transmural pressure without changing the perfusion pressure. This elicits a strong myogenic response and significant decrease in blood flow.

As Fig. 11.7 shows, the myogenic response is robust but clearly does not deliberately maintain constant blood flow. For the specific case of changing arterial pressure, Johnson proposed that myogenic autoregulation can occur when arterial smooth muscle incorporates a sensor responsive to changing wall tension coupled in series with the contractile elements and moderate feedback

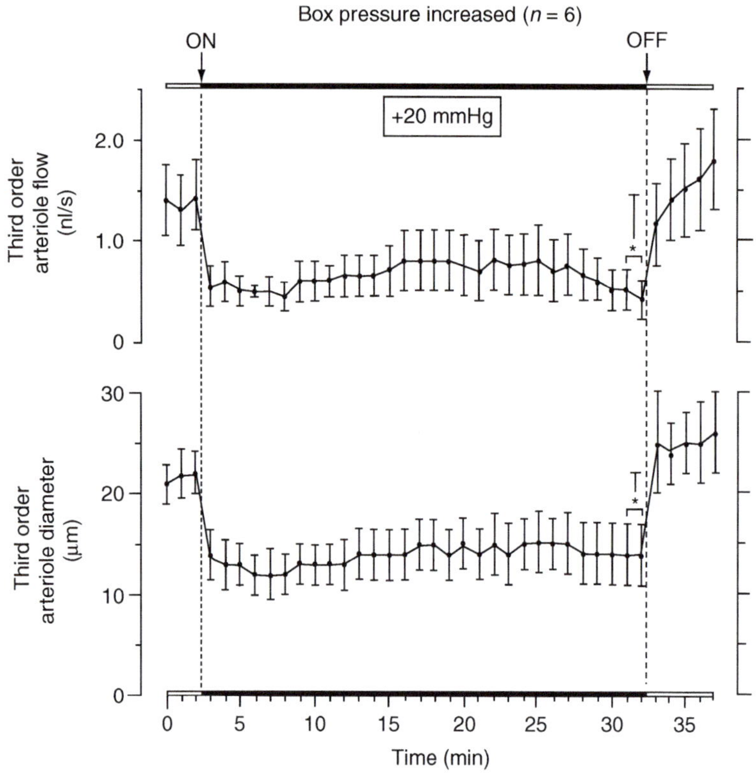

Fig. 11.7 Rat cremaster arteriole constriction and decrease in blood flow in response to increased transmural pressure at constant perfusion pressure [12]

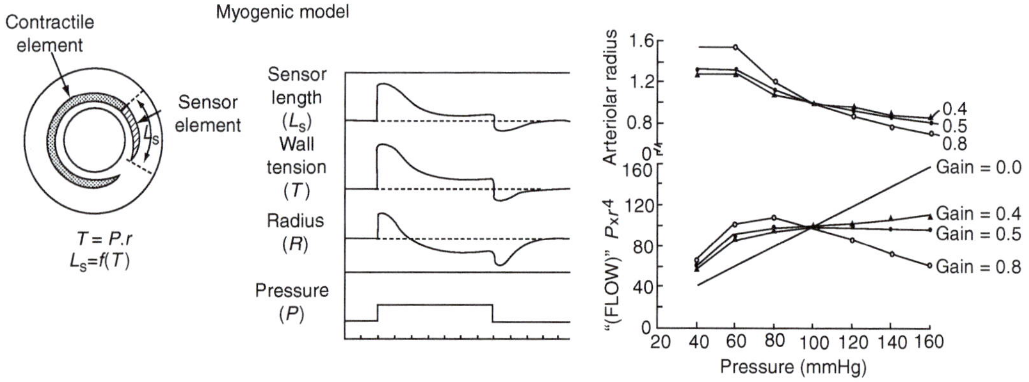

Fig. 11.8 Myogenic model of autoregulation. Model assumes arterial resistance vessels have a sensor element coupled in series with the contractile element that responds to changes in wall tension as defined by the law of Laplace (tension equals the transmural pressure times the radius, $T = P \cdot r$). The model addresses a conceptual problem with myogenic autoregulation, i.e., if the arterial wall is stretched by an increase in pressure, the muscle cannot simply contract back to its original length, since that would return the vessel to its original radius, resistance would be unchanged, and flow would increase. For flow to remain constant despite an increase in pressure, the muscle fibers need to shorten to less than their prestretched length, so that the radius is less than control and resistance increases. With moderate feedback gain, the model predicts autoregulatory flow behavior. If the gain is too high, however, the model predicts "superregulation" where flow decreases in response to increased perfusion pressure [13, 16]

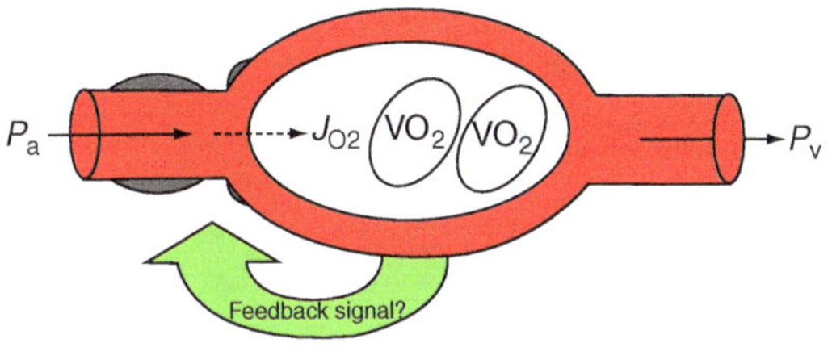

Fig. 11.9 Metabolic local control. Over the autoregulatory range of perfusion pressure ($P_a - P_v$, arterial pressure minus venous pressure), arterial resistance and the number of perfused capillaries are modulated by a feedback signal linked to oxygen delivery (J_{O2}) and parenchymal metabolism (VO_2)

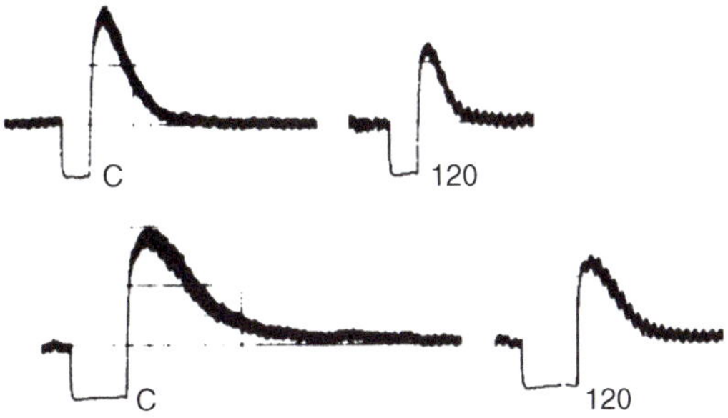

Fig. 11.10 Coronary reactive hyperemia. Blood flow overshoot in canine circumflex artery after release of 15 s (*top*) and 30 s (*bottom*) arterial occlusions before (*left*) and 120 min after (*right*) administration of the nonselective adenosine antagonist, theophylline. Magnitude of postocclusion blood flow overshoot increases with occlusion duration and is blunted by adenosine blockade [20]

gain (Fig. 11.8) [13, 16]. Such a regulatory feedback loop would permit the vessel radius adjustments necessary to maintain flow when pressure changes (e.g., if arterial pressure increases, the arterial contraction must decrease the radius below control to maintain blood flow constant). Johnson also noted that in terms of homeostasis, the myogenic mechanism is better suited to regulating tissue capillary hydrostatic pressure than blood flow (e.g., if arterial or venous pressure rises, arterial myogenic vasoconstriction in both cases would tend to preserve capillary hydrostatic pressure).

11.3.2 Metabolic Local Control

The basic premise of metabolic local control is that tissues regulate their blood flow to insure the delivery of nutrients and removal of waste in accordance with their metabolic requirements (Fig. 11.9) [17]. The metabolic hypothesis assumes communication between parenchymal cells and the smooth muscle cells controlling the tissue vascular resistance (arterioles) and capillary flow distribution (precapillary sphincters and pericytes) [18, 19]. Because most tissues utilize aerobic metabolism, the convective delivery of oxygen by blood flow to the tissue is often considered the regulated variable. If oxygen delivery decreases (e.g., due to a fall in arterial pressure) or oxygen demand increases (e.g., increased neuronal activity), the parenchymal cells produce a vasodilatory signal that increases tissue blood flow and capillary perfusion such that oxygen delivery is again matched to oxygen demand. Conversely, if oxygen delivery exceeds demand, the parenchymal cells decrease production of the vasodilatory signal until delivery and demand are again matched. There are numerous vasodilator candidates linked to metabolism that can act as the feedback signal (e.g., adenosine, CO_2, H+, lactate, etc.), and it is likely that all participate to a variable extent depending on the tissue. Because it is a vasoconstrictor, oxygen can also modulate local resistance in accordance with metabolic demand.

Several lines of evidence support the metabolic local control hypothesis: reactive hyperemia (Fig. 11.10), functional hyperemia (Fig. 11.11), modulation of pressure-flow autoregulation by metabolic stimulation (Fig. 11.12), and hypoxic hyperemia (Fig. 11.13).

11.3.3 Flow-Mediated Vasodilation

In vitro and in vivo studies of large and small arteries show that flow elicits endothelium-dependent vasodilation (Fig. 11.14) [24–28].

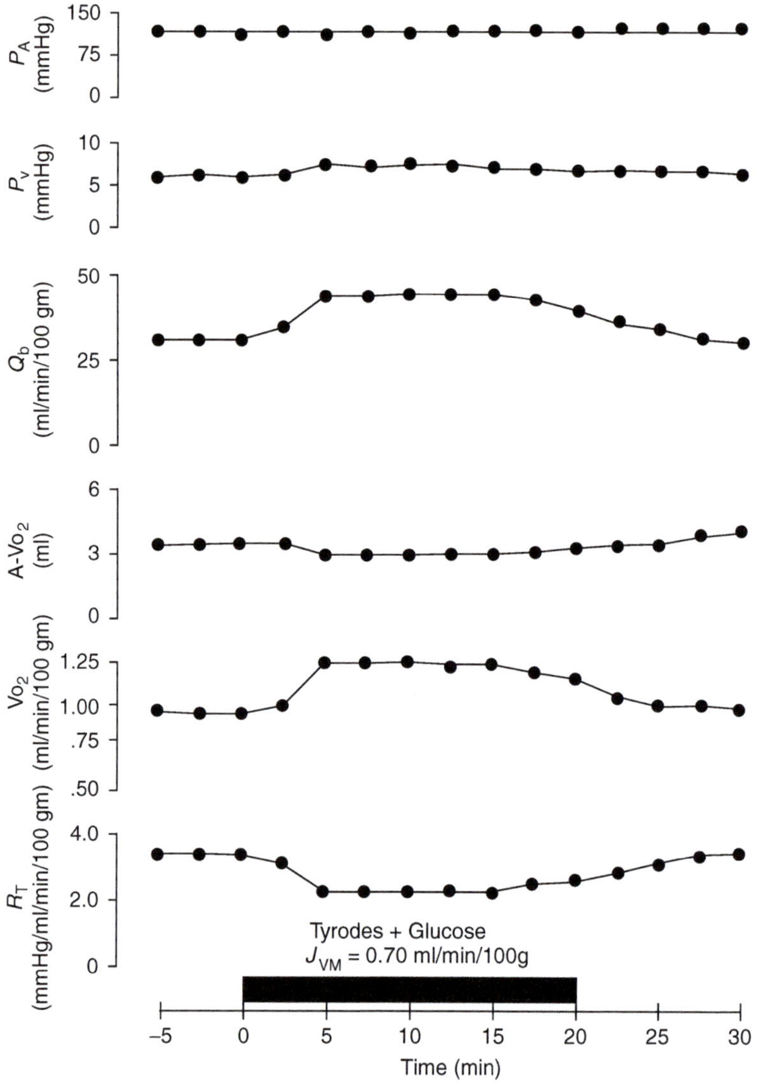

Fig. 11.11 Intestinal functional hyperemia. Blood flow (Q_b) in an isolated loop of feline ileum before and after filling the lumen with a glucose solution. Oxygen consumption (VO_2) increased during glucose absorption, which was achieved by increased Q_b rather than increased arteriovenous oxygen extraction (A-VO_2). The increased Q_b was due to a decrease in vascular resistance (R_t) since arterial (P_a) and venous (P_v) pressures were unaltered [21]

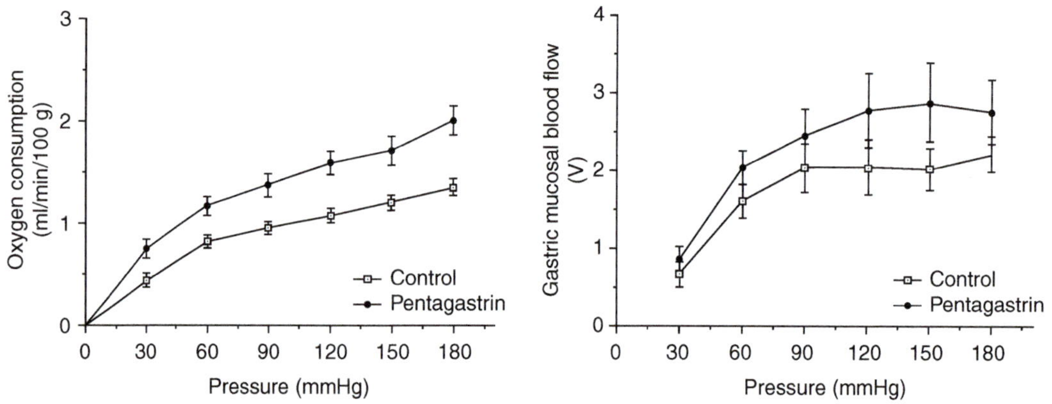

Fig. 11.12 Gastric mucosal autoregulation and metabolic activity. Effect of changing perfusion pressure on total oxygen consumption and mucosal blood flow in a pressure-perfused canine stomach preparation before and during pentagastrin-stimulated acid secretion. Increased oxygen consumption was associated with an upward shift in the mucosal pressure-flow relationship [22]

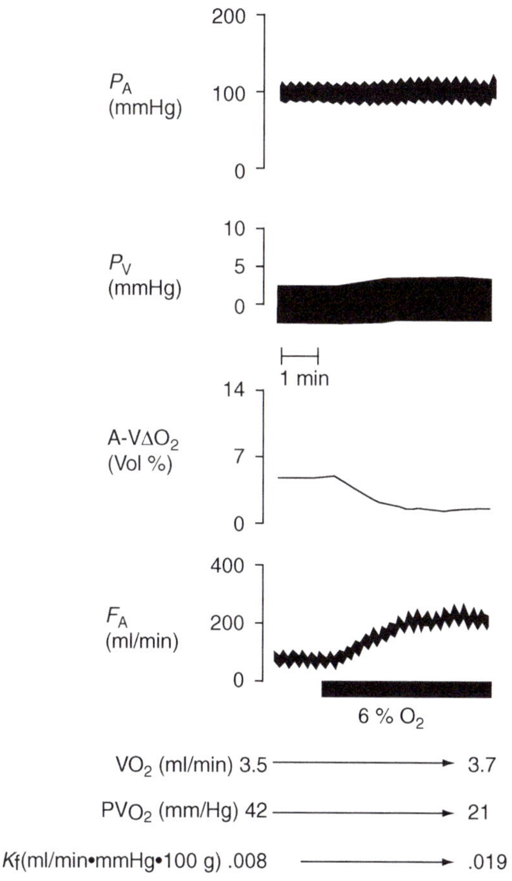

Fig. 11.13 Hypoxic vasodilation. Systemic hypoxia elicits increased hind limb skeletal muscle blood flow (F_A) in a denervated dog preparation [23]

The response appears to be mediated by shear stress exerted on the endothelial cells by the velocity and viscosity of blood moving within the vessel lumen (Fig. 11.15). The response is inhibited by indomethacin (Fig. 11.15) and nitric oxide synthase inhibitors, indicating that endothelial release of vasoactive prostaglandins and nitric oxide play a role in the response [29, 30]. The role of flow-mediated vasodilation in local control of tissue blood flow is complex since it has the potential to be inherently unstable (i.e., an increase in flow elicits a vasodilation that causes a further increase in flow). However, the robustness of the response varies with location in the arterial tree and is likely modulated by metabolic and myogenic local control mechanisms.

11.3.4 Flow Control by Intercellular Conduction

One potential integrating mechanism for the various local control mechanisms is intercellular communication along the arterial tree [31]. The evidence for this mechanism is the rapid propagation of a focal vasodilation elicited by iontophoretic application of acetylcholine (Fig. 11.16) [32]. The vasodilation spreads from one region of

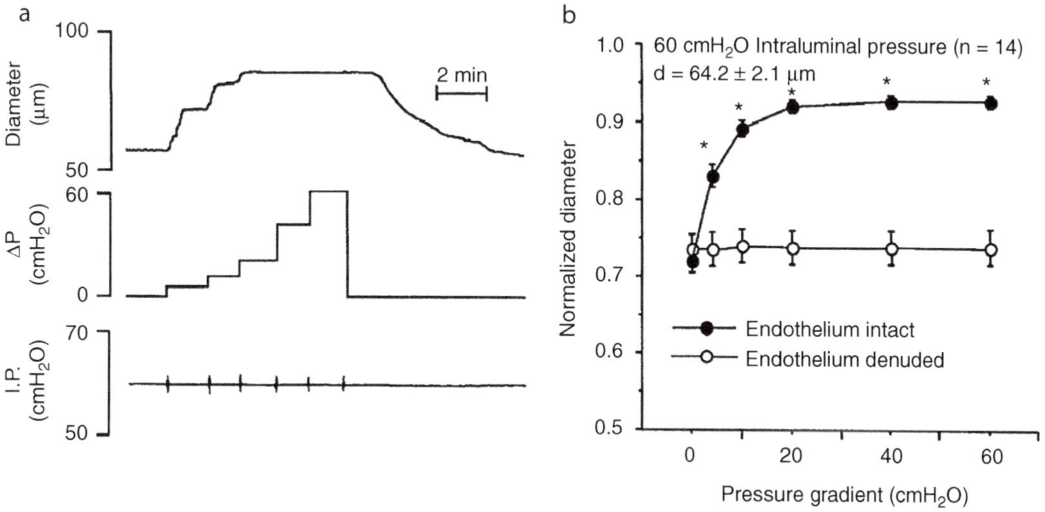

Fig. 11.14 Flow-mediated vasodilation. (**a**) Isolated pig coronary arterioles dilate in response to flow increase caused by increased perfusion pressure (ΔP) while holding the midpoint intraluminal pressure constant. (**b**) Flow-induced vasodilation is abolished after removal of endothelium [28]

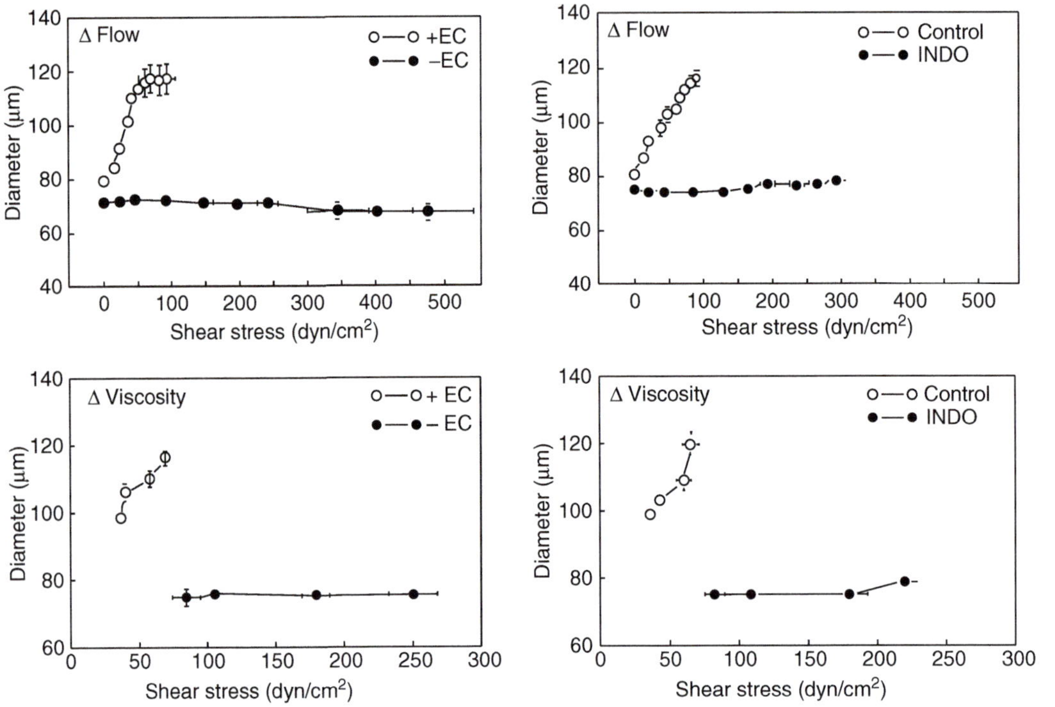

Fig. 11.15 Flow-mediated vasodilation and shear stress. Isolated rat cremaster arterioles dilate in response to shear stress increased by raising flow (*top*) or viscosity (*bottom*). Shear stress-induced dilatation is eliminated by endothelial removal (*left*) or indomethacin (*right*) [29]

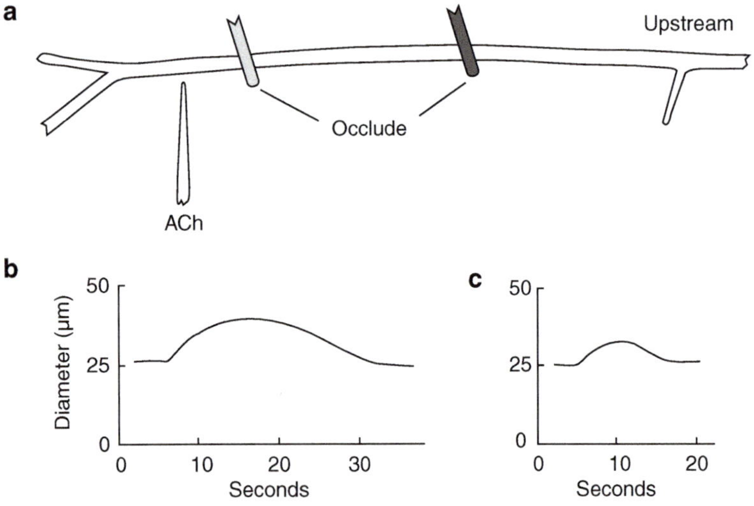

Fig. 11.16 Propagated vasodilation. In hamster cheek pouch arterioles (**a**), a point application of acetylcholine (ACh) causes vasodilation (**b**) that propagates upstream past a double occlusion (**c**) indicative of cell-to-cell communication along the arterial tree [32]

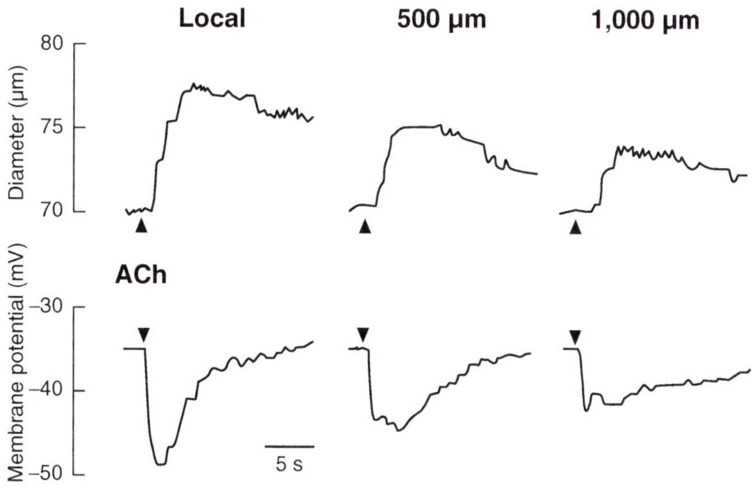

Fig. 11.17 During propagated vasodilation, there is a corresponding hyperpolarization of the endothelial membrane potential [33]

an artery or arteriole to another. The fact that the vasodilation spreads past an upstream occlusion indicates that flow-mediated vasodilation is not involved. The propagated vasodilation is associated with a hyperpolarization of the endothelial cells [33] (Fig. 11.17).

11.4 Ocular Local Control

The study of local control mechanisms in the ocular circulations is difficult because of the eye's unique anatomy and the limitations of current blood flow measuring technology. Nonetheless, there are clear examples of local control phenomena in each of the ocular circulations.

11.4.1 Optic Nerve Head (ONH)

Numerous studies show evidence of pressure-flow autoregulation in the prelaminar ONH (Fig. 11.18). The autoregulatory behavior occurs in different species with blood flow measured with different techniques and perfusion pressure varied by changing IOP or blood pressure [34–37]. However, the local control mechanisms responsible for this behavior are unclear, and all may contribute, though metabolic and myogenic seem most likely.

There are several pieces of evidence in favor of metabolic autoregulation. First, the optic nerve head undergoes a robust functional hyperemic response during retinal illumination with a flickering light stimulus (Fig. 11.19) [38]. The functional hyperemia is linked to metabolism since it occurs in conjunction with a decrease in tissue PO_2 indicative of increased oxygen consumption and a rise in interstitial potassium concentration consistent with increased neuronal activity [39]. There is also an increase in local NO production, which presumably contributes, but the functional hyperemia persists despite inhibition of nitric oxide synthase [40]. The response is also modulated by exogenous adenosine, but it is unclear whether endogenous adenosine is involved [40]. Second, the pressure-flow relationship is shifted upward during metabolic stimulation (Fig. 11.20) [41]. Interestingly, the breakpoint in the pressure-flow relationship occurs at a higher perfusion pressure than in control. This may be due to decreased action potentials arriving from the retina due to inadequate retinal perfusion, or the vasodilatory reserve may be exhausted sooner at the higher metabolic rate (i.e., the ONH arterioles have a maximum achievable diameter, all of which is available to respond to decreased perfusion pressure under control conditions, but less is available when doubly challenged by increased metabolism and decreased perfusion pressure).

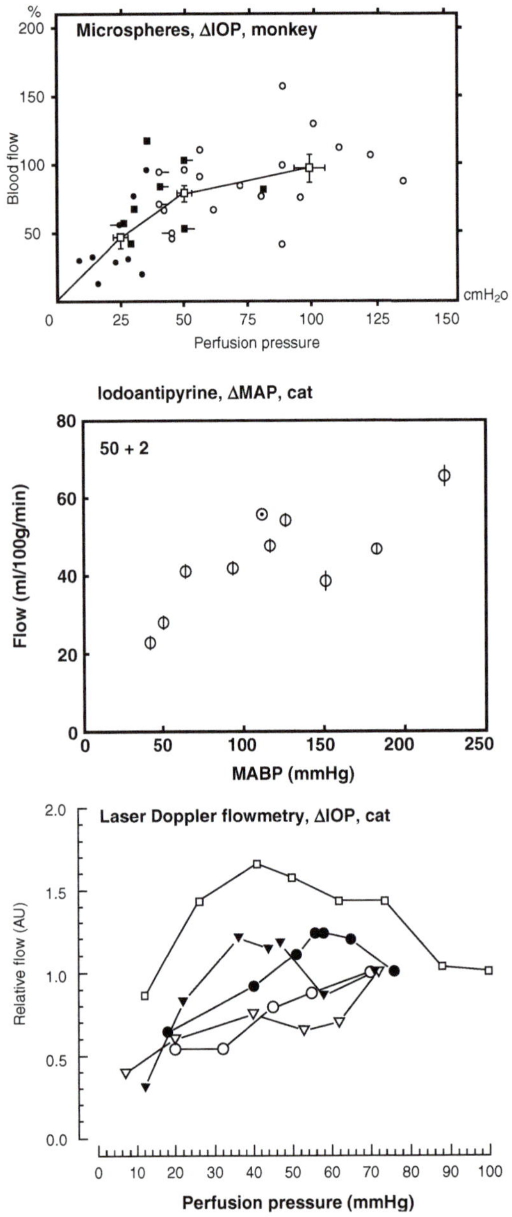

Fig. 11.18 Autoregulation of optic nerve head blood flow reported in different species with different blood flow measuring techniques and different perfusion pressure perturbations [34–36]

Third, hyperoxia decreases ONH blood flow and hypoxia increases it (Fig. 11.21) [42]. Lastly, the ONH appears to undergo a reactive hyperemia following brief periods of ischemia, at least in some species (Fig. 11.22) [41]. It seems to occur in cats, monkeys, and humans [43], but the experiments were not designed specifically to study reactive hyperemia (i.e., the ischemic period duration was not varied systematically or performed at different levels of metabolic activity). These pieces of evidence are consistent with metabolic local control in the ONH.

There is also evidence against metabolic involvement in ONH local control, at least in the rabbit (Fig. 11.23). In that species, a 10-min period of relative ischemia did not elicit a reactive hyperemic response, although there was clear evidence of autoregulation (i.e., the relative decrease in ONH velocity was much smaller than the relative decrease in perfusion pressure) [45]. Moreover, the speed of the autoregulatory response in the rabbit to a step-decrease in perfusion pressure was quite rapid (<5 s), which seems fast for the accumulation of vasoactive metabolites and perhaps more in keeping with a myogenic response (though the flicker response, which is clearly metabolic, is also quite fast) [46]. The difference in the reactive hyperemic responses in the rabbit versus the cat and monkey may reflect species differences, but more studies designed specifically to identify the local control mechanisms responsible for ONH autoregulation are clearly needed.

11.4.2 Choroid

In contrast to optic nerve head autoregulation, the evidence regarding choroidal autoregulation is less consistent. The advent of the microsphere technique in the eye [47] in the early 1970s stimulated several studies of choroidal blood flow responses to changing perfusion pressure, typically by raising IOP. In some cases, the results indicated no choroidal autoregulation, as in the cat study by Alm and Bill, while the results in another cat study by Weiter et al. indicated the choroid has autoregulatory ability (Fig. 11.24) [48, 49]. In another study by Alm and Bill, the results in primates were ambiguous (Fig. 11.25) [50]. Accounting for their use of femoral arterial pressure as an index of ophthalmic artery

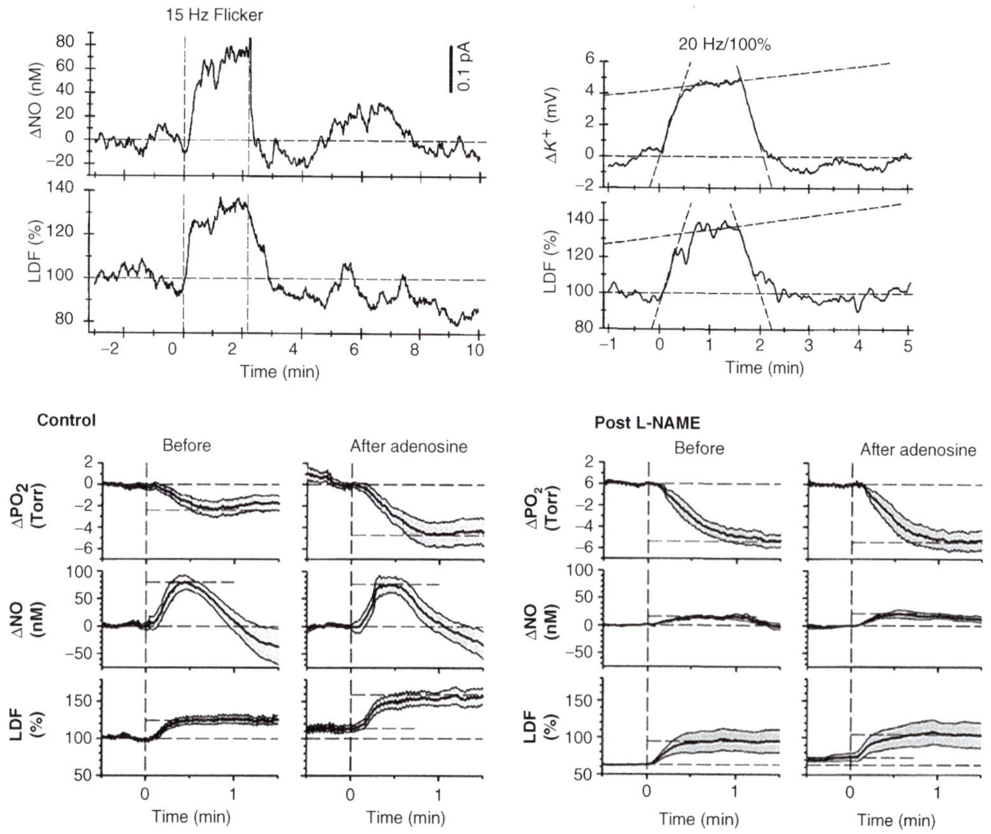

Fig. 11.19 Functional hyperemia in the optic nerve head. (*Top, left*) Retinal flicker stimulation in the anesthetized cat increases ONH nitric oxide (NO) and ONH blood flow measured by LDF. (*Top, right*) Flicker also increases cat ONH potassium levels and ONH blood flow. (*Bottom*) Retinal flicker responses are also modulated by adenosine and the nitric oxide synthase inhibitor L-NAME [39, 40, 44]

pressure, the authors noted "If the line connecting the values for the two eyes is extrapolated and then intercepts either the positive flow axis or the positive pressure axis below 25 cmH_2O [18.4 mmHg] this suggests a vasodilation in the eye with reduced perfusion pressure.... Although some lines point toward the positive flow axis the mean result does not indicate any marked reduction in vascular resistance in response to a reduction in perfusion pressure." Using the authors' definition, roughly half the animals autoregulated and the other half did not (Fig. 11.25). Given the lack of clear evidence of autoregulation, the authors concluded that the choroid was passive. Other studies came to a similar conclusion.

Stronger evidence of choroidal autoregulation came later from a series of studies in rabbits using a fiber-optic-based laser Doppler flowmeter (Fig. 11.26) [6, 51, 52]. In those studies, the efficacy of autoregulation depended on the method used to vary the perfusion pressure, i.e., autoregulation occurred over a wider perfusion pressure range when arterial pressure was manipulated without controlling IOP than when IOP was raised at a constant MAP. It was also noted that the efficacy of choroidal autoregulation depended on the type of anesthesia, i.e., autoregulation occurred under pentobarbital anesthesia but was largely abolished by Nembutal, which contains 40% propylene glycol and 10% alcohol [53]. Methodological issues such as these and others

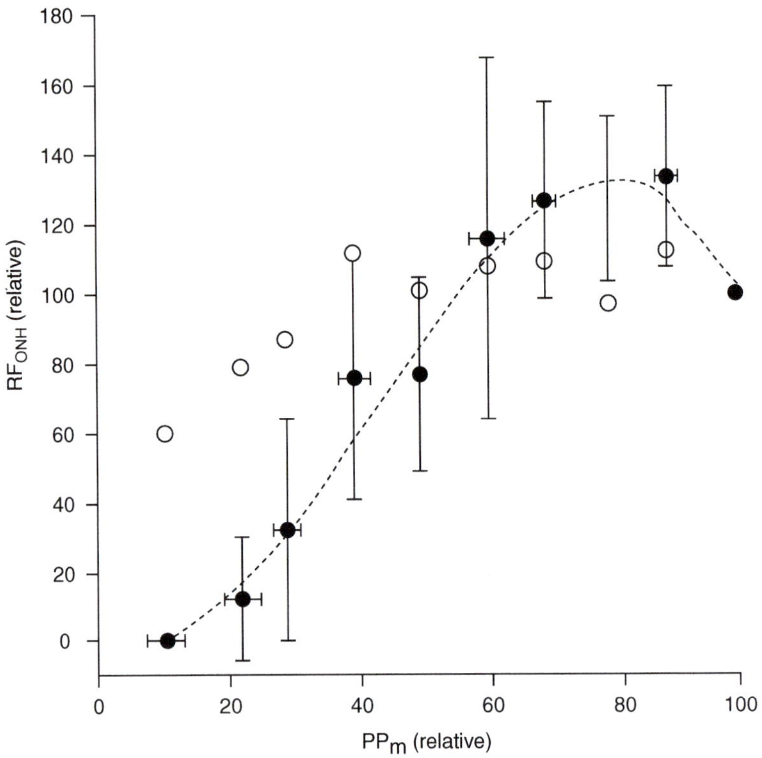

Fig. 11.20 ONH autoregulation and metabolic activity. Feline ONH pressure-flow relationship is broader during flicker (*closed circles*) than without (*open circles*) flicker (RF$_{ONH}$: ONH blood flow response measured by LDF) [41]

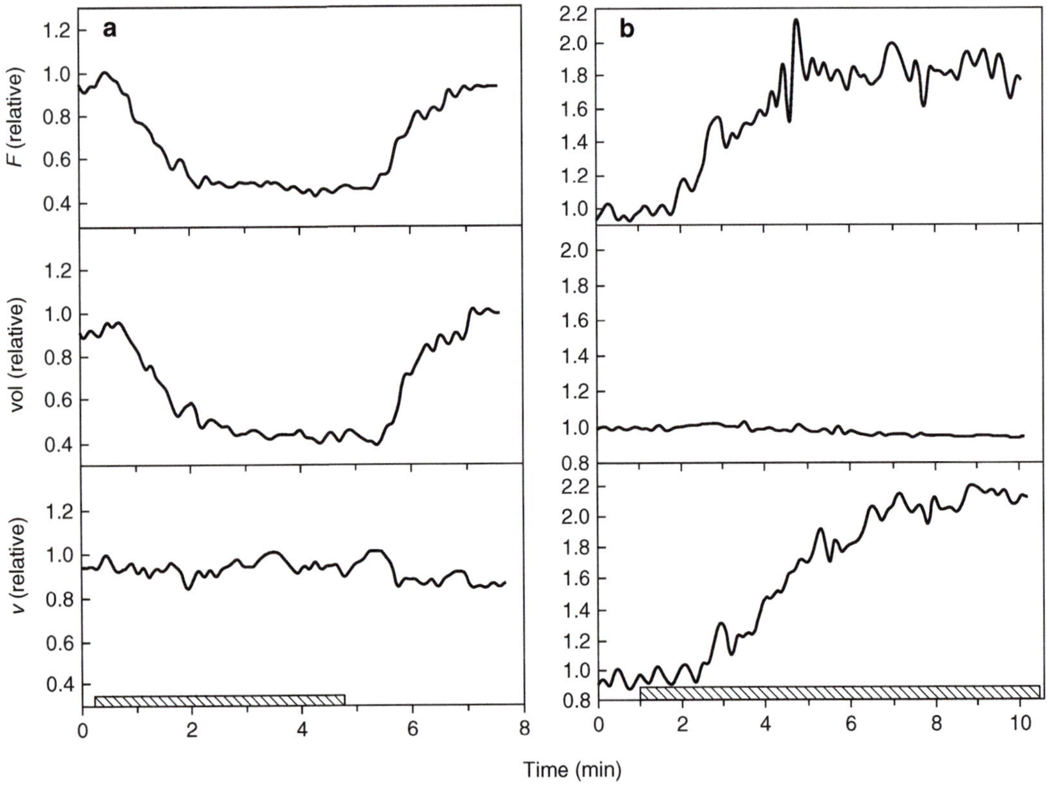

Fig. 11.21 Cat ONH blood flow (*F*) decreases in response to hyperoxia (**a**) and increases in response to hypoxia (**b**) [42]

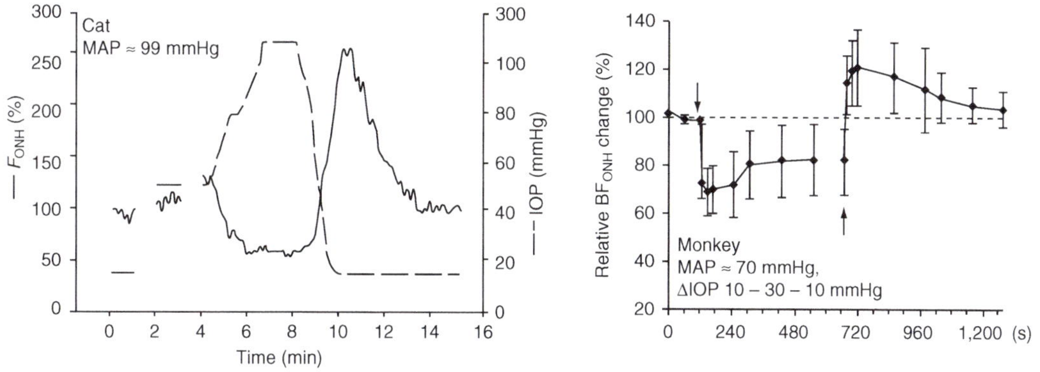

Fig. 11.22 ONH reactive hyperemia following periods of raised IOP in the cat (*left*) and monkey (*right*) [37, 41]

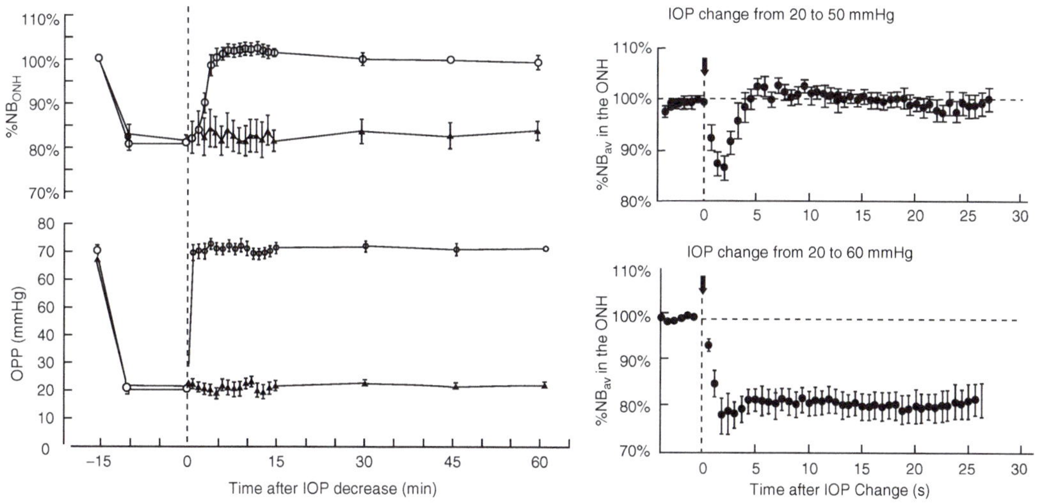

Fig. 11.23 Observations inconsistent with metabolic local control in the ONH. (*Left*) Rabbit fails to exhibit reactive hyperemia after period of ischemia. (*Right*) Rabbit ONH autoregulatory response to raised IOP is very rapid [45, 46]

may account for the discrepant findings in the literature regarding choroidal autoregulation.

Unlike the optic nerve and retina, the choroid is richly innervated. Consequently, an interplay of local and neural control mechanisms is inherent in the choroidal response to changing perfusion pressure and designating the response as autoregulation (i.e., due solely to local control) may not be appropriate. For that reason, some authors prefer the term "baroregulation" when all regulatory mechanisms are intact and choroidal blood flow does not change with perfusion pressure. However, it is possible to minimize the neural control contribution by systemic ganglionic blockade. In rabbits, this causes a slight downward shift in the choroidal pressure-flow relationship, but choroidal blood flow remains pressure-independent until perfusion pressure falls below approximately 40 mmHg [52]. This result indicates that the choroid is capable of autoregulation (Fig. 11.27).

Metabolic local control does not seem to play a role in choroidal autoregulation, since the choroid does not respond markedly to retinal metabolic stimulation by flicker, or to hyperoxia or hypercapnia, though it does respond to hypocapnia (Fig. 11.28) [54, 55]. The choroid also fails to undergo a reactive hyperemic response to brief

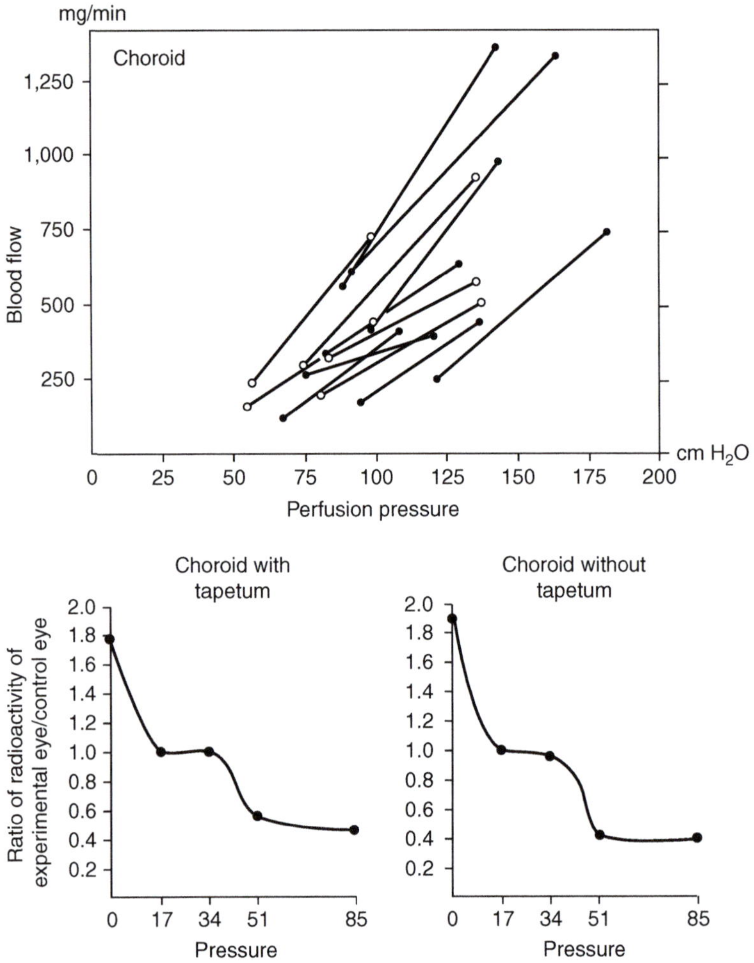

Fig. 11.24 Early microsphere evidence of passive (*left*) and autoregulating (*right*) choroid in cats in response to uniocular raised IOP [48, 49]

(Fig. 11.29) or long periods of ischemia [56]. Choroidal blood flow also does not increase (and may even decrease) when retinal oxygen consumption increases in the dark such that the inner photoreceptor layer becomes nearly anoxic (see below). The apparent absence of choroidal metabolic local control may be due to the high flow rate in the choroid and its anatomical organization. Metabolic local control is thought to depend on metabolic vasodilators reaching the resistance vessels. The choroid itself has little parenchymal metabolism, so the source of metabolic vasodilators would be the RPE and photoreceptors. The sheetlike choriocapillaris design is thought to optimize oxygen delivery and waste removal for the RPE and photoreceptors. If efficient, the waste removal function of the choriocapillaris would preclude vasodilators from the RPE and photoreceptors from reaching the choroidal arterioles. This is one plausible explanation for the lack of choroidal metabolic local control.

There is some evidence for choroidal myogenic local control, at least in the rabbit [51, 57]. To see if a myogenic mechanism could account for autoregulatory-like behavior in the rabbit choroid, Kiel and Shepherd created a mathematical model incorporating myogenic control of choroidal arterial resistance and passive (Starling resistor) control of choroidal venous resistance. The model simulated pressure-flow relationship

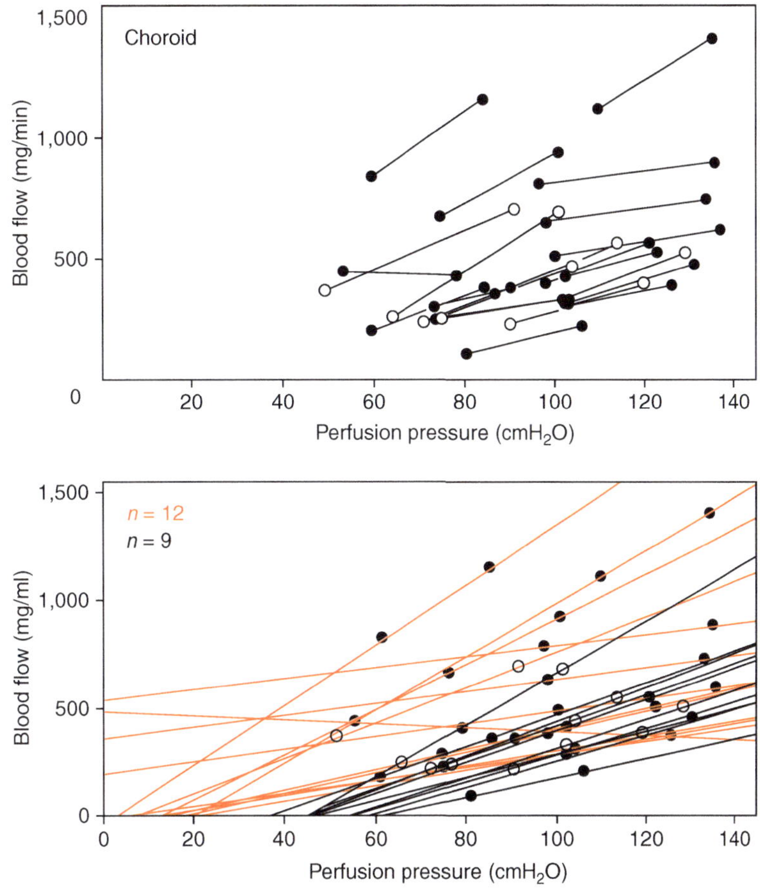

Fig. 11.25 Binocular microsphere measurements of choroidal blood flow in primates with IOP elevated in one eye. Lines connect the data points for eye pairs in each monkey. Autoregulation threshold is at perfusion pressure of 25 cmH_2O; lines intersecting the *x*-axis below that threshold indicate autoregulation, and lines intersecting above that threshold indicate passive behavior (*Top*: original figure[50], *Bottom*: figure redrawn with linear regression lines) [5]

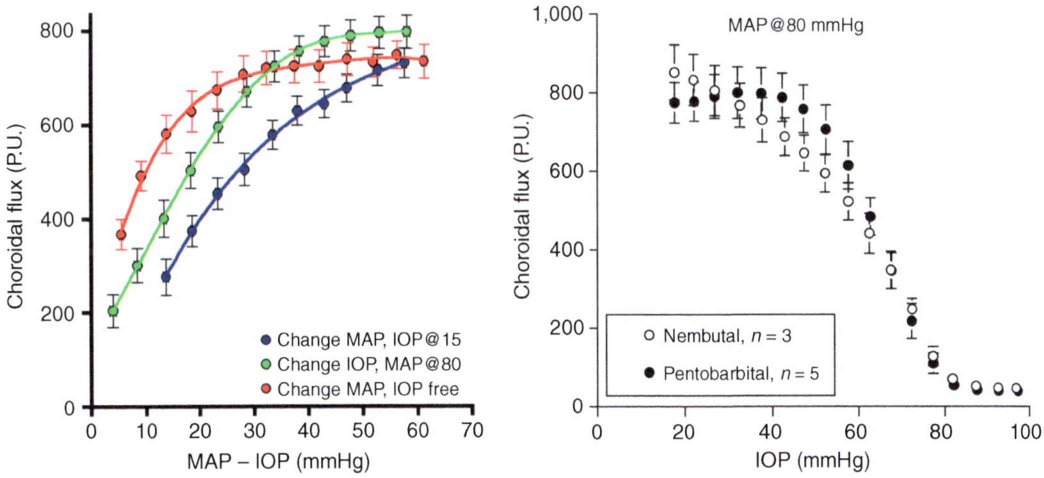

Fig. 11.26 Rabbit choroidal autoregulation depends on how perfusion pressure is manipulated (*left*) and type of anesthesia (*right*) [6, 53]

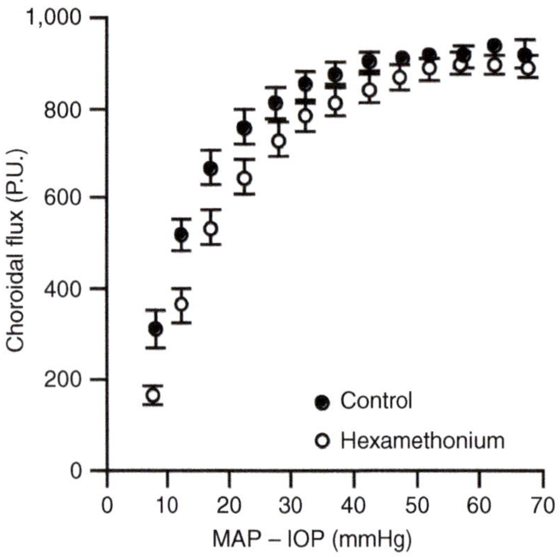

Fig. 11.27 Choroidal autoregulation in the rabbit occurs in the absence of neural control [52]

was similar to that found in vivo, suggesting that a myogenic mechanism could account for the autoregulatory-like behavior. Moreover, the model predicted decreased choroidal blood flow in response to increased perfusion pressure in conditions conducive to heightened wall tension, and similar responses were found in vivo. Thus, it appears myogenic local control may be operative in the choroid (Fig. 11.30).

As noted earlier, from a function perspective, the myogenic mechanism is better suited to regulating capillary hydrostatic pressure than blood flow. If so, a myogenic mechanism in the choroid could help to stabilize blood volume during changes in arterial pressure and thereby minimize arterial pressure-dependent fluctuations in IOP. Consistent with this idea, normally the IOP rises a few mmHg in response to a 30–40-mmHg mechanically induced increase in arterial pressure, but if vascular control is blocked with a systemic vasodilator, the IOP response to a similar arterial pressure increase is much greater [57]. This IOP protective response occurs during autonomic blockade, so it has a local mechanism. If the arterial hypertension is neurally or humorally mediated, the local mechanism may augment the choroidal vasoconstriction relative to that occurring throughout the systemic circulation [58] (Fig. 11.31).

A function proposed for the choroidal circulation that may have a local component is retinal temperature regulation. Given the small amount of metabolically active tissue (i.e., heat generating) relative to the size of the globe, the temperatures of the retina and the eye are primarily determined by the blood-borne convective heat delivery from the core and heat dissipation from the exposed ocular surface. As the largest of the ocular circulations, the choroid provides most of the convective heat delivery to the eye. In an environment cooler than core body temperature, the thermal gradient favors ocular heat loss; conversely, in a hot environment, the thermal gradient favors ocular heat gain. An additional source of heat gain is light absorption in the pigmented RPE and choroid. Given these thermodynamic parameters, it follows that eye temperature will fall if ocular blood flow is reduced in a cold environment, or rise if the environment is hot. If the temperature outside the eye is the same as core body temperature, stopping ocular blood flow will not change retinal temperature.

Parver et al. reported results from two monkeys in which raising IOP above systolic blood pressure decreased the temperature of a thermistor inside a 23-gauge needle inserted in the retina-choroid by 1–1.2°C [59]. The temperatures of the environment, the saline drip on the cornea, and the infusate used to raise IOP were not provided so the thermal gradients are unclear. Given that the retina-choroid temperatures were measured at the macula where the thermal gradient should be dominated by the temperature of the orbit, the magnitude and speed (i.e., "not less than one minute after alteration of the pressure") of the retina-choroid temperature decreases suggest that the thermal gradient to the exposed anterior eye surface must have been quite large. In the same two monkeys, exposure of the cornea to 1.09 mW/cm^2 of light caused an increase in retina-choroid temperature of ≈0.9°C when the IOP was set at 20 mmHg, which increased another ≈0.8°C when IOP was raised above systolic blood pressure.

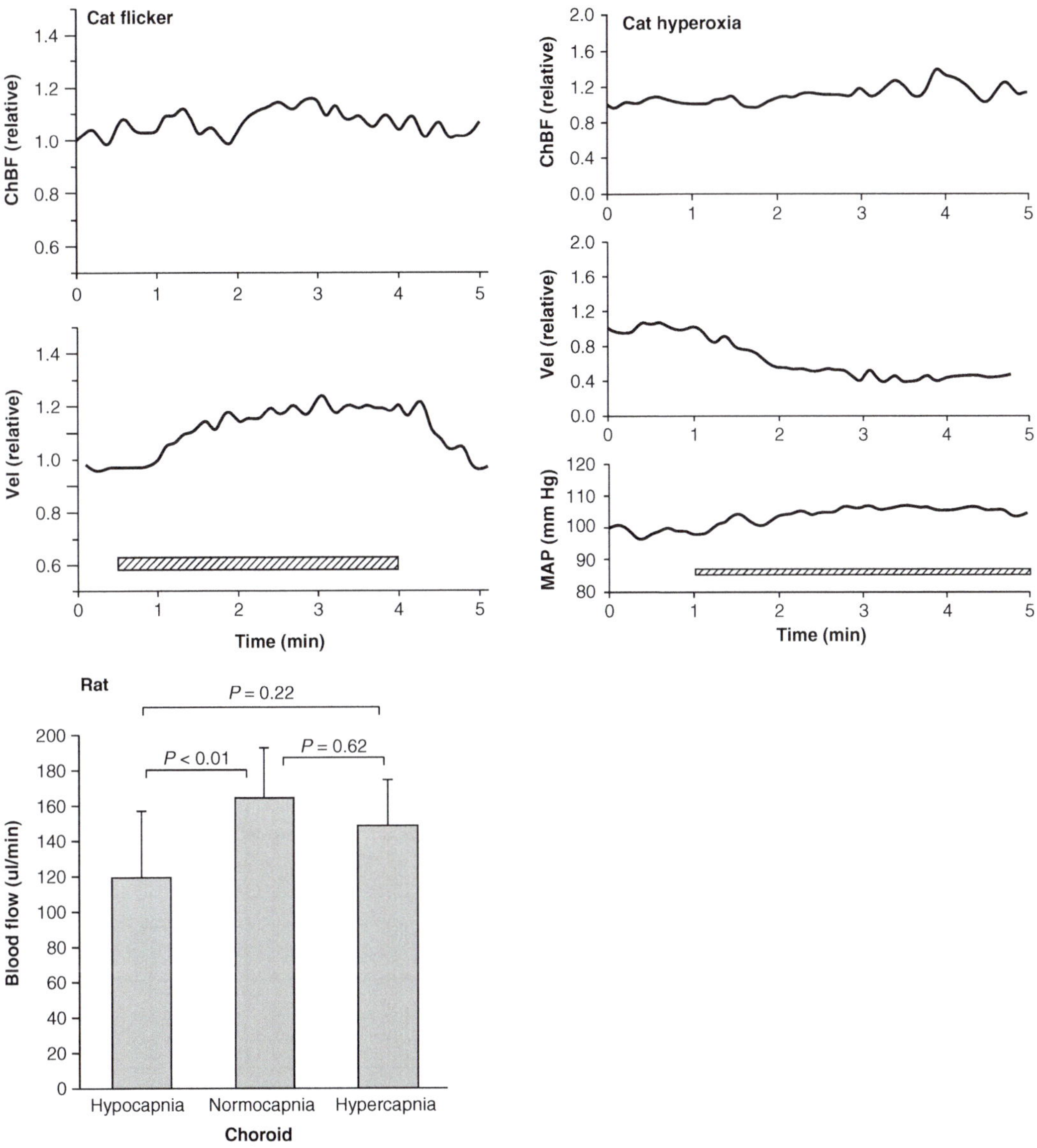

Fig. 11.28 Little choroidal blood flow response to flickering light stimulation (*left*), hyperoxia (*middle*), or hypercapnia (*right*) [54, 55]

Based on these results, the authors suggested "that the high flow choroidal circulation normally functions to stabilize the temperature environment of the retinal pigment epithelium and outer retinal layers." [59] A problem with this suggestion is that the temperature of choroidal blood is set by the core body temperature, which can vary significantly. For example, walking for 30 min in a cool environment followed by walking for 1 h in a hot environment can raise core body temperature by 3.5°C (Fig. 11.32) [60]. Thus, in addition to the thermal gradient mentioned earlier, the ability of the choroid to stabilize retinal temperature depends greatly on whole-body thermoregulation. Moreover, the 0.9°C increase in retina-choroid temperature upon light exposure

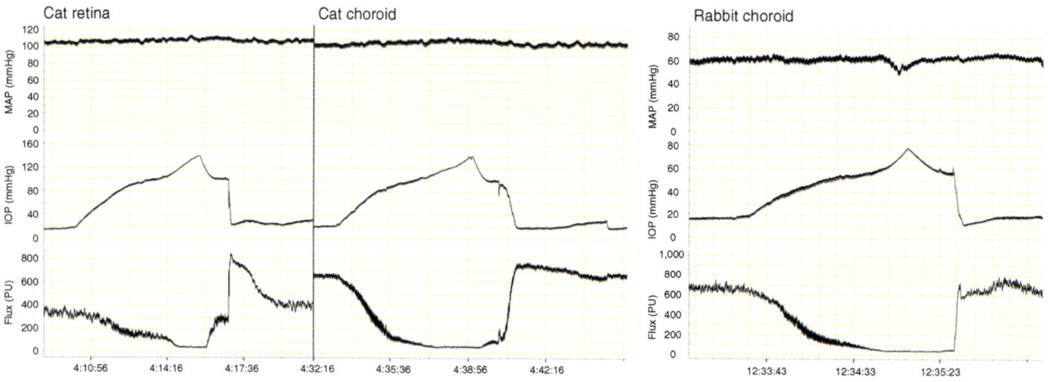

Fig. 11.29 Lack of choroidal reactive hyperemia in cat and rabbit [5]

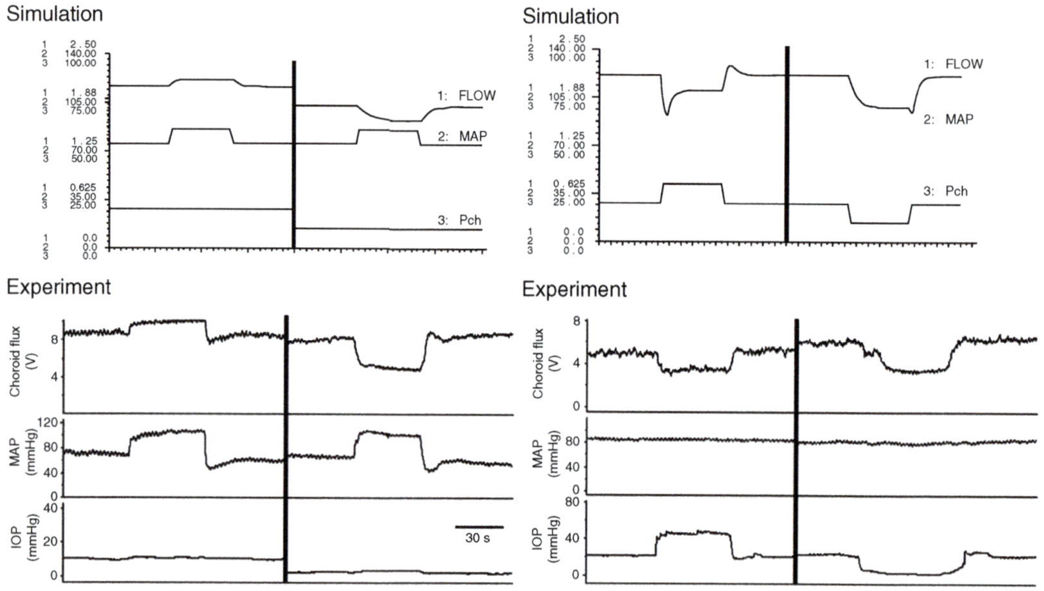

Fig. 11.30 Evidence of choroidal myogenic local control [51]

with IOP at 20 mmHg suggests that choroidal blood flow does a poor job of stabilizing retinal temperature. Indeed, some of that increase in temperature may have been a light-induced reflex, since Parver et al. found that light applied to the contralateral eye increased ipsilateral retina-choroid and scleral temperature as well as an index of choroidal blood flow; the reflex presumably also works when the ipsilateral eye is light-exposed [61]. Such a reflex suggests that choroidal blood flow is not regulated to maintain retinal temperature.

11.4.3 Retina

Evidence of retinal autoregulation comes from different species using various blood flow measuring techniques and methods of perfusion pressure manipulation (Fig. 11.33). Retinal

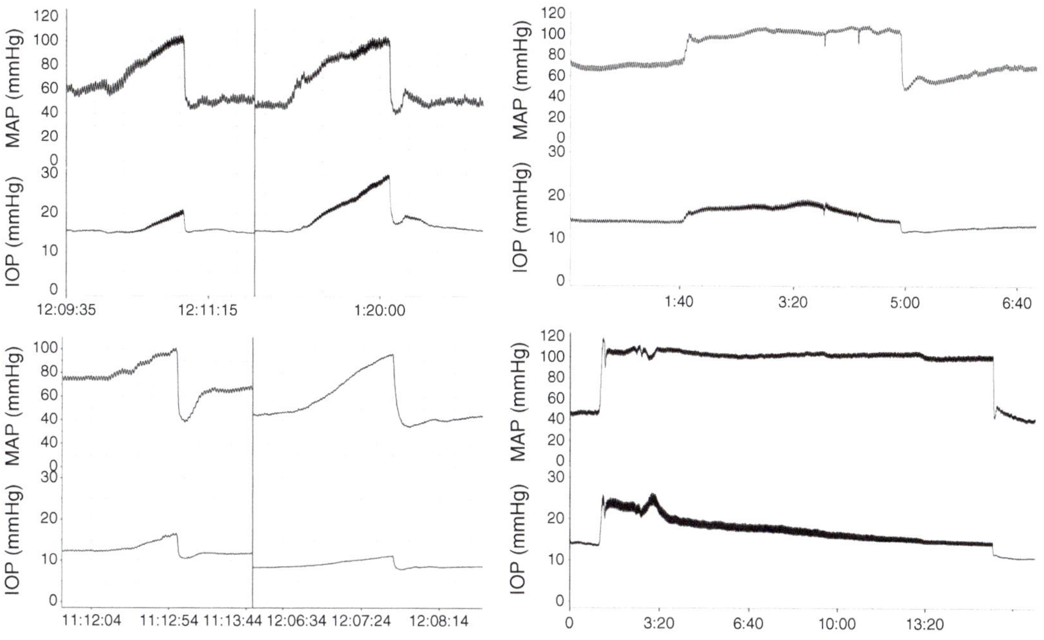

Fig. 11.31 Choroidal myogenic mechanism may protect the eye from arterial pressure-dependent changes in IOP (time in seconds) [57]

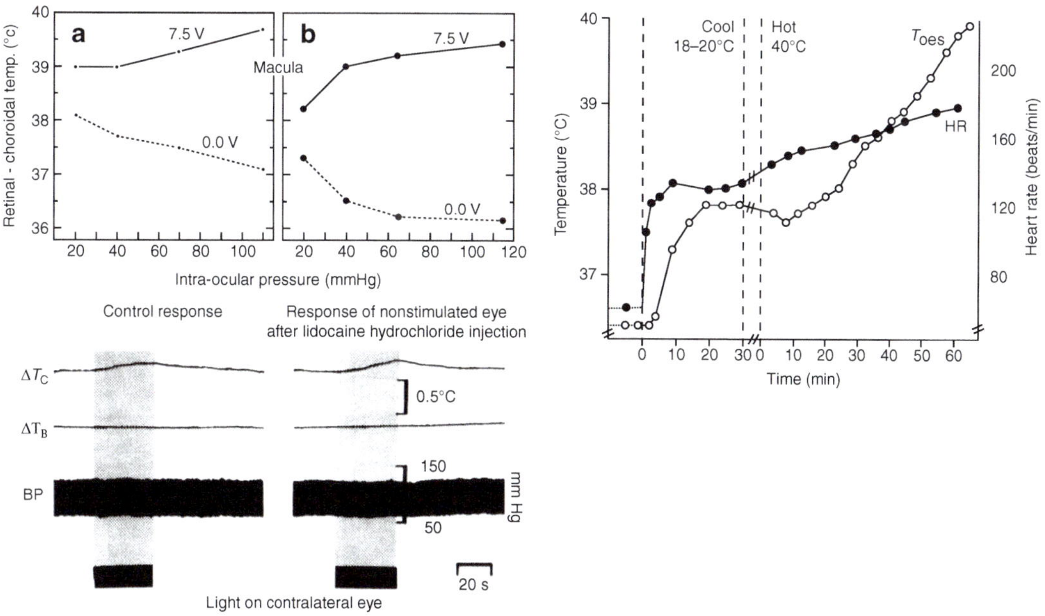

Fig. 11.32 Choroidal thermoregulation. (*Upper left*) Retinal temperature increases as choroidal blood flow is reduced by raising IOP in two monkey eyes when retina is illuminated with a lamp powered at 7.5 V and decreased when the lamp is off. (*Upper right*) Core body temperature range during walking in a cool and hot environment. (*Lower left*) Primate conjunctival temperature (ΔT_C) increase during illumination of contralateral eye is unaffected by retrobulbar anesthesia, while blood pressure (BP) and core body temperature (ΔT_B) remain unchanged [59–61]

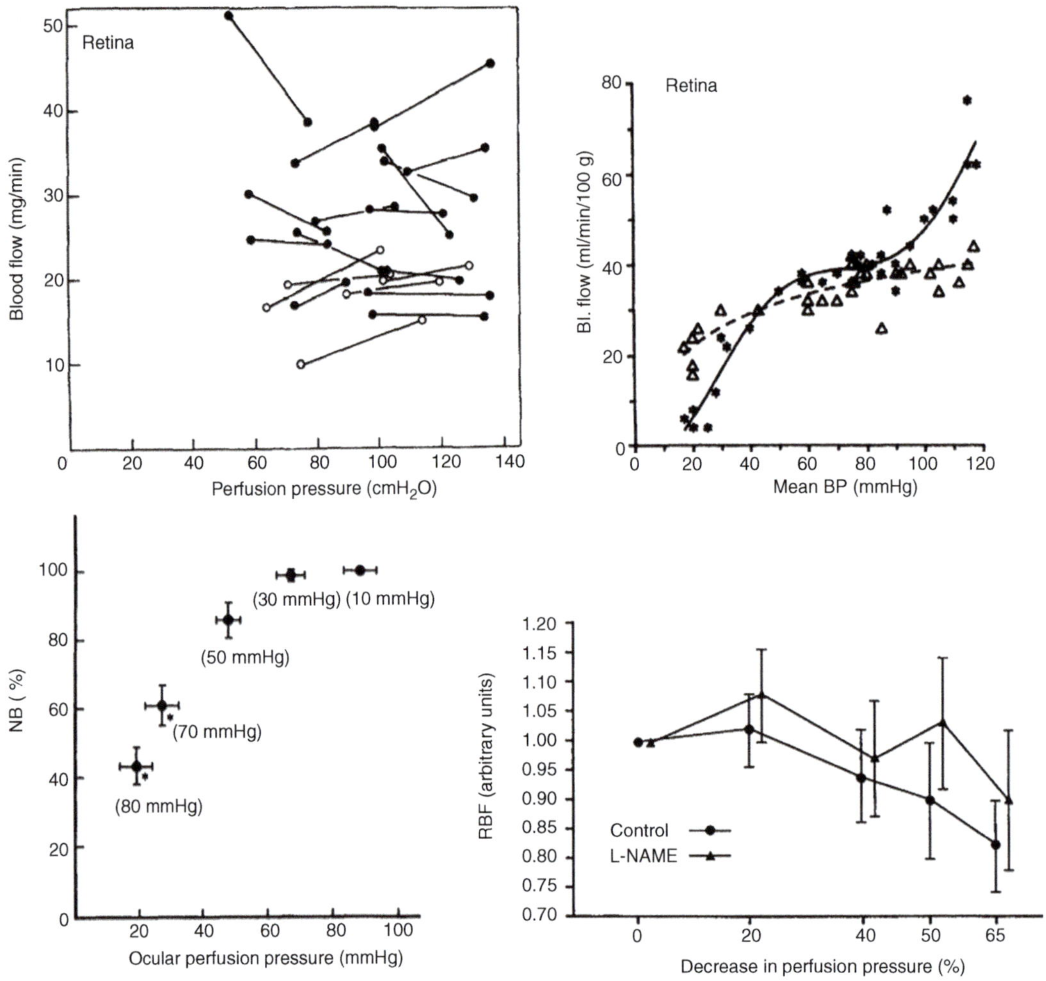

Fig. 11.33 Retinal blood flow autoregulation in monkey (*top left*), piglet before (*circles*) and after (*triangles*) ibuprofen (*top right*), rabbit (*bottom left*) and cat before and after L-NAME [48, 50, 62–64]

autoregulation is perhaps not surprising given the lack of autonomic innervation and the metabolic needs of the retina. However, local control in the retina is by no means simple. In many species, retinal nutrient delivery and waste removal are provided by retinal and choroidal circulations, but some species have a negligible retinal circulation (e.g., rabbits) and others have none (e.g., guinea pigs), and the fovea region of primates lacks retinal vessels despite its high density of metabolically active photoreceptors. Clearly, the link between retinal perfusion and metabolism is complex and varies by species and location, which makes understanding metabolic local control difficult. The negative visual consequences of retinal edema underscore the likely importance of myogenic local control, but this mechanism is difficult to study in the in vivo retina though it is evident under in vitro conditions (Fig. 11.4). The contributions of flow-mediated vasodilation and intercellular conduction are even harder to study and less well understood. Thus, while the evidence for retinal metabolic local control predominates, the other forms of local control may contribute as well.

A counterintuitive phenomenon in the retina is that its oxygen consumption increases in the dark due to increased Na/K ATPase activity

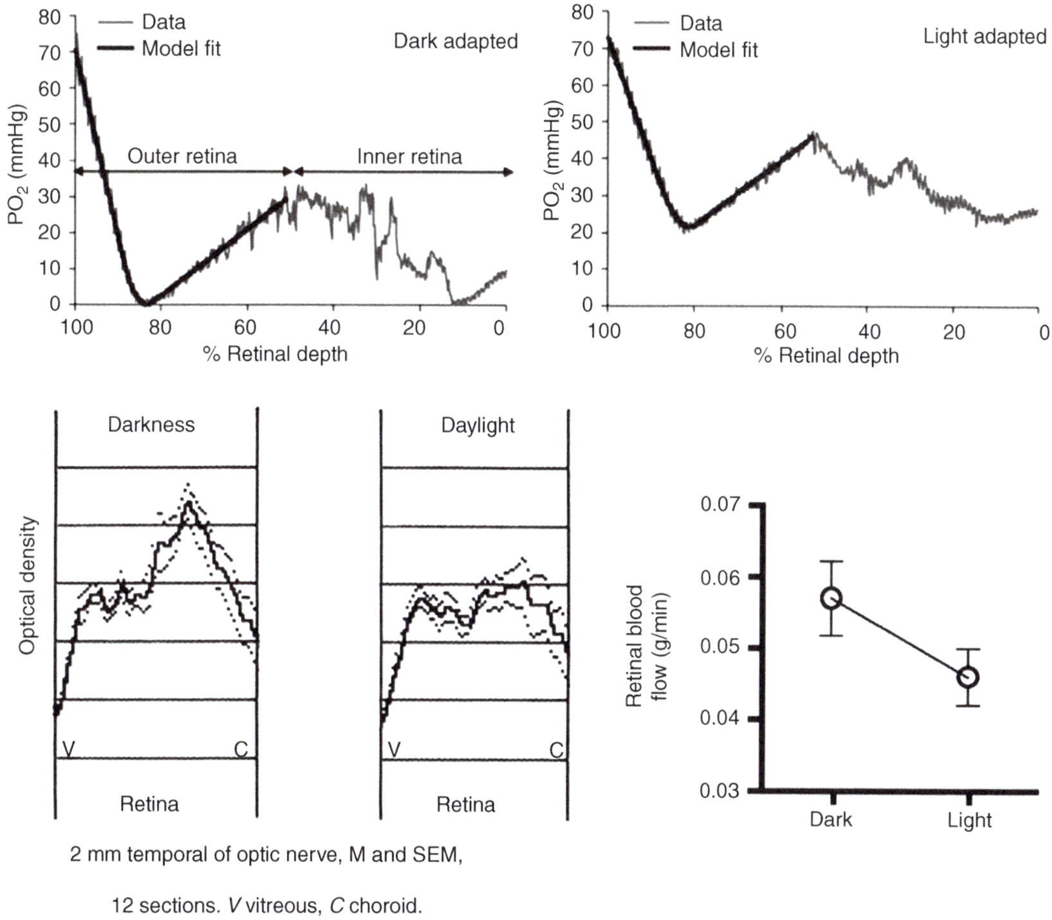

Fig. 11.34 Retinal metabolism and blood flow increase in darkness [68, 69] (*lower right* graph data from Bill and Sperber [69])

[65, 66]. In species with a dual retinal blood supply (i.e., retinal and choroidal circulations), the dark-stimulated increase in oxygen consumption is sufficient to lower the PO_2 of the photoreceptor inner segments to near zero (Fig. 11.34) [67, 68]. There is a corresponding increase in glucose consumption in approximately same location in the dark, and retinal blood flow is also higher in the dark [69]. This behavior appears to be an example of functional hyperemia, even though the metabolic action is occurring in the outer retina while the blood flow action is in the inner retina.

A clearer example of functional hyperemia is the retinal response to flickering light stimulation. In this case, the increase in retinal blood flow is associated with increased oxygen consumption indicated by the arteriovenous oxygen difference in paired retinal arteries and veins as well as increased glucose consumption in the inner retina [69–71]. Interestingly, there appears to be the greatest increase in retinal blood flow in the area with the highest density of ganglion cells, consistent with a link between metabolic demand and perfusion [71]. Additional evidence indicating significant retinal metabolic local control include the increase in retinal blood flow in response to hypoxia [72] and hypercapnia [73] as well as the decrease in blood flow in response to hyperoxia [74, 75] (Fig. 11.36) and the reactive hyperemia

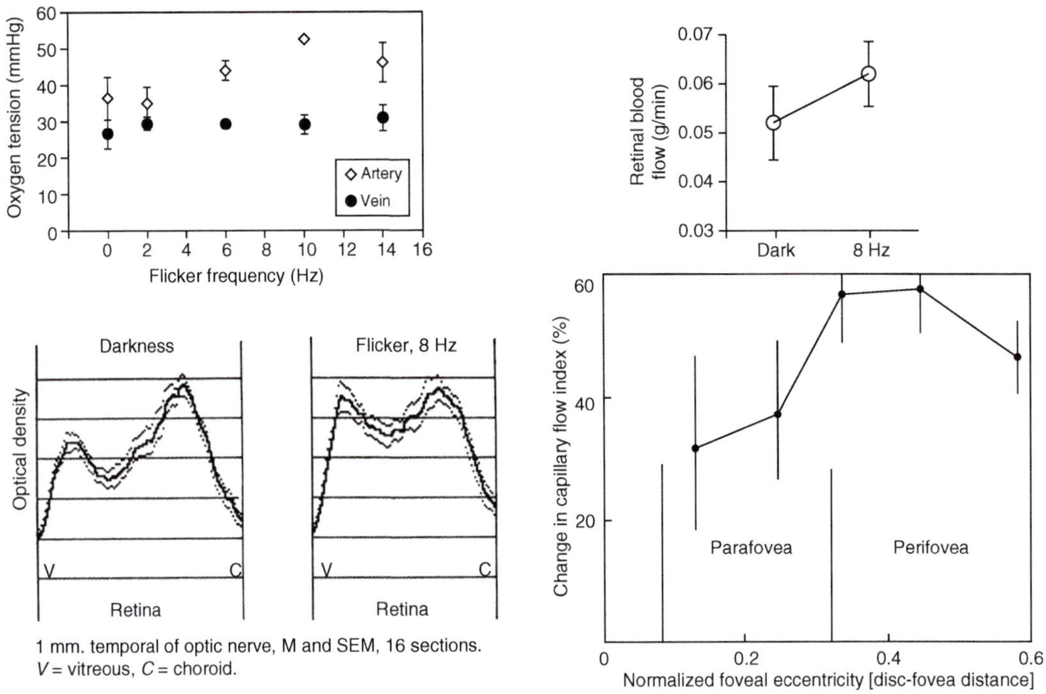

Fig. 11.35 Flickering light stimulation increases retinal arteriovenous oxygen difference (*top left*) and inner retinal glucose uptake (*bottom left*) and elicits a retinal functional hyperemia (*right top* and *bottom*) [69–71] (*upper right graph* data from Bill and Sperber [69])

after brief (Fig. 11.29) and long periods of ischemia (Fig. 11.37) [56]. Surprisingly, the retinal pressure-flow relationships under light- and dark-adapted conditions have not been determined, but it seems likely that the dark-adapted curve would be shifted upward due to the heightened oxygen consumption. Similarly, reactive hyperemic responses to brief ischemic periods would be expected to be larger in the dark (Fig. 11.35).

11.4.4 Ciliary Body

While the retina has essentially one function (processing light information for the central nervous system) with a dual blood supply, the ciliary body has two functions (accommodation and aqueous secretion) with a single blood supply. Distinguishing between ciliary muscle and secretory tissue is fairly straightforward by histology, but difficult in vivo, and measuring the blood flow to the two tissue types is even harder. Microsphere measurements at normal and high IOP suggest that both ciliary muscle and ciliary processes autoregulate, but the measurements vary widely, perhaps due to the small number of spheres that can be trapped in such a small amount of tissue (Fig. 11.38).

Laser Doppler flowmetry measurements in the rabbit also show evidence of ciliary autoregulation. However, it is also clear that the ciliary circulation is under neural control since ganglionic blockade caused an upward and leftward shift in the pressure-flow relationship indicative of tonic neuroconstrictor tone (Fig. 11.39). This is an interesting example of the interplay between neural and local control. Ciliary blood flow provides the oxygen and nutrient delivery to support aqueous production, and ciliary blood flow can

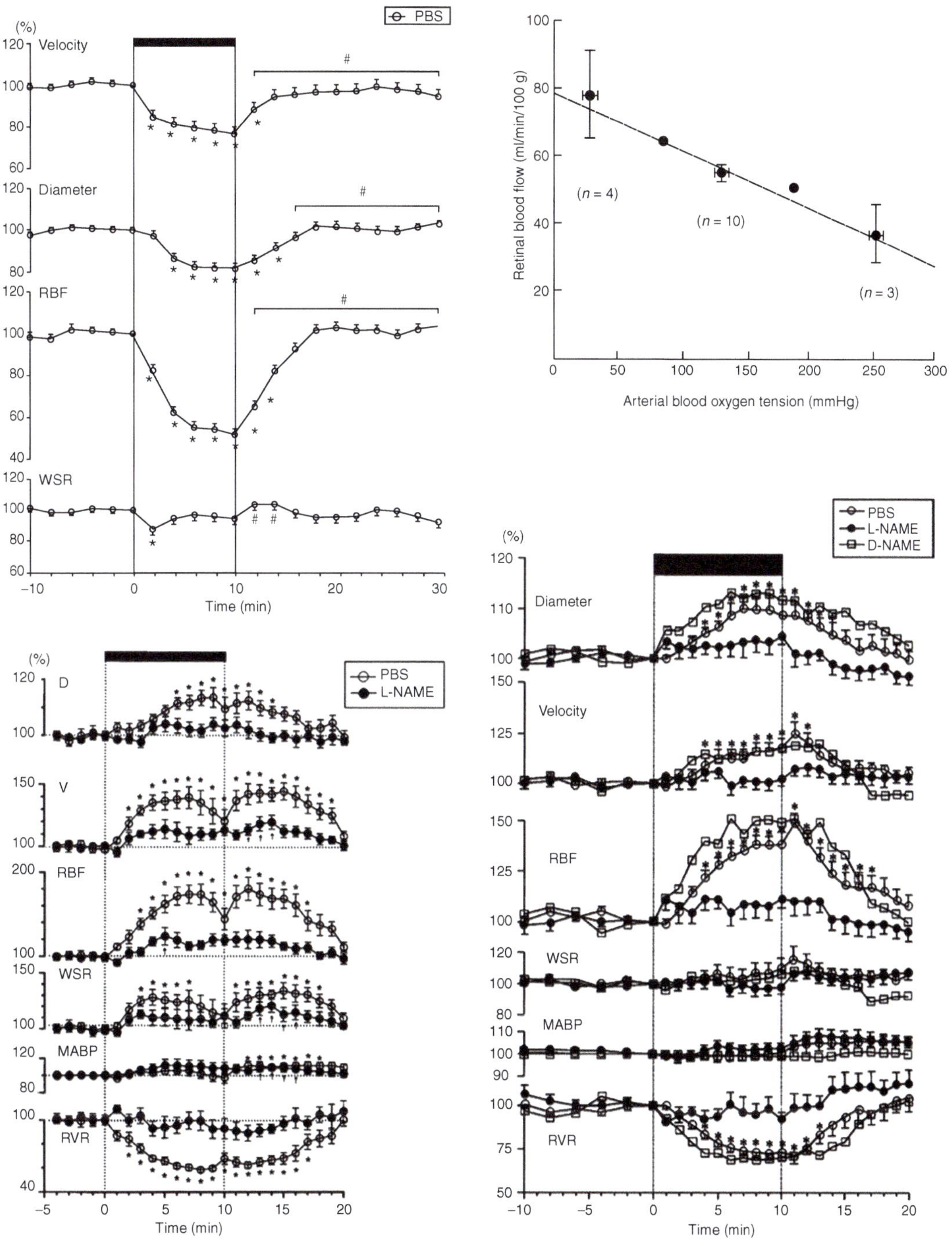

Fig. 11.36 Feline retinal blood flow responses in cats consistent with metabolic local control. Retinal perfusion decreases during hyperoxia (*top left* and *right*) and increases during hypoxia (*bottom left*) and hypercapnia (*bottom right*) [72–75]

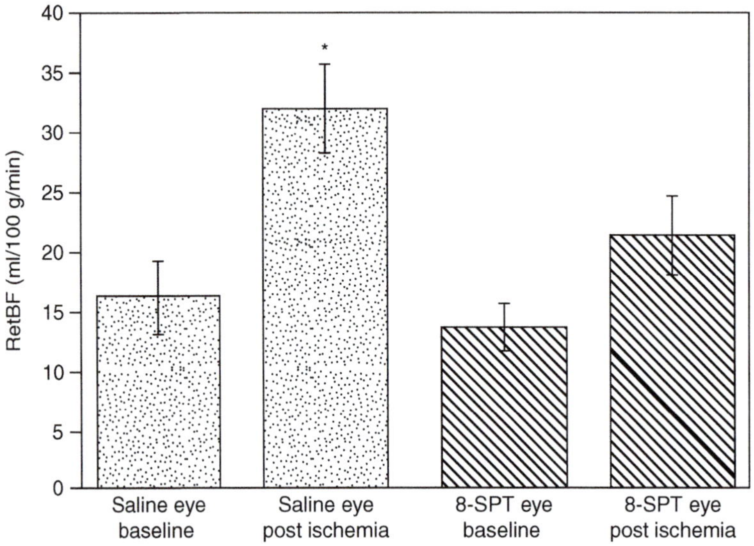

Fig. 11.37 Reactive hyperemia response in rat retina after 1 h of ischemia is attenuated by adenosine receptor blockade with 8-sulfophenyltheophylline (*8-SPT*) [56]

decrease approximately 30% before aqueous production is compromised under control conditions. It may be this perfusion reserve that makes possible the constrained autoregulatory range seen when both neural and local controls are operative. The local mechanisms involved in ciliary autoregulation have not been investigated systematically, but the author has rarely seen flow behavior suggestive of metabolic local control. For example, ciliary blood flow does not undergo a reactive hyperemic response following periods of ischemia (Fig. 11.40).

11.4.5 Iris

There are few published studies of iris blood flow in animals. Microsphere measurements in monkeys suggest the iris autoregulates, but the measurements were quite variable (Fig. 11.40). Laser speckle flowgraphy measurements in pigmented rabbits suggest little iris autoregulatory ability (Fig. 11.41). In humans, iris blood flow also does not seem to be autoregulated, but curiously, there is evidence of iridial reactive hyperemia [77]. Given the limited number of studies, definitive comments on iris local control require additional data.

11.5 Caveats

Defining local control behavior and identifying the underlying mechanisms and their interactions are challenging even in more accessible vascular beds, but it is particularly hard in the ocular circulations. Local control clearly occurs in the eye, and further study will broaden our knowledge of its role in normal ocular function and disease. Until then, attempts to manipulate ocular local control pharmacologically should be done cautiously. If "vascular dysregulation" does in fact contribute to a pathologic process, giving a vasodilator or vasoconstrictor that overrides any remaining local control may be detrimental in the long term.

11.6 Summary for the Clinician

Normal ocular metabolism and function depend on an appropriate blood supply for oxygen and nutrient delivery, tissue hydration, and waste

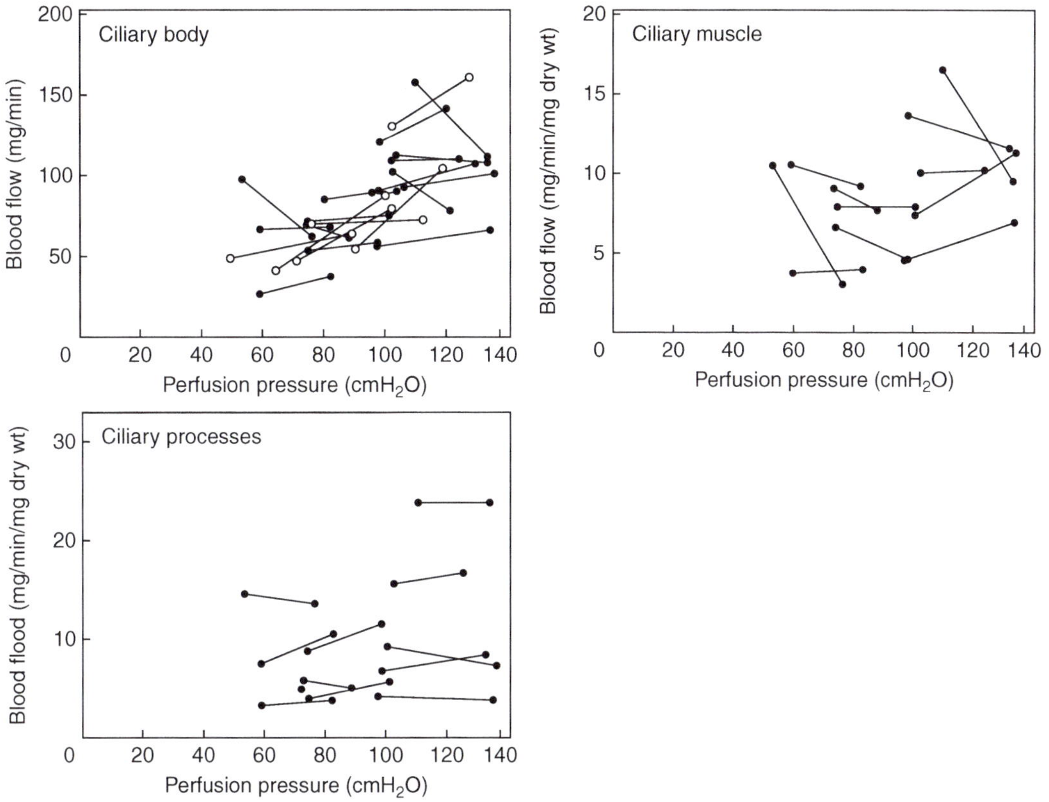

Fig. 11.38 Feline ciliary autoregulation [50]

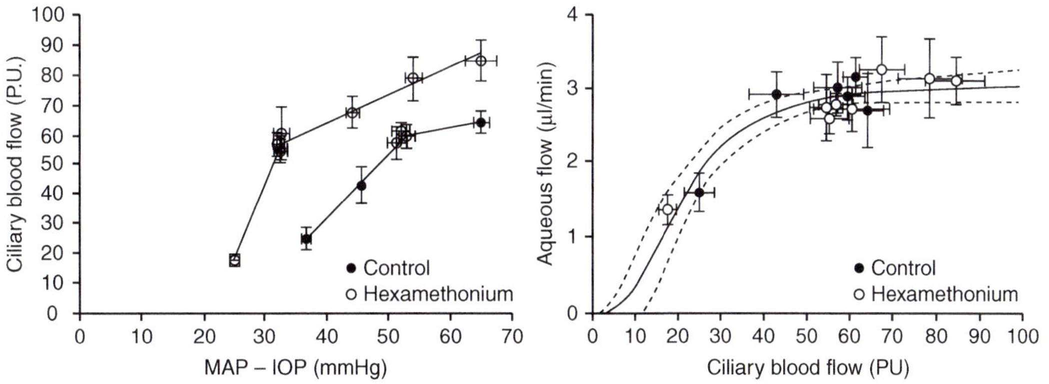

Fig. 11.39 Pressure-flow relation in rabbit ciliary body shifts up and to the left after ganglionic block with hexamethonium, but the relation between ciliary blood flow and aqueous flow is unaltered (Adapted from Reitsamer and Kiel [76])

removal. The adequacy of ocular blood flow is determined by physical factors (the arterial pressure and IOP) and the interplay between neurohumoral and local control mechanisms. Determining which local mechanisms are operative is difficult, but their potential contributions to ocular health, disease, and treatment should always be a consideration.

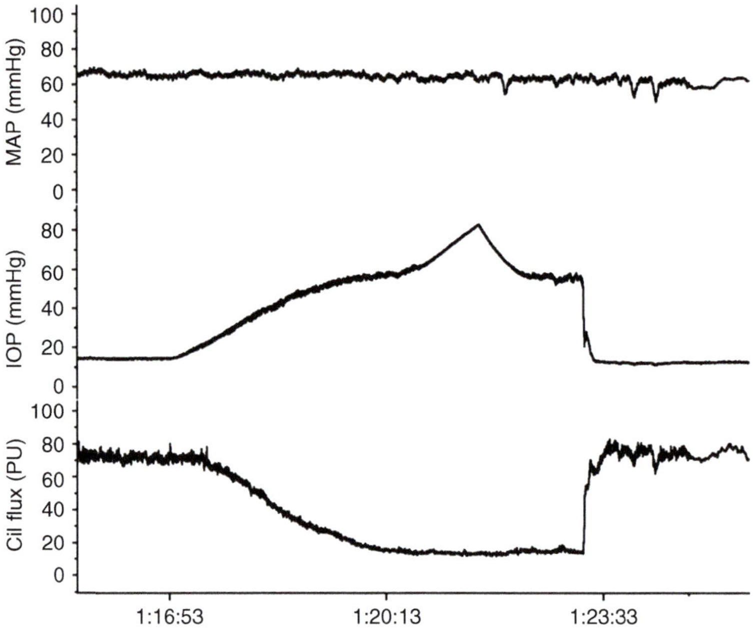

Fig. 11.40 No reactive hyperemia after ischemia in rabbit ciliary body suggests little metabolic local control [5]

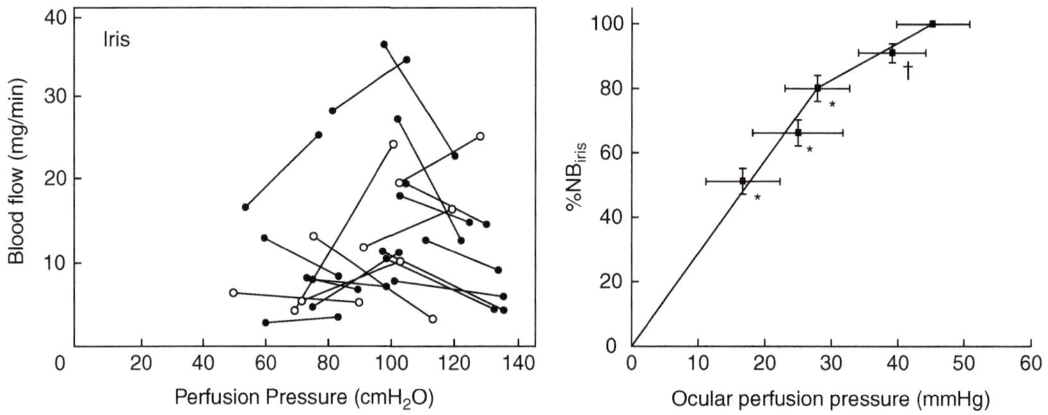

Fig. 11.41 Evidence of iris autoregulation as perfusion pressure is decreased by raising IOP in the monkey (*left*) and rabbit (*right*) [50, 78]

References

1. Patterson SW, Starling EH (1914) On the mechanical factors which determine the output of the ventricles. J Physiol 48:357–379
2. Moses RA (1963) Hydrodynamic model eye. Ophthalmologica 146:137–142
3. Fry DL, Thomas LJ, Greenfield JC (1980) Flow in collapsible tubes. In: Patel DJ, Vaishnav RN (eds) Basic hemodynamics and its role in disease processes. University Park Press, Baltimore, pp 407–424
4. Maepea O (1992) Pressures in the anterior ciliary arteries, choroidal veins and choriocapillaris. Exp Eye Res 54:731–736
5. Kiel JW (2011) The Ocular Circulation. In: Granger DN, Granger J (eds) Colloquium Series in Integrated Systems Physiology: from Molecule to function. www.morganclaypool.com, pp 1-68
6. Kiel JW, van Heuven WAJ (1995) Ocular perfusion pressure and choroidal blood flow in the rabbit. Invest Ophthalmol Vis Sci 36:579–585
7. Baylis WM (1902) On the local reactions of the arterial wall to changes of internal pressure. J Physiol (London) 28:220–231

8. Hein TW, Rosa RH, Yuan Z, Roberts E, Kuo L (2010) Divergent roles of nitric oxide and Rho kinase in vasomotor regulation of human retinal arterioles. Invest Ophthalmol Vis Sci 51(3):1583–1590
9. Davis MJ, Sikes PJ (1990) Myogenic responses of isolated arterioles: test for a rate-sensitive mechanism. Am J Physiol 259:H1890–H1900
10. Falcone JC, Davis MJ, Meininger GA (1991) Endothelial independence of myogenic response in isolated skeletal muscle arterioles. Am J Physiol 260:H130–H135
11. Loutzenhiser R, Bidani A, Chilton L (2002) Renal myogenic response: kinetic attributes and physiological role. Circ Res 90:1316–1324
12. Meininger GA, Mack CA, Fehr KL, Bohlen HG (1987) Myogenic vasoregulation overrides local metabolic control in resting rat skeletal muscle. Circ Res 60:861–870
13. Johnson PC (1980) The myogenic response. In: Bohr D, Somlyo A, Sparks H, Geiger S (eds) Handbook of physiology: the cardiovascular system. American Physiological Society, Bethesda, pp 409–442
14. Rubanyi GM (1993) Mechanoreception by the vascular wall. Futura Publishing Co, Mount Kisco
15. Wiederhielm CA, Bouskela E, Heald R, Black L (1979) A method for varying arterial and venous pressures in intact, unanesthetized mammals. Microvasc Res 18:124–128
16. Johnson PC, Intaglietta M (1976) Contributions of pressure and flow sensitivity to autoregulation in mesenteric arterioles. Am J Physiol 231:1686–1698
17. Gaskell WH (1877) On the changes of the blood stream through stimulation of their nerves. J Anat 11:360–404
18. Granger HJ, Goodman AH, Granger DN (1973) Intrinsic metabolic regulation of blood flow, O_2 extraction and tissue O_2 delivery in dog skeletal muscle. Adv Exp Med Biol 37A:451–456
19. Granger HJ, Shepherd AP (1973) Intrinsic microvascular control of tissue oxygen delivery. Microvasc Res 5:49–72
20. Gidday JM, Esther JW, Ely SW, Rubio R, Berne RM (1990) Time-dependent effects of theophylline on myocardial reactive hyperaemias in the anaesthetized dog. Br J Pharmacol 100:95–101
21. Valleau JD, Granger DN, Taylor AE (1979) Effect of solute-coupled volume absorption on oxygen consumption in cat ileum. Am J Physiol 236:E198–E203
22. Kiel JW, Riedel GL, Shepherd AP (1987) Local control of canine gastric mucosal blood flow. Gastroenterology 93: 1041–1053
23. Granger HJ, Goodman AH, Granger DN (1976) Role of resistance and exchange vessels in local microvascular control of skeletal muscle oxygenation in the dog. Circ Res 38:379–385
24. Schretzenmayr A (1933) Über kreislaufregulatorische Vorgänge an den grossen Arterien bei der Muskelarbeit. Pfluegers Arch Ges Physiol 232:743–748
25. Lie M, Sejersted OM, Kiil F (1970) Local regulation of vascular cross section during changes in femoral arterial blood flow in dogs. Circ Res 27:727–737
26. Hilton SM (1959) A peripheral arterial conducting mechanism underlying dilatation of the femoral artery and concerned in functional vasodilatation in skeletal muscle. J Physiol 149:93–111
27. Holtz J, Forstermann U, Pohl U, Giesler M, Bassenge E (1984) Flow-dependent, endothelium-mediated dilation of epicardial coronary arteries in conscious dogs: effects of cyclooxygenase inhibition. J Cardiovasc Pharmacol 6:1161–1169
28. Kuo L, Davis MJ, Chilian WM (1990) Endothelium-dependent, flow-induced dilation of isolated coronary arterioles. Am J Physiol 259:H1063–H1070
29. Koller A, Sun D, Kaley G (1993) Role of shear stress and endothelial prostaglandins in flow- and viscosity-induced dilation of arterioles in vitro. Circ Res 72:1276–1284
30. Stepp DW, Nishikawa Y, Chilian WM (1999) Regulation of shear stress in the canine coronary microcirculation. Circulation 100:1555–1561
31. Figueroa XF, Duling BR (2009) Gap junctions in the control of vascular function. Antioxid Redox Signal 11:251–266
32. Segal SS, Duling BR (1986) Flow control among microvessels coordinated by intercellular conduction. Science 234:868–870
33. Dora KA, Xia J, Duling BR (2003) Endothelial cell signaling during conducted vasomotor responses. Am J Physiol Heart Circ Physiol 285:H119–H126
34. Geijer C, Bill A (1979) Effects of raised intraocular pressure on retinal, prelaminar, laminar, and retrolaminar optic nerve blood flow in monkeys. Invest Ophthalmol Vis Sci 18:1030–1042
35. Weinstein JM, Duckrow RB, Beard D, Brennan RW (1983) Regional optic nerve blood flow and its autoregulation. Invest Ophthalmol Vis Sci 24:1559–1565
36. Shonat RD, Wilson DF, Riva CE, Cranstoun SD (1992) Effect of acute increases in intraocular pressure on intravascular optic nerve head oxygen tension in cats. Invest Ophthalmol Vis Sci 33:3174–3180
37. Liang Y, Downs JC, Fortune B et al (2009) Impact of systemic blood pressure on the relationship between intraocular pressure and blood flow in the optic nerve head of nonhuman primates. Invest Ophthalmol Vis Sci 50:2154–2160
38. Riva CE, Logean E, Falsini B (2005) Visually evoked hemodynamical response and assessment of neurovascular coupling in the optic nerve and retina. Prog Retin Eye Res 24:183–215
39. Buerk DG, Riva CE, Cranstoun SD (1995) Frequency and luminance-dependent blood flow and K^+ ion changes during flicker stimuli in cat optic nerve head. Invest Ophthalmol Vis Sci 36:2216–2227
40. Buerk DG, Riva CE (2002) Adenosine enhances functional activation of blood flow in cat optic nerve head during photic stimulation independently from nitric oxide. Microvasc Res 64:254–264
41. Riva CE, Cranstoun SD, Petrig BL (1996) Effect of decreased ocular perfusion pressure on blood flow and the flicker-induced flow response in the cat optic nerve head. Microvasc Res 52:258–269

42. Riva CE, Harino S, Petrig BL, Shonat RD (1992) Laser Doppler flowmetry in the optic nerve. Exp Eye Res 55:499–506
43. Riva CE, Hero M, Titze P, Petrig B (1997) Autoregulation of human optic nerve head blood flow in response to acute changes in ocular perfusion pressure. Graefes Arch Clin Exp Ophthalmol 235: 618–626
44. Buerk DG, Riva CE, Cranstoun SD (1996) Nitric oxide has a vasodilatory role in cat optic nerve head during flicker stimuli. Microvasc Res 52:13–26
45. Takayama J, Tomidokoro A, Tamaki Y, Araie M (2005) Time course of changes in optic nerve head circulation after acute reduction in intraocular pressure. Invest Ophthalmol Vis Sci 46:1409–1419
46. Takayama J, Tomidokoro A, Ishii K et al (2003) Time course of the change in optic nerve head circulation after an acute increase in intraocular pressure. Invest Ophthalmol Vis Sci 44:3977–3985
47. O'Day DM, Fish MB, Aronson SB, Coon A, Pollycove M (1971) Ocular blood flow measurement by nuclide labeled microspheres. Arch Ophthalmol 86:205–209
48. Alm A, Bill A (1972) The oxygen supply to the retina, II. Effects of high intraocular pressure of increased arterial carbon dioxide tension on uveal & retinal blood flow in cats. Acta Physiol Scand 84:306–319
49. Weiter JJ, Schachar A, Ernest JT (1973) Control of intraocular blood flow. I. Intraocular pressure. Invest Ophthalmol 12:327–334
50. Alm A, Bill A (1973) Ocular and optic nerve blood flow at normal and increased intraocular pressures in monkeys (macaca irus): a study with radioactively labeled microspheres including flow determinations in brain and some other tissues. Exp Eye Res 15:15–29
51. Kiel JW, Shepherd AP (1992) Autoregulation of choroidal blood flow in the rabbit. Invest Ophthalmol Vis Sci 33:2399–2410
52. Kiel JW (1999) Modulation of choroidal autoregulation in the rabbit. Exp Eye Res 69:413–429
53. Kiel JW, Lovell MO (1996) Adrenergic modulation of choroidal blood flow in the rabbit. Invest Ophthalmol Vis Sci 37:673–679
54. Riva CE, Cranstoun SD, Mann RM, Barnes GE (1994) Local choroidal blood flow in the cat by laser Doppler flowmetry. Invest Ophthalmol Vis Sci 35:608–618
55. Wang L, Grant C, Fortune B, Cioffi GA (2008) Retinal and choroidal vasoreactivity to altered $PaCO_2$ in rat measured with a modified microsphere technique. Exp Eye Res 86:908–913
56. Roth S (1995) Post-ischemic hyperemia in the cat retina: the effects of adenosine receptor blockade. Curr Eye Res 14:323–328
57. Kiel JW (1994) Choroidal myogenic autoregulation and intraocular pressure. Exp Eye Res 58:529–544
58. Bill A, Linder M, Linder J (1977) The protective role of ocular sympathetic vasomotor nerves in acute arterial hypertension. Bibl Anat 16:30–35
59. Parver LM, Auker C, Carpenter DO (1980) Choroidal blood flow as a heat dissipating mechanism in the macula. Am J Ophthalmol 89:641–646
60. Nielsen B, Savard G, Richter EA, Hargreaves M, Saltin B (1990) Muscle blood flow and muscle metabolism during exercise and heat stress. J Appl Physiol 69:1040–1046
61. Parver LM, Auker CR, Carpenter DO, Doyle T (1982) Choroidal blood flow II. Reflexive control in the monkey. Arch Ophthalmol 100:1327–1330
62. Chemtob S, Beharry K, Rex J et al (1991) Ibuprofen enhances retinal and choroidal blood flow autoregulation in newborn piglets. Invest Ophthalmol Vis Sci 32:1799–1807
63. Tamaki Y, Araie M, Kawamoto E, Eguchi S, Fujii H (1994) Noncontact, two-dimensional measurement of retinal microcirculation using laser speckle phenomenon. Invest Ophthalmol Vis Sci 35: 3825–3834
64. Harino S, Nishimura K, Kitanishi K, Suzuki M, Reinach P (1999) Role of nitric oxide in mediating retinal blood flow regulation in cats. J Ocul Pharmacol Ther 15:295–303
65. Zuckerman R, Weiter JJ (1980) Oxygen transport in the bullfrog retina. Exp Eye Res 30:117–127
66. Linsenmeier RA, Steinberg RH (1984) Effects of hypoxia on potassium homeostasis and pigment epithelial cells in the cat retina. J Gen Physiol 84: 945–970
67. Linsenmeier RA (1986) Effects of light and darkness on oxygen distribution and consumption in the cat retina. J Gen Physiol 88:521–542
68. Birol G, Wang S, Budzynski E, Wangsa-Wirawan ND, Linsenmeier RA (2007) Oxygen distribution and consumption in the macaque retina. Am J Physiol Heart Circ Physiol 293:H1696–H1704
69. Bill A, Sperber GO (1990) Control of retinal and choroidal blood flow. Eye 4:319–325
70. Shakoor A, Blair NP, Mori M, Shahidi M (2006) Chorioretinal vascular oxygen tension changes in response to light flicker. Invest Ophthalmol Vis Sci 47:4962–4965
71. Kiryu J, Asrani S, Shahidi M, Mori M, Zeimer R (1995) Local response of the primate retinal microcirculation to increased metabolic demand induced by flicker. Invest Ophthalmol Vis Sci 36:1240–1246
72. Nagaoka T, Sakamoto T, Mori F, Sato E, Yoshida A (2002) The effect of nitric oxide on retinal blood flow during hypoxia in cats. Invest Ophthalmol Vis Sci 43:3037–3044
73. Sato E, Sakamoto T, Nagaoka T et al (2003) Role of nitric oxide in regulation of retinal blood flow during hypercapnia in cats. Invest Ophthalmol Vis Sci 44:4947–4953
74. Izumi N, Nagaoka T, Sato E et al (2008) Role of nitric oxide in regulation of retinal blood flow in response to hyperoxia in cats. Invest Ophthalmol Vis Sci 49: 4595–4603

75. Stefansson E, Wagner HG, Seida M (1988) Retinal blood flow and its autoregulation measured by intraocular hydrogen clearance. Exp Eye Res 47: 669–678
76. Reitsamer HA, Kiel JW (2003) Relationship between ciliary blood flow and aqueous production in rabbits. Invest Ophthalmol Vis Sci 44:3967–3971
77. Chamot SR, Movaffaghy A, Petrig BL, Riva CE (2000) Iris blood flow response to acute decreases in ocular perfusion pressure: a laser Doppler flowmetry study in humans. Exp Eye Res 70:107–112
78. Tomidokoro A, Araie M, Tamaki Y, Tomita K (1998) In vivo measurement of iridial circulation using laser speckle phenomenon. Invest Ophthalmol Vis Sci 39:364–371

Neural Control of Ocular Blood Flow

12

Anton Reiner, Malinda E.C. Fitzgerald, and Chunyan Li

Core Messages

- Neural regulation of ocular blood flow is mediated by parasympathetic, sympathetic and sensory innervation of ocular and orbital blood vessels.
- The parasympathetic, and sympathetic inputs to ocular vessels are controlled by central nervous system circuitry that is responsive to interoceptive and exteroceptive stimuli.
- The parasympathetic input to choroid appears to maintain high ocular blood flow during low systemic blood pressure, and may also mediate increased blood flow in response to increased retinal activity.
- The sympathetic input to choroid appears to prevent excessive ocular blood flow during systemic hypertension.
- Impaired autonomic control of choroidal blood flow leads to retinal pathology and dysfunction.
- Impairments in the neural control of choroidal blood flow occur with aging, and various ocular or systemic diseases such as glaucoma, age-related macular degeneration, hypertension and diabetes, and such impairments in neural control of choroidal blood flow may contribute to retinal pathology and dysfunction in these diseases and conditions.

12.1 Overview of Ocular Blood Supplies and Their Neural Control

The eye possesses several major vascular beds, including those of the retina, the optic nerve, the ciliary body, and the iris. The retina itself has two vascular supplies in most mammalian species, the choroidal vasculature and the vessels of the inner retina [37]. The choroid accounts for >85% of the blood supply to the retina and all of the blood supply to the retinal pigment epithelium (RPE) and photoreceptors [33, 37]. These blood vessels are innervated by parasympathetic, sympathetic, and sensory nerve fibers, and they are thus under neural control. The blood supply to the choroid arises from the ophthalmic branch of the internal carotid artery, via the long and short ciliary arteries (Fig. 12.1). At one time, it was

A. Reiner, Ph.D. (✉) • C. Li, M.D., Ph.D.
Department of Anatomy and Neurobiology, University of Tennessee Health Science Center, 855 Monroe Ave., Memphis, TN 38163, USA
e-mail: areiner@utmem.edu; chli@uthsc.edu

M.E.C. Fitzgerald, Ph.D., FARVO
Department of Anatomy and Neurobiology and Ophthalmology, University of Tennessee Health Science Center, 855 Monroe Ave., Memphis, TN 38163, USA

Department of Biology, Christian Brothers University, Memphis, TN, USA
e-mail: malinda@cbu.edu

L. Schmetterer, J.W. Kiel (eds.), *Ocular Blood Flow*,
DOI 10.1007/978-3-540-69469-4_12, © Springer-Verlag Berlin Heidelberg 2012

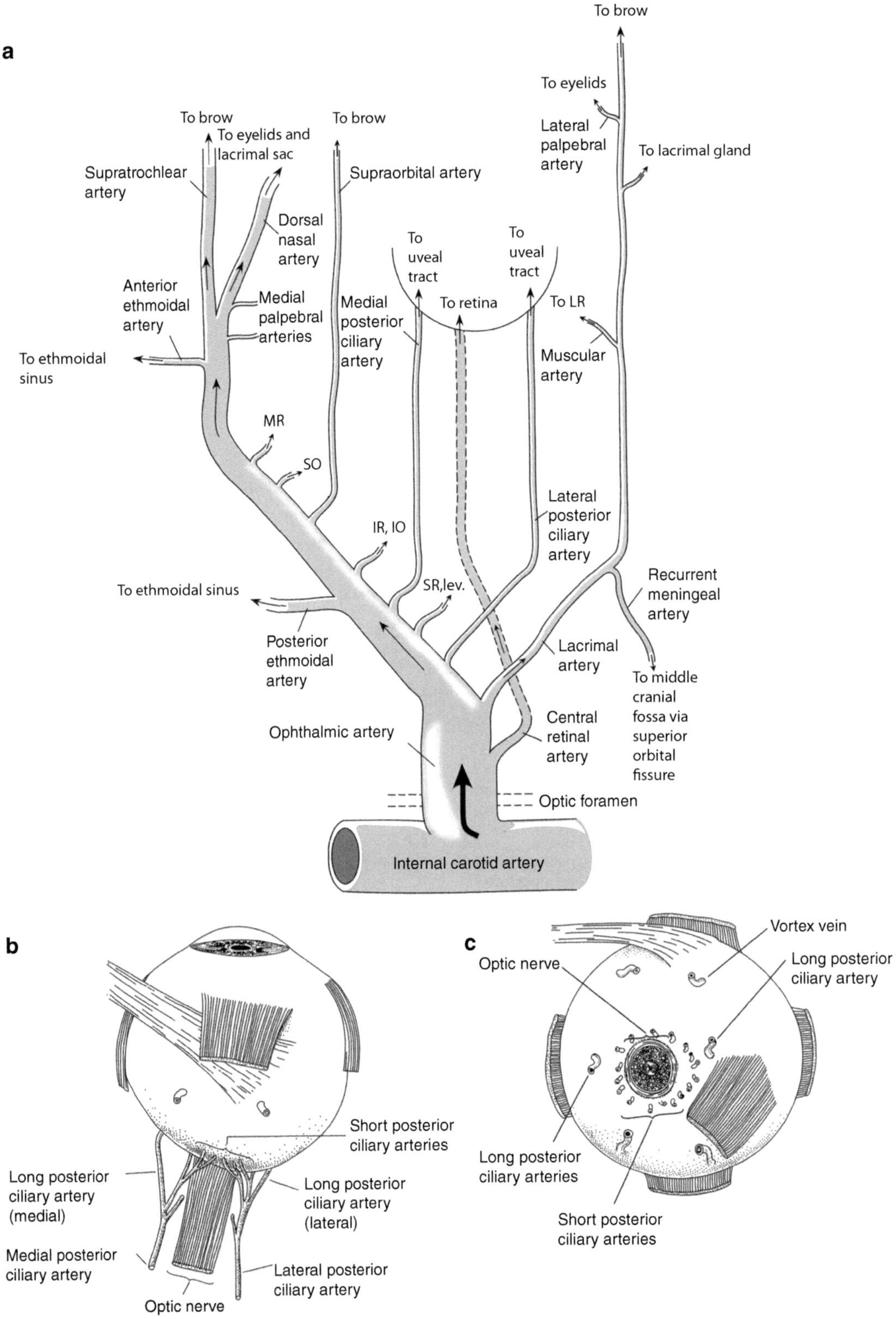
a
To brow
To eyelids and lacrimal sac
To brow
Supratrochlear artery
Supraorbital artery
Dorsal nasal artery
To eyelids
Lateral palpebral artery
To lacrimal gland
To brow
To uveal tract
To uveal tract
To retina
To LR
Anterior ethmoidal artery
Medial palpebral arteries
Medial posterior ciliary artery
Muscular artery
To ethmoidal sinus
MR
SO
IR, IO
SR,lev.
Lateral posterior ciliary artery
Recurrent meningeal artery
To ethmoidal sinus
Posterior ethmoidal artery
Lacrimal artery
To middle cranial fossa via superior orbital fissure
Ophthalmic artery
Central retinal artery
Optic foramen
Internal carotid artery
b
Short posterior ciliary arteries
Long posterior ciliary artery (medial)
Long posterior ciliary artery (lateral)
Medial posterior ciliary artery
Lateral posterior ciliary artery
Optic nerve
c
Vortex vein
Optic nerve
Long posterior ciliary artery
Long posterior ciliary arteries
Short posterior ciliary arteries

thought that choroidal blood flow was so great that it did not need to be regulated [33, 37]. As will be discussed below, despite the high flow rate in the choroid, it clearly is and does need to be adaptively controlled. The remainder of the blood supply to the retina is via the central retinal artery, whose branches radiate from the optic nerve head onto the inner retinal surface and then give rise to branches that penetrate into the retina through the depth of the inner nuclear layer, and supply blood to the inner layers of the retina (approximately its inner two-thirds) [10]. In some mammalian groups, the retinal vessels are sparse (such as rabbits), and in some groups of vertebrates with large eyes and acute vision, retinal vessels are altogether lacking (e.g., birds). Birds do possess a highly vascularized comb-shaped structure called the pecten that extends into the vitreous and may diffuse nutrients to the inner retina via the vitreous [234]. The retinal vessels in mammals and the pecten vessels in birds are themselves not innervated, but the central retinal artery is innervated by parasympathetic, sympathetic, and sensory nerve fibers, as discussed in more detail below. The central retinal artery (which arises from the ophthalmic artery) is also the blood supply of the optic nerve, and the postlaminar part of the optic nerve head in mammals additionally receives contributions from the short ciliary arteries that supply the choroid (Fig. 12.2). The same is true of the laminar part of the optic nerve. By contrast, the prelaminar part of the optic nerve head is supplied by the central retinal artery and by the choroid, the latter of which is as well innervated by parasympathetic, sympathetic, and sensory nerve fibers. Thus, the blood supply to the optic nerve and optic nerve head is under neural control, and flow to the retinal and pecten vessels may be under indirect neural control via the control exerted at the level of the central retinal artery, short ciliary arteries, and choroid. The ciliary body and iris receive their blood supply from the long posterior ciliary arteries (which arise from the ophthalmic artery) and the anterior ciliary artery branches from the blood supply to the extraocular muscles (Fig. 12.3), and the feeding vessels and arteries within both ciliary body and iris are innervated and subject to autonomic and sensory control, as discussed below [37]. The iris and ciliary body are components of the uveal tract of the eye, to which the choroid also belongs. Finally, the extraocular parts of the major orbital vessels (ophthalmic and the long and short ciliary arteries) also are typically innervated by parasympathetic, sympathetic, and trigeminal nerve sensory fibers, as are the blood vessels supplying the various orbital glands.

12.2 Neural Control of Optic Nerve and Retinal Blood Flow

The ophthalmic artery in rat is innervated by: (1) parasympathetic fibers containing the vasodilatory neuropeptide vasoactive intestinal polypeptide (VIP) and containing the neuronal form of nitric oxide synthase (nNOS), which is responsible for synthesizing the gaseous neurotransmitter and vasodilator nitric oxide (NO); (2) sympathetic nerve fibers containing the enzymes for

Fig. 12.1 Schematic of ophthalmic artery in human and its arterial branches for the right eye. Schematic (**a**) is from Oyster [266], Fig. 6.8. Blood is supplied to the eye by the central retinal artery, the posterior ciliary arteries, and the rectus muscle branches that give rise to the ciliary arteries. Other orbital arteries pass through the orbit to supply structures outside the orbit, some of which also give rise to intraorbital branches in passing (such as glandular branches). Schematics (**b** and **c**) show the posterior ciliary artery and its branches in human (Redrawing based on [266], Fig. 6.15). The medial and the lateral posterior ciliary arteries typically arise from the ophthalmic. Small branches from the posterior ciliary arteries that penetrate the sclera around the optic nerve are the short posterior ciliary arteries, while the main posterior ciliary arteries continue as the medial and lateral long posterior ciliary arteries. *MR* medial rectus, *SO* superior oblique, *IR* inferior rectus, *IO* inferior oblique, *SR* superior rectus, *LR* lateral rectus, *lev* levator palpebrae (Drawings by N. Guley)

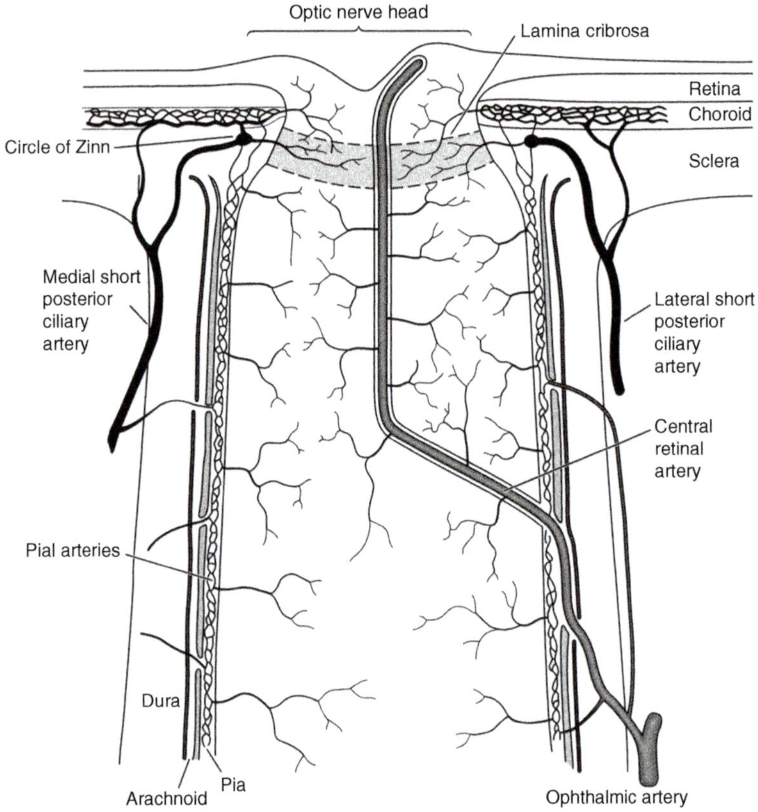

Fig. 12.2 Schematic of the central retinal artery, and the optic nerve and nerve head blood supply in human (Redrawing based on [266], Fig. 6.23). The central core of the optic nerve is supplied and drained by the central retinal artery and vein, respectively. The more peripheral portions of the nerve and nerve head are supplied by the short posterior ciliary arteries and their branches. The laminar part of the optic nerve is also supplied from central retinal artery and the short posterior ciliary arteries. By contrast, the prelaminar part of the optic nerve head is supplied from the central retinal artery and the choroid (Drawings by N. Guley)

producing noradrenaline and the vasoconstrictory neuropeptide Y (NPY); and (3) sensory nerve fibers derived from the trigeminal nerve co-containing the neuropeptides substance P (SP) and calcitonin gene-related neuropeptide (CGRP) [27, 76]. All of the nNOS+ parasympathetic fibers but not all of the VIP+ fibers arise from the pterygopalatine ganglion (PPG, also known as the sphenopalatine ganglion), which receives its preganglionic input from the superior salivatory nucleus of the facial motor complex, and the NPY+ sympathetic fibers arise from the superior cervical ganglion [76]. The PPG nerve fibers to the orbit appear to be cholinergic as well [71]. The facial parasympathetic input to orbital and ocular vessels in both mammals and birds actually arises from a set of distributed ganglia along the course of the preganglionic nerve to the PPG (the greater petrosal in mammals) and along the course of the plexus of postganglionic nerves that enter the eye (Fig. 12.4), with the main ganglion located behind the eye [63, 194, 357]. In mammals, it is the largest of these at the rear of the orbit that is commonly referred to as the PPG, but for simplicity, we will refer to all of these as the PPG. The sympathetic nerve fibers innervating the ophthalmic and ciliary arteries constrict these vessels by releasing noradrenaline, which acts on the vascular smooth muscle via alpha-adrenergic receptors [367]. In general, adrenergic receptors on ocular blood vessels are alpha-adrenergic [37]. SP and CGRP are potent vasodilators that work, in part, by stimulating vasodilator release from the vascular endothelium, notably NO [163]. The innervation of the ophthalmic artery and its derived ocular branches is similar in other mammalian species, and in avian species as well [63, 346].

The central retinal artery, which arises from the ophthalmic artery in mammals, and its small

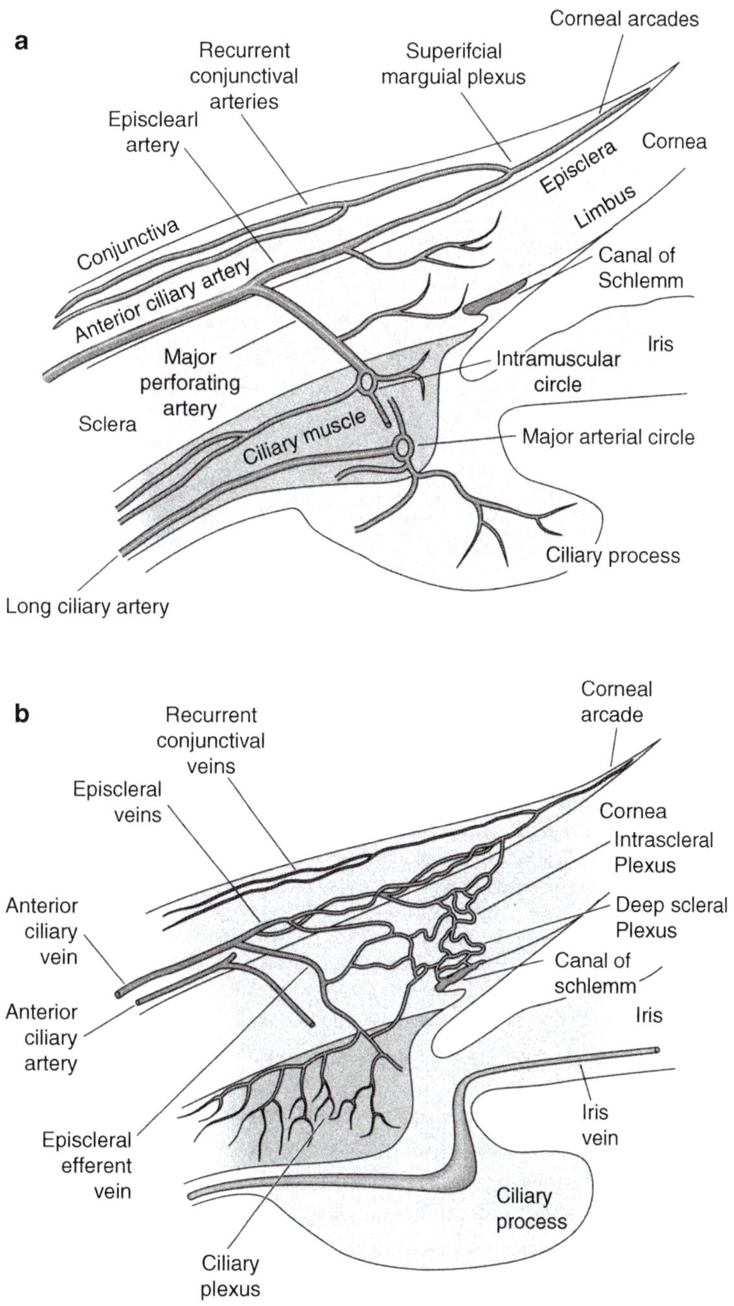

Fig. 12.3 Schematic of anterior uveal vessels in human (Redrawing based on [266], Figs. 6.12a and 6.14b). Schematic (**a**) depicts the arterial supply, while (**b**) depicts the venous drainage. After passing over the rectus muscles, the anterior ciliary arteries branch. One set, the episcleral arteries, remains on the ocular surface, while the other set perforates the sclera and travels to the ciliary body. The major arterial circle of the ciliary body gives rise to the arteries to the iris and the ciliary processes. The episcleral efferent veins receive blood from the ciliary muscle (the ciliary plexus), the limbus (the intrascleral plexus), the conjunctiva, and the cornea arcades (Drawings by N. Guley)

branches within the optic nerve behind the lamina cribrosa have been shown to be innervated by VIP+ and nNOS+ facial parasympathetic, NPY+ noradrenergic sympathetic, and CGRP+ trigeminal sensory nerve fibers in diverse primate, carnivore, lagomorph, and rodent species [27, 37, 73, 197, 403] (Laties and Jacobowitz 1966). The small vessels of the optic nerve behind laminar cribrosa also have VIP+ fibers, but none of these parasympathetic nerve fibers enter the eye [403]. Similarly,

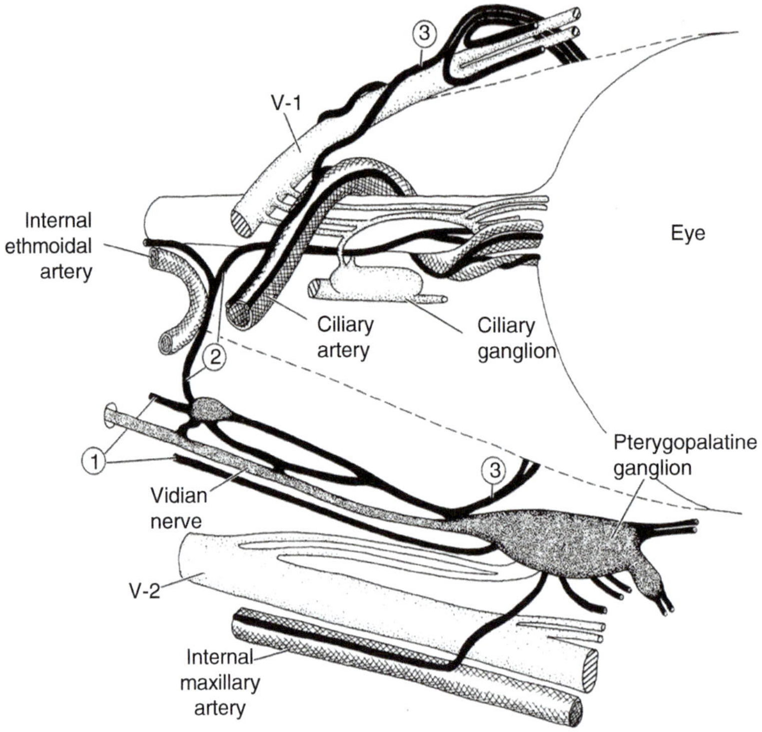

Fig. 12.4 Figure 1 modified from Lin et al. [202], showing the PPG (and one of its constituent accessory microganglia) and its nerve pathways to eye in cats. The rete mirabile is cut away to allow visualization of the fiber pathways of the PPG. The PPG nerves are represented by *thick black lines*, and the plexus of fine nerve fibers in arterial walls is indicated by *cross-hatching*. The numbers indicate each of the three pathways from the PPG to the eye: 1 – nerves that traverse the rete mirabile and travel to the eye on blood vessels and as small nerves, 2 – nerves traveling along the inferior optic nerve sheath and into the eye via the ciliary nerves and arteries, and 3 – nerves traveling with terminal branches of the nasociliary nerve, and to the eye via the ciliary nerves and ciliary arteries. The *thin dashed lines* show the limits of the extraocular muscle cone, which is omitted for simplicity. *V-1* ophthalmic nerve, *V-2* maxillary nerve; Vidian nerve – nerve of the pterygoid canal. Note the Vidian nerve is formed by the union of the facial preganglionics of the greater petrosal nerve and the superior cervical postganglionics of the deep petrosal nerve

in monkeys and rats, the central retinal artery (and its smaller offshoot vessels behind the lamina cribrosa) possesses sensory trigeminal fibers co-containing SP/CGRP+ but not its extension into the eye as the retinal vasculature [192, 403]. The choroidal contributions to the prelaminar part of the optic nerve also are innervated by parasympathetic, sympathetic, and sensory nerve fibers. Bergua et al. [27] have further shown for rat that NOS+ fibers are present on blood vessels of the optic nerve head. These anatomical findings imply neural control of blood flow in the optic nerve and in the pre- and postlaminar parts of the optic nerve head. While the central retinal artery enters the eye and ramifies into the retinal vessels, the general view is that innervation of the central retinal artery does not extend into or beyond the optic nerve head itself [137]. Thus, retinal vessels are not directly controlled by autonomic or trigeminal nerve fibers. Retinal vessels, however, appear to be similar to vessels that enter the brain parenchyma – they do not need to be neurally controlled since local metabolic coupling may help them match blood flow to local retinal need [83, 403] (Takano et al. 1995). Moreover, retinal blood flow has been shown to compensate well for reductions in ocular perfusion pressure (a phenomenon called autoregulation), down to perfusion pressure as low as about 30 mmHg, in rabbits, cats, pigs, and monkeys [10]. The retinal autoregulation is presumably effected by local myogenic or metabolic mechanisms. Optic nerve head blood flow has also been shown to be regulated in response to ocular perfusion pressure variation in monkeys and humans [10]. The possibility that retinal vessels in at least some species are innervated to some small extent is raised by the finding of Bergua et al. [27] that some retinal vessels are innervated in rabbits.

Consistent with facial nerve innervation of optic nerve blood vessels, facial nerve activation in monkeys and cats was found to yield increased optic nerve blood flow, but retinal blood flow did not increase with such activation [252]. The increases in optic nerve blood flow could be prevented with blockade of transmission through the PPG by the neuronal nicotinic receptor antagonist hexamethonium (a general blocker of transmission through autonomic ganglia). Nilsson [255] reported, however, that facial nerve stimulation in rabbits does increase blood flow in retinal vessels, and this effect could be blocked with NOS inhibition. Thus, PPG-mediated increases in central retinal artery flow can have an impact on retinal flow in at least some species under some conditions. Consistent with parasympathetic control of the central retinal artery blood flow, Toda et al. [368] showed that activation of the perivascular nerve plexus on the central retinal artery in dogs (with topical nicotine) results in NO-dependent vasodilation. Endothelial NO release appears, in part, to underlie the role of NO in optic nerve head vasodilation [350]. Further evidence shows that optic nerve head blood flow is also strongly modulated by NO. For example, Luksch et al. [211] showed that systemic inhibition of NO production in humans decreases optic nerve head blood flow (as measured by laser Doppler flowmetry or LDF). Similarly, inhibitors of NOS such as LNAME diminish optic nerve head blood flow in rabbits [120, 350] and reduce visually evoked potentials [120]. Conversely, oral administration of the NO donor ISMO (5-isosorbide mononitrate) increases optic nerve head blood flow in humans [128]. Basal sympathetic vasoconstrictory tone of optic nerve head vessels is evidenced by the finding that alpha-adrenergic blockade with bunazosin increases optic nerve head blood flow in rabbits [120].

Optic nerve and optic nerve head blood flow is regulated in response to ocular perfusion pressure, and flicker-induced retinal activity [80, 246, 294, 389]. Such regulation is important potentially in the etiology of glaucoma, since autoregulation of optic nerve head blood flow (maintenance of stable flow despite fluctuation in perfusion pressure) is defective in primary open angle glaucoma and normotensive glaucoma [137]. It may be that endothelially or neurally derived factors such as NO or noradrenaline play a role in the normal vascular autoregulation observed in the optic nerve and optic nerve head during fluctuations in ocular perfusion pressure. Studies in rabbits [264], in fact, show that the vasodilatory optic nerve head vascular autoregulation in response to perfusion pressure reductions caused by IOP elevation and the increased optic nerve head flow caused in cats by flicker are mediated by NO. Given the possible contributory role of defective optic nerve head blood flow and its regulation in primary open angle glaucoma and normotensive glaucoma [137], the above-noted vasodilatory effect of ISMO on optic nerve head vessels may explain the benefit of nitrate therapy in glaucomatous humans [418].

Endothelially derived NO is involved in basal regulation of retinal vessel tone in rabbits, cats, and dogs – blood flow being diminished when NO production is blocked by NOS inhibition [69, 247, 313]. Further consistent with a role of NO in retinal blood flow, Goldstein et al. [119] reported that NO relaxes retinal arteries in humans and increases retinal blood flow, as does the NO donor ISMO [151]. NO relaxes vascular smooth muscle by activating soluble guanylate cyclase, resulting in production of cGMP, which acts on smooth muscle as a relaxant and is broken down by phosphodiesterase-5 [119]. Consistent with this mechanism of action, inhibitors of phosphodiesterase-5 such as Viagra (Sildenafil) dilate retinal arterioles and increase retinal blood flow [233, 275]. Benedito et al. [24] have demonstrated that bovine retinal arteries show endothelium-dependent relaxation in response to acetylcholine. Studies in piglets show that endothelium-dependent vasodilatory effect of acetylcholine on retinal vessels appears to involve endothelial release of NO [115], and similar studies on bovine retinal vessels show the cholinergic stimulation of NO release is mediated by muscarinic receptors on retinal endothelial cells [148]. Retinal vessels in humans, cats, pigs, and cows not only have muscarinic receptors but also contain choline acetyltransferase (ChAT), the enzyme needed for synthesis of acetylcholine [82]. Thus, retinal vessels themselves may provide the acetylcholine involved in endothelium NO-dependent retinal vessel dilation. Retinal vessels also relax and dilate in response to nicotine (Toda et al. 1994), due to stimulation of NO release from endothelium. Thus, retina vessels may locally regulate blood flow via cholinergic elicitation of endothelial NO formation. Riva et al. [291], however, did not find evidence in cat that intravenous acetylcholine increases retinal flow. They did find that retinal flow increases in response to diffuse luminance flicker, a stimulus inducing heightened inner retinal neural activity and metabolic demand. The flicker-induced increase in retinal blood flow may occur by local metabolic coupling of the retina and vasculature. Others have also observed increased retinal blood flow with diffuse luminance flicker in various mammalian groups [109, 177, 302, 314]. The retinal vasculature also shows vasodilatory compensation for reduced ocular perfusion pressure [123, 290], and NO may be involved in this effect. For example, LNAME blockade of retinal NO production in piglets hinders retinal blood flow compensation for decreased perfusion pressure caused by increased IOP [156]. The work of Okuno et al. [264] shows, similarly, that vasodilatory retinal vascular autoregulation in response to perfusion pressure reductions caused by IOP manipulation in rabbits is mediated by NO. By contrast, NO does not appear to mediate retinal blood flow increases in newborn piglets in response to hypoxia and hypercapnia, or in response to very low perfusion pressure (40 mmHg) [115]. Note, however, retinal vascular autoregulation in newborn animals is immature, and not typical of regulation in adult animals [135]. Consistent with this caveat, endothelially derived NO appears to be involved in increasing retinal blood flow during hypoxia or hypercapnia in adult cats [247, 301].

Stimulation of sympathetic input to the eye has generally been reported to have no noteworthy effect on retinal blood flow [6, 9, 10]. Intravenous delivery of the sympathetic nervous system co-transmitter NPY also has no reported effect on retinal blood flow [254]. Moreover, sympathetic denervation of the eye has little effect on retinal blood flow in monkeys, suggesting that the sympathetic nervous system has no major direct or indirect effect on blood flow in monkey retinal vessels [230]. Alm and Bill [6], however, have shown that stimulation of the cervical sympathetic chain does lower vitreal oxygen tension near the retina in cats, which they interpreted as suggesting that sympathetic activation does reduce retinal blood flow somewhat, at least in cats. Consistent with some possible sympathetic effects on retinal blood flow, direct application of NPY to isolated bovine retinal arteries causes contraction [280]. Retinal vessels show compensatory vasoconstriction to increased systemic perfusion pressure, and adrenergic mechanisms appear involved in this effect [159, 160]. Note that while ocular vessels utilize alpha-adrenergic receptors in their responses to sympathetic nervous system activation, their responses to beta-adrenergic blockade are of relevance as well, since such blockers are used to lower IOP as a glaucoma treatment. Unfortunately, the literature is inconsistent regarding whether topical administration of beta-adrenergic blockers such as timolol affects flow in retinal vessels, with some studies in humans finding no effect on retinal or optic nerve head blood flow [405], and others reporting that topical timolol does increase retinal blood flow in humans [124, 220, 288]. This unresolved issue is of importance for understanding whether the beneficial effect of timolol in treating glaucoma stems from its inhibitory effect on aqueous production, or from improving ocular blood flow.

Although retinal vessels receive no CGRP+ sensory fibers, bovine and dog retinal vessels have been shown to dilate in response to application of CGRP [43, 261, 279], indicating they possess CGRP receptors and can respond to blood-borne or retinal diffusion-carried CGRP. Retinal vessels also relax and dilate in response to SP [178], via stimulation of NO release from endothelium. The retina itself could be the source of any SP or CGRP having effects on the retinal vasculature in vivo.

12.3 Neural Control of Iridial and Ciliary Body Blood Flow

The iris contains the pupillary control musculature, while the ciliary body contains the muscles of accommodation. The ciliary body is continuous with the ciliary processes, which are rich in arteries and are responsible for production of aqueous humor, whose abundance regulates IOP. Outflow of aqueous humor through the trabecular meshwork at the limbus of the eye, a portal through which aqueous humor returns to the venous side of the vasculature, also contributes to IOP and may be regulated by autonomic and sensory nerve fibers. The ciliary body and iris in mammals typically receive their blood supply from the long posterior ciliary arteries and the anterior ciliary artery branches from the blood supply to the extraocular muscles (Fig. 12.3). The long posterior ciliary arteries (typically one medial and one lateral) arise from the posterior ciliary arteries, which themselves arise from the ophthalmic artery behind the eye. The posterior ciliary arteries give rise to many short posterior ciliary arteries that penetrate the eye at its posterior pole just around the optic nerve, and enter the choroid (Fig. 12.1). The long posterior ciliary arteries pierce the sclera only slightly more anteriorly and course within the choroid on the medial and lateral sides of the eye to the ciliary body, where they give rise to the blood supply to the muscles of accommodation and the ciliary processes [10, 266]. The anterior ciliary arteries typically give rise to an annular anastomosis within the ciliary body that gives off radially oriented blood vessels to the iris, as well as additional branches to the ciliary processes. A venous network in the ciliary body drains the ciliary body and iris to the anterior ciliary veins and the episcleral veins (Fig. 12.3). The ophthalmic artery and posterior and anterior ciliary arteries and their branches within iris and ciliary body are innervated, in some places heavily, as are nonvascular muscular structures of the iris and ciliary body. For example, as noted above, the ophthalmic artery in rats is innervated by NOS+ and VIP+ cholinergic fibers from the PPG, NPY+ noradrenergic fibers from the superior cervical ganglion, and SP+ and CGRP+ sensory fibers from the trigeminal nerve [65, 76, 346]. Consistent with PPG influences on the ophthalmic artery, Bakken et al. [17] showed that the pig ophthalmic artery dilates to VIP and acetylcholine, with the cholinergic response eliminated by removal of endothelium, and with acetylcholine and VIP being synergistic in their effects. Similarly, Wang et al. [387] showed that cholinergic vasodilation of ophthalmic arteries in dogs involves endothelial NO release. Consistent with sympathetic influences on the ophthalmic artery, noradrenaline administration or cranial sympathetic nerve stimulation constricts the ophthalmic artery in dogs [260]. Similarly, the long posterior ciliary arteries have been shown to be under alpha-adrenergic sympathetic vasoconstrictory control and NO- and VIP-mediated and muscarinic cholinergic parasympathetic dilatory control [263, 265, 348], as well as vasorelaxant control by SP+ and CGRP+ fibers [265]. The NO-mediated control of the long posterior ciliary artery is regulated by cholinergic control of endothelial NO release, as well as by direct release of NO from the PPG innervation [407]. Blood flow in the long posterior ciliary artery in cats is not affected by cervical sympathetic nerve section, implying minimal resting sympathetic tone [186].

The ciliary body and iris are under the control of two sets of parasympathetic fibers, those from the PPG and those from the ciliary ganglion.

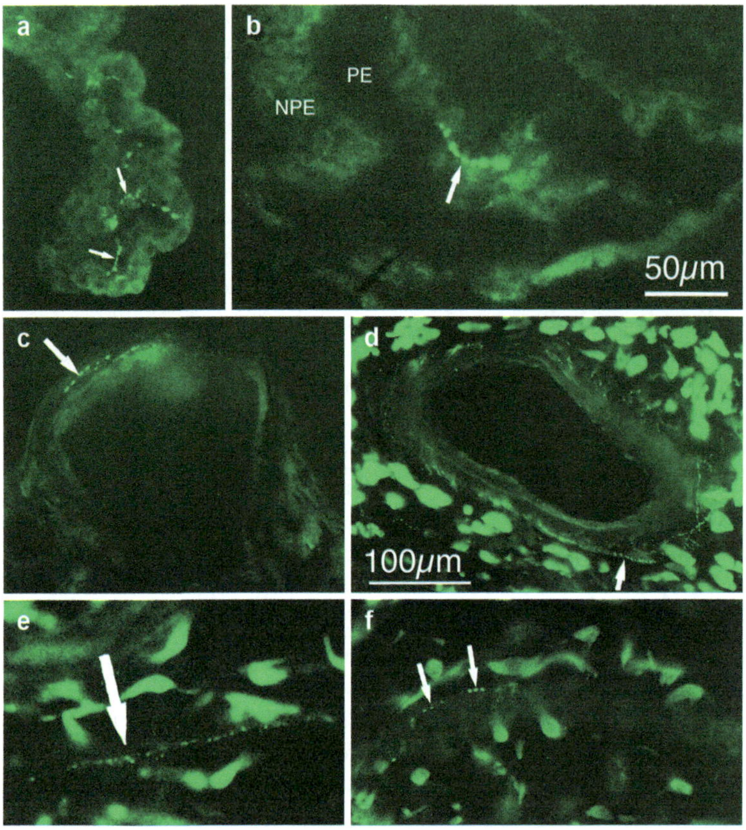

Fig. 12.5 Images showing innervation of various structures of the anterior uvea, all modified from Stone [340]. Image (**a**) shows VIP+ nerve fibers in the ciliary processes of rat (Fig. 3a from [340]). Image (**b**) shows VIP+ nerve fibers in the ciliary processes of guinea pig (Fig. 3b from [340]). Image (**c**) shows VIP+ nerve fibers on a large blood vessel of guinea pig iris (Fig. 4a from [340]). Image (**d**) shows VIP+ nerve fibers on a large blood vessel in cat iris (Fig. 4b from [340]). Image (**e**) shows VIP+ nerve fibers in the iris stroma of rhesus monkey (Fig. 4e from [340]). Image (**f**) shows VIP+ nerve fibers on a longitudinal blood vessel in the iris root of rhesus monkey (Fig. 4f from [340]). *Arrows* in each image show some of the labeled fibers. Magnification is the same in images (**a**–**c**, **e** and **f**). *NPE* nonpigmented epithelium, *PE* pigmented epithelium

The latter are thought to mainly or only serve to control pupil constrictor muscles and accommodation muscles, and any (presumably indirect) impact they have on IOP and blood flow within iris and ciliary body is discussed below. The PPG input to the iris and ciliary body appears to be largely to the vasculature [47] and is discussed in this paragraph. The iris stroma and its vessels, the blood vessels and musculature of the ciliary body, and/or the blood vessels of the ciliary processes have been shown to be innervated by VIP+ fibers from the PPG in guinea pigs, rats, squirrels, cats, monkeys, and humans (Fig. 12.5) [319, 344, 345, 359, 376, 379]. VIP+ nerve fibers have also been observed in the trabecular meshwork in humans [345]. The blood vessels of the ciliary processes in guinea pig and rat but not cat and monkey have also been reported to possess VIP+ innervation [344, 379]. The PPG is also the source of a neuronal NOS+ innervation of vessels in the iris, ciliary body, ciliary processes, and limbus, and all neuronal NOS+ PPG neurons have been found to also contain VIP in rats, but not all VIP+ neurons contain nNOS [11, 398]. The PPG tends to be highly enriched in nNOS neurons in mammals [11], with 70% of PPG neurons being nNOS+ in humans [122]. In pigs at least, the anterior uvea is also innervated by VIP+ and nNOS+ neurons associated with the ciliary nerves as they course through the choroid to the anterior uvea [222].

Consistent with the PPG as a source of ciliary body and iris vascular innervation, the iridial and ciliary body vasculature in cats, rabbits, and monkeys does show a flow increase with activation of the preganglionic input to the PPG via facial nerve stimulation [252]. This vasodilation is blockable with the ganglionic blocker hexamethonium. Moreover, Bill et al. [31] showed that ganglionic blockade with hexamethonium slightly reduced basal iris and ciliary body blood flow in rabbit, presumably by diminishing basal activation of PPG neurons by their preganglionic input from the facial motor complex. The increased iridial and ciliary body blood flow with facial nerve stimulation is likely to be mediated by VIP or NO released from PPG fibers in anterior uvea, with the stimulation parameters used by Nilsson et al. [252].

Nilsson [255] demonstrated that the increases in blood flow in the ciliary body and iris in rabbits with facial nerve stimulation could largely be prevented at low stimulation frequencies (2 Hz) by NOS inhibition but not at higher frequencies (5 Hz). Nilsson suggested that another vasodilator released from PPG terminals other than NO might mediate the vasodilation at the higher of the two stimulation frequencies – presumably VIP. Consistent with this, Nilsson (2000) reported that the increased blood flow in the iris and ciliary body in cats occurring with facial nerve stimulation at 5 Hz was only minimally reduced by the nonselective NOS inhibitor N^{W}-nitro-l-arginine (LNA). Combined treatment with LNA and the nonselective muscarinic antagonist atropine also had little inhibitory effect on the anterior uveal blood flow increase with facial nerve stimulation. These results suggest neither neuronally derived NO release nor endothelially derived NO release driven by acetylcholine plays a noteworthy role in the vasodilatory action of the PPG input to anterior uvea in cats at a 5-Hz firing rate, again implicating VIP as the major dilator at this firing rate. At 10-Hz stimulation, LNA or atropine alone was largely ineffective in blocking the anterior uveal blood flow increase with facial nerve stimulation. At 10 Hz, however, LNA and atropine together were effective at nearly completely blocking the facial nerve-evoked vasodilation in the iris and ciliary body. This result is puzzling in light of the differential effects of facial nerve stimulation frequencies in rabbits and suggests that VIP is not the major vasodilator at 10 Hz in lacrimal gland in cats. Rather, both neurally derived and endothelially derived NO, the later driven by cholinergic action, are both seemingly involved in the anterior uveal increase with facial nerve stimulation at 10 Hz in cats, and the doses of LNA and atropine used were apparently not able to block the 10-Hz vasodilation when used singly.

Consistent with a role of NO in control of iris and ciliary body blood flow, Deussen et al. [69] reported that NOS inhibition with LNAME in dogs decreased basal anterior uveal blood flow about 50%, despite causing about a 20% increase in mean arterial blood pressure. Similarly, Seligsohn and Bill [313] reported that LNAME decreased iris and ciliary body blood flow in rabbit by 40–60%. Similarly, Jacot et al. [156] showed in piglet that systemic NOS inhibition with LNAME reduces anterior uveal blood flow. The purpose of facial nerve control of anterior uveal blood flow is uncertain. The anterior uvea in cats and monkeys but not rabbits has been shown to compensate well for reductions in ocular perfusion pressure caused by elevation of IOP or reduction of systemic blood pressure [10]. It may be that the facial input contributes to this regulation and thereby aids in maintaining stable blood flow to the anterior uvea regardless of momentary fluctuations caused by variations in ocular perfusion pressure. As we will discuss in the section on choroidal blood flow, the facial input may be especially involved in compensation for systemic hypotension. Iridial blood flow has, however, been reported to not show autoregulation in humans when perfusion pressure is reduced by increasing IOP [51].

Since the PPG innervates both vessels involved in aqueous production and the territory involved in aqueous outflow, the impact of the PPG innervation of anterior uvea on IOP has been of interest. This interest was initiated by the finding of Ruskell [297] that PPGectomy in monkeys led to a long-lasting diminution of IOP. Conversely, PPG activation leads to increased IOP in monkey [252]. While these effects might stem from effects on choroidal blood flow (whose volume affects IOP), the possibility of more direct effects cannot be ignored. Nilsson and Bill [251] showed that intravenous administration of VIP in rabbits yielded increased choroidal blood flow and increased IOP but no change in anterior uveal blood flow. By contrast, intracameral administration of VIP vasodilated vessels in iris and ciliary body but had no effect on IOP, pupil diameter, or blood-retinal barrier. Thus, increased choroidal blood flow can be associated with increased IOP, while increased anterior uveal blood flow by itself need not increase IOP. Nilsson et al. [253], however, showed that there might be species differences in the impact of PPG input to the anterior uvea on IOP. They showed that intracameral VIP in monkeys can increase IOP by increasing aqueous production and increasing venous outflow resistance. NO production by PPG terminals on ocular vessels also exerts an effect on IOP as well. For example, in rabbits, inhibition of NO production reduces long posterior ciliary artery blood flow (and presumably ciliary process blood flow as well) and thereby reduces aqueous production (Kiel et al. 2001). NO also yields trabecular meshwork relaxation and thus increased outflow [391]. This effect on outflow may be why treatments that potentiate NO action by preventing its breakdown or treatments that cause release of NO tend to decrease IOP [21]. Finally, M3 muscarinic receptors are present in human trabecular meshwork [130], and acetylcholine dilates ciliary process vessels in cats and rabbits [382]. It may be that the PPG input to these structures can act via cholinergic mechanisms to influence aqueous production and outflow, at least in some species. A role of the facial input in the nocturnal rise in IOP is possible [35], given the apparent input that the facial preganglionic neurons receive from the SCN nucleus region, as discussed below in more detail in the section on facial regulation of choroidal blood flow control.

Ciliary ganglion parasympathetic cholinergic fibers innervate iris sphincter and ciliary body muscles and thereby cause pupil constriction and accommodation, respectively, when activated. Consistent with cholinergic influences in these muscular structures, M1, M2, and M3 muscarinic receptors have been reported in human iris, ciliary body, and ciliary processes, and M3 receptors have been reported in the trabecular meshwork [130, 131]. Stimulation of the ciliary ganglion input has, however, been reported to also have unexpected effects on anterior uveal blood flow. For example, Stjernschantz et al. (1973) found that intracranial stimulation of the oculomotor nerve (and thus the preganglionic input to the ciliary ganglion) decreased blood flow in the rabbit iris, ciliary body, and ciliary processes. They suggested that mechanical compressive effects on blood vessels due to ciliary body and iris muscle contraction might be the basis of the diminished blood flow. Consistent with this interpretation, eye illumination sufficient to cause pupil constriction in rabbits is accompanied by reduced iris blood flow, and both the pupil constriction and the decreased blood flow can be blocked with the peripheral muscarinic blocker biperiden [336]. Similarly, Bill et al. [31] showed that ganglionic blockade with hexamethonium eliminated and muscarinic blockade with biperiden reduced the iris and ciliary body blood flow decrease obtained in rabbits with intracranial oculomotor nerve stimulation. Sympathectomy in rabbits showed that the anterior uveal vasoconstriction with oculomotor nerve stimulation was not due to inadvertent activation of sympathetic input to the anterior uvea [8]. Thus, the reduced anterior uveal blood flow with oculomotor nerve stimulation in rabbits is mediated by the input of the oculomotor nerve to the ciliary ganglion (which arises specifically from the nucleus of Edinger-Westphal). The same research group later showed that the uniformly reduced anterior uveal blood flow with oculomotor nerve stimulation might be unique to rabbits, because both oculomotor nerve stimulation and activation of the cholinergic input from the ciliary ganglion caused iris vasoconstriction and ciliary body vasodilation in cats and monkeys [37, 337]. The effects in cats but not monkeys could be blocked with the muscarinic blocker atropine. Consistent with the possibility that reduced blood flow in the iris in cats is caused by direct cholinergic vascular mechanisms, acetylcholine and ciliary ganglion activation were both found to vasoconstrict anterior uveal vessels in arterially perfused cat eyes [214, 216]. In contrast to the blood flow-reducing effects of activation of the ciliary ganglion input to the anterior uvea in rabbits, the specific action of acetylcholine on ciliary process vessels in both cats and rabbits has been reported to be dilatory [382]. It may be this effect is mediated by vascular muscarinic receptors postsynaptic to PPG cholinergic input. Consistent with cholinergic vasodilatory effects mediated in some parts of the anterior uvea in some species, Alm et al. [7] showed that the cholinesterase inhibitor neostigmine or the muscarinic agonist pilocarpine applied to the cornea caused pupil constriction and increased ciliary body, ciliary process, and iris blood flow in monkeys. The most consistent interpretation of these diverse results is that the mechanical effects of prominent pupil constriction caused by activation of the oculomotor – ciliary ganglion input to the iris sphincter muscle compress iris blood vessels and cause diminished iridial blood flow in rabbits, cats, and monkeys. The above-noted study by Alm et al. [7] may have observed increased iris blood flow with cholinergic agents because the pupil constriction was too mild for its mechanical effects to override the vasodilatory action of muscarinic activation of vessels. By contrast, the influence of the ciliary ganglion input to the ciliary body and processes varies among species. In rabbits, the vasoconstrictory effect is likely to be mechanical, since cholinergic vasodilatory mechanisms are in place. In monkeys and cats, the vasodilatory effect could be direct, or the mechanical effects may somehow diminish vascular resistance.

Activation of the ciliary ganglion input to the anterior uvea also has an impact on IOP, with stimulation of the preganglionic neurons of the nucleus of Edinger-Westphal (EW) of the oculomotor nuclear complex in rabbits and cats causing an IOP rise [113, 303]. In Gherezghiher et al. [113], the stimulation was sufficient to produce pupil constriction and a 35% rise in IOP that was blockable by hexamethonium. The basis of this effect is uncertain, but the mechanical effects of pupil constriction on outflow or a vasodilatory effect on uveal vessels have been raised as a possibility [113, 303]. An effect on IOP other than by the transitory pupil constriction is implied by the finding that ciliary ganglionectomy causes an IOP drop in cats and monkeys [57, 78]. Pharmacological stimulation of accommodation with pilocarpine, however, increases outflow in monkey and human eye, apparently by a mechanical effect on the trabecular meshwork mediated by the pull of the muscles of accommodation on the scleral spur, as well as by a direct effect on the outflow pathway [77, 102]. This finding implies that the accommodative effect of ciliary ganglion activation does not account for the increased IOP with ciliary ganglion activation. Both aqueous production and outflow, however, were reportedly increased in cats and rabbits by oculomotor nerve or ciliary ganglion stimulation [215, 216, 229]. Whether ciliary ganglion input directly or indirectly mediates these effects is uncertain. In any event, these results suggest that IOP increases caused by oculomotor nerve or ciliary ganglion activation stem from pupil constriction and perhaps greater aqueous production than outflow. It may also be that choroidal blood flow increases contribute to IOP rise with oculomotor nerve or ciliary ganglion activation. As will be discussed below, however, the evidence that oculomotor nerve activation increases choroidal blood flow is not definitive.

Among the nerve fibers innervating iris, ciliary body and the aqueous outflow path are sympathetic nerve fibers from the superior cervical ganglion. These nerve fibers contain adrenaline and NPY [39, 240], and they have been demonstrated in diverse mammalian species, including rat, monkey, cat, and rabbit [195, 356]. The iris also has sympathetic innervation in chicks [175]. Although some of the sympathetic input to iris ends on the iris dilator muscle, some has also been shown to end on iris vessels in rat [143], and they vasoconstrict vessels by alpha-adrenergic mechanisms [142]. Alm and Bill [6] showed that cervical sympathetic stimulation decreases iris, ciliary body, and ciliary process blood flow in cats. This effect is blocked by the alpha-adrenergic antagonist phentolamine [6], and pharmacological studies show that sympathetic constriction of ciliary body arteries is mediated by alpha 2a adrenoreceptor mechanisms [393]. Additionally, intravenous NPY decreases iridial flow by 30% and ciliary body flow by 50% [254]. The sympathetic input to the ciliary body and processes also influences aqueous production. Chronic sympathetic stimulation in rabbits decreases aqueous production (by decreasing blood flow presumably) but increases outflow resistance, resulting in an initial short-term decline in IOP and an eventual gradual return to basal IOP levels [22]. Sympathetic beta-adrenergic mechanisms acting at the level of the ciliary processes are involved in increased aqueous production. In rabbits, it has been shown that both an outflow resistance increase mediated by alpha-adrenergic mechanisms and an aqueous inflow increase mediated by beta-adrenergic mechanisms cause the nocturnal IOP rise [205, 406].

Sensory fibers from the trigeminal ganglion that contain SP and CGRP also innervate the ciliary body and vessels, ciliary processes, iris stroma and vessels, limbal vessels, and/or the trabecular meshwork in guinea pigs, rats, squirrels, rabbits, pigs, cats, monkeys, and humans [20, 23, 74, 143, 198, 202, 224, 236, 319, 342, 343, 346, 347, 359–363, 380]. Double-label studies show that the SP and CGRP typically co-occur in single fibers in these regions [193]. SP and CGRP are common in neurons of the trigeminal ganglion [380], and trigeminal ganglionectomy or transection of the ophthalmic nerve (plus maxillary nerve) eliminates SP and CGRP from the anterior uvea and limbal vessels [362, 363, 380]. Consistent with its SP+ input, rat and rabbit iris possess SP receptors [68], and consistent with its CGRP+ input, the iris and ciliary body in pig, guinea pig, and cat possess CGRP receptors [138]. Sensory fibers such as those of the trigeminal nerve send a central message of hot, cold, pain, or touch and can elicit ocular reflexes, such as blinking and tearing in response to their activation [23, 103]. Peripheral fibers can also participate in antidromic responses in which they release SP and CGRP and cause local irritation responses, which include a vascular component [23]. For example, stimulation of the ophthalmic nerve in rabbits causes blood flow increases in the iris and ciliary body, IOP increases, increased extravasation of albumin in the iris and ciliary body, increased albumin in the aqueous, and pupil constriction [338]. Ocular irritation or injection of CGRP into the eye of rabbits increases iridial and ciliary body blood flow and causes blood-retinal barrier breakdown, IOP rise, and pupil constriction [188]. The SP released from the ophthalmic nerve endings does not appear to contribute to the blood-retinal barrier breakdown, but the CGRP does [13, 36]. The ocular irritation response involves edema and vasodilation to wash away irritants. In monkeys and cats, however, trigeminal nerve stimulation or SP and/or CGRP injection into the eye causes lesser effects on iridial and ciliary body blood flow and blood-aqueous barrier integrity than they do in rabbits [10, 37, 262]. SP and CGRP also exert an effect on aqueous production and aqueous outflow in monkeys, cats, and rabbits, although the nature of the effects and the mechanisms underlying them may vary among species [12, 189, 262, 340, 377]. For example, an effect on IOP can be mediated by a vasodilatory effect of trigeminal sensory fibers on the arteries of the ciliary processes and/or the episcleral veins of the outflow channel.

12.4 Neural Control of Blood Flow in Orbital Glands

The orbit also contains various glandular structures that are responsible for lubricating the cornea and washing away small debris that might injure the cornea. These structures include the lacrimal gland (which is located laterally in the orbit and secretes tears that lubricate and moisturize the cornea), the Meibomian glands of the tarsal plates of the eyelids (which secrete an oily fluid that coats the cornea and limits dehydration), goblet cells of the conjunctival fornices (which secrete mucin to aid in limiting corneal dehydration), and the Harderian gland, which is a sebaceous gland that acts as an accessory to the lacrimal gland in most mammalian species and is very prominent in birds, where it is located medially in the orbit and is larger than the lacrimal gland [63, 266, 273]. The blood supply to the lacrimal gland arises from the ophthalmic artery via the lacrimal artery, while the blood supply to the Harderian gland arises from a more medial and posterior branch of the ophthalmic artery (in birds, it arises from the ophthalmotemporal artery). The blood supply to the Meibomian glands and goblet cells is via branches of the palpebral arteries to the eyelids. Secretion from these glands is under neural control, as is blood flow to and within these glands.

For example, Ruskell has shown that the lacrimal gland in primates is innervated by parasympathetic and sympathetic nerve fibers. In an early study [300], he showed parasympathetic terminals from the PPG to the lacrimal gland in rabbits. In a later study, he showed that rami from the PPG (presumably secretomotor) and perivascular nerves traveling on the lacrimal artery both enter the lacrimal gland in humans [299]. The PPG neurons projecting to the lacrimal gland appear to arise from a different part of the PPG than those to the iris [357]. Ruskell [296] showed sympathetic terminals that arise from the superior cervical ganglion on arteries, veins, and capillaries in the lacrimal gland of monkeys. Ven der Werf and coworkers [18, 381] confirmed that the PPG and superior cervical ganglion innervate the lacrimal gland in monkey using retrograde labeling methods, and they also showed trigeminal sensory innervation. The PPG and superior cervical ganglion innervation of lacrimal gland has been confirmed by immunolabeling, retrograde labeling, and/or anterograde labeling for diverse rodent species as well, including guinea pig, rat, and/or mouse [19, 71, 72, 311, 357, 376]. The PPG fibers in rodents are cholinergic and contain VIP and nNOS as well, and they and the sympathetic input end near acinar cells, as well as on blood vessels. In mouse at least, parasympathetic and sympathetic fibers innervate different parts of the mouse lacrimal gland and thus different secretory cells [71, 72].

The autonomic innervation of the lacrimal gland regulates both blood flow and tear secretion in the gland. Reflecting its secretomotor role, transection of preganglionic input to the PPG has been shown to cause dry eye in rabbits [372]. Cholinergic mechanisms are involved in the secretomotor role of PPG input to the lacrimal gland [231]. Additionally, VIP released from PPG terminals has a secretomotor role in lacrimal gland function, since VIP causes lacrimal gland tear secretion in rats, rabbits, and pigs [68, 321, 341]. Nilsson [256] showed that activation of PPG input to the eye via facial nerve stimulation also increases blood flow in rabbit lacrimal gland. The increase at 2-Hz stimulation was nearly completely blocked by NOS inhibition, but only reduced at 5 Hz, implying a greater role for NO at low stimulation frequencies and a greater role for VIP at high frequencies. Subsequently, Nilsson (2000) provided further evidence for frequency-dependent roles of NO, VIP, and cholinergic mechanisms in vasodilation in lacrimal glands from studies in cats. Facial nerve stimulation at 5 Hz yielded increases in blood flow in lacrimal gland that could be greatly reduced by NOS inhibition alone (with LNA), and further reduced by combined NOS inhibition and muscarinic blockade with atropine. This result suggests a prominent role of NO in lacrimal gland vasodilation at 5 Hz, involving both NO release from PPG terminals and acetylcholine-evoked NO release from endothelium. The fact that LNA alone did not entirely block lacrimal gland vasodilation at 5 Hz suggests some role of VIP at this frequency as well. At 10 Hz, facial nerve stimulation-evoked increases in glandular blood flow were greatly attenuated but not completely blocked by combined LNA and atropine. Moreover, the facial nerve stimulation-evoked increases in glandular blood flow at 10 Hz were no more reduced by combined LNA and atropine than by NOS inhibition alone, and atropine alone did not reduce the vasodilation at all. These results suggest a role of neurally derived NO in vasodilation in lacrimal gland at 10 Hz, as well as a role for an additional vasodilator that is presumably VIP. Yasui et al. [402] showed that 20-Hz facial nerve stimulation yielded both blood flow increases and tear secretion from lacrimal gland in cats. The tear secretion at this frequency was greatly dependent on muscarinic cholinergic mechanisms since it was blocked with scopolamine, while the blood flow increase was not blocked by scopolamine. Thus, lacrimal vasodilation mediated by the PPG input at a 20-Hz activation frequency may occur via VIP but not via acetylcholine-evoked release of endothelially derived NO, and secretion may be mainly cholinergic in its basis. Given the prominent role of muscarinic mechanisms in lacrimal secretion, the contributions of VIP and NO to such secretion may occur (at least in part) via their effects on blood flow. Finally, consistent with regionally differential parasympathetic and sympathetic innervation of mouse lacrimal gland, both beta-adrenergic and cholinergic agonists produce increased glandular secretion, but in different parts of the mouse lacrimal gland [71, 72].

The Harderian gland receives cholinergic innervation in mammals [273], on blood vessels, secretory cells, and secretory tubules. VIP+ fibers from the PPG have been reported in mammalian Harderian gland, near blood vessels, and acini [161, 388]. VIP+ and SP+ fibers have been reported in avian Harderian gland, and SP+ fibers have been reported in rabbit Harderian gland [47, 63, 384]. In pigeons, blood vessels to the Harderian gland have been shown to be also richly innervated with nNOS+ nerve fibers that arise from the PPG and the distributed microganglia of the orbit that appear to be part of the avian PPG system [63]. The acini of the avian Harderian gland appear to be innervated by both VIP+ and nNOS+ fibers as well [63, 384]. The function of the innervation of the Harderian gland is presumed to be similar to that in the lacrimal gland. Nilsson [255] showed that facial nerve activation of PPG input to the eye increases blood flow in rabbit Harderian gland, with NO mediating responses to 2-Hz stimulation, and NO plus a second vasodilator (presumably VIP) mediating responses to 5-Hz stimulation. Subsequently, Nilsson (2000) showed a role of NO (possibly both neural and endothelial) and presumably VIP in vasodilation in Harderian gland at a 5-Hz facial nerve stimulation frequency in cats, while neural NO and VIP appeared to mediate vasodilation at a 10-Hz stimulation frequency.

The acini of the Meibomian glands and eyelid blood vessels have been shown to be innervated by VIP+ cholinergic parasympathetic fibers from the PPG, by NPY+ noradrenergic sympathetic fibers from the superior cervical ganglion, and by SP+ and CGRP+ trigeminal fibers in diverse mammalian groups, including rats, pigs, monkeys, and humans [55, 75, 176, 209, 210, 312, 320, 321]. The distribution of PPG neurons innervating the Meibomian glands overlaps that of PPG neurons innervating iris [357], choroid [65, 200], and cerebral vasculature [352]. Sympathetic fibers are more abundant on blood vessels of the Meibomian glands and eyelid, while parasympathetic fibers are more numerous near secretory cells [200]. LeDoux et al. [200] used pseudorabies virus transneuronal retrograde labeling to identify the central cholinergic preganglionic neurons that control the Meibomian glands. They found them to reside within the lateral aspect of the superior salivatory nucleus (SSN) of the facial motor complex of the pons. The role of the nerve fibers innervating the Meibomian glands has not been extensively studied, but they are likely to play roles in blood flow control and regulation of secretion [55]. The work of Nilsson [255] in rabbits and cats discussed above indicates that the SSN-facial nerve-PPG circuit mediates vasodilation within the eyelid, with NO (possibly both neural and endothelial) being the main vasodilator at a lower firing frequency, and NO (neural and endothelial) and VIP involved at a higher firing frequency.

Several studies have shown that goblet cells have VIP+ parasympathetic fibers, and TH+ and DBH+ (containing dopamine beta-hydroxylase) sympathetic fibers, and CGRP+ sensory fibers near them in mouse, rats, and human [67, 312, 70]. Diebold et al. (2001) also showed M3 muscarinic receptors and beta-adrenergic receptors on goblet cells in mouse and human, while [289] demonstrated M2 and M3 receptors and VIPR2 receptors on goblet cells in rats. Ten Tusscher et al. [357] confirmed the PPG projection to goblet cells by retrograde labeling in rats. VIP has been shown to cause goblet cell mucus secretion [168, 289] as has the cholinergic agonist carbachol.

12.5 Neural Control of Choroidal Blood Flow

12.5.1 Importance of the Choroid

The choroid accounts for about 85% of the blood supply to the retina and all of the blood supply to the retinal pigment epithelium (RPE) and photoreceptors [33, 37]. Moreover, the choroid is the major or exclusive vascular supply for both outer and inner retina in regions lacking retinal vessels (such as the macula in primates) and in animal groups in which retinal vessels are sparse (rabbits) or absent (birds). The importance of the choroid is emphasized by the rapid photoreceptor loss that occurs with severely diminished choroidal blood flow [58, 110, 111, 226]. Photoreceptors are particularly dependent on the choroid because of their high metabolism and continuous outer segment renewal [37, 408]. One specialization that helps the choroid meet retinal needs is the high choroidal blood flow [37]. Although it was once thought that choroidal blood flow (ChBF) so exceeds retinal requirements as to eliminate need for its regulation, it now seems likely that: (1) the high ChBF is essential as the driving force for diffusion of oxygen and nutrients through Bruch membrane, the retinal pigment epithelium (RPE), and the depth of the outer neural retina, and to the inner retina in the case of retinal regions devoid of retinal vessels (or in the cases of species devoid of or poor in retinal vessels); (2) ChBF must be stably matched to the thermal and metabolic needs of the retina; and (3) disturbances in ChBF result in impairments in retinal function [33, 37, 61, 325, 364, 399, 400, 401]. Thus, the ability of ChBF to respond adaptively to variations in retinal need imposed by changes in illumination level and retinal activity, and to maintain a stable high flow rate despite fluctuations in bodily state and time of day, is likely to be important for the health and proper functioning of the RPE and outer neural retina.

12.5.2 Choroidal Innervation: Overview of Anatomy

Retinal blood vessels are located directly within the retina itself and are therefore able to respond to the local concentrations of carbon dioxide and oxygen, and to autoregulate blood flow accordingly, as typical of most vascular beds [35, 37], or by metabolic coupling, as in the brain [83, 150, 355]. In contrast, the choroid and its capillary bed, the choriocapillaris, are separated from the outer retina by Bruch membrane and the RPE. This barrier and the high blood flow rate in the choroid make ChBF largely unable to be regulated by vasogenic metabolites derived from the outer retina [35]. Although the choroid is not regulated by local retinal metabolic factors, it is under neural control. Three major types of nerve fibers innervate the choroid in mammals and birds (Fig. 12.6): (1) parasympathetic fibers arising from the PPG that co-contain neuronal NOS and acetylcholine [63, 298, 346, 398], (2) noradrenergic sympathetic fibers from the superior cervical ganglion that also contain NPY [129, 346], and (3) sensory fibers from the trigeminal ganglion co-containing SP and CGRP [318, 346]. In addition, cholinergic parasympathetic fibers to the choroid also arise from the ciliary ganglion in birds, and perhaps some mammal groups as well [62, 182, 221, 286, 335, 346] [232]. The nNOS+ and VIP+ PPG fibers have a vasodilatory influence on choroidal vessels and increase ChBF [35–37], while nor adrenergic fibers have a vasoconstrictory action that decreases ChBF [35–37, 183]. The input to the choroid from the ciliary ganglion also has a vasodilatory influence, mediated by muscarinic elicitation of endothelial NO release [64, 88, 412–414]. The sensory input has a vasodilatory influence as well. The anatomy and function of these inputs in birds and mammals is detailed in the following sections.

12.5.3 Facial Nucleus Parasympathetic Input

12.5.3.1 Peripheral Anatomy of Facial Circuitry for Control of ChBF – Mammals

The distributed orbital neurons composing the PPG profusely innervate the choroid in all mammalian groups studied, including rats, guinea pigs, cats, squirrels, monkeys, and humans, with postganglionic axons from the PPG entering the choroid directly, on ciliary vessels, and by joining the short ciliary nerves arising from the ciliary ganglion (Fig. 12.4) [19, 202, 297, 298, 319, 344, 357, 359, 376, 378]. Some pass through the ciliary ganglion on their way to the choroid [19]. The PPG neurons innervating choroid in rats are found within the proximal part of the PPG and scattered along the greater petrosal nerve itself, in which the preganglionic fibers to the PPG

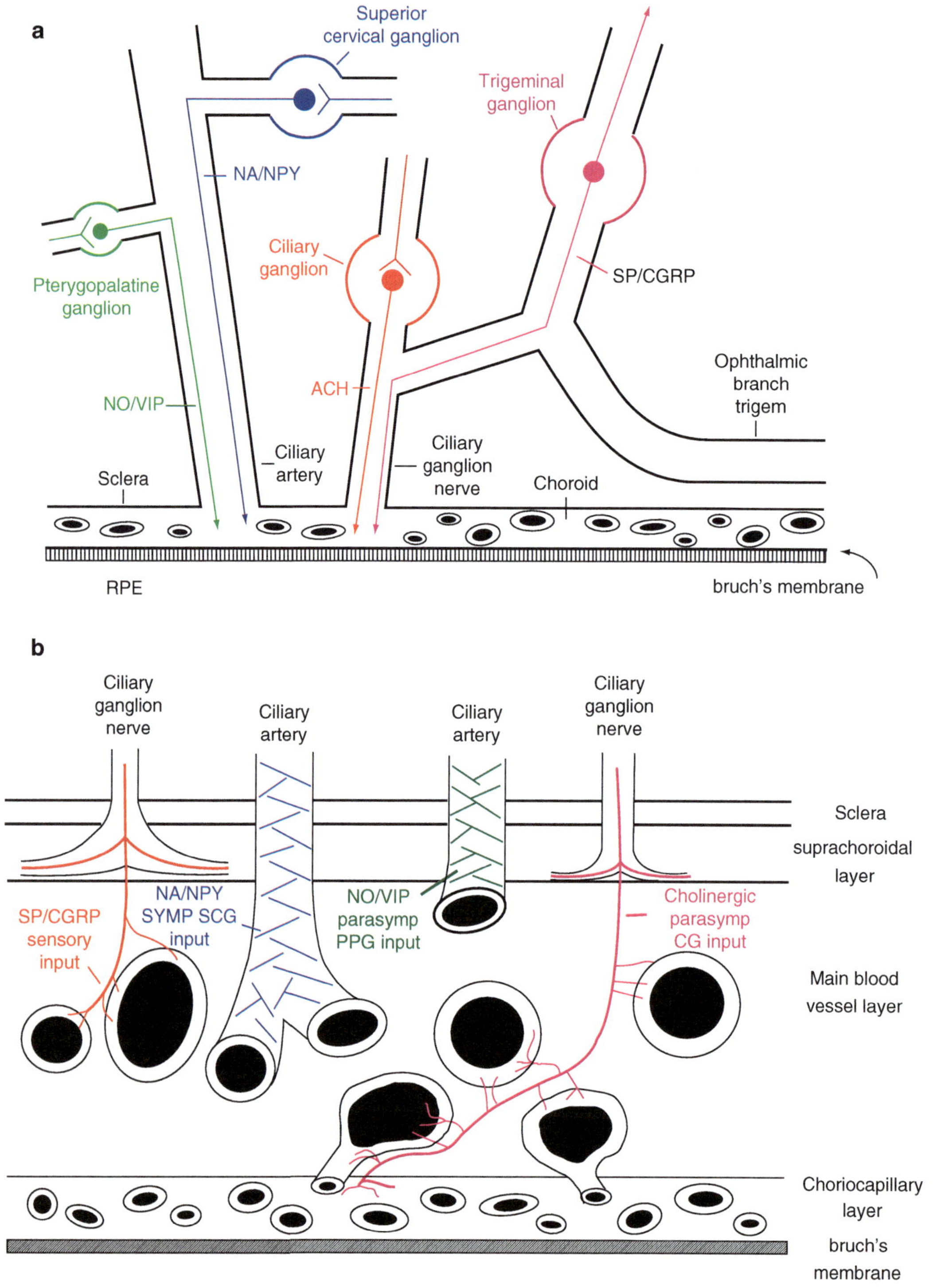

Fig. 12.6 Schematics illustrating the ganglia innervating choroid (**a**) and the route fibers from each ganglion take to enter the choroid (**b**). This organization and neurochemistry for the choroidal innervation shown is true of both birds and mammals

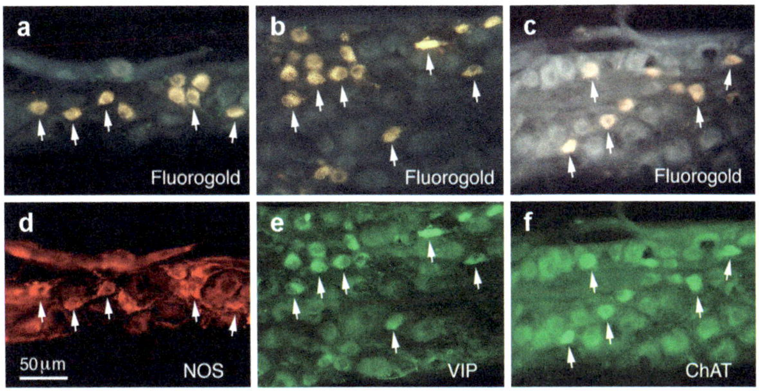

Fig. 12.7 A series of images of fluorescent labeling showing that neurons of the rat PPG that project to the choroid contain NOS, VIP, and ChAT. Image (**a**) shows neurons of the PPG that had been retrogradely labeled by intrachoroidal FG injection into the temporal sector of the choroid, and image (**d**) shows NOS immunolabeling in this same field of view. NOS is present in most of the FG-labeled PPG neurons, as indicated by *arrows* for some of the double-labeled neurons. Image (**b**) shows neurons of the PPG that were retrogradely labeled by the same intrachoroidal FG injection, and image (**e**) shows VIP+ neurons in this same field. Note that VIP is also present in most of the FG-labeled PPG neurons (some indicated by *arrows*). Image (**c**) shows neurons of the PPG that had been retrogradely labeled by the same intrachoroidal FG injection as in (**a** and **b**), while image (**f**) shows ChAT immunolabeling in the same field. Note that ChAT is also present in most of the FG-labeled PPG neurons (some indicated by *arrows*). All images are at the same magnification

travel [65]. Small accessory ganglia near the ciliary ganglion may be part of the PPG system as well, since in rats, they project to the eye and are VIP+ [194, 346, 357]. PPG neurons are VIP+, and their extirpation eliminates VIP and VIP+ fibers from cat choroid [47, 376]. The PPG is also the source of a profuse nNOS+ innervation of choroid, and all nNOS+ PPG neurons have been found to also contain VIP in rats (Fig. 12.7) [65, 398]. Additionally, PPG neurons and their input to choroid appear to be cholinergic as well [65, 134, 344, 359, 378]. As evidenced by VIP immunolabeling, PPG input to choroid ends on arteries, veins, and melanocytes (Fig. 12.8) [19]. The distribution within PPG of neurons innervating choroid overlaps those innervating the iris [357], Meibomian glands [200], and cerebral vasculature [342]. Thus, it is uncertain if the PPG neurons controlling the choroid exclusively subserve the choroid. Based on the number of retrogradely labeled neurons we observed in the PPG and greater petrosal nerve after intrachoroidal fluorogold injection, and based on the extent of the fluorogold injections, it appears that at least 200–300 PPG neurons innervate choroid in rats [65]. Additional PPG neurons may innervate orbital vessels feeding into choroid and thereby exert a further influence on ChBF [346, 370].

12.5.3.2 Central Anatomy of Facial Circuitry for Control of ChBF – Mammals

The PPG in mammals receives its preganglionic parasympathetic input from the superior salivatory nucleus (SSN) subdivision of the facial nucleus motor complex (SSN) [59, 357]. We have used transneuronal retrograde pathway tracing in rats with the Bartha K strain of pseudorabies virus (PRV) to identify the SSN neurons innervating those PPG neurons projecting to choroid, and to identify central neurons projecting to that part of SSN [65]. Double-labeling immunofluorescence <75 h after intrachoroidal virus injection (i.e., at relatively short survival times) showed that the PRV+ neurons in SSN were cholinergic, confirming their preganglionic identity (Fig. 12.9). Moreover, the choroidal SSN neurons appeared to be slightly more rostral and medial in location (Fig. 12.10) than those observed after injection of PRV into other peripheral PPG targets, such as the Meibomian glands [200] and the lacrimal gland [373], indicating choroidal control may be local-

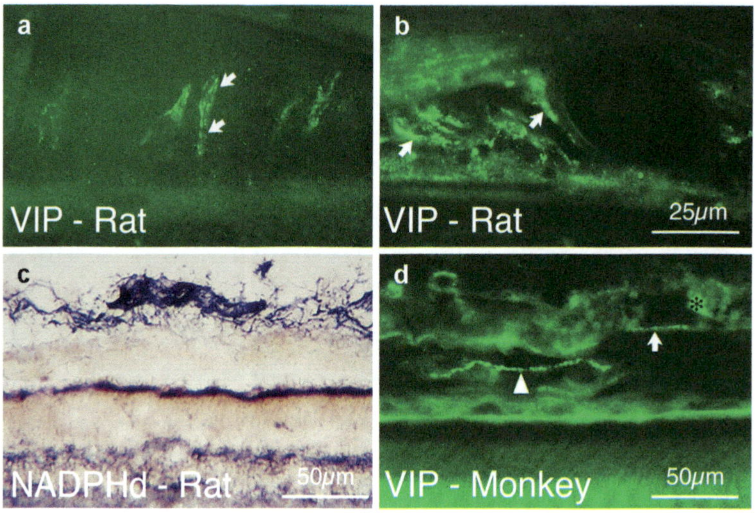

Fig. 12.8 Images showing PPG input to the choroid in mammals. Images (**a** and **b**) show thin and beaded VIP+ fibers in and around choroidal vessels in rat *(arrows)*. Image (**c**) shows a dense plexus of NADPHd+ nerve fibers within the rat choroid. NADPHd is a marker for NOS fibers, and (**c**) thus shows that the rat choroid is rich in NOS+ innervation. These fibers arise from the PPG and co-contain VIP. Image (**d**) shows VIP+ nerve fibers adjacent to a large vessel cut longitudinally *(arrow)* and also more superficially *(arrowhead)* in the rhesus monkey choroid (Modified from Stone [344]). An autofluorescent red blood cell is marked by a *black asterisk* in (**d**). Magnification in (**a**) same as in (**b**)

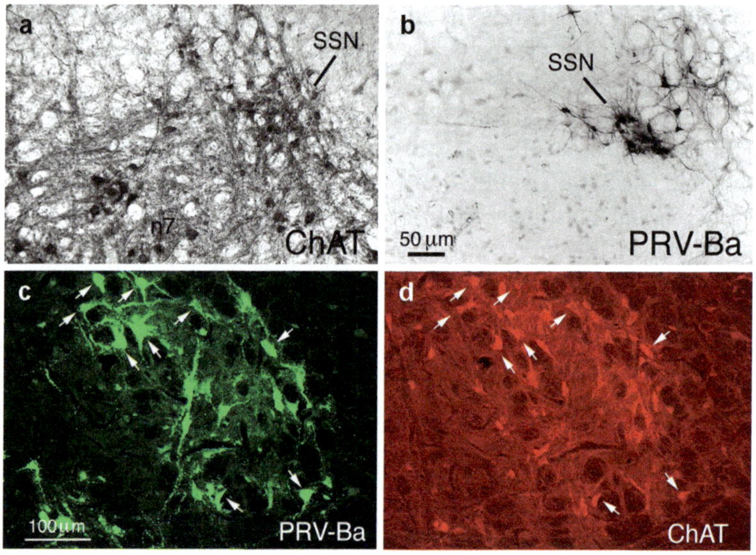

Fig. 12.9 A pair of images of transverse sections of the SSN at its rostral level (**a**, **b**), showing neurons within rat SSN that are preganglionic to choroidal neurons of the PPG. Image (**a**) shows a section from a normal rat immunolabeled for choline acetyltransferase (*ChAT*), a marker of cholinergic neurons, while image (**b**) shows neurons in SSN transneuronally labeled 63 h after a pseudorabies injection into the choroid. The pair of images shows that SSN and facial motor nucleus (n7) neurons are ChAT+, and neurons regulating choroid are restricted to more ventromedial SSN. The magnification is the same in images (**a** and **b**). Images (**c** and **d**) show images of a single field of view of the SSN from tissue double labeled with immunofluorescence for pseudorabies (**c**) and ChAT (**d**), from an animal that survived 65 h after unilateral virus injection into the choroid. The *arrows* indicate neurons within the SSN that were labeled for PRV from choroid and were cholinergic. These results show that the PRV+ neurons within the SSN labeled transneuronally from the choroid were cholinergic preganglionic neurons, which represent a subset of SSN neurons. Images (**c** and **d**) are at the same magnification. All four fields are of the right side of the brain, with medial to the left and dorsal to the top

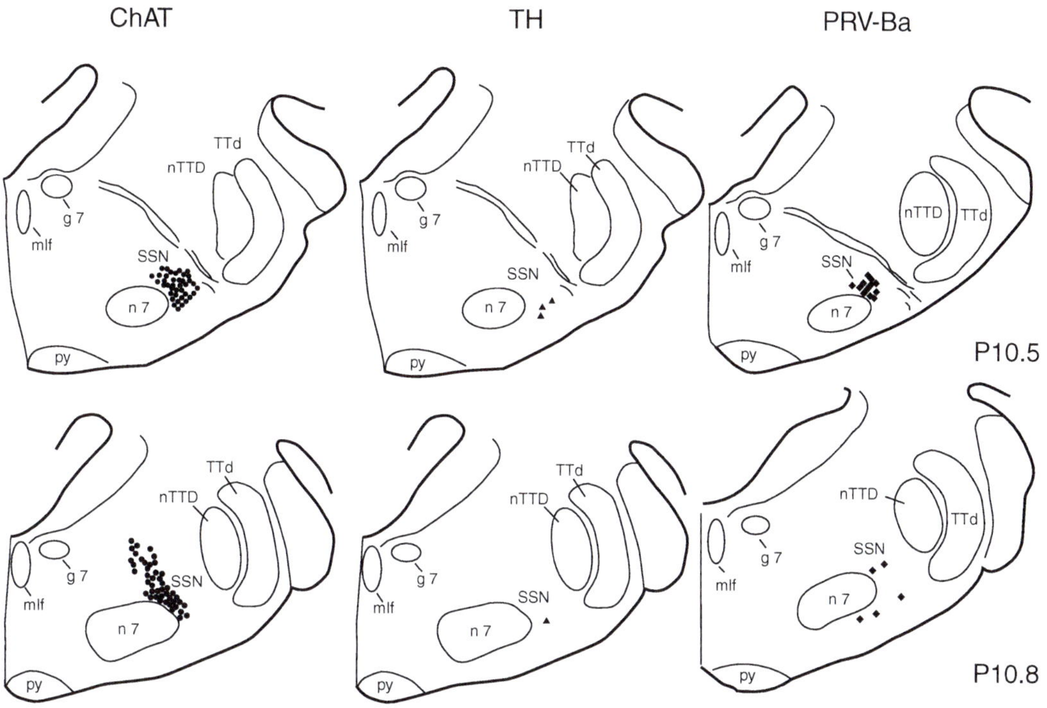

Fig. 12.10 Camera lucida drawings of the right side of the brain in a rostral (*top row*)-to-caudal (*bottom row*) set of transverse sections showing the location of the rat SSN (cholinergic neurons immunolabeled for ChAT illustrated by *dots* in the *left column*), noradrenergic neurons immunolabeled for TH (illustrated by *triangles* in the *middle column*), and SSN neurons transneuronally retrogradely labeled from the choroid with PRV (illustrated by *diamonds* in the *right column*). The numbers at the *right* indicate the rostrocaudal level of each row in terms of the distance behind the skull suture Bregma. The first two columns show that the cholinergic neurons of SSN and the TH+ neurons of the A5 cell group have only slight overlap, with the SSN neurons largely lying caudomedial to A5. The PRV neurons transneuronally retrogradely labeled from the choroid were restricted to the rostral and ventromedial SSN (at P10.5). *g7* genu of the facial nerve, *mlf* medial longitudinal fasciculus, *n7* facial motor nucleus, *N7* facial nerve root, *nTTd* nucleus of the descending trigeminal tract, *py* pyramidal tract, *SSN* superior salivatory nucleus, *TTd* descending trigeminal tract, *A5* nor adrenergic cell group 5

ized to a select subpopulation of SSN neurons. The choroidal SSN neurons commonly were also NOS+, and the majority of NOS+ neurons within the SSN were PRV+, suggesting NOS to be a marker of choroidal neurons within rat SSN. With longer post intrachoroidal injection times, higher-order labeling of neurons projecting directly or polysynaptically to SSN became evident (Fig. 12.11). The dorsal part of the paraventricular nucleus (PVN) of the hypothalamus notably contained PRV+ neurons bilaterally (more heavily ipsilaterally). The input of PVN to choroidal SSN is of interest, because PVN is responsive to systemic blood pressure (BP) and plays a role in maintaining stable systemic BP [16, 190, 276, 397], and the PVN region exerts a vasodilatory influence on cerebral blood flow (CBF) [117]. With similar survival times, PRV+ neurons were also observed in the caudolateral part of the nucleus of the solitary tract (NTS) known to receive aortic baroreceptor input via the vagus nerve and respond to BP fluctuation [56, 116, 295]. Like the PVN, the nucleus of the solitary tract exerts a vasodilatory influence on cerebral blood flow [248] and has been shown to project directly to the SSN [2]. Additionally, the NTS projects directly and indirectly via the lateral parabrachial region (PBL) to the PVN [48, 118]. The parabrachial region is known to be part of the circuitry for relaying BP-related information from NTS to PVN, and we found in our PRV studies that it too projects to SSN [137, 157]. These

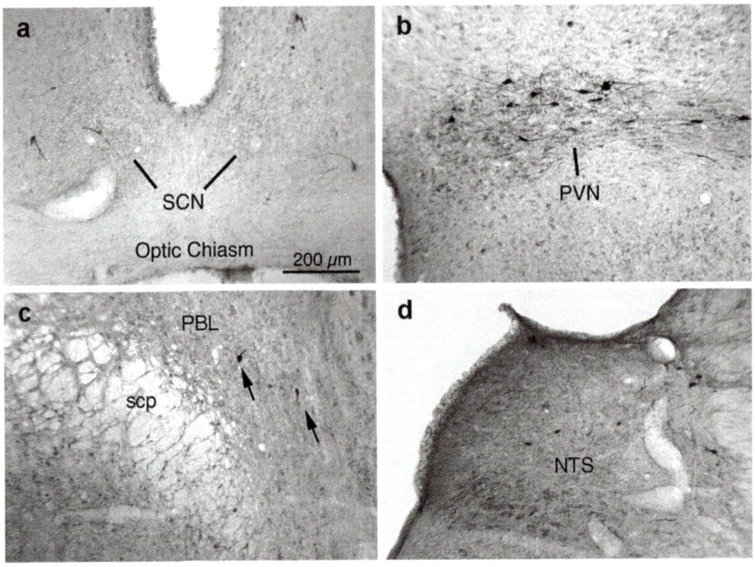

Fig. 12.11 Images of PRV-labeled neurons from a rat that had survived 69 h after an intrachoroidal injection of virus into the right eye, with the labeled neurons detected by peroxidase-antiperoxidase immunolabeling with diaminobenzidine. All images are of coronal sections. Image (**a**) shows a midline view of the suprachiasmatic nucleus (*SCN*) region of the hypothalamus. Note that a few PRV+ neurons that send dendrites toward and into the SCN are present on both sides of the brain. Image (**b**) shows a higher-power view of the right paraventricular nucleus of the hypothalamus (*PVN*). Note the many PRV+ neurons in PVN. Image (**c**) shows PRV+ neurons (*arrows*) in the lateral parabrachial nucleus (*PBL*) lateral to the superior cerebellar peduncle (*scp*), on the right side of the brain. Image (**d**) shows PRV+ neurons in the rostral nucleus of the solitary tract (*NTS*), on the right side of the brain. The magnification is the same in all images

findings in rat suggest that brain regions involved in sensing and controlling systemic BP may influence PPG neurons innervating the choroid.

The phenomenon of blood flow stability over a range of systemic BPs has been well documented for the cerebral vasculature, and cerebral compensation for low BP is thought to maintain cerebral function and prevent ischemic cerebral injury [272]. At the same time, the PVN, NTS, and PBL may drive cerebral blood flow increases in response to low systemic BP, and these sites can drive peripheral vasoconstriction as well [48, 118, 276]. The combination of increased vasodilation in the cerebral vasculature combined with peripheral vasoconstriction would work together to maintain a stable CBF despite BP declines. It may be that the inputs of PVN, NTS, and PBL to choroidal SSN play a similar role in maintaining choroidal blood flow during normal or accidental bouts of low systemic BP. Such circuitry could serve to maintain high ChBF during episodes of decreased systemic BP, such as might occur normally during inactivity, rest, or sleep, or traumatically due to injury and blood loss. Reflexive PPG-mediated choroidal vasodilation could be important for maintaining metabolic support of the retina during the bouts of low BP, as well as for preventing retinal ischemia during extremes of the BP-lowering events [33, 35]. While ChBF stability during BP variation has been reported in humans and rabbits [169, 171, 292, 293], it is unknown if the SSN-PPG circuit mediates ChBF stability with BP declines. The circuitry data we obtained using PRV transneuronal labeling in rats suggest that the PPG input to choroid may be involved in such a function. It may be, therefore, that neurogenic and myogenic mechanisms both contribute to stabilize both choroidal and cerebral blood flow during downward fluctuations in systemic BP, as discussed further in the following section on the function of the SSN-PPG circuit.

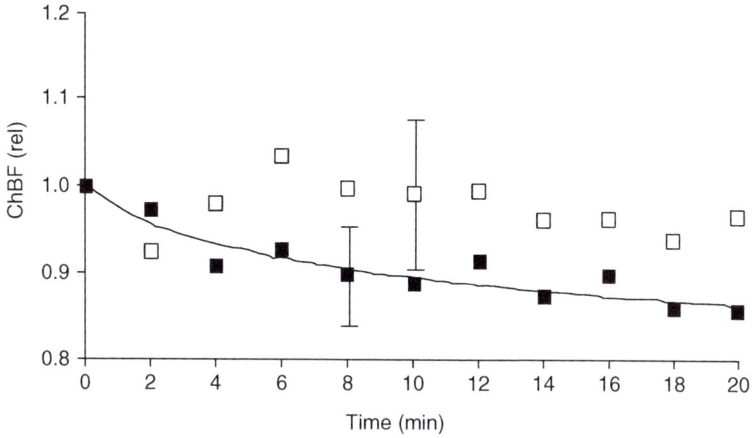

Fig. 12.12 Illustration (Fig. 2) redrawn from Longo et al. (2000), showing the mean response of ChBF for eight humans during 20 min of room light and during 20 min of darkness. Note that ChBF rapidly increased with light onset and remained consistently higher in the light than it was in the dark. Note also that ChBF declined progressively during the 20-min exposure to darkness. A log regression revealed that the decline was significantly linear with time. Typical *error bars* are shown

The SSN innervates two cranial parasympathetic ganglia, the PPG and the submandibular ganglion, with the submandibular preganglionic neurons situated dorsal to those innervating the PPG [357]. As noted above, the PPG innervates the choroid, orbital vessels, the lacrimal gland, and Meibomian glands [200, 357], as well as nonorbital tissues such as the cerebral vasculature and the nasal mucosa [385]. Given that PVN region stimulation or NTS stimulation increases cerebral blood flow [56, 117, 190, 248, 295], and given that PVN and the NTS innervate choroidal neurons of the SSN, the possibility exists that the same populations of PPG neurons and SSN neurons that regulate choroidal blood flow also regulate cerebral blood flow. The fact that the cerebrum and eye are both neural tissues with high metabolic requirements calling for a stable blood supply is supportive of this possibility.

Because of the evidence that retinal illumination can increase ChBF via central circuits in birds, as discussed in more detail in the section on ciliary ganglion innervation of the choroid [88, 90, 315], because of the evidence that light or transition from dark to light increases ChBF in rabbits, monkeys, and humans (Fig. 12.12) [35, 100, 149, 206 269, 270, 364,] and because of the evidence that flicker-modulated regulation of blood flow occurs in the short ciliary arteries [415], visual inputs to choroidal SSN neurons are of interest. Retinal activity-dependent regulation of ChBF, mediated by retinal input to central autonomic circuitry, may serve as a replacement for the direct metabolic coupling evident between retina and the inner retinal circulation, which is not possible for the choroidal circulation due to its position behind the retina. While photoreceptor metabolism is high in the dark and declines in the light [45, 61, 204, 229, 364, 383, 386], flickering light (which may better simulate a changing visual scene than constant illumination) results in considerable increase in inner retinal metabolism with little change in photoreceptor metabolism [35, 45]. Retinal activity-mediated increases in ChBF may be especially important for driving nutrients and oxygen to inner retina under these circumstances, again for species poor in retinal vessels, and for retinal regions poor in retinal vessels in those species otherwise possessing a well-developed retinal vasculature. Our anatomical data with pseudorabies virus transneuronal retrograde labeling suggest the possibility that light-mediated increases in ChBF may occur via the facial parasympathetic circuit in mammals.

For example, we have observed that neurons surrounding the retinorecipient hypothalamic cell group the suprachiasmatic nucleus (SCN) project to SSN, suggesting a direct projection from retinorecipient peri-SCN neurons to SSN choroidal neurons (Fig. 12.11). Visual input to SSN may also come via a visual input to PVN neurons projecting to SSN. The retina and SCN are both known to project to ipsilateral PVN in rodents [140, 409], and it is possible the PVN target neurons include those that project to choroidal neurons of SSN. Finally, the olivary pretectal nucleus, which also receives retinal input, may also be a source of visual input to SSN or to the nucleus of Edinger-Westphal (which projects to ciliary ganglion) [86, 227, 274]. In any event, the finding that ChBF changes in human macula attending dark–light transitions can be blocked by NOS inhibition is consistent with the possible involvement of the PPG input to thet choroid in light-mediated control of ChBF [149]. Note that an SCN input to choroidal neurons of PVN would provide not only a neural substrate for visual control of ChBF but also potentially impose a circadian rhythm on ChBF [244], which would seem plausible given the demonstrated circadian rhythms in disk shedding, IOP, ocular length, choroidal thickness, photoreceptor function, and outer retinal metabolism [38, 41, 187, 204, 250, 267, 408]. Finally, while light-mediated increases in ChBF have been demonstrated in many studies in mammals, light-mediated ChBF increases have not been observed in some prior studies in rabbits and newborn piglets [332, 336], nor were ChBF increases in response to diffuse luminance flicker observed in cats [291].

12.5.3.3 Physiological Studies of Facial Parasympathetic Control of ChBF – Mammals

The influence of facial parasympathetic control of choroidal blood flow in mammals has been studied pharmacologically with drugs that affect or mimic the actions of the peripheral vasodilators used by the PPG (VIP, NO, and acetylcholine) and by stimulation of the preganglionic neurons at the level of the facial nerve or the SSN. The latter studies have also examined the role of VIP, NO, and acetylcholine in mediating the effects of facial nerve or SSN stimulation on ChBF. In early studies, Ruskell showed that severing PPG input to the choroid yielded diminished IOP [297]. While the basis of this effect was uncertain at the time, it is now thought that reduced ChBF caused by loss of basal PPG vasodilatory influences on the choroid is the basis of the lessened IOP. This interpretation is consistent with the slight rise in IOP in monkey caused by facial nerve stimulation [252]. Numerous researchers have subsequently carried out detailed studies on the physiology and pharmacology of PPG input to the choroid.

Intravenous delivery of VIP in rabbits increases ChBF and IOP [251]. Other studies in mammals have shown that NO plays a role in maintaining basal ChBF [185, 69, 211, 218] and that NO mediates the cholinergic neurogenic vasodilation on choroidal vessels [185, 69, 114, 255, 326]. For example, Deussen et al. [69] showed that LNAME increases mean arterial pressure by 20% and yet decreases ChBF by 40–50% in dogs. Similarly, NO inhibition substantially decreases basal ChBF in humans [211, 212], rabbits [313], piglets [156], and rats [166, 185, 259]. Two LDF studies in cats showed that intravenous acetylcholine increases ChBF despite its systemic hypotensive effect [218, 291]. Mann et al. [218] showed that the ChBF increase with intravenous acetylcholine in cats was attenuated 50% by NOS inhibition with LNA despite a 40% increase in blood pressure. [256] showed in cats that nonselective NOS inhibition or atropine blockade of muscarinic cholinergic control of the choroid both reduce ChBF. These results and the evidence that NOS inhibition with LNAME yields endothelium-dependent vasoconstriction of pig ophthalmic arteries [17, 235, 387] suggest that at least part of the effect of NO on basal choroidal tone might be mediated by endothelial release of NO [401]. Neurogenic NO-mediated relaxation of the intraocular part of the bovine long posterior ciliary arteries was, however, found to be driven by neurally derived NO dependent, but not be endothelial NO dependent [392]. These results suggest a role in uveal blood flow control by PPG cholinergic elicitation of endothelial NO release, as well as by direct release of NO from PPG terminals on choroidal vessels. The ophthalmic artery and its short and long ciliary branches supplying the choroid are also innervated by VIP+, nNOS+, and cholinergic (cholinesterase-containing) nerves of PPG origin, and they show NO-mediated relaxation in response to transmural activation of these nerves or to cholinergic stimulation [17, 132, 133, 258, 366, 369, 401, 410]. Thus, PPG cholinergic stimulation of endothelial NO release, as well as direct neural release of NO and VIP, acting at the level of the vessels supplying the choroid also may affect ChBF.

Activation of preganglionic input to the PPG by facial nerve stimulation or by stimulation of the preganglionic neurons within SSN increases ChBF in mammals. For example, facial nerve stimulation in rabbits increases ChBF and also yields a slight IOP rise [339]. A later study reported that facial nerve activation in cats, rabbits, or monkeys yields a choroidal blood flow increase that is blockable with ganglionic blockade with hexamethonium, and a slight IOP rise was seen due to the ChBF increase in monkeys [252]. Nilsson [255] showed that 2-Hz facial nerve stimulation in rabbits increases ChBF, with the increase reduced 50% by NOS inhibition, while the ChBF increase to 5-Hz stimulation was minimal. A role of some vasodilator system in addition to NO (presumably) was implied by the finding that the ChBF increase at 5-Hz facial nerve stimulation was only minimally suppressed by NOS inhibition. In a subsequent study addressing this finding in cats, Nilsson (2000) reported that the ChBF increases with 5-Hz facial nerve stimulation could be completely blocked with systemic NOS inhibition with LNA, but the ChBF increases with 10-Hz stimulation were reduced but not blocked by LNA alone or atropine alone, and more greatly reduced by LNA and atropine together. The results in rabbits and cats suggest that the different vasoactive substances released by PPG terminals play greater roles at different nerve firing frequencies – more so muscarinic mechanisms driving endothelially derived NO, and perhaps neurally derived NO as well, at low frequencies, and more so VIP, cholinergic release of endothelially derived NO, and perhaps neurally derived NO at higher frequencies. Similarly, studies on submandibular ganglion and otic ganglion control of blood flow and salivary secretion in the submandibular and parotid glands, respectively, suggest that low-frequency firing of parasympathetic postganglionic fibers may preferentially yield vasodilation and salivation via acetylcholine release and a muscarinic cholinergic action that stimulates endothelial NO release, while high-frequency firing of parasympathetic postganglionic fibers may preferentially yield vasodilation and salivation by VIP release [81, 213, 219].

Activation of the SSN neurons projecting to those PPG neurons innervating choroid also yields vortex vein and choroidal vessel vasodilation in rats [326], as measured by transcleral LDF, yielding 200% increases above baseline (Fig. 12.13). Since the retinal vasculature is meager at these LDF sites [417], the SSN-elicited blood flow increases in rats must be attributable to ChBF increases (Koss 1988) [326]. Intravenous administration of the general NOS inhibitor LNAME or the nNOS selective inhibitor TRIM significantly decreased basal ChBF and vortex vein blood flow in rats and attenuated the ChBF increases yielded by SSN stimulation. These results employing the proven selective nNOS inhibitor TRIM indicate that activation of preganglionic input to PPG increases ChBF and this effect may involve NO release from intrachoroidal (as well as possibly from orbital) PPG fibers. The ciliary arteries show acetylcholine-mediated vasodilation in monkeys, suggesting a role for cholinergic endothelial NO-mediated vasodilation in control of ChBF as well [370]. One previous study, however, reported that atropine did not block SSN-elicited choroidal vasodilation in rat [326], but it is uncertain if this was the case because of the stimulation frequencies examined or because of the atropine dose used. In chronically sympathectomized rats, cholinergic PPG mechanisms are revealed, as M3 muscarinic receptor antagonists and atropine block ChBF increases with SSN stimulation [327]. This result is consistent with the finding that PPG neurons

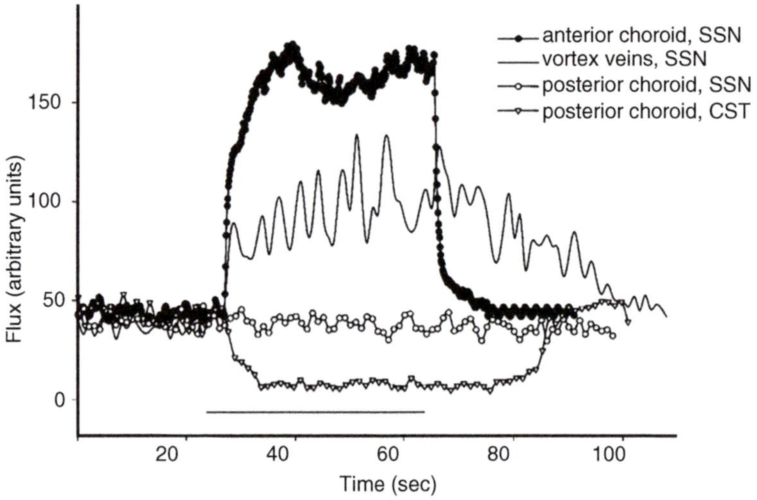

Fig. 12.13 Figure 1 redrawn from Steinle et al. [322] showing a laser Doppler blood flow record from the anterior choroid, posterior choroid, and vortex veins of rats during stimulation of either the SSN or the cervical sympathetic trunk (*CST*). The *bar* indicates the duration of the SSN (20 Hz) or the CST (12-Hz stimulation). Note that the CST stimulation decreased ChBF in posterior choroid, while SSN stimulation increased flow in the anterior choroid and vortex veins

also appear to be cholinergic [63, 354] and muscarinic agonists yield choroidal vasodilation [114, 218].

12.5.3.4 Choroidal Autoregulation and the PPG Input to Choroid – Mammals

The issue of choroidal autoregulation (i.e., compensation for ocular perfusion pressure changes so as to maintain ChBF near basal levels) has been somewhat controversial, since earlier studies reported that ChBF decreased linearly with reductions in choroidal perfusion pressure caused by acute hemorrhage or increased IOP [4, 10, 33, 37]. These observations had, in fact, led to a dogma that while cerebral blood flow and retinal blood flow do autoregulate to maintain stale flow despite reduced ocular perfusion pressure, choroidal blood flow does not. It became increasingly evident from subsequent studies, however, that ChBF does compensate for perfusion pressure declines. For example, some autoregulation with IOP elevations was noted in cats [97, 390], rabbits [54], and humans [292]. Detailed studies in rabbits have shown that when ocular perfusion pressure is experimentally reduced by lowering BP rather than by raising IOP, stable ChBF over a blood pressure (BP) range of 40–50 below basal BP is observed [169]. The compensation was hypothesized to stem from myogenic mechanisms [171]. In a later study, Kiel [173] noted that both NOS inhibition and ganglionic blockade diminished the ChBF compensation to systemic hypotension in rabbit, implying some involvement of neurogenic vasodilatory mechanisms in the compensation. We refer to the blood flow compensation for reduced systemic BP as baroregulation. By contrast, Jacot et al. [156] showed in piglet that ChBF compensation to perfusion pressure changes caused by IOP manipulation do not involve NO release, and thus involve different mechanisms that are involved in choroidal baroregulation. It is now evident than it is unlikely that ChBF would not show compensation for reduced ocular perfusion pressure, given the potentially adverse consequences of either supranormal ChBF or subnormal ChBF on retinal health and function. Without autoregulation, high BP would yield an ocular perfusion pressure resulting in excessively high ChBF, causing fluid accumulation in retina and deficient exchange of wastes and nutrients between retina and choroid [35, 170]. Similarly, without autoregulation, low BP would yield an ocular perfusion pressure resulting in low ChBF, causing retinal hypoxia and impaired retinal function [325, 399, 400].

Given the input of hypothalamic and solitary nucleus blood pressure-sensitive sites to the choroidal neurons of the SSN [65, 147, 154, 323], at least part of the choroidal compensation to BP declines may be mediated by the SSN-PPG circuit. Consistent with this possibility, prior studies have suggested that at least part of the compensation of cerebral blood flow for declines in BP may be mediated by the PPG [121, 245]. Moreover, systemic hypotension does not activate sympathetic input to the choroid, while it does cause peripheral vasoconstriction [34]. Thus, the eye (like the brain) is a privileged tissue during systemic hypotension. Nonetheless, Linder [203] found that facial nerve stimulation increased choroidal blood flow in hypotensive rabbit, but facial nerve section did not reduce choroidal blood flow in hypotensive or normotensive rabbit, suggesting little contribution of the facial nerve system to hypotensive or normotensive tone in anesthetized rabbit. These results argue against the notion that the SSN-PPG circuit to the choroid participates in ChBF compensation for low systemic blood pressure, at least in rabbits. Given the anatomical evidence for BP-sensitive inputs to the SSN, however, further studies are needed to assess the contribution of the SSN-PPG circuit in mammals to ChBF baroregulation.

Note that the choroidal circulation does not appear to regulate (decrease) to high oxygen levels but does regulate (increase) in response to high CO_2. For example, ChBF is unaltered in response to breathing 100% oxygen in humans [112, 167, 291] but is increased by breathing carbogen (95% O_2 – 5% CO_2) [112]. Similarly, hypercarbia increases ChBF in newborn piglets, cats, sheep, and baboons [5, 238, 332, 395]. High CO_2 is also known to increase cerebral blood flow as well [395]. The mechanism of the increased blood flow with hypercapnia is uncertain. Schmetterer et al. [305] reported that NO is involved in hypercapnia-mediated increases in blood flow in the human ophthalmic artery, raising the possibility that the same is true for the choroid. Cyclooxygenase products do not appear to mediate the hypercapnic ChBF increase in newborn piglets [331]. Note that some newborn mammals such as piglets [333] but not sheep [243] show a ChBF decrease to breathing 100% O_2 – the increase in piglets also does not appear to be mediated by cyclooxygenase products. Whether vasodilatory PPG input plays a role in hypercapnic ChBF increases is unknown.

12.5.3.5 Peripheral Anatomy of Facial Circuitry for Control of ChBF – Birds

The pigeon PPG consists of an interconnected series of three to four microganglia of about 50–200 neurons each and numerous lesser microganglia (Figs. 12.14 and 12.15) [63]. The main microganglia of the PPG network in pigeons lie along the superior aspect of the Harderian gland. Neurons of all of these microganglia are extremely rich in VIP and nNOS, and moderate in ChAT (and thus make acetylcholine), and the majority co-contain VIP and nNOS (Fig. 12.16). In pigeons and chickens, the PPG has been shown to innervate choroidal vessels, as well as orbital vessels supplying the choroid [63, 85]. Axons containing VIP and nNOS extend from the PPG network to perivascular fiber plexi on orbital blood vessels [63]. These orbital vessels, many of which enter the choroid posteriorly and nasally, are a conduit by which PPG postganglionic fibers reach the choroid (Fig. 12.15). Within the choroid, VIP+ and nNOS+ fibers are widely scattered but sparse, and most abundant in nasal choroid. These results suggest that PPG neurons in birds use VIP and NO, and also possibly acetylcholine, to exert vasodilatory control over blood flow to and within the avian choroid. A few VIP+ and nNOS+ neurons were also observed in the choroid. In some avian groups, such as ducks, many more intrinsic choroidal neurons co-containing VIP and nNOS have been reported [26, 306, 307, 309], as described here in more detail in a later section.

12.5.3.6 Central Anatomy of Facial Circuitry for Control of ChBF – Birds

Several studies have suggested that the preganglionic neurons innervating the PPG in birds reside in the superior salivatory nucleus in a similar brainstem location as in mammals [108, 228, 310]. Schroedl et al. [310] recently carried out a detailed anatomical study on the localization of the avian SSN (Fig. 12.17). ChAT+ neurons in brainstem were retrogradely labeled via the radix autonomica of the facial nerve, which conveys preganglionic axons from the SSN to the PPG. The SSN neurons were located dorsolateral to somatic facial

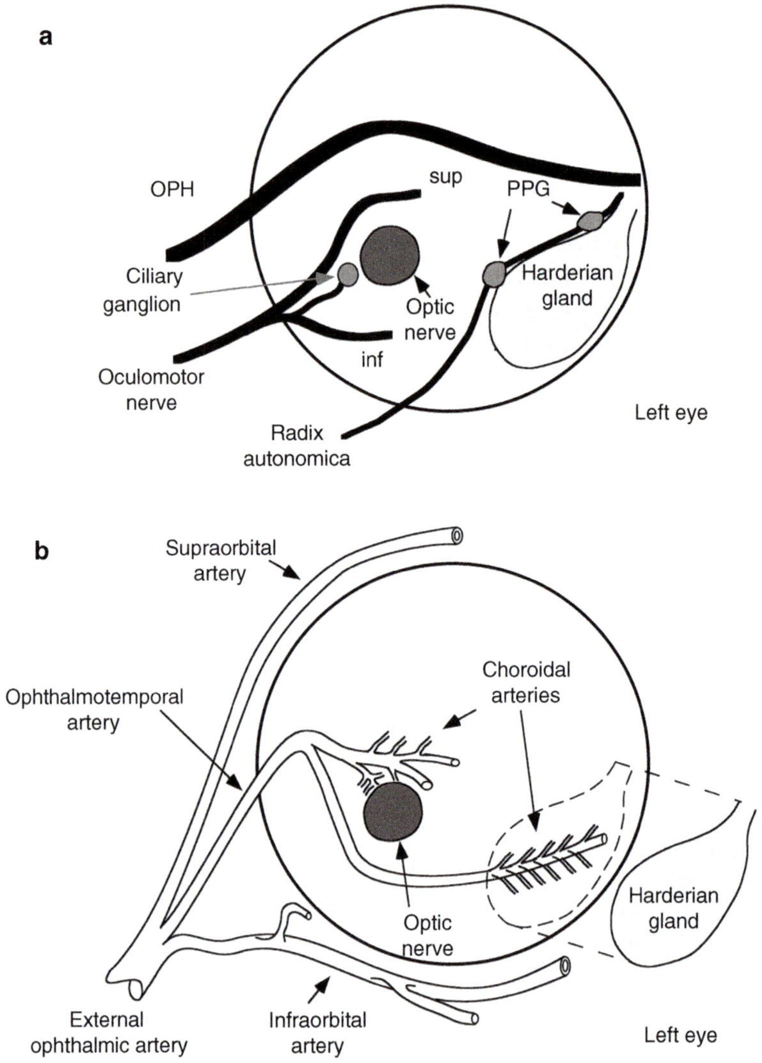

Fig. 12.14 Schematic illustrations of the major ocular nerves (**a**) and vessels (**b**) and their relationship to the Harderian gland in birds, both as viewed from the posterior aspect of the left eye. Schematic (**a**) shows the course and relative locations of several major orbital nerves, as well as the locations of the ciliary ganglion (*CG*) and a simplified version of the PPG system of microganglia. A more detailed version of the PPG is shown in Fig. 12.15. Schematic (**b**) illustrates the origin of the ophthalmotemporal artery from the external ophthalmic artery (which is itself a branch of the internal carotid) and its orbital course along the left eye. Note the course of the ophthalmotemporal artery along the temporal, posterior, and nasal poles of the eye, and note that it gives rise to choroidal arteries throughout its course. It also gives rise to additional muscular and glandular branches. The ophthalmotemporal artery is accompanied by a vein of the same name whose major branches are somewhat different from those of the artery. Superior is to the top and nasal to the right in both schematics. *inf* inferior branch of oculomotor nerve, *OPH* ophthalmic nerve, *sup* superior branch of oculomotor nerve

motoneurons, as they are in mammals. As in mammals, the SSN region receives input from the nucleus of the solitary tract [14], the parabrachial region [394], and the SSN [180]. As in mammals, the parabrachial region receives input from the nucleus of the solitary tract, which receives baroreceptive input [28, 164]. Thus, as in mammals, the avian SSN-PPG circuit may be responsive to

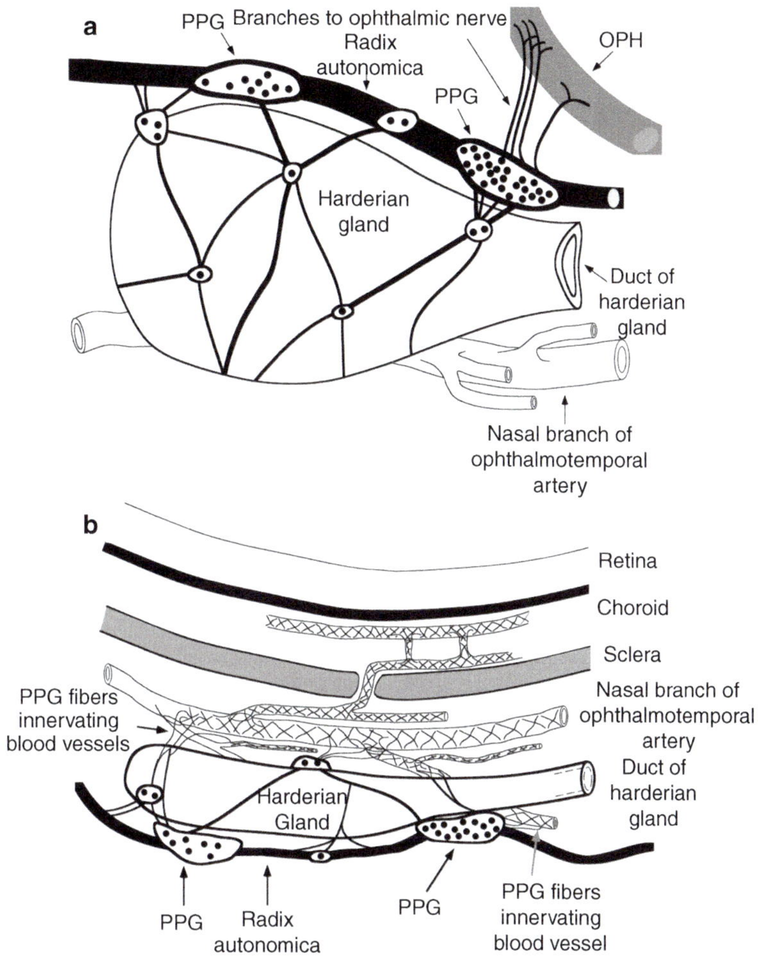

Fig. 12.15 Image (**a**) provides a schematic view of the left Harderian gland and associated PPG plexus, as seen from the nasal side (i.e., the side facing the orbit). The ophthalmic nerve (*OPH*) is shown as coursing superior to the gland and receiving fibers from the PPG plexus. The nasal branch of the ophthalmotemporal artery is shown behind the gland. The two major PPG microganglia are located along the superior aspect of the Harderian gland and are indicated by *arrows*. The more rostral of these two is typically referred to as "the PPG" in many previous published works. NADPH-diaphorase neurons within the various microganglia are shown as *solid circles*. Image (**b**) provides a superior view of the Harderian gland and the ophthalmotemporal artery between it and the eye. Nerve fibers on the artery and its branches to the choroid are illustrated

baroreceptor information and thus regulate ChBF as a function of systemic blood pressure.

12.5.3.7 Physiological Studies of Facial Parasympathetic Control of ChBF – Bird

In unpublished studies, we used transcleral LDF to measure ChBF in pigeons while systematically electrically stimulating brainstem in the vicinity of the facial motor nucleus, focusing on the region of small cholinergic neurons between the two motoneuron pools comprising the SSN in chickens that had been shown to project to the PPG [108, 310]. We found that the region of the SSN of birds was effective for eliciting ChBF increases (100% or more) without significant concomitant systemic BP increases. The NOS inhibitor 7-nitroindazole (7NI) greatly attenuated (about 50%) the ChBF increases that could be elicited from this region, consistent with an involvement

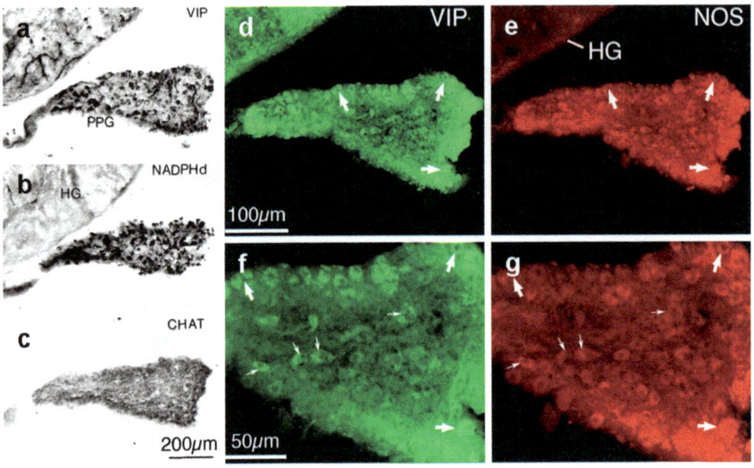

Fig. 12.16 Photomicrographs of immunolabeling in the main ganglion making up the avian PPG network. Images (**a**–**c**) show three adjacent sections of the main PPG, labeled for VIP (**a**), NADPHd (**b**), and ChAT (**c**). The ganglion is rich in VIP+ and NADPHd+ perikarya, but poorer in ChAT perikarya. The ChAT perikarya are embedded within the ChAT+ neuropil of this PPG microganglion. The axon bundle shown to the extreme left in each photomicrograph contains VIP+, NADPHd+, and ChAT+ axons. The Harderian gland (*HG*) is shown to the *upper left* in all three photomicrographs. Images (**d**–**g**) show single fields of view of the main PPG microganglion in frontal sections, double-labeled by immunofluorescence for VIP and nNOS. As can be seen in image pair (**d** and **e**), and a higher-power view of part of the same field (**f** and **g**), numerous individual perikarya in the gland are labeled for VIP and nNOS. The *large arrows* in (**d** and **e**) indicate three such double-labeled neurons, and these same neurons are indicated in (**f** and **g**) by *large arrows*. In addition, a number of other neurons labeled for both VIP and nNOS are indicated in (**f** and **g**) by *small arrows*. Magnification the same in (**a**–**c**). Magnification the same in (**d** and **e**). Magnification the same in (**f** and **g**)

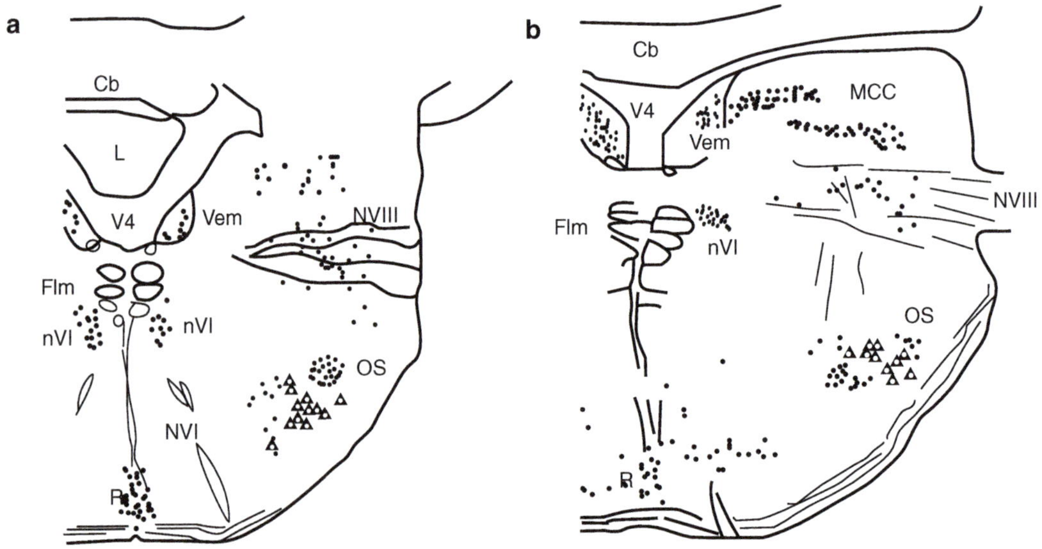

Fig. 12.17 Schematic images from Fig. 9 of Schrodl et al. [310]. The schematics show a mapping of the right side of the brainstem in a rostral (**a**) to caudal (**b**) pair of transverse sections. Preganglionic parasympathetic neurons of the SSN, as identified by *retrograde* tracing, are indicated by *open triangles*. *Black dots* represent motoneurons of somatic facial motor nucleus nerve VII. *Cb* cerebellum, *Flm* fasciculus longitudinalis medialis, *L* lingula, *MCC* nucleus magnocellularis cochlearis, *nVI* nucleus abducens, *NVI* nervus abducens, *NVIII* nervus vestibulocochlearis, *OS* nucleus olivaris, *V4* fourth ventricle, *Vem* nucleus vestibularis medialis, *R* raphe nucleus

of PPG NOS+ neurons in the elicited increases. Note that 7NI has widely been referred to as a selective nNOS inhibitor. This conclusion was at odds with the published evidence that 7NI is a potent inhibitor of both nNOS and eNOS *in vitro*, but its isoform selectivity in vivo was hypothesized to be to due to uptake by neurons but not endothelial cells [243]. The evidence for *in vivo* selectivity was that 7NI reportedly did not produce hypertension. This claim was, however, consistently based on a meager number of animals in the studies claiming selectivity, with the result being that a small hypertensive effect was ignored as insignificant. In our own studies, we showed that 7NI does induce a clear pressor effect in rats when the group size is adequate for statistical power, that this pressor effect is peripherally mediated, and that 7NI does inhibit endothelium-dependent cholinergic vasodilation, all suggesting that the selectivity of 7NI for nNOS over eNOS was overstated [414]. Thus, 7NI is not nNOS selective, as others have now also come to recognize [3, 15].

12.5.3.8 Choroidal Autoregulation and the PPG Input to Choroid – Birds

As part of our interest in the significance of the facial parasympathetic control of ChBF, we investigated if ChBF in pigeons showed baroregulation (i.e., compensation for perfusion pressure changes caused by BP changes so as to maintain ChBF near basal levels). In one line of study [287], we determined whether pigeons can compensate for an acute decrease in arterial BP and maintain stable ChBF (Fig. 12.18). ChBF was measured using transcleral LDF in anesthetized pigeons, and a stable decrease in arterial BP was produced by blood withdrawal from the brachial artery. The ChBF response to the acute BP drop was assessed by calculating the gain factor (Gf). A Gf of zero means that the ChBF change is completely proportional to the arterial BP change, without compensation (i.e., without baroregulation). A Gf of 1 indicates complete stability of the ChBF despite the change in BP (i.e., perfect baroregulation). During the withdrawal itself, BP decreased rapidly, as did ChBF. The ChBF decline, however, was typically not as great as the arterial BP decline and showed recovery once arterial BP had stabilized at its lower level. For postwithdrawal BP above 40 mmHg, the Gf by 1 min after blood withdrawal was 0.4–0.5, indicating that the decline in ChBF was proportionally less than the decline in arterial BP. When the arterial BP declined to a level at or below 40 mmHg, the Gf was about 0, suggesting that the ChBF could not compensate when arterial BP was this low. While these results were consistent with the dogma that ChBF declines when BP declines, they nonetheless confirmed that ChBF does significantly compensate for BP declines. In a second line of study [287], we determined if ChBF baroregulation occurred during spontaneous BP fluctuation (Fig. 12.18). We found that ChBF compensated well for arterial BP declines to the 50–80 mmHg range and for increases above 130 mmHg (with the Gf approaching 1), but poorly just above and below basal ABP. Additionally, compensation failed below a BP of about 40–45 mmHg. Our studies therefore show that significant ChBF baroregulation does occur in pigeons when the BP deviates +30 or −15 from basal BP. While baroregulation is poor with BP deflections just above and below baseline, these small BP changes exert only a small effect on ChBF, and thus the ChBF stays near baseline notwithstanding the poor baroregulation within this range (Fig. 12.18). Our finding that NOS inhibition with LNAME eliminates ChBF baroregulation with blood withdrawal-produced systemic hypotension suggests a possible role of NO-mediated neurogenic mechanisms in the process [287]. The observation that a specific threshold BP change must occur for significant ChBF baroregulation to be observed is consistent with a neural mechanism, as has been reported to play a role in cerebral blood flow baroregulation [121, 249]. ChBF baroregulation would prevent underperfusion-related ischemia or overperfusion-related edema and impaired fluid-tissue exchange in the outer retina [35, 162, 170], and thus seemingly be an important ocular homeostatic mechanism. For the cerebral vasculature, in which baroregulation is also evident, both an intrinsic vascular smooth muscle myogenic mechanism (which acts to maintain vessel wall stretch within

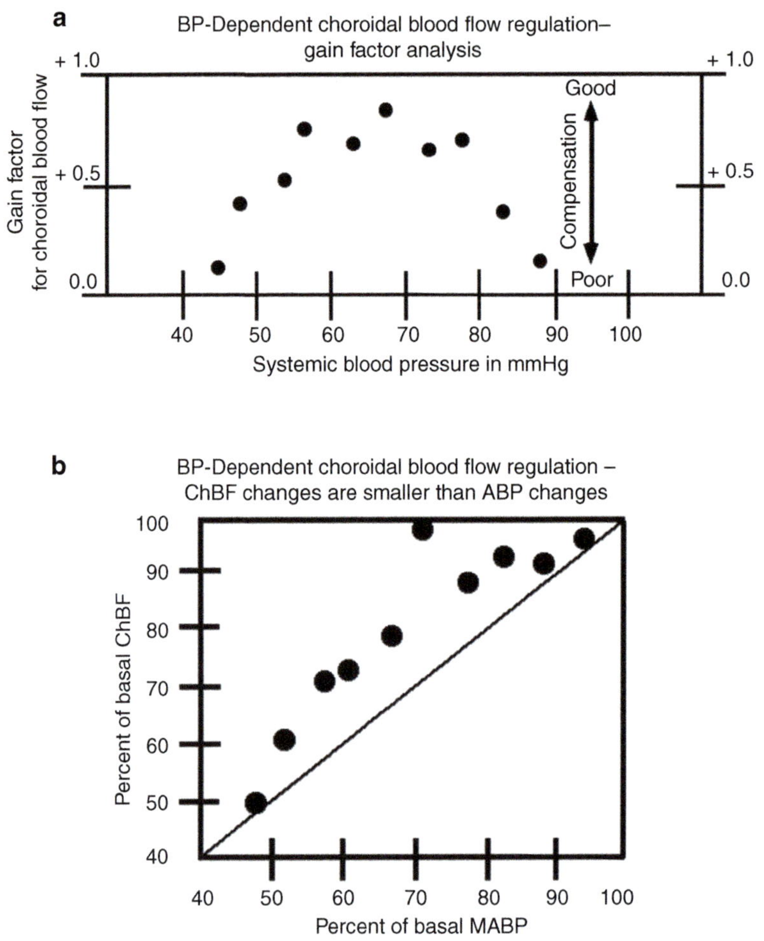

Fig. 12.18 In graph (**a**), the ChBF response to BP fluctuation was assessed by calculating the gain factor (Gf) as follows, Gf=1, (dChBF/bChBF)/(dMABP/bMABP), where dChBF and dMABP are the change in basal (b) ChBF and mean (M) arterial BP after blood withdrawal. A Gf of zero means that the ChBF change is completely proportional to the arterial BP (ABP) change, without compensation (i.e., without baroregulation). A Gf of 1 indicates complete stability of the ChBF despite the change in BP (i.e., perfect baroregulation). For the 25 birds analyzed in a study of the behavior of ChBF during spontaneous fluctuation in ABP, the mean Gf for ChBF (±SEM) for each 5 mmHg step in ABP was determined and graphed as a function of the corresponding ABP. As can be seen, the Gf values for ChBF were 0.50–0.75 (good compensation) over the 50–80 mmHg range and were consistently positive over nearly the entire ABP range examined. Below a BP of 45 mmHg compensation, however, appeared to fail. Graph (**b**) shows the results for the same 25 birds, with mean ABP for each 5 mmHg step in pressure expressed as a percent of basal ABP. The mean ChBF at each 5 mmHg step was also expressed as a percent of basal ChBF, and each ChBF value (±SEM) was then graphed as a function of the corresponding ABP. The *diagonal line* indicates the expected ChBF value if ChBF passively changed with declining ABP. As can be seen, ChBF values were consistently higher than would be predicted if ChBF passively followed ABP. ChBF remained near or above 90% of basal over an ABP range of 70–100% of basal ABP. Even from 50% to 70% of basal ABP, ChBF showed prominent compensation

a preferred range) and a PPG-mediated neurogenic mechanism have been proposed to contribute to baroregulation to low systemic BP [121, 153, 272]. In addition to a choroidal neurogenic mechanism, a myogenic regulation has also been proposed to play a role in choroidal baroregulation [169, 170]. Note that ocular perfusion pressure can be modulated by head position due to hydrostatic forces, without an alteration of systemic blood pressure. Orthostatic modulation of perfusion pressure requires a myogenic mechanism for autoregulation, since baroreceptors do not detect fluctuations in choroidal perfusion pressure caused by such a process.

12.5.4 Oculomotor Nucleus Parasympathetic Input

12.5.4.1 Ciliary Ganglion Circuitry – Mammals

As will be discussed below, birds also possess a parasympathetic input to the choroid via the ciliary ganglion, which receives its preganglionic input from the nucleus of Edinger-Westphal (EW) of the oculomotor nuclear complex (Figs. 12.19 and 12.20) [284]. While avian choroid receives both ciliary ganglion and PPG input, the ciliary ganglion input is the more prominent parasympathetic input. In mammals, the PPG gives rise to a massive choroidal innervation, but it has remained uncertain if the ciliary ganglion also innervates the choroid. Various lines of evidence suggest it does, but none of the evidence is unambiguous. This evidence is reviewed here.

The ciliary ganglion innervates the muscles of accommodation and the sphincter muscle of the iris. The ciliary ganglion postganglionic nerves to these structures travel via the short ciliary nerves, which penetrate the posterior globe, and course to the front of the eye, to reach these muscle groups. In birds, two distinct neuron types are present in the ciliary ganglion, a smaller one innervating choroid called choroidal neurons and a larger one innervating the muscles of accommodation and the sphincter muscle of the iris called ciliary neurons (Fig. 12.19) [232, 284, 285]. In mammals, no such distinctly different neuron types have been reported in the ciliary ganglion, and no distinct subdivision of EW not subserving pupil constriction or accommodation and thus potentially subserving choroidal blood flow regulation has been identified [107, 225]. May [221], however, noted numerous historical references in the literature to branching of the short ciliary nerves from the ciliary ganglion to choroidal blood vessels. The ciliary ganglion neurons are cholinergic, and Imai [152] identified cholinergic nerves by acetylcholinesterase staining that enter the choroid via the short ciliary nerves and ramify on choroidal blood vessels in albino rats and rabbits. Imai noted a similar result in pigeons, and in pigmented rabbits and rats, in which the nerves were more difficult to discern due to the choroidal pigmentation. Imai [152] also noted cholinergic nerves in monkey choroid. Using biochemical methods, Mindel and Mittag [239] confirmed the presence of ChAT in the choroid of rabbits, cows, and humans, consistent with the presence of cholinergic nerves in choroid. In all of these instances of short ciliary nerve branching to choroidal blood vessels, however, an interpretation difficulty is posed by the fact that some PPG postganglionic branches join and travel with the short ciliary nerves, and others travel through the ciliary ganglion to emerge in the short ciliary nerves (Fig. 12.4) [194, 202, 297, 357]. Thus, the cholinergic fibers in the short ciliary nerves described in some prior studies might be of PPG origin.

Additional evidence for ciliary ganglion control of choroid in mammals involves the evidence for an input from SCN to EW in mammals [274]. This input resembles that from SCN to EW in birds (see below) and could thus be to a part of an EW circuit mediating control of choroidal blood flow. If the facial parasympathetic input to choroid mainly mediates systemic BP-dependent control of ChBF, an additional circuit would in principle be needed for retinal activity-dependent control of ChBF. The SCN input to a putative ChBF-control part of EW would provide such a mechanism, and as discussed below, there is evidence for both EW and ciliary ganglion involvement in choroidal vasodilation in mammals.

12.5.4.2 Function of the EW-Ciliary Ganglion Circuit – Mammals

As noted above in the section on the PPG, there is evidence that ChBF can be increased by muscarinic cholinergic activation. Additionally, Alm et al. [7] showed that preventing acetylcholine breakdown by corneal administration of the

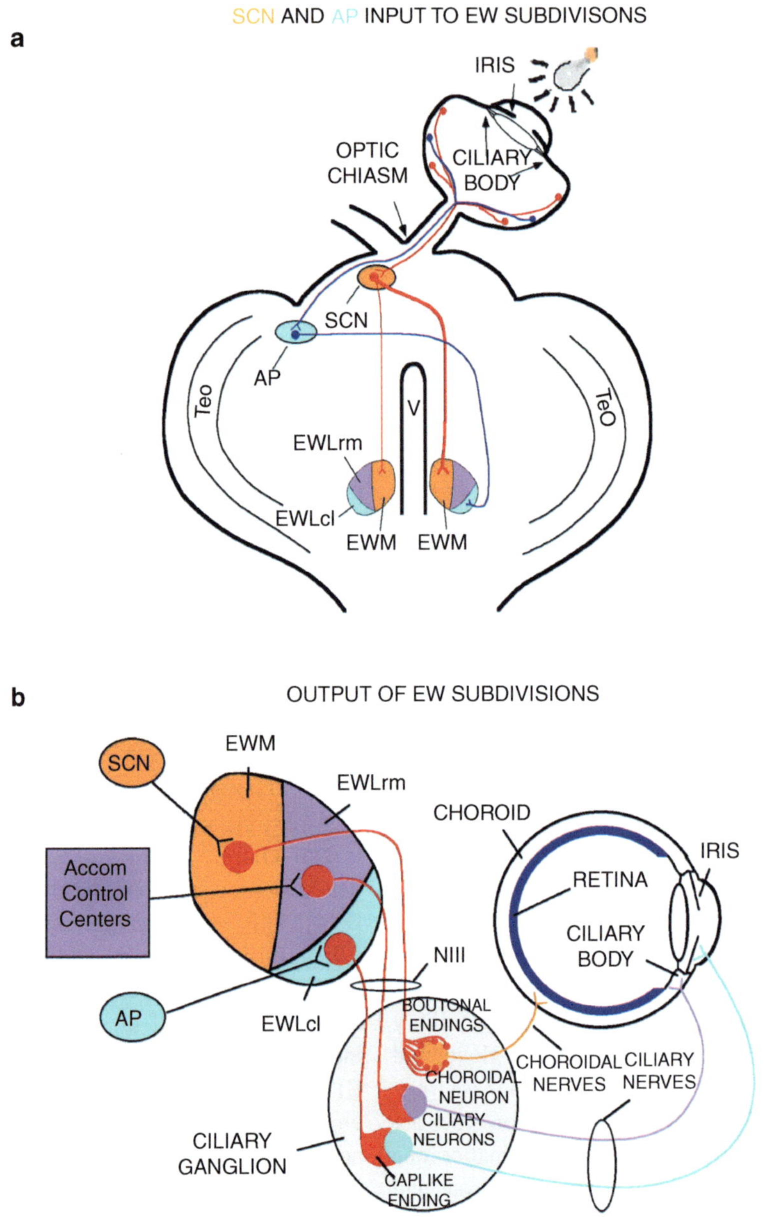
a
SCN AND AP INPUT TO EW SUBDIVISONS
IRIS
OPTIC CHIASM
CILIARY BODY
SCN
AP
Teo
V
TeO
EWLrm
EWLcl
EWM
EWM
b
OUTPUT OF EW SUBDIVISIONS
SCN
EWM
EWLrm
Accom Control Centers
AP
EWLcl
NIII
CHOROID
RETINA
IRIS
CILIARY BODY
BOUTONAL ENDINGS
CHOROIDAL NEURON
CILIARY NEURONS
CAPLIKE ENDING
CILIARY GANGLION
CHOROIDAL NERVES
CILIARY NERVES

cholinesterase inhibitor neostigmine or ocular muscarinic activation by corneal administration of the agonist pilocarpine causes a slight increase in ChBF, despite the distance from the choroid. These various cholinergic effects might be mimicking the actions of vessels postsynaptic to cholinergic ciliary ganglion input. Consistent with this possibility, [335] showed that intracranial stimulation of the oculomotor nerve in rabbits increased blood flow in the choroid. They were surprised by this effect and speculated that perhaps their stimulation had inadvertently activated the nearby trigeminal nerve. Bill et al. [31] showed that ganglionic blockade with hexamethonium or peripheral muscarinic blockade with biperiden slightly reduced the vasodilatory effect of intracranial stimulation of the oculomotor nerve in rabbits on the choroid. Hexamethonium also reduced basal ChBF. In a later study, Stjernschantz and Bill [337] reported that intracranial stimulation of the oculomotor nerve now failed to yield a ChBF increase in rabbits, cats, or monkeys. It was uncertain why the ChBF increase in rabbits was no longer evident. Similarly, Bill [29] stimulated the ciliary ganglion input to the choroid in rabbits and cats and collected ocular blood efflux to measure flow. He reported no increase in ChBF with ciliary ganglion activation, but method sensitivity might be an issue. Gherezghiher et al. [113] reported that oculomotor nerve stimulation in cats increased IOP. This effect could be blocked by hexamethonium, and they suggested the IOP increase might have been due to an increase in ChBF. Consistent with these results, ciliary ganglion removal causes an IOP drop [61, 85] in cats and monkeys. Nakanome et al. [250] found that upon short ciliary nerve stimulation in cats at a point just before the eye and distal to the ciliary ganglion, capsaicin-sensitive increases in ChBF were obtained. The capsaicin sensitivity indicates that the ChBF increase was largely mediated by trigeminal sensory fibers that had joined the short ciliary nerves just prior to scleral penetration. By contrast, stimulation of the short ciliary nerves just after their emergence from the ciliary ganglion yielded a prominent capsaicin-insensitive vasodilation in the choroid that was parasympathetic in origin (sensory fibers being absent at this point). While this is again consistent with a ciliary ganglion innervation of the choroid, the possibility remains that it was PPG fibers traveling with the short ciliary nerves that were responsible for the evoked choroidal vasodilation (see Fig. 12.4).

12.5.4.3 Ciliary Ganglion Circuitry – Birds

In birds, we have identified the major central components of the circuit regulating ChBF via the ciliary ganglion [62, 64, 104, 284, 285]. The central components of this circuit are (arbitrarily beginning with the right eye): the right retina – the left suprachiasmatic nucleus (SCN) – the right (and to a lesser extent the left) medial subdivision of the nucleus of Edinger-Westphal

Fig. 12.19 Schematized horizontal views of the midbrain and the eye (**a**) and of EW, the ciliary ganglion and the eye (**b**) showing the circuitry in pigeon for the bisynaptic retinal pathways to the nucleus of Edinger-Westphal (EW) that drive ChBF increases and pupil constriction. The pathway shown in (**a**) with *red lines* depicts the crossed projection from retinal ganglion cells to the suprachiasmatic nucleus (*SCN*) that, in turn, has a bilateral (but greater contralateral than ipsilateral) projection to medial EW (*EWM*), which controls ChBF via its ipsilateral projection to choroidal neurons of the ciliary ganglion, as depicted in (**b**). The pathway depicted with *blue lines* in (**a**) shows a crossed projection from retinal ganglion cells to area pretectalis (*AP*), which then projects to the contralateral caudolateral part of lateral EW (*EWLcl*), which controls the pupillary light reflex (PLR) via an ipsilateral projection to pupilloconstrictive neurons of the ciliary ganglion, as depicted in (**b**). The lower schematic (**b**) details the peripheral circuitry controlling ChBF and PLR, with EW, the ciliary ganglion (CG), and the eye, all in horizontal view. The EW projects ipsilaterally via the oculomotor nerve to the CG, where EWM input terminates with boutonal endings on choroidal neurons that project to choroidal blood vessels. Projections from both the rostromedial part of lateral EW (*EWLrm*) and from EWLcl terminate with cap-like endings on ciliary neurons that project to the ciliary body and the iris, and control accommodation and the PLR, respectively. The subdivisions of EW are color-coded in (**a**) and (**b**), and the projections of each to the eye via the ciliary ganglion in (**b**) are as well. *EWL* lateral subdivision of the nucleus of Edinger-Westphal, *LRF* lateral reticular formation, *TeO* optic tectum

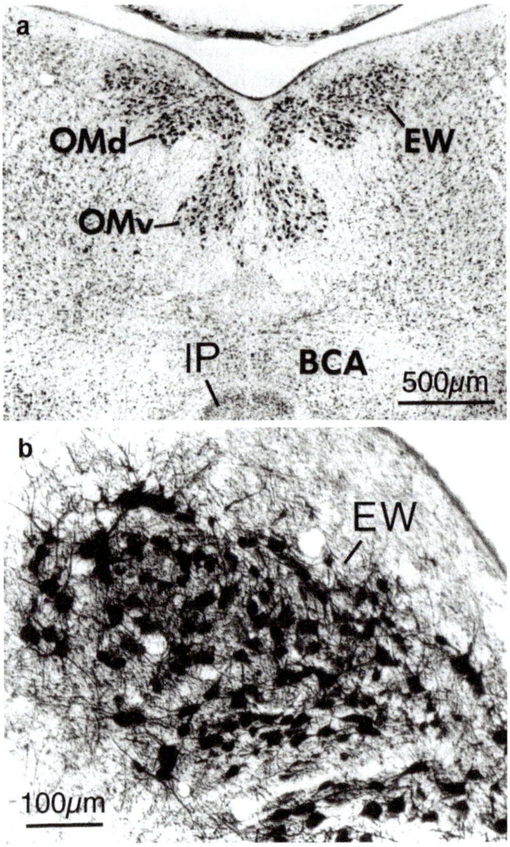

Fig. 12.20 Low-power photomicrograph of a cresyl-violet-stained transverse section showing the location of EW in the pigeon midbrain (**a**), higher-power photomicrograph of EW on the left side of the brain (**b**) in a section that had been labeled using the peroxidase-antiperoxidase procedure for the localization of choline acetyltransferase (ChAT). Most EW neurons in pigeons are also enriched in the neuropeptides substance P and enkephalin (see figure 12.21). *BCA* brachium conjunctivum ascendens, *IP* interpeduncular nucleus, *OMd* dorsal subdivision of the oculomotor nucleus, *OMv* ventral subdivision of the oculomotor nucleus

(EWM) (Figs. 12.19 and 12.20). Neurons of EWM, in turn, innervate those neurons of the ipsilateral ciliary ganglion that give rise to an extensive cholinergic innervation of ipsilateral choroidal blood vessels. By contrast, neurons of lateral EW (EWL) innervate the ciliary neurons of the ciliary ganglion, some of which innervate the iris sphincter muscle and the rest of which innervate the muscles of accommodation. Those ciliary ganglion neurons controlling iris receive their input from caudolateral EWL, which in turn receives its input from a retinorecipient pretectal cell group, called area pretectalis in birds, of the contralateral side of the brain [105, 284]. Area pretectalis receives mainly contralateral retinal input. The more rostromedial part of EWL controls accommodation and receives input from the reticular midbrain, which appears to receive input from the arcopallium, a motor area of the telencephalon [106, 284]. Stimulation of the pigeon arcopallium yields accommodation [284]. Based on recent analysis of the hypothalamic circadian control system in birds, [49, 50] have suggested that the mammalian SCN is equivalent to two hypothalamic regions in birds. A medial region termed the mSCN in birds appears to be responsible for circadian rhythms but receives only meager retinal input, while the region projecting to EWM has been termed the vSCN [49, 50]. These two regions are interconnected, and together they have the neurochemistry, connections, and function of the mammalian SCN. For these reasons, we will henceforth term the part of avian SCN controlling ChBF the vSCN. The anatomy of the vSCN-EWM-ciliary ganglion circuit suggests that it might be involved in the light-regulated control of ChBF, and in studies described below, we have shown this is the case.

Neurons of EW are cholinergic and possess AMPA-type glutamate receptors, implying the inputs from SCN, AP, and reticular formation are in large part glutamatergic (Fig. 12.20) [232, 285, 371]. Choroidal neurons release acetylcholine (Fig. 12.21) [232], and cholinergic fibers from the ciliary ganglion are widespread in avian choroid (Fig. 12.22) [62]. In pigeons, they are most abundant in the superior and temporal ocular quadrants [62]. A published histochemical study claimed that choroidal neurons of the ciliary ganglion in bird were also rich in NOS [351]. This claim was based on histochemical staining for NADPH-diaphorase (NADPHd), an indirect marker for nitric oxide synthase (NOS). We sought to determine if the NADPHd labeling was truly attributable to nNOS, by using immunolabeling for nNOS and NOS-selective blockers of the NADPHd histochemical stain. Our results

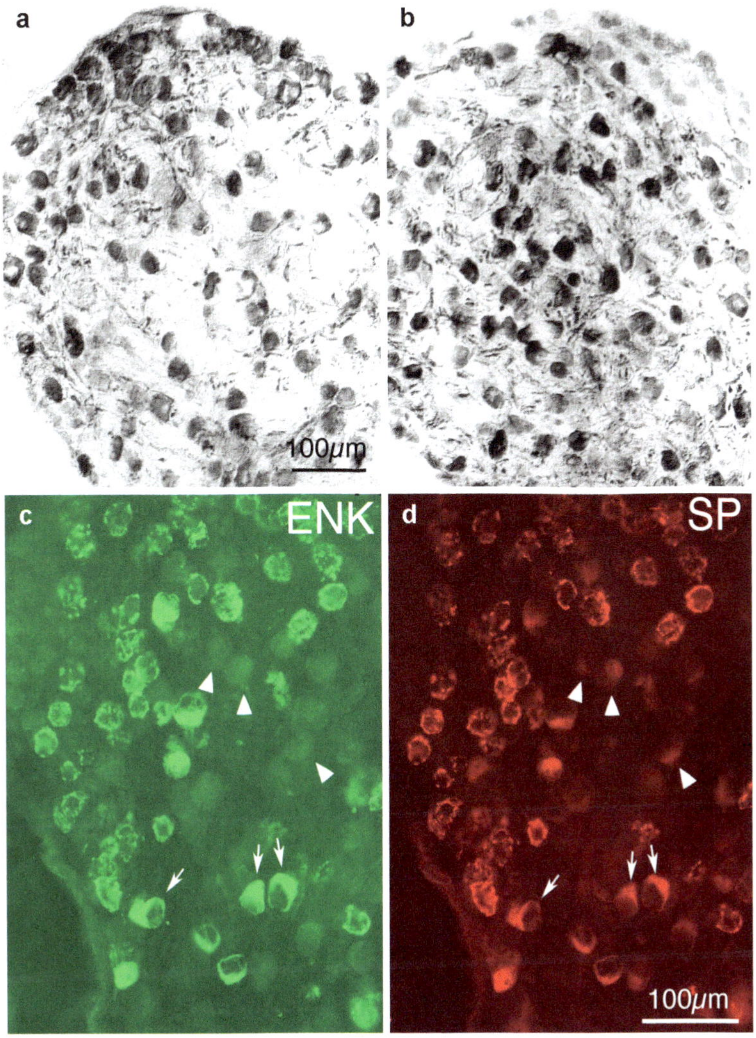

Fig. 12.21 Images (**a** and **b**) show two sections through the pigeon ciliary ganglion immunolabeled for ChAT. Note that many neurons and the preganglionic terminals on them in the ganglion are immunolabeled, with ciliary neurons and their cap-like endings tending to be more intensely labeled than the choroidal neurons and their boutonal endings. Same magnification in (**a**) as in (**b**). Images (**c** and **d**) show a pair of images illustrating the presence of preganglionic SP+ (substance P-containing) and ENK+ (enkephalinergic) boutonal endings on choroidal neurons (upper part of ciliary ganglion) and SP+ and ENK+ cap-like endings on ciliary neurons (*arrows*) as revealed by immunofluorescence double labeling. Images (**c** and **d**) show a single field of view from a section that had been double labeled for SP and ENK, which labeled preganglionic endings due to the presence of SP and ENK in most EW neurons. Note that not all cap-like endings contain both SP and ENK (*arrowheads*). Same magnification in (**c**) as in (**d**)

revealed that NOS is found only at a low level in some ciliary neurons and is absent from choroidal neurons. The latter finding is consistent with the fact that NOS+ and NADPHd+ fibers are much scarcer in choroid than are the cholinergic fibers demonstrably arising from ciliary ganglion [62, 64], and those NOS+/NADPHd+ fibers that are present are likely to arise from the PPG [63]. The prior report that NADPHd is abundant in choroidal neurons of the pigeon CG [351], thus, appears to have been based, regrettably, in a failure to distinguish between specific and nonspecific NADPHd staining. Consistent with the cholinergic projection of the ciliary ganglion to the eye [152, 232, 285], Fischer et al. [84] found that M2, M3, and M4 receptors are present in chick choroid and ciliary body by Western blots and immunolabeling.

12.5.4.4 Function of vSCN-EWM-Ciliary Ganglion Circuit – Birds

Our anatomical data suggest that the vSCN-EWM circuit participates in the light-regulated control of ChBF [284], and in a series of studies, we have shown that this is the case. For example, we showed that electrical stimulation of EWM increases ChBF in the ipsilateral eye, as measured by laser Doppler flowmetry (LDF)

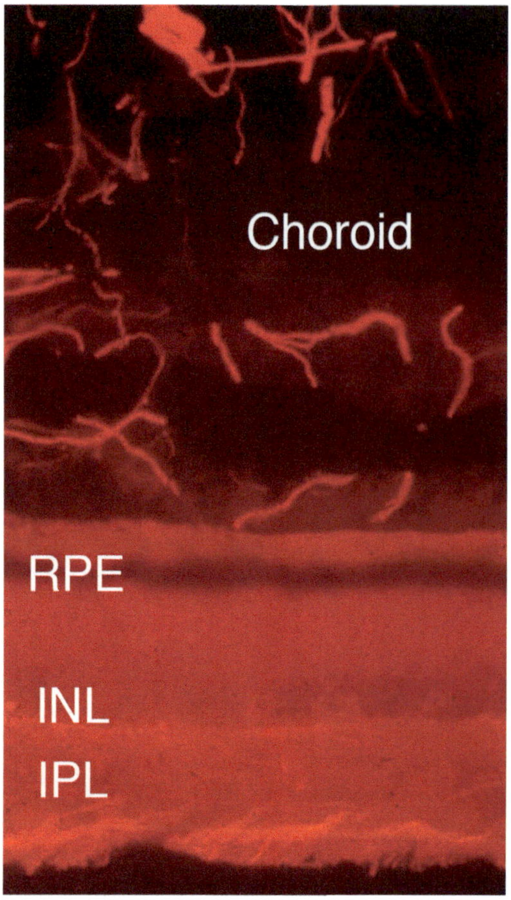

Fig. 12.22 Photomicrograph of a field of view from a section immunolabeled according to immunofluorescence procedures for the 3A10 neurofilament antigen that detects cholinergic ciliary ganglion input to the choroid. The image shows 3A10+ labeled fibers in the choroid of the superior ocular quadrant of a normal pigeon eye. *INL* inner nuclear layer, *IPL* inner plexiform layer, *RPE* retinal pigment epithelium

(Fig. 12.23) [88]. The increases are driven by increases in choroidal volume (vasodilation) and are not accompanied by increases in systemic BP. In Fitzgerald et al. [90], we also showed that activation by electrical stimulation of what we now call vSCN yields increases in ChBF in the opposite eye, while retinal illumination yields ChBF increases in the illuminated eye, as would be predicted by the layout of the vSCN-EWM circuit (Fig. 12.23). Furthermore, the vSCN- and light-elicited increases could be blocked reversibly by lidocaine injection into the EWM ipsilateral to the recorded eye (Fig. 12.23). Control studies confirmed that the light-elicited increases were not artifactually generated by transocular illumination of the LDF probe. In a study on chicks, we showed that severing the ciliary nerves permanently dilates the pupil and causes increased ChBF [315]. We interpreted the ChBF increase to stem from the increased illumination falling on the retina due to the chronically dilated pupil. An opaque occluder that diminished light entry in chick eyes with severed ciliary nerves eliminated the ChBF increase [316]. These results confirmed that the vSCN-EWM-ciliary ganglion circuit regulates ChBF in response to retinal illumination and/or activity. Thus, a major natural stimulus activating this circuit is the pattern and intensity of retinal illumination [90], and perhaps flicker may as well [317]. Such reflexive responses to increased retinal illumination or increased complexity of the patterns imaged on the retina may be adaptive, since such retinal activation alters the metabolic and/or thermal demands on the retina [35, 204, 325].

In a 1996 study [412], we evaluated the effects of two NOS inhibitors, 7NI and LNAME, on the increases in ChBF elicited by electrical stimulation of EWM in pigeons, as detected using transcleral LDF. We found that 7NI (which at the time we thought was nNOS selective but now know is not) and LNAME each attenuated the EWM-evoked response by about 80% (Fig. 12.24). In light of our findings that choroidal neurons do not contain NOS or make NO [64], the results of Zagvazdin et al. [412] indicate that endothelially derived NO must mediate EWM-elicited choroidal vasodilation. Since choroidal neurons of the avian ciliary ganglion do release acetylcholine [232], and since acetylcholine is known to stimulate endothelial NO release [242], we studied the role of muscarinic cholinergic mechanisms in ciliary ganglion-mediated ChBF increases in pigeon [413]. Using LDF and atropine as well as selective blockers of the M3-type muscarinic receptor (4-diphenylacetoxy-*N*-methylpiperedine, 4DAMP) and the M2-type muscarinic receptor (himbacine), we found that atropine and the M3-type muscarinic receptor blockade greatly (by about 90%) inhib-

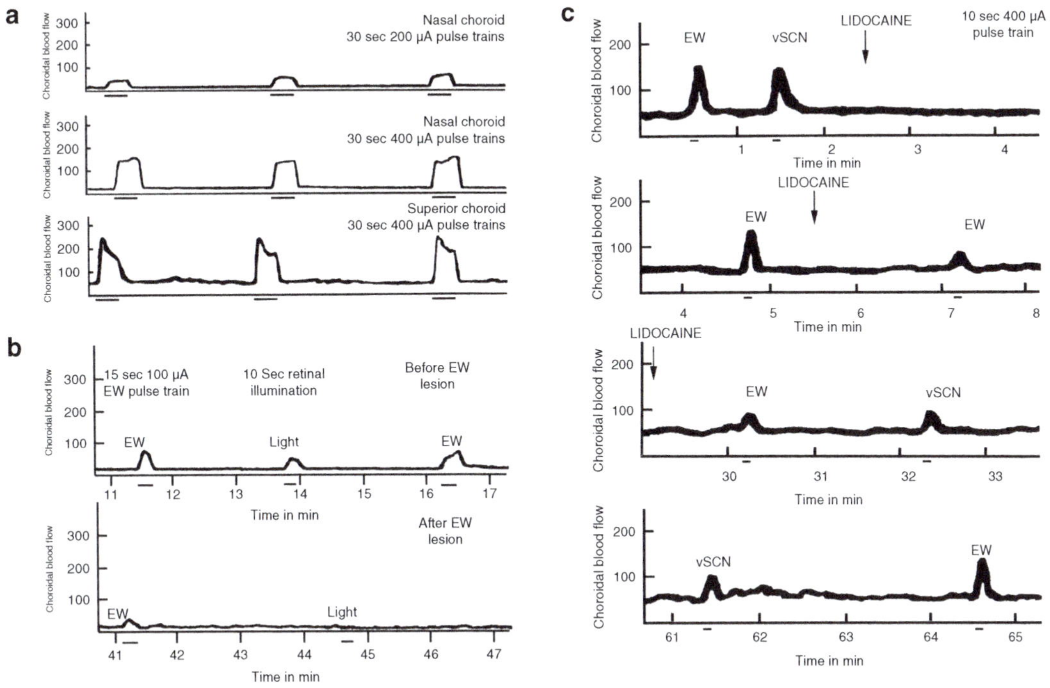

Fig. 12.23 Chart records showing the effects SCN-EWM circuit activation on ChBF in pigeon, with ChBF measured using laser Doppler flowmetry and expressed in relative units referred to as blood flow units. The records in (**a**) show that stimulation of EWM yields blood flow increases in superior and nasal choroid, with the increases being stimulus duration and stimulus amplitude dependent. The records in (**b**) show right eye ChBF responses to right EW activation or right retinal illumination. Note that both elicit ChBF increases and that the light-mediated response is prevented by an EWM lesion, showing that it is mediated via EWM. The records in (**c**) show right eye ChBF responses to right EWM or left vSCN stimulation. Note both elicit clear ChBF increases that are stimulus duration dependent, and note that transiently inactivating EW with lidocaine reversibly diminishes both the ChBF increases obtained with vSCN or EW stimulation. These overall results are consistent with the interpretation that the vSCN-EWM circuit in birds is involved in retinal activity-dependent increases in ChBF

ited EWM-evoked increases in ChBF (Fig. 12.24), while M2-type receptor inhibition increased ChBF by about 100%. Based on our findings that the ciliary ganglion input to choroid does not synthesize NO but inhibitors of NO production do block EWM-evoked choroidal vasodilation, it seems likely that the M3 receptors acted on by 4DAMP are present on choroidal endothelial cells, and the ciliary ganglion thereby mediates choroidal vasodilation via M3 cholinergic stimulation of endothelial NO release. In contrast, M2 muscarinic receptors may play a presynaptic role in downregulating EWM-evoked parasympathetic cholinergic vasodilation in avian choroid. Inhibiting them thus would be a means to potentiate ChBF. Our finding that EW lesions significantly diminish ChBF in pigeons indicates that basal activity in the preganglionic input to the ciliary ganglion is needed to maintain basal choroidal tone [90], and our finding that severing the choroidal nerves from the choroidal neurons of the ciliary ganglion to the choroid in chickens vastly diminishes ChBF supports the view that ciliary ganglion input prominently controls basal choroidal tone [315].

12.5.4.5 Studies of the Importance of ChBF Control by the vSCN-EWM-Ciliary Ganglion Circuit – Birds

While large reductions in ChBF lead to severe photoreceptor loss [111, 226], even slight reduc-

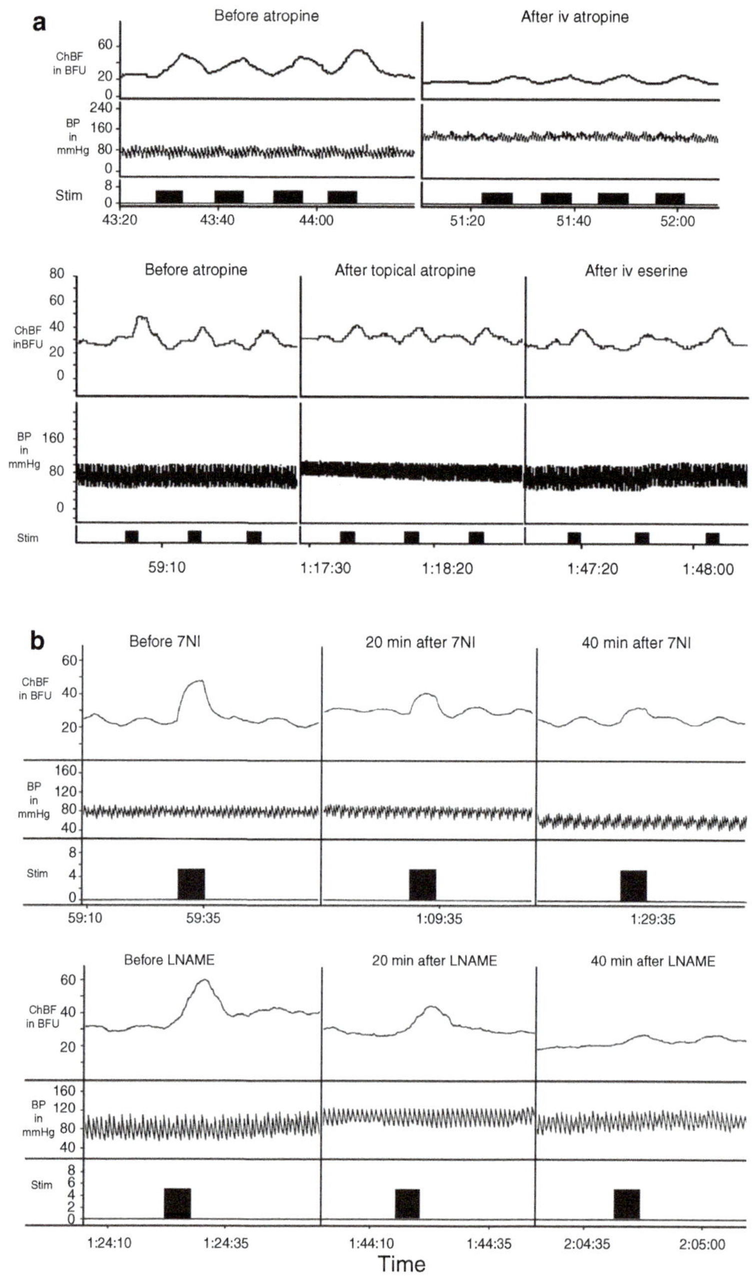

Fig. 12.24 The chart records in (**a**) show examples of the effect of intravenous (1 mg/kg) and topical (4–5 drops of 5% solution) atropine, a nonselective muscarinic antagonist, on the increase in ChBF evoked by electrical stimulation of EWM and on systemic arterial blood pressure (BP) in pigeons. Eserine (100 μg/kg iv), an acetylcholine esterase inhibitor, was administered 30 min after atropine administration. CBF was measured by laser Doppler flowmetry and is presented as relative blood flow units (BFU). The electrical stimulation (Stim) was applied as three–four 5–10 s trains with 5 s (iv atropine) and 40 s (topical atropine) intervals between trains. Note that atropine produced an eserine-reversible decrease in the EWM-evoked response and an elevation in systemic BP. The chart records in (**b**) show examples of the effect of nitric oxide synthase inhibition by 7NI and LNAME on the increase in ChBF after 5 s electrical stimulation of EWM. The EWM stimulation was not associated with arterial BP changes and was diminished after the injection of 7NI (50 mg/kg administered ip) or LNAME (30 mg/kg). Note the reduction in baseline choroidal blood flow and the increase in blood pressure after LNAME. Time (expressed as h/min/s) from the start of data acquisition is shown on the *x*-axis

tions in ChBF hinder the ability of oxygen and nutrients to reach the outer retina, and such ChBF reductions thereby have rapid and deleterious functional consequences for the outer retina [325, 399, 400]. ChBF is also vital for supporting the inner retina in those retinal regions poor in retinal vessels [37, 169]. Given that ChBF is likely to be regulated by the nervous system so as to match ChBF to retinal need, it is also likely that adaptive neural regulation of ChBF is important for the long-term health of the retina. We have specifically addressed this issue in studies on the avian vSCN-EWM-ciliary ganglion system in which we destroyed EWM. Such lesions reduce basal ChBF to 50–75% of normal in the ipsilateral eye and block adaptive ChBF regulation by the vSCN-EWM-ciliary ganglion circuit (e.g., light-mediated ChBF increases) [90]. In pigeon eyes affected by EWM lesions, we have found evidence of retinal functional disturbance and pathology, including (1) increased glial fibrillary acidic protein (GFAP) in retinal Müller cells [87, 180] and (2) losses in behaviorally assessed visual acuity [145]. In the former studies, we examined the effects of EWM lesions on the health of the retina (as assessed by GFAP immunolabeling) in birds housed under normal circadian lighting conditions [174]. We found that the GFAP increases in Müller cells following EWM destruction are progressive up to 24 weeks and occur preferentially in superior/temporal retina, which is heavily innervated by the ciliary ganglion in pigeons (Fig. 12.25). After 24 months, GFAP expression begins to diminish, but the GFAP upregulation is still evident 1 year after the lesion. In our behavioral studies, we examined the effects of EWM lesions on visual acuity in pigeons [145]. Bilateral lesions of EWM were made electrolytically, and visual acuity for high-contrast, square-wave gratings was determined behaviorally about 1 year later and compared to that in a group of pigeons that had received sham lesions of EW about 1 year prior to acuity testing (Fig. 12.26). Because lesions targeting EWM invariably result in damage to the adjoining EWL, two additional control groups were studied. In one control group, bilateral lesions in area pretectalis (AP), which innervates the pupillary control part of EWL and thereby controls pupillary constriction [284], were made, and the effects on visual acuity determined about 1 year later. In the second additional control group, the effects of acute accommodative and pupillary dysfunction on acuity were studied in cyclopleged pigeons. The mean acuities of birds with AP lesions (9.1 ± 1.4 cycles/degree) and sham lesions (7.1 ± 1.5 cycles/degree) were not significantly different from normal. In contrast, pigeons with lesions that completely destroyed EW bilaterally showed visual acuity (2.7 ± 0.1 cycles/degree) that was well below the acuity of the sham- and AP-lesion control groups. The acuity of the cycloplegic pigeons (4.8 ± 0.3 cycles/degree) and one pigeon with a nearly complete bilateral EWL but a unilateral EWM lesion (6.4 cycles/degree) indicated that only about half of the loss with a bilateral EW lesion could be attributed to accommodative dysfunction. Thus, bilateral destruction of EWM led to a loss in visual acuity, suggesting that disruption of adaptive neural regulation of ChBF causes retinal injury that impairs vision.

Our lesion studies, therefore, support the notion that interrupting neural control of ChBF by the vSCN-EWM-ciliary ganglion circuit is harmful for the retina. It seems likely therefore that retinopathy would also ensue from central or peripheral damage to facial, sensory, or sympathetic circuits controlling ChBF [277]. The precise nature of the retinal injury and the circumstances under which the impaired neural control of ChBF might be especially harmful, however, are uncertain and would depend on the precise role that the damaged circuit plays in supporting ocular health. It is possible that the retinal impairments observed with EWM lesions that disable parasympathetic control of ChBF stem from ChBF insufficiency that renders the retina chronically hypoxic and ischemic (Fig. 12.27) [45, 87, 399, 400]. Impaired parasympathetic control of ChBF may also result in harmful accumulation of waste products in the outer retina or an inadequate nutrient supply for outer retina renewal (e.g., amino acids, sugars, and fats) [141, 199]. Regardless of the basis of the retinal damage that occurs with disturbed ciliary ganglion-mediated control of ChBF in birds, it is likely that these same potentially damaging processes are

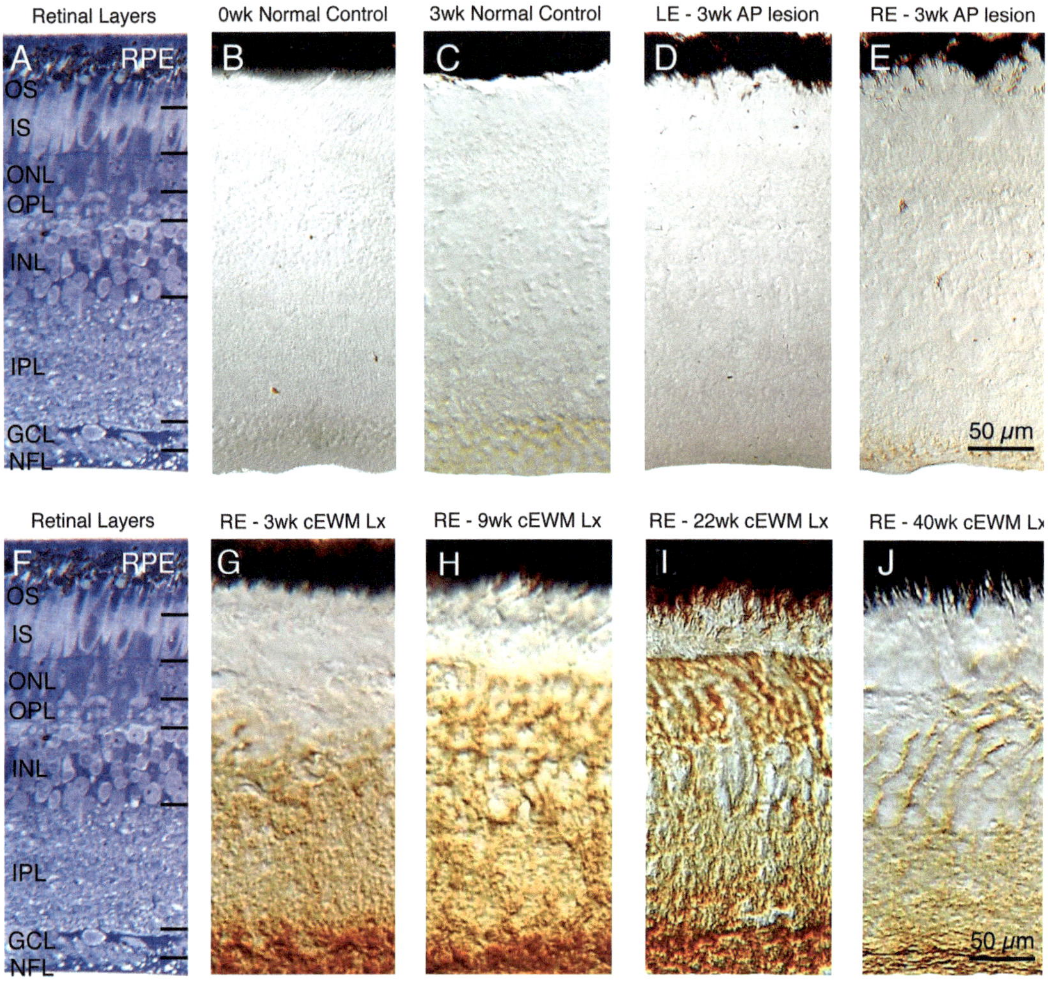

Fig. 12.25 Images of representative GFAP-immunolabeled sections through the superior-central retina of normal pigeons (**b**, **c**), a pigeon with a left area pretectalis (AP) lesion (**d**, **e**), a pigeon that survived for 3 weeks with a complete right EW lesion (**g**), a pigeon that survived for 9 weeks with a complete right EW lesion (**h**), a pigeon that survived for 22 weeks with a complete right EW lesion (**i**), and a pigeon that survived for 40 weeks with a complete right EW lesion (**j**), all housed in a 12-h moderate light/12-h dark cycle. Images (**a**) and (**f**) show a 1-mm-thick toluidine blue-stained plastic section of pigeon retina, with the different retinal layers delimited by hash marks, and the hash mark between the outer nuclear layer and the inner segment layer located at the outer limiting membrane. Retinal sections from a normal pigeon never housed in an individual cage show no GFAP immunolabeling (**b**), while the retinal section from a normal pigeon housed for 3 weeks in an individual cage shows slight GFAP immunostaining of the nerve fiber layer (*NFL*) and ganglion cell layer (*GCL*). The lesion of left AP eliminated the pupil light reflex and chronically dilated the pupil of the right eye. In the left eye, GFAP immunolabeling is weak and does not extend beyond the NFL, while GFAP immunolabeling in the right eye fills Müller cell processes into the GCL. The view of the right eye (**g**) of a bird 3 weeks after a complete lesion of both the right EWM and EWL shows GFAP immunolabeling fills Müller cell processes into the inner nuclear layer (*INL*). The image of the right eye (**h**) of a bird 9 weeks after a complete lesion of right EWM and EWL shows GFAP immunolabeling extends through the outer plexiform layer (*OPL*). The image of the right (**i**) eye of a bird 22 weeks after a complete lesion of EWM and EWL shows GFAP immunolabeling extends through the outer plexiform layer to the outer limiting membrane (OLM). Finally, the image of the right eye (**j**) of a pigeon 40 weeks after a complete lesion of EWM and EWL shows GFAP immunolabeling filled the Müller cell processes through the INL. *GCL* ganglion cell layer, *INL* inner nuclear layer, *IPL* inner plexiform layer, *IS* inner segment layer, *NFL* nerve fiber layer, *ONL* outer nuclear layer, *OS* outer segment, *RPE* retinal pigment epithelium. Magnification the same in all images

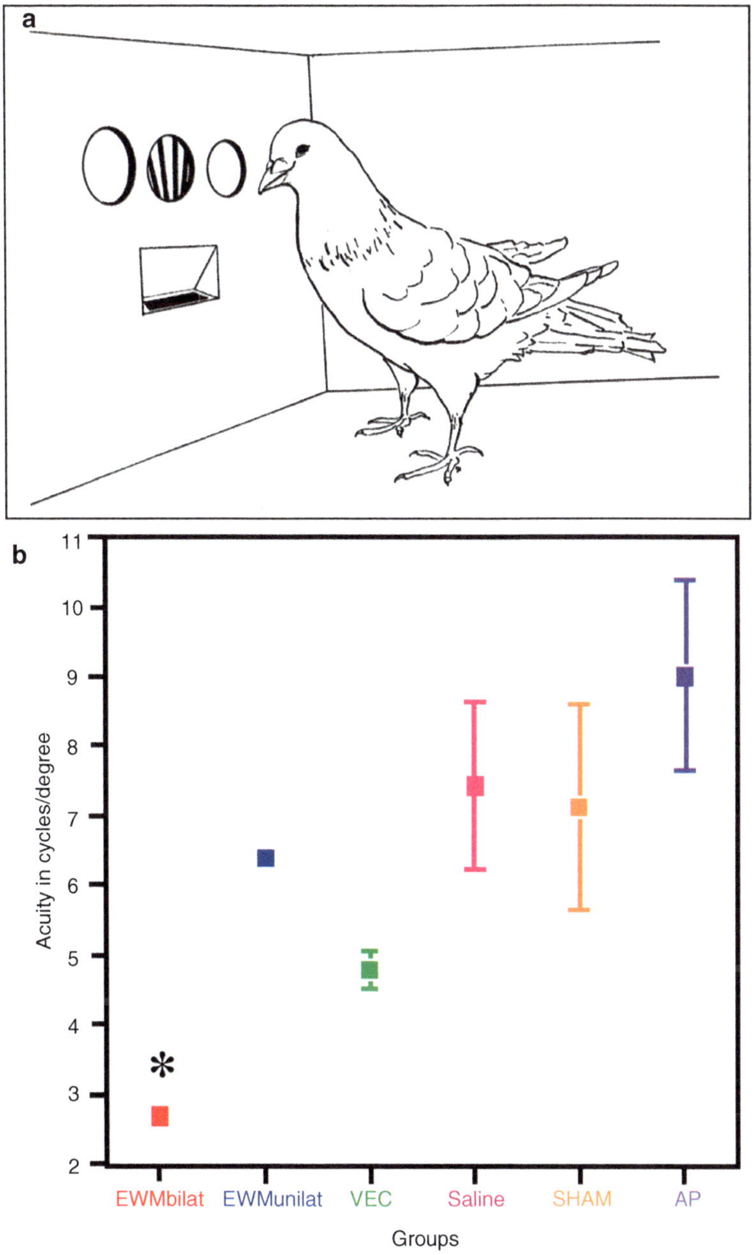

Fig. 12.26 Image (**a**) shows a schematic representation of the training chamber used to measure visual acuity in pigeons, using a discriminative behavioral task. Image (**b**) shows the mean acuity data (±SEM) for the complete EW-lesion birds, the partial EW-lesion bird, the AP-lesion birds, the sham-lesion birds, the cycloplegic birds treated with saline (*Saline*), and the same cycloplegic birds treated with vecuronium (*Vec*). Note that the mean data for the three birds with the complete bilateral destruction (*EWMbilat*) are graphed separately from the data for the one bird with the bilateral EWL but unilateral EWM lesion (*EWMunilat*). The SEM is too small for the former birds to be evident, and no SEM can be calculated for the single unilateral EW bird. The visual acuity for the bilaterally EW-lesioned birds was significantly poorer than for the five other categories of birds, among which there were no statistically significant differences

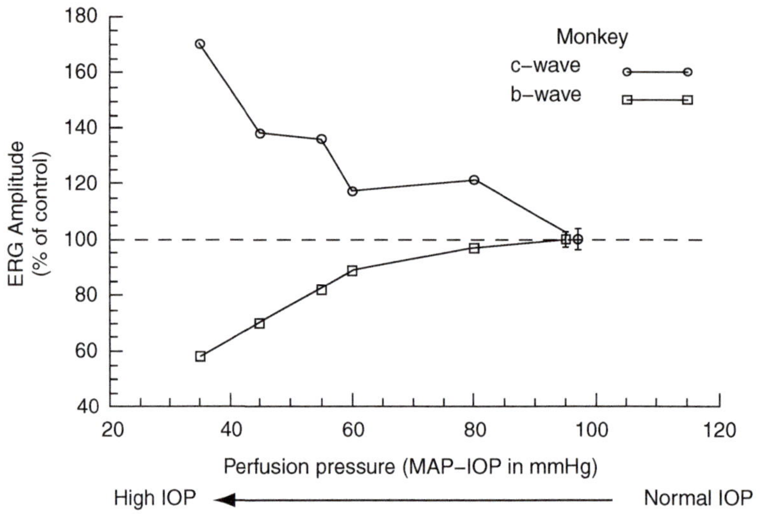

Fig. 12.27 Effect of ocular perfusion pressure on the c-wave and the b-wave of the ERG in cat, expressed as a percent of the control amplitude (Fig. 7 from [399]). The perfusion pressure was manipulated by increasing the IOP, which is known to reduce ChBF, with only modest autoregulation by the choroid in response to the reduced perfusion pressure [169]. Note that with decreasing perfusion pressure, abnormalities in both the c-wave and the b-wave become evident, especially below a perfusion pressure of 60 mmHg

normally held in check by intact adaptive ciliary ganglion-mediated control of ChBF.

12.5.5 Sympathetic Superior Cervical Ganglion Input

Sympathetic noradrenergic nerve fibers from the superior cervical ganglion innervate the choroid in mammals [346] and birds [129]. In birds and mammals, the innervation is to blood vessels, and in birds to the smooth muscle of the choroidal stroma as well. Mammalian groups in which sympathetic innervation of the choroid has been demonstrated by catecholamine fluorescence or immunolabeling include rats, guinea pigs, rabbits, cats, and monkeys (Fig. 12.28) [73, 89, 179, 201, 217, 359]. These nerve fibers utilize noradrenaline as a neurotransmitter, and they thus contain the enzymes involved in its synthesis (such as tyrosine hydroxylase and dopamine beta-hydroxylase). The sympathetic nerve fibers from the superior cervical ganglion travel to the choroid via orbital blood vessels or by joining the ophthalmic nerve [322]. Consistent with the sympathetic innervation of choroid and vessels supplying the choroid, cervical sympathetic stimulation in rats, rabbits, cats, and monkeys increases uveal resistance and decreases ChBF (Figs. 12.10 and 12.29) [1, 6, 9, 10, 29, 291, 326]. The choroidal vasoconstriction caused by direct administration of noradrenaline or by activation of sympathetic nerves to the choroid is mediated by alpha-adrenergic receptors [1, 37, 114, 165, 183, 184, 326]. Kawarai and Koss [165] specifically showed that alpha1-adrenoreceptors mediate sympathetic vasoconstriction in the rat choroid. Blockers of beta-adrenergic receptors, by contrast, have been shown in pig to be only marginally effective in dilating the short posterior ciliary arteries [44], and thus unlikely to have a significant vascular role in sympathetic choroidal control. The sympathetic co-transmitter NPY does, however, have a role in ChBF regulation, since intravenous NPY in rabbits decreases ChBF by 50% [254]. NPY may be responsible for that part of choroidal sympathetic vasoconstriction that is not blocked by alpha-adrenergic receptor antagonists [10]. NPY appears to particularly contribute to choroidal vasoconstriction with high sympathetic nerve firing rates, and noradrenaline with low firing rates [36].

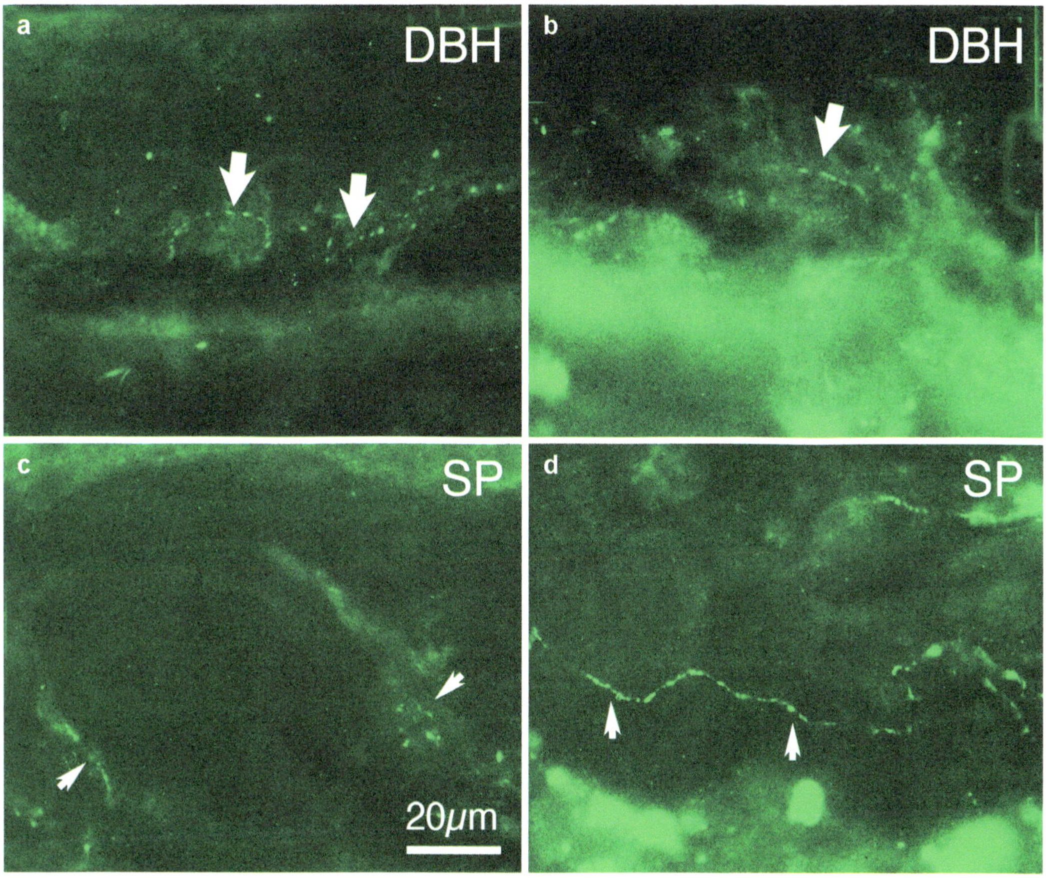

Fig. 12.28 Images (**a** and **b**) show sympathetic nerve fibers with varicosities in rat choroid immunolabeled for dopamine beta-hydroxylase (*DBH*). Images (**c** and **d**) show sensory nerve fibers with varicosities in rat choroid immunolabeled for substance P (*SP*). All images at the same magnification *arrows* indicate labeled axons and terminals

Consistent with a role of alpha-adrenergic receptors in choroidal control, Kiel and Lovell [172] reported that alpha-adrenoreceptor block increased ChBF in rabbits, implying thus also a level of basal sympathetic tone in anesthetized rabbit choroid. Similarly, Chou et al. [52] reported ChBF increased at low ocular perfusion pressures in rabbit 1 week after sympathetic denervation of choroid. Such a manipulation also produced adrenergic supersensitivity [53]. Other investigators have, however, questioned the presence of significant basal sympathetic tone in ChBF at normal systemic blood pressure in awake animals [34, 37]. Consistent with this, Zhan et al. [416] reported that 24 h after sympathetic denervation ChBF was unchanged in rabbits – there was thus no effect on resting tone. This issue was more extensively investigated by Chou et al. [54] in rabbits using LDF. They found that with reductions in perfusion pressure caused by increased IOP, ChBF remained stable until a perfusion pressure of <55 mmHg, at which point ChBF was proportional to perfusion pressure. In rabbits with either unilateral or bilateral cervical sympathectomy, ChBF declines were not as severe with extremely low perfusion pressure as in normal animals. Thus, sympathetic input does exert some tone on choroidal vessels that is evident at low perfusion pressures, at least in some species under some conditions.

Bill [37] has suggested that the sympathetic innervation of choroid becomes activated with

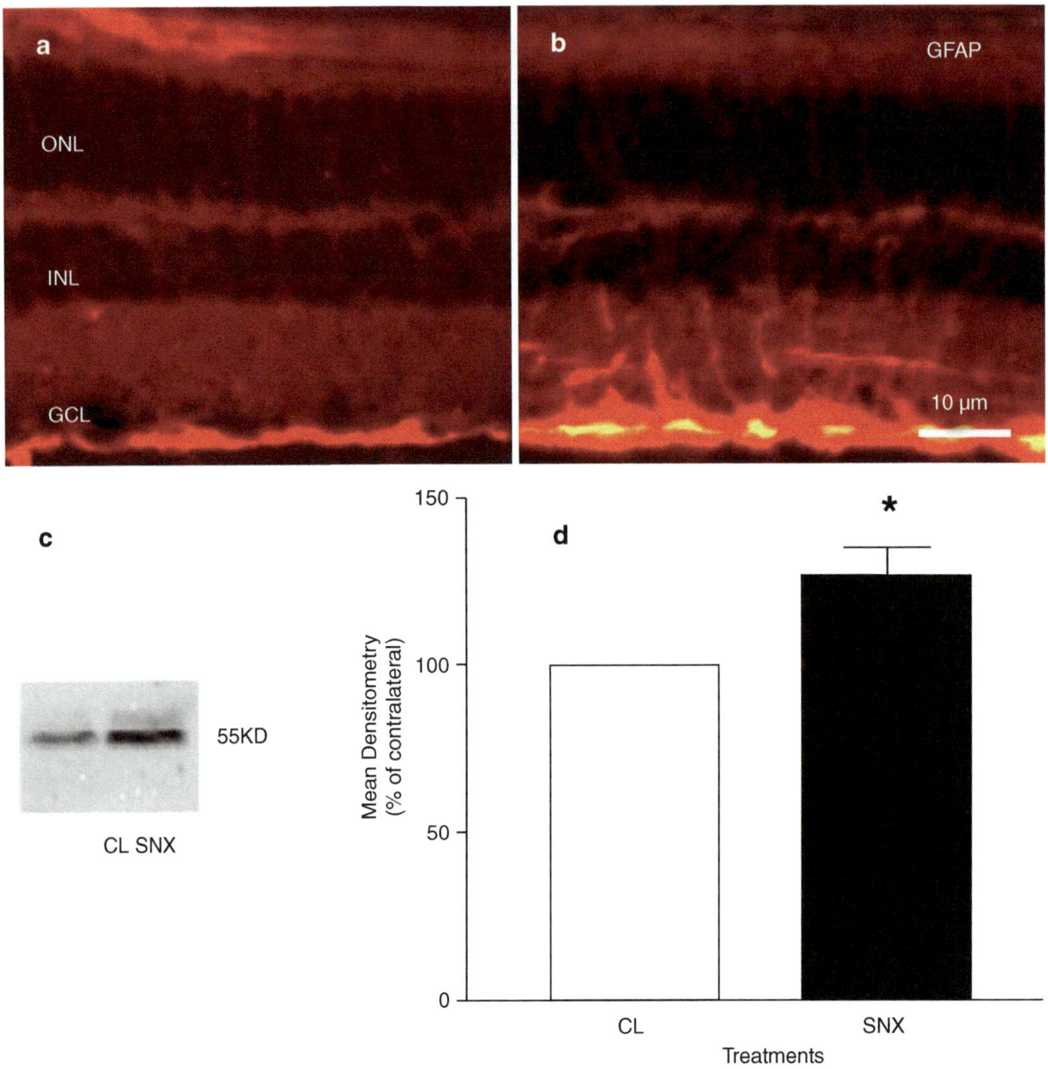

Fig. 12.29 Images and graphs showing retinal changes in rats after sympathectomy (From Steinle et al. [329]). Images (**a** and **b**) (Fig. 5 from [329]) show immunofluorescent GFAP labeling in the contralateral (**a**) and sympathectomized (**b**) retina. Greater GFAP immunostaining is observed in Müller cells in the sympathectomized retina. Images (**c** and **d**) (Fig. 6 from [329]) show representative results from a Western blot for GFAP protein levels for control (*CL*) and sympathectomized (*SNX*) retina (**c**) and densitometric analysis of Western blot data for four pairs of control and sympathectomized retinas (**d**). In (**d**), the results are presented as a percent of control. The *asterisk* indicates a significant difference between experimental and control. *GCL* ganglion cell layer, *INL* inner nuclear layer, *IPL* inner plexiform layer. Magnification the same in both (**a** and **b**)

high systemic blood pressure, serving to vasoconstrict the choroid to prevent the overperfusion that would otherwise occur with the increased perfusion pressure caused by the elevated systemic BP. Naturally occurring increases in systemic blood pressure can occur during stress or heightened activity levels. Sustained elevations of ChBF would cause increased IOP, breakdown of blood-retinal barriers, and edema and be harmful for retinal health and function. Such ocular overperfusion and/or vascular leakiness has, in fact, been demonstrated in sympathectomized rabbits with systemic blood pressure elevation caused by aortic clamping [30, 34] and in sympathectomized monkeys subjected to systolic hypertension [79]. Several studies in humans have shown that the choroid vasoconstricts and thereby compensates for exercise-induced increases in systemic blood pressure [208, 292]. Fuchsjäger-Mayerl et al. [101] reported that endothelial release of the vasoconstrictor endothelin-1 plays a role in this effect. It may be that central baroreceptor-responsive circuitry acting via the sympathetic input to choroid contributes to choroidal baroregulation to high systemic BP. Longo et al. [207] noted, however, that orthostatic increases in ocular perfusion pressure caused by posture change (moving to a supine position) are not compensated for by choroidal vasoconstriction, as would be expected since such changes in perfusion pressure are not accompanied by increases in systemic pressure that would activate aortic baroreceptors.

Steinle et al. [328] investigated the consequences of loss of sympathetic tone on the rat choroid. They reported vascular remodeling after cervical sympathetic transection – choroidal arteries and veins were larger and more numerous than in normal rats. No changes in vessel abundance or ChBF were yet observed 2 days after the sympathetic transection, but by 6 weeks, vessels were much more numerous, vessel area was increased, and ChBF was fourfold increased. Steinle and Lashbrook [330] noted elevation of angiogenic factors in choroid after cervical sympathetic transection, and they suggested that these might contribute to increased vessel abundance. Vessel dilation might contribute to increased vessel size as well. Steinle et al. [329] noted that there was a significant reduction (30%) in photoreceptor cell numbers in sympathectomized rat eyes. This loss appeared to be due to apoptosis, since there was a doubling in apoptotic photoreceptor cell numbers after sympathectomy. The photoreceptor loss in sympathectomized eyes resulted in reduced width of the retinal outer nuclear layer. Increased Müller cell immunostaining for GFAP spanning the ganglion cell layer and inner nuclear layer was also noted after sympathectomy (Fig. 12.29). These results suggest that loss of sympathetic innervation causes significant changes to the physiology of the choroid that are adverse for retinal health.

12.5.6 Trigeminal Sensory Input

Sensory nerve fibers from the trigeminal ganglion co-containing SP and CGRP innervate the choroid in mammals and birds (Fig. 12.28) [60, 318, 346]. These largely are branches of the ophthalmic nerve, but some that reach the orbit arise from the maxillary nerve. The sensory nerve fibers typically join the meshwork of nerves behind the eye, with one prominent branch entering the ciliary ganglion (Fig. 12.10). Nerve fibers reach the choroid by traveling with the short ciliary nerves and on blood vessels of the orbit. Among mammals, SP+ and CGRP+ fibers to the choroid have been observed in rats [89], guinea pigs [193], monkeys [347], and humans [343, 347]. Sensory fibers such as those of the trigeminal nerve send a central message of hot, cold, pain, or touch and can elicit ocular reflexes, such as blinking and tearing in response to their activation [23, 103]. It is possible that they also participate in temperature-dependent ChBF reflexes. Peripheral fibers can also participate in antidromic responses in which they release SP and CGRP and cause local responses, which include a vascular component [23]. Consistent with an antidromic action of the SP+ input, rat and rabbit choroid possess SP receptors [68], and consistent with an antidromic action of the CGRP+ input, the choroid in pigs, guinea pigs, and monkeys has been shown to possess CGRP receptors [138]. The neuropeptides SP and CGRP are vasodilators, and their release from intrachoroidal trigeminal sensory fibers would be expected to act on ChBF [35–37, 318, 346]. Consistent with this, stimulation of the ophthalmic nerve in rabbit increases ChBF [338] and is associated with uveal release of SP [34]. Orbital vessels too are sites at which sensory fibers can affect the blood supply to the choroid. For example, Bakken et al. [17] showed that the pig ophthalmic artery dilates to CGRP. Additionally, trigeminal fibers have SP+ and CGRP+ terminals in the PPG, which could be a basis of sensory-autonomic vascular reflexes in choroid [353]. Our work in chicks suggests that the sensory input to the choroid may be involved in temperature-dependent regulation of ChBF (i.e., retinal thermoregulation) [318], which may be important for retinal homeostasis [268]. Reduced ocular blood flow caused by elevated IOP has been shown to result in reduced scleral and retinal temperature [32]. Heating an eye in which blood flow has been reduced by increased IOP (thereby reducing ocular perfusion pressure and ChBF) results in an exaggerated rise in ocular temperature, showing that choroidal blood flow can be involved in ocular cooling as well [32]. Regulation of ocular temperature has been shown to be important for the normal physiological functioning of the retina [146, 241].

12.5.7 Intrinsic Choroidal Neurons

Studies over the past 25 years have proven that the choroid in many mammalian and avian species contains neuronal cells that possess various neuroactive substances typical of parasympathetic PPG neurons. Although intrachoroidal neurons were initially reported by silver staining methods in the nineteenth century [308], Terenghi et al. [359] first noted them for their neuroactive substance content (VIP) in guinea pig choroid after colchicine treatment (which prevented transport of the neuroactive substances out of the perikarya). Shortly thereafter, Miller et al. [237] and Stone et al. [345] saw VIP+ neurons in human choroid. This phenomenon was subsequently more extensively studied by several groups. Using NADPHd histochemistry, which detects neurons using NOS to synthesize NO [25] reported the presence of many isolated or grouped NADPHd+ neurons in human choroid, with individual neurons being about 30 μm in perikaryal size. Flügel et al. [94] made similar observations of human choroid and reported that VIP was typically present in the NOS+ intrachoroidal neurons. The ganglion cells tended to be most numerous in the central retina around the human fovea. No such neurons were, however, observed in afoveate species such as rats, rabbits, tree shrews, cats, pigs, or owl monkeys [94, 95]. Flügel-Koch et al. [95] reported that NOS+ and VIP+ intrachoroidal ganglion cells are also present in cynomolgus monkeys, which have a fovea centralis. Since these neurons were absent from primate and nonprimate species lacking a fovea centralis, these authors suggested that the plexus was associated with choroidal blood flow control in the region of the fovea. The intrachoroidal ganglion cells in humans give rise to processes that can be observed to join the perivascular network of NADPHd+ and VIP+ fibers in the choroid, and in some cases observed to end directly on arteries. The intrachoroidal ganglion cells in humans have been noted to receive terminals, some of which contain NOS or VIP [94, 223]. Schrödl et al. [308] reported on the connectivity of the intrinsic choroidal neurons in human eye. Using immunolabeling for nNOS or VIP, as well as single-cell filling, they reported that intrachoroidal neurons send processes to other ganglion cells, as well as to vascular and nonvascular smooth muscle of the choroid. CGRP+ sensory boutons were also observed ending on intrachoroidal neurons. The evidence of sensory input to intrachoroidal ganglion cells suggests they have a role in reflexive modulation of ChBF in response to sensory stimuli such as heat or cold detected by the sensory fibers.

Intrinsic choroidal neurons are also present in birds and of a similar size (30 μm) and neurochemistry (nNOS+ and VIP+) to those in mammals [26, 65, 309]. These neurons give rise to processes that contact choroidal blood vessels, and they receive CGRP+ contacts from sensory axons [306]. The evidence of sensory input to intrachoroidal ganglion cells again suggests they have a role in reflexive modulation of ChBF in response to sensory stimuli such as heat or cold. Intrachoroidal ganglion neurons in birds also receive sympathetic nerve terminals, indicating their sympathetic modulation as well [307]. The intrachoroidal ganglion neurons vary among avian groups in their abundance, with the highest abundance among studied groups being in goose, the next highest in turkeys and chickens, and the lowest in ducks [309]. Their spatial distribution also varied among studied avian groups, presumably in relation to retinal high acuity specializations.

As in mammals, the intrachoroidal ganglion neurons are also found associated with the ciliary nerves in their course to the anterior uvea [26]. It is possible that the intrachoroidal ganglion cells play a role in the repositioning of the focal plane of the retina that is caused by increased ChBF and retinal thickening in chicks recovering from myopia [92].

12.5.8 Disturbed Neural Control of Choroidal Blood Flow in Aging and Retinal Disease

12.5.8.1 Effect of Aging on Retina and Choroid

Choroidal blood flow and its adaptive regulation can be impaired by aging (Fig. 12.30). For example, reductions in basal ChBF occur in the macula of humans as they age [271, 283]. Moreover, reductions in ChBF in excess of what aging alone would predict have been observed beginning in the early stages of age-related macular degeneration [42, 98, 99, 125, 126, 281], the leading cause of blindness in humans older than 65 years [136]. Adaptive regulation of ChBF also is impaired in aged humans, since compensatory responses of ChBF to fluctuations in systemic blood pressure are abnormal in elderly humans [278]. These various findings raise the possibility that abnormalities in macular ChBF and/or in its adaptive regulation contribute to normal age-related declines in retinal degeneration (Fig. 12.30). The mechanisms responsible for the reduced ChBF and impaired adaptive regulation with age are uncertain. Loss and narrowing of submacular choroidal vessels in normal aged eyes and in atrophic AMD eyes has been reported [125, 282, 324]. Such changes may contribute to the reduction in basal ChBF seen with aging and in early AMD (Fig. 12.30). To examine the possible role of changes in neuronal control of ChBF, we used immunolabeling to assess the impact of age on parasympathetic innervation of human choroid, using VIP immunolabeling. Our results indicated a significant age-related decline in VIP-positive nerve fibers and vessel diameter in the submacular choroid in disease-free human donor eyes (Fig. 12.31) [155]. These findings suggest that a decline in the neural control of ChBF and vessel diameter may explain reductions in ChBF and its adaptive control observed with aging. Impaired neural control of ocular and choroidal blood flow may occur in other conditions as well. For example, production of the vasodilator NO by the endothelium of ophthalmic, ciliary and retinal vessels and the optic nerve head can be impaired in hypertension, hypercholesterolemia, and diabetes [119, 133, 137].

As part of an effort to characterize the effects of aging on the retina and choroid, we carried out detailed studies in pigeon [91, 93] that sought to: (1) determine if choroidal and outer retinal deterioration occurs as a function of aging in a nonhuman species that possesses both rod and cone photoreceptors (i.e., pigeons), and might therefore be widespread concomitants of aging, and (2) obtain data that might shed light on the relationship between age-related retinal and choroidal changes. In a sample of 64 pigeons ranging in age over much of the pigeon life span (0.5–20 years), we measured diverse ocular parameters by physiological or histological means, including: ChBF (by LDF); choriocapillary vessel abundance (by LM histology); acuity (by behavioral methods); and photoreceptor abundance (by LM histology). Statistical methods were used to ascertain the pattern of age-related changes and determine ages at which or by which significant changes occurred in specific parameters. In the sample of 53 birds for which we had obtained visual acuity and/or photoreceptor data, we observed a prominent stepwise decline of about 20% in photoreceptor abundance at the age of 4.7 years (Fig. 12.32), followed by lesser decline thereafter, and a curvilinear decline in acuity (with half the decline having occurred by the age of 5 years). The period of prominent photoreceptor loss (4–6 years of age) coincided with ages during which about 10% of photoreceptors appeared to be showing degenerative changes. For the sample of 45 birds for which we measured choroidal parameters, choriocapillary vessel abundance showed a highly curvilinear decline with age and at least half of this decline had occurred by the age of 3 years (Fig. 12.33). ChBF showed an abrupt decline of about 20% at

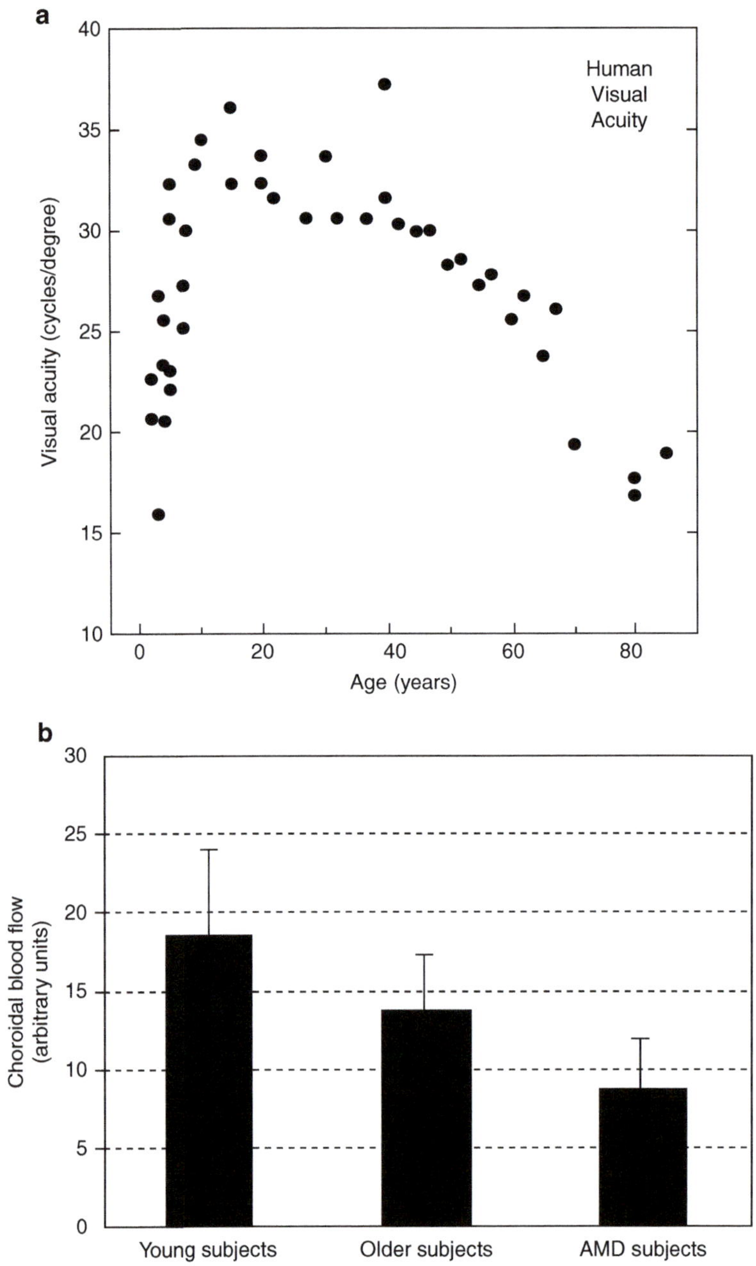

Fig. 12.30 Graphs showing visual acuity changes in humans over the life span (**a**), and basal choroidal blood flow in young humans, aged humans, and humans with age-related macular degeneration (AMD) (**b**). Graph (**a**) shows the decline in human visual acuity with advancing age (Fig. 3 redrawn from [145]). In (**b**), the blood flow values are taken form Grunwald et al. [125, 126] and are expressed in arbitrary units based on laser Doppler flowmetry measurements. Note that basal ChBF declines with age, and individuals with AMD have even lower basal ChBF than aged humans without AMD

the age of 4 years and a further 20% decline thereafter. Our results clearly show that ChBF and choroidal vascularity decline significantly with age in pigeons, as do acuity and photoreceptor abundance. Our statistical analyses suggest that there is a positive relationship between choroidal and visual functions in pigeons, and that prominent choroidal vascular decline precedes visual

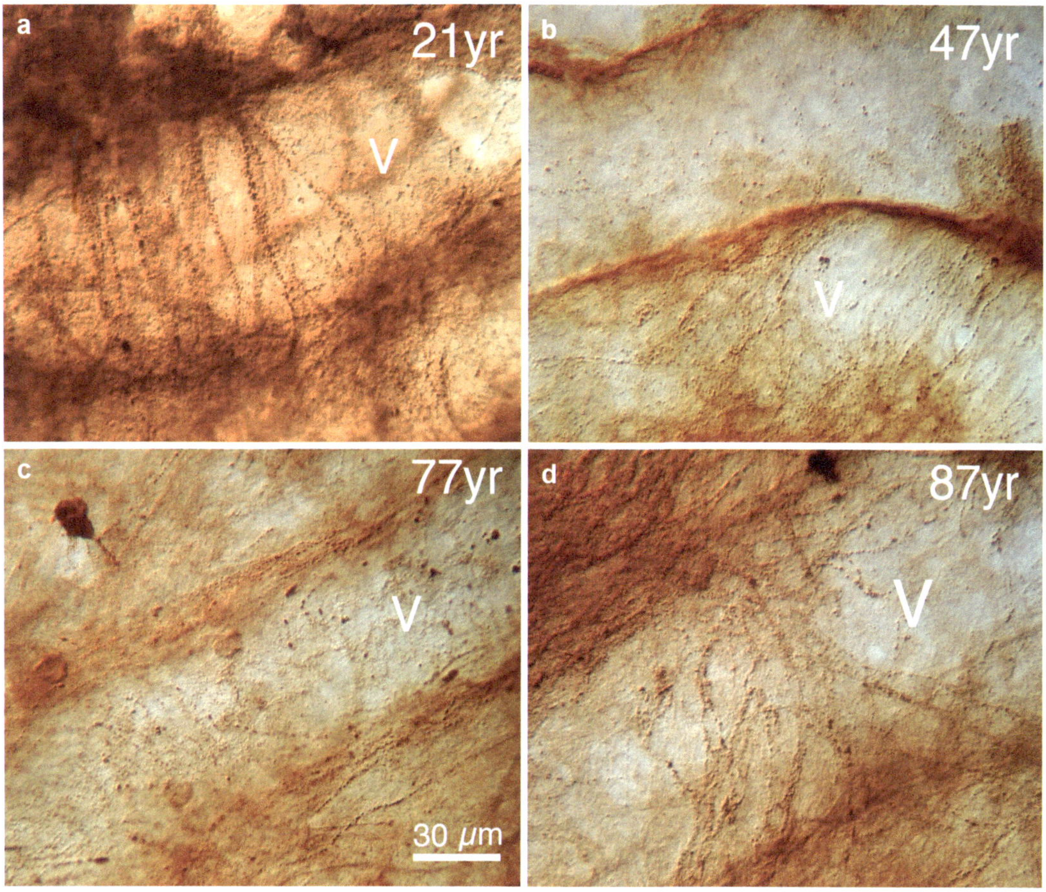

Fig. 12.31 Examples of VIP immunolabeling of nerve fibers on choroidal vessels in humans of differing age. The view of the choroidal specimens shows representative submacular choroidal vessels (*V*), surrounded by stroma. The abundant beaded striae running across the vessel lumens are the VIP-positive fibers of the choroidal vessel. Note the generally lesser amount of VIP-positive fibers in elderly compared with young eyes. Note also the VIP+ intrachoroidal neuron in (**c**). Magnification the same in all images

decline as pigeons age (Fig. 12.34). Thus, our findings are consistent with the view that age-related decline in choroidal function might contribute to age-related vision loss in pigeons. Note that the possibility exists that retinal degeneration is the primary event and diminished ChBF the consequence of diminished retinal need. Nonetheless, outer retinal degeneration does not invariably lead to diminished ChBF, as evidenced in the Abyssinian cat [257].

In further studies, we sought to determine if age-related changes in parasympathetic regulation of ChBF could contribute to age-related choroidal and retinal decline [93]. To this end, we used immunohistochemical methods to detect choroidal nerve fibers from the ciliary ganglion immunolabeled for ChAT or a neurofilament-associated antigen in fixed cryostat sections of the eye in 0.5- to 20-year-old pigeons. Additionally, transcleral LDF was used to measure basal ChBF, light-evoked ChBF increases and EWM-evoked ChBF increases in the superior choroid to assess the functional integrity of the vSCN-EWM-ciliary ganglion circuit. We observed a marked age-related linear decline in the ciliary ganglion innervation of the choroid (Fig. 12.33). Moreover, we observed pronounced declines in ciliary ganglion-mediated control of ChBF. The decline in EWM control of ChBF was pronounced by 5 years of age, and half of the

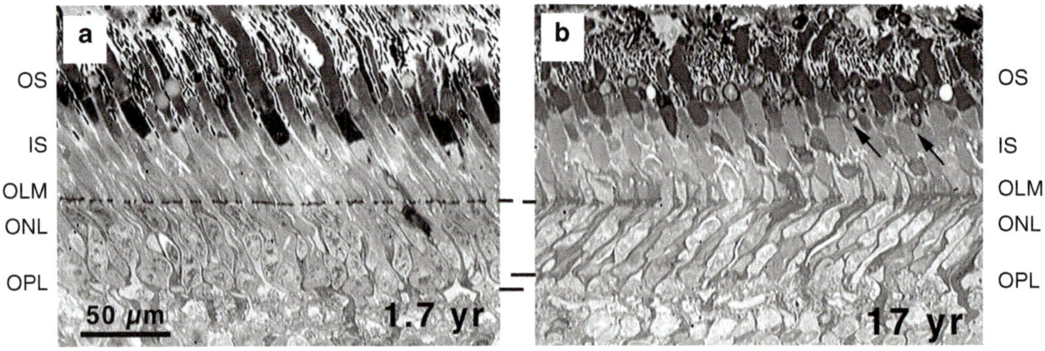

Fig. 12.32 Light micrographic images of retinal sections from a 1.7-year-old pigeon (**a**) and a 17-year-old pigeon (**b**) showing two extremes in the abundance of photoreceptor cell bodies. The 1.7-year-old animal had 47 photoreceptors per 100 μm, while the 17-year-old animal had 32 photoreceptors per 100 μm. In addition, the thickness of the outer nuclear layer (*ONL*) is less in the older bird (note lines demarcating ONL – outer plexiform layer (*OPL*) border in both images). Age-related photoreceptor pathology was also evidenced by loss of outer segments (*OS*) and vitread displacement of some lipid droplets from their normal position (*arrows*). *IS* inner segment, *OLM* outer limiting membrane, *ONL* outer nuclear layer, *OPL* outer plexiform layer

functional decline occurred by about 2–3 years of age. Thus, significant loss in choroidal vascularity and innervation appear to lead to impaired basal and adaptive parasympathetic ChBF control early in the life span of pigeons (Fig. 12.34), which is consistent with the notion they contribute to age-related vascular insufficiency and attendant age-related damage to the retina. These overall results for pigeons suggest that aging may deleteriously affect the retina, in part, by impairing ChBF and its neural control. The apparent impact of age-related loss of ciliary ganglion input to the choroid in birds is consistent with the impact we see of EWM lesions in younger birds.

12.5.8.2 Effect of Disease on Retina and Choroid

Based on the effects of interruption of the vSCN-EWM-ciliary ganglion circuit in birds, it seems plausible that diseases that compromise ChBF and its neural control might have retinal dysfunction and pathology as their consequences. Several conditions in humans and animals that involve autonomic neuropathy or endothelial cell pathology (the latter of which would affect endothelial-dependent neural control of ChBF), in fact, have diminished ChBF and outer retinal pathology as concomitants, including aging [404], age-related macular degeneration [99], chronic hypertension [133, 374], insulin-dependent diabetes [195, 305] and glaucoma [127, 158, 191, 244]. Disturbances in ChBF and/or its neural regulation may also be involved in ischemic outer retinal disease and myopic retinopathy [40, 110, 286]. Consistent with this, impaired regulation of ChBF in response to exercise-induced increases in systemic BP (and thus ocular perfusion pressure) is observed in inactive central serous chorioretinopathy [365]. Adverse lifestyle choices may also deleterious affect adaptive neural regulation of ChBF, since smokers show impaired ChBF compensation for exercise-induced ocular perfusion pressure increase [396]. The specific link between parasympathetic ChBF impairment and retinopathy in aging is confirmed by our studies in pigeons [91, 93]. The above-described work of Steinle and coworkers in rats shows a similar importance of sympathetic control of ChBF. Thus, the available data are consistent with the view that disturbances in the maintenance of basal neurogenic choroidal tone and/or adaptive ChBF neural control could be contributing factors to the retinal declines seen in humans and other species in aging, hypertension, diabetes, glaucoma, ischemic retinal disease, and myopia. Autonomic neuropathy in diabetes may in particular contribute to overperfusion and leakiness of ocular vessels in diabetes [35]. Clearly, other mechanisms may also contribute to retinal damage in these conditions, such as edema, retinal detachment, and neovascularization

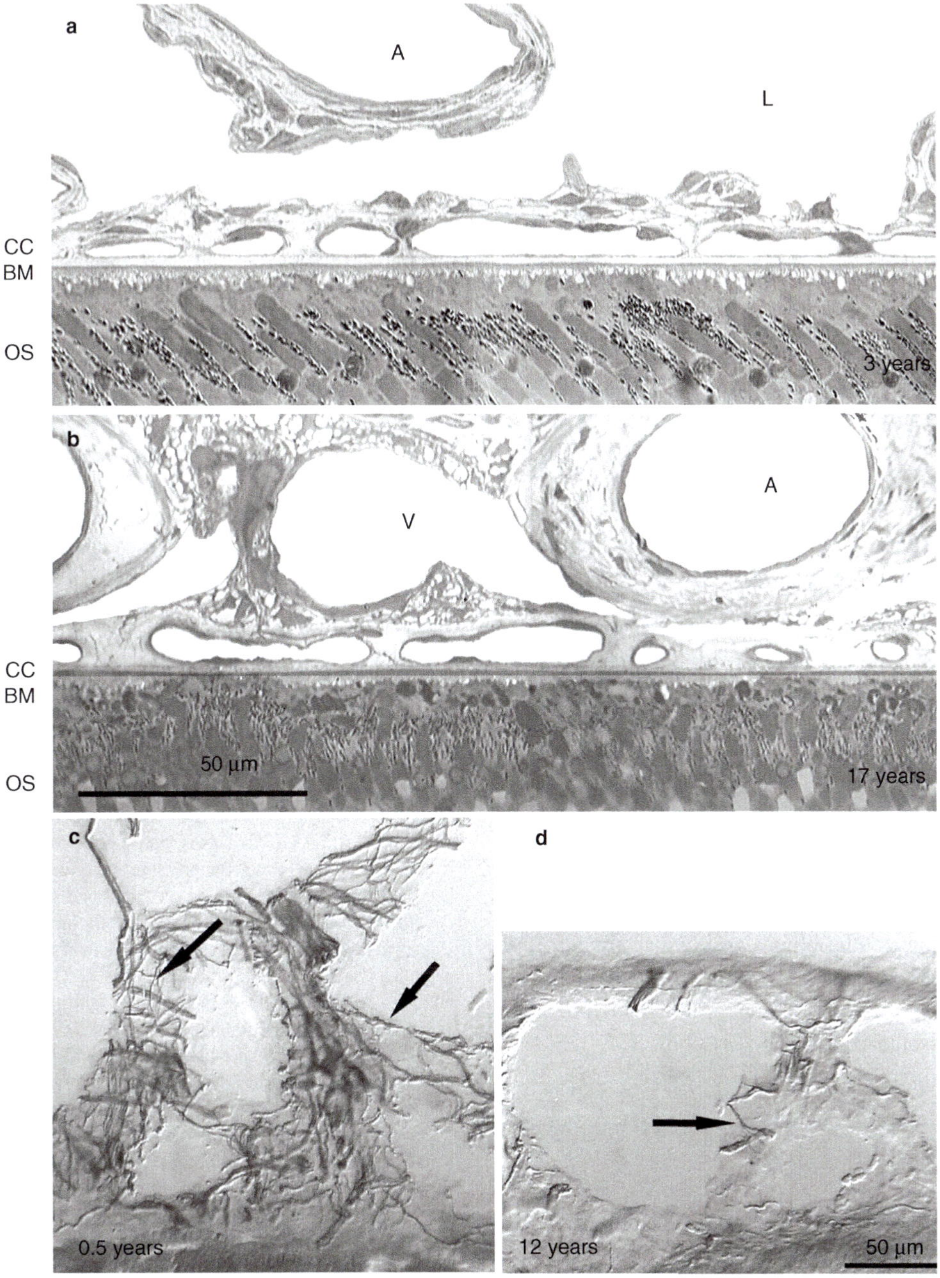

Fig. 12.33 Images showing the decline in choroidal vascularity and innervation with age in pigeons. Images (**a** and **b**) show a pair of LM images that illustrate the age-related decline in choriocapillary vessel abundance, showing choroid from a 3-year-old animal that had a choriocapillary coverage of 86% (**a**), with a mean intercapillary distance (ICD) of 7.26 mm, and a choroid from a 17-year-old pigeon that had a capillary coverage of 48% and a mean ICD of 13.15 mm (**b**). Images (**c** and **d**) illustrate the abundance of intrachoroidal nerve fibers of ciliary ganglion origin from a representative 0.5-year-old pigeon (**c**) and a representative 12-year-old pigeon (**d**). The nerve fibers were immunolabeled by the peroxidase-antiperoxidase method using antibodies against the 3A10 neurofilament-associated antigen. These images typify the age-related decline in nerve fiber abundance in pigeon choroid. Magnification same in (**a**) as in (**b**), and in (**c**) as in (**d**). *A* artery, *L* lacunae, *CC* choriocapillaris, *BM* Bruch membrane, *OS* outer segment, *V* vein

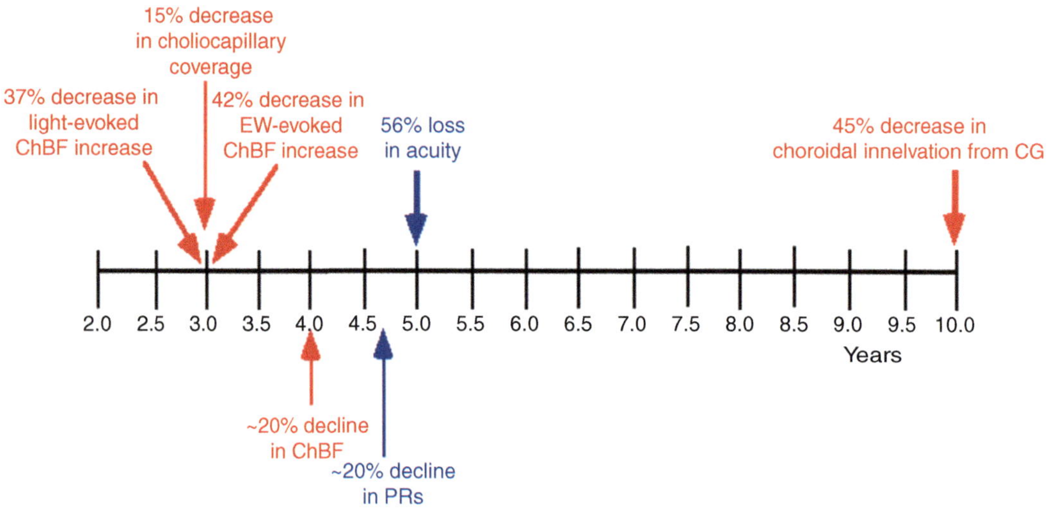

Fig. 12.34 A time line summarizing the ages by which or at which significant change in visual and choroidal parameters occurred in our sample of pigeons. The age by which 50% loss had occurred for the retinal illumination and EWM-evoked ChBF responses was around 3 years (2.9 and 3.16, respectively). Despite the early decline in the function of the EWM-ciliary ganglion circuit, the age by which 50% of the loss in the ciliary ganglion innervation of the choroid was not until 10 years. These results suggest that the functional decline in this circuit precedes the actual intrachoroidal fiber loss. These findings are compared with findings for basal ChBF, choriocapillary vessel abundance, photoreceptor loss and acuity decline and choroidal parameters. The half loss point for choriocapillary vessels was also at 3 years, but the significant drop in photoreceptor abundance and the 50% for acuity did not occur until after the major declines in the various choroidal vascular and functional parameters

[96, 368, 375]. Nonetheless, impaired neural regulation of ChBF may contribute to degeneration of neural retina, as well as facilitate (in conjunction with other etiological factors) the pathological changes in the RPE or Bruch membrane that further edema, retinal detachment, and neovascularization [96, 181, 277, 374, 375].

References

1. Abe S, Karita K, Izumi H et al (1995) Increased and decreased choroidal blood flow elicited by cervical sympathetic nerve stimulation in the cat. Jpn J Physiol 45:347–353
2. Agassandian K, Fazan VP, Adanina V et al (2002) Direct projections from the cardiovascular nucleus tractus solitarii to pontine preganglionic parasympathetic neurons: a link to cerebrovascular regulation. J Comp Neurol 452:242–254
3. Alderton WK, Cooper CE, Knowles RG (2001) Nitric oxide synthases: structure, function and inhibition. Biochem J 357:593–615
4. Alm A, Bill A (1970) Blood flow and oxygen extraction in the cat uvea at normal and high intraocular pressures. Acta Physiol Scand 80:19–28
5. Alm A, Bill A (1972) The oxygen supply to the retina. II. Effects of high intraocular pressure and of increased arterial carbon dioxide tension on uveal and retinal blood flow in cats. A study with radioactively labelled microspheres including flow determinations in brain and some other tissues. Acta Physiol Scand 84:306–319
6. Alm A, Bill A (1973) The effect of stimulation of the cervical sympathetic chain on retinal oxygen tension and on uveal, retinal and cerebral blood flow in cats. Acta Physiol Scand 88:84–94
7. Alm A, Bill A, Young FA (1973) The effects of pilocarpine and neostigmine on the blood flow through the anterior uvea in monkeys. A study with radioactively labelled microspheres. Exp Eye Res 15: 31–36
8. Alm A, Stjernschantz J, Bill A (1976) Effects of oculomotor nerve stimulation on ocular blood flow in rabbits after sympathetic denervation. Exp Eye Res 23:609–613
9. Alm A (1977) The effect of sympathetic stimulation on blood flow through t, e uvea, retina and optic nerve in monkeys (*Macacca irus*). Exp Eye Res 25:19–24
10. Alm A (1992) Ocular circulation. In: Hart WM (ed) Adler's physiology of the eye: clinical application. Mosby, St Louis, pp 198–227
11. Alm P, Uvelius B, Ekstrom J et al (1995) Nitric oxide synthase-containing neurons in rat parasympathetic, sympathetic and sensory ganglia: a comparative study. Histochem J 27:819–831

12. Almegard B, Andersson SE (1990) Outflow facility in the monkey eye: effects of calcitonin gene-related peptide, cholecystokinin, galanin, substance P and capsaicin. Exp Eye Res 51:685–689
13. Andersson SE, Bill A (1989) Effects of intravenous calcitonin gene-related peptide (CGRP) and substance P on the blood-aqueous barrier in the rabbit. Acta Physiol Scand 135:349–357
14. Arends JJ, Wild JM, Zeigler HP (1988) Projections of the nucleus of the tractus solitarius in the pigeon (*Columba livia*). J Comp Neurol 278:405–429
15. Ayajiki K, Fujioka H, Okamura T et al (2001) Relatively selective neuronal nitric oxide synthase inhibition by 7-nitroindazole in monkey isolated cerebral arteries. Eur J Pharmacol 423:179–183
16. Badoer E, Merolli J (1998) Neurons in the hypothalamic paraventricular nucleus that project to the rostral ventrolateral medulla are activated by haemorrhage. Brain Res 791:317–320
17. Bakken IJ, Vincent MB, Sjaavaag I et al (1995) Vasodilation in porcine ophthalmic artery: peptide interaction with acetylcholine and endothelial dependence. Neuropeptides 29:69–75
18. Baljet B, VanderWerf F (2005) Connections between the lacrimal gland and sensory trigeminal neurons: a WGA/HRP study in the cynomolgous monkey. J Anat 206:257–263
19. Beckers HJ, Klooster J, Vrensen GF et al (1993) Facial parasympathetic innervation of the rat choroid, lacrimal glands and ciliary ganglion. An ultrastructural pterygopalatine tracing and immunohistochemical study. Ophthalmic Res 25:319–330
20. Beckers HJ, Klooster J, Vrensen GF et al (1993) Substance P in rat corneal and iridal nerves: an ultrastructural immunohistochemical study. Ophthalmic Res 25:192–200
21. Behar-Cohen FF, Goureau O, D'Hermies F et al (1996) Decreased intraocular pressure induced by nitric oxide donors is correlated to nitrite production in the rabbit eye. Invest Ophthalmol Vis Sci 37:1711–1715
22. Belmonte C, Bartels SP, Liu JH et al (1987) Effects of stimulation of the ocular sympathetic nerves on IOP and aqueous humor flow. Invest Ophthalmol Vis Sci 28:1649–1654
23. Belmonte C, Garcia-Hirschfeld J, Gallar J (1997) Neurobiology of ocular pain. Prog Retin Eye Res 16:117–156
24. Benedito S, Prieto D, Nielsen PJ et al (1991) Role of the endothelium in acetylcholine-induced relaxation and spontaneous tone of bovine isolated retinal small arteries. Exp Eye Res 52:575–579
25. Bergua A, Jünemann A, Naumann GO (1993) NADPH-D-reaktive chorioidale Ganglienzellen beim Menschen. Klin Monatsbl Augenheilkd 203:77–82
26. Bergua A, Mayer B, Neuhuber WL (1996) Nitrergic and VIPergic neurons in the choroid and ciliary ganglion of the duck *Anis carina*. Anat Embryol (Berl) 193:239–248
27. Bergua A, Schrodl F, Neuhuber WL (2003) Vasoactive intestinal and calcitonin gene-related peptides, tyrosine hydroxylase and nitrergic markers in the innervation of the rat central retinal artery. Exp Eye Res 77:367–374
28. Berk ML, Smith SE, Mullins LA (1993) Distribution, parabrachial region projection, and coexistence of neuropeptide and catecholamine cells of the nucleus of the solitary tract in the pigeon. J Comp Neurol 327:416–441
29. Bill A (1962) Autonomic nervous control of uveal blood flow. Acta Physiol Scand 56:70–81
30. Bill A, Linder J (1976) Sympathetic control of cerebral blood flow in acute arterial hypertension. Acta Physiol Scand 96:114–121
31. Bill A, Stjernschantz J, Alm A (1976) Effects of hexamethonium, biperiden and phentolamine on the vasoconstrictive effects of oculomotor nerve stimulation in rabbits. Exp Eye Res 23:615–622
32. Bill A, Sperber G, Ujiie K (1983) Physiology of the choroidal vascular bed. Int Ophthalmol 6:101–107
33. Bill A (1985) Some aspects of the ocular circulation. Friedenwald lecture. Invest Ophthalmol Vis Sci 26:410–424
34. Bill A, Nilsson SF (1985) Control of ocular blood flow. J Cardiovasc Pharmacol 7(Suppl 3):S96–S102
35. Bill A, Sperber GO (1990) Control of retinal and choroidal blood flow. Eye (Lond) 4(Pt 2):319–325
36. Bill A (1991) The 1990 Endre Balazs Lecture. Effects of some neuropeptides on the uvea. Exp Eye Res 53:3–11
37. Bill A (1984) The circulation in the eye. In: Renkin EM, Michel CC (eds) Handbook of physiology: the cardiovascular system IV: microcirculation part 2. Waverly Press, Baltimore, pp 1001–1035
38. Birch DG (1987) Diurnal rhythm in the human rod ERG: retinitis pigmentosa. Invest Ophthalmol Vis Sci 28:2042–2048
39. Bjorklund H, Hokfelt T, Goldstein M et al (1985) Appearance of the noradrenergic markers tyrosine hydroxylase and neuropeptide Y in cholinergic nerves of the iris following sympathectomy. J Neurosci 5:1633–1640
40. Blacharski P (1988) Pathological progressive myopia. In: Newsome DA (ed) Retinal dystrophies and degenerations. Raven Press, New York, pp 257–269
41. Bok D (1985) Retinal photoreceptor-pigment epithelium interactions. Friedenwald lecture. Invest Ophthalmol Vis Sci 26:1659–1694
42. Boker T, Fang T, Steinmetz R (1993) Refractive error and choroidal perfusion characteristics in patients with choroidal neovascularization and age-related macular degeneration. Ger J Ophthalmol 2:10–13
43. Boussery K, Delaey C, Van de Voorde J (2005) The vasorelaxing effect of CGRP and natriuretic peptides in isolated bovine retinal arteries. Invest Ophthalmol Vis Sci 46:1420–1427
44. Braakman R, van der Linden P, Sipkema P (1999) Effects of topical beta-blockers on the diameter of the isolated porcine short posterior ciliary artery. Invest Ophthalmol Vis Sci 40:370–377
45. Braun RD, Linsenmeier RA, Goldstick TK (1995) Oxygen consumption in the inner and outer retina of the cat. Invest Ophthalmol Vis Sci 36:542–554

46. Buijs RM, la Fleur SE, Wortel J et al (2003) The suprachiasmatic nucleus balances sympathetic and parasympathetic output to peripheral organs through separate preautonomic neurons. J Comp Neurol 464:36–48
47. Butler JM, Ruskell GL, Cole DF et al (1984) Effects of VIIth (facial) nerve degeneration on vasoactive intestinal polypeptide and substance P levels in ocular and orbital tissues of the rabbit. Exp Eye Res 39:523–532
48. Calarescu FR, Ciriello J, Caverson MM et al (1984) Functional neuroanatomy of ventral pathways controlling the circulation. In: Kochen TA, Guthrie CP (eds) Hypertension and the brain. Futura Publications, New York, pp 3–21
49. Cantwell EL, Cassone VM (2006) Chicken suprachiasmatic nuclei: II. Autoradiographic and immunohistochemical analysis. J Comp Neurol 499:442–457
50. Cantwell EL, Cassone VM (2006) Chicken suprachiasmatic nuclei: I. Efferent and afferent connections. J Comp Neurol 496:97–120
51. Chamot SR, Movaffaghy A, Petrig BL et al (2000) Iris blood flow response to acute decreases in ocular perfusion pressure: a laser Doppler flowmetry study in humans. Exp Eye Res 70:107–112
52. Chou P, Lu DW, Chen JT (2000) Bilateral superior cervical ganglionectomy increases choroidal blood flow in the rabbit. Ophthalmologica 214:421–425
53. Chou PI, Lu DW, Chen JT (2001) Adrenergic supersensitivity of rabbit choroidal blood vessels after sympathetic denervation. Curr Eye Res 23:352–356
54. Chou PI, Lu DW, Chen JT (2002) Effect of sympathetic denervation on rabbit choroidal blood flow. Ophthalmologica 216:60–64
55. Chung CW, Tigges M, Stone RA (1996) Peptidergic innervation of the primate meibomian gland. Invest Ophthalmol Vis Sci 37:238–245
56. Ciriello J (1983) Brainstem projections of aortic baroreceptor afferent fibers in the rat. Neurosci Lett 36:37–42
57. Colasanti BK, Powell SR (1985) Effect of delta 9-tetrahydrocannabinol on intraocular pressure after removal of autonomic input. J Ocul Pharmacol 1:47–57
58. Collier RH (1967) Experimental embolic ischemia of the choroid. Arch Ophthalmol 77:683–692
59. Contreras RJ, Gomez MM, Norgren R (1980) Central origins of cranial nerve parasympathetic neurons in the rat. J Comp Neurol 190:373–394
60. Corvetti G, Pignocchino P, Sisto Daneo L (1988) Distribution and development of substance P immunoreactive axons in the chick cornea and uvea. Basic Appl Histochem 32:187–192
61. Cringle SJ, Yu DY, Alder V et al (1999) Light and choroidal PO_2 modulation of intraretinal oxygen levels in an avascular retina. Invest Ophthalmol Vis Sci 40:2307–2313
62. Cuthbertson S, White J, Fitzgerald ME et al (1996) Distribution within the choroid of cholinergic nerve fibers from the ciliary ganglion in pigeons. Vision Res 36:775–786
63. Cuthbertson S, Jackson B, Toledo C et al (1997) Innervation of orbital and choroidal blood vessels by the pterygopalatine ganglion in pigeons. J Comp Neurol 386:422–442
64. Cuthbertson S, Zagvazdin YS, Kimble TD et al (1999) Preganglionic endings from nucleus of Edinger-Westphal in pigeon ciliary ganglion contain neuronal nitric oxide synthase. Vis Neurosci 16:819–834
65. Cuthbertson S, LeDoux MS, Jones S et al (2003) Localization of preganglionic neurons that innervate choroidal neurons of pterygopalatine ganglion. Invest Ophthalmol Vis Sci 44:3713–3724
66. Dartt DA, Baker AK, Vaillant C et al (1984) Vasoactive intestinal polypeptide stimulation of protein secretion from rat lacrimal gland acini. Am J Physiol 247:G502–G509
67. Dartt DA, McCarthy DM, Mercer HJ et al (1995) Localization of nerves adjacent to goblet cells in rat conjunctiva. Curr Eye Res 14:993–1000
68. Denis P, Fardin V, Nordmann JP et al (1991) Localization and characterization of substance P binding sites in rat and rabbit eyes. Invest Ophthalmol Vis Sci 32:1894–1902
69. Deussen A, Sonntag M, Vogel R (1993) L-arginine-derived nitric oxide: a major determinant of uveal blood flow. Exp Eye Res 57:129–134
70. Diebold Y, Ríos JD, Hodges RR, Rawe I, Dartt DA (2001) Presence of nerves and their receptors in mouse and human conjunctival goblet cells. Invest Ophthalmol Vis Sci 42:2270–2282
71. Ding C, Walcott B, Keyser KT (2001) Neuronal nitric oxide synthase and the autonomic innervation of the mouse lacrimal gland. Invest Ophthalmol Vis Sci 42:2789–2794
72. Ding C, Walcott B, Keyser KT (2003) Sympathetic neural control of the mouse lacrimal gland. Invest Ophthalmol Vis Sci 44:1513–1520
73. Ehinger B (1966) Adrenergic nerves to the eye and to related structures in man and the cynomolgus monkey. Invest Ophthalmol 5:42–52
74. Ehinger B, Sundler F, Tervo K et al (1983) Substance P fibres in the anterior segment of the rabbit eye. Acta Physiol Scand 118:215–218
75. Elsas T, Edvinsson L, Sundler F et al (1994) Neuronal pathways to the rat conjunctiva revealed by retrograde tracing and immunocytochemistry. Exp Eye Res 58:117–126
76. Elsas T, Uddman R, Sundler F (1996) Vasoactive intestinal peptide- and nitric oxide synthase-containing nerve fibers in the rat ophthalmic artery have different origins. Graefes Arch Clin Exp Ophthalmol 234:125–130
77. Erickson KA, Schroeder A (2000) Direct effects of muscarinic agents on the outflow pathways in human eyes. Invest Ophthalmol Vis Sci 41:1743–1748
78. Erickson-Lamy KA, Kaufman PL (1988) Effect of cholinergic drugs on outflow facility after ciliary ganglionectomy. Invest Ophthalmol Vis Sci 29:491–494

79. Ernest JT (1977) The effect of systolic hypertension on rhesus monkey eyes after ocular sympathectomy. Am J Ophthalmol 84:341–344
80. Falsini B, Riva CE, Logean E (2002) Flicker-evoked changes in human optic nerve blood flow: relationship with retinal neural activity. Invest Ophthalmol Vis Sci 43:2309–2316
81. Fazekas A, Gazelius B, Edwall B et al (1987) VIP and noncholinergic vasodilatation in rabbit submandibular gland. Peptides 8:13–20
82. Ferrari-Dileo G, Davis EB, Anderson DR (1989) Biochemical evidence for cholinergic activity in retinal blood vessels. Invest Ophthalmol Vis Sci 30:473–477
83. Filosa JA, Bonev AD, Straub SV et al (2006) Local potassium signaling couples neuronal activity to vasodilation in the brain. Nat Neurosci 9:1397–1403
84. Fischer AJ, McKinnon LA, Nathanson NM et al (1998) Identification and localization of muscarinic acetylcholine receptors in the ocular tissues of the chick. J Comp Neurol 392:273–284
85. Fischer AJ, Stell WK (1999) Nitric oxide synthase-containing cells in the retina, pigmented epithelium, choroid, and sclera of the chick eye. J Comp Neurol 405:1–14
86. Fite KV (1985) Pretectal and accessory optic visual nuclei of fish, amphibia, and reptiles: theme and variations. Brain Behav Evol 26:71–90
87. Fitzgerald ME, Vana BA, Reiner A (1990) Evidence for retinal pathology following interruption of neural regulation of choroidal blood flow: Muller cells express GFAP following lesions of the nucleus of Edinger-Westphal in pigeons. Curr Eye Res 9:583–598
88. Fitzgerald ME, Vana BA, Reiner A (1990) Control of choroidal blood flow by the nucleus of Edinger-Westphal in pigeons: a laser Doppler study. Invest Ophthalmol Vis Sci 31:2483–2492
89. Fitzgerald ME, Caldwell RB, Reiner A (1992) Vasoactive intestinal polypeptide-containing nerve fibers are increased in abundance in the choroid of dystrophic RCS rats. Curr Eye Res 11:501–515
90. Fitzgerald ME, Gamlin PD, Zagvazdin Y et al (1996) Central neural circuits for the light-mediated reflexive control of choroidal blood flow in the pigeon eye: a laser Doppler study. Vis Neurosci 13:655–669
91. Fitzgerald ME, Tolley E, Frase S et al (2001) Functional and morphological assessment of age-related changes in the choroid and outer retina in pigeons. Vis Neurosci 18:299–317
92. Fitzgerald ME, Wildsoet CF, Reiner A (2002) Temporal relationship of choroidal blood flow and thickness changes during recovery from form deprivation myopia in chicks. Exp Eye Res 74:561–570
93. Fitzgerald ME, Tolley E, Jackson B et al (2005) Anatomical and functional evidence for progressive age-related decline in parasympathetic control of choroidal blood flow in pigeons. Exp Eye Res 81:478–491
94. Flügel C, Tamm ER, Mayer B et al (1994) Species differences in choroidal vasodilative innervation: evidence for specific intrinsic nitrergic and VIP-positive neurons in the human eye. Invest Ophthalmol Vis Sci 35:592–599
95. Flügel-Koch C, Kaufman P, Lütjen-Drecoll E (1994) Association of a choroidal ganglion cell plexus with the fovea centralis. Invest Ophthalmol Vis Sci 35:4268–4272
96. Frank RN (1988) Studies in diabetic retinopathy. In: Tso MOM (ed) Retinal disease: biomedical foundations & clinical management. Lippincott Co., Philadelphia, pp 165–180
97. Friedman E (1970) Choroidal blood flow. Pressure-flow relationships. Arch Ophthalmol 83:95–99
98. Friedman E, Ivry E, Glynn R et al (1989) Increased scleral rigidity and age-related macular degeneration. Ophthalmology 196:104–108
99. Friedman E, Krupsky S, Lane AM et al (1995) Ocular blood flow velocity in age-related macular degeneration. Ophthalmology 102:640–646
100. Fuchsjäger-Mayrl G, Polska E, Malec M et al (2001) Unilateral light–dark transitions affect choroidal blood flow in both eyes. Vision Res 41:2919–2924
101. Fuchsjäger-Mayrl G, Luksch A, Malec M et al (2003) Role of endothelin-1 in choroidal blood flow regulation during isometric exercise in healthy humans. Invest Ophthalmol Vis Sci 44:728–733
102. Gabelt BT, Crawford K, Kaufman PL (1991) Outflow facility and its response to pilocarpine decline in aging rhesus monkeys. Arch Ophthalmol 109:879–882
103. Gallar J, Acosta MC, Belmonte C (2003) Activation of scleral cold thermoreceptors by temperature and blood flow changes. Invest Ophthalmol Vis Sci 44:697–705
104. Gamlin PD, Reiner A, Karten HJ (1982) Substance P-containing neurons of the avian suprachiasmatic nucleus project directly to the nucleus of Edinger-Westphal. Proc Natl Acad Sci USA 79:3891–3895
105. Gamlin PD, Reiner A, Erichsen JT et al (1984) The neural substrate for the pupillary light reflex in the pigeon (*Columba livia*). J Comp Neurol 226:523–543
106. Gamlin PD, Reiner A (1991) The Edinger-Westphal nucleus: sources of input influencing accommodation, pupilloconstriction, and choroidal blood flow. J Comp Neurol 306:425–438
107. Gamlin PDR (2000) Functions of the Edinger-Westphal Nucleus. In: Burnstock G, Sillito AM (eds) Nervous control of the eye. Harwood Academic Publishers, Newark, pp 117–154
108. Ganchrow D, Gentle MJ, Ganchrow JR (1987) Central distribution and efferent origins of facial nerve branches in the chicken. Brain Res Bull 19:231–238
109. Garhöfer G, Zawinka C, Resch H et al (2004) Diffuse luminance flicker increases blood flow in major retinal arteries and veins. Vision Res 44:833–838
110. Gaudric A, Coscas G, Bird AC (1982) Choroidal ischemia. Am J Ophthalmol 94:489–498
111. Gay AJ, Goldor H, Smith M (1964) Chorioretinal vascular occlusions with latex spheres. Invest Ophthalmol 3:647–656

112. Geiser MH, Riva CE, Dorner GT et al (2000) Response of choroidal blood flow in the foveal region to hyperoxia and hyperoxia-hypercapnia. Curr Eye Res 21:669–676
113. Gherezghiher T, Hey JA, Koss MC (1990) Parasympathetic nervous control of intraocular pressure. Exp Eye Res 50:457–462
114. Gherezghiher T, Okubo H, Koss MC (1991) Choroidal and ciliary body blood flow analysis: application of laser Doppler flowmetry in experimental animals. Exp Eye Res 53:151–156
115. Gidday JM, Zhu Y (1995) Nitric oxide does not mediate autoregulation of retinal blood flow in newborn pig. Am J Physiol 269:H1065–H1072
116. Godino A, Giusti-Paiva A, Antunes-Rodrigues J et al (2005) Neurochemical brain groups activated after an isotonic blood volume expansion in rats. Neuroscience 133:493–505
117. Golanov EV, Christensen JR, Reis DJ (2001) Neurons of a limited subthalamic area mediate elevations in cortical cerebral blood flow evoked by hypoxia and excitation of neurons of the rostral ventrolateral medulla. J Neurosci 21:4032–4041
118. Goldstein DS, Kopin IJ (1990) The autonomic nervous system and catecholamines in normal blood pressure control and in hypertension. In: Laragh JH, Brenner BM (eds) Hypertension: pathophysiology, diagnosis, and management. Raven Press, New York, pp 711–747
119. Goldstein IM, Ostwald P, Roth S (1996) Nitric oxide: a review of its role in retinal function and disease. Vision Res 36:2979–2994
120. Goto W, Oku H, Okuno T et al (2003) Amelioration by topical bunazosin hydrochloride of the impairment in ocular blood flow caused by nitric oxide synthase inhibition in rabbits. J Ocul Pharmacol Ther 19:63–73
121. Gotoh F, Tanaka K (1988) Regulation of cerebral blood flow. In: Bruyn GW, Vinken PJ (eds) Handbook of clinical neurology. Elsevier, Amsterdam, pp 47–77
122. Gottanka J, Kirch W, Tamm ER (2005) The origin of extrinsic nitrergic axons supplying the human eye. J Anat 206:225–229
123. Grunwald JE, Riva CE, Kozart DM (1988) Retinal circulation during a spontaneous rise of intraocular pressure. Br J Ophthalmol 72:754–758
124. Grunwald JE (1991) Effect of two weeks of timolol maleate treatment on the normal retinal circulation. Invest Ophthalmol Vis Sci 32:39–45
125. Grunwald JE, Hariprasad SM, DuPont J (1998) Effect of aging on foveolar choroidal circulation. Arch Ophthalmol 116:150–154
126. Grunwald JE, Hariprasad SM, DuPont J et al (1998) Foveolar choroidal blood flow in age-related macular degeneration. Invest Ophthalmol Vis Sci 39:385–390
127. Grunwald JE, Piltz J, Hariprasad SM et al (1998) Optic nerve and choroidal circulation in glaucoma. Invest Ophthalmol Vis Sci 39:2329–2336
128. Grunwald JE, Iannaccone A, DuPont J (1999) Effect of isosorbide mononitrate on the human optic nerve and choroidal circulations. Br J Ophthalmol 83:162–167
129. Guglielmone R, Cantino D (1982) Autonomic innervation of the ocular choroid membrane in the chicken: a fluorescence-histochemical and electron-microscopic study. Cell Tissue Res 222:417–431
130. Gupta N, Drance SM, McAllister R et al (1994) Localization of M3 muscarinic receptor subtype and mRNA in the human eye. Ophthalmic Res 26: 207–213
131. Gupta N, McAllister R, Drance SM et al (1994) Muscarinic receptor M1 and M2 subtypes in the human eye: QNB, pirenzipine, oxotremorine, and AFDX-116 in vitro autoradiography. Br J Ophthalmol 78:555–559
132. Haefliger IO, Flammer J, Lüscher TF (1993) Heterogeneity of endothelium-dependent regulation in ophthalmic and ciliary arteries. Invest Ophthalmol Vis Sci 34:1722–1730
133. Haefliger IO, Flammer J, Beny JL et al (2001) Endothelium-dependent vasoactive modulation in the ophthalmic circulation. Prog Retin Eye Res 20:209–225
134. Hara H, Hamill GS, Jacobowitz DM (1985) Origin of cholinergic nerves to the rat major cerebral arteries: coexistence with vasoactive intestinal polypeptide. Brain Res Bull 14:179–188
135. Hardy P, Varma DR, Chemtob S (1997) Control of cerebral and ocular blood flow autoregulation in neonates. Pediatr Clin North Am 44:137–152
136. Hawkins BS, Bird A, Klein R et al (1999) Epidemiology of age-related macular degeneration. Mol Vis 5:26
137. Hayreh SS (2001) Blood flow in the optic nerve head and factors that may influence it. Prog Retin Eye Res 20:595–624
138. Heino P, Oksala O, Luhtala J et al (1995) Localization of calcitonin gene-related peptide binding sites in the eye of different species. Curr Eye Res 14:783–790
139. Herbert H, Moga MM, Saper CB (1990) Connections of the parabrachial nucleus with the nucleus of the solitary tract and the medullary reticular formation in the rat. J Comp Neurol 293:540–580
140. Hermes ML, Buijs RM, Renaud LP (1996) Electrophysiology of suprachiasmatic nucleus projections to hypothalamic paraventricular nucleus neurons. Prog Brain Res 111:241–252
141. Herron WL, Riegel BW, Myers OE et al (1969) Retinal dystrophy in the rat – a pigment epithelial disease. Invest Ophthalmol 8:595–604
142. Hill CE, Gould DJ (1995) Modulation of sympathetic vasoconstriction by sensory nerves and nitric oxide in rat irideal arterioles. J Pharmacol Exp Ther 273:918–926
143. Hirai R, Tamamaki N, Hukami K et al (1994) Ultrastructural analysis of tyrosine hydroxylase-, substance P-, and calcitonin gene-related peptide-immunoreactive nerve fibers in the rat iris. Ophthalmic Res 26:169–180

144. Hodos W (1991). Animal models of life-span development. In The Changing Visual System: Maturation and Aging in Central Nervous System, ed. Bagnoli, P. & Hodos, W., pp. 21–32. New York: Plenum Press
145. Hodos W, Miller RF, Ghim MM et al (1998) Visual acuity losses in pigeons with lesions of the nucleus of Edinger-Westphal that disrupt the adaptive regulation of choroidal blood flow. Vis Neurosci 15: 273–287
146. Horiguchi M, Miyake Y (1991) Effect of temperature on electroretinograph readings during closed vitrectomy in humans. Arch Ophthalmol 109:1127–1129
147. Hosoya Y, Matsushita M, Sugiura Y (1984) Hypothalamic descending afferents to cells of origin of the greater petrosal nerve in the rat, as revealed by a combination of retrograde HRP and anterograde autoradiographic techniques. Brain Res 290:141–145
148. Hoste AM, Andries LJ (1991) Contractile responses of isolated bovine retinal microarteries to acetylcholine. Invest Ophthalmol Vis Sci 32:1996–2005
149. Huemer KH, Garhofer G, Aggermann T et al (2007) Role of nitric oxide in choroidal blood flow regulation during light/dark transitions. Invest Ophthalmol Vis Sci 48:4215–4219
150. Iadecola C, Nedergaard M (2007) Glial regulation of the cerebral microvasculature. Nat Neurosci 10:1369–1376
151. Iannaccone AE, DuPont J, Grunwald JE (2000) Human retinal hemodynamics following administration of 5-isosorbide mononitrate. Curr Eye Res 20:205–210
152. Imai K (1977) Cholinergic innervation of the choroid. Ophthalmic Res 9:194–200
153. Ishitsuka T, Iadecola C, Underwood MD et al (1986) Lesions of nucleus tractus solitarii globally impair cerebrovascular autoregulation. Am J Physiol 251:H269–H281
154. Ito H, Seki M (1998) Ascending projections from the area postrema and the nucleus of the solitary tract of *Suncus murinus*: anterograde tracing study using *Phaseolus vulgaris* leucoagglutinin. Okajimas Folia Anat Jpn 75:9–31
155. Jablonski MM, Iannaccone A, Reynolds DH et al (2007) Age-related decline in VIP-positive parasympathetic nerve fibers in the human submacular choroid. Invest Ophthalmol Vis Sci 48:479–485
156. Jacot JL, O'Neill JT, Scandling DM et al (1998) Nitric oxide modulation of retinal, choroidal, and anterior uveal blood flow in newborn piglets. J Ocul Pharmacol Ther 14:473–489
157. Jhamandas JH, Harris KH, Petrov T, Krukoff TL (1992) Characterization of the parabrachial nucleus input to the hypothalamic paraventricular nucleus in the rat. J Neuroendocrinol 4:461–471
158. James CB, Smith SE (1991) Pulsatile ocular blood flow in patients with low tension glaucoma. Br J Ophthalmol 75:466–470
159. Jandrasits K, Polak K, Luksch A et al (2001) Effects of atropine and propranolol on retinal vessel diameters during isometric exercise. Ophthalmic Res 33: 185–190
160. Jeppesen P, Sanye-Hajari J, Bek T (2007) Increased blood pressure induces a diameter response of retinal arterioles that increases with decreasing arteriolar diameter. Invest Ophthalmol Vis Sci 48:328–331
161. Johansson O, Lundberg JM (1981) Ultrastructural localization of VIP-like immunoreactivity in large dense-core vesicles of 'cholinergic-type' nerve terminals in cat exocrine glands. Neuroscience 6:847–862
162. Johnson PC (1980) The myogenic response. In: Bohr DF, Somlyo AP, Sparks HV (eds) Handbook of physiology: the cardiovascular system: vascular smooth muscle. Waverly Press, Baltimore, pp 409–442
163. Kahl BF, Reid TW (1995) Substance P and the eye. Prog Retin Eye Res 14:473–504
164. Katz DM, Karten HJ (1979) The discrete anatomical localization of vagal aortic afferents within a catecholamine-containing cell group in the nucleus solitarius. Brain Res 171:187–195
165. Kawarai M, Koss MC (1998) Sympathetic vasoconstriction in the rat anterior choroid is mediated by alpha1-adrenoceptors. Eur J Pharmacol 363:35–40
166. Kelly PA, Buckley CH, Ritchie IM et al (1998) Possible role for nitric oxide releasing nerves in the regulation of ocular blood flow in the rat. Br J Ophthalmol 82:1199–1202
167. Kergoat H, Faucher C (1999) Effects of oxygen and carbogen breathing on choroidal hemodynamics in humans. Invest Ophthalmol Vis Sci 40:2906–2911
168. Kessler TL, Dartt DA (1994) Neural stimulation of conjunctival goblet cell mucous secretion in rats. Adv Exp Med Biol 350:393–398
169. Kiel JW, Shepherd AP (1992) Autoregulation of choroidal blood flow in the rabbit. Invest Ophthalmol Vis Sci 33:2399–2410
170. Kiel JW (1994) Choroidal myogenic autoregulation and intraocular pressure. Exp Eye Res 58:529–543
171. Kiel JW, van Heuven WA (1995) Ocular perfusion pressure and choroidal blood flow in the rabbit. Invest Ophthalmol Vis Sci 36:579–585
172. Kiel JW, Lovell MO (1996) Adrenergic modulation of choroidal blood flow in the rabbit. Invest Ophthalmol Vis Sci 37:673–679
173. Kiel JW (1999) Modulation of choroidal autoregulation in the rabbit. Exp Eye Res 69:413–429
174. Kimble TD, Fitzgerald ME, Reiner A (2006) Sustained upregulation of glial fibrillary acidic protein in Muller cells in pigeon retina following disruption of the parasympathetic control of choroidal blood flow. Exp Eye Res 83:1017–1030
175. Kirby ML, Diab IM, Mattio TG (1978) Development of adrenergic innervation of the iris and fluorescent ganglion cells in the choroid of the chick eye. Anat Rec 191:311–319
176. Kirch W, Horneber M, Tamm ER (1996) Characterization of Meibomian gland innervation in the cynomolgus monkey (*Macaca fascicularis*). Anat Embryol (Berl) 193:365–375
177. Kiryu J, Asrani S, Shahidi M et al (1995) Local response of the primate retinal microcirculation to

increased metabolic demand induced by flicker. Invest Ophthalmol Vis Sci 36:1240–1246
178. Kitamura Y, Okamura T, Kani K et al (1993) Nitric oxide-mediated retinal arteriolar and arterial dilatation induced by substance P. Invest Ophthalmol Vis Sci 34:2859–2865
179. Klooster J, Beckers HJ, Ten Tusscher MP et al (1996) Sympathetic innervation of the rat choroid: an autoradiographic tracing and immunohistochemical study. Ophthalmic Res 28:36–43
180. Korf HW (1984) Neuronal organization of the avian paraventricular nucleus: intrinsic, afferent, and efferent connections. J Exp Zool 232:387–395
181. Korte GE, Burns MS, Bellhorn RW (1989) Epithelium-capillary interactions in the eye: the retinal pigment epithelium and the choriocapillaris. Int Rev Cytol 114:221–248
182. Koss MC, Kuhlman L, Gherezghiher T (1991) IOP and ChBF control by the EW nucleus in cats. Invest Ophthalmol Vis Sci Suppl 32:1186
183. Koss MC, Gherezghiher T (1993) Adrenoceptor subtypes involved in neurally evoked sympathetic vasoconstriction in the anterior choroid of cats. Exp Eye Res 57:441–447
184. Koss MC (1994) Adrenoceptor mechanisms in epinephrine-induced anterior choroidal vasoconstriction in cats. Exp Eye Res 59:715–722
185. Koss MC (1998) Role of nitric oxide in maintenance of basal anterior choroidal blood flow in rats. Invest Ophthalmol Vis Sci 39:559–564
186. Koss MC (1999) Analysis of blood flow in the long posterior ciliary artery of the cat. Invest Ophthalmol Vis Sci 40:800–804
187. Krishna R, Mermoud A, Baerveldt G et al (1995) Circadian rhythm of intraocular pressure: a rat model. Ophthalmic Res 27:163–167
188. Krootila K, Uusitalo H, Palkama A (1988) Effect of neurogenic irritation and calcitonin gene-related peptide (CGRP) on ocular blood flow in the rabbit. Curr Eye Res 7:695–703
189. Krootila K, Uusitalo H, Palkama A (1991) Intraocular and cardiovascular effects of calcitonin gene-related peptide (CGRP)-I and -II in the rabbit. Invest Ophthalmol Vis Sci 32:3084–3090
190. Krukoff TL, Mactavish D, Jhamandas JH (1997) Activation by hypotension of neurons in the hypothalamic paraventricular nucleus that project to the brainstem. J Comp Neurol 385:285–296
191. Kubota T, Jonas JB, Naumann GO (1993) Decreased choroidal thickness in eyes with secondary angle closure glaucoma. An aetiological factor for deep retinal changes in glaucoma? Br J Ophthalmol 77:430–432
192. Kumagai N, Yuda K, Kadota T et al (1988) Substance P-like immunoreactivity in the central retinal artery of the rabbit. Exp Eye Res 46:591–596
193. Kuwayama Y, Stone RA (1987) Distinct substance P and calcitonin gene-related peptide immunoreactive nerves in the guinea pig eye. Invest Ophthalmol Vis Sci 28:1947–1954
194. Kuwayama Y, Grimes PA, Ponte B et al (1987) Autonomic neurons supplying the rat eye and the intraorbital distribution of vasoactive intestinal polypeptide (VIP)-like immunoreactivity. Exp Eye Res 44:907–922
195. Langham ME, Grebe R, Hopkins S et al (1991) Choroidal blood flow in diabetic retinopathy. Exp Eye Res 52:167–173
196. Laties AM, Jacobowitz D (1966) A comparative study of the autonomic innervation of the eye in monkey, cat, and rabbit. Anat Rec 156:383–395
197. Laties AM (1967) Central retinal artery innervation. Absence of adrenergic innervation to the intraocular branches. Arch Ophthalmol 77:405–409
198. Laties AM, Stone RA, Brecha NC (1981) Substance P-like immunoreactive nerve fibers in the trabecular meshwork. Invest Ophthalmol Vis Sci 21:484–486
199. LaVail MM (1981) Analysis of neurological mutants with inherited retinal degeneration. Friedenwald lecture. Invest Ophthalmol Vis Sci 21:638–657
200. LeDoux MS, Zhou Q, Murphy RB et al (2001) Parasympathetic innervation of the meibomian glands in rats. Invest Ophthalmol Vis Sci 42:2434–2441
201. Li H, Grimes P (1993) Adrenergic innervation of the choroid and iris in diabetic rats. Curr Eye Res 12:89–94
202. Lin T, Grimes PA, Stone RA (1988) Nerve pathways between the pterygopalatine ganglion and eye in cats. Anat Rec 222:95–102
203. Linder J (1981) Effects of facial nerve section and stimulation on cerebral and ocular blood flow in hemorrhagic hypotension. Acta Physiol Scand 112:185–193
204. Linsenmeier RA (1986) Effects of light and darkness on oxygen distribution and consumption in the cat retina. J Gen Physiol 88:521–542
205. Liu JH, Dacus AC, Bartels SP (1991) Adrenergic mechanism in circadian elevation of intraocular pressure in rabbits. Invest Ophthalmol Vis Sci 32:2178–2183
206. Longo A, Geiser M, Riva CE (2000) Subfoveal choroidal blood flow in response to light-dark exposure. Invest Ophthalmol Vis Sci 41:2678–2683
207. Longo A, Geiser MH, Riva CE (2004) Posture changes and subfoveal choroidal blood flow. Invest Ophthalmol Vis Sci 45:546–551
208. Lovasik JV, Kergoat H, Riva CE et al (2003) Choroidal blood flow during exercise-induced changes in the ocular perfusion pressure. Invest Ophthalmol Vis Sci 44:2126–2132
209. Luhtala J, Palkama A, Uusitalo H (1991) Calcitonin gene-related peptide immunoreactive nerve fibers in the rat conjunctiva. Invest Ophthalmol Vis Sci 32:640–645
210. Luhtala J, Uusitalo H (1991) The distribution and origin of substance P immunoreactive nerve fibres in the rat conjunctiva. Exp Eye Res 53:641–646
211. Luksch A, Polak K, Beier C et al (2000) Effects of systemic NO synthase inhibition on choroidal and

optic nerve head blood flow in healthy subjects. Invest Ophthalmol Vis Sci 41:3080–3084
212. Luksch A, Polska E, Imhof A et al (2003) Role of NO in choroidal blood flow regulation during isometric exercise in healthy humans. Invest Ophthalmol Vis Sci 44:734–739
213. Lundberg JM, Anggard A, Fahrenkrug J (1982) Complementary role of vasoactive intestinal polypeptide (VIP) and acetylcholine for cat submandibular gland blood flow and secretion. Acta Physiol Scand 114:329–337
214. Macri FJ (1971) Vasoconstriction produced in the iris-ciliary body of the cat eye by stimulation of local ganglion-like receptors. Invest Ophthalmol 10:581–588
215. Macri FJ, Cevario SJ (1975) Ciliary ganglion stimulation. II. Neurogenic, intraocular pathway for excitatory effects on aqueous humor production and outflow. Invest Ophthalmol 14:471–475
216. Macri FJ, Cevario SJ (1975) Ciliary ganglion stimulation. I. Effects on aqueous humor inflow and outflow. Invest Ophthalmol 14:28–33
217. Malmfors T (1965) The adrenergic innervation of the eye as demonstrated by fluorescence microscopy. Acta Physiol Scand 65:259–267
218. Mann RM, Riva CE, Stone RA et al (1995) Nitric oxide and choroidal blood flow regulation. Invest Ophthalmol Vis Sci 36:925–930
219. Mansson B, Ekstrom J (1991) On the non-adrenergic, non-cholinergic contribution to the parasympathetic nerve-evoked secretion of parotid saliva in the rat. Acta Physiol Scand 141:197–205
220. Martin XD, Rabineau PA (1989) Vasoconstrictive effect of topical timolol on human retinal arteries. Graefes Arch Clin Exp Ophthalmol 227:526–530
221. May CA (1997) Description and function of the ciliary nerves – some historical remarks on choroidal innervation. Exp Eye Res 65:1–5
222. May CA, Fuchs AV, Scheib M et al (2002) Characterization of nitrergic neurons in the porcine and human ciliary nerves. Invest Ophthalmol Vis Sci 43:581–586
223. May CA, Neuhuber W, Lutjen-Drecoll E (2004) Immunohistochemical classification and functional morphology of human choroidal ganglion cells. Invest Ophthalmol Vis Sci 45:361–367
224. May CA, Skorski LM, Lutjen-Drecoll E (2005) Innervation of the porcine ciliary muscle and outflow region. J Anat 206:231–236
225. May PJ, Reiner AJ, Ryabinin AE (2008) Comparison of the distributions of urocortin-containing and cholinergic neurons in the perioculomotor midbrain of the cat and macaque. J Comp Neurol 507:1300–1316
226. McCulloch C (1988) Choroideremia and other choroidal atrophies. In: Newsome DA (ed) Retinal dystrophies and degenerations. Raven Press, New York, pp 285–295
227. McKenna OC, Wallman J (1985) Accessory optic system and pretectum of birds: comparisons with those of other vertebrates. Brain Behav Evol 26:91–116
228. Medina L, Reiner A (1994) Distribution of choline acetyltransferase immunoreactivity in the pigeon brain. J Comp Neurol 342:497–537
229. Medrano CJ, Fox DA (1995) Oxygen consumption in the rat outer and inner retina: light- and pharmacologically-induced inhibition. Exp Eye Res 61: 273–284
230. Menage MJ, Robinson JC, Kaufman PL et al (1994) Retinal blood flow after superior cervical ganglionectomy: a laser Doppler study in the cynomolgus monkey. Br J Ophthalmol 78:49–53
231. Meneray MA, Fields TY, Bennett DJ (1997) Gs and Gq/11 couple vasoactive intestinal peptide and cholinergic stimulation to lacrimal secretion. Invest Ophthalmol Vis Sci 38:1261–1270
232. Meriney SD, Pilar G (1987) Cholinergic innervation of the smooth muscle cells in the choroid coat of the chick eye and its development. J Neurosci 7:3827–3839
233. Metelitsina TI, Grunwald JE, DuPont JC et al (2006) Effect of viagra on retinal vein diameter in AMD patients. Exp Eye Res 83:128–132
234. Meyer DB (1977) The avian eye and its adaptations. In: Crescitelli F (ed) Handbook of sensory physiology. Springer, Berlin, pp 549–611
235. Meyer P, Flammer J, Luscher TF (1993) Endothelium-dependent regulation of the ophthalmic microcirculation in the perfused porcine eye: role of nitric oxide and endothelins. Invest Ophthalmol Vis Sci 34:3614–3621
236. Miller A, Costa M, Furness JB et al (1981) Substance P immunoreactive sensory nerves supply the rat iris and cornea. Neurosci Lett 23:243–249
237. Miller AS, Coster DJ, Costa M et al (1983) Vasoactive intestinal polypeptide immunoreactive nerve fibres in the human eye. Aust J Ophthalmol 11:185–193
238. Milley JR, Rosenberg AA, Jones MD Jr (1984) Retinal and choroidal blood flows in hypoxic and hypercarbic newborn lambs. Pediatr Res 18:410–414
239. Mindel JS, Mittag TW (1976) Choline acetyltransferase in ocular tissues of rabbits, cats, cattle, and man. Invest Ophthalmol 15:808–814
240. Mione MC, Cavanagh JF, Lincoln J et al (1990) Long-term chemical sympathectomy leads to an increase of neuropeptide Y immunoreactivity in cerebrovascular nerves and iris of the developing rat. Neuroscience 34:369–378
241. Mizota A, Adachi-Usami E (2002) Effect of body temperature on electroretinogram of mice. Invest Ophthalmol Vis Sci 43:3754–3757
242. Moncada S, Palmer RM, Higgs EA (1991) Nitric oxide: physiology, pathophysiology, and pharmacology. Pharmacol Rev 43:109–142
243. Moore PK, Wallace P, Gaffen Z et al (1993) Characterization of the novel nitric oxide synthase inhibitor 7-nitro indazole and related indazoles: antinociceptive and cardiovascular effects. Br J Pharmacol 110:219–224
244. Moore RY (1983) Organization and function of a central nervous system circadian oscillator: the

suprachiasmatic hypothalamic nucleus. Fed Proc 42:2783–2789
245. Morita Y, Hardebo JE, Bouskela E (1995) Influence of cerebrovascular sympathetic, parasympathetic, and sensory nerves on autoregulation and spontaneous vasomotion. Acta Physiol Scand 154:121–130
246. Movaffaghy A, Chamot SR, Petrig BL et al (1998) Blood flow in the human optic nerve head during isometric exercise. Exp Eye Res 67:561–568
247. Nagaoka T, Sakamoto T, Mori F et al (2002) The effect of nitric oxide on retinal blood flow during hypoxia in cats. Invest Ophthalmol Vis Sci 43: 3037–3044
248. Nakai M, Ogino K (1984) The relevance of cardiopulmonary-vascular reflex to regulation of the brain vessels. Jpn J Physiol 34:193–197
249. Nakanome Y, Karita K, Izumi H et al (1995) Two types of vasodilatation in cat choroid elicited by electrical stimulation of the short ciliary nerve. Exp Eye Res 60:37–42
250. Nickla DL, Wildsoet C, Wallman J (1998) Visual influences on diurnal rhythms in ocular length and choroidal thickness in chick eyes. Exp Eye Res 66:163–181
251. Nilsson SF, Bill A (1984) Vasoactive intestinal polypeptide (VIP): effects in the eye and on regional blood flows. Acta Physiol Scand 121:385–392
252. Nilsson SF, Linder J, Bill A (1985) Characteristics of uveal vasodilation produced by facial nerve stimulation in monkeys, cats and rabbits. Exp Eye Res 40:841–852
253. Nilsson SF, Sperber GO, Bill A (1986) Effects of vasoactive intestinal polypeptide (VIP) on intraocular pressure, facility of outflow and formation of aqueous humor in the monkey. Exp Eye Res 43:849–857
254. Nilsson SF (1991) Neuropeptide Y (NPY): a vasoconstrictor in the eye, brain and other tissues in the rabbit. Acta Physiol Scand 141:455–467
255. Nilsson SF (1996) Nitric oxide as a mediator of parasympathetic vasodilation in ocular and extraocular tissues in the rabbit. Invest Ophthalmol Vis Sci 37:2110–2119
256. Nilsson SF (2000) The significance of nitric oxide for parasympathetic vasodilation in the eye and other orbital tissues in the cat. Exp Eye Res 70:61–72
257. Nilsson SF, Maepea O, Alm A et al (2001) Ocular blood flow and retinal metabolism in abyssinian cats with hereditary retinal degeneration. Invest Ophthalmol Vis Sci 42:1038–1044
258. Nyborg NC, Nielsen PJ (1990) The level of spontaneous myogenic tone in isolated human posterior ciliary arteries decreases with age. Exp Eye Res 51:711–715
259. O'Brien C, Kelly PA, Ritchie IM (1997) Effect of chronic inhibition of nitric oxide synthase on ocular blood flow and glucose metabolism in the rat. Br J Ophthalmol 81:68–71
260. Ohkubo H, Chiba S (1989) Vascular responses of ophthalmic arteries to exogenous and endogenous norepinephrine. Exp Eye Res 48:539–547
261. Okamura T, Ayajiki K, Kangawa K et al (1997) Mechanism of adrenomedullin-induced relaxation in isolated canine retinal arteries. Invest Ophthalmol Vis Sci 38:56–61
262. Oksala O, Stjernschantz J (1988) Increase in outflow facility of aqueous humor in cats induced by calcitonin gene-related peptide. Exp Eye Res 47:787–790
263. Okubo H, Gherezhiher T, Koss MC (1990) Long posterior ciliary arterial blood flow and systemic blood pressure. Invest Ophthalmol Vis Sci 31: 819–826
264. Okuno T, Oku H, Sugiyama T et al (2002) Evidence that nitric oxide is involved in autoregulation in optic nerve head of rabbits. Invest Ophthalmol Vis Sci 43:784–789
265. Overend J, Wilson WS, Martin W (2005) Biphasic neurogenic vasodilatation in the bovine intraocular long posterior ciliary artery: involvement of nitric oxide and an additional unidentified neurotransmitter. Br J Pharmacol 145:1001–1008
266. Oyster CW (1999) The human eye: structure and function. Sinauer Associates, Sunderland
267. Papastergiou GI, Schmid GF, Riva CE et al (1998) Ocular axial length and choroidal thickness in newly hatched chicks and one-year-old chickens fluctuate in a diurnal pattern that is influenced by visual experience and intraocular pressure changes. Exp Eye Res 66:195–205
268. Parver LM, Auker C, Carpenter DO (1980) Choroidal blood flow as a heat dissipating mechanism in the macula. Am J Ophthalmol 89:641–646
269. Parver LM, Auker CR, Carpenter DO et al (1982) Choroidal blood flow II. Reflexive control in the monkey. Arch Ophthalmol 100:1327–1330
270. Parver LM, Auker CR, Carpenter DO (1983) Choroidal blood flow. III. Reflexive control in human eyes. Arch Ophthalmol 101:1604–1606
271. Pauleikhoff D, Chen JC, Chisholm IH et al (1990) Choroidal perfusion abnormality with age-related Bruch's membrane change. Am J Ophthalmol 109:211–217
272. Paulson OB, Strandgaard S, Edvinsson L (1990) Cerebral autoregulation. Cerebrovasc Brain Metab Rev 2:161–192
273. Payne AP (1994) The harderian gland: a tercentennial review. J Anat 185(Pt 1):1–49
274. Pickard GE, Smeraski CA, Tomlinson CC et al (2002) Intravitreal injection of the attenuated pseudorabies virus PRV Bartha results in infection of the hamster suprachiasmatic nucleus only by retrograde transsynaptic transport via autonomic circuits. J Neurosci 22:2701–2710
275. Polak K, Wimpissinger B, Berisha F et al (2003) Effects of sildenafil on retinal blood flow and flicker-induced retinal vasodilatation in healthy subjects. Invest Ophthalmol Vis Sci 44:4872–4876
276. Porter JP, Brody MJ (1986) A comparison of the hemodynamic effects produced by electrical stimulation of subnuclei of the paraventricular nucleus. Brain Res 375:20–29

277. Potts AM (1966) An hypothesis on macular disease. Trans Am Acad Ophthalmol Otolaryngol 70: 1058–1062
278. Pournaras CJ, Logean E, Riva CE et al (2006) Regulation of subfoveal choroidal blood flow in age-related macular degeneration. Invest Ophthalmol Vis Sci 47:1581–1586
279. Prieto D, Benedito S, Nielsen PJ et al (1991) Calcitonin gene-related peptide is a potent vasodilator of bovine retinal arteries in vitro. Exp Eye Res 53:399–405
280. Prieto D, Simonsen U, Nyborg NC (1995) Regional involvement of an endothelium-derived contractile factor in the vasoactive actions of neuropeptide Y in bovine isolated retinal arteries. Br J Pharmacol 116:2729–2737
281. Prunte C, Niesel P (1988) Quantification of choroidal blood-flow parameters using indocyanine green video-fluorescence angiography and statistical picture analysis. Graefes Arch Clin Exp Ophthalmol 226:55–58
282. Ramrattan RS, van der Schaft TL, Mooy CM et al (1994) Morphometric analysis of Bruch's membrane, the choriocapillaris, and the choroid in aging. Invest Ophthalmol Vis Sci 35:2857–2864
283. Ravalico G, Toffoli G, Pastori G et al (1996) Age-related ocular blood flow changes. Invest Ophthalmol Vis Sci 37:2645–2650
284. Reiner A, Karten HJ, Gamlin PDR et al (1983) Parasympathetic ocular control: functional subdivisions and circuitry of the avian nucleus of Edinger-Westphal. Trends Neurosci 6:140–145
285. Reiner A, Erichsen JT, Cabot JB et al (1991) Neurotransmitter organization of the nucleus of Edinger-Westphal and its projection to the avian ciliary ganglion. Vis Neurosci 6:451–472
286. Reiner A, Shih YF, Fitzgerald ME (1995) The relationship of choroidal blood flow and accommodation to the control of ocular growth. Vision Res 35:1227–1245
287. Reiner A, Zagvazdin Y, Fitzgerald ME (2003) Choroidal blood flow in pigeons compensates for decreases in arterial blood pressure. Exp Eye Res 76:273–282
288. Richard G, Weber J (1987) Der Einfluss der Betablocker Timolol und Pindolol auf die retinale Hamodynamik–eine videoangiographische Studie. Klin Monatsbl Augenheilkd 190:34–39
289. Ríos JD, Zoukhri D, Rawe IM, Hodges RR, Zieske JD, Dartt DA (1999) Immunolocalization of muscarinic and VIP receptor subtypes and their role in stimulating goblet cell secretion. Invest Ophthalmol Vis Sci 40:1102–1111
290. Riva CE, Grunwald JE, Petrig BL (1986) Autoregulation of human retinal blood flow. An investigation with laser Doppler velocimetry. Invest Ophthalmol Vis Sci 27:1706–1712
291. Riva CE, Cranstoun SD, Mann RM et al (1994) Local choroidal blood flow in the cat by laser Doppler flowmetry. Invest Ophthalmol Vis Sci 35: 608–618
292. Riva CE, Titze P, Hero M et al (1997) Choroidal blood flow during isometric exercises. Invest Ophthalmol Vis Sci 38:2338–2343
293. Riva CE, Titze P, Hero M et al (1997) Effect of acute decreases of perfusion pressure on choroidal blood flow in humans. Invest Ophthalmol Vis Sci 38:1752–1760
294. Riva CE, Falsini B, Logean E (2001) Flicker-evoked responses of human optic nerve head blood flow: luminance versus chromatic modulation. Invest Ophthalmol Vis Sci 42:756–762
295. Rogers RF, Paton JF, Schwaber JS (1993) NTS neuronal responses to arterial pressure and pressure changes in the rat. Am J Physiol 265: R1355–R1368
296. Ruskell GL (1967) Vasomotor axons of the lacrimal glands monkeys and the ultrastructural identification of sympathetic terminals. Z Zellforsch Mikrosk Anat 83:321–333
297. Ruskell GL (1970) An ocular parasympathetic nerve pathway of facial nerve origin and its influence on intraocular pressure. Exp Eye Res 10:319–330
298. Ruskell GL (1971) Facial parasympathetic innervation of the choroidal blood vessels in monkeys. Exp Eye Res 12:166–172
299. Ruskell GL (2004) Distribution of pterygopalatine ganglion efferents to the lacrimal gland in man. Exp Eye Res 78:329–335
300. Ruskell GL (1965) The orbital distribution of the sphenopalatine ganglion in the rabbit. In: Rohen J (ed) The structure of the eye. Schattauer, Stutgart, pp 323–339
301. Sato E, Sakamoto T, Nagaoka T et al (2003) Role of nitric oxide in regulation of retinal blood flow during hypercapnia in cats. Invest Ophthalmol Vis Sci 44:4947–4953
302. Scheiner AJ, Riva CE, Kazahaya K et al (1994) Effect of flicker on macular blood flow assessed by the blue field simulation technique. Invest Ophthalmol Vis Sci 35:3436–3441
303. Schmerl E, Steinberg B (1949) The role of ciliary and superior cervical ganglia in ocular tension. Am J Ophthalmol 32:947–90
304. Schmetterer L, Findl O, Strenn K et al (1997) Role of NO in the O2 and CO2 responsiveness of cerebral and ocular circulation in humans. Am J Physiol 273:R2005–R2012
305. Schmetterer L, Findl O, Fasching P et al (1997) Nitric oxide and ocular blood flow in patients with IDDM. Diabetes 46:653–658
306. Schrödl F, Schweigert M, Brehmer A et al (2001) Intrinsic neurons in the duck choroid are contacted by CGRP-immunoreactive nerve fibres: evidence for a local pre-central reflex arc in the eye. Exp Eye Res 72:137–146
307. Schrödl F, Tines R, Brehmer A et al (2001) Intrinsic choroidal neurons in the duck eye receive sympathetic input: anatomical evidence for adrenergic modulation of nitrergic functions in the choroid. Cell Tissue Res 304:175–184

308. Schrödl F, De Laet A, Tassignon MJ et al (2003) Intrinsic choroidal neurons in the human eye: projections, targets, and basic electrophysiological data. Invest Ophthalmol Vis Sci 44: 3705–3712
309. Schrödl F, De Stefano ME, Reese S et al (2004) Comparative anatomy of nitrergic intrinsic choroidal neurons (ICN) in various avian species. Exp Eye Res 78:187–196
310. Schrödl F, Brehmer A, Neuhuber WL et al (2006) The autonomic facial nerve pathway in birds: a tracing study in chickens. Invest Ophthalmol Vis Sci 47:3225–3233
311. Segade LA, Quintanilla JS (1990) Distribution of postganglionic parasympathetic fibers originating in the pterygopalatine ganglion in the maxillary and ophthalmic nerve branches of the trigeminal nerve; HRP and WGA-HRP study in the guinea pig. Brain Res 522:327–332
312. Seifert P, Spitznas M (1999) Vasoactive intestinal polypeptide (VIP) innervation of the human eyelid glands. Exp Eye Res 68:685–692
313. Seligsohn EE, Bill A (1993) Effects of NG-nitro-L-arginine methyl ester on the cardiovascular system of the anaesthetized rabbit and on the cardiovascular response to thyrotropin-releasing hormone. Br J Pharmacol 109:1219–1225
314. Shakoor A, Blair NP, Mori M et al (2006) Chorioretinal vascular oxygen tension changes in response to light flicker. Invest Ophthalmol Vis Sci 47:4962–4965
315. Shih YF, Fitzgerald ME, Reiner A (1993) Effect of choroidal and ciliary nerve transection on choroidal blood flow, retinal health, and ocular enlargement. Vis Neurosci 10:969–979
316. Shih YF, Fitzgerald ME, Reiner A (1994) The effects of choroidal or ciliary nerve transection on myopic eye growth induced by goggles. Invest Ophthalmol Vis Sci 35:3691–3701
317. Shih YF, Lin SY, Huang JK et al (1997) The choroidal blood flow response after flicker stimulation in chicks. J Ocul Pharmacol Ther 13: 213–218
318. Shih YF, Fitzgerald ME, Cuthbertson SL et al (1999) Influence of ophthalmic nerve fibers on choroidal blood flow and myopic eye growth in chicks. Exp Eye Res 69:9–20
319. Shimizu Y (1982) Localization of neuropeptides in the cornea and uvea of the rat: an immunohistochemical study. Cell Mol Biol 28:103–110
320. Simons E, Smith PG (1994) Sensory and autonomic innervation of the rat eyelid: neuronal origins and peptide phenotypes. J Chem Neuroanat 7:35–47
321. Singh J, Adeghate E, Burrows S et al (1994) Protein secretion and the identification of neurotransmitters in the isolated pig lacrimal gland. Adv Exp Med Biol 350:57–60
322. Smith PG, Fan Q (1996) Sympathetic nerve trajectories to rat orbital targets: role of connective tissue pathways. J Comp Neurol 365:69–78
323. Spencer SE, Sawyer WB, Wada H et al (1990) CNS projections to the pterygopalatine parasympathetic preganglionic neurons in the rat: a retrograde transneuronal viral cell body labeling study. Brain Res 534:149–169
324. Spraul CW, Lang GE, Grossniklaus HE (1996) Morphometric analysis of the choroid, Bruch's membrane, and retinal pigment epithelium in eyes with age-related macular degeneration. Invest Ophthalmol Vis Sci 37:2724–2735
325. Steinberg RH (1987) Monitoring communications between photoreceptors and pigment epithelial cells: effects of "mild" systemic hypoxia. Friedenwald lecture. Invest Ophthalmol Vis Sci 28:1888–1904
326. Steinle JJ, Krizsan-Agbas D, Smith PG (2000) Regional regulation of choroidal blood flow by autonomic innervation in the rat. Am J Physiol Regul Integr Comp Physiol 279:R202–R209
327. Steinle JJ, Smith PG (2000) Presynaptic muscarinic facilitation of parasympathetic neurotransmission after sympathectomy in the rat choroid. J Pharmacol Exp Ther 294:627–632
328. Steinle JJ, Pierce JD, Clancy RL et al (2002) Increased ocular blood vessel numbers and sizes following chronic sympathectomy in rat. Exp Eye Res 74:761–768
329. Steinle JJ, Lindsay NL, Lashbrook BL (2005) Cervical sympathectomy causes photoreceptor-specific cell death in the rat retina. Auton Neurosci 120:46–51
330. Steinle JJ, Lashbrook BL (2006) Cervical sympathectomy regulates expression of key angiogenic factors in the rat choroid. Exp Eye Res 83:16–23
331. Stiris T, Suguihara C, Hehre D et al (1992) Effect of cyclooxygenase inhibition on retinal and choroidal blood flow during hypercarbia in newborn piglets. Pediatr Res 31:127–130
332. Stiris TA, Hall C, Christensen T et al (1991) Effect of different phototherapy lights on retinal and choroidal blood flow. Dev Pharmacol Ther 17:70–78
333. Stiris TA, Suguihara C, Flynn J et al (1996) Effects of the cyclooxygenase inhibitor ibuprofen on retinal and choroidal blood flow during hyperoxia in newborn piglets. Biol Neonate 69:101–108
334. Stjernschantz J (1976) Effect of parasympathetic stimulation on intraocular pressure, formation of the aqueous humour and outflow facility in rabbits. Exp Eye Res 22:639–645
335. Stjernschantz J, Alm A, Bill A (1976) Effects of intracranial oculomotor nerve stimulation on ocular blood flow in rabbits: modification by indomethacin. Exp Eye Res 23:461–469
336. Stjernschantz J, Alm A, Bill A (1977) Cholinergic and aminergic control of uveal blood flow in rabbits. Bibl Anat 16:42–46
337. Stjernschantz J, Bill A (1979) Effect of intracranial stimulation of the oculomotor nerve on ocular blood flow in the monkey, cat, and rabbit. Invest Ophthalmol Vis Sci 18:99–103
338. Stjernschantz J, Geijer C, Bill A (1979) Electrical stimulation of the fifth cranial nerve in rabbits: effects on ocular blood flow, extravascular albumin

content and intraocular pressure. Exp Eye Res 28:229–238
339. Stjernschantz J, Bill A (1980) Vasomotor effects of facial nerve stimulation: noncholinergic vasodilation in the eye. Acta Physiol Scand 109:45–50
340. Stjernschantz J, Sears M, Stjernschantz L (1981) Intraocular effects of substance P in the rabbit. Invest Ophthalmol Vis Sci 20:53–60
341. Stolze HH, Sommer HJ (1985) Influence of secretagogues on volume and protein pattern in rabbit lacrimal fluid. Curr Eye Res 4:489–492
342. Stone RA, Laties AM, Brecha NC (1982) Substance P-like immunoreactive nerves in the anterior segment of the rabbit, cat and monkey eye. Neuroscience 7:2459–2468
343. Stone RA, Kuwayama Y (1985) Substance P-like immunoreactive nerves in the human eye. Arch Ophthalmol 103:1207–1211
344. Stone RA (1986) Vasoactive intestinal polypeptide and the ocular innervation. Invest Ophthalmol Vis Sci 27:951–957
345. Stone RA, Tervo T, Tervo K et al (1986) Vasoactive intestinal polypeptide-like immunoreactive nerves to the human eye. Acta Ophthalmol (Copenh) 64:12–18
346. Stone RA, Kuwayama Y, Laties AM (1987) Regulatory peptides in the eye. Experientia 43:791–800
347. Stone RA, McGlinn AM (1988) Calcitonin gene-related peptide immunoreactive nerves in human and rhesus monkey eyes. Invest Ophthalmol Vis Sci 29:305–310
348. Su EN, Alder VA, Yu DY et al (1994) Adrenergic and nitrergic neurotransmitters are released by the autonomic system of the pig long posterior ciliary artery. Curr Eye Res 13:907–917
349. Su WW, Cheng ST, Hsu TS et al (2006) Abnormal flow-mediated vasodilation in normal-tension glaucoma using a noninvasive determination for peripheral endothelial dysfunction. Invest Ophthalmol Vis Sci 47:3390–3394
350. Sugiyama T, Oku H, Ikari S et al (2000) Effect of nitric oxide synthase inhibitor on optic nerve head circulation in conscious rabbits. Invest Ophthalmol Vis Sci 41:1149–1152
351. Sun W, Erichsen JT, May PJ (1994) NADPH-diaphorase reactivity in ciliary ganglion neurons: a comparison of distributions in the pigeon, cat, and monkey. Vis Neurosci 11:1027–1031
352. Suzuki N, Hardebo JE, Owman C (1988) Origins and pathways of cerebrovascular vasoactive intestinal polypeptide-positive nerves in rat. J Cereb Blood Flow Metab 8:697–712
353. Suzuki N, Hardebo JE, Owman C (1989) Trigeminal fibre collaterals storing substance P and calcitonin gene-related peptide associate with ganglion cells containing choline acetyltransferase and vasoactive intestinal polypeptide in the sphenopalatine ganglion of the rat. An axon reflex modulating parasympathetic ganglionic activity? Neuroscience 30: 595–604
354. Suzuki N, Hardebo JE, Owman C (1990) Origins and pathways of choline acetyltransferase-positive parasympathetic nerve fibers to cerebral vessels in rat. J Cereb Blood Flow Metab 10:399–408
355. Takano T, Tian GF, Peng W et al (2006) Astrocyte-mediated control of cerebral blood flow. Nat Neurosci 9:260–267
356. Ten Tusscher MP, Klooster J, van der Want JJ et al (1989) The allocation of nerve fibres to the anterior eye segment and peripheral ganglia of rats. II. The sympathetic innervation. Brain Res 494:105–113
357. Ten Tusscher MP, Klooster J, Baljet B et al (1990) Pre- and post-ganglionic nerve fibres of the pterygopalatine ganglion and their allocation to the eyeball of rats. Brain Res 517:315–323
358. Terenghi G, Polak JM, Allen JM, Zhang SQ, Unger WG, Bloom SR. (1983) Neuropeptide Y-immunoreactive nerves in the uvea of guinea pig and rat. Neurosci Lett 42:33–38
359. Terenghi G, Polak JM, Probert L et al (1982) Mapping, quantitative distribution and origin of substance p- and VIP-containing nerves in the uvea of guinea pig eye. Histochemistry 75:399–417
360. Terenghi G, Polak JM, Ghatei MA et al (1985) Distribution and origin of calcitonin gene-related peptide (CGRP) immunoreactivity in the sensory innervation of the mammalian eye. J Comp Neurol 233:506–516
361. Tervo K, Tervo T, Eranko L et al (1981) Immunoreactivity for substance P in the Gasserian ganglion, ophthalmic nerve and anterior segment of the rabbit eye. Histochem J 13:435–443
362. Tervo K, Tervo T, Eranko L et al (1982) Effect of sensory and sympathetic denervation on substance P immunoreactivity in nerve fibres of the rabbit eye. Exp Eye Res 34:577–585
363. Tervo K, Tervo T, Eranko L et al (1982) Substance P-immunoreactive nerves in the human cornea and iris. Invest Ophthalmol Vis Sci 23:671–674
364. Tillis TN, Murray DL, Schmidt GJ et al (1988) Preretinal oxygen changes in the rabbit under conditions of light and dark. Invest Ophthalmol Vis Sci 29:988–991
365. Tittl M, Maar N, Polska E et al (2005) Choroidal hemodynamic changes during isometric exercise in patients with inactive central serous chorioretinopathy. Invest Ophthalmol Vis Sci 46:4717–4721
366. Toda M, Okamura T, Azuma I et al (1997) Modulation by neurogenic acetylcholine of nitroxidergic nerve function in porcine ciliary arteries. Invest Ophthalmol Vis Sci 38:2261–2269
367. Toda M, Okamura T, Ayajiki K et al (1999) Neurogenic vasoconstriction as affected by cholinergic and nitroxidergic nerves in dog ciliary and ophthalmic arteries. Invest Ophthalmol Vis Sci 40:1753–1760
368. Toda N, Ayajiki K, Yoshida K et al (1993) Impairment by damage of the pterygopalatine ganglion of nitroxidergic vasodilator nerve function in canine cerebral and retinal arteries. Circ Res 72:206–213

369. Toda N, Kitamura Y, Okamura T (1995) Functional role of nerve-derived nitric oxide in isolated dog ophthalmic arteries. Invest Ophthalmol Vis Sci 36:563–570
370. Toda N, Toda M, Ayajiki K et al (1998) Cholinergic nerve function in monkey ciliary arteries innervated by nitroxidergic nerve. Am J Physiol 274: H1582–H1589
371. Toledo CA, Britto LR, Pires RS et al (2002) Interspecific differences in the expression of the AMPA-type glutamate receptors and parvalbumin in the nucleus of Edinger-Westphal of chicks and pigeons. Brain Res 947:122–130
372. Toshida H, Nguyen DH, Beuerman RW et al (2007) Evaluation of novel dry eye model: preganglionic parasympathetic denervation in rabbit. Invest Ophthalmol Vis Sci 48:4468–4475
373. Tóth IE, Boldogkoi Z, Medveczky I et al (1999) Lacrimal preganglionic neurons form a subdivision of the superior salivatory nucleus of rat: transneuronal labelling by pseudorabies virus. J Auton Nerv Syst 77:45–54
374. Tso MOM, Jampol LM (1990) Hypertensive retinopathy, choroidopathy & optic neuropathy of hypertensive ocular disease. In: Laragh JH, Brenner BM (eds) Hypertension: pathophysiology, diagnosis and management. Raven Press, New York, pp 433–465
375. Tso MOM (1988) Photic injury to the retina and pathogenesis of age-related macular degeneration. In: Tso MOM (ed) Retinal diseases: biomedical foundations and clinical management. Lippincott Co., Philadelphia, pp 187–214
376. Uddman R, Alumets J, Ehinger B et al (1980) Vasoactive intestinal peptide nerves in ocular and orbital structures of the cat. Invest Ophthalmol Vis Sci 19:878–885
377. Unger WG, Terenghi G, Ghatei MA et al (1985) Calcitonin gene-related polypeptide as a mediator of the neurogenic ocular injury response. J Ocul Pharmacol 1:189–199
378. Uusitalo H, Lehtosalo J, Palkama A et al (1984) Vasoactive intestinal polypeptide (VIP)-like immunoreactivity in the human and guinea-pig choroid. Exp Eye Res 38:435–437
379. Uusitalo H, Lehtosalo JI, Palkama A (1985) Vasoactive intestinal polypeptide (VIP)-immunoreactive nerve fibers in the anterior uvea of the guinea pig. Ophthalmic Res 17:235–240
380. Uusitalo H, Krootila K, Palkama A (1989) Calcitonin gene-related peptide (CGRP) immunoreactive sensory nerves in the human and guinea pig uvea and cornea. Exp Eye Res 48:467–475
381. van der Werf F, Baljet B, Prins M et al (1996) Innervation of the lacrimal gland in the cynomolgous monkey: a retrograde tracing study. J Anat 188(Pt 3): 591–601
382. van Pinxteren PC, van Alphen GW (1984) Acetylcholine can exert two opposite effects on uveal flow in isolated arterially perfused eyes of cats and rabbits. Curr Eye Res 3:1001–1006
383. Voaden MJ, Hussain AA, Taj M et al (1983) Light and retinal metabolism. Biochem Soc Trans 11:679–681
384. Walcott B, Sibony PA, Keyser KT (1989) Neuropeptides and the innervation of the avian lacrimal gland. Invest Ophthalmol Vis Sci 30:1666–1674
385. Walters BB, Gillespie SA, Moskowitz MA (1986) Cerebrovascular projections from the sphenopalatine and otic ganglia to the middle cerebral artery of the cat. Stroke 17:488–494
386. Wang L, Kondo M, Bill A (1997) Glucose metabolism in cat outer retina. Effects of light and hyperoxia. Invest Ophthalmol Vis Sci 38:48–55
387. Wang Y, Okamura T, Toda N (1993) Mechanisms of acetylcholine-induced relaxation in dog external and internal ophthalmic arteries. Exp Eye Res 57: 275–281
388. Webb SM, Puig-Domingo ML, Viader M et al (1992) Harderian gland peptides. In: Webb SM, Hoffman RA, Puig-Domingo ML, Reiter RJ (eds) Harderian glands: porphyrin metabolism: behavioral and endocrine effects. Springer, Berlin, pp 235–243
389. Weinstein JM, Duckrow RB, Beard D et al (1983) Regional optic nerve blood flow and its autoregulation. Invest Ophthalmol Vis Sci 24:1559–1565
390. Weiter JJ, Schachar RA, Ernest JT (1973) Control of intraocular blood flow. I. Intraocular pressure. Invest Ophthalmol 12:327–331
391. Wiederholt M, Sturm A, Lepple-Wienhues A (1994) Relaxation of trabecular meshwork and ciliary muscle by release of nitric oxide. Invest Ophthalmol Vis Sci 35:2515–2520
392. Wiencke AK, Nilsson H, Nielsen PJ et al (1994) Nonadrenergic noncholinergic vasodilation in bovine ciliary artery involves CGRP and neurogenic nitric oxide. Invest Ophthalmol Vis Sci 35:3268–3277
393. Wikberg-Matsson A, Simonsen U (2001) Potent alpha(2A)-adrenoceptor-mediated vasoconstriction by brimonidine in porcine ciliary arteries. Invest Ophthalmol Vis Sci 42:2049–2055
394. Wild JM, Arends JJ, Zeigler HP (1990) Projections of the parabrachial nucleus in the pigeon (*Columba livia*). J Comp Neurol 293:499–523
395. Wilson TM, Strang R, MacKenzie ET (1977) The response of the choroidal and cerebral circulations to changing arterial PCO2 and acetazolamide in the baboon. Invest Ophthalmol Vis Sci 16:576–580
396. Wimpissinger B, Resch H, Berisha F et al (2003) Effects of isometric exercise on subfoveal choroidal blood flow in smokers and nonsmokers. Invest Ophthalmol Vis Sci 44:4859–4863
397. Wyss JM, Oparil S, Chen YF (1990) The role of the central nervous system in hypertension. In: Laragh JH, Brenner BM (eds) Hypertension: pathophysiology, diagnosis, and management. Raven Press, New York, pp 679–701
398. Yamamoto R, Bredt DS, Snyder SH et al (1993) The localization of nitric oxide synthase in the rat eye and related cranial ganglia. Neuroscience 54: 189–200

399. Yancey CM, Linsenmeier RA (1988) The electroretinogram and choroidal PO2 in the cat during elevated intraocular pressure. Invest Ophthalmol Vis Sci 29:700–707
400. Yancey CM, Linsenmeier RA (1989) Oxygen distribution and consumption in the cat retina at increased intraocular pressure. Invest Ophthalmol Vis Sci 30:600–611
401. Yao K, Tschudi M, Flammer J et al (1991) Endothelium-dependent regulation of vascular tone of the porcine ophthalmic artery. Invest Ophthalmol Vis Sci 32:1791–1798
402. Yasui T, Karita K, Izumi H et al (1997) Correlation between vasodilatation and secretion in the lacrimal gland elicited by stimulation of the cornea and facial nerve root of the cat. Invest Ophthalmol Vis Sci 38:2476–2482
403. Ye XD, Laties AM, Stone RA (1990) Peptidergic innervation of the retinal vasculature and optic nerve head. Invest Ophthalmol Vis Sci 31:1731–1737
404. Yoneya S, Amano H, Mori K et al (1995) Indocyanine green angiography of the choroid in young and aged eyes. Invest Ophthalmol Vis Sci Suppl 36:187
405. Yoshida A, Feke GT, Ogasawara H et al (1991) Effect of timolol on human retinal, choroidal and optic nerve head circulation. Ophthalmic Res 23:162–170
406. Yoshitomi T, Gregory DS (1991) Ocular adrenergic nerves contribute to control of the circadian rhythm of aqueous flow in rabbits. Invest Ophthalmol Vis Sci 32:523–528
407. Yoshitomi T, Ishikawa H, Hayashi E (2000) Pharmacological effects of pilocarpine on rabbit ciliary artery. Curr Eye Res 20:254–259
408. Young RW (1978) The daily rhythm of shedding and degradation of rod and cone outer segment membranes in the chick retina. Invest Ophthalmol Vis Sci 17:105–116
409. Youngstrom TG, Weiss ML, Nunez AA (1987) A retinal projection to the paraventricular nuclei of the hypothalamus in the Syrian hamster (*Mesocricetus auratus*). Brain Res Bull 19:747–750
410. Yu DY, Alder VA, Su EN et al (1992) Relaxation effects of diltiazem, verapamil, and tolazoline on isolated cat ophthalmociliary artery. Exp Eye Res 55:757–766
411. Yu DY, Cringle SJ, Su EN (2005) Intraretinal oxygen distribution in the monkey retina and the response to systemic hyperoxia. Invest Ophthalmol Vis Sci 46:4728–4733
412. Zagvazdin Y, Sancesario G, Wang YX et al (1996) Evidence from its cardiovascular effects that 7-nitroindazole may inhibit endothelial nitric oxide synthase in vivo. Eur J Pharmacol 303:61–69
413. Zagvazdin Y, Fitzgerald ME, Reiner A (2000) Role of muscarinic cholinergic transmission in Edinger-Westphal nucleus-induced choroidal vasodilation in pigeon. Exp Eye Res 70:315–327
414. Zagvazdin YS, Fitzgerald ME, Sancesario G et al (1996) Neural nitric oxide mediates Edinger-Westphal nucleus evoked increase in choroidal blood flow in the pigeon. Invest Ophthalmol Vis Sci 37:666–672
415. Zeitz O, Mayer J, Hufnagel D, Praga R, Wagenfeld L, Galambos P, Wiermann A, Rebel C, Richard G, Klemm M (2009) Neuronal activity influences hemodynamics in the paraoptic short posterior ciliary arteries: a comparison between healthy and glaucomatous subjects. Invest Ophthalmol Vis Sci 50: 5846–5850
416. Zhan GL, Lee PY, Ball DC et al (2002) Time dependent effects of sympathetic denervation on aqueous humor dynamics and choroidal blood flow in rabbits. Curr Eye Res 25:99–105
417. Zhang HR (1994) Scanning electron-microscopic study of corrosion casts on retinal and choroidal angioarchitecture in man and animals. Prog Retin Eye Res 13:243–270
418. Zurakowski D, Vorwerk CK, Gorla M et al (1998) Nitrate therapy may retard glaucomatous optic neuropathy, perhaps through modulation of glutamate receptors. Vision Res 38:1489–1494

13 Endothelial and Adrenergic Control

Gerhard Garhöfer and Leopold Schmetterer

Core Messages

- In the recent years it has become clear that the vasular endothelium plays a major role in the regulation of vascular tone. Thus, intact blood flow regulation is largely dependent on a functioning vascular endothelium and endothelium derived substances play a substantial role in regulation vascular tone in health and diseases. This chapter will summarize our knowledge on the role of the vascular endothelium in blood flow regulation.

Within the last 20 years, it has become clear that the endothelium plays a key role in maintaining vascular tone within all vascular beds of the human body. Since the first in vitro study showing the obligatory role of the endothelium in mediating acetylcholine-induced vasodilatation in 1980 [66], the study of endothelium-derived vasoactive substances has become an important research area. Nowadays, it is assumed that the endothelium produces a large variety of vasodilators and vasoconstrictors. Only if there is a balance between the production of endothelium-derived vasodilators and endothelium-derived vasoconstrictors is a vessel under normal physiological tone (Fig. 13.1). This also holds true for the eye, where numerous in vitro animal and human studies have proven the concept of endothelial control of blood flow in the ocular vascular systems.

G. Garhöfer, M.D. (✉)
Department of Clinical Pharmacology,
Medical University of Vienna,
Waehringer Guertel 18-20, Vienna A-1090, Austria
e-mail: gerhard.garhoefer@meduniwien.ac.at

L. Schmetterer, Ph.D.
Department of Clinical Pharmacology,
Center of Medical Physics and Biomedical Engineering, Medical University of Vienna,
Waehringer Guertel 18-20, Vienna A-1090, Austria
e-mail: leopold.schmetterer@meduniwien.ac.at

13.1 Nitric Oxide

In their original experiment, Furchgott and Zawadzki proposed the existence of a potent endothelium-derived relaxing factor (EDRF), which was, however, not identified. In the following years, it became clear that this EDRF is nitric oxide (NO), produced from the amino acid l-arginine, with cyclic GMP as a second messenger [105, 184, 185, 198].

Three distinct isoforms of nitric oxide synthase (NOS), which are products of different genes, are used to produce NO. NOS is required to oxidize the guanidine group of l-arginine in a process involving five electrons. The three isoforms of NOS are termed NOS1, NOS2, and NOS3. In older textbooks and articles, NOS1 was termed neuronal NOS, NOS2 was termed inducible NOS, and NOS3 was termed endothelial NOS. NOS1 and NOS3 were characterized as constitutive and

L. Schmetterer, J.W. Kiel (eds.), *Ocular Blood Flow*,
DOI 10.1007/978-3-540-69469-4_13, © Springer-Verlag Berlin Heidelberg 2012

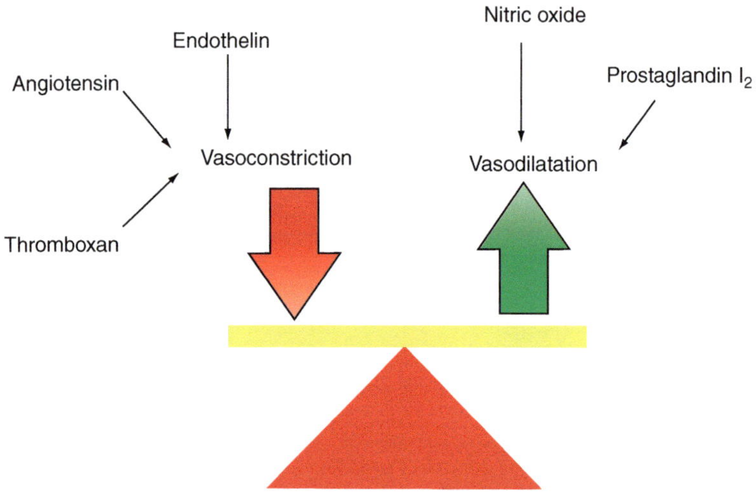

Fig. 13.1 The endothelium plays a key role in maintaining basal vascular tone. A balance between the production of endothelium-derived vasodilators and vasoconstrictors is required

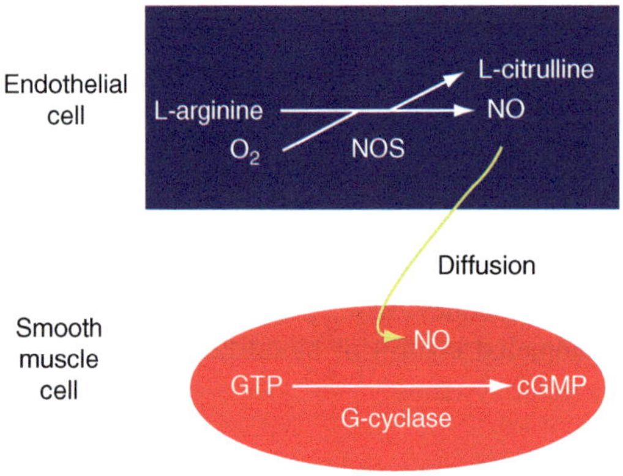

Fig. 13.2 Biosynthesis of nitric oxide. For details please see text

NOS2 as inducible. However, the notion that all NOS isoforms are regulated dynamically required a new nomenclature. When NOS1 and NOS3 are activated, NO is produced via the calcium/calmodulin complex, when NOS2 is activated NO production is independent of calcium. Large amounts of NO via NOS-2 are produced in the presence of immunological and inflammatory stimuli (Fig. 13.2). The NOS gene family shares similar compositions with each other: All have two domains: N-terminal half of heme-oxygenase domain, with tetrahydrobiopterin, heme- and arginine-binding sites, and C-terminal half of P-450 reductase domain with the positions of recognition sites for NADPH, as well as for flavin mononucleotide (FMN) and flavin adenosine dinucleotide (FAD). The NOS1, NOS 2, and NOS3 genes have been mapped on the chromosomes 12q24, 17q11.2, and 7q35-q36, respectively.

Much of our knowledge of the role of NO in the control of blood flow is based on experiments using NOS inhibitors. NOS inhibition can be achieved by L-arginine analogs such as N^G-monomethyl-L-arginine (L-NMMA), N^G-nitro-L-arginine methyl ester (L-NAME) and N^G-nitro-L-arginine (L-NA). They are competitive inhibitors of NOS and not specific for any of the isoforms. Short-term effects of L-NMMA, L-NAME, or L-NA can be reversed by excess doses of l-arginine. To study the role of the different isoforms of NO, a number of specific inhibitors were also employed. 7-Nitroindazole (7-NINA) is the most widely used inhibitor of NOS1, but at higher dosages, it also inhibits NOS3 in cerebral arteries [11]. In earlier studies, aminoguanidine was used as a specific inhibitor of NOS2, although this drug exerts a variety of other pharmacological actions including inhibition of advanced glycation end products. Inhibitors of NOS are also produced endogenously. Among the identified endogenous inhibitors, asymmetric dimethylarginine (ADMA) appears to be the most important. Plasma levels of ADMA are increased in a variety of vascular diseases including end-stage renal disease, hypertension, hypercholesterolemia, atherosclerosis, and diabetes [21]. In these diseases, it appears to contribute to endothelial dysfunction. Whether endogenous inhibitors also play a role in ocular vascular disease is not established.

Due to its small size and its specific properties NO is an ubiquitous messenger throughout the human body. NO is soluble in tissues and can easily diffuse across membranes like other small molecule gases such as O_2, CO_2, or CO. Nitric oxide has a very short half life of only a few seconds. Accordingly, NO production is regulated at the level of biosynthesis because it cannot be stored in vivo. Also related to the small size of the molecule is its ability to diffuse over large distances up to several hundreds of micrometers. Hence, a single NO molecule can affect numerous cells adjacent to the location of its production despite the short half-life.

Nitric oxide has a key role in the maintenance of vascular tone in humans [246] and is a major regulator of systemic blood pressure [99]. NO also exerts a variety of other physiological and pathophysiological effects, which are not directly related to the control of vascular tone and blood flow. In the eye, this includes processes related to signal transduction, neurotransmission, neurodegeneration, and oxidative stress. A more detailed discussion of these effects is, however, beyond the scope of this chapter.

All three types of NOS were identified in the eye. As in other tissues, staining of cells for NADPH diaphorase activity has been widely used to determine the regional distribution of NOS because of its high sensitivity. The different isoforms of NOS can, however, not be distinguished. Immuncytochemistry and fluorescence methods overcome this problem, employing specific monoclonal and polyclonal antibodies. Molecular biology based methods including measurement of mRNA expression by reverse transcription polymerase chain reaction, measurement of protein expression using SDS-polyacrylamide gel electrophoresis and western blotting, or in situ hybridization were also used to characterize NOS in ocular tissues.

In the retina, NOS was found in amacrine and ganglion cells [37, 43, 128] retinal pigment epithelium [70, 71], Müller cells [127], photoreceptors and nerve fibers in the inner and outer plexiform layers [256]. As in other vascular beds, NOS is also present in the endothelium of retinal vessels [155, 240] and in retinal capillary endothelial cells and pericytes [31]. As expected, NOS3 was also identified in the endothelium of optic nerve head and choroidal blood vessels [35, 61, 155, 165].

Under physiological conditions, NO is continuously produced in the endothelium to ensure that vessels are under constant vasodilator tone. Numerous in vitro, animal, and human experiments indicate that this is also the case in the ocular vasculature. In isolated porcine ophthalmic and ciliary arteries, inhibition of NOS with L-NMMA induces dose-dependent contraction [83, 260]. NO also relaxes the contractile tone of retinal bovine pericytes [85]. Given that the ratio of pericytes to endothelial cells in the retinal microvasculature is extremely high (approximately 1:1), this indicates a major role for NO also in the smaller retinal vessels, where most of the resistance to flow occurs. In the isolated perfused porcine eye, NOS inhibition decreases ocular blood flow and increases vascular resistance [152].

In animal and human studies, evidence for a reduction in ocular blood flow after NOS inhibition has been accumulated for all vascular beds of the eye (Fig. 13.3). Using a variety of different techniques, unequivocal data have been presented for the choroid, the optic nerve head, the ciliary body, and the iris [47, 73, 118, 130, 132, 142, 148, 208, 232]. Data have also been generated that the effects of the NOS inhibitor L-NMMA in the human choroid is reversible by administration of high-dose l-arginine indicating the vasoconstrictor effect is specific to the NO pathway. In contrast, some [45, 49, 50, 52, 211], but not all studies [47, 182, 183, 238] indicated that NOS inhibition also reduces retinal blood flow. One human study reported a dose-dependent vasoconstrictor effect of L-NMMA on retinal arterial and arterial vessel diameters after systemic administration [52]. The negative results obtained were all collected with the radioactive microsphere technique and are most likely related to the limitations of this technique in assessing retinal blood flow [204]. Only few data are available for specific NOS inhibitors. In rats, L-NAME, but not 7-NINA, increased blood pressure. Both drugs, however, decreased ocular blood flow, suggesting a role for NOS1 in the maintenance of basal vascular tone [117]. A variety of other studies reported, however, that

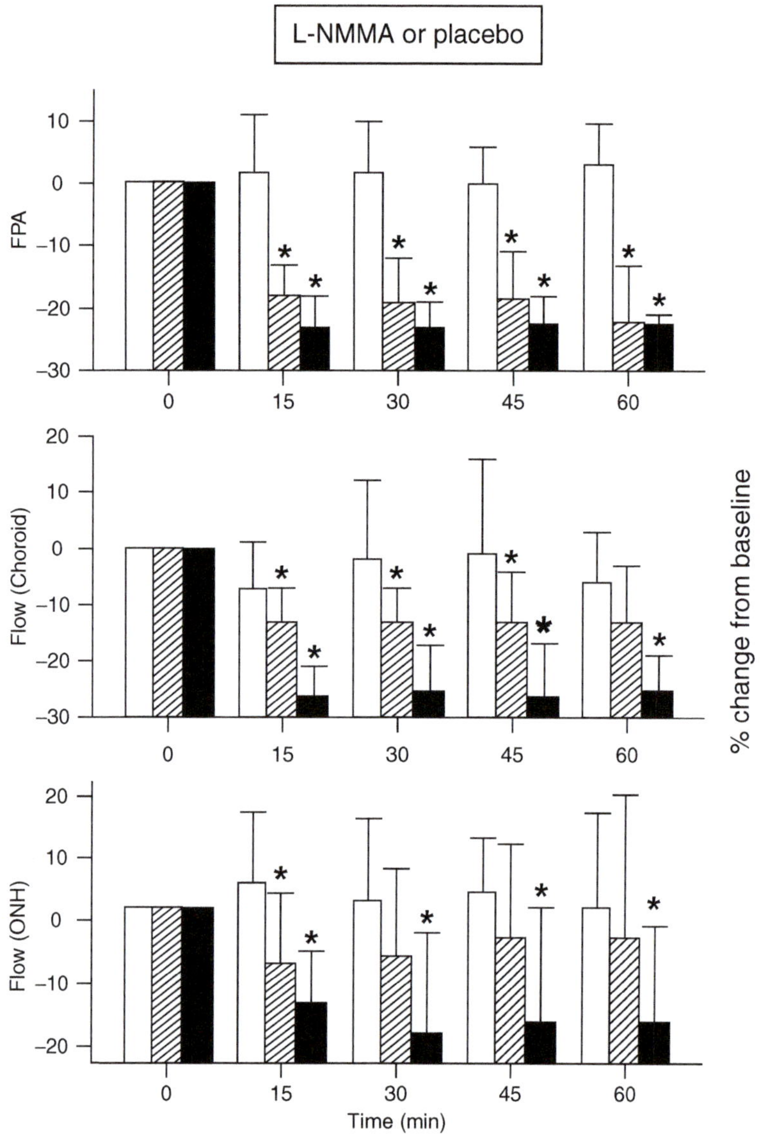

Fig. 13.3 Effect of the NO synthase inhibitor L-NMMA on ocular blood flow parameters in young healthy volunteers. Percent change in fundus pulsation amplitude (FPA); blood flow in the choroid, FLOW (Choroid); and blood flow in the optic nerve head, FLOW (ONH) after administration of L-NMMA (hatched bars: 3 mg/kg over 5 minutes followed by 30 μg/kg per minute over 55 minutes; solid bars: 6 mg/kg over 5 minutes followed by 60 μg/kg per minute over 55 minutes) or placebo (hollow bars). Data are presented as mean ± SD (n = 12). Asterisks indicate significant effects of L-NMMA versus baseline as calculated from the absolute values

7-NINA does not affect retinal or choroidal blood flow under physiological conditions in the cat [169], rat [132] or pigeon [263].

Nitric oxide is also a key candidate for regulating ocular blood flow during changes in perfusion pressure. Obviously, vascular resistance decreases when blood flow is kept constant during a decrease in perfusion pressure and increases during an increase in perfusion pressure. According to the work of [118] employing laser Doppler flowmetry in the rabbit, NO is a key candidate to control vascular tone during changes in perfusion pressure (Fig. 13.4). On the other hand, Koss [133] failed to detect an effect of a NO synthase inhibitor on the choroidal pressure-flow relationship in the cat during changes in perfusion pressure. In humans, NO synthase inhibition alters the ocular perfusion pressure/choroidal flow relationship during an increase in perfusion pressure induced by isometric exercise [143] (Fig. 13.5). Whether this truly indicates a role for NO in human choroidal autoregulation is, however, unclear because the neural input to the choroid cannot be eliminated in a human experiment. NOS inhibition also modulated the response of ONH blood flow assessed with hydrogen clearance to an increase in IOP

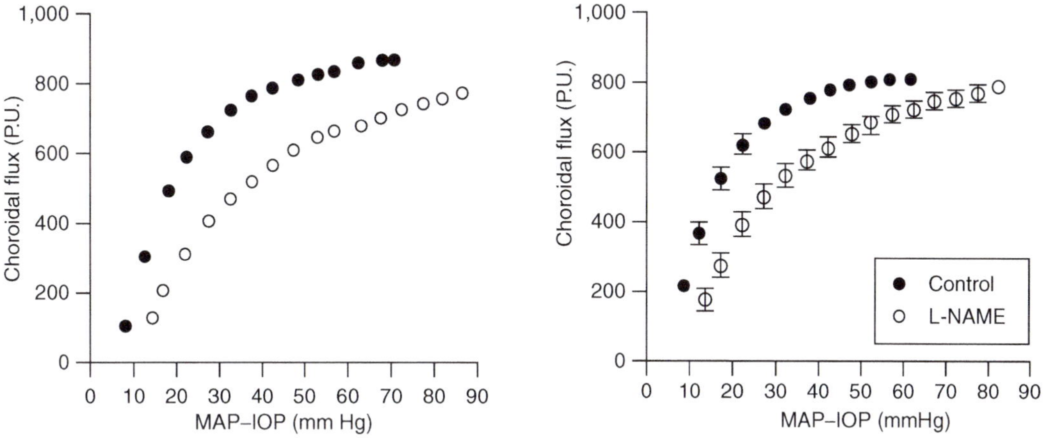

Fig. 13.4 Effect of the nitric oxide synthase inhibitor L-NAME on choroidal pressure/flow relationships in the rabbit. The left graph show tracings as obtained in a single rabbit. The right graph shows means ± SDs as obtained from a group of animals

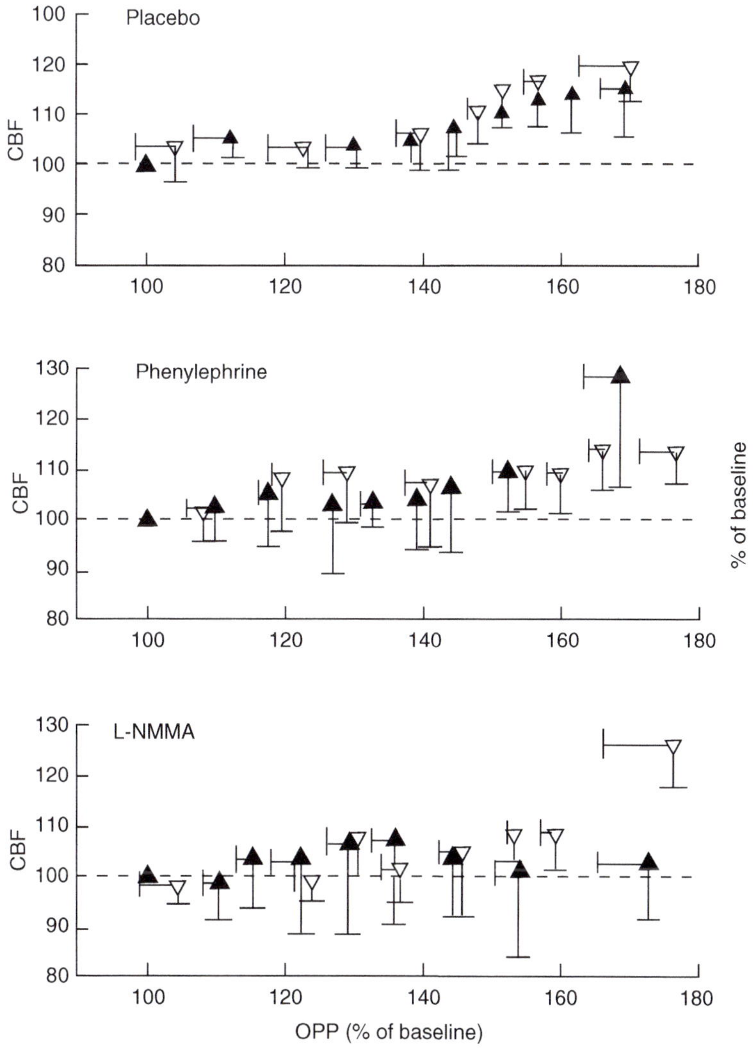

Fig. 13.5 Choroidal Pressure–flow relationship using the categorized ocular perfusion pressue (OPP) and choroidal blood flow (CBF) values during isometric exercise. Relative data were sorted into groups of nine values each, according to ascending OPPs. The first period of squatting was performed without drug administration (baseline; open down triangles). The second squatting period was performed during administration of placebo, L-NMMA, or PE (solid up triangles). The means and the lower limits of the 95% confidence intervals are shown (n = 12). The dotted line indicates 100% of baseline

[177]. Data that NO is involved in retinal and choroidal autoregulation have also been generated in newborn animals [89, 109]. Such data are of major relevance for the understanding of diseases such as retinopathy of permaturity, but most likely cannot be applied to adults.

Nitric oxide has also been shown to play a major role in mediating the ocular vasodilator effects to a variety of agonists as well as in response to changes in perfusion pressure. As in many other vascular beds, removal of the vascular endothelium in isolated bovine retinal as well as in human retinal and ophthalmic arteries is associated with a significant reduction of acetylcholine-induced relaxation [16, 83, 260]. In canine ophthalmic and retinal arteries, relaxations to acetylcholine were, however, endothelium-independent indicating significant species differences [240, 248]. Data for NO-dependent relaxation of bradykinin are more consistent and were collected for isolated ocular vessels from different species [83, 84, 260, 266] and in the isolated perfused porcine eye [152]. The list of agonists that appear to induce vasodilatation via NO-dependent mechanisms in the ocular vasculature is long and includes substance P [123], low-dose arginine [175], and a selective antidiuretic desmopressin [241], dipyridamole [154]. The evidence for histamine- and insulin-induced vasodilatation as well as the NO dependence of these effects will be discussed later in this chapter.

Some, but not all, experiments indicate that NO also interacts with the changes in ocular blood flow during changes in pO_2. In retinal pericytes, hypoxia amplified relaxations to the NO donor sodium nitroprusside, but hypoxia alone did not influence pericyte basal tone [86]. In vivo NOS inhibition did not affect choroidal blood flow after hyperoxia in humans [206] but reduced the retinal blood flow response to hypoxia in the cat [162], indicating that NO contributes to hypoxia-induced vasodilatation. As in the brain, the vasodilator response to CO_2 is significantly reduced by NO-synthase inhibitors in the retina [203] and choroid [206] indicating NO-dependence of hypercapnia-induced effects. In the cat retina, this effect can also be achieved with 7-NINA, suggesting a role for neuronal NOS [203].

A major role for NO was discovered for vasodilator effects after neural stimulation. This topic is discussed in some detail in another chapter of this book. In isolated dog ophthalmic arteries, relaxation by nicotine or electrical neural stimulation was abolished by NO-synthase inhibition and restored by adding high-dose l-arginine [239]. This is in keeping with data showing NO release from the autonomic system in the posterior ciliary arteries [225, 253]. The presence of nerves releasing NO was also shown for choroidal arterioles by measuring membrane potentials with the microelectrode technique [97]. This is in good agreement with a variety of in vivo studies. In the primate, electrical stimulation of the pterygopalatine or geniculate ganglion dilates the ophthalmic artery. This effect was abolished by L-NA and restored by high-dose l-arginine, proving NO-dependence [10]. Neurally derived NO also plays a major role in the choroidal blood flow increase caused by stimulation of the Edinger-Westphal nucleus in pigeons [263] or facial nerve stimulation in the cat [169].

Flicker light stimulation in the miniature pig increases NO concentrations in preretinal vitreous humor [49]. This is in keeping with data in cats, showing increased NO levels in the vitreous humor near the optic nerve head [25]. In addition, these experiments revealed that the flicker-induced increase in ONH blood flow measured with laser Doppler flowmetry is reduced but not abolished by NOS inhibition, which is in good agreement with microsphere experiments [130]. Data are also available for humans, indicating that L-NMMA blunts the retinal vasodilator response to flicker stimulation [51]. Nitric oxide synthase inhibition also modulates the response of human choroidal blood flow to a light/dark transition [103]. The physiological relevance of the blood flow decrease in the choroid during a light/dark transition is largely unknown. It appears, however, that this is a neurally mediated effect [65] as it is in the pigeon [60].

The role of NO in the development of ocular perfusion abnormalities in vascular diseases of the eye is discussed in other articles of this book. Alterations in the l-arginine/NO system affecting blood flow regulation have, however, been proven in patients with diabetes [205] and glaucoma [193]. This makes the l-arginine/NO system a major candidate for therapeutic interventions in ocular vascular disease. Indeed, high-dose

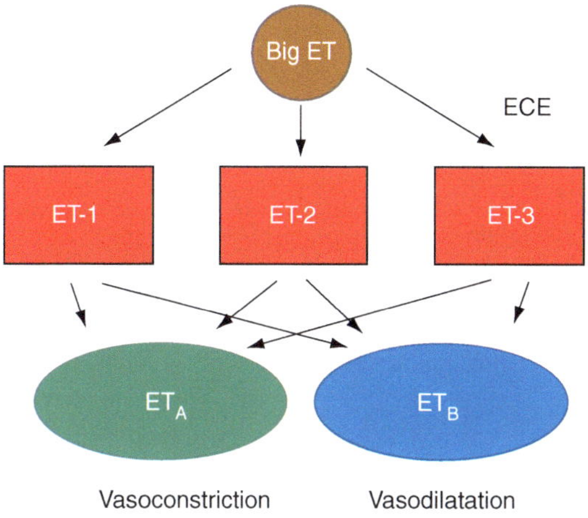

Fig. 13.6 Endothelin pathway. For details please see text

l-arginine increases choroidal and optic nerve head blood flow in healthy humans [42, 67]. Nitrates and sodium nitroprusside appear to dilate retinal vessels [81, 190], but as with all systemically administered drugs, the pronounced effects on blood pressure and pulse rate need to be considered [102]. To prevent these systemic effects, topical administration of NO donors may be effective. Development of nitric oxide-donating prostaglandin F2-alpha analogs for the treatment of glaucoma may represent such an approach.

13.2 Endothelins

Endothelin (ET) was first described in 1988 and the potent vasoconstrictor effects were immediately recognized [259]. Three structurally different ET isoforms, namely ET-1, ET-2, and ET-3 have been identified. Among the three ET isopeptides, the 21–amino acid peptide ET-1 is considered the most important isoform in controlling vascular tone. Endothelin-1 is formed from pre-pro-ET-1 via a 39–amino acid intermediate named big ET-1. Big ET-1 is processed to ET-1 by a family of ET-converting enzymes (ECEs, Fig. 13.6). Under physiological conditions, ET-1 is produced in small concentrations in endothelial cells. Endothelin-1 primarily acts as an autocrine/paracrine mediator and circulating levels of ET-1 are small. Under pathophysiological conditions however, the production may be stimulated in a large number of different cell types, including endothelial cells, vascular smooth muscle cells and inflammatory cells such as macrophages and leukocytes. This may significantly increase local ET-1 production reflected in elevated ET-1 plasma levels [200].

The biological effects of ET-1 are mediated by two pharmacologically different receptor subtypes, ET_A and ET_B receptors. In the vasculature, the ET_A receptor is mainly located on vascular smooth muscle cells and mediates potent vasoconstriction. The ET_{B1} receptor is primarily located on endothelial cells. Stimulation of the endothelial ET_{B1} receptor induces vasodilatation due to the release of NO and prostacyclin. Stimulation of the ET_{B2} receptor located at the vascular smooth muscle cell results in direct vasoconstriction. The net effect of ET-1 is therefore determined by receptor localization and the ratio between ET_A and ET_B receptors, which may vary considerably among vascular beds. Generally, very low doses of ET-1 cause vasodilatation, whereas with high concentrations of the peptide, vasoconstriction dominates.

Early studies identified ET-1 in cultured bovine retinal endothelial cells [234], the iris, and ciliary body [54, 146]. Immunoreactivity to ET-1 and ET-3 appears to be present in all ocular tissues except the cornea, where no immunoreactivity to ET-3 is found. Particularly, high levels are observed in the choroid [30]. Production of ETs in the eye is not limited to endothelial cells, but also includes nonpigmented ciliary epithelial cells [137, 196] and retinal pigmented epithelial cells [164], Endothelin receptors are present at cornea, iris, and ciliary processes [181]. In the posterior pole of the eye, ET_A-like receptor-binding sites are localized at retinal and choroidal blood vessels, whereas the ET_B-like receptor-binding sites can be found at neural and glial tissues of the retina [145]. In the rat retina, the ratio of ET_B to ET_A receptors is as high as 35:65 [44].

ET-1 causes a dose-dependent contraction of the human ophthalmic artery [83]. When the in vitro effects of ET-1 are compared among porcine ophthalmic and posterior ciliary artery, it becomes obvious that the vasoconstrictor effects are the more pronounced the smaller the vessels [84]. In the isolated perfused porcine eye, ET-1 induced early vasodilatation followed by

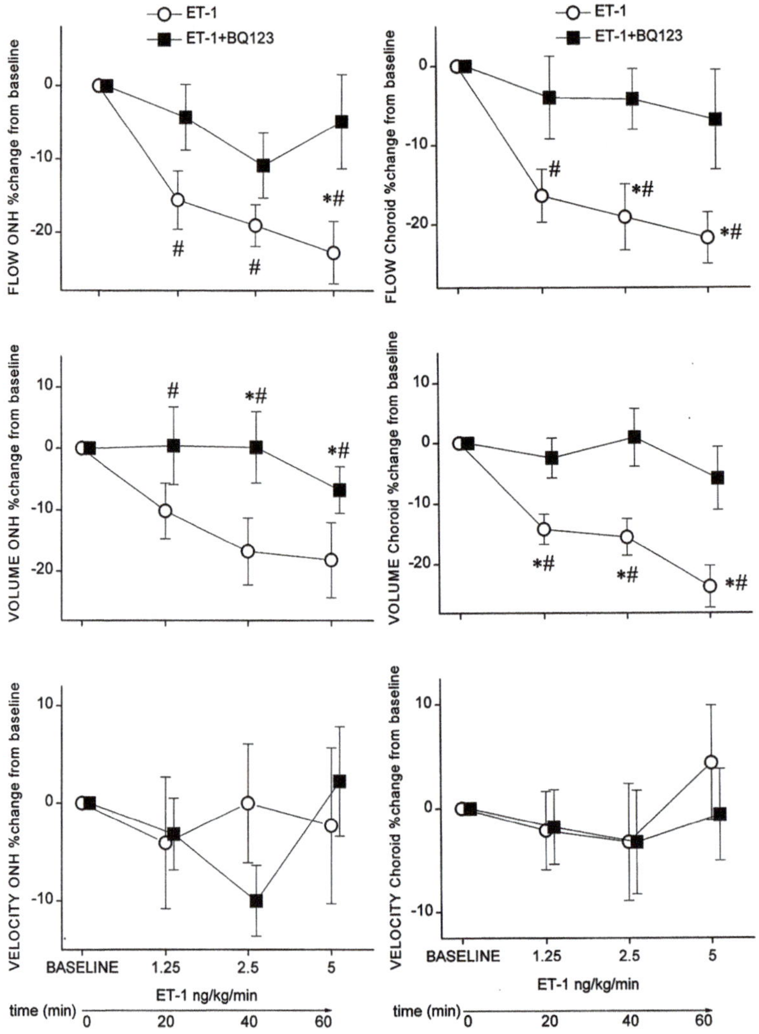

Fig. 13.7 Effects of endothelin-1 (ET-1) and the endothelinA receptor antagonist BQ-123 on optic nerve head (left panel) and choroidal blood flow (right panel). Percentage change from baseline of red blood cell flow, volume, and velocity as measured with laser Doppler flowmetry during administration of ET-1 in stepwise increasing doses of 1.25, 2.5, and 5 ng/kg · min (open circles); coinfusion of ET-1 in stepwise increasing doses with BQ123 at 60 μg/kg · min (solid squres). Data are presented as means ± SEM (n = 12). Statistical significance: #, versus baseline; *, between treatments

pronounced vasoconstriction. ET-3 caused similar, but less pronounced effects. The effects of both peptides were abolished by a specific ET_A receptor antagonist, proving the major role of this receptor subtype in mediating vasoconstriction [152].

In an early in vivo study in cats, using the radioactive microsphere technique intravitreal injection of ET-1 caused a reduction in retinal but not in ciliary body, iris, or choroid blood flow [74]. Intravitreal injection of ET-1 into the rabbit eye, however, decreased choroidal and optic nerve head blood flow as assessed with hydrogen clearance [229] and induced vasoconstriction of the iris–ciliary body microvasculature [230]. A decrease in optic nerve head blood flow was also seen after intravenous administration of ET-1 in the rabbit [229] and the cat [170]. In the retina, both ET-1 and ET-3 induced a reduction in retinal blood flow, with the former being more potent [26]. In humans, ET-1 dose-dependently reduced ocular fundus pulsation amplitude [207]. In addition, intravenous infusion of ET-1 in healthy subjects induces a decrease in choroidal, optic nerve head and retinal blood flow but does not affect retinal arterial or venous diameters [191, 192] (Fig. 13.7). These effects of exogenous ET-1 are reversed with the specific ET_A receptor antagonist BQ-123, indicating that they are primarily mediated via the ET_A receptor subtype. BQ-123 alone, however, did not affect retinal, optic nerve head, or choroidal blood flow but increased blood velocities in the ophthalmic artery. Again, one needs to consider that the concentrations of both, ET-1 and BQ-123 may differ at the levels of the endothelial cell and the smooth muscle cell, respectively.

A major role for the ET system has also been proposed in ocular blood flow regulation. In the rabbit, ET-1 induced a downward shift of the pressure/flow curve [119]. A downward shift of the pressure-flow relationship was also observed with a specific ET_B receptor antagonist. By contrast, neither a specific ET_A receptor antagonist nor a dual ET_A/ET_B receptor antagonist affected choroidal pressure/flow curves. The vasoconstriction seen with ET-1 in the rabbit choroid was reduced with the ET_A receptor antagonist and enhanced with the ET_B receptor antagonist. The results with the ET_B receptor antagonist indicate that endogenous ET-1 produced under physiological conditions controls vascular tone. Hence, ET_B receptor blockade withdraws ET-1 associated vasodilator tone. If this is true, one would, however, also assume that an ET_A receptor induces vasodilatation because it withdraws vasoconstrictor tone. Most likely, this is related to a lower ET-1 concentration at the smooth muscle cell as compared to the endothelial cell. These rabbit data are in good agreement with results in healthy subjects. During isometric exercise, the increase in choroidal blood flow associated with the increase in ocular perfusion pressure is significantly more pronounced in the presence than in the absence of the ET_A receptor antagonist BQ-123 [63]. This indicates that the regulatory vasoconstrictor response to an increase in ocular perfusion pressure is at least partially mediated by ET-1. Data have also been generated for the retina. The vasoconstrictor response of a retinal artery during isometric exercise is also blunted by BQ123, again indicating a role for ET-1 [144]. ET-1 has also been implicated to play a role in the vasoconstrictor response to hyperoxia. Data in newborn pigs, adult rats, and humans indicate that ET_A receptor mediated vasoconstriction contributes to oxygen-induced blood flow reductions but do not fully account for them [41, 233, 267].

ET-1 is also assumed to play a major role in the regulation of contractile tone of retinal pericytes. ET receptors are present at retinal pericytes [234]. ET-1 induces a pronounced release of free intracellular Ca^{++} in the pericyte indicating that the peptide is an important agonist for pericyte contraction in vivo [197]. This is compatible with data showing that ET-1 contracts the tone of pericytes [29] and regulates intercellular communication within pericyte-containing microvessels [116]. Whereas it has been hypothesized that ET-1 regulates microvascular retinal blood flow via this effect on pericyte contractility, an in vivo confirmation of this mechanism is lacking.

Due to its potent vasoconstrictor effects, animal models of chronic ocular ischemia have been developed using the ET system. Repeated intravitreal administration of ET-1 in the rabbit causes enlargement and excavation of the disk cup without increasing IOP [176]. In several species, a model using continuous release of ET-1 to the optic nerve head via osmotically driven minipumps was developed. This leads to a selective loss of retinal ganglion cells [33, 38, 179] at doses that reduce optic nerve head blood flow by 30–50% [33, 38]. Whether the loss of retinal ganglion cells in this model is a consequence, optic nerve head ischemia is, however, unknown. Endothelins also appear to have a role in the control of IOP, but this topic is beyond the scope of the present chapter, and the reader is referred to recent reviews on the involvement of the ET system in glaucoma [32, 261].

Hence, the ET system has been implicated in the therapy of ocular vascular disease. As in other vascular beds, the vasoconstrictor effects of ET-1 can be reduced or even abolished with Ca^{++} channel blockers [153, 224, 229, 244].

13.3 Arachidonic Acid Metabolites

Arachidonic acid metabolites play an important and complex role in the regulation of ocular blood flow. Arachidonic acid is liberated from membrane-bound phospholipids, usually by the action of phospholipases. Once released, arachidonic acid may be metabolized by four distinct pathways: (1) cyclooxygenase (COX, also known as prostaglandin endoperoxidase synthase), (2) lipoxygenase, (3) P450 epoxygenase, and (4)

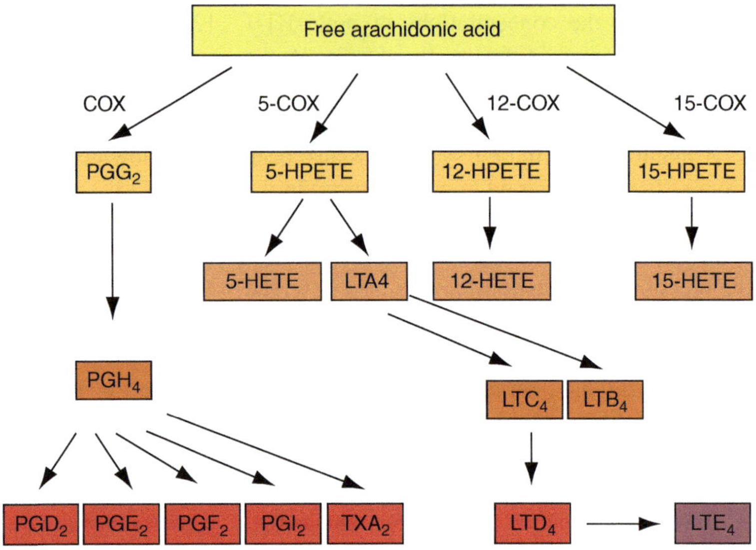

Fig. 13.8 Prostanoid synthesis from the arachidonic acid pathway

nonenzymatic isoprostane biosynthesis. In this chapter, the focus is directed toward the pathway catalyzed by cyclooxygenase because it results in numerous important vasoactive metabolites including prostaglandins (PGDs), thromboxane A2 (TXA2), and prostacyclin (PGI2, Fig. 13.8). Two COX isoforms, COX-1 and COX-2, are the key regulatory enzymes of this pathway. COX-1 is a constitutive enzyme whose expression is only developmentally regulated. Hence, the metabolites produced via COX-1 are considered to regulate vascular tone under physiological conditions. COX-2 is an inducible enzyme in most tissues but is expressed in high concentrations in response to growth factors or cytokines. COX-2 therefore plays a major role in prostaglandin production during inflammatory states.

The effects of prostanoids released as a consequence of the processes outlined above are mediated by specific cell membrane spanning G protein-coupled prostanoid receptors. Five major subdivisions of receptors have been defined pharmacologically. These correspond to each of the COX metabolites: DP for PGD2, EP for PGE2, FP for PGF2, IP for PGI2, and TP for TXA2. Most of these receptors exist in several isoforms.

Interestingly, relatively little information is available about the role of arachidonic acid metabolites in the control of ocular blood flow. The retinal microvasculature contains all prostanoid receptors with the exception of EP4 [1] whereas the choroidal vasculature includes the EP4 receptor subtype [2].

Prostanoids are locally produced in ocular tissues [114], but their synthesis is higher in the perinatal than in the adult eye [91]. Accordingly, a specific interest was directed toward the role of prostaglandins in ocular blood flow regulation in newborn animals. Evidence has accumulated that prostanoids are involved in the processes that lead to retinopathy of prematurity, particularly in those related to oxidative stress [91]. In the newborn animal, they are involved in blood flow regulation during changes in blood gases and ocular perfusion pressure but do not play a major role in the control of basal blood flow [34, 88, 267]. In adult animals, PGE_2 is produced during an increase in perfusion pressure and appears to mediate in part the autoregulatory vasoconstrictor response [88]. In newborn animals, however, PGE_2 induces choroidal vasodilatation [1]. Generally, it appears that immature animals respond more to vasodilator PGs and less to vasoconstrictor PGs than adult animals [91].

The data are, however, controversial. For the anterior uvea, data have been accumulated indicating vasodilator effects of PGE_1 and PGE_2 and to a lesser degree of $PGF_{2\alpha}$ [76, 221]. By contrast, $PGF_{2\alpha}$ appears to have little effect on retinal and choroidal blood flow [9, 76, 223]. At very high doses, $PGF_{2\alpha}$ induces constriction of isolated retinal, ciliary, and ophthalmic arteries [168, 174, 226], an effect which may be mediated via the TP receptor. Evidence for ocular vasoconstrictor effects of high doses of $PGF_{2\alpha}$ have also been obtained in experiments using the isolated perfused rat eye [227]. PGE_1 induces retinal vasodilatation after intravitreal administration in the pig [195]. Recent data in the rat indicate that intravenous PGE_2 act as a potent vasodilator in retinal and choroidal circulation, with the EP_2 receptor subtype playing a major role in this effect [160]. In cats, no effects of PGE_1 were observed on optic nerve head blood flow as assessed with laser Doppler flowmetry [125]. By contrast, an increase of blood flow was seen with low-dose liposomal PGE_1 but not with high-dose liposomal PGE_1, most likely because the higher dose induced a pronounced decrease in systemic blood pressure. In healthy humans, alprostadil, a synthetic form of the naturally occurring PGE_1, did not affect retinal or choroidal blood flow after intravenous administration [53]. In patients with intermittent claudication, a 21-week treatment with PGE_1 increased blood velocities in the central retinal and the ophthalmic artery, but this study was not masked [219].

Specific interest has also been directed toward the role of $PGF_{2\alpha}$ and the prostaglandin FP receptor in the control of ocular blood flow because the ester analog of prostaglandin $F_{2\alpha}$ latanoprost is used clinically to reduce intraocular pressure in glaucoma. Animal and human studies have provided little evidence for an effect of topical latanoprost or bimatoprost on blood flow at the posterior pole of the eye [3, 8, 106, 111, 167, 212, 222, 223]. Some studies in humans do, however, indicate that latanoprost increases pulsatile ocular blood flow in patients with glaucoma [68, 139, 218]. In addition, a reduction in resistive index in the central retinal and ophthalmic artery was seen with latanoprost and travoprost but not with bimatoprost [135]. Given the potent ocular hypotensive effect of latanoprost, one needs, however, to be careful to interpret such data as evidence for choroidal vasodilatation. On the other hand, a set of experiments in rabbits, monkeys, and humans was performed using laser speckle technology [108], reporting increased optic nerve head blood flow after latanoprost independently of its ocular hypotensive effects. The authors speculated that latanoprost reaching the posterior pole of the eye stimulates the release of endogenous prostaglandins.

The most important vasodilator metabolite of the arachidonic acid pathway is PGI2. There is evidence that PGI2 plays a major role in NO-induced vasorelaxation in the retina and choroid of the piglet [90]. In addition, it is involved in the early vasodilator response to endothelin-1 in the isolated perfused pig eye [152]. In addition, prostacyclin increases retinal blood flow after intravitreal and intravenous administration in the rat [98, 159, 160] mediated via the IP receptor subtype.

A large set of experiments on vessel caliber in the microvasculature associated with human retinal tissues grafted into the hamster cheek pouch membrane was published by Spada and coworkers [217]. In these experiments, a selective DP receptor antagonist induced dose-dependent vasodilatation. By contrast, a selective TP receptor antagonist induced dose-dependent vasoconstriction. By contrast, $PGF_{2\alpha}$, a selective FP agonist, PGE_2, agonists for the EP receptor as well as agonists selective for EP_2 and EP_3, a selective IP agonist, and a thromboxane mimetic induced little effect on vascular tone.

Some data are also available from inhibitors of COX. Both isoforms of COX are reversibly inhibited by nonsteroidal anti-inflammatory drugs (NSAIDs), but differences in the selectivity exist among different drugs. The mechanism of action of aspirin is through the irreversible acetylation of the cyclooxygenase site of the enzyme, without affecting peroxidase activity. This is compatible with the finding that COX-1 is completely inhibited by aspirin acetylation, whereas COX-2 is still able to convert arachidonic acid to 15-R-hydroxyeicosatetraenoic acid (15-HETE) after aspirin treatment. In the recent years, specific COX-2 inhibitors were developed as an alternative for NSAIDs with less pronounced side effects, but no data are available for this class of drugs in the eye.

It was recognized early that indomethacin may reduce retinal blood flow in rabbits [19].

Close arterial administration of indomethacin induces vasoconstriction of retinal arterioles in pigs [23]. This is in good agreement with data showing that indomethacin reduces optic nerve head oxygen tension [187]. In healthy humans, systemic administration of indomethacin is associated with a pronounced decrease in choroidal and retinal blood flow [252]. Whether these effects of indomethacin are, however, related to the effects of COX inhibition is doubtful, because ibuprofen and naproxen do not change retinal or choroidal blood flow [186]. This hypothesis only recently gained support by the observation that indomethacin, but not ibuprofen, diclofenac, ketoprofen, a parecyclo-oxygenase-2 inhibitor, and lornoxicam affected optic nerve head oxygen tension [171]. Only few data are available for ocular hemodynamic effects of aspirin. In a study involving eight patients with diabetes, an improvement of retinal blood flow was however observed after 14 days of aspirin treatment [59].

As mentioned above, lipoxygenase and P450 epoxygenase can also convert arachidonic acid. The role of the resulting metabolites such as hydroperoxyeicosatetraenoic acids (HPETEs), HETEs, and epoxyeicosatrienoic acid (EETs) in controlling ocular perfusion is not well characterized. In the newborn pig, however, the P 450 epoxygenase metabolite 20-HETE appears to mediate part of the vasoconstrictor response to hyperoxia [267].

In the literature unoprostone has also been discussed as a prostaglandin analog. Unoprostone isopropyl, however, is a synthetic docosanoid that has been introduced for the treatment of glaucoma. It is a compound with a 22-carbon chain and resembles the naturally occurring oxygenated metabolites of the docosahexaenoic and the docosatetraenoic acids. The latter are 22-carbon polyunsaturated omega-3 fatty acids. In the rabbit, topical unoprostone induces an increase in iris blood flow as evidenced with two different methods [242] and improves optic nerve head perfusion [172, 228]. In healthy subjects and normal-tension glaucoma patients, laser speckle studies indicates that topical unoprostone is capable of increasing blood flow to the posterior pole of the eye [121, 129, 147, 236]. A study using laser Doppler flowmetry does not confirm these results but reported that unoprostone is capable of reversing the vasoconstrictor effects of exogenous ET-1 [194]. A study using pulsatile ocular blood flow also did not find an effect of topical unoprostone on ocular blood flow [126]. In patients with normal-tension glaucoma, a 7-day treatment with unoprostone did not affect choroidal or optic nerve blood flow as assessed by means of laser Doppler flowmetry [14].

13.4 Adrenergic Control

Given that alpha receptor agonists as well as beta-receptor blockers have been shown to effectively reduce IOP and are therefore widely used in the therapy of glaucoma, the effect of adrenergic drugs on ocular blood flow has attracted much interest.

13.5 Alpha Receptors

Alpha receptors have been indentified in several ocular tissues, and it is well known today that both alpha-1 and alpha-2 adrenoceptors exist in the human eye. However, the situation is complicated by the fact that for both, alpha-1 and alpha-2 adrenoceptors, several different subtypes exist. In particular, at least three different alpha-1 adrenoceptors have been detected (alpha-1a, alpha-1b, and alpha-1c) [156], and there is compelling evidence that alpha-1 receptors play also an important role in the eye. More specifically, alpha-1a and alpha-1b receptors were found in the pig retina. In the rabbit, alpha-1a and alpha1b receptors were detected in the ciliary body, iris and, retina, whereas in the choroid, only alpha-1a adrenoceptors were detected [255]. In addition, alpha-1 adrenoceptors have also been found in both the glia and the neurons of the rabbit retina [180].

Using the radioligand technique, alpha-2 adrenoceptors have been found in the iris and the ciliary body of rabbits [158]. In human eyes, alpha-2 adrenoceptors have been detected in the iris, the ciliary epithelium, the ciliary muscle, and the choroid [151]. As for the alpha-1 adrenoceptors, in mammals, at least three distinct alpha-2 receptors exist, namely, alpha-2a, alpha-2b, and alpha-2c adrenoceptors. Using the radioligand technique in pig eyes, evidence has been provided that the only alpha adrenoceptor subtype present in the choroid, ciliary body, and iris is alpha-2a, while the retina contains both alpha-2a and alpha-2c adrenoceptors [254].

Several attempts have been made to investigate the role of the adrenergic system in the control of ocular blood flow. Back in 1972, Alm et al. investigated the role of exogenous administration of noradrenaline on retinal blood flow in experimental animals [4]. However, the authors did not observe an effect of noradrenaline on retinal blood flow. Similar experiments have also been performed in human subjects. Dollery et al. investigated the effect of exogenous-administered noradrenaline and reported vasoconstriction of retinal vessels [48]. However, given that only retinal vessel diameter, not retinal blood flow was reported, the results remain difficult to interpret. More recently, the effect of noradrenaline on retinal blood flow in healthy subjects was determined by a combined measurement of retinal vessel diameters and red blood cell velocity. Whereas, as expected, noradrenaline induced a systemic hypertensive response, no changes in retinal hemodynamic parameters were observed after administration of noradrenaline [110]. From these experiments, the authors of the latter study conclude that adrenergic system does not play a major role in retinal blood flow regulation.

Kitanishi et al. have investigated the effect of systemic administration of noradrenaline on optic nerve head blood flow and choroidal blood flow in anesthetized cats [124]. Whereas no change was observed in the optic nerve head, the authors found a pronounced increase in choroidal blood flow after administration of noradrenaline. In contrast, a decrease in choroidal blood flow, as measured with laser Doppler flowmetry was noted when noradrenaline was administered in the same species [69].

Unfortunately, all experiments investigating catecholamines in vivo are hampered by the fact that these substances, when administered systemically in vivo, cause considerable change in systemic hemodynamic parameters. Due to the autoregulatory properties of the retinal vascular bed, these changes in ocular perfusion pressure may also induce autoregulatory responses of the retinal vasculature, which complicates the interpretation of the results and may at least partially explain the contradictory results.

In addition, as mentioned above, the distribution of alpha adrenoceptors is complex and may vary from species to species. For example, Thörig and Bill have shown that selective alpha-2 adrenoceptor agonists reduce choroidal blood flow in a rabbit model [237]. In the cat choroid, the administration of adrenaline has been shown to induce vasoconstriction, predominantly mediated by alpha-2 adrenoceptors [131].

Systemic administration of clonidine, an alpha-2 agonist, constricts episcleral and conjunctival vessels but does not affect retinal vessel diameters of rabbits [13]. Experimental evidence from isolated vessels, however, did not show a significant change in diameter of canine internal, external ophthalmic and ciliary arteries in response to clonidine [173]. In addition, the effect of intravenously administered clonidine on retinal and choroidal circulation has been evaluated in healthy subjects in a randomized, double-masked, placebo-controlled, two-way crossover study. The authors of the latter study report a considerable decrease of retinal and choroidal blood flow after administration of the alpha-2 adrenoceptor agonist [251].

13.6 Topical Administration

13.6.1 Clonidine

Interestingly, alpha-2 agonists may also alter ocular hemodynamics when administered topically. Using fundus pulsation amplitude measurements to assess pulsatile choroidal blood flow, evidence has been provided that single topical doses of clonidine decrease choroidal and optic nerve head blood flow [209]. Weigert et al. have investigated the effect of the two topically administered alpha-2 agonists clonidine and brimonidine on choroidal blood flow during isometric exercise in healthy subjects [250]. For this purpose, two drops of either clonidine or brimonidine were administered in the subjects' study eyes. During a 6-min squatting period,

Table 13.1 Examples of beta-blocker currently used in ophthalmology

Nonselective beta-blocker	β1-selective beta-blocker
Timolol	Betaxolol
Carteolol	
Levobunolol	

choroidal blood flow regulation was assessed by the means of laser Doppler flowmetry. Interestingly, the two alpha agonists led to decreased choroidal blood flow during squatting, when compared to baseline conditions. The authors of the study concluded that clonidine and brimonidine induce changes in choroidal blood flow regulation, even after single administration [250].

13.6.2 Brimonidine

The effect of topical brimonidine tartrate 0.2% on retinal capillary blood flow has been evaluated in patients with ocular hypertension using confocal scanning laser Doppler flowmetry. Although brimonidine considerably reduced intraocular pressure, this double-masked, randomized, placebo-controlled trial did not detect a change in retinal blood flow [28]. This is in keeping with the results of other studies. Using CDI, no change in retrobulbar blood velocity was observed after topical administration of brimonidine BID for 2 weeks [113].

13.6.3 Beta Receptors

Beta receptors have been identified in several ocular tissues, such as the retina and the choroid [24, 72]. Beta-blockers are a class of drugs, which bind to beta receptors, but do not activate them. Depending on whether beta-blockers bind to beta-1, beta-2, or both receptors, one can distinguish between nonselective beta-blockers, such as timolol, or selective beta-blockers such as betaxolol, which binds with a stronger affinity to beta-1 receptors (Table 13.1).

13.6.4 Timolol

Given that timolol is widely used as an IOP-lowering drug, extensive research has been performed to determine the effect of timolol on ocular blood flow. After a single dose, both timolol and betaxolol showed pronounced constriction of rabbit ciliary arteries. Interestingly, after 7 weeks of daily treatment, tolerance significantly reduced the response to betaxolol whereas the response to timolol was unchanged [247].

Inconsistent results have been reported on the effect of timolol on the perfusion of the anterior segment of the eye. Whereas some authors found a decrease of ciliary body perfusion after topical treatment with timolol [249], others failed to find an effect of timolol at a concentration of 0.5% on anterior segment perfusion [112]. Both experiments were performed in rabbit eye models.

Unfortunately, the same holds true also for the posterior segment perfusion. Chiou and Chen have investigated the effects of antiglaucoma drugs on ocular blood flow in ocular hypertensive rabbits [36]. The authors found that all nonspecific beta-adrenergic blockers such as timolol, levobunolol, or metipranolol decreased retinal and choroidal blood flow in their model of hypertensive rabbits measured with microsphere technique [36]. These results are in keeping with another experiment investigating the effect of the beta-adrenoceptor antagonists timolol and carteolol in an arterially perfused bovine eye model, again revealing a decrease of perfusion parameters [157]. Other experiments, however, failed to find a decrease in ocular perfusion parameter or even found an increase. For example, Tomidokoro et al. evaluated the effects of topical carteolol or timolol on choroidal blood flow by the means of laser speckle technique, but did not find an effect of both drugs on choroidal blood flow [243]. Another experiment revealed that L-timolol reduced significantly retinal blood flow, whereas d-timolol induced a slight increase in retinal blood flow, without affecting systemic blood pressure or heart rate [258].

The short-term effect of the topically delivered nonselective beta-blockers, timolol and carteolol, and the selective beta-blocker betaxolol has been investigated using the hydrogen clearance method. However, no significant effect was observed in response to timolol treatment [202]. Likewise, a 30-day treatment with timolol maleate 0.5% did not produce any optic nerve vasomotor effect in the rabbit eye [178].

Liu et al. investigated the vascular resistance to flow in the ophthalmic artery of rabbits receiving topical treatment with timolol [140]. As a measure of vascular resistance, Pourcelot's resistive index based on the peak systolic velocity and the end-diastolic velocity assessed by CDI was chosen. The authors found a dose-dependent increase in resistive index after topical treatment with timolol at a concentration of 1% and 3.3%.

A different approach was used by Kiel and Patel [120]. Whereas the above-mentioned studies have focused on the effect of beta-receptor antagonists on baseline ocular circulation, the authors investigated the pressure-flow relationship in the rabbit choroid. For this purpose, pentobarbital-anesthetized rabbits were instrumented with hydraulic occluders on the aorta and inferior vena cava to control mean arterial pressure and two vitreous cannulas to control and measure intraocular pressure. Then, choroidal blood flow was measured by laser Doppler flowmetry during changes in mean arterial pressure induced by the hydraulic occluders in the presence of either timolol or betaxolol. Whereas timolol induced a slight increase in choroidal vascular resistance, none of the drugs changed the pressure-flow relationship, indicating that none of the drug alters the autoregulatory properties of the choroid.

13.6.5 Human Studies

A large number of experiments investigating the effect of timolol on ocular blood flow in healthy as well as in glaucoma subjects have been performed. Unfortunately, the experiments in humans are inconsistent and sometime contradicting. Yoshida et al. have investigated the effect of timolol on pulsatile ocular blood flow [262]. In this double-masked, randomized, placebo-controlled study, the effect of topical timolol maleate 0.5% on the retinal, choroidal, and optic nerve head circulation was investigated [262]. The authors report that although choroidal blood flow decreased after treatment, no statistically significant change was observed in the other hemodynamic parameters. However, whether the lack of statistical significance is related to the small sample size or can be explained by a missing effect of the drug in these vascular beds remains unclear.

Haefliger et al. have used the Heidelberg retina flowmeter (HRF) to assess the hemodynamic effects of either a single topical ocular instillation of timolol 0.5% or betaxolol 0.5% [87] and found decreased HRF parameters at the papilla of healthy volunteers after a single instillation of timolol. Retrobulbar hemodynamics have been assessed using CDI in patients with normal-tension glaucoma in order to measure retrobulbar blood velocities and resistance index [93]. Either timolol or betaxolol were administered for 1 month in a double-masked crossover design, with a 3-week washout before each drug. These results of this study indicate that, at least in the investigated group of patients with normal-tension glaucoma, timolol lowers intraocular pressure without altering retrobulbar hemodynamics [93].

The retinal circulation of eyes with ocular hypertension has been the focus of a study using laser Doppler velocimetry in combination with fundus photography to determine retinal blood flow [77]. The data of this study revealed, that treatment with timolol maleate 0.5% increased retinal blood flow in patients with ocular hypertension [77]. The authors attributed this effect to the increase in ocular perfusion pressure caused by the strong intraocular pressure lowering effect of timolol. The effect of a more prolonged therapy with timolol 0.5% on ocular blood flow was investigated by the same group of authors [78]. In this study, 15 healthy volunteers received topical treatment for 2 weeks. Again, volumetric blood flow was determined by a combination of bidirectional laser Doppler velocimetry and fundus photography. Similar to their first report, the authors found an increase in retinal blood flow after treatment with timolol [78].

Pulsatile ocular blood flow as assessed with laser interferometry after administration of topical antiglaucoma treatment was the focus of another study. The authors of this double-blind,

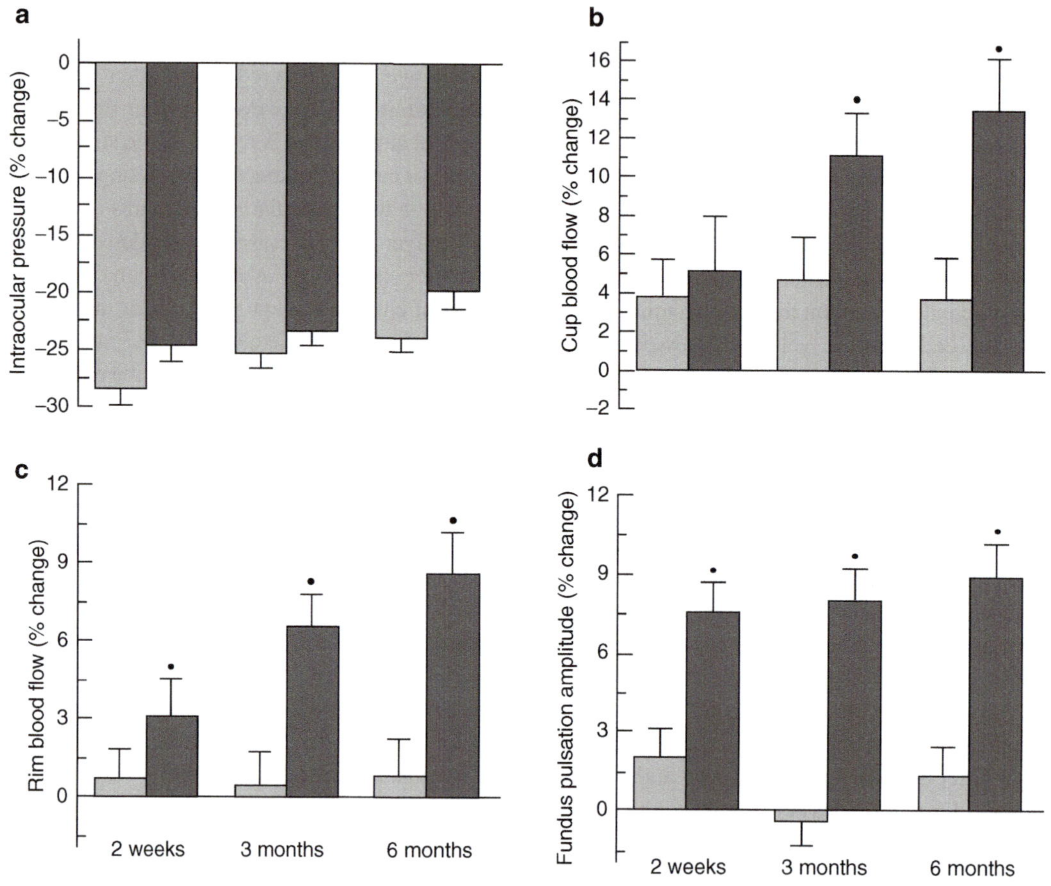

Fig. 13.9 Effects of dorzolamide (darker bars) and timolol (lighter bars) on (**a**) intraocular pressure, (**b**) cup blood flow, (**c**) rim blood flow, and (**d**) fundus pulsation amplitude. Asterisks indicate significant effects of dorzolamide compared with timolol

placebo-controlled, randomized crossover study report that timolol as well as the beta-blocking agents befunolol and metipranolol reduced pulsatile choroidal blood flow, even after single topical administration [209]. In contrast, no hemodynamic effect was observed after treatment with betaxolol or levobunolol.

The most of the above mentioned studies investigated the hemodynamic effect of timolol on healthy subjects. Although the data gained from these studies teach us about the drug effect and the physiology of blood flow regulation, the more important question might be whether timolol also alters blood flow in patients with glaucoma. Fuchsjäger-Mayrl et al. investigated ocular hemodynamics in patients with primary open-angle glaucoma and patients with ocular hypertension [64]. For this purpose, 140 glaucoma patients were included in a controlled, randomized, double-masked study in two parallel groups. Patients were randomized to receive either timolol or dorzolamide for a 6 month period. To assess ocular hemodynamics, pulsatile choroidal blood flow was measured using laser interferometric measurement of fundus pulsation amplitude. Scanning laser Doppler flowmetry was used to measure blood flow in the temporal neuroretinal rim and the cup of the optic nerve head. The study revealed that although the effect of both drugs on IOP was comparable, dorzolamide increased the optic nerve head and choroidal blood flow after a 6-month treatment period whereas timolol had no effect (Fig. 13.9).

For patients with glaucoma, the reported ocular hemodynamic effects of timolol are inconsistent. Optic nerve head perfusion as measured by the Heidelberg retina flowmeter has been investigated in patients with primary open-angle glaucoma and compared to a healthy control group [141]. A treatment period of 3 weeks was scheduled for both groups. The authors did not find a change in optic nerve head blood flow data after treatment with timolol.

Several studies investigating pulsatile choroidal blood flow in patients with glaucoma or ocular hypertensives found unchanged POBF after treatment with timolol. Morsman et al. randomized 33 ocular hypertensive patients to receive either timolol, levobunolol, or betaxolol in one eye [161]. Although the authors found a tendency toward a decreased puslatile choroidal blood flow, this effect did not reach statistical significance. These results are in good agreement with the data of another group investigating pulsatile ocular blood flow in patients with chronic open-angle glaucoma [245].

Although several reports also focused on the effect of timolol on retrobulbar blood flow, the majority of the studies did not find a significant change in retrobulbar hemodynamics caused by timolol treatment. Harris et al. did not find a timolol effect of on retrobulbar flow velocities as determined by color Doppler imaging in patients with normal-tension glaucoma [93]. This is in keeping with the results of several other reports that failed to find substantial hemodynamic changes in the retrobulbar vessels after timolol treatment [55, 92, 167].

In contrast, treatment with timolol was found to decrease vascular resistance in a group of glaucoma patients, whereas no change was found in the ocular hypertensive group [17]. No change was observed in blood flow velocities in the ophthalmic artery after treatment. The authors have speculated that the different timolol effect between the glaucoma group and the ocular hypertensive group may be related to a potentially defective autoregulation in the ocular hypertensive group [17].

13.6.6 Betaxolol

In contrast to timolol, betaxolol is a selective beta-1 receptor blocker, originally developed to reduce systemic side effects such as bronchospasm. As for timolol, much effort has been spent to investigate the hemodynamic effects of betaxolol. In a rabbit model, 30 days of repeated application of betaxolol 0.5% did not produce any observable effect in optic nerve head blood flow [178]. In contrast, Araie and colleagues have found a small but significant increase in tissue blood velocity in the iris and optic nerve head as measured with the laser speckle technique [6]. Instead of measuring only static choroidal blood flow, Kiel and Patel focused on the choroidal pressure-flow relationship measured before and after topical treatment with betaxolol [120]. Betaxolol did not induce a change in baseline choroidal blood flow or in the pressure flow relationship in this rabbit model.

13.6.7 Human Studies

Various methods have been used to investigate the effects of betaxolol on ocular blood flow in humans. Perimacular hemodynamic parameters have been studied after a single administration of betaxolol in normal subjects [92]. Two hours after topical administration of the drug, no change in perimacular leukocyte velocity or density was observed. Data obtained with the Heidelberg retina flowmeter focused on topical treatment with either timolol or betaxolol on optic nerve head blood flow. In contrast to timolol, which showed a slight decrease in optic nerve head blood flow, betaxolol did not show any effect on HRF parameters [87]. Similarly, Schmetterer et al. did not observe an effect of a single topical dose of betaxolol on pulstile choroidal blood flow in healthy volunteers [209].

In contrast to a single topical instillation, long-term betaxolol instillation revealed different effects. The effect of topical betaxolol on tissue circulation in the human optic nerve head has been investigated using the laser speckle technique [235]. The authors report a small but significant increase in tissue blood velocity in the human optic nerve head after topical instillation of betaxolol twice daily for 3 weeks. However, whether this increase is of clinical relevance has yet to be shown.

In addition, several studies have focused on the effect of topical betaxolol in patients with glaucoma or ocular hypertension. Again the data reported about betaxolol and its effect on ocular blood flow are not consistent. Steigerwalt et al. investigated the effect of betaxolol in patients with ocular hypertension [220]. The authors show that topical betaxolol as well as topical carteolol and timolol led to an increase in the flow velocity of the central retinal artery, indicating but not finally proving an increase in blood flow [220].

However, given that data about long-term treatment may be more relevant for glaucoma patients, several studies have focused on the hemodynamic effects of long-term betaxolol treatment. Carenini et al. have investigated whether a 12-month treatment with either topical timolol or betaxolol may influence pulsatile ocular blood flow in patients with primary open-angle glaucoma [27]. Although treatment with betaxolol and timolol showed similar reductions of the IOP, timolol decreased pulsatile ocular blood flow over the 12-month observation period, whereas no change was observed in betaxolol-treated patients.

Using the color Doppler imaging technique, a 1-month drug treatment with either timolol or betaxolol was tested in a double-masked two-way crossover design in patients with normal-tension glaucoma [93]. The authors found a decrease in resistance index after betaxolol treatment, whereas no effect was observed in the timolol group. However, whether a decrease in resistance index as found in the latter study can really be interpreted as an increase in blood flow remains unclear.

The effect of betaxolol was also investigated in a selected group of glaucoma patients, namely in patients who exhibited ocular vasospasm [55]. Patients were evaluated for blood flow velocity of the retrobulbar vessels using color Doppler imaging and resistance indices were calculated for the ophthalmica artery, the central retinal, and temporal posterior ciliary arteries. Measurements were done before and after a 4-week treatment with either timolol or betaxolol. Treatment with timolol did not induce a change in hemodynamic parameters. Surprisingly, the authors found that after betaxolol treatment resistance index significantly decreased in the ophthalmic artery, whereas an increase in resistance index was observed in both the central retinal and temporal posterior ciliary arteries. Whether these data can be applied correspondingly to a typical glaucoma patient without signs of vasospasm remains unclear.

The hemodynamic effect of topical treatment with either timolol, betaxolol, carteolol, or levobunolol drops was investigated using the CDI technique in patients with primary open-angle glaucoma [5]. Retrobulbar ocular blood flow velocity was measured using the color Doppler imaging method in the ophthalmic artery, central retinal artery, and temporal posterior ciliary arteries. The authors report that in the timolol group, an increase in resistance index values of the posterior ciliary artery was observed. In the betaxolol group, resistance index decreased in the central retinal artery and the posterior ciliary artery, whereas in the carteolol group, there was a significant decrease only in the central retinal artery. Finally, no change in retrobulbar hemodynamics was observed in the levobunolol group.

13.6.8 Levobunolol

Although not as widely used as timolol or betaxolol, the hemodynamic effects of the nonselective beta-adrenoreceptor antagonist levobunolol have been the focus of several experiments. Arend et al. report that levobunolol, as the other beta-adrenoceptor blockers tested increased blood velocities in retinal and epipapillary capillaries [7]. In this study, blood velocities were measured by digital image analysis of scanning laser fluorescein angiograms. None of the drugs tested induced a change in vessel diameters but increased epipapillary and macular papillary blood velocities.

Data about the effect of levobunolol 0.5% is also available from a randomized, double-masked, placebo-controlled study investigating puslatile ocular blood flow in patients with glaucoma and healthy volunteers [22]. The authors found that pulsatile ocular blood flow was increased after treatment with levobunolol in both the glaucoma group as well as in the healthy subjects.

In contrast, other reports did not find a hemodynamic effect of levobunolol. Harris et al. failed to find an effect in perimacular blood flow assessed with the blue-field technique 2 h after topical drug administration [92]. This is also in keeping with a randomized, placebo-controlled study investigating the effect of levobunolol in healthy subjects by the means of fundus pulsation amplitudes in the macula and the optic disk [209]. The authors report that application of a single drop of levobunolol did not change the measured ocular hemodynamic parameters. These results seem also to hold true for the retinal circulation. Leung et al. have investigated the effect of topical administered levobunolol on retinal blood flow by a combined measurement of retinal vessel diameters and red blood cell velocity as determined with laser Doppler velocimetry [138]. As with the blue-field technique, there was no statistically significant effect on calculated volumetric blood flow rate after administration of levobunolol. The same technique has been used to investigate within 1 week of topical treatment with levobunolol on retinal blood flow [20]. In contrast to the latter data reported of single instillation, 1 week treatment with levobunolol lead to a slight but statistically significant increase in volumetric retinal blood flow.

13.6.9 Carteolol

Carteolol is a non cardio-selective beta-adrenergic blocking agent. In contrast to other beta-blockers such as timolol or betaxolol, cartelolol possesses intrinsic sympathomimetic activity, which might at least theoretically provide some reduced potential for systemic effects.

A variety of studies have been performed to study the blood flow effects of carteolol in animal models as well as in humans. In a rabbit model, topical instillation of carteolol induced a significant reduction of choroidal and retinal blood flow [202], whereas intravenous administration of carteolol was found to result in an increase in blood flow parameters [231].

Yamazaki et al. found an increase in pulsatile ocular blood flow after installation of carteolol [257]. In contrast to these findings, no changes in retinal haemodynmics were reported using the bluefield technique to assess white blood cell flux [92] and combined measurement of red blood cell velocity and retinal vessel diameters [79]. Both studies were performed in healthy subjects. Steigerwalt et al. have investigated the effect of carteolol 2% on blood flow velocity of retrobulbar vessel after topical administration of the drug in patients with ocular hypertension [220]. The authors found a significant increase in blood velocity in the central retinal artery after carteolol administration in this group of patients.

13.6.10 Serotonin

Serotonin (5-hydroxytryptamine, 5-HT) is a monoamino neurotransmitter and an important tissue hormone in several organ systems. However, serotonin also has vasoactive functions. Unfortunately, these vasoactive properties are complex and differ considerably depending on the vascular bed investigated. Given that at least 19 different serotonin receptors have been indentified so far, this complex nature of effects may be in part attributed to the large number of serotonin receptors existing [166]. For example, it has been shown in the heart that serotonin can cause vasodilation of coronary arterial vessels with a diameter smaller than 100 μm while causing constriction of larger coronary artery segments causing a net increase in coronary blood flow in the healthy heart [136]. The authors have hypothesized that this difference in the effect of serotonin on the vessels can be attributed to the different serotonin receptors involved.

Currently, knowledge of the effects of serotonin in the ocular circulation is sparse. One study investigated the effect of serotonin on the ocular circulation with the microsphere method in a monkey model [57]. In this experiment, exogenous administration of serotonin did not significantly affect retinal or choroidal blood flow. However, in contrast, the same experiment showed that in arteriosclerotic monkeys, administration of serotonin induces pronounced vasoconstriction, which in turn lead to a reduction of retinal and choroidal blood flow [57]. Although the exact reason for this effect remains unclear, the authors hypothesized that this effect can be attributed to a pronounced release of vasoconstrictor components including serotonin and thromboxane by the platelets aggregated in arteriosclerotic plaques. This hypothesis

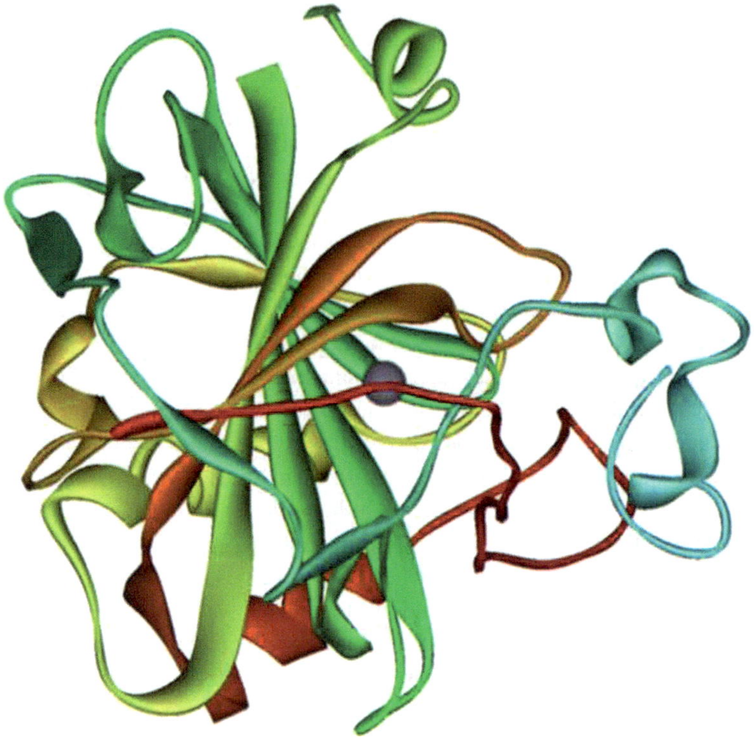

Fig. 13.10 Ribbon diagram of human carbonic anhydrase II

is also supported by another experiment from the same group. In this experiment, the vasoconstrictor response of the retinal circulation to administration of serotonin was examined in a primate model [58]. Monkeys with arteriosclerotic lesions, induced by an atherogenic diet, showed a pronounced vasoconstrictor response to serotonin. However, after an 18-month normal diet, the monkeys showed regression of the atherosclerotic lesions, which also led to an abolished response to serotonin [58].

Hayreh et al. have studied the effect of exogenously administered serotonin on the ocular circulation in a monkey model [101]. For this purpose, serotonin in a dose of 40 μg/kg/min was administered intravenously in 18 arteriosclerotic and 5 normal cynomolgus monkeys. To evaluate changes in the ocular fundus, fundus photography and fluoroscein angiography was performed under basal condition and during serotonin infusion. Whereas the authors observed no changes in normal monkeys, serotonin produced a transient occlusion or delayed filling of both the central retinal artery and the posterior ciliary artery in arteriosclerotic animals [101]. Based on these results, the authors have speculated that ischemic episodes in the eye such as amaurosis fugax or retinal arterial occlusions could be due to vasospasm induced by serotonin released by platelet aggregation in atherosclerotic vessels [100]. However, the study also revealed a marked interindividual as well as interocular variability in the rate and site of susceptibility to vasospasm, not related to generally accepted risk factors for atherosclerosis such as cholesterol. Thus, the hypothesis that serotonin is involved in ischemic events in the eye has to be proven in further experiments.

13.7 Carbonic Anhydrase Inhibitors

Carbonic anhydrase inhibitors (CAIs) are a class of drugs that suppress the activity of carbonic anhydrase, an enzyme necessary to catalyze the rapid conversion of carbon dioxide to bicarbonate and protons. Given that this reaction is ubiquitous throughout the body, several isoforms of the CA (Fig. 13.10) can be found, from which CA-I,

$$H_2O + CO_2 \rightleftharpoons HCO_3^- + H^+$$

Fig. 13.11 The reaction catalyzed by carbonic anhydrase

CA-II, and CA-IV are considered to be most important ones in the human eye. Today, CAIs are widely used to decrease intraocular pressure. The use of CAIs to lower intraocular pressure is based on the finding made 50 years ago that the inhibition of CA leads to a pronounced reduction in aqueous humor production and thus in a decrease in intraocular pressure [15]. This has been explained by the fact that aqueous humor secretion depends on the production of bicarbonate (HCO_3^-), catalyzed by carbonic anhydrase II, which is located in the ciliary epithelium. CA induces the conversion of CO_2 to H_2CO_3, which in turn dissociates into HCO_3^- and H^+ (Fig. 13.11). Consequently, inhibition of CA leads to decreased production of bicarbonate and to decreased intraocular pressure.

However, blockade of CA has also additional impact on the vasculature. In particular, the shift of the equilibrium from bicarbonate to CO_2 due to carboanhydrase inhibition may lead to increased tissue pCO_2 and to lower tissue pH. This is of special importance because local pCO_2 and tissue pH play a role in local blood flow regulation. Based on these theoretical considerations, it has been hypothesized that inhibition of carboanhydrase may lead to vasodilatation and increased blood flow. Consequently, many studies have focused on the effect of systemic and local CAIs on ocular blood flow.

13.8 Acetazolamide

The first CAI widely used in ophthalmology was acetazolamide. However, beside its strong intraocular pressure lowering effect, the chronic treatment is accompanied by considerable side effects such as paresthesias, malaise, or hypokalemia in the majority of patients [188]. The obvious approach to limit the systemic side effects by topical administration of the drug was precluded by the fact that both topical administration and the subconjunctival injection of acetazolamide failed to lower the IOP of rabbits [62, 75].

First evidence of an effect on ocular blood flow caused by systemic administration of acetazolamide was derived from animal experiments in a model of ocular hypertensive rabbits. By the means of the microsphere method, it was observed that administration of acetazolamide leads to a pronounced increase in retinal and choroidal blood flow [36]. These preliminary results have been confirmed by several other studies in different species, including humans. Rassam et al. investigated the effect of intravenously administered acetazolamide on ocular blood flow in healthy subjects [199]. Acetazolamide was administered intravenously at a dose of 500 mg, and retinal blood flow was calculated based on measurements of red cell velocity using laser Doppler velocimetry and vessel diameter measurement using computerized digital image analysis of fundus photographs. The data indicate a significant increase in both retinal blood flow and retinal vessel diameters 60 min after drug injection [199].

To assess perimacular retinal blood flow, Grunwald et al. have measured leukocyte flow with the blue-field system before and after ingestion of 500 mg acetazolamide in a randomized, placebo-controlled study [80]. However, the authors did not find a significant change in leukocyte flow after drug administration.

Differing results for the effect of acetazolamide on retrobulbar blood flow as assessed by the color Doppler technique have been reported. Harris et al. did not find a significant effect of 1,000 mg of acetazolamide administered orally on peak systolic, end-diastolic velocities or resistance index in the ophthalmic or central retinal arteries [95]. These results are in contrast to the findings of Dallinger et al., who also focused on the effects of acetazolamide on retrobulbar blood flow [40]. This study indicated that intravenous administration of acetazolamide increases mean blood flow velocity in the middle cerebral artery and ophthalmic artery in a dose dependent manner [40] (Fig. 13.12).

These results have been confirmed by another study of the same group [122] who found an

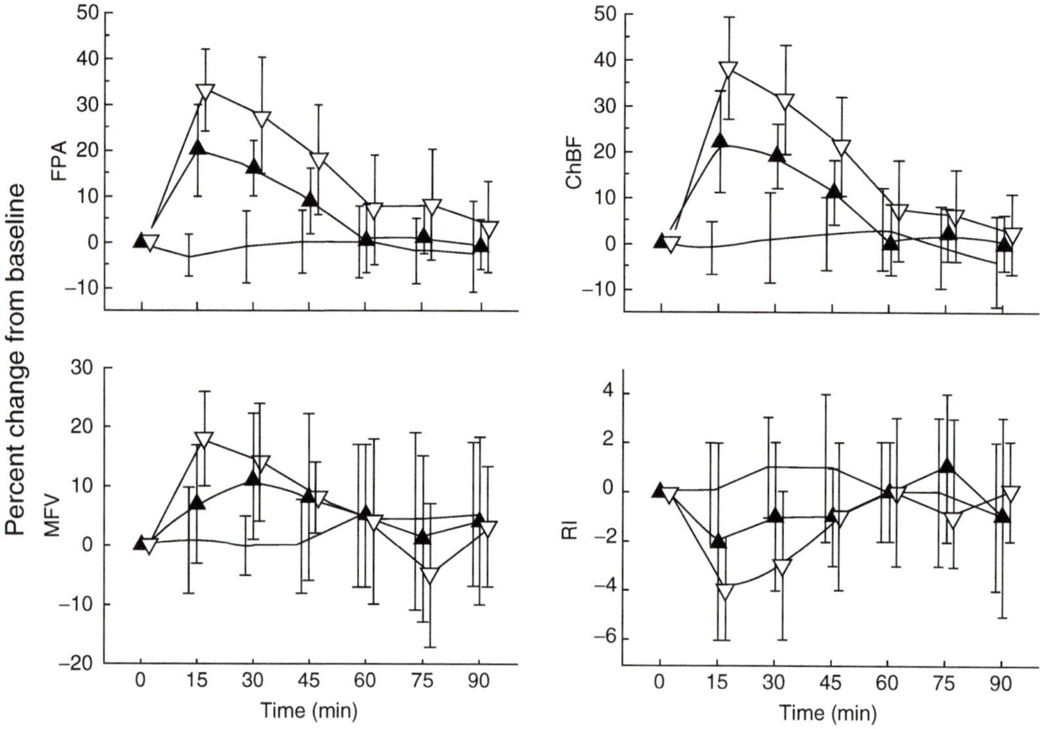

Fig. 13.12 Effects of acetazolamide (1000 mg = open triangles, 500 mg = black triangles) or placebo (no symbols) on fundus pulsation amlitude, choroidal blood flow (ChBF), resistance index and mean flow velocity in the ophthalmic artery

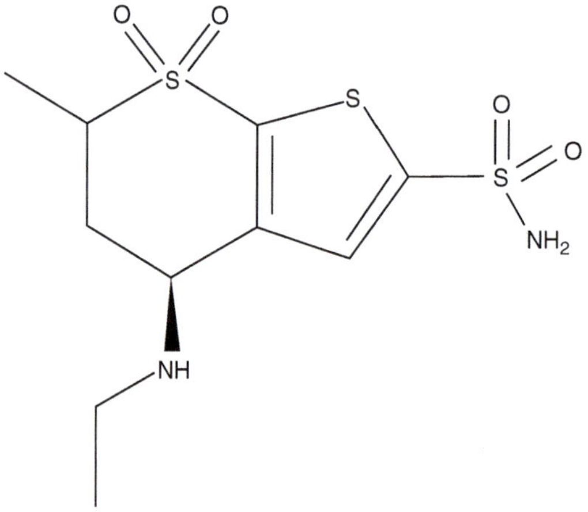

Fig. 13.13 Dorzolamide

increase in blood velocity after administration of the drug. The reason for the differing results is however still unclear but may be related to different routes of drug administration used in the studies. Whereas in the work of Harris et al., acetazolamide was administered orally, the latter studies used intravenously administration of the study drug, which in turn may lead to higher plasma drug concentrations.

The work of Dallinger et al. also indicates that acetazolamide increases fundus pulsation amplitude as measured with laser interferometery, which gives an estimate of pulsatile choroidal blood flow [40]. These results are in keeping with other studies investigating the effect of acetazolamide on choroidal blood flow [122].

13.9 Dorzolamide

Dorzolamide hydrochloride is a water-soluble inhibitor of carbonanhydrase (Fig. 13.13). From a chemical view, it consists of a heterocyclic thienothiopyran resulting in an increased lipophilicity compared to acetazolamide [188]. Because of its better penetration through the cornea, dorzolamide is used as a topical CA-inhibitor to lower intraocular pressure. Topical dorzolamide leads to a pronounced decrease in intraocular pressure [215] due to a strong reduction in aqueous humor production [107].

The hemodynamic effect of dorzolamide was the focus of a several studies in healthy subjects and glaucoma patients that reached differing conclusions about the ability of dorzolamide to increase ocular hemodynamics, but most studies indicate vasodilator effects. The reason for the differing results is still a matter of controversy. This may be related to the variety of methods used for assessing blood flow and the differences in the ocular vascular beds under study. Furthermore, given that dorzolamide is mainly used in glaucoma patients, a large number of studies have been performed in glaucoma patients. However, it has to be considered that glaucoma is a multifactorial disease and drug effects may be different depending on the type of glaucoma.

13.10 Retrobulbar Blood Flow

As one of the first studies in humans, Harris et al. investigated the effect of topical dorzolamide blood velocity in four retrobulbar vessels (nasal and temporal posterior ciliary, central retinal and ophthalmic artery) and on retinal arteriovenous passage time with SLO [94]. For this purpose, two drops 2% dorzolamide or placebo were instilled in a group of 11 healthy volunteers and ocular hemodynamic parameters were assessed at baseline and after drug administration in a double-masked, balanced study. As measured 2 h after drug administration, no difference in retrobulbar hemodynamic parameters was observed. However, the same study revealed accelerated retinal arteriovenous passage of fluorescein as well as an increase in capillary velocity in the optic nerve head, both variables indicating but not proving an increase in blood flow [94].

The same group investigated the effect of dorzolamide on ocular hemodynamics in patients with glaucoma. Again, blood flow velocities in retrobulbar vessels were measured with the CDI technique, as well as retinal arteriovenous passage time and retinal arterial/venous diameters with SLO [96]. Included were 18 patients with normal-tension glaucoma, treated for 4 weeks with 2% topical dorzolamide after a washout phase and compared to a placebo group. Measurements were made at baseline and 2 and 4 weeks after start with dorzolamide treatment, respectively. In agreement with their previous results, the authors did not find a change in retrobulbar hemodynamic parameters in response to dorzolamide [96]. However, although no changes in retinal vessel diameters were observed, an increased retinal arteriovenous passage time was observed.

Data about the effect of dorzolamide on retrobulbar blood flow are also available from Matinez et al. The authors have investigated the effect of dorzolamide on 26 patients with open-angle glaucoma compared to a control group consisting of 13 normal eyes [150]. All eyes underwent CDI measurements of all major retrobulbar vessels. In keeping with the results of Harris et al., dorzolamide did not change peak systolic velocities of the ophthalmic artery and the central retinal artery. However, in contrast, Martinez et al. found an increase of end-diastolic velocity and a decrease of resistance index [150]. The reason for these differing results is not entirely clear but may be related to the fact that patients with different types of glaucoma have been included in these studies.

Zeitz et al. have investigated retrobulbar blood flow in patients with normal-tension glaucoma and in patients with primary open-angle glaucoma in two different experiments. In the first experiment, peak systolic and end-diastolic blood flow velocities in the short posterior ciliary artery were assessed by color Doppler imaging in 42 patients with normal-tension glaucoma [264]. Measurements were done shortly before and after a 1-month treatment with latanoprost, bimatoprost, or dorzolamide. Whereas no changes were observed in the latanoprost and the bimatoprost groups, dorzolamide accelerated peak systolic blood flow velocities [264]. In the second experiments, CDI measurements were performed in patients with primary open-angle glaucoma. However, in this study, no changes in blood flow velocities were detected after the application of dorzolamide [265].

Bergstrand et al. investigated the effect of a 6-week period of topical dorzolamide administration two times a day versus timolol in a double-masked study [18]. As outcome variables, the authors assessed retrobulbar blood flow velocities using CDI and scanning laser ophthalmoscope fluorescein angiograms to measure retinal hemodynamics. Whereas both drugs induced a significant decrease in intraocular pressure, none of the blood flow parameters, retrobulbar or retinal, changed significantly in response to the treatment. Based on these results, the authors have concluded that none of the administered drugs induced a measurable vascular effect in previously untreated glaucoma eyes [18].

Recently a comparative analysis of the effects of brimonidine and dorzolamide on ocular blood flow velocity in patients with newly diagnosed primary open-angle glaucoma has been performed [216]. Forty-four patients with newly diagnosed primary open-angle glaucoma were included in this prospective, comparative, randomized clinical study. Ocular retrobulbar hemodynamics were assessed by CDI. Patients were randomly assigned to receive either brimonidine 0.2% or dorzolamide 2% for a 3-month treatment period. Whereas an increase in peak systolic velocity in the central retinal artery was observed in both treatment groups, no changes were observed in the other measured retrobulbar vessels.

13.11 Retinal Blood Flow

The effect of topically administered dorzolamide on retinal blood flow of healthy volunteers was investigated using the laser Doppler techniques. Combined measurement of bidirectional laser Doppler velocimetry and monochromatic fundus photography to assess blood velocity and retinal vessel size, respectively, was performed in a major temporal vein in healthy subjects [82]. Measurements were performed before and 2 h after a single drop instillation of dorzolamide 2% in a double-masked randomized study. The authors did not find a statistically significant effect of dorzolamide on retinal blood flow in healthy subjects.

Arend et al. have investigated retinal hemodynamics using fluorescein angiography by means of a scanning laser ophthalmoscope in a prospective, randomized, crossover study in a group of newly diagnosed primary open-angle glaucoma patients [8]. Retinal hemodynmaics were assessed at baseline and after a 4-week treatment period. Arteriovenous passage times as well as peripapillary retinal arterial and venous calibers were assessed from SLO angiograms, using digital image processing. The results of the study revealed that dorzolamide treatment shortened the arteriovenous passage time compared to baseline, whereas no changes in vessel diameters were observed [8]. Although these findings may suggest an increase in retinal blood flow, the findings do not necessarily prove an increase in blood flow due to limitation of the technique.

Using a Retinal Vessel Analyzer, retinal vessel diameters were measured before and after 4 weeks of treatment with dorzolamide in 12 previously untreated patients with primary open-angle glaucoma [163]. Beside the expected pronounced drop in intraocular pressure, the authors observed an increased in retinal vessel diameters after dorzolamide treatment. However, whether this can be interpreted as a mechanical response to the decrease in IOP or as a vasodilatation caused by CAI remains unclear. Furthermore, the interpretation of the latter study is limited by the fact that only vessel diameters were measured, and therefore, no conclusion about retinal blood flow can be made.

The short-time effect of topical dorzolamide on retinal blood flow has been investigated using the Canon laser Doppler flowmeter [56]. Retinal blood flow was measured before and 1 h after topical administration of the study drug in a group of 17 healthy volunteers. The authors did not find a significant change in retinal vessel caliber, red blood cell velocity, or blood flow in the measured vessels.

Costagliola et al. focused on the effect of dorzolamide on peripapillary retinal hemodynamics in juvenile primary open-angle glaucoma patients previously treated with timolol [39]. Measurements with laser scanning flowmetry were performed at baseline and after coadministration of dorzolamide. The study revealed an increase of retinal blood flow as assessed with HRF after coadministration of dorzolamide, accompanied with an increase in ocular perfusion pressure.

13.12 Choroidal and Optic Nerve Head Blood Flow

Several studies focused on the effect of dorzolamide on choroidal blood flow using different methodological approaches. Using measurements of ocular pulse amplitude as an estimate for pulsatile choroidal blood flow, the effect of dorzolamide was investigated in primary open-angle glaucoma patients with elevated intraocular pressure and non-glaucomatous controls after topical administration of the drug [210]. Measurements were done before and after 2 days of treatment. The results of this study indicate that dorzolamide increases ocular pulse amplitude in both groups, which may be interpreted as an increase in choroidal blood flow [210].

The same group also focused on the effect of dorzolamide on optic nerve head blood flow. In this double-masked, randomized clinical trial, dorzolamide eye drops were applied to both eyes of 15 healthy subjects three times daily for 3 days and compared to a control group receiving placebo [189]. Scanning laser Doppler flowmetry using the Heidelberg retina flowmeter and laser Doppler flowmetry using the Oculix instrument were used to measure optic nerve head blood flow. Although the authors found the expected decrease in intraocular pressure, no significant change in optic nerve head blood flow was observed [189]. The authors have attributed these results to the strong autoregulatory capacity of the optic nerve head tissue.

The effect of dorzolamide and timolol on retinal and optic nerve head blood flow has been investigated in a prospective randomized double-masked clinical trial [64]. For this purpose, 140 patients with POAG or OHT were included in two parallel groups from which 70 patients were randomized to receive timolol and 70 to receive dorzolamide for a period of 6 months. Scanning laser Doppler flowmetry was used to measure blood flow in the temporal neuroretinal rim and the cup of the optic nerve head. The pulsatile component of choroidal blood flow was assessed using laser interferometric measurement of fundus pulsation amplitude [64].

Whereas the effects of both drugs on IOP and ocular perfusion pressure were comparable between dorzolamide and timolol, dorzolamide, but not timolol, increased blood flow in the temporal neuroretinal rim and the cup of the optic nerve head. Furthermore, dorzolamide but not timolol increased fundus pulsation amplitude as measured with laser interferometry. Based on these results, the authors have concluded that dorzolamide, but not timolol, augments blood flow in the optic nerve head and choroid after 6 months of treatment with dorzolamide. Whether this effect is also beneficial for patients in terms of delayed progression of the disease remains to be investigated.

Siesky et al. have performed a meta-analysis for all studies of topical carbonic anhydrase inhibitors including dorzolamide on ocular blood flow [214]. Based on the meta-analysis, the authors have concluded that topical administration of dorzolamide increases ocular blood flow velocities in the retinal circulation, central retinal, and short posterior ciliary arteries, whereas no consistent effect of the drug on the ophthalmic artery was observed [214].

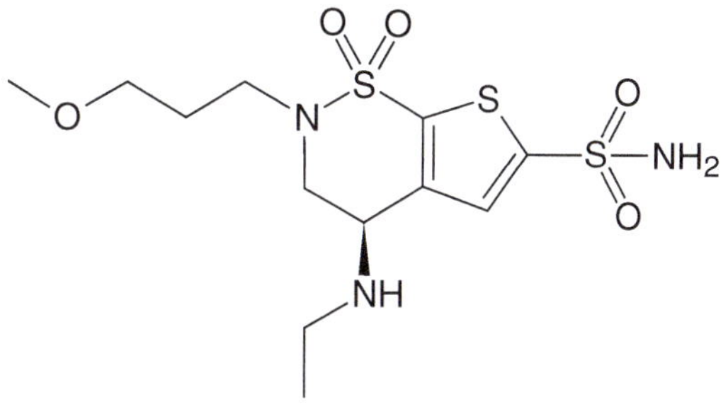

Fig. 13.14 Brinzolamide

13.13 Brinzolamide

Brinzolamide (Fig. 13.14) is a potent inhibitor of CA, with a special affinity to CA-II, which is involved in aqueous humor production. From a chemical view, brinzolamide also belongs to the sulfonamide family. However, brinzolamide is more lipophilic than either dorzolamide or acetazolamide at physiologic pH, which favors its ability to cross cell membranes and may allow for a better penetration through the cornea. Comparable to dorzolamide, topical administration of brinzolamide leads to a pronounced decrease in intraocular pressure due to a reduction in aqueous humor secretion. An overview of the chemical and pharmacological properties of brinzolamide has been published elsewhere [46].

The first evidence of a blood flow effect of brinzolamide came from animal experiments. Using a three-way crossover study design, brinzolamide, dorzolamide, and placebo were administered to tranquilized Dutch-belted rabbits. Laser Doppler measurements on the optic nerve head revealed that both CAIs increased optic nerve head blood flow [12].

In a preliminary report, optic nerve head and retinal hemodymanics were measured in 25 healthy subjects as well as in 72 glaucoma patients, from which 24 eyes were treated with betaxolol, 24 with brinzolamide, and 24 treated with brimonidine [201]. Optic nerve head blood flow was measured using scanning laser Doppler flowmetry. However, the authors did not find any evidence for a blood flow increasing effect with any of the administered drugs. In contrast, a decrease in optic nerve head blood flow was observed in the brinzolamide group [201].

The effect of topical brinzolamide on retinal capillary blood flow was measured by Lester et al. in 20 patients with primary open-angle glaucoma [104]. The Heidelberg retina flowmeter was used to measure blood flow at baseline and after 1 month of treatment. An increase in retinal blood flow was observed in the optic nerve head tissue. The effect of topically administered brinzolamide on retinal vessel diameters was assessed with a retinal vessel analyzer 2 h after administration [134]. This study, however, did not find a statistically significant change in retinal vessel sized after drug administrations. It has to be considered, however, that this study was only capable of detecting short-term effects and that retinal vessel diameters alone are not a good measure of retinal blood flow.

The effect of brinzolamide on retrobulbar hemodynamics has been investigated using the CDI technique. In a study by Kaup et al., 30 healthy volunteers were randomized to receive either brinzolamide or placebo during a 2-week double-masked treatment period [115]. Peak systolic and end-diastolic velocities of the ophthalmic artery, the central retinal artery, and the short posterior ciliary arteries were determined, and resistive indices were calculated. In addition, the arteriovenous passage time and retinal vessel diameters were measured by digital image based analysis of video fluorescein angiograms. Administration of brinzolamide resulted in a significant reduction of arteriovenous passage time while leaving retinal vessel diameters unchanged. This was interpreted by the authors as evidence for an increase of retinal blood flow, but does, however, not finally prove this effect. In contrast to the data for the retinal circulation, no effect of the brinzolamide treatment on retrobulbar hemodynamics was observed.

The results of the latter study were confirmed recently in a randomized, crossover, double-masked experiment. Fifteen patients with open-angle glaucoma were evaluated at baseline and after 3 months treatment with either brinzolamide or dorzolamide [213]. Ocular blood flow was assessed using scanning laser Doppler flowmetry and retrobulbar blood flow with color Doppler imaging. Whereas scanning laser Doppler flowmetry suggested an increase in retinal blood flow, no significant changes were found in CDI measurements of retrobulbar blood velocity during either of the two treatments [213].

Another report on the long-term treatment effects of brinzolamide and dorzolamide on retrobulbar blood velocity was published recently [149]. In this prospective, evaluator-masked study, 146 patients with primary open-angle glaucoma were randomized to receive either dorzolamide 2% or brinzolamide 1% BID, each added to timolol 0.5%. Retrobulbar hemodynamic parameters were assessed in the ophthalmic artery, the central retinal artery, and the short posterior ciliary arteries by the means of CDI

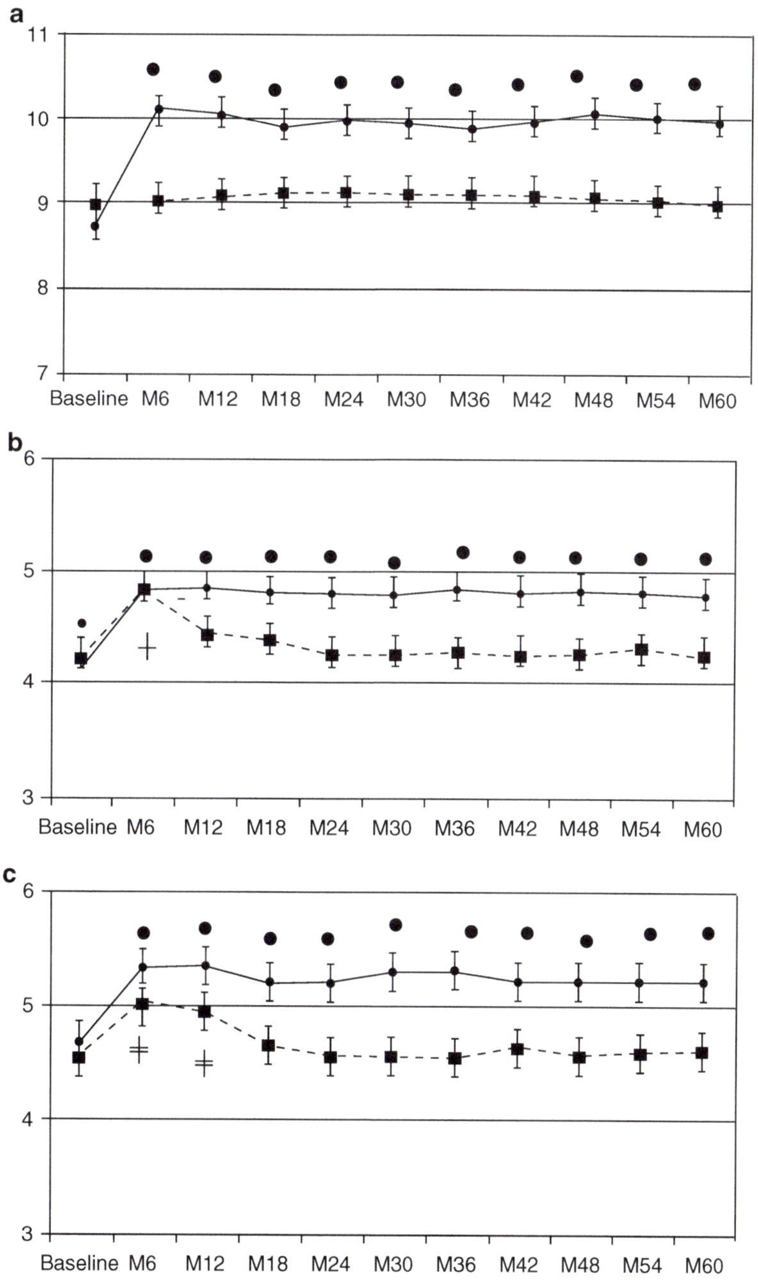

Fig. 13.15 Mean end-diastolic velocities in the ophthalmic artery (**a**), central retinal artery (**b**) and short ciliary artery (**c**) starting at baseline up to month 60. The solid line with solid rhombus indicates the dorzolamide/timolol treatment group, whereas the dotted line with solid squares corresponds to the brinzolamide/timolol treatment group. Black dots indicates significant effects of dorzolamide/timolol compared with baseline and brinzolamide/timolol. ≠ Indicates significant effects of brinzolamide/timolol compared with baseline ($P < 0.001$). + Indicates no significant differences between dorzolamide/timolol and brinzolamide/timolol treatment groups

every 6 months, with a total observation period of 60 months. Interestingly, the authors found that dorzolamide significantly increased the end-diastolic velocity in the ophthalmic artery, whereas the brinzolamide group did not show any significance change in retrobulbar blood flow at the end of the 60 months observation period (Fig. 13.15) [149] It has, however, to be mentioned that most of the blood supplied by the OA does not go into the eye.

In summary, most but not all studies indicate an increase in ocular blood flow after administration of brinzolamide. However, further studies are needed to evaluate whether the ocular hemodynamic properties have a beneficial effect for patients under treatment.

References

1. Abran D et al (1995) Characterization and ontogeny of PGE2 and PGF2 alpha receptors on the retinal vasculature of the pig. Prostaglandins 50:253–267
2. Abran D et al (1997) Characterization and regulation of prostaglandin E2 receptor and receptor-coupled functions in the choroidal vasculature of the pig during development. Circ Res 80:463–472
3. Akarsu C et al (2004) Short-term effect of latanoprost on ocular circulation in ocular hypertension. Clin Experiment Ophthalmol 32:373–377
4. Alm A (1972) Effects of norepinephrine, angiotensin, dihydroergotamine, papaverine, isoproterenol, histamine, nicotinic acid, and xanthinol nicotinate on retinal oxygen tension in cats. Acta Ophthalmol (Copenh) 50:707–719
5. Altan-Yaycioglu R et al (2001) The effects of beta-blockers on ocular blood flow in patients with primary open angle glaucoma: a color Doppler imaging study. Eur J Ophthalmol 11:37–46
6. Araie M, Muta K (1997) Effect of long-term topical betaxolol on tissue circulation in the iris and optic nerve head. Exp Eye Res 64:167–172
7. Arend O et al (1998) The acute effect of topical beta-adrenoreceptor blocking agents on retinal and optic nerve head circulation. Acta Ophthalmol Scand 76:43–49
8. Arend O et al (2003) Evaluation of retinal haemodynamics and retinal function after application of dorzolamide, timolol and latanoprost in newly diagnosed open-angle glaucoma patients. Acta Ophthalmol Scand 81:474–479
9. Astin M et al (1994) Role of nitric oxide in PGF2 alpha-induced ocular hyperemia. Exp Eye Res 59:401–407
10. Ayajiki K et al (2000) Evidence for nitroxidergic innervation in monkey ophthalmic arteries in vivo and in vitro. Am J Physiol Heart Circ Physiol 279:H2006–H2012
11. Ayajiki K et al (2001) Relatively selective neuronal nitric oxide synthase inhibition by 7-nitroindazole in monkey isolated cerebral arteries. Eur J Pharmacol 423:179–183
12. Barnes GE et al (2000) Increased optic nerve head blood flow after 1 week of twice daily topical brinzolamide treatment in Dutch-belted rabbits. Surv Ophthalmol 44(Suppl 2):S131–S140
13. Baurmann H et al (1986) Simultaneous study of ocular vascular reactions of the anterior and posterior segments following clonidine and propranolol in normotensive and hypertensive experimental animals. Klin Monbl Augenheilkd 189:467–476
14. Beano F et al (2001) An evaluation of the effect of unoprostone isopropyl 0.15% on ocular hemodynamics in normal-tension glaucoma patients. Graefes Arch Clin Exp Ophthalmol 239:81–86
15. Becker B (1954) Decrease in intraocular pressure in man by a carbonic anhydrase inhibitor, diamox; a preliminary report. Am J Ophthalmol 37:13–15
16. Benedito S et al (1991) Role of the endothelium in acetylcholine-induced relaxation and spontaneous tone of bovine isolated retinal small arteries. Exp Eye Res 52:575–579
17. Bergstrand IC et al (2001) Timolol increased retrobulbar flow velocities in untreated glaucoma eyes but not in ocular hypertension. Acta Ophthalmol Scand 79:455–461
18. Bergstrand IC et al (2002) Dorzolamide and ocular blood flow in previously untreated glaucoma patients: a controlled double-masked study. Acta Ophthalmol Scand 80:176–182
19. Bill A (1979) Effects of indomethacin on regional blood flow in conscious rabbits – a microsphere study. Acta Physiol Scand 105:437–442
20. Bloom AH et al (1997) Effect of one week of levobunolol HCl 0.5% on the human retinal circulation. Curr Eye Res 16:191–196
21. Boger RH, Bode-Boger SM (2000) Asymmetric dimethylarginine, derangements of the endothelial nitric oxide synthase pathway, and cardiovascular diseases. Semin Thromb Hemost 26:539–545
22. Bosem ME et al (1992) Short-term effects of levobunolol on ocular pulsatile flow. Am J Ophthalmol 114:280–286
23. Brazitikos PD et al (1993) Microinjection of L-lactate in the preretinal vitreous induces segmental vasodilation in the inner retina of miniature pigs. Invest Ophthalmol Vis Sci 34:1744–1752
24. Bruinink A et al (1986) Catecholaminergic binding sites in cat retina, pigment epithelium and choroid. Exp Eye Res 43:147–151
25. Buerk DG et al (1996) Nitric oxide has a vasodilatory role in cat optic nerve head during flicker stimuli. Microvasc Res 52:13–26
26. Bursell SE et al (1995) The in vivo effect of endothelins on retinal circulation in nondiabetic and diabetic rats. Invest Ophthalmol Vis Sci 36:596–607
27. Carenini AB et al (1994) Differences in the longterm effect of timolol and betaxolol on the pulsatile ocular blood flow. Surv Ophthalmol 38(Suppl):S118–S124
28. Carlsson AM et al (2000) The effect of brimonidine tartrate on retinal blood flow in patients with ocular hypertension. Am J Ophthalmol 129:297–301
29. Chakravarthy U et al (1992) The effect of endothelin 1 on the retinal microvascular pericyte. Microvasc Res 43:241–254
30. Chakravarthy U et al (1994) Immunoreactive endothelin distribution in ocular tissues. Invest Ophthalmol Vis Sci 35:2448–2454
31. Chakravarthy U et al (1995) Nitric oxide synthase activity and expression in retinal capillary endothelial cells and pericytes. Curr Eye Res 14:285–294
32. Chauhan BC (2008) Endothelin and its potential role in glaucoma. Can J Ophthalmol 43:356–360
33. Chauhan BC et al (2004) Model of endothelin-1-induced chronic optic neuropathy in rat. Invest Ophthalmol Vis Sci 45:144–152
34. Chemtob S et al (1991) Differences in the effects in the newborn piglet of various nonsteroidal antiinflammatory drugs on cerebral blood flow but not on cerebrovascular prostaglandins. Pediatr Res 30:106–111

35. Chen Z et al (1998) Histochemical mapping of NADPH-diaphorase in monkey and human eyes. Curr Eye Res 17:370–379
36. Chiou GC, Chen YJ (1993) Effects of antiglaucoma drugs on ocular blood flow in ocular hypertensive rabbits. J Ocul Pharmacol 9:13–24
37. Chun MH et al (1999) Light and electron microscopical analysis of nitric oxide synthase-like immunoreactive neurons in the rat retina. Vis Neurosci 16: 379–389
38. Cioffi GA et al (1995) An in vivo model of chronic optic nerve ischemia: the dose-dependent effects of endothelin-1 on the optic nerve microvasculature. Curr Eye Res 14:1147–1153
39. Costagliola C et al (2007) Effect of 2% dorzolamide on retinal blood flow: a study on juvenile primary open-angle glaucoma patients already receiving 0.5% timolol. Br J Clin Pharmacol 63:376–379
40. Dallinger S et al (1998) Effects of acetazolamide on choroidal blood flow. Stroke 29:997–1001
41. Dallinger S et al (2000) Endothelin-1 contributes to hyperoxia-induced vasoconstriction in the human retina. Invest Ophthalmol Vis Sci 41:864–869
42. Dallinger S et al (2003) Vasodilator effects of L-arginine are stereospecific and augmented by insulin in humans. Am J Physiol Endocrinol Metab 284:1106–1111
43. Dawson TM et al (1991) Nitric oxide synthase and neuronal NADPH diaphorase are identical in brain and peripheral tissues. Proc Natl Acad Sci USA 88:7797–7801
44. de Juan JA et al (1995) Identification of endothelin receptor subtypes in rat retina using subtype-selective ligands. Brain Res 690:25–33
45. Delles C et al (2004) Impaired endothelial function of the retinal vasculature in hypertensive patients. Stroke 35:1289–1293
46. DeSantis L (2000) Preclinical overview of brinzolamide. Surv Ophthalmol 44(Suppl 2):S119–S129
47. Deussen A et al (1993) L-arginine-derived nitric oxide: a major determinant of uveal blood flow. Exp Eye Res 57:129–134
48. Dollery CT et al (1963) The response of normal retinal blood vessels to angiotensin and noradrenaline. J Physiol 165:500–507
49. Donati G et al (1995) Nitric oxide controls arteriolar tone in the retina of the miniature pig. Invest Ophthalmol Vis Sci 36:2228–2237
50. Donati G et al (1997) Decreased nitric oxide production accounts for secondary arteriolar constriction after retinal branch vein occlusion. Invest Ophthalmol Vis Sci 38:1450–1457
51. Dorner GT et al (2003) Hyperglycemia affects flicker-induced vasodilation in the retina of healthy subjects. Vision Res 43:1495–1500
52. Dorner GT et al (2003) Nitric oxide regulates retinal vascular tone in humans. Am J Physiol Heart Circ Physiol 285:H631–H636
53. Dorner GT et al (2007) Effects of pentoxifylline and alprostadil on ocular hemodynamics in healthy humans. Invest Ophthalmol Vis Sci 48:815–819
54. Eichhorn M, Lutjen-Drecoll E (1993) Distribution of endothelin-like immunoreactivity in the human ciliary epithelium. Curr Eye Res 12:753–757
55. Evans DW et al (1999) Primary open-angle glaucoma patients characterized by ocular vasospasm demonstrate a different ocular vascular response to timolol versus betaxolol. J Ocul Pharmacol Ther 15:479–487
56. Faingold D et al (2004) Assessment of retinal hemodynamics with the Canon laser blood flowmeter after a single dose of 2% dorzolamide hydrochloride eyedrops. Can J Ophthalmol 39:506–510
57. Faraci FM et al (1989) Atherosclerosis potentiates constrictor responses of cerebral and ocular blood vessels to thromboxane in monkeys. Stroke 20:242–247
58. Faraci FM et al (1991) Dietary treatment of atherosclerosis abolishes hyperresponsiveness of retinal blood vessels to serotonin in monkeys. Stroke 22:1405–1408
59. Feke GT et al (1996) Retinal blood flow increases following short-term aspirin usage in type I diabetics with no or minimal retinopathy. Ophthalmic Res 28:108–116
60. Fitzgerald ME et al (1996) Central neural circuits for the light-mediated reflexive control of choroidal blood flow in the pigeon eye: a laser Doppler study. Vis Neurosci 13:655–669
61. Flugel C et al (1994) Species differences in choroidal vasodilative innervation: evidence for specific intrinsic nitrergic and VIP-positive neurons in the human eye. Invest Ophthalmol Vis Sci 35:592–599
62. Foss RH (1955) Local application of diamox; an experimental study of its effect on the intraocular pressure. Am J Ophthalmol 39:336–337
63. Fuchsjager-Mayrl G et al (2003) Role of endothelin-1 in choroidal blood flow regulation during isometric exercise in healthy humans. Invest Ophthalmol Vis Sci 44:728–733
64. Fuchsjager-Mayrl G et al (2005) Effect of dorzolamide and timolol on ocular blood flow in patients with primary open angle glaucoma and ocular hypertension. Br J Ophthalmol 89:1293–1297
65. Fuchsjäger-Mayrl G et al (2001) Unilateral light–dark transitions affect choroidal blood flow in both eyes. Vision Res 41(22):2919–2924
66. Furchgott RF, Zawadzki JV (1980) The obligatory role of endothelial cells in the relaxation of arterial smooth muscle by acetylcholine. Nature 288:373–376
67. Garhofer G et al (2005) Intravenous administration of L-arginine increases retinal and choroidal blood flow. Am J Ophthalmol 140:69–76
68. Georgopoulos GT et al (2002) The short-term effect of latanoprost on intraocular pressure and pulsatile ocular blood flow. Acta Ophthalmol Scand 80:54–58
69. Gherezghiher T et al (1991) Choroidal and ciliary body blood flow analysis: application of laser Doppler flowmetry in experimental animals. Exp Eye Res 53:151–156
70. Goureau O et al (1993) Differential regulation of inducible nitric oxide synthase by fibroblast growth

factors and transforming growth factor beta in bovine retinal pigmented epithelial cells: inverse correlation with cellular proliferation. Proc Natl Acad Sci USA 90:4276–4280
71. Goureau O et al (1994) Human retinal pigmented epithelial cells produce nitric oxide in response to cytokines. Biochem Biophys Res Commun 198:120–126
72. Grajewski AL et al (1991) Beta-adrenergic responsiveness of choroidal vasculature. Ophthalmology 98:989–995
73. Granstam E, Granstam SO (1999) Regulation of uveal and retinal blood flow in STZ-diabetic and non-diabetic rats; involvement of nitric oxide. Curr Eye Res 19:330–337
74. Granstam E et al (1992) Ocular effects of endothelin-1 in the cat. Curr Eye Res 11:325–332
75. Grant WM, Trotter RR (1954) Diamox (acetazoleamide) in treatment of glaucoma. AMA Arch Ophthalmol 51:735–739
76. Green K et al (1985) Ocular blood flow after experimental alkali burns and prostaglandin administration. Arch Ophthalmol 103:569–571
77. Grunwald JE (1990) Effect of timolol maleate on the retinal circulation of human eyes with ocular hypertension. Invest Ophthalmol Vis Sci 31:521–526
78. Grunwald JE (1991) Effect of two weeks of timolol maleate treatment on the normal retinal circulation. Invest Ophthalmol Vis Sci 32:39–45
79. Grunwald JE, Delehanty J (1992) Effect of topical carteolol on the normal human retinal circulation. Invest Ophthalmol Vis Sci 33:1853–1856
80. Grunwald JE, Zinn H (1992) The acute effect of oral acetazolamide on macular blood flow. Invest Ophthalmol Vis Sci 33:504–507
81. Grunwald JE et al (1997) Effect of chronic nitrate treatment on retinal vessel caliber in open-angle glaucoma. Am J Ophthalmol 123:753–758
82. Grunwald JE et al (1997) Effects of dorzolamide hydrochloride 2% on the retinal circulation. Acta Ophthalmol Scand 75:236–238
83. Haefliger IO et al (1992) Nitric oxide and endothelin-1 are important regulators of human ophthalmic artery. Invest Ophthalmol Vis Sci 33:2340–2343
84. Haefliger IO et al (1993) Heterogeneity of endothelium-dependent regulation in ophthalmic and ciliary arteries. Invest Ophthalmol Vis Sci 34:1722–1730
85. Haefliger IO et al (1994) Relaxation of retinal pericyte contractile tone through the nitric oxide-cyclic guanosine monophosphate pathway. Invest Ophthalmol Vis Sci 35:991–997
86. Haefliger IO et al (1997) Effect of oxygen on relaxation of retinal pericytes by sodium nitroprusside. Graefes Arch Clin Exp Ophthalmol 235:388–392
87. Haefliger IO et al (1999) Modulation of Heidelberg Retinal Flowmeter parameter flow at the papilla of healthy subjects: effect of carbogen, oxygen, high intraocular pressure, and beta-blockers. Surv Ophthalmol 43(Suppl 1):S59–S65
88. Hardy P et al (1994) Free radicals in retinal and choroidal blood flow autoregulation in the piglet: interaction with prostaglandins. Invest Ophthalmol Vis Sci 35:580–591
89. Hardy P et al (1996) Nitric oxide in retinal and choroidal blood flow autoregulation in newborn pigs: interactions with prostaglandins. Pediatr Res 39:487–493
90. Hardy P et al (1998) A major role for prostacyclin in nitric oxide-induced ocular vasorelaxation in the piglet. Circ Res 83:721–729
91. Hardy P et al (2000) Oxidants, nitric oxide and prostanoids in the developing ocular vasculature: a basis for ischemic retinopathy. Cardiovasc Res 47:489–509
92. Harris A et al (1995) Acute effect of topical (beta-adrenergic antagonists on normal perimacular hemodynamics. J Glaucoma 4:36–40
93. Harris A et al (1995) Retrobulbar arterial hemodynamic effects of betaxolol and timolol in normal-tension glaucoma. Am J Ophthalmol 120:168–175
94. Harris A et al (1996) Effects of topical dorzolamide on retinal and retrobulbar hemodynamics. Acta Ophthalmol Scand 74:569–572
95. Harris A et al (1996) Acetazolamide and CO_2: acute effects on cerebral and retrobulbar hemodynamics. J Glaucoma 5:39–45
96. Harris A et al (1999) Dorzolamide, visual function and ocular hemodynamics in normal-tension glaucoma. J Ocul Pharmacol Ther 15:189–197
97. Hashitani H et al (1998) Neuroeffector transmission in arterioles of the guinea-pig choroid. J Physiol 510(Pt 1):209–223
98. Hata Y et al (2000) Retinal expression, regulation, and functional bioactivity of prostacyclin-stimulating factor. J Clin Invest 106:541–550
99. Haynes WG et al (1993) Inhibition of nitric oxide synthesis increases blood pressure in healthy humans. J Hypertens 11:1375–1380
100. Hayreh SS (1999) Retinal and optic nerve head ischemic disorders and atherosclerosis: role of serotonin. Prog Retin Eye Res 18:191–221
101. Hayreh SS et al (1997) Serotonin-induced constriction of ocular arteries in atherosclerotic monkeys. Implications for ischemic disorders of the retina and optic nerve head. Arch Ophthalmol 115:220–228
102. Hessemer V, Schmidt KG (1996) Effect of the vasodilator isosorbide dinitrate on ocular circulation. Klin Monbl Augenheilkd 208:188–195
103. Huemer KH et al (2007) Role of nitric oxide in choroidal blood flow regulation during light/dark transitions. Invest Ophthalmol Vis Sci 48:4215–4219
104. Iester M et al (2004) Retinal peripapillary blood flow before and after topical brinzolamide. Ophthalmologica 218:390–396
105. Ignarro LJ et al (1987) Endothelium-derived relaxing factor produced and released from artery and vein is nitric oxide. Proc Natl Acad Sci USA 84:9265–9269
106. Inan UU et al (2003) The effects of latanoprost and brimonidine on blood flow velocity of the retrobulbar vessels: a 3-month clinical trial. Acta Ophthalmol Scand 81:155–160
107. Ingram CJ, Brubaker RF (1999) Effect of brinzolamide and dorzolamide on aqueous humor flow in human eyes. Am J Ophthalmol 128:292–296
108. Ishii K et al (2001) Effects of topical latanoprost on optic nerve head circulation in rabbits, monkeys, and humans. Invest Ophthalmol Vis Sci 42:2957–2963

109. Jacot JL et al (1998) Nitric oxide modulation of retinal, choroidal, and anterior uveal blood flow in newborn piglets. J Ocul Pharmacol Ther 14:473–489
110. Jandrasits K et al (2002) Effect of noradrenaline on retinal blood flow in healthy subjects. Ophthalmology 109:291–295
111. Januleviciene I et al (2004) A comparison of the effects of dorzolamide/timolol fixed combination versus latanoprost on intraocular pressure and pulsatile ocular blood flow in primary open-angle glaucoma patients. Acta Ophthalmol Scand 82:730–737
112. Jay WM et al (1984) Effect of topical epinephrine and timolol on ocular and optic nerve blood flow in phakic and aphakic rabbit eyes. Curr Eye Res 3:1199–1202
113. Jonescu-Cuypers CP et al (2001) Effect of brimonidine tartrate on ocular hemodynamics in healthy volunteers. J Ocul Pharmacol Ther 17:199–205
114. Kass MA, Holmberg NJ (1979) Prostaglandin and thromboxane synthesis by microsomes of rabbit ocular tissues. Invest Ophthalmol Vis Sci 18:166–171
115. Kaup M et al (2004) Effects of brinzolamide on ocular haemodynamics in healthy volunteers. Br J Ophthalmol 88:257–262
116. Kawamura H et al (2002) Endothelin-induced changes in the physiology of retinal pericytes. Invest Ophthalmol Vis Sci 43:882–888
117. Kelly PA et al (1998) Possible role for nitric oxide releasing nerves in the regulation of ocular blood flow in the rat. Br J Ophthalmol 82:1199–1202
118. Kiel JW (1999) Modulation of choroidal autoregulation in the rabbit. Exp Eye Res 69:413–429
119. Kiel JW (2000) Endothelin modulation of choroidal blood flow in the rabbit. Exp Eye Res 71:543–550
120. Kiel JW, Patel P (1998) Effects of timolol and betaxolol on choroidal blood flow in the rabbit. Exp Eye Res 67:501–507
121. Kimura I et al (2005) Effect of topical unoprostone isopropyl on optic nerve head circulation in controls and in normal-tension glaucoma patients. Jpn J Ophthalmol 49:287–293
122. Kiss B et al (1999) Acetazolamide-induced cerebral and ocular vasodilation in humans is independent of nitric oxide. Am J Physiol 276:R1661–R1667
123. Kitamura Y et al (1993) Nitric oxide-mediated retinal arteriolar and arterial dilatation induced by substance P. Invest Ophthalmol Vis Sci 34:2859–2865
124. Kitanishi K et al (1997) Optic nerve head and choroidal circulation measured by laser Doppler flowmetry in response to intravenous administration of noradrenaline. Nippon Ganka Gakkai Zasshi 101:215–219
125. Kitanishi K et al (2001) Liposomal prostaglandin E1 enhances optic nerve head blood flow in cats. J Ocul Pharmacol Ther 17:115–122
126. Kitaya N et al (1997) Effect of timolol and UF-021 (a prostaglandin-related compound) on pulsatile ocular blood flow in normal volunteers. Ophthalmic Res 29:139–144
127. Kobayashi M et al (2000) Nitric oxide synthase expression in ischemic rat retinas. Jpn J Ophthalmol 44:235–244
128. Koistinaho J et al (1993) NADPH-diaphorase (nitric oxide synthase)-reactive amacrine cells of rabbit retina: putative target cells and stimulation by light. Neuroscience 57:587–597
129. Kojima S et al (1997) Effect of topically applied isopropyl unoprostone on microcirculation in the human ocular fundus evaluated with a laser speckle microcirculation analyser. Nippon Ganka Gakkai Zasshi 101:605–610
130. Kondo M et al (1997) The role of nitric oxide in hyperaemic response to flicker in the retina and optic nerve in cats. Acta Ophthalmol Scand 75:232–235
131. Koss MC (1994) Adrenoceptor mechanisms in epinephrine-induced anterior choroidal vasoconstriction in cats. Exp Eye Res 59:715–722
132. Koss MC (1998) Role of nitric oxide in maintenance of basal anterior choroidal blood flow in rats. Invest Ophthalmol Vis Sci 39:559–564
133. Koss MC (1999) Functional role of nitric oxide in regulation of ocular blood flow. Eur J Pharmacol 374:161–174
134. Kothy P, Hollo G (2001) Does glaucoma medication influence the diameter of the retinal arteriole in the human eye? (A pilot study using the retinal vessel analyser). Acta Physiol Hung 88:281–292
135. Koz OG et al (2007) Comparison of the effects of travoprost, latanoprost and bimatoprost on ocular circulation: a 6-month clinical trial. Acta Ophthalmol Scand 85:838–843
136. Lamping KG et al (1989) Nonuniform vasomotor responses of the coronary microcirculation to serotonin and vasopressin. Circ Res 65:343–351
137. Lepple-Wienhues A et al (1992) Endothelin-like immunoreactivity in the aqueous humour and in conditioned medium from cultured ciliary epithelial cells. Curr Eye Res 11:1041–1046
138. Leung M, Grunwald JE (1997) Short-term effects of topical levobunolol on the human retinal circulation. Eye (Lond) 11(Pt 3):371–376
139. Liu CJ et al (2002) Effect of latanoprost 0.005% and brimonidine tartrate 0.2% on pulsatile ocular blood flow in normal tension glaucoma. Br J Ophthalmol 86:1236–1239
140. Liu JH et al (2007) Resistance to blood flow in the rabbit ophthalmic artery after topical treatment with timolol. J Ocul Pharmacol Ther 23:103–109
141. Lubeck P et al (2001) Effect of timolol on anterior optic nerve blood flow in patients with primary open-angle glaucoma as assessed by the Heidelberg retina flowmeter. J Glaucoma 10:13–17
142. Luksch A et al (2000) Effect of systemic NO synthase inhibition on choroidal and optic nerve head blood flow in healthy subjects. Invest Ophthalmol Vis Sci 41:3080–3084
143. Luksch A et al (2003) Role of NO in choroidal blood flow regulation during isometric exercise in healthy humans. Invest Ophthalmol Vis Sci 44:734–739
144. Luksch A et al (2006) ETa-receptor blockade, but not ACE inhibition, blunts retinal vessel response during isometric exercise. Am J Physiol Heart Circ Physiol 290:H1693–H1698
145. MacCumber MW, D'Anna SA (1994) Endothelin receptor-binding subtypes in the human retina and choroid. Arch Ophthalmol 112:1231–1235

146. MacCumber MW et al (1991) Ocular effects of the endothelins. Abundant peptides in the eye. Arch Ophthalmol 109:705–709
147. Makimoto Y et al (2002) Long-term effect of topically applied isopropyl unoprostone on microcirculation in the human ocular fundus. Jpn J Ophthalmol 46:31–35
148. Mann RM et al (1995) Nitric oxide and choroidal blood flow regulation. Invest Ophthalmol Vis Sci 36:925–930
149. Martinez A, Sanchez-Salorio M (2009) A comparison of the long-term effects of dorzolamide 2% and brinzolamide 1%, each added to timolol 0.5%, on retrobulbar hemodynamics and intraocular pressure in open-angle glaucoma patients. J Ocul Pharmacol Ther 25:239–248
150. Martinez A et al (1999) Dorzolamide effect on ocular blood flow. Invest Ophthalmol Vis Sci 40:1270–1275
151. Matsuo T, Cynader MS (1992) Localization of alpha-2 adrenergic receptors in the human eye. Ophthalmic Res 24:213–219
152. Meyer P et al (1993) Endothelium-dependent regulation of the ophthalmic microcirculation in the perfused porcine eye: role of nitric oxide and endothelins. Invest Ophthalmol Vis Sci 34:3614–3621
153. Meyer P et al (1995) Effects of calcium channel blockers on the response to endothelin-1, bradykinin and sodium nitroprusside in porcine ciliary arteries. Exp Eye Res 60:505–510
154. Meyer P et al (1996) Effect of dipyridamole on vascular responses of porcine ciliary arteries. Curr Eye Res 15:387–393
155. Meyer P et al (1999) Localization of nitric oxide synthase isoforms in porcine ocular tissues. Curr Eye Res 18:375–380
156. Michel MC et al (1995) Classification of alpha 1-adrenoceptor subtypes. Naunyn Schmiedebergs Arch Pharmacol 352:1–10
157. Millar JC et al (1995) Drug effects on intraocular pressure and vascular flow in the bovine perfused eye using radiolabelled microspheres. J Ocul Pharmacol Ther 11:11–23
158. Mittag TW, Tormay A (1985) Adrenergic receptor subtypes in rabbit iris-ciliary body membranes: classification by radioligand studies. Exp Eye Res 40:239–249
159. Mori A et al (2007) Intravenously administered vasodilatory prostaglandins increase retinal and choroidal blood flow in rats. J Pharmacol Sci 103:103–112
160. Mori A et al (2007) Stimulation of prostanoid IP and EP(2) receptors dilates retinal arterioles and increases retinal and choroidal blood flow in rats. Eur J Pharmacol 570:135–141
161. Morsman CD et al (1995) The effect of topical beta-adrenoceptor blocking agents on pulsatile ocular blood flow. Eye (Lond) 9(Pt 3):344–347
162. Nagaoka T et al (2002) The effect of nitric oxide on retinal blood flow during hypoxia in cats. Invest Ophthalmol Vis Sci 43:3037–3044
163. Nagel E et al (2005) Dorzolamide influences the autoregulation of major retinal vessels caused by artificial intraocular pressure elevation in patients with POAG: a clinical study. Curr Eye Res 30:129–137
164. Narayan S et al (2003) Endothelin-1 synthesis and secretion in human retinal pigment epithelial cells (ARPE-19): differential regulation by cholinergics and TNF-alpha. Invest Ophthalmol Vis Sci 44: 4885–4894
165. Neufeld AH et al (1997) Nitric oxide synthase in the human glaucomatous optic nerve head. Arch Ophthalmol 115:497–503
166. Nichols DE, Nichols CD (2008) Serotonin receptors. Chem Rev 108:1614–1641
167. Nicolela MT et al (1996) A comparative study of the effects of timolol and latanoprost on blood flow velocity of the retrobulbar vessels. Am J Ophthalmol 122:784–789
168. Nielsen PJ, Nyborg NC (1989) Calcium antagonist-induced relaxation of the prostaglandin-F2 alpha response of isolated calf retinal resistance arteries. Exp Eye Res 48:329–335
169. Nilsson SF (2000) The significance of nitric oxide for parasympathetic vasodilation in the eye and other orbital tissues in the cat. Exp Eye Res 70:61–72
170. Nishimura K et al (1996) Effects of endothelin-1 on optic nerve head blood flow in cats. J Ocul Pharmacol Ther 12:75–83
171. Noergaard MH et al (2008) Indomethacin decreases optic nerve oxygen tension by a mechanism other than cyclo-oxygenase inhibition. Br J Ophthalmol 92:126–130
172. Ohashi M et al (2007) Effects of topical travoprost and unoprostone on optic nerve head circulation in normal rabbits. Curr Eye Res 32:743–749
173. Ohkubo H, Chiba S (1987) Pharmacological analysis of vasoconstriction of isolated canine ophthalmic and ciliary arteries to alpha-adrenoceptor agonists. Exp Eye Res 45:263–270
174. Ohkubo H, Chiba S (1987) Responses of isolated canine ophthalmic and ciliary arteries to vasoactive substances. Jpn J Ophthalmol 31:627–634
175. Okamura T et al (1997) Receptor subtypes involved in relaxation and contraction by arginine vasopressin in canine isolated short posterior ciliary arteries. J Vasc Res 34:464–472
176. Oku H et al (1999) Experimental optic cup enlargement caused by endothelin-1-induced chronic optic nerve head ischemia. Surv Ophthalmol 44(Suppl 1): S74–S84
177. Okuno T et al (2002) Evidence that nitric oxide is involved in autoregulation in optic nerve head of rabbits. Invest Ophthalmol Vis Sci 43:784–789
178. Orgul S et al (1995) Optic nerve vasomotor effects of topical beta-adrenergic antagonists in rabbits. Am J Ophthalmol 120:441–447
179. Orgul S et al (1996) An endothelin-1 induced model of optic nerve ischemia in the rabbit. Invest Ophthalmol Vis Sci 37:1860–1869
180. Osborne NN (1993) Neuromediators and their receptors (adrenergic and endothelin types) in the eye. Therapie 48:549–558

181. Osborne NN et al (1993) Endothelin receptors in the cornea, iris and ciliary processes. Evidence from binding, secondary messenger and PCR studies. Exp Eye Res 56:721–728
182. Ostwald P et al (1995) Effect of nitric oxide synthase inhibition on blood flow after retinal ischemia in cats. Invest Ophthalmol Vis Sci 36:2396–2403
183. Ostwald P et al (1997) Adenosine receptor blockade and nitric oxide synthase inhibition in the retina: impact upon post-ischemic hyperemia and the electroretinogram. Vision Res 37:3453–3461
184. Palmer RM et al (1987) Nitric oxide release accounts for the biological activity of endothelium-derived relaxing factor. Nature 327:524–526
185. Palmer RM et al (1988) L-arginine is the physiological precursor for the formation of nitric oxide in endothelium-dependent relaxation. Biochem Biophys Res Commun 153:1251–1256
186. Parys-Van Ginderdeuren R et al (1992) Dissociation between prostaglandin levels and blood flow to the retina and choroid in the newborn pig after nonsteroidal antiinflammatory drugs. Invest Ophthalmol Vis Sci 33:3378–3384
187. Pedersen DB et al (2004) Indomethacin lowers optic nerve oxygen tension and reduces the effect of carbonic anhydrase inhibition and carbon dioxide breathing. Br J Ophthalmol 88:1088–1091
188. Pfeiffer N (1997) Dorzolamide: development and clinical application of a topical carbonic anhydrase inhibitor. Surv Ophthalmol 42:137–151
189. Pillunat LE et al (1999) Effect of topical dorzolamide on optic nerve head blood flow. Graefes Arch Clin Exp Ophthalmol 237:495–500
190. Polak K et al (2000) Evaluation of the Zeiss retinal vessel analyser. Br J Ophthalmol 84:1285–1290
191. Polak K et al (2001) Effect of endothelin and BQ123 on ocular blood flow parameters in healthy subjects. Invest Ophthalmol Vis Sci 42:2949–2956
192. Polak K et al (2003) Regulation of human retinal blood flow by endothelin-1. Exp Eye Res 76:633–640
193. Polak K et al (2007) Altered nitric oxide system in patients with open-angle glaucoma. Arch Ophthalmol 125:494–498
194. Polska E et al (2002) Partial antagonism of endothelin 1-induced vasoconstriction in the human choroid by topical unoprostone isopropyl. Arch Ophthalmol 120:348–352
195. Pournaras C et al (1978) Studies on the role of prostaglandins in the regulation of retinal blood flow. Exp Eye Res 26:687–697
196. Prasanna G et al (1998) Regulation of endothelin-1 in human non-pigmented ciliary epithelial cells by tumor necrosis factor-alpha. Exp Eye Res 66:9–18
197. Ramachandran E et al (1993) Effects of endothelin on cultured bovine retinal microvascular pericytes. Invest Ophthalmol Vis Sci 34:586–595
198. Rapoport RM et al (1983) Endothelium-dependent relaxation in rat aorta may be mediated through cyclic GMP-dependent protein phosphorylation. Nature 306:174–176
199. Rassam SM et al (1993) The effect of acetazolamide on the retinal circulation. Eye (Lond) 7(Pt 5):697–702
200. Rubanyi GM, Polokoff MA (1994) Endothelins: molecular biology, biochemistry, pharmacology, physiology, and pathophysiology. Pharmacol Rev 46:325–415
201. Sampaolesi J et al (2001) Antiglaucomatous drugs effects on optic nerve head flow: design, baseline and preliminary report. Int Ophthalmol 23:359–367
202. Sato T et al (2001) Short-term effect of beta-adrenoreceptor blocking agents on ocular blood flow. Curr Eye Res 23:298–306
203. Sato E et al (2003) Role of nitric oxide in regulation of retinal blood flow during hypercapnia in cats. Invest Ophthalmol Vis Sci 44:4947–4953
204. Schmetterer L, Polak K (2001) Role of nitric oxide in the control of ocular blood flow. Prog Retin Eye Res 20:823–847
205. Schmetterer L et al (1997) Nitric oxide and ocular blood flow in patients with IDDM. Diabetes 46: 653–658
206. Schmetterer L et al (1997) Role of NO in the O_2 and CO_2 responsiveness of cerebral and ocular circulation in humans. Am J Physiol 273:R2005–R2012
207. Schmetterer L et al (1997) Effects of endothelin-1 (ET-1) on ocular hemodynamics. Curr Eye Res 16:687–692
208. Schmetterer L et al (1997) The effect of systemic nitric oxide-synthase inhibition on ocular fundus pulsations in man. Exp Eye Res 64:305–312
209. Schmetterer L et al (1997) Effects of antiglaucoma drugs on ocular hemodynamics in healthy volunteers. Clin Pharmacol Ther 61:583–595
210. Schmidt KG et al (1998) Topical carbonic anhydrase inhibition increases ocular pulse amplitude in high tension primary open angle glaucoma. Br J Ophthalmol 82:758–762
211. Seligsohn EE, Bill A (1993) Effects of NG-nitro-L-arginine methyl ester on the cardiovascular system of the anaesthetized rabbit and on the cardiovascular response to thyrotropin-releasing hormone. Br J Pharmacol 109:1219–1225
212. Seong GJ et al (1999) Effects of 0.005% latanoprost on optic nerve head and peripapillary retinal blood flow. Ophthalmologica 213:355–359
213. Siesky B et al (2008) A comparative study of the effects of brinzolamide and dorzolamide on retinal oxygen saturation and ocular microcirculation in patients with primary open-angle glaucoma. Br J Ophthalmol 92:500–504
214. Siesky B et al (2009) Literature review and meta-analysis of topical carbonic anhydrase inhibitors and ocular blood flow. Surv Ophthalmol 54:33–46
215. Silver LH (2000) Dose–response evaluation of the ocular hypotensive effect of brinzolamide ophthalmic suspension (Azopt). Brinzolamide Dose–response Study Group. Surv Ophthalmol 44(Suppl 2):S147–S153
216. Simsek T et al (2006) Comparative analysis of the effects of brimonidine and dorzolamide on ocular blood flow velocity in patients with newly diagnosed

primary open-angle glaucoma. J Ocul Pharmacol Ther 22:79–85
217. Spada CS et al (2002) Vascular activities of prostaglandins and selective prostanoid receptor agonists in human retinal microvessels. Exp Eye Res 75: 155–163
218. Sponsel WE et al (2002) Comparative effects of latanoprost (Xalatan) and unoprostone (Rescula) in patients with open-angle glaucoma and suspected glaucoma. Am J Ophthalmol 134:552–559
219. Steigerwalt RD Jr et al (2001) Ocular and orbital blood flow velocity in patients with peripheral vascular disease and diabetes treated with intravenous prostaglandin E1. J Ocul Pharmacol Ther 17:529–535
220. Steigerwalt RD Jr et al (2001) Ocular and retrobulbar blood flow in ocular hypertensives treated with topical timolol, betaxolol and carteolol. J Ocul Pharmacol Ther 17:537–544
221. Stjernschantz J et al (1989) Vasodynamic and angiogenic effects of eicosanoids in the eye. Prog Clin Biol Res 312:155–170
222. Stjernschantz J et al (1999) Effect of latanoprost on regional blood flow and capillary permeability in the monkey eye. Arch Ophthalmol 117:1363–1367
223. Stjernschantz J et al (2000) Microvascular effects of selective prostaglandin analogues in the eye with special reference to latanoprost and glaucoma treatment. Prog Retin Eye Res 19:459–496
224. Strenn K et al (1998) Reversal of endothelin-1-induced ocular hemodynamic effects by low-dose nifedipine in humans. Clin Pharmacol Ther 63:54–63
225. Su EN et al (1994) Adrenergic and nitrergic neurotransmitters are released by the autonomic system of the pig long posterior ciliary artery. Curr Eye Res 13:907–917
226. Su EN et al (1994) Effects of extracellular pH on agonist-induced vascular tone of the cat ophthalmociliary artery. Invest Ophthalmol Vis Sci 35:998–1007
227. Su EN et al (1995) Altered vasoactivity in the early diabetic eye: measured in the isolated perfused rat eye. Exp Eye Res 61:699–711
228. Sugiyama T, Azuma I (1995) Effect of UF-021 on optic nerve head circulation in rabbits. Jpn J Ophthalmol 39:124–129
229. Sugiyama T et al (1993) Effect of endothelin-1 on ocular circulation. Nippon Ganka Gakkai Zasshi 97:678–682
230. Sugiyama K et al (1996) The effects of intravitreally injected endothelin-1 on the iris-ciliary body microvasculature in rabbits. Curr Eye Res 15:633–637
231. Sugiyama T et al (1998) The effect of continuous and intravenous application of carteolol chloride on tissue blood flow in the rabbit optic nerve head. Nippon Ganka Gakkai Zasshi 102:662–666
232. Sugiyama T et al (2000) Effect of nitric oxide synthase inhibitor on optic nerve head circulation in conscious rabbits. Invest Ophthalmol Vis Sci 41:1149–1152
233. Takagi C et al (1996) Endothelin-1 action via endothelin receptors is a primary mechanism modulating retinal circulatory response to hyperoxia. Invest Ophthalmol Vis Sci 37:2099–2109
234. Takahashi K et al (1989) Production of endothelin 1 by cultured bovine retinal endothelial cells and presence of endothelin receptors on associated pericytes. Diabetes 38:1200–1202
235. Tamaki Y et al (1999) Effect of topical betaxolol on tissue circulation in the human optic nerve head. J Ocul Pharmacol Ther 15:313–321
236. Tamaki Y et al (2001) Effect of topical unoprostone on circulation of human optic nerve head and retina. J Ocul Pharmacol Ther 17:517–527
237. Thorig L, Bill A (1986) Effects of B-HT 920 in the eye and on regional blood flows in anaesthetized and conscious rabbits. Curr Eye Res 5:565–573
238. Tilton RG et al (1999) Role for nitric oxide in the hyperpermeability and hemodynamic changes induced by intravenous VEGF. Invest Ophthalmol Vis Sci 40:689–696
239. Toda N et al (1995) Functional role of nerve-derived nitric oxide in isolated dog ophthalmic arteries. Invest Ophthalmol Vis Sci 36:563–570
240. Toda N et al (1995) Mechanisms underlying endothelium-independent relaxation by acetylcholine in canine retinal and cerebral arteries. J Pharmacol Exp Ther 274:1507–1512
241. Toda M et al (1998) Desmopressin-induced dog ciliary artery relaxation. Eur J Pharmacol 344: 197–201
242. Tomidokoro A et al (1998) In vivo measurement of iridial circulation using laser speckle phenomenon. Invest Ophthalmol Vis Sci 39:364–371
243. Tomidokoro A et al (1999) Effects of topical carteolol and timolol on tissue circulation in the iris and choroid. Curr Eye Res 18:381–390
244. Toriu N et al (2001) Effects of lomerizine, a novel Ca^{2+} channel blocker, on the normal and endothelin-1-disturbed circulation in the optic nerve head of rabbits. J Ocul Pharmacol Ther 17:131–149
245. Trew DR, Smith SE (1991) Postural studies in pulsatile ocular blood flow: II. Chronic open angle glaucoma. Br J Ophthalmol 75:71–75
246. Vallance P et al (1989) Effects of endothelium-derived nitric oxide on peripheral arteriolar tone in man. Lancet 2:997–1000
247. Van Buskirk EM et al (1990) Ciliary vasoconstriction after topical adrenergic drugs. Am J Ophthalmol 109:511–517
248. Wang Y et al (1993) Mechanisms of acetylcholine-induced relaxation in dog external and internal ophthalmic arteries. Exp Eye Res 57:275–281
249. Watanabe K, Chiou GC (1983) Action mechanism of timolol to lower the intraocular pressure in rabbits. Ophthalmic Res 15:160–167
250. Weigert G et al (2007) Effects of topical clonidine versus brimonidine on choroidal blood flow and intraocular pressure during squatting. Invest Ophthalmol Vis Sci 48:4220–4225
251. Weigert G et al (2007) Intravenous administration of clonidine reduces intraocular pressure and alters ocular blood flow. Br J Ophthalmol 91:1354–1358
252. Weigert G et al (2008) Effect of unspecific inhibition of cyclooxygenase by indomethacin on retinal and

choroidal blood flow. Invest Ophthalmol Vis Sci 49:1065–1070
253. Wiencke AK et al (1994) Nonadrenergic noncholinergic vasodilation in bovine ciliary artery involves CGRP and neurogenic nitric oxide. Invest Ophthalmol Vis Sci 35:3268–3277
254. Wikberg-Matsson A et al (1996) Characterization of alpha 2-adrenoceptor subtypes in the porcine eye: identification of alpha 2A-adrenoceptors in the choroid, ciliary body and iris, and alpha 2A- and alpha 2C-adrenoceptors in the retina. Exp Eye Res 63:57–66
255. Wikberg-Matsson A et al (2000) Characterization of alpha(1)-adrenoceptor subtypes in the eye. Exp Eye Res 70:51–60
256. Yamamoto R et al (1993) The localization of nitric oxide synthase in the rat eye and related cranial ganglia. Neuroscience 54:189–200
257. Yamazaki S, Baba H (1993) Acute effect of topical carteolol on ocular pulsatile volume change. Acta Ophthalmol (Copenh) 71:760–764
258. Yan HY, Chiou GC (1987) Effects of L-timolol, D-timolol, haloperidol and domperidone on rabbit retinal blood flow measured with laser Doppler method. Ophthalmic Res 19:45–48
259. Yanagisawa M et al (1988) A novel peptide vasoconstrictor, endothelin, is produced by vascular endothelium and modulates smooth muscle Ca^{2+} channels. J Hypertens Suppl 6:S188–S191
260. Yao K et al (1991) Endothelium-dependent regulation of vascular tone of the porcine ophthalmic artery. Invest Ophthalmol Vis Sci 32:1791–1798
261. Yorio T et al (2002) Endothelin: is it a contributor to glaucoma pathophysiology? J Glaucoma 11: 259–270
262. Yoshida A et al (1991) Effect of timolol on human retinal, choroidal and optic nerve head circulation. Ophthalmic Res 23:162–170
263. Zagvazdin YS et al (1996) Neural nitric oxide mediates Edinger-Westphal nucleus evoked increase in choroidal blood flow in the pigeon. Invest Ophthalmol Vis Sci 37:666–672
264. Zeitz O et al (2005) Effects of glaucoma drugs on ocular hemodynamics in normal tension glaucoma: a randomized trial comparing bimatoprost and latanoprost with dorzolamide [ISRCTN18873428]. BMC Ophthalmol 5:6
265. Zeitz O et al (2006) Volumetric colour Doppler imaging: a useful tool for the determination of ocular blood flow in glaucoma patients? Eye (Lond) 20: 668–673
266. Zhu P et al (1997) Relaxation by bradykinin in porcine ciliary artery. Role of nitric oxide and K(+)-channels. Invest Ophthalmol Vis Sci 38:1761–1767
267. Zhu Y et al (1998) Mechanisms of hyperoxia-induced reductions in retinal blood flow in newborn pig. Exp Eye Res 67:357–369

Retinal Ischemia in Relation to Glaucoma and Neuroprotection

14

Neville N. Osborne

Core Messages

- Ischemia occurs in many ocular diseases when the blood delivery does not meet the cellular energy demands of the tissue. Ischemia has characteristics associated with hypoxia/anoxia but hypoxia/anoxia does not imply ischemia.
- It is proposed that a cause for the initiation of glaucoma to the retina is a mild ischemic insult to the optic nerve head region. In contrast, a complete loss of blood supply occurs in ischemic optic neuropathy.
- In glaucoma, ganglion cell mitochondria are affected by lack of an optimum blood supply resulting in oxidative stress so making them susceptible to additional insults like light. A more drastic reduction of optic nerve head blood supply will result in a complete loss of ganglion cell mitochondrial function as in ischemic optic neuropathy.
- A mild reduction of blood supply to dividing retinal glial cells results in them becoming activated releasing substances to support and enhance neuronal function. However, as glaucoma progresses the continuous release and accumulation of such substances in the extracellular space might reach high levels so becoming toxic to the already fragile ganglion cells.
- It is therefore suggested that as glaucoma progresses the combined elevated levels of substances released from glial cells together with light impinging on ganglion cell intra-axonal mitochondria triggers their demise at different times. Thus, the receptor profile and number of mitochondria associated in individual ganglion cells might be major factors at determining when they die after initiation of the disease.

14.1 Introduction

The word ischemia was coined by Virchow, who combined the Greek iskho, meaning "I hold back," with háima, meaning "blood". Hence, ischemia refers to a pathological situation involving an inadequacy (not necessarily a complete lack of) blood flow to a tissue, with failure to meet cellular energy demands. Ischemia should be distinguished from anoxia (a complete lack of oxygen) and hypoxia (a reduction in oxygen): ischemia always has a component of

N.N. Osborne, Ph.D., D.Sc.
Fundación de Investigación Oftalmológica,
Avda. Doctores Fernández-Vega 34,
E-33012 Oviedo, Asturias, Spain

Nuffield Laboratory of Ophthalmology,
Oxford University,
Headley Way, Oxford OX3 9DU, UK
e-mail: neville.osborne@eye.ox.ac.uk

L. Schmetterer, J.W. Kiel (eds.), *Ocular Blood Flow*,
DOI 10.1007/978-3-540-69469-4_14, © Springer-Verlag Berlin Heidelberg 2012

Ocular diseases where retinal ischemia is implicated
(retina blood supplies – retinal, choroidal, optic nerve)

1. Occlusive retinal arterial and arteriolar diseases	Embolism Atherosclerosis Vasospasm Vasculitis Microaneurysms Hemoglobinopathies Behcet disease
2.Capillary diseases	Congenital retinal telangiectasis Diabetes mellitus X-ray irradiation Sickle-cell diseases Incontinentia pigmenti
3. Occlusive venous diseases	Central retinal vein occlusion Branch retinal vein occlusion Glaucoma
4. Miscellaneous	Retinopathy of prematurity Myocardial infarction Ocular trauma/subarachnoid Hemorrhage Retinal detachment Viral infections e.g., HIV Bacterial infections e.g., syphilis Hypercoagulable stress Takayasu's disease Homocystinuria Fabry's disease Atrial myxoma Fibromuscular dysplasia involving the ophthalmic artery Intravenous drug abuse Cervical trauma Multiple sclerosis

Fig. 14.1 Diseases associated with the retina where ischemia is implicated. Ischemia can be global affecting all blood supplies to the retina or focal where defined blood supplies associated with different parts of the retina and/or optic nerve head. Moreover, ischemia can be complete or incomplete as well as being associated with reperfusion or not

hypoxia/anoxia, but hypoxia/anoxia does not imply ischemia. For example, the retina may become hypoxic at high altitudes, producing loss of vision, but it is not ischemic. Similarly, anemia (generally a reduction, rather than complete absence of hemoglobin) is always a component of ischemia but not vice versa.

Ischemia deprives a tissue of three requirements: oxygen, metabolic substrates and removal of waste products. The loss of these requirements will initially lower homeostatic responses and, with time, will induce injury to the tissue. If withheld for a sufficiently long time, the tissue will die (an infarct). The molecular events that accompany ischemia and exactly what constitutes a "sufficiently long time" vary and depend on different factors. The retina being a thin tissue surrounded by the vitreous humor is therefore more resistant than deeper brain structures to a persistent complete reduction of blood flow.

Ischemia plays a part in a number of ocular pathologies (Fig. 14.1). For example, occlusion of central retinal vein or artery is associated with acute retinal ischemia and rapid visual loss [23]. In other conditions which gradually develop, such as diabetic retinopathy and glaucoma, chronic retinal ischemia (not a complete but a reduced blood supply) is a pathogenic feature of the disease [10, 34, 72]. Ischemic challenges also contribute to the pathogenesis of less prevalent conditions to the retina, such as sickle cell retinopathy, radiation retinopathy, and retinopathy of prematurity [24, 61]. Moreover, ischemic challenges also occur in the aging retina which probably results in neuronal loss [70]. The complexity of mechanisms involved in causing ischemic damage to central nervous tissue is suggested by the variety of ways by which the process can be attenuated experimentally (Table 14.1).

14.2 Retinal Ischemia Basic Mechanisms

Mammalian retinal ischemia results in irreversible morphological and functional changes. These are the consequence of depleted ATP stores, due to deprivation of both glucose and oxygen, though

Table 14.1 This table documents the number of ways by which insults of ischemia to central nervous tissue can be attenuated in various animal studies

Maintain retinal energy supply
Decrease extracellular glutamate levels
Block excitotoxic depolarization (e.g., blockade of NMDA, AMPA/kainate receptors)
Activate inhibitory receptors (e.g., GABA, glycine receptor agonists)
Block calcium entry (e.g., voltage-sensitive calcium channels)
Block axonal sodium channels
Inhibit influx of chloride ions
Enhance intracellular glutamate catabolism
Inhibit free radical production
Inhibit nitric oxide production
Prevent lipid peroxidation
Prevent apoptotic cell death pathways
Block mitochondrial permeability transition
Prevent inflammation
Prevent oxidative stress
Compensate for compromised axonal flow of neurotrophins (e.g., BDNF, NGF, NT-4/5)
Growth factors (e.g., bFGF, EGF, CNTF)
Induction of heat shock proteins

transient loss of these substrates is not immediately lethal. The cell death is the result of an extremely complex (not completely understood) cascade of biochemical responses initiated by energy failure. The tissue damage and functional deficits that follow periods of transient ischemia reflect the combined effects of several, often interrelated, pathophysiological pathways. These result in drastic changes in ion movements, neurotransmitter levels, and metabolites.

It is useful to consider that stroke lesions consist of a densely ischemic focus, the ischemic core, surrounded by a better perfused area, the ischemic penumbra [27]. Cells in the focus are usually doomed unless reperfusion is quickly instituted. In contrast, penumbral cells may remain viable for several hours and can be saved by reperfusion or by drugs that prevent the infarction extending into the penumbral zone [27].

During the past decade, a considerable amount of experimental work has been devoted to the elucidation of the mechanisms of ischemic neuronal injury, but there is still much debate over the underlying processes causing the injury. One area of interest is the role of glutamate and aspartate, whose extracellular concentrations increase markedly during ischemia [12]. It is generally accepted that glutamate release during the early phase of brain ischemia triggers events leading to irreversible injury not only in those areas in which oxygen supply is critically reduced but also in regions of seemingly less disturbed energy metabolism, i.e. the penumbra of focal ischemia. However, there is still much controversy over the extent of their role in the pathophysiology of neural ischemia. Moreover, neuronal bodies (gray matter) and axons (white matter) are not affected in precisely the same way by ischemia [53].

It is important to remember that much of the work studying ischemic neuronal and axonal damage has been carried out on tissues derived from defined brain regions, so care is important when drawing comparisons with what happens in the retina. In particular, the way photoreceptors respond to ischemia in light and dark conditions may be unique. Moreover, the thinness of the retina, its well-defined blood systems and the large glycogen supply associated with the Müller cells provide the tissue with a unique energy supply when compared with the brain, so making the response of tissues to ischemia not the same.

14.3 Oxidative Stress

Oxidative stress is imposed on cells as a result of one of three factors: (1) an increase in oxidant generation, (2) a decrease in antioxidant protection, or (3) a failure to repair oxidative damage. Cell damage is induced by reactive oxygen species (ROS). ROS are either free radicals, reactive anions containing oxygen atoms or molecules containing oxygen atoms that can either produce free radicals or are chemically activated by them. Examples are hydroxyl radical, superoxide, hydrogen peroxide and peroxynitrite. The main source of ROS in vivo is aerobic respiration, although ROS are also produced by peroxisomal β-oxidation of fatty acids, microsomal cytochrome P450 metabolism of xenobiotic compounds, stimulation of phagocytosis by pathogens or lipopolysaccharides, arginine

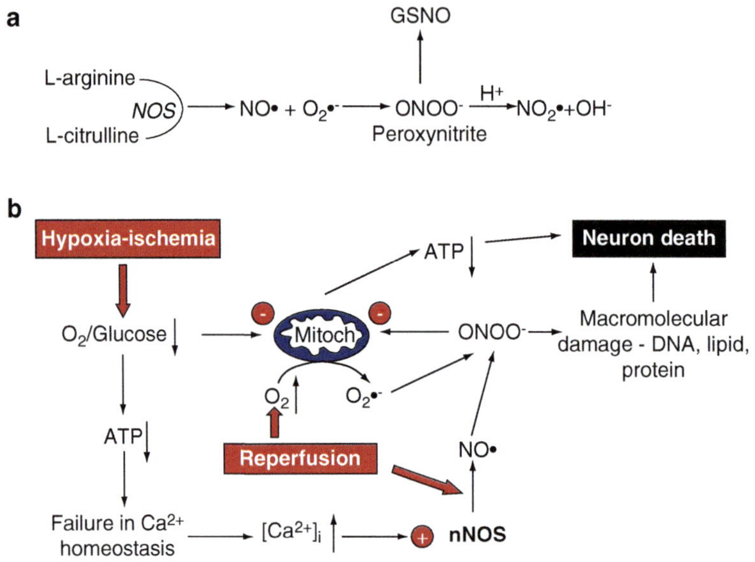

Fig. 14.2 Production and reactions of nitrogen-derived radicals. (**a**) Nitric oxide synthase (NOS) produces nitric oxide radicals in the conversion of l-arginine to l-citrulline. This compound acts as an important homeostatic modulating agent under physiological conditions via its vasodilator, antioxidant, antiplatelet, and antineutrophil actions. In the event that superoxide radicals are present ($\cdot O_2^-$), usually as a result of rapid tissue reoxygenation subsequent to an ischemic event, peroxynitrite is formed which rapidly decomposes to highly reactive oxidant species that can cause tissue injury. Under physiological conditions, there is a critical balance between cellular concentrations of NO, $\cdot O_2^-$ and superoxide dismutase activity which favor NO production. In pathological conditions such as reperfusion following an ischemic event, the formation of $ONOO^-$ is favored. The latter compound can be rapidly detoxified if it is combined with reduced glutathione (*GSH*) to form *S*-nitrosoglutathione (*GSNO*), but this depends upon the cellular antioxidant defense system being functional, and this is generally overwhelmed during tissue reperfusion. (**b**) Reactions and formation of nitric oxide radicals as a result of ischemia-reperfusion. The combined production of nitric-oxide-derived radicals and failure of cellular antioxidant defense will lead to widespread macromolecular damage and cell death

metabolism and tissue-specific enzymes. Under normal conditions, ROS are cleared from the cell by the action of superoxide dismutase (SOD), catalase or glutathione (GSH) peroxidase. The main damage to cells results from the ROS-induced alteration of macromolecules such as polyunsaturated fatty acids in membrane lipids, essential proteins and DNA. Additionally, oxidative stress and ROS have been implicated in retinal ischemic disease states.

14.4 The Role of Free Radicals in Retinal Ischemia (Fig. 14.2)

Many cascades generated by glutamate and glucose/oxygen deprivation result in the formation of free radicals [51], and it has been proposed that free radicals are important mediators in damage caused by retinal ischemia [3, 46]. Reperfusion injury after ischemia appears paradoxical, but oxygen-derived and other free radicals are principally formed when reduced compounds, which accumulate during ischemia, are reoxidized. There is evidence that this free radical burst, produced during the early stage of reperfusion, overwhelms normal cellular antioxidant defense mechanisms, causing oxidative stress and a variety of types of tissue injury [21].

There are many ways in which free ROS can be formed during ischemia-reperfusion, but the burst of superoxide radicals ($\cdot O_2^-$) which occurs during the early stage of reperfusion is thought to occur by the following pathway. During ischemia, degradation of ATP leads to the formation of hypoxanthine, and increases in intracellular calcium in neurones activate the Ca^{2+}-dependent protease calpain. Calpain converts xanthine dehydrogenase into xanthine oxidase, and upon reperfusion, the latter enzyme oxidizes the accumulated hypoxanthine to uric acid resulting in the release of $\cdot O_2^-$. These two molecules react by the Haber-Weiss mechanism to yield the highly toxic hydroxyl radical ($\cdot$OH). This reaction is catalyzed by iron, which is released from its protein-bound stores at the low pH generated during ischemia. In addition, $\cdot O_2^-$ interacts with nitric oxide (NO$\cdot$), which is produced in considerable amounts following ischemia, leading to the formation of peroxynitrite, nitroxyl radical, and eventually $\cdot$OH [21]. It is not just from the mitochondrial systems of neuronal cells that free radicals are generated, activation of glial cells and infiltrating leukocytes release inflammatory mediators, such as arachidonic acid, nitric oxide, and cytokines, which all play major roles in the formation of free radicals following ischemia.

An early indication that excessive free radical formation may be detrimental to the retina was the finding that iron-ascorbate perfusion rapidly attenuates the isolated rat retina's b-wave and causes lipid peroxidation [13]. However, the first studies to provide evidence for the involvement of free radicals in retinal damage after ischemia were performed by Szabo and colleagues who showed that administration of superoxide dismutase (SOD) affords protection against ischemia-induced histological damage [68] and the ionic imbalance that occurs in the reperfusion period [66]. The work of Szabo and colleagues showed, albeit indirectly, that the superoxide free radical is generated in such quantities during retinal ischemia-reperfusion that the normal endogenous levels of superoxide dismutase expressed by the retina are overwhelmed and are unable to protect the tissue from oxidative damage caused by this radical. Subsequent studies supported the involvement not only of superoxide in ischemic injury to the retina but also of hydrogen peroxide and the hydroxyl radical (e.g., [1, 35, 58]). Furthermore, the inability of endogenous free radical quenching mechanisms to cope with the demand posed following ischemia is illustrated by the capacity of a variety of free radical scavengers, such as extract of *Ginkgo biloba*, epigallocatechin gallate (extract from green tea), α-lipoic acid, vitamin E, thioredoxin (an ascorbic acid derivative), mannitol, and the iron chelator desferrioxamine to protect the retina from ischemia-reperfusion injury [46].

Although it can be inferred from all of these studies that there is an elevated level of free radicals in the retina after ischemia, direct measurement of free radical formation has only been performed by Muller and coworkers [35], who showed increased free radical formation both during and after ischemia, and by Szabo et al. [65] who subjected diabetic retinas to ischemia and documented increased levels of free radicals during the reperfusion phase.

As previously mentioned, xanthine oxidase has been established as an important source of oxygen free radicals in certain ischemia-reperfusion injuries, and there is evidence for a similar action in the retina. Xanthine oxidase activity has been reported to increase fivefold within 10 min of reperfusing the ischemic rat retina, while concentrations of hypoxanthine and xanthine, respectively the substrate and product of xanthine oxidase, increase in a time-related fashion following ischemia-reperfusion [1, 59] and enhance free radical formation [75]. Furthermore, administration of allopurinol or oxypurinol, blockers of xanthine oxidase, both result in significant improvement of the ERG after ischemia [49, 59]. These combined data suggest that xanthine-oxidase-mediated processes contribute to the functional anomalies of retinal ischemia. However, there is limited evidence to suggest that xanthine oxidase may not be an important source of free radicals after retinal ischemia. Faberowski et al. [16] reported that allopurinol provided no significant protection against ischemia produced by transient ligation of the optic nerve, while Szabo et al. [67] found that allopurinol was only neuroprotective when administered in combination with extract of *Ginkgo biloba*. The reported variations may be due to either the allopurinol doses used or the manner of its administration, or to the methods of evaluating injury.

One potential source of free radicals in the retina following ischemia is polymorphonuclear leukocytes. In the brain, neutrophils oxidize NADPH to generate superoxide and are important free radical donors during and after focal ischemia [30]. Agents that prevent the accumulation or activation of neutrophils are protective [28]. In the retina, infiltration of neutrophils occurs during the early phase of reperfusion, probably in response to increased levels of cytokines and free radical formation. Although there is no direct evidence for increased free radical formation from neutrophils, blocking leukocyte accumulation has been shown to afford protection to the ischemic retina [73].

Oxygen-derived free radicals cause extensive cellular damage in the brain. One of the main mechanisms by which this occurs is by attacking unsaturated fatty acids, which leads to lipid peroxidation of membranes. This will result in loss of membrane fluidity, cell swelling, oedema, and feed-forward production of more free radicals. Many additional mechanisms of damage have been ascribed to free radicals generated during ischemia-reperfusion, including attacking sulphhydryl protein bonds, which leads to the destruction of amino acids and polypeptide chains; fragmentation of DNA molecules, which leads to activation of poly(ADP-ribose) polymerase; activation of cytokines and NF-κB, which are likely to be instrumental in upregulations of iNOS and COX-2 and subsequent release of glutamate; and effects on Ca^{2+} homeostasis [28]. Interestingly, a clear association between neuronal cell death and formation of free radicals and lipid peroxides in the retina subjected to ischemia-reperfusion has only recently been demonstrated [8]. In all of these studies, ischemia-reperfusion induced free radical formation, lipid peroxidation, and neuronal injury and were largely preventable by administration of free radical scavengers [8, 63] and a novel metal chelate [1].

14.5 Ischemia in Glaucoma (Figs. 14.3 and 14.4)

Glaucoma is defined as an optic neuropathy associated with characteristic changes in the optic nerve [54, 55]. Patients suffer a gradual loss of vision that can span a 30–40-year period, with retinal ganglion cells slowly, but constantly, dying. The disease is often associated with raised intraocular pressure, but this does not occur in approximately one sixth of all glaucoma patients [55]. Thus, glaucoma can be viewed as a neurodegenerative disease of the retinal ganglion cells.

It has been proposed that in most cases of glaucoma, ischemia initiates the disease by affecting solely components in the optic nerve head of the retina [17, 47]. In this particular region of the eye, minute blood vessels exist, and when intraocular pressure is increased, perfusion pressure in these vessels is reduced [19, 39]. In laboratory studies, it can be shown that a sustained elevation of intraocular pressure causes glaucoma-like pathology in animals [46]. We also know that various risk factors often associated with glaucoma subjects (e.g., hemorrhaging, inability to autoregulate, hypotension, hypertension) relate to the vasculature [17–19]. Thus, a body of evidence suggests that the initiation of glaucoma is caused by an ischemic or other metabolic insult to the optic nerve head region. It should be noted that ischemia is defined here as a pathological situation involving a degree of inadequacy (but not a lack) in the blood flow with a subsequent failure to meet appropriate cellular energy demands. Moreover, it is possible that at least in some glaucoma patients, ischemia to the optic nerve head is not constant but intermittent, and this would result in additional oxidative stress, reperfusion injury [46]. Fluctuations in intraocular pressure causing changes of blood flow to the optic nerve head would theoretically cause both ischemia and reperfusion injury.

It has been proposed that ischemia to the optic nerve head initiates a process that eventually results in ganglion cells dying at different rates [17, 45, 47] in glaucoma. Tissue components

Fig. 14.4 Hypothetical cascade of events that lead to ganglion cells dying at different times during the lifetime of a glaucoma patient. The initial insult in the optic nerve head region affects all components (astrocytes, microglia, Müller cell end feet, ganglion cell axons, lamina cribrosa) resulting in ganglion cells existing at a reduced energetic state and an elevation of various substances (glutamate, TNFα, NO, endothelin, d-serine, potassium) in the extracellular space caused by activated astrocyte and microglia cells. These substances together with light impinging on mitochondria in ganglion cell axons eventually cause apoptotic insults to the energetically compromised ganglion cells at different times, depending on the axonal length and receptor profile of individual ganglion cells

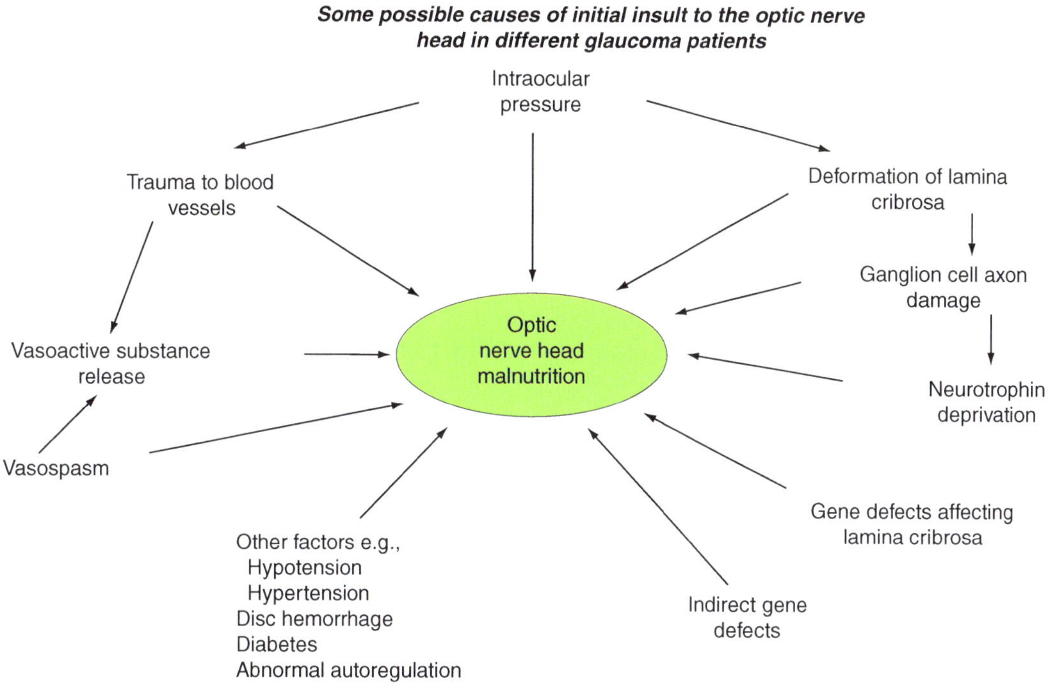

Fig. 14.3 Potential causes for an ischemic insult to the optic nerve head as might occur in glaucoma

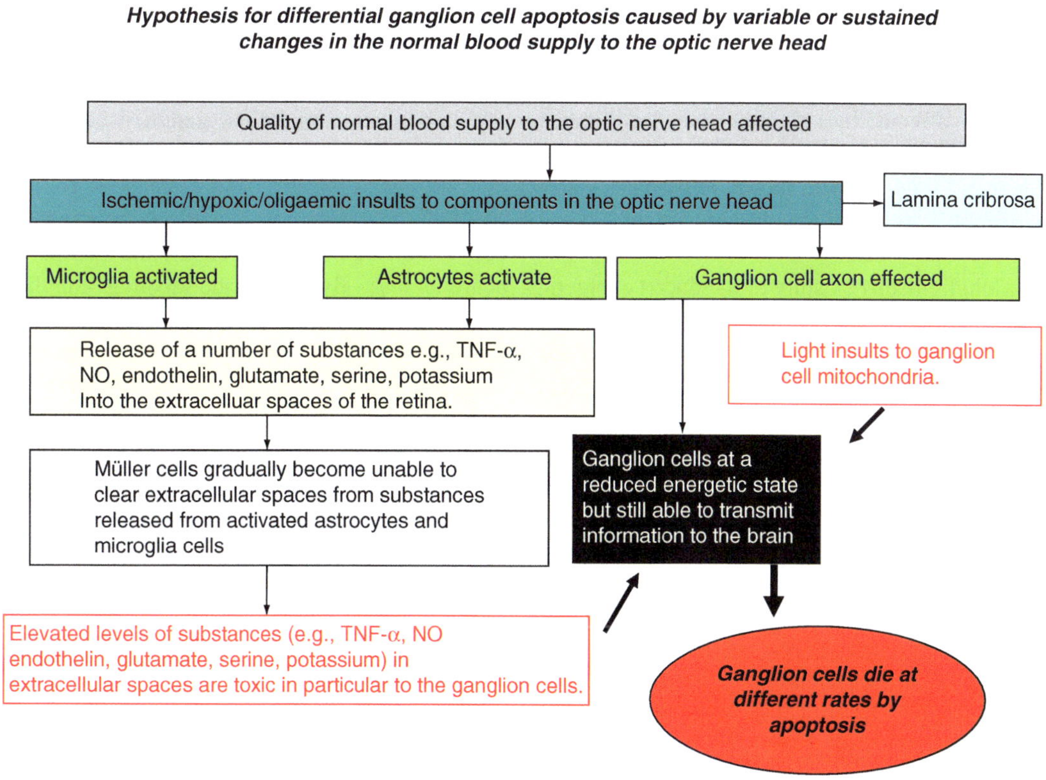

likely to be affected by ischemia will include the lamina cribrosa, ganglion cell axons, astrocytes, microglia, and blood vessels. Moreover, altered neurotrophin delivery to the retina may occur [50]. The activation of both microglia and astrocytes is likely to result in a release of various factors into the extracellular space. These will include substances such as TGFβ, TNFα, nitric oxide, d-serine, prostaglandins, endothelin, and glutamate [42, 47]. Laboratory studies using cultures of microglia or astrocytes have added support to this notion [22, 71]. With the progress of time, it is likely that other areas of the retina will eventually become secondarily affected by these extracellular factors. In particular, Müller cells are likely to be stimulated, in an attempt to maintain a homeostatic extracellular environment. It can be postulated that eventually the extracellular environment will gradually become contaminated because normal Müller cell function will be insufficient. Thus, a gradual rise in the extracellular level of substances released from activated glial cells will occur. Some of these substances (e.g., glutamate, prostaglandins, endothelin, and nitric oxide) are likely to be toxic to neurones depending on the receptor profile of a particular neurone. Other substances like GABA are likely to act as survival factors [47, 48]. Since ganglion cells are affected by ischemia in the optic nerve head region, these neurones will exist at a reduced energetic state and will therefore be particularly susceptible to a rise of any contaminant in the extracellular space.

Ganglion cells existing at a reduced energetic state might also be susceptible to light entering the globe. Ganglion cell axons within the globe are laden with mitochondria [6, 45], and recent studies show that light as entering the globe is able to trigger neuronal death by acting directly on mitochondria [26, 44, 45]. Our working hypothesis for explaining why ganglion cells die at differential rates in glaucoma is therefore as follows. An initial ischemic insult to the optic nerve head results in ganglion cells existing at a reduced energetic state. This makes them susceptible to secondary insults caused by an increase in extracellular "contaminants" caused by ischemia to astrocytes and microglia as well as light impinging on their intra-axonal mitochondria. The rate of death of individual ganglion cells is in turn dependent on their receptor profile and number of intra-axonal mitochondria [44, 45]. For a summary of events, see Fig. 14.4.

14.6 Animal Studies Relating Ischemia, Glaucoma, and Neuroprotection

14.6.1 Retinal Ischemia

There is a large body of evidence linking ischemia with ganglion cell death in situ [46]. Since retinal ganglion cells express both NMDA- and AMPA/kainate-type ionotropic glutamate receptors [4, 52], they are at risk from the elevation of extracellular glutamate that is known to follow retinal ischemic incidences [29]. Retinal ischemia can be experimentally induced in a number of ways, although the relevance of such physiological insults to glaucoma remains debatable. In situ, retinal ischemia can be caused by raising the IOP above the systolic blood pressure [43, 62] or by vascular ligation [2, 33]. Table 14.1 refers to a number of studies that describe ganglion cell death after different experimental paradigms of retinal ischemia.

Quantification of retinal ganglion cell death following different periods of raised IOP in adult rats led Sellés-Navarro and colleagues [62] to conclude that there was a direct relationship between the duration of insult and the quantity of neurone loss. By retrograde-labeling ganglion cells with fluorogold, they showed that after 120 min of high IOP, up to 95% of ganglion cells were lost between 5 and 30 days following the insult. Subsequent studies have revealed that the mode of death most often associated with glaucomatous loss of ganglion cells is apoptosis [31]. It has also been demonstrated that apoptosis accounts for a significant percentage of the death of these neurones following ischemia [5]. These data provide more circumstantial evidence for the role of ischemia in the process of ganglion neuronal death during glaucoma.

Studies in our laboratory have utilized another technique to follow the effect of ischemia upon ganglion cells. By analyzing selective ganglion cell markers such as the surface-associated antigen, Thy-1, or the structural protein neurofilament-light (NF-L) in rat retinal sections after ischemia, it is possible to directly determine the effects of the insult upon these cells. It is found that the immunoreactive protein and mRNA levels for Thy-1 and NF-L are much reduced after ischemia-reperfusion [9].

It is important to note that the actual experimental paradigm of retinal ischemia will relate directly to the pathological outcome. For example, in most of the studies that can detect significant ganglion cell loss, ischemia is achieved by raising the IOP above the systolic blood pressure (Table 14.1). In contrast, occlusion of retinal blood vessels (e.g., clamping of carotid and/or vertebral arteries) causes a different pattern of histological damage to the retina. Photoreceptors initially survive while ganglion cells are affected, but the former cells die some months later, due to a loss of the pupillary response leading to increased retinal light exposure [64]. These data suggest that various factors are associated with the outcome of retinal ischemia. These may include the specific effects on different ocular blood supplies and any associated inflammatory responses. Thus, the choice of experimental paradigm used to initiate retinal ischemia is critical in determining the exact relationship between ischemia, glaucoma, and ganglion cell death.

14.6.2 Ischemia and Oxidative Stress (Figs. 14.2 and 14.5)

Oxygen free radicals and other reactive oxygen species (ROS) can react detrimentally with most macromolecular constituents of the cell and lead to protein modification, lipid peroxidation, and nucleic acid breakdown (e.g., [28]). It is for this reason that cells have developed natural antioxidant defense mechanisms, which include the use of enzymes (catalase, glutathione reductase, glutathione peroxidase, superoxide dismutase) and other compounds (ascorbate, uric acid, α-tocopherol, glutathione). In ischemia-reperfusion, oxygen free radicals are generated in excess of the natural cellular antioxidant defense systems, and cellular damage or destruction can therefore occur. As a consequence, any form of treatment that can enhance the natural cellular antioxidant defense system will have a neuroprotective action in retinal ischemia. This has indeed been demonstrated to be the case in experimental studies. Elevation of IOP to cause retinal ischemia in the rabbit generates oxygen-derived free radicals as well as retinal injury that can be attenuated by administration of the antioxidant compound dimethylthiourea or other antioxidant treatments. Protection of the retina after ischemia-reperfusion has also been demonstrated by the use of a variety of antioxidants as discussed above.

14.6.3 Ischemia and Excitotoxicity (Figs. 14.5 and 14.6)

Reports have accumulated to suggest that glutamate toxicity, resulting either from ischemia or from excessive stimulation of ionotropic glutamate receptors (excitotoxicity), may contribute to ganglion cell death in glaucoma [48]. Dreyer et al. [14] described high levels of glutamate in the vitreous from glaucoma patients, and it has been shown that there is a similar increase in this neurotransmitter in the rabbit vitreous following retinal ischemia [69].

Evidence is also available to support the notion that ischemic or excitotoxic loss of ganglion cells is mediated predominantly via NMDA receptors [46]. Dreyer et al. [15] found that large (magnocellular) ganglion cells were more sensitive to intraocular injection of NMDA or glutamate than small (parvocellular) ganglion cells. This is an interesting finding since it has been reported that large ganglion cells are the most sensitive to destruction in glaucoma [56].

Systemic or intraocular administration of MK-801 or memantine, specific NMDA antagonists, can protect against many of the destructive effects of either NMDA or experimental retinal ischemia upon ganglion neurones. However, since glial/neuronal take-up of glutamate within the retina is achieved highly efficiently, it is surprising that glutamate itself caused ganglion cell death at such low concentrations (26–34 μM) in this paradigm. It is evident that a number of

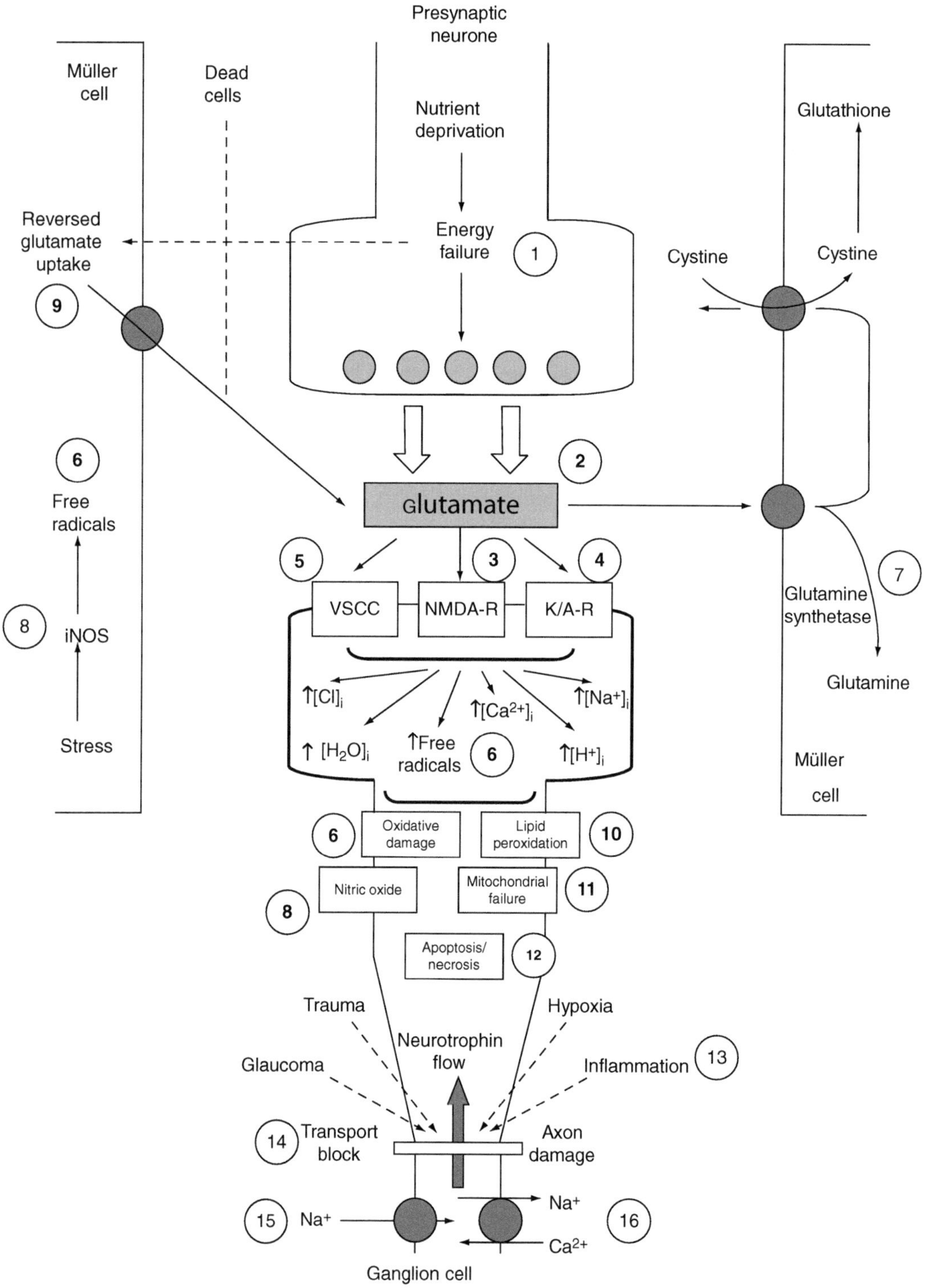

sub-populations of ganglion cells exist, each with their own repertoire of glutamate and other receptor types. It is likely that the number and/or classes of receptor expressed by an individual or sub-population of cell(s) will determine whether a particular ganglion cell can withstand the effects of the excessive release of neurotransmitters known to follow an ischemic incident.

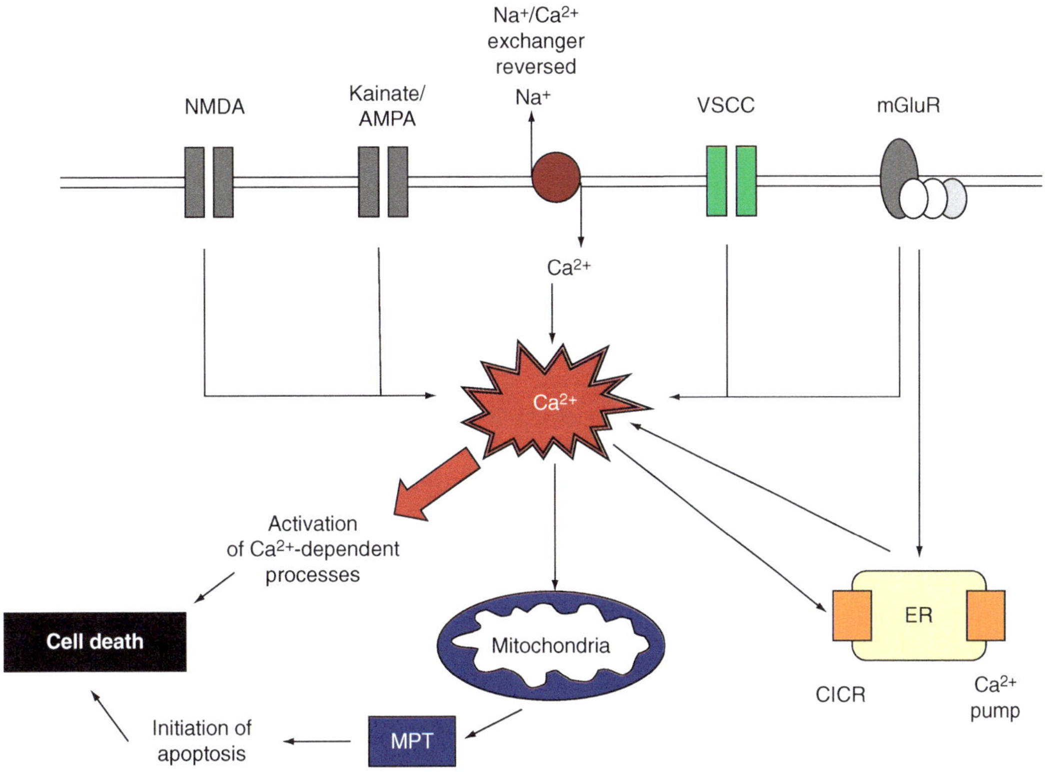

Fig. 14.6 The central role of calcium in ischemic neuronal death processes. As a result of ischemia, intracellular calcium levels can increase by several mechanisms: neurotransmitter (e.g., glutamate)-gated channels (e.g., the ionotropic glutamate receptors of the NMDA or AMPA/kainate types), by reversal of sodium-calcium exchange, through voltage-sensitive calcium channels (*VSCC*) or by stimulation of metabotropic neurotransmitter receptors (e.g., mGluRs) which can release calcium from intracellular stores. The increased intracellular levels of calcium result in activation of many enzymes which are dependent on this cation for full activity (e.g., protein kinase C, calpains, nitric oxide synthase), and these enzymes produce changes which lead to the breakdown of cellular homeostatic mechanisms and structural integrity to cause cell death. Furthermore, when calcium levels are increased in the cytoplasm, mitochondria attempt to take up this cation, but these organelles are also overcome by ischemia-induced elevations, and this leads to mitochondrial permeability transition, cytoplasmic release of cell-death-promoting factors such as cytochrome C and the initiation of apoptotic cell death

14.6.4 Glaucoma and Neurotrophic Support (Fig. 14.5)

It has been hypothesized that withdrawal of trophic factor support is one of the primary causes of ganglion cell death in glaucoma [38]. It is known that all neurones are dependent upon certain peptide factors (growth factors, cytokines, neurotrophic factors, neurotrophins) to reach their targets during growth and development of nervous tissues. It is also known that neurones rely upon a constant source of such factors throughout their lifetime in order to maintain normal functioning. There are a variety of such factors in mammalian

Fig. 14.5 Diagrammatic representation outlining some putative targets for intervention in ganglion cell death. 1 – Prevention of energy failure; 2 – reduction of extracellular glutamate levels; 3 – blockade of NMDA-type ionotropic glutamate receptors (*NMDA-R*); 4 – inhibition of non-NMDA, kainate/AMPA (*K/A-R*) ionotropic glutamate receptors; 5 – blockade of voltage-sensitive calcium channels (VSCC); 6 – scavenging of free radicals; 7 – enhancement of glial glutamate metabolism; 8 – prevention of detrimental effects of nitric oxide; 9 – inhibition of glutamate carriers; 10 – prevention of lipid peroxidation; 11 – prevention of mitochondria failure; 12 – apoptotic pathway inhibitors; 13 – prevention of inflammation; 14 – maintenance of neurotrophin support; 15 – inhibition of axonal sodium channels; 16 – inhibition of sodium-calcium exchanger nervous tissues, but the neurotrophin that is most commonly linked with ganglion neurones is brain-derived neurotrophic factor (BDNF). BDNF is released by target neurones in the brain. Ganglion cell axons will synapse with these neurones, take up the neurotrophin, and axonally transport it back in a retrograde manner to the retina. During development, after the first BDNF arrives in the retina in this way, all ganglion cells become dependent upon its presence in order to survive. Thus, if anything occurs to compromise the transport of BDNF to the retina, ganglion neurones are thought to die, mainly by apoptosis.

This fact can be supported experimentally: removal of the ganglion cell axonal brain target (superior colliculus) from neonatal Wistar rats results in a rapid loss of ganglion cells [11]. Transection of the adult rat optic nerve also leads to degeneration of 80% or more of the ganglion cells. This is presumed to occur through the physical effects of ganglion cell axotomy and the associated cessation in retrograde transport of neurotrophins. Application of BDNF or neurotrophin-4/5 to the retina under these conditions can significantly prevent ganglion cell degeneration and indeed can promote regeneration of the axons intraretinally [60]. In an elegant study by Castillo et al. [7], conditioned medium from astrocytes that had been genetically modified to produce elevated levels of BDNF was added to rat retinal cultures. In such cultures, the excessive levels of BDNF led to a significant (15-fold) survival of Thy-1 immunoreactive ganglion cells as compared with control cultures.

14.6.5 Inflammation (Fig. 14.5)

By hypothesizing that ganglion cells die by apoptosis in glaucomatous optic neuropathy, one would be led to conclude that there would be no associated inflammation, since this type of cell death involves phagocytic destruction of the cellular corpses by surrounding or invading phagocytes. Indeed, clinical evidence appears to support this notion. The disappearance of ganglion cell bodies and a thinning of the inner retina in glaucoma 211 appear not to be associated with oedema or inflammation. Neufeld and colleagues [36], for example, could find no evidence for any inflammatory response associated with glaucoma. He showed that the level of inducible cyclooxygenase (COX)-2-immunoreactivity in the optic nerve of glaucoma eyes was unaffected. COX enzymes are involved in inflammatory responses by influencing the biosynthesis of eicosanoids. Interestingly, we have shown that when NMDA is injected into the rat eye, COX-2 mRNA is upregulated, suggesting that inflammation may in fact be associated with excitotoxicity in vivo.

The generation of nitric oxide (endothelium-derived relaxing factor) is often associated with both a stimulation of ionotropic glutamate receptors and with inflammation. Injection of the nitric oxide donor *S*-nitroso-*N*-acetyl-dl-penicillamine (SNAP, final concentration, 200 nmol) into the vitreous humor of albino rabbits causes marked degenerative changes that include death of large numbers of ganglion cells [41]. These data show that ganglion cells are susceptible to attack by nitric oxide, and if this compound is generated by invading macrophages or microglia during a post-traumatic inflammatory response, then ganglion cell layer degeneration will result.

Microglia exist in the optic nerve head in significant amounts, and these cells become activated when the optic nerve is transected [32] or when this region is subjected to ischemia. Microglial cells have now been shown to release nerve growth factor (NGF) in vitro which can directly kill retinal neurones [20]. Thus, it is possible that during ischemia, ganglion cells can be induced to die not only by excessive release of substances from retinal neurones but also through a release of mediators such as NGF, from microglia. It seems likely that the release of substances from microglia is associated with inflammation. Thus, if ischemia is associated with glaucoma, as seems likely, the suggestion is made that some degree of retinal inflammation will actually result.

14.6.6 Role of Mitochondria (Fig. 14.6)

It is expected that if cells gradually accumulate calcium as during ischemia [25], then mitochondrial failure will result. This is because mitochondria will sequester intracellular calcium as levels become elevated [37]. This leads to mitochondrial membrane depolarization and a consequent production of reactive oxygen species [25]. Mitochondrial dysfunction has been reported to occur as a result of ischemia, excitotoxicity, inflammation, oxidative stress, and loss of neurotrophic support resulting in cell death.

Apoptosis appears to be responsible for retinal ganglion cell death during development and to a lesser extent after axotomy or ischemia [40]. Mitochondria have been implicated in controlling signal transduction pathways in apoptosis, mainly due to their retaining a preserved intact morphology throughout the death process. This relates to the disease process in Leber's hereditary optic neuropathy (LHON), where genetic mutations in mitochondrial complex I lead specifically to gradual ganglion cell death by apoptosis [6]. Although no reports exist that describe mitochondrial dysfunction leading to ganglion cell atrophy on a large scale or as a result of glaucoma, this would be expected should such organelles malfunction.

It has been suggested that apoptosis can be triggered by the opening of mitochondrial megachannels, in the process of mitochondrial permeability transition (MPT) [74]. The link between MPT and the activation of the cytosolic proteolytic cascade resulting in apoptotic death is complicated, but, for example, it has been reported that a soluble mitochondrial protein, apoptosis-inducing factor (AIF), is released through the open mitochondrial megapores to cause nuclear apoptosis. This can be blocked by overexpression of Bcl-2, which is a protein found associated with the outer mitochondrial membrane.

As stated previously, optic nerve transection leads to apoptotic degeneration of retinal ganglion cells. In transgenic mice that overexpress the Bcl-2 gene, approximately 65% of ganglion cells survive this insult up to 3.5 months following the insult. In contrast, non-transgenic control mice had less than 10% of their original number of ganglion cells remaining after this period of time. Similar mice that are transgenic for Bcl-2 do not lose as many neurones during normal developmental ganglion cell death: they have up to 50% more of these neurones. These mice also have a much reduced infarct volume subsequent to ischemia induced by middle cerebral artery occlusion. Thus, the Bcl-2 protein, which protects against the destructive effects of mitochondrial failure, can prevent the type of ganglion cell death that is thought to be associated with glaucoma.

14.7 Neuroprotection Specifically Related to Glaucoma

The cause of visual loss in glaucoma is due to death of retinal ganglion cells. Their differential demise with time is proposed to be elicited by a combination of ischemia to their axons in the optic nerve head region and chemicals released from activated microglia and astrocytes, as well as light impinging on their intra-axonal mitochondria. Clearly, reducing the amount of light entering the eye is one way forward. Also, drugs targeted to reduce the release of chemicals from activated glia, the impact of ischemia to the optic nerve head region (vasoconstrictors) and the negative effects of raised extracellular chemicals to ganglion cells are required. It is unlikely however that a single substance with a defined mode of action will achieve all these aims. It is therefore proposed that for neuroprotection in glaucoma to become a reality, a spectrum of drug actions is required which might be achieved by a cocktail of defined substances or drugs that display multiple mechanisms of actions. Such drugs should be taken orally to allow penetration to the retina with minimum side effects. Present research suggests that extracts from variety of naturally occurring substances like green tea and *Ginkgo* [34, 57] have characteristics that might be exploited for such use.

Acknowledgement NNO acknowledges with thanks the support (Cátedra de Biomedicina) from the Fundación BBVA in Spain.

References

1. Banin E, Berenshtein E, Kitrossky N, Pe'er J, Chevion M (2000) Gallium-desferrioxamine protects the cat retina against injury after ischemia and reperfusion. Free Radic Biol Med 28:315–323
2. Barnett NL, Osborne NN (1995) Prolonged bilateral carotid artery occlusion induces electrophysiological and immunohistochemical changes to the rat retina without causing histological damage. Exp Eye Res 61:83–90
3. Bonne C, Muller A, Villain M (1998) Free radicals in retinal ischemia. Gen Pharmacol 30:275–280

4. Brandstatter JH, Hartveit E, Sassoe Pognetto M, Wassle H (1994) Expression of NMDA and high-affinity kainate receptor subunit mRNAs in the adult rat retina. Eur J Neurosci 6:1100–1112
5. Buchi ER, Suivaizdis I, Fu J (1991) Pressure-induced retinal ischemia in rats: an experimental model for quantitative study. Ophthalmologica 203:138–147
6. Carelli V, Ross-Cisneros FC, Sadun AA (2004) Mitochondrial dysfunction as a cause of optic neuropathies. Prog Retin Eye Res 23:53–89
7. Castillo B Jr, del Cerro M, Breakefield XO, Frim DM, Barnstable CJ, Dean DO, Bohn MC (1994) Retinal ganglion cell survival is promoted by genetically modified astrocytes designed to secrete brain-derived neurotrophic factor (BDNF). Brain Res 647:30–36
8. Celebi S, Dilsiz N, Yilmaz T, Kukner AS (2002) Effects of melatonin, vitamin E and octreotide on lipid peroxidation during ischemia-reperfusion in the guinea pig retina. Eur J Ophthalmol 12:77–83
9. Chidlow G, Osborne NN (2003) Rat retinal ganglion cell loss caused by kainate, NMDA and ischemia correlates with a reduction in mRNA and protein of Thy-1 and neurofilament light. Brain Res 963: 298–306
10. Clarke M, Dodson PM (2007) PKC inhibition and diabetic microvascular complications. Best Pract Res Clin Endocrinol Metab 21:573–586
11. Cui Q, Harvey AR (1995) At least two mechanisms are involved in the death of retinal ganglion cells following target ablation in neonatal rats. J Neurosci 15:8143–8155
12. Davalos A, Castillo J, Serena J, Noya M (1997) Duration of glutamate release after acute ischemic stroke. Stroke 28:708–710
13. Doly M, Braquet P, Bonhomme B, Meyniel G (1984) Effects of lipid peroxidation on the isolated rat retina. Ophthalmic Res 16:292–296
14. Dreyer EB, Zurakowski D, Schumer RA, Podos SM, Lipton SA (1996) Elevated glutamate levels in the vitreous body of humans and monkeys with glaucoma. Arch Ophthalmol 114:299–305
15. Dreyer E, Pan Z, Storm S, Lipton S (1994) Greater sensitivity of larger retinal ganglion cells to NMDA-mediated cell death. Neuroreport 5:629–631
16. Faberowski N, Stefansson E, Davidson RC (1989) Local hypothermia protects the retina from ischemia. A quantitative study in the rat. Invest Ophthalmol Vis Sci 30:2309–2313
17. Flammer J, Mozaffarieh M (2007) What is the present pathogenetic concept of glaucomatous optic neuropathy? Surv Ophthalmol 52(Suppl 2):S162–S173
18. Flammer J, Haefliger IO, Orgül S, Resink T (1999) Vascular dysregulation: a principal risk factor for glaucomatous damage? J Glaucoma 8:212–219
19. Flammer J, Orgül S (1998) Optic nerve blood-flow abnormalities in glaucoma. Prog Retin Eye Res 17: 267–289
20. Frade JM, Barde YA (1998) Microglia-derived nerve growth factor causes cell death in the developing retina. Neuron 20:35–41
21. Gilgun Sherki Y, Rosenbaum Z, Melamed E, Offen D (2002) Antioxidant therapy in acute central nervous system injury: current state. Pharmacol Rev 54:271–284
22. Hernandez MR, Pena JD (1997) The optic nerve head in glaucomatous optic neuropathy. Arch Ophthalmol 115:389–395
23. Hayreh SS (2005) Prevalent misconceptions about acute retinal vascular occlusive disorders. Prog Retin Eye Res 24:493–519
24. Kaur C, Foulds WS, Ling EA (2008) Blood-retinal barrier in hypoxic ischaemic conditions: basic concepts, clinical features and management. Prog Retin Eye Res 27:622–647
25. Kristian T, Siesjo BK (1998) Calcium in ischemic cell death. Stroke 29:705–718
26. Lascaratos G, Ji D, Osborne NN (2007) Visible light affects mitochondrial function and induces neuronal death in retinal cell cultures. Vision Res 47:1191–1201
27. Leker RR, Shohami E (2002) Cerebral ischemia and trauma-different etiologies yet similar mechanisms: neuroprotective opportunities. Brain Res Rev 39:55–73
28. Lipton P (1999) Ischemic cell death in brain neurons. Physiol Rev 79:1431–1568
29. Louzada Junior P, Dias JJ, Santos WF, Lachat JJ, Bradford HF, Coutinho NJ (1992) Glutamate release in experimental ischaemia of the retina: an approach using microdialysis. J Neurochem 59:358–363
30. Matsuo Y, Kihara T, Ikeda M, Ninomiya M, Onodera H, Kogure K (1995) Role of neutrophils in radical production during ischemia and reperfusion of the rat brain: effect of neutrophil depletion on extracellular ascorbyl radical formation. J Cereb Blood Flow Metab 15:941–947
31. McKinnon SJ (1997) Glaucoma, apoptosis and neuroprotection. Curr Opin Ophthalmol 8:28–37
32. Moore S, Thanos S (1996) The concept of microglia in relation to central nervous system disease and regeneration. Prog Neurobiol 48:441–460
33. Mosinger JL, Olney JW (1989) Photothrombosis-induced ischemic neuronal degeneration in the rat retina. Exp Neurol 105:110–113
34. Mozaffarieh M, Grieshaber MC, Orgül S, Flammer J (2008) The potential value of natural antioxidative treatment in glaucoma. Surv Ophthalmol 53:479–505
35. Muller A, Pietri S, Villain M, Frejaville C, Bonne C, Culcas M (1997) Free radicals in rabbit retina under ocular hyperpressure and functional consequences. Exp Eye Res 64:637–643
36. Neufeld A, Hernandez M, Gonzalez M, Geller A (1997) Cycooxygenase-1 and cyclooxygenase-2 in the human optic nerve head. Exp Eye Res 65:739–745
37. Nicholls DG, Budd SL (2000) Mitochondria and neuronal survival. Physiol Rev 80:315–360
38. Nickells R (1996) Retinal ganglion cell death in glaucoma; the how, the why and the maybe. J Glaucoma 5:345–356
39. Nicolela MT (2008) Clinical clues of vascular dysregulation and its association with glaucoma. Can J Ophthalmol 43:337–341

40. Okouchi M, Ekshyyan O, Maracine M, Aw TT (2007) Neuronal apoptosis and neurodegeneration. Antioxid Redox Signal 9:1059–1095
41. Oku H, Yamaguchi H, Sugiyama T, Kojima S, Ota M, Azuma I (1997) Retinal toxicity of nitric oxide released by administration of a nitric oxide donor in the albino rabbit. Invest Ophthalmol Vis Sci 38: 2540–2544
42. Osborne NN (2009) Recent clinical findings with memantine should not mean that the idea of neuroprotection in glaucoma is abandoned. Acta Ophthalmol 87(4):450–454
43. Osborne NN, Herrera AJ (1994) The effect of experimental ischaemia and excitatory amino acid agonists on the GABA and serotonin immunoreactivities in the rabbit retina. Neuroscience 59:1071–1081
44. Osborne NN, Li GY, Ji D, Mortiboys HJ, Jackson S (2008) Light affects mitochondria to cause apoptosis to cultured cells: possible relevance to ganglion cell death in certain optic neuropathies. J Neurochem 105:2013–2028
45. Osborne NN, Lascaratos G, Bron AJ, Chidlow G, Wood JP (2006) A hypothesis to suggest that light is a risk factor in glaucoma and the mitochondrial optic neuropathies. Br J Ophthalmol 90:237–241
46. Osborne NN, Casson RJ, Wood JP, Chidlow G, Graham MM, Melena J (2004) Retinal ischemia: mechanisms of damage and potential therapeutic strategies. Prog Retin Eye Res 23:91–147
47. Osborne NN, Melena J, Chidlow G, Wood JPM (2001) A hypothesis to explain ganglion cell death caused by vascular insults at the optic nerve head: possible implication for the treatment of glaucoma. Br J Ophthalmol 85:1252–1259
48. Osborne NN, Ugarte M, Chao M, Chidlow G, Bae JH, Wood JPM, Nash MS (1999) Neuroprotection in relation to retinal ischemia and relevance to glaucoma. Surv Ophthalmol 43(Suppl 1):102–128
49. Peachey NS, Green DJ, Ripps H (1993) Ocular ischemia and the effects of allopurinol on functional recovery in the retina of the arterially perfused cat eye. Invest Ophthalmol Vis Sci 34:58–65
50. Pease ME, McKinnon SJ, Quigley HA, Kerrigan-Baumrind LA, Zack DJ (2000) Obstructed axonal transport of BDNF and its receptor TrkB in experimental glaucoma. Invest Ophthalmol Vis Sci 41: 764–774
51. Pellegrini Giampietro DE, Cherici G, Alesiani M, Carla V, Moroni F (1990) Excitatory amino acid release and free radical formation may cooperate in the genesis of ischemia-induced neuronal damage. J Neurosci 10:1035–1041
52. Peng YW, Blackstone CD, Huganir RL, Yau KW (1995) Distribution of glutamate receptor subtypes in the vertebrate retina. Neuroscience 66:483–497
53. Petty MA, Wettstein JG (1999) White matter ischaemia. Brain Res Rev 31:58–64
54. Quigley HA (1999) Neuronal death in glaucoma. Prog Retin Eye Res 18:39–57
55. Quigley HA (2005) Glaucoma: macrocosm to microcosm the Friedenwald lecture. Invest Ophthalmol Vis Sci 46:2662–2670
56. Quigley HA, Dunkelberger GR, Green WR (1988) Chronic human glaucoma causing selectively greater loss of large optic nerve fibers. Ophthalmology 95:357–363
57. Rich R (2005) Complimentary therapy for the treatment of glaucoma: a perspective. Ophthalmol Clin North Am 18:597–609
58. Rios L, Cluzel J, Vennat JC, Menerath JM, Doly M (1999) Comparison of intraocular treatment of DMTU and SOD following retinal ischemia in rats. J Ocul Pharmacol Ther 15:547–556
59. Roth S, Park SS, Sikorski CW, Osinski J, Chan R, Loomis K (1997) Concentrations of adenosine and its metabolites in the rat retina/choroid during reperfusion after ischemia. Curr Eye Res 16:875–885
60. Sawai H, Clarke DB, Kittlerova P, Bray GM, Aguayo AJ (1996) Brain-derived neurotrophic factor and neurotrophin-4/5 stimulate growth of axonal branches from regenerating retinal ganglion cells. J Neurosci 16:3887–3894
61. Schmidt D (2008) The mystery of cotton-wool spots – a review of recent and historical descriptions. Eur J Med Res 13:231–266
62. Selles Navarro I, Villegas Perez MP, Salvador Silva M, Ruiz Gomez JM, Vidal SM (1996) Retinal ganglion cell death after different transient periods of pressure-induced ischemia and survival intervals. A quantitative in vivo study. Invest Ophthalmol Vis Sci 37:2002–2014
63. Shibuki H, Katai N, Yodoi J, Uchida K, Yoshimura N (2000) Lipid peroxidation and peroxynitrite in retinal ischemia-reperfusion injury. Invest Ophthalmol Vis Sci 41:3607–3614
64. Stevens WD, Fortin T, Pappas BA (2002) Retinal and optic nerve degeneration after chronic carotid ligation: time course and role of light exposure. Stroke 33:1107–1112
65. Szabo ME, Droy Lefaix MT, Doly M (1997) Direct measurement of free radicals in ischemic/reperfused diabetic rat retina. Clin Neurosci 4:240–245
66. Szabo ME, Droy Lefaix MT, Doly M, Braquet P (1992) Ischaemia- and reperfusion-induced Na^+, K^+, Ca^{2+} and Mg^{2+} shifts in rat retina: effects of two free radical scavengers, SOD and EGB 761. Exp Eye Res 55:39–45
67. Szabo ME, Droy Lefaix MT, Doly M, Braquet P (1993) Modification of ischemia/reperfusion-induced ion shifts (Na^+, K^+, Ca^{2+} and Mg^{2+}) by free radical scavengers in the rat retina. Ophthalmic Res 25: 1–9
68. Szabo ME, Droy Lefaix MT, Doly M, Carre C, Braquet P (1991) Ischemia and reperfusion-induced histologic changes in the rat retina. Demonstration of a free radical-mediated mechanism. Invest Ophthalmol Vis Sci 32:1471–1478
69. Tamai K, Toumoto E, Majima A (1997) Local hypothermia protects the retina from ischaemic injury in vitrectomy. Br J Ophthalmol 81:789–794

70. Takagi H (2007) Aging and retinal vascular diseases. Nippon Ganka Gakkai Zasshi 111:207–230
71. Tezel G, Wax MB (2000) Increased production of tumor necrosis factor-alpha by glial cells exposed to simulated ischemia or elevated hydrostatic pressure induces apoptosis in cocultured retinal ganglion cells. J Neurosci 20:8693–8700
72. Toda N, Nakanishi-Toda M (2007) Nitric oxide: ocular blood flow, glaucoma, and diabetic retinopathy. Prog Retin Eye Res 26:205–238
73. Tsujikawa A, Ogura Y, Hiroshiba N, Miyamoto K, Kiryu J, Tojo SJ, Miyasaka M, Honda Y (1999) Retinal ischemia-reperfusion injury attenuated by blocking of adhesion molecules of vascular endothelium. Invest Ophthalmol Vis Sci 40:1183–1190
74. Uchino H, Kuroda Y, Morota S, Hirabayashi G, Ishii N, Shibasaki F, Ikeda Y, Hansson MJ, Elmer E (2008) Probing the molecular mechanisms of neuronal degeneration: importance of mitochondrial dysfunction and calcineurin activation. J Anesth 22:253–262
75. Zhang H, Agardh CD, Agardh E (1995) Increased catalase levels and hypoxanthine-enhanced nitro-blue tetrazolium staining in rat retina after ischemia followed by circulation. Curr Eye Res 14:47–54

Part IV

Disease

Ocular Blood Flow in Diabetes: Contribution to the Microvascular Lesions of Diabetic Retinopathy

15

Tim M. Curtis and Tom A. Gardiner

Core Messages

- Diabetic retinopathy is a leading cause of vision loss in the working population of developed countries.
- Changes in retinal haemodynamics have been proposed to play a key role in the initiation and progression of diabetic retinopathy.
- Substantial evidence suggests that there is an early reduction in retinal perfusion prior to the onset of diabetic retinopathy followed by a gradual increase in blood flow as the disease progresses.
- Two major mechanisms have been proposed to explain how hyperglycaemia decreases retinal blood flow in early diabetes, namely, protein kinase C (PKC) activation and ion channel dysfunction in the contractile mural cells of retinal microvessels.
- The functional reduction in retinal blood flow observed during early diabetic retinopathy may be additive or synergistic to pro-inflammatory changes, leukostasis and vaso-occlusion and thus may be intimately linked to the progressive ischaemic hypoxia and increased blood flow associated with later stages of the disease.
- A unifying haemodynamic framework is presented that explains how changes in retinal perfusion may contribute to the microvascular lesions and vision loss in diabetic retinopathy.
- Large-scale prospective studies are currently needed to determine whether retinal blood flow measurements may be useful as a surrogate end point for clinical drug trials in diabetic retinopathy.

15.1 Introduction

Diabetes mellitus is a condition of chronic hyperglycaemia and is currently classified into two main forms. Type 1 diabetes is due primarily to autoimmune-mediated destruction of pancreatic-β-cell islets, resulting in absolute insulin deficiency. People with type 1 diabetes are usually younger than 30 years old at diagnosis and are dependent on continuing supplemental insulin. Type 2 diabetes is a metabolic disorder normally of middle-life characterised by insulin resistance and/or abnormal insulin secretion, either of which may predominate. People with type 2 diabetes

T.M. Curtis, Ph.D. (✉)
Centre for Vision and Vascular Sciences School of Medicine, Dentistry and Biomedical Sciences, The Queen's University of Belfast, Institute of Clinical Science - Block A, Royal Victoria Hospital, Grosvenor Road, Belfast, Northern Ireland BT12 6BA, UK
e-mail: t.curtis@qub.ac.uk

T.A. Gardiner, Ph.D.
Centre for Biomedical Sciences Education, School of Medicine, Dentistry and Biomedical Sciences, The Queen's University of Belfast, Whitla Medical Building, 97 Lisburn Road, Belfast, Northern Ireland BT9 7B, UK
e-mail: t.gardiner@qub.ac.uk

L. Schmetterer, J.W. Kiel (eds.), *Ocular Blood Flow*,
DOI 10.1007/978-3-540-69469-4_15, © Springer-Verlag Berlin Heidelberg 2012

are not usually dependent on exogenous insulin but may require it for control of blood glucose levels if this is not achieved with diet alone or with oral hypoglycaemic agents.

Diabetes is associated with a number of complications which share an aetiology that is, at least in part, vascular. As a consequence of microvascular pathology, diabetes is a leading cause of blindness, end-stage renal disease and a variety of debilitating neuropathies. Macrovascular complications manifest themselves as accelerated atherosclerosis resulting in an increased risk of myocardial infarction, stroke and limb amputation. Diabetes is increasing at a startling rate and consequently imposes an ever-increasing burden on health-care authorities in both developed and developing countries. The global figure of people with diabetes is set to rise from the present estimate of 150–220 million in 2010 and 300 million in 2025 [13, 110]. Most cases will be of type 2 diabetes, which is strongly associated with sedentary lifestyle and obesity [199].

Retinopathy is one of the most common microvascular complications of diabetes [10]. After 20 years of diabetes, nearly all patients with type 1 diabetes will have at least some retinopathy. Approximately 80% of type 2 diabetic patients who require insulin and 50% type 2 diabetic patients who do not require insulin will have retinopathy after 20 years [111]. Although the pathogenic basis of diabetic retinopathy is not wholly understood at the cellular and molecular level, large-prospective clinical studies have demonstrated a strong relationship between time-averaged mean levels of glycaemia and the rate of development of retinopathy in both type 1 and type 2 diabetes [1, 3].

Diabetic retinopathy is traditionally regarded as disease of the intra-retinal microvessels and is generally classified into two main clinical forms: non-proliferative diabetic retinopathy and proliferative diabetic retinopathy. Mild non-proliferative diabetic retinopathy (background retinopathy) and severe proliferative retinopathy represent different ends of a spectrum of the same disease process. Non-proliferative diabetic retinopathy is characterised by a complex array of vasodegenerative lesions within the retinal microvascular bed, including thickening of capillary basement membranes [77], pericyte and vascular smooth muscle cell dropout [81, 123], microaneurysms [34] and capillary occlusion and acellularity [67]. Visual impairment normally occurs in the later stages of diabetic retinopathy with the development of macula oedema as a direct consequence of blood-retinal barrier breakdown. In the proliferative phase of the disease, there is an abnormal growth of new blood vessels (retinal neovascularisation) that give rise to sight-threatening complications such as vitreous haemorrhage and tractional retinal detachment [76]. Arguably, both macular oedema and retinal neovascularisation occur as a result of increasing inner retinal ischaemia and hypoxia-driven secretion of vascular endothelial growth factor (VEGF). Sight-threatening diabetic retinopathy can be treated or contained to some extent by laser photocoagulation or vitreoretinal surgery, but this is often at the expense of functional retina and visual performance [79]. The use of VEGF-blocking agents has shown efficacy in both of these sight-threatening sequelae, but the underlying vascular insufficiency may compromise retinal cell survival [148].

While there is no doubt that hyperglycaemia is the primary insult in the pathogenesis of diabetic retinopathy [1, 3], haemodynamic factors have also been implicated in the development of this condition. This chapter begins by providing an overview of the haemodyamic changes that occur in the retina during the development of diabetic retinopathy. Attention is also given to the pathophysiological mechanisms that have been proposed to underlie these alterations. The remainder of the chapter is then devoted to describing how retinal blood flow changes in diabetes may contribute the sight-threatening lesions of diabetic retinopathy. From these discussions, it is clear that treatments designed to normalise retinal blood flow during early and long-term diabetes may provide a novel means of delaying the onset and progression of this devastating condition.

15.2 Retinal Blood Flow in Diabetes

The earliest suggestion that retinal blood flow may be disrupted in diabetes came from studies performed over 60 years ago showing that dilation of retinal veins is common in persons with diabetes [24, 187]. Later, Skovborg and coworkers reported that the calibre of retinal arteries is also increased in diabetic subjects [170]. The first direct empirical evidence of altered retinal blood flow in persons with diabetes came from Kohner et al. in the mid-1970s coinciding with the development of methods for the measurement of mean retinal circulation time from fluorescein angiograms [118]. Their results showed that retinal blood flow was increased in diabetic subjects with absent or mild retinopathy but not in those with moderate or severe diabetic retinopathy. Since then, the haemodynamics of the retina in persons with diabetes, as well as diabetic animal models, has attracted considerable research interest. Several comprehensive review articles have appeared in recent years that provide an overview of retinal perfusion abnormalities in the different stages of diabetic retinopathy [46, 49, 163]. There is significant discrepancy between the results obtained in the various clinical and experimental studies of ocular blood flow in diabetes. This may be attributable to the variety of techniques used to measure retinal haemodynamics, the use of differing sites to measure retinal blood flow and the fact that in many studies, relatively little attention has been paid to the demographic and metabolic parameters (blood glucose, lipids, insulin, blood pressure, diabetes duration, etc.) of the study cohorts. Although there are some conflicting reports, the majority of studies suggest that in patients with a short duration of diabetes (<5 years), there is a constriction of the major arteries and arterioles [115, 192], and retinal blood flow is decreased [29, 36, 73]. With longer durations of diabetes and the presence of clinical retinopathy, arterial vessels begin to dilate, and bulk retinal blood flow increases in proportion to the severity of retinopathy [47, 72, 74, 91, 152].

15.3 Retinal Hypoperfusion

A decrease in retinal blood flow is one of the earliest abnormalities observed in the diabetic retina. Bertram et al. [29] measured the arteriovenous passage time in patients with type 1 diabetes using video fluorescein angiography (VFA). The arteriovenous passage time is defined as the interval between the first influx of fluorescein into a retinal artery and its first appearance in the corresponding retinal vein. They found that the arteriovenous passage time was increased in diabetic patients with no retinopathy, indicative of reduced retinal blood flow. Similarly, Bursell et al. [36] reported that the mean circulation time, a parameter closely related to the arteriovenous passage time, is increased in patients with type 1 diabetes and no retinopathy. Laser Doppler techniques have also been used to quantify blood flow in the major retinal vessels of diabetic patients with no retinopathy. Feke et al. [73] found that arterial blood speeds were reduced by ~30% in type 1 diabetic patients prior to the appearance of overt diabetic retinopathy. Several studies have investigated perimacular capillary perfusion in early diabetic retinopathy using confocal scanning laser ophthalmoscopy. A reduction in flow velocities has been observed in both type 1 and type 2 diabetic patients with absent or mild retinopathy [16, 17, 191]. Impairment of retinal blood flow has also been reported in diabetic rodents up to 12-week disease duration [37, 49, 101].

15.3.1 Mechanisms of Hypoperfusion

15.3.1.1 Glycaemic Control

Studies in both humans and animals have suggested that hypoperfusion in diabetes is closely associated with poor glycaemic control. Indeed, several reports have shown that retinal perfusion is inversely correlated with HbA_{1c} in patients with both type 1 and type 2 diabetes [28, 29, 47]. Furthermore, normal retinal haemodynamics have been reported in well-controlled diabetic patients ($HbA_{1c} \leq 7.5\%$) with no or minimal retinopathy, even after several years of diabetes [86, 133]. In experimental studies, normalisation of retinal blood flow in streptozotocin (STZ)-induced diabetic rats has been accomplished by primary intervention with insulin therapy [48, 101]. Reversal of abnormal retinal haemodynamics in diabetic rats has also been reported using acarbose, an α-glucosidase inhibitor that reduces blood glucose concentration [178]. As outlined below, two major mechanisms have been proposed to explain how hyperglycaemia decreases retinal blood flow in early diabetes, namely, protein kinase C (PKC) activation and ion channel dysfunction in the contractile mural cells of retinal microvessels.

15.3.1.2 Protein Kinase C

Diabetes causes an increase in diacylglycerol (DAG) concentrations in vascular tissues associated with diabetic complications, including the retina [121, 167]. The molecular species of these pathophysiological DAGs are consistent with them being produced by the shunting of excess glucose through the de novo synthesis pathway [194]. These in turn activate several conventional and novel isoforms of PKC, including PKCα, PKCβ, PKCδ and PKCε [121]. PKC activation in diabetes may also arise through oxidative stress or increased concentrations of free fatty acids [56]. In the diabetic rat retina, the PKCβII isoform is preferentially activated [167]. There is now good evidence to suggest that PKC activation contributes to the impaired retinal blood flow observed in experimental and human diabetes: Intravitreal injection of phorbol dibutyrate, a PKC activator, or R59949, a DAG kinase inhibitor that elevates total retinal DAG levels, have been shown to decrease retinal blood flow in non-diabetic rats [38]. Also ruboxistaurin, a specific PKCβ inhibitor, can improve retinal blood flow in diabetic animals [103], and diabetic PKCβ knockout mice have been reported to exhibit no abnormalities in retinal blood flow [49]. In clinical studies, increases in the mean retinal circulation time were ameliorated by ruboxistaurin in type 1 and type 2 diabetic patients with no or very mild diabetic retinopathy [9].

Activation of PKC contracts retinal arterioles by sensitisation of the contractile apparatus to Ca^{2+} (Fig 15.1) [55], and this represents the most likely mechanism through which PKC activation in diabetes impairs retinal blood flow. Another mechanism by which PKC has been proposed to enhance microvascular constriction in the diabetic retina is through up-regulation of the vasoconstrictor peptide, endothelin-1(Et-1) [177, 197]. Increased Et-1 immunoreactivity has been reported in the retina of STZ-induced diabetic rats and the spontaneously diabetic BB/W rat [39, 40, 43]. The increased expression of Et-1 is thought to be partly due to the activation of the β and δ isoforms of PKC [151]. However, it should be stressed that the effects of elevated Et-1 in the retina during diabetes are offset by an increased resistance of the retinal microvessels to this vasoconstrictor [56], and therefore, it may not play a primary role in mediating arteriolar dysfunction in this disease.

15.3.1.3 Ion Channel Dysfunction

Ion channels in plasma membranes of retinal vascular smooth muscle cells and capillary pericytes play a central role in the regulation of vascular tone and blood flow in the retina [157, 165]. There is growing evidence that disruption of ion channel function may contribute to retinal vasoconstriction and decreased retinal blood flow during early diabetes.

The retinal arterioles constitute the main site of local blood flow regulation within the retinal microvascular tree [102]. Retinal arteriolar smooth muscle cells express several kinds of plasma membrane ion channels, including voltage-gated K^+ channels [137], large-conductance Ca^{2+}-activated K^+ channels (BK channels) [136], Ca^{2+}-activated Cl^- channels [135] and L-type Ca^{2+} channels [135, 164], all of which may be involved in the regulation of retinal vascular tone [165]. Among these channels, BK channels are known to play a critical role because their blockade with the specific inhibitor Penitrem A causes vasoconstriction in pressurised, isolated retinal vessels [136]. Normally, the opening of these channels in response to localised intracellular Ca^{2+} transients (Ca^{2+} sparks) results in spontaneous outward currents (STOCs), which cause membrane hyperpolarisation. This closes voltage-dependent Ca^{2+} channels, which decreases Ca^{2+} influx, thereby leading to vasorelaxation. BK channels are composed of pore-forming α-subunits and accessory β-subunits [117]. The β1 subunit is preferentially expressed in vascular smooth muscle [33], and this subunit increases the sensitivity of the BK channels to Ca^{2+} [52]. Recent studies from our own laboratory suggest that impairment of BK

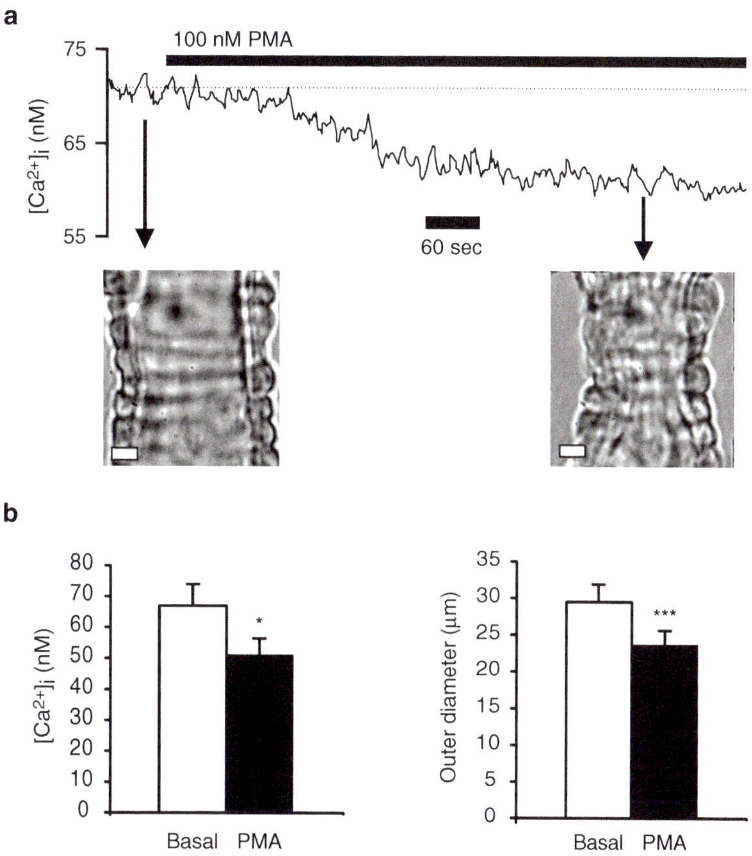

Fig. 15.1 PKC activation induces retinal arteriolar vasoconstriction by sensitising the contractile apparatus to Ca^{2+}. (**a**) An original recording of $[Ca^{2+}]_i$ and vessel diameter in a rat retinal arteriole exposed to the PKC activator phorbol myristate acetate (*PMA*). PMA reduced basal $[Ca^{2+}]_i$ but caused the arteriole to constrict (*photomicrographs below*). *Scale bars* = 5 µm on photos (From Curtis et al. [55] with permission). (**b**) *Bar charts* showing mean $[Ca^{2+}]_i$ and outer vessel diameter for arterioles before and 5 min after exposure to 100 nM PMA ($n = 9$). *Error bars* are SEMs; * and *** denote $p < 0.05$ and $p < 0.001$, respectively

channel activity may contribute to retinal hypoperfusion in early diabetes [136]. We found that STOCs were smaller (Fig 15.2a), but the amplitude of Ca^{2+} sparks was larger in retinal arteriolar myocytes from STZ-induced diabetic rats. This was explained by a reduced Ca^{2+} sensitivity of the BK channels to Ca^{2+} associated with a decreased abundance of the β1-subunit at the mRNA and protein level (Fig. 15.2b). This effect is early in onset, occurring within 1 month of diabetes induction [138]. It appears, therefore, that this downregulation decreases BK channel activity in retinal arteriolar smooth muscle cells, reducing negative feedback by intracellular Ca^{2+} and promoting contraction. This effect also appears to be highly selective for BK channels since no changes were observed when other conductances were compared between diabetic and age-matched control tissues [138].

While arterioles play a major role in regulating blood flow, recent studies have suggested that retinal perfusion may also be actively regulated at the capillary level. Retinal capillaries are richly endowed with contractile pericytes on their abluminal surface, and in vitro work has shown that these cells can modulate capillary luminal diameters in response to retinal neurotransmitters [155, 193]. Electrophysiological recordings have shown that retinal pericytes express a range of functional ion channels, including inward rectifier potassium channels (K_{IR} channels) [134, 157]. K_{IR} channels are believed to play an important role in establishing the resting membrane potential, and thereby the contractile tone of retinal pericytes. Diabetes has been reported to reduce outward K_{IR} currents in pericytes located at proximal sites within the retinal microvascular network [134]. These diabetes-induced changes seem to be mediated by the polyamine, spermine, which is elevated in the diabetic eye [147]. Consistent with the reduced K^+ efflux through K_{IR} channels, proximal pericytes on diabetic

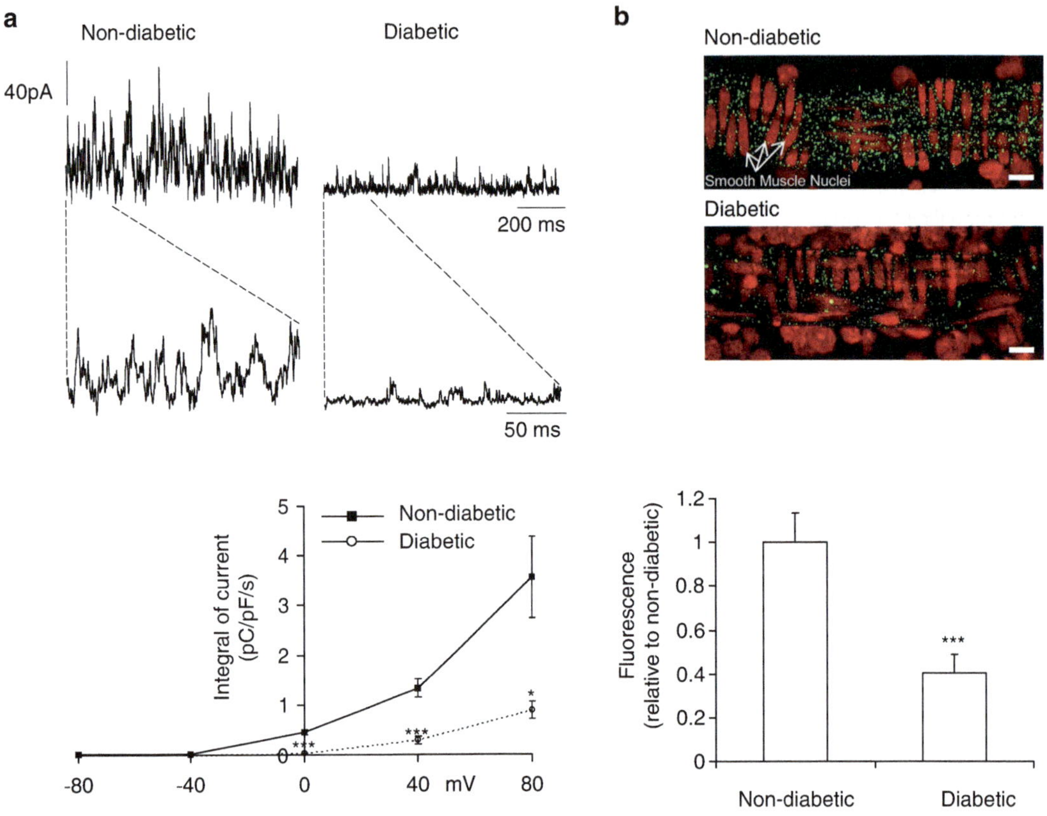

Fig. 15.2 BKβ1 downregulation in diabetic retinopathy. (**a**) *Top panel*, whole-cell patch-clamp recordings of STOC activity in a non-diabetic and diabetic retinal arteriole at a holding potential of +40 mV. *Bottom panel graph* showing the mean integrated current density versus voltage for STOCs from non-diabetic ($n=8$) and diabetic ($n=8$) retinal arterioles. * and *** denote $p < 0.05$ and $p < 0.001$. (**b**) *Top*, confocal images of non-diabetic and diabetic retinal arterioles embedded within retinal flatmount preparations and labelled with anti-BKβ1 antibody (*green*) and propidium iodide (*red*: nuclear label). Labelling of the circular smooth muscle is reduced in the tissue from the diabetic animal. *Bottom*, summary data showing statistically significant reduction in anti-BKβ1 fluorescence for diabetic samples ($n=6$ retinas, 30 vessels) relative to non-diabetics ($n=6$ retinas, 25 vessels) (From McGahon et al. [136] with permission)

retinal capillaries are more depolarised. This depolarisation increases Ca^{2+} influx through voltage-dependent Ca^{2+} channels, and the resulting increase in pericyte tone may contribute to the retinal hypoperfusion observed early in the course of diabetes.

15.4 Retinal Hyperperfusion

A transition from decreased to increased retinal blood flow is observed with progression to overt background and pre-proliferative diabetic retinopathy. Using fluoroangiographic methods, Cunha-Vaz et al. observed a small increase in retinal blood flow of diabetic patients with minimal background retinopathy, whilst a larger increase was observed in patients with more advanced stages of background retinopathy and macular oedema [53]. These findings were later substantiated by Yoshida et al., who reported that mean retinal circulation times decreased with the progression of background diabetic retinopathy [198]. Laser Doppler velocimetry studies have also confirmed that retinal perfusion is increased in diabetic patients with background retinopathy: Grunwald et al. observed that in patients with background retinopathy and poor glycaemic control, total venous cross section and volumetric blood flow was increased [91]. In another study by Patel et al., retinal hyperperfusion was found to correlate with the severity of retinopathy: Retinal blood flow was significantly increased by 30% in those with background retinopathy, by 50% in pre-proliferative patients and by over 60% in patients with proliferative diabetic retinopathy [152]. The precise nature of the retinal perfusion abnormalities observed in patients with proliferative retinopathy seems to depend on the specific pathological features present. For example, proliferative patients with severe ischemia have smaller volumetric flow rates than those with mild capillary non-perfusion [89]. These findings may be reconciled on the basis that capillary density is known to correlate inversely with vascular resistance in various tissues.

15.4.1 Mechanisms of Hyperperfusion: A Link to Hypoperfusion, Tissue Hypoxia and Retinal Leukostasis?

At present, the exact mechanisms underlying the change from hypo- to hyperperfusion as diabetic retinopathy progresses are not well understood. However, it has been suggested that a gradual increase in the levels of tissue hypoxia during the course of the disease may play an important role [49]. As outlined below, the hypoperfusion observed prior to the appearance of retinopathy may establish a functional and therefore reversible hypoxia that contributes to the progressive increase in irreversible hypoxia by promoting leukocyte-mediated retinal vascular occlusion and capillary cell death.

Tissue hypoxia occurs at a very early stage in the diabetic retina which may, at least in part, be due to the initial impairment of retinal blood flow [58, 189]. In addition, alterations in the oxygen affinity of glycosylated haemoglobin [64] and increases in the oxygen consumption of the diabetic retina [12] may also contribute. Direct measurements of intra-retinal PO_2 in diabetic cats have demonstrated that inner retinal hypoxia is present even before capillary dropout is clinically observable [128]. Further evidence of retinal hypoxia relatively early in the disease has been provided by studies showing that hyperoxia improves contrast sensitivity [97] and colour vision [59] in patients with absent or minimal diabetic retinopathy. As diabetic retinopathy progresses, tissue hypoxia may be exacerbated by non-perfusion and degeneration of the retinal capillaries. Evidence from experimental studies suggests that leukocyte adhesion and entrapment (leukostasis) is a major cause of vascular occlusion and capillary cell death in the early stages of diabetic retinopathy [108]. The adhesion and accumulation of leukocytes in the retinal vasculature begins as early as 1 week following the onset of experimental diabetes and results in endothelial cell apoptosis through a FasL-mediated mechanism [105, 106]. Multiple studies have shown that inhibition of retinal leukostasis by using selective pharmacological agents or in transgenic animals protects against the occlusion and dropout of capillaries in the diabetic retina [45, 108]. The pathogenic mechanisms that have been implicated in diabetic retinal leukostasis include hyperglycaemia-mediated up-regulation of endothelial cell adhesion molecules [140], an increase in the white blood cell count [182] and a decrease in the deformability of leukocytes [154]. The early decrease in retinal blood flow may also predispose the retinal microcirculation to enhanced leukostasis since the reduced hydrodynamic forces will make it easier for leukocytes to firmly adhere to the vessel wall [30]. It is also conceivable that the early hypoperfusion/tissue hypoxia may act together with high glucose to induce the up-regulation of adhesion molecules on the surface of the endothelial cells. Retinal hypoxia has been shown to increase the expression of pro-inflammatory cytokines, including tumour necrosis factor (TNF)-α [80], which is known to up-regulate the expression of retinal endothelial cell adhesion molecules such as intracellular adhesion molecule (ICAM)-1 and vascular cell adhesion molecule (VCAM)-1 [44].

As in most vascular beds, tissue hypoxia elicits retinal vasodilatation leading to an increase in retinal blood flow [165]. In the diabetic retina, when tissue hypoxia reaches a certain threshold, it may override the direct vasoconstrictive effects of diabetes, thereby instigating the shift to retinal hyperperfusion. Among the proposed mechanisms to explain retinal hypoxic vasodilatation are the release of metabolic factors from the surrounding neural tissues and the production of vasoactive agents from the endothelium [60]. Several mediators have been suggested to play a role in retinal hypoxic vasodilatation, including lactate [32, 82, 99, 195], adenosine [84], VEGF [47], retinal-derived relaxing factor [60] and nitric oxide (NO) [144]. Although a number of these factors are known to be elevated in the diabetic retina [156], their relative contribution to the retinal hyperperfusion observed in the later stages of diabetic retinopathy has yet to be determined.

15.4.2 Retinal Autoregulation in Diabetes

As the retinal vasculature lacks autonomic innervation [124], modulation of retinal blood flow is dependent on local control mechanisms, such as pressure and metabolic autoregulation [62].

Pressure autoregulation refers to the innate ability of the retina to keep blood flow constant despite physiological changes in systemic arterial blood pressure. For example, when blood pressure is increased during exercise, or dips nocturnally, then retinal vessels may constrict or dilate, respectively, to maintain a constant supply of nutrients to the retina [65]. Pressure autoregulatory responses in the retina are thought to be myogenic in origin [61], whereby the smooth muscle of the retinal arteries and arterioles responds to increased intraluminal pressure/cell stretch by contracting, thereby increasing vascular resistance. Previous studies have shown that pressure autoregulation of retinal blood flow is impaired in patients with early background retinopathy. For example, Rassam et al. studied changes in retinal haemodynamics in response to an increase in blood pressure (elicited by intravenous infusion of tyramine) in normotensive control subjects and diabetic patients under conditions of low (<10 mmol l^{-1}) and high glucose (>15 mmol l^{-1}) [159]. They found that retinal blood flow remained constant in non-diabetic subjects until the mean arterial pressure (MAP) was raised by 40%. However, a 30% pressure elevation increased retinal blood flow in well-controlled diabetic patients, and only a 15% increase in MAP was necessary to increase retinal perfusion in the same patients under hyperglycaemic conditions. Abnormal pressure autoregulation in the retinal vasculature of diabetic patients has also been demonstrated in response to increased intraocular pressure [169].

Metabolic autoregulation describes the adjustment of blood flow to closely match retinal metabolic demand. Several metabolic mediators have been proposed to play a role including oxygen, carbon dioxide, K^+, H^+, lactate and adenosine [160]. Hyperoxia exerts a potent vasoconstrictor effect on the retinal vasculature, and this response has been used as a method for determining metabolic autoregulatory function in the retina during diabetes. Grunwald et al. examined the effect of pure oxygen breathing on retinal blood flow in controls and diabetic patients with varying degrees of retinopathy. Hyperoxia produced a decrease in retinal blood flow of 61% in control subjects, 53% in patients with no retinopathy, 38% in those with background retinopathy and 24% in patients with proliferative retinopathy [92]. Similarly, Patel et al. showed that the oxygen reactivity of the retinal circulation was impaired in diabetic patients with background retinopathy under conditions of normoglycaemia and reduced further in the same patients when their blood glucose levels were elevated [153]. In agreement with a reduction in the oxygen reactivity of the retinal circulation, it has been reported that diabetic dogs exhibit a larger increase in preretinal oxygen tension than normal dogs when given 100% oxygen to breath [69]. More recently, Berkowitz and colleagues have demonstrated similar findings in diabetic patients [184] and experimental rodent models [27] using MRI-based imaging.

Since autoregulatory mechanisms are active under normal physiological conditions, impairment of retinal vascular autoregulation may contribute to the retinal hyperperfusion observed in the later stages of diabetic retinopathy [120].

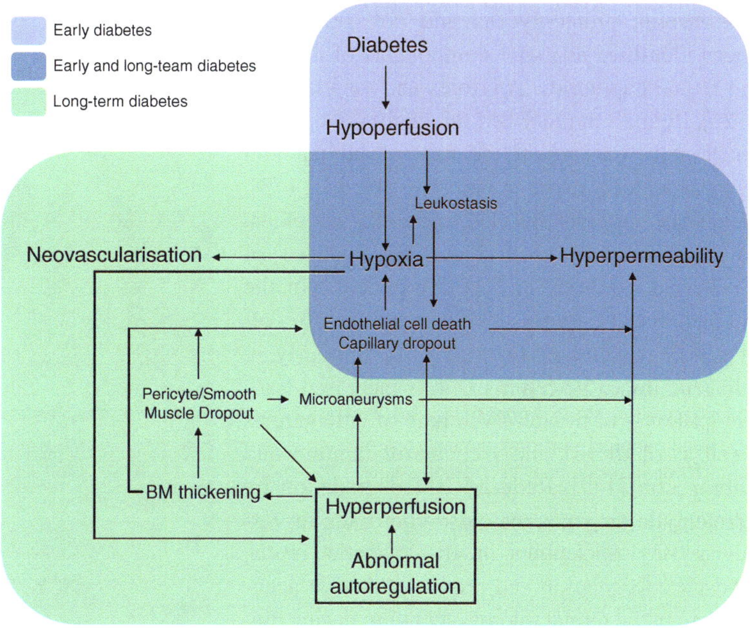

Fig. 15.3 A unifying haemodynamic model for the pathogenesis of diabetic retinopathy

However, the loss of retinal blood flow autoregulation most likely results from the progressive tissue hypoxia that occurs as the disease develops. For instance, a number of studies have shown that panretinal photocoagulation, which increases oxygenation of the inner retinal layers [35], markedly improves the retinal vascular response to hyperoxia in diabetic patients with proliferative retinopathy [92, 93].

15.5 Retinal Blood Flow and the Long-Term Lesions of Diabetic Retinopathy

We have proposed a model to explain how alterations in retinal blood flow may lead to the initiation and progression of diabetic retinopathy (Fig. 15.3). In particular, an emphasis has been placed on interrelationships between retinal blood flow and the evolution of the microvascular lesions that are characteristic of the disease. While retinal hypoperfusion may be an initiating event in the pathogenesis of diabetic retinopathy, considerable evidence suggests that retinal hyperperfusion may play an important role in accelerating microvascular damage in the later stages of the disease. The importance of retinal hyperperfusion in diabetic retinopathy is suggested by clinical observations that raised intraocular pressure [186] and moderate stenosis of the carotid artery [152], which reduce or normalise retinal blood flow, protect against the development of advanced stages of the disease. In contrast, conditions such as hypertension [85] and pregnancy [107], which increase retinal blood flow, are associated with a faster disease progression. Significantly, it has also been shown that patients with mild diabetic retinopathy who fail to demonstrate an improvement in haemodynamic indices through normalisation of blood glucose tend to show a more rapid progression of the disease on follow-up [90]. In the following section of this chapter, we evaluate the microvascular changes that are manifest in the retinal microvasculature during long-term diabetes and consider how alterations in retinal blood flow may contribute to the pathology observed.

15.5.1 Basement Membrane Thickening

The vascular basement membrane (BM) provides structural support and influences the growth, function and survival of the endothelial and mural cells of retinal microvessels [42]. Laminin, fibronectin, proteoglycans and collagens have been identified as major components of the BM in retinal capillaries, arterioles and venules [70, 126]. Retinal capillary BM thickening is a prominent and characteristic feature of human and experimental diabetic retinopathy [41, 131, 173]. Evidence suggests that BM thickening develops, at least in part, as a direct consequence of increased synthesis of BM components. In the human diabetic retina, mRNA expression of collagen IV is increased twofold [162]. Similarly, in diabetic mice, the retina exhibits increased transcript levels of the α1 (IV) chain of collagen, as well as of the β1 and γ chains of laminin and fibronectin [149]. Reduced BM degradation by proteolytic enzymes may also contribute to vascular BM thickening in the diabetic retina. Advanced glycation end products (AGEs) accumulate in the retinal microvasculature during diabetes [175], and these adducts precipitate enhanced cross-linking in vascular BMs with associated reductions in BM degradability and abnormal vessel elasticity [143]. Indeed, inhibition of AGE formation during diabetes can protect against BM thickening in diabetic rats [78].

Increased blood flow may also be a significant factor in retinal capillary BM expansion during long-term diabetes [120]: Retinal hyperperfusion, together with an increase in blood viscosity [171], tends to increase shear stress to the vessel wall. Shear stress is known to regulate endothelial cell gene expression through activation of multiple intracellular signalling cascades. Increased shear stress increases endothelial cell expression of several BM components including collagen, laminin and fibronectin [180]. In addition, increased shear stress may act to preserve the integrity of the microvascular wall by inducing the expression of inhibitors of matrix degrading proteases [139]. Morphometric analysis of the retinal vasculature from diabetic rats and dogs supports the concept that increased blood flow may play a role in retinal capillary BM thickening. Shear stress levels are higher on the arterial side of the retinal microcirculation [145], and in diabetes, BM thickening is more pronounced in capillaries that are situated in close vicinity to the retinal arteries and arterioles (Fig. 15.4) [14, 173].

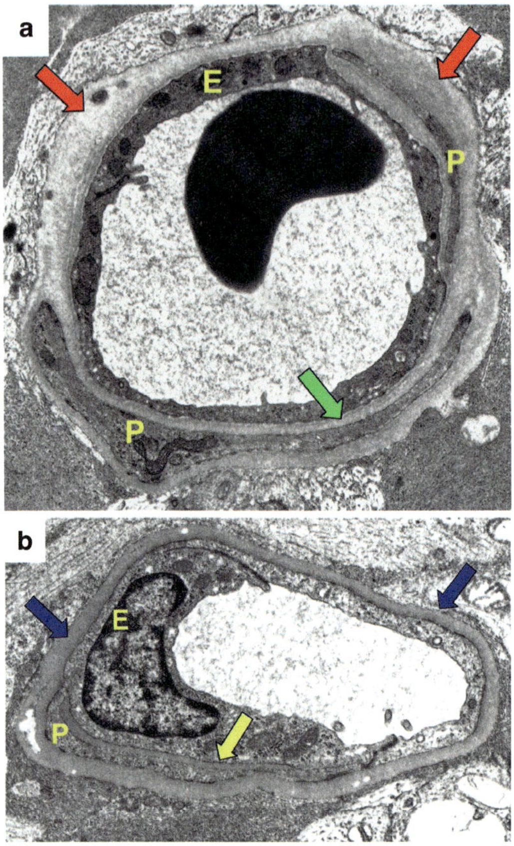

Fig. 15.4 (**a**) Arterial capillary from a 5-year diabetic dog retina shows greater and more irregular basement membrane (*BM*) thickening (*red arrows*) than in a venous capillary from the same retina (**b**). The BM between the endothelial cell (*E*) and pericyte processes (*P*) is characteristically thinner than at other cellular interfaces (*green arrow*). Pericyte coverage is greater in vessel *A*, consistent with its proximity to the arterial circulation. (**b**) Venous capillary from the same 5-year diabetic dog retina depicted in (**a**) shows less basement membrane (*BM*) thickening (*blue arrows*) than in the arterial capillary. Once again, the BM between the endothelial cell (*E*) and pericyte process (*P*) is characteristically thinner than at other locations (*yellow arrow*)

15.5.2 Pericyte and Smooth Muscle Cell Death

Although pericytes are present in all vascular beds, they are thought to be particularly important in the retina where the ratio of these cells to vascular endothelial cells is greater than in any other microvascular bed [23]. The selective loss of pericytes from the capillary/venular wall is a

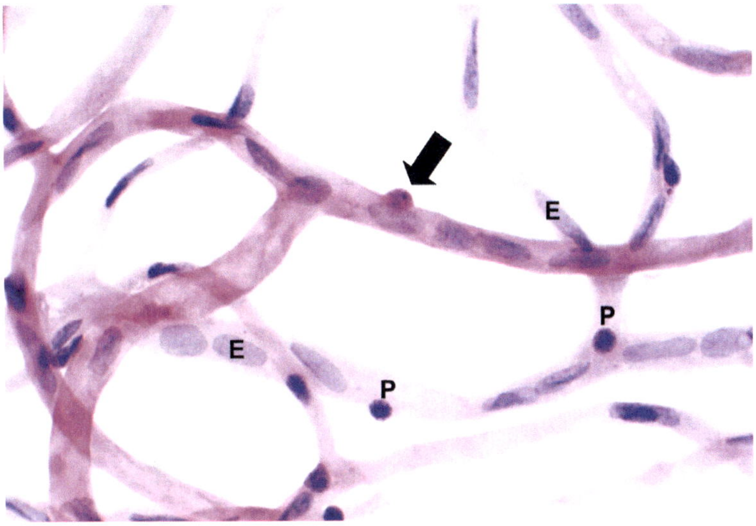

Fig. 15.5 Trypsin digest of retinal vasculature of 5-year diabetic dog. A pericyte "ghost" (*arrow*) in the wall of a small venule stains red with the periodic acid-Schiff technique. Normal pericyte (*P*) and endothelial cell nuclei (*E*) stain blue with haematoxylin

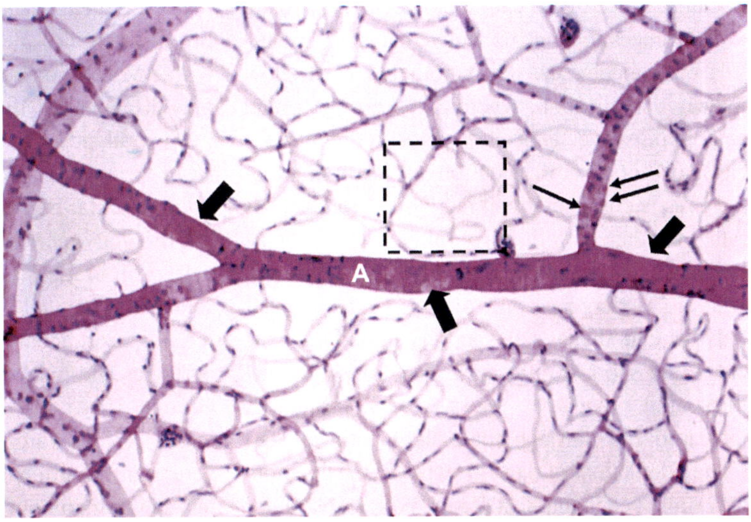

Fig. 15.6 Trypsin digest of retinal vasculature from 5-year diabetic dog shows widespread loss of arteriolar smooth muscle (*SM*) cells (*thick arrows*). SM cell "ghosts" stain red with the periodic acid-Schiff technique (*thin arrows*). Several contiguous capillaries adjacent to the major arteriole (*A*) are completely acellular (*dashed box*)

histopathological hallmark of diabetic retinopathy [50, 51], and pericyte dropout can be detected by the presence of "ghosts" in trypsin digest (Fig. 15.5) and transmission electron microscopy preparations. Pericyte ghosts represent pockets of cell debris within the BM pocket at sites of pericyte cell death. In parallel with pericyte loss, there is also a selective degeneration of vascular smooth muscle cells in the retinal arteries and arterioles (Fig. 15.6) [79, 81]. In trypsin digests, vascular smooth muscle cell loss is associated with vasodilatation, which is often focal in nature, but may also be more widespread across the entire length the vessel. The dropout of retinal pericytes and smooth muscle cells may severely aggravate the retinal haemodynamic abnormalities observed in long-term diabetes and probably explains why retinal blood flow changes appear to be irreversible in some diabetic patients with established retinopathy [90].

The mechanisms underlying pericyte and smooth muscle cell death are complex and not fully elucidated. However, the loss of both cell types is probably linked as they have a common ontogeny [19], display similar morphological characteristics and share common survival factors [79]. It seems unlikely that retinal hyperperfusion directly contributes to pericyte and smooth muscle cell death during long-term diabetes, but it may act indirectly as a consequence of BM thickening. A close communication between endothelial cells and pericytes/vascular smooth muscle cells is vital for the maintenance of vessel integrity and function. Endothelial cells release platelet-derived growth factor B (PDGF-B) for pericyte/vascular smooth muscle cell recruitment and survival [26, 31, 127], while pericytes/vascular smooth muscle cells express vascular endothelial growth factor and angiopoietin-1 which enhance the survival and integrity of the endothelium [19, 57, 176, 185]. BM thickening may limit communication between endothelial cells and pericytes/vascular smooth muscle cells and thus contribute to accelerated vascular cell death and vessel instability in the diabetic retina. The reason why the mural cells are lost in advance of the endothelial cells may relate to differences in their replicative capacity. In vessels of diabetic rats, the endothelial cells show increased replication, whereas the pericytes and smooth muscle cells display almost no replicative capacity [166]. The distribution of mitotic endothelial cells is consistent with increased cell turnover rather than uncontrolled proliferation [141]. Thus, it appears that in vivo both the mural and endothelial cells undergo premature cell death, but that only the endothelial cells can make good the deficit, in the short term at least. In addition to BM thickening, oxidative stress, AGE formation, increased flux through the polyol pathway and retinal depletion of key growth/survival factors have all been implicated in pericyte/vascular smooth muscle cell apoptosis in the diabetic retinal vasculature [66].

15.5.3 Microaneurysms

Capillary microaneurysms are normally the earliest clinically recognisable feature of diabetic retinopathy. Although retinal microaneurysms have also been described in other non-diabetic conditions (e.g. hypertension, retinal branch vein occlusions and leukaemia), they appear with the greatest frequency in diabetic retinopathy [20]. Ophthalmoscopically, microaneurysms appear as small, dark red spots, varying between 10 and 100 μm in diameter. On fluorescein angiography, microaneurysms are manifest as hyperfluorescent spots that fade in the later phases of the angiogram. Trypsin digest preparations of post-mortem diabetic eyes show that microaneurysms may be classified into several distinct morphological forms, ranging from thin-walled cellular types to dense, acellular, hyalinised forms [83]. Ultrastructurally, early stage microaneurysms show an extensive accumulation of large numbers of monocytes and polymorphonuclear cells [174]. The recruitment of inflammatory cells may damage the endothelium, and it is apparent that later-stage microaneurysms are without an endothelial cell lining [174]. Loss of endothelium-derived anticoagulants may account for the observed accumulation of red blood cells and microthrombi that are typical of these microvascular abnormalities. With time, microaneurysms become filled with BM components and lipid and consequently may disappear on fluorescein angiography [100].

Several possible mechanisms of microaneurysm formation have been proposed including microvascular BM damage, changes in intraluminal rheology, pericyte cell death and endothelial cell proliferation [79]. Of these mechanisms, selective pericyte loss appears to be critical since retinal microaneurysms invariably lack viable pericytes and only occur in capillary beds where pericyte loss is established [196]. However, the loss of pericytes alone is unlikely to cause microaneurysm formation since diabetic retinal capillaries can remain perfused and viable without a pericyte covering [174]. Retinal arteriolar vasodilatation and hyperperfusion is thought to be associated with an increase in capillary pressure [120, 183], which may account for capillary wall stretching at weak points and subsequent microaneurysm formation. The role of intraluminal pressure in the pathogenesis of microaneurysms is supported by the observation that these lesions tend to occur largely on the arterial side of the capillary bed in both diabetic patients and long-term animal models such as diabetic dogs (Fig. 15.7) [79, 174]. It is intriguing to note that microaneurysms do not develop in diabetic rodents. This may be because the rodent retinal vasculature, unlike that in humans and dogs,

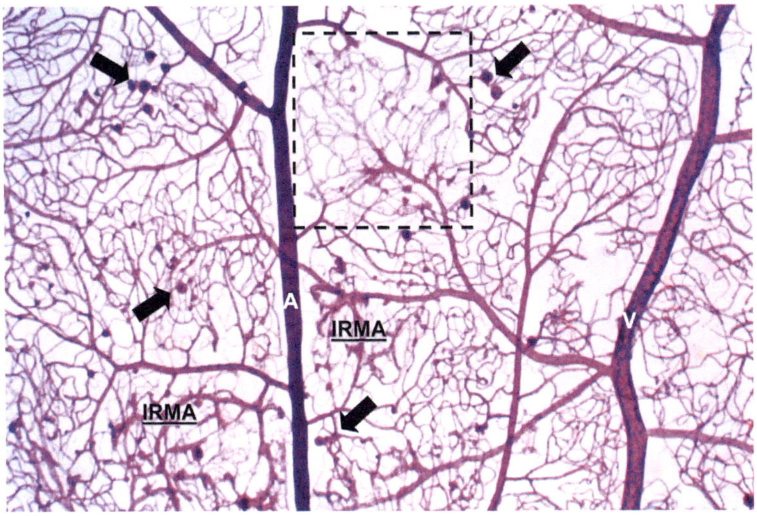

Fig. 15.7 Low-magnification image of a trypsin digest from a type 2 diabetic patient with long-standing (>15 years) non-proliferative diabetic retinopathy. Note that microaneurysms are predominantly located on capillaries adjacent to pre-capillary arterioles (*arrows*). Faintly stained acellular capillaries underlie many regions of the capillary bed, although confluent regions tend to have a peri-arteriolar location and may be bridged by intra-retinal microvascular abnormalities (*IRMA*), as in the lower part of the image or not, as in the *boxed region*. Artery (*A*), vein (*V*)

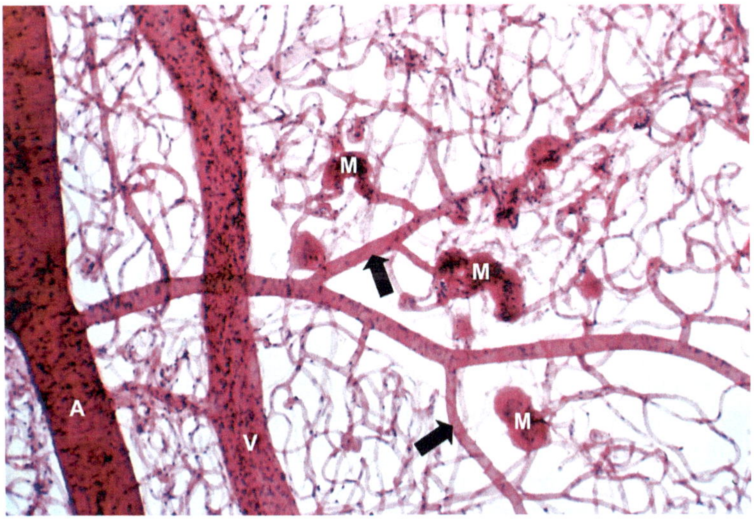

Fig. 15.8 Low-magnification image of a trypsin digest from a type 2 diabetic patient with long-standing (>15 years) non-proliferative diabetic retinopathy. Several microaneurysms (*M*) lie immediately downstream of pre-capillary arterioles and upstream of acellular capillaries. Smooth muscle cell loss is evident in the arteriolar branches (*arrows*). First-order retinal arteriole (*A*) and major retinal vein (*V*)

contains pre-capillary sphincters [150] which may render the capillary beds less susceptible to hypertensive injury. In diabetic patients, microaneurysms tend to cluster upstream of large areas of capillary non-perfusion, suggesting that these lesions may contribute to downstream capillary occlusion in the diabetic retina (Fig. 15.8) [79]. Consistent with this, microaneurysms have been identified as important predictive lesions for the progression of diabetic retinopathy [116].

15.5.4 Capillary Acellularity

Diabetes appears to greatly accelerate the death and renewal of retinal microvascular endothelial cells [142, 166]. However, with prolonged exposure to the diabetic milieu, endothelial cells may exhaust their replicative lifespan, as occurs with all somatic cells when their Hayflick limit is exceeded [98, 129]. In addition, recent experimental and clinical studies have suggested circulating bone-marrow-derived endothelial progenitor cells are dysfunctional in diabetes, thereby leading to inadequate endothelial cell regeneration and vascular repair [71]. These changes may eventually lead to the formation of acellular capillaries (empty tubes of basement membrane devoid of an endothelial covering and mural pericytes, Figs. 15.6, 15.7, and 15.8), which are a universal finding in retina from long-term diabetic animal models and post-mortem specimens [79]. Through careful comparison of fluorescein angiograms with retinal trypsin digest preparations, it is clear that acellular capillary beds correspond clinically with areas of retinal non-perfusion [119].

As described earlier, changes in retinal blood flow during diabetes may contribute to the degeneration of retinal capillaries by promoting retinal leukostasis, vascular BM thickening and microaneurysm formation. It has also been postulated that increased shear stress resulting from increased blood flow in the later stages of diabetic retinopathy may directly provoke retinal endothelial cell death and capillary occlusion [49, 120]. Physiological levels of shear stress are known to suppress endothelial cell apoptosis through activation of pro-survival signalling pathways such as the PI3/Akt pathway [63]. In contrast, pathologically high levels of shear stress have been reported to induce endothelial cell death [125]. Interestingly, there is geographic variability in the distribution of acellular capillaries (and microaneurysms) within the diabetic retina. In both dogs and humans, capillary acellularity is significantly more prevalent within the temporal retina than in the nasal retina [109, 179] and progresses from the arterial to venous side of the retinal circulation as the disease progresses [21]. These findings may provide some support for the notion that retinal haemodynamics contribute to acellular capillary formation in the diabetic retina since the blood flow rate in temporal retinal vessels is known to be much greater than that in nasal vessels [158], and flow-related stress is likely to be greater in the juxta-arteriolar capillary beds.

15.6 Retinal Blood Flow and Vision Loss in Diabetic Retinopathy

15.6.1 Diabetic Macular Oedema

Diabetic macular oedema (DME) is a manifestation of diabetic retinopathy that produces loss of central vision. DME is characterised by hyperpermeability of the retinal blood vessels leading to swelling of the neural retina within the macular region. Although DME can develop at virtually any stage of diabetes, it is most prevalent during the later phases of the disease. In the Wisconsin Epidemiologic Study of diabetic retinopathy, DME was discovered in 2–6% of patients with mild non-proliferative retinopathy, 20–63% of patients with moderate to severe non-proliferative retinopathy and 70–74% of patients with proliferative retinopathy [112]. In fluorescein angiograms, DME is classified as being either focal or diffuse. Focal DME is caused by microaneurysms that leak fluid into the retina, while diffuse DME is characterised by generalised leakage from dilated capillaries within the macular area [87]. Untreated, over one-half of diabetic patients with DME will lose two or more lines of visual acuity within 2 years [75]. The Early Treatment Diabetic Retinopathy Study (ETDRS) demonstrated that focal laser photocoagulation in the macula provides benefit to patients with DME [2, 6]. However, the principal effect of laser photocoagulation was to reduce further visual loss rather than to improve visual acuity. As a result, considerable research effort has been directed towards a better understanding of the molecular pathogenesis of DME with a view to identifying new therapeutic strategies for the treatment of this debilitating condition.

The disruption of retinal blood flow may contribute both directly and indirectly to the development of diabetic maculopathy (Fig. 15.3). According to Starling's law, fluid distribution across the retinal capillary beds depends upon the hydrostatic and oncotic pressure gradients. The hydrostatic pressure gradient is created by blood pressure and acts to push fluid out of the capillaries into the tissue interstitial spaces. The oncotic pressure gradient is caused by the larger concentration of proteins in the capillaries and acts to draw interstitial fluid into the capillaries. When the hydrostatic pressure gradient and oncotic pressure gradient are in balance, there is no net movement of fluid between the capillaries and the tissue compartment. In the retina, it is particularly important that these forces balance one another out since there are no lymphatic channels present within this tissue to drain excess fluid from the interstitium. In diabetic retinopathy, DME probably arises as a consequence of both increases in capillary hydrostatic pressures and a reduction in the oncotic pressure gradient induced by inner blood-retinal barrier breakdown and the extravasation of plasma proteins. As described earlier (see Sect. 15.5.3), the hydrostatic pressure in the retinal capillaries is increased during long-term diabetes due to retinal arteriolar vasodilatation and increased retinal blood flow. Work from Stefansson and colleagues has shown that a progressive dilatation of the retinal arterioles occurs prior to the development of DME [122] and that regression of DME following laser photocoagulation is associated with retinal vasoconstriction [88]. Further evidence that DME may result from increased hydrostatic pressure within the retina comes from studies showing that the risk of diffuse DME is >3-fold greater in patients with hypertension [132] and that reducing blood pressure may protect against this condition [4, 5]. The reduction in the oncotic pressure gradient in the diabetic retina may be primarily attributed to hypoxia-induced up-regulation of vasopermeability factors, especially VEGF [146]. The inner blood-retinal barrier is formed by tight junctions between adjacent endothelial cells of the retinal vessels, which effectively blocks paracellular vascular permeability (i.e. the transport of substances between cells). Several tight junction proteins, including transmembrane proteins, such as occludin, the claudin family and the membrane associated protein zonula occludens, contribute to the formation of the blood-retinal barrier [68]. VEGF is believed to increase retinal vascular permeability, at least in part, through a pathway involving PKC-dependent phosphorylation of occludin [95]. Several clinical studies have recently demonstrated the potential of anti-VEGF therapies in the treatment of patients with DME [18, 22, 54, 96, 146].

15.6.2 Proliferative Diabetic Retinopathy

Proliferative diabetic retinopathy involves the formation of new blood vessels that develop from the venous side of the retinal circulation [83]. The new vessels break through the inner limiting membrane and grow either along the retinal surface or directly out into the vitreous cavity. Ultrastructurally, the newly formed vessels consist of a thin-walled endothelium ensheathed by a basement membrane and vascular pericytes [94, 190]. Unlike normal retinal vessels, the endothelial cells of pre-retinal vessels are fenestrated [188] and presumably contribute to the development of DME or its persistence in patients with proliferative diabetic retinopathy. The neovascular process introduces fibroblasts/myofibroblasts into the vitreous cavity, and consequently, some of the new vessels become enveloped by a thick and densely fibrous connective tissue layer [25]. This fibrous tissue contributes to the formation of firm adhesions at the posterior hyaloid membrane (interface between retina and vitreous body) and may eventually contract as it matures. Such traction can result in pre-retinal or vitreous haemorrhages or tractional retinal detachment leading to sudden visual loss. Proliferative retinopathy occurs in approximately 50% of patients with type 1 diabetes [113] and in about 15% of patients with type 2 diabetes who have the disease for 25 years [114].

It is widely accepted that ischaemic hypoxia constitutes the main driving force responsible for the development of proliferative diabetic retinopathy. Although several hypoxia-regulated cytokines and growth factors have been implicated in the pathogenesis of proliferative diabetic retinopathy, most attention has focused on the role of VEGF. The vitreous VEGF concentration is elevated by ~20-fold in patients with active proliferative retinopathy compared to those with quiescent proliferative retinopathy [8]. Furthermore, it has been shown that intraocular injection of VEGF can stimulate neovascularisation in experimental animals [181], while interventions to block VEGF synthesis and intravitreal injection of soluble VEGF-receptor chimeric proteins can suppress this neovascularisation [11, 161]. Panretinal laser photocoagulation markedly reduces retinal VEGF levels [130]. Several recent studies have demonstrated that intraocular delivery of anti-VEGF agents can cause a rapid regression of retinal neovascularisation and an improvement in visual acuity in patients with proliferative diabetic retinopathy [7, 22, 104, 168, 172], the use of these agents is not without risks: VEGF represents the major survival factor for vascular endothelium and has been shown to act as a neuroprotectant in the retina [148]. Therefore, it is arguable that prolonged VEGF blockade may compromise the surviving intra-retinal circulation and exacerbate intra-retinal ischaemia. Alternatively, or in addition to vascular injury, VEGF-blocking agents may precipitate the death of retinal neurons and glia, which are known to be stressed and vulnerable to the diabetic milieu [15].

15.7 Conclusions

Diabetic retinopathy is a highly complex and multifactorial disease that results from a range of interactions between metabolic and haemodynamic factors within the eye. Although there has been some controversy regarding the exact nature of the retinal haemodynamic changes that occur in diabetes, strong evidence suggests that there is an early reduction in retinal perfusion prior to the onset of diabetic retinopathy followed by a gradual increase in blood flow as the disease progresses. We have proposed a unifying haemodynamic framework that not only explains how the biphasic changes in retinal blood flow observed in diabetes may be interrelated, but also how these changes may contribute directly to the development of retinal microangiopathy and vision loss in diabetes. This work may act as a primer to facilitate future studies aimed at better understanding the role of retinal haemodynamics in the pathogenesis of diabetic retinopathy. The development of new treatments for diabetic retinopathy has been hampered by the slow time course over which the disease progresses. Retinal blood flow measurements could be useful as a surrogate end point for clinical drug trials in diabetic retinopathy. However, this will require that retinal blood flow changes are fully validated as a prognostic biomarker for the downstream microvascular complications of diabetic retinopathy. This may only be achieved through large-scale, long-term, prospective studies that specifically focus on this issue.

15.8 Summary for the Clinician

Diabetic retinopathy is the leading cause of visual impairment and blindness in adults of working age in developed countries. During early diabetes (<5 years), there is a reduction in retinal blood flow that may contribute to a low-grade, chronic inflammation of the retinal microvasculature resulting in retinal vascular occlusion and capillary cell death. As the disease progresses, an irreversible hypoxia is established which triggers a switch to retinal hyperperfusion and the development of a range of microvascular lesions characteristic of diabetic retinopathy, including BM thickening, pericyte and smooth muscle cell dropout, microaneurysms and capillary occlusion/acellularity. Careful analysis of these lesions in terms of their temporal appearance, ultrastructural features and spatial localisation within the retina strongly supports the concept that alterations in retinal blood flow contribute to the vasodegenerative stages of diabetic retinopathy which eventually lead to vision loss due to clinically significant macular oedema and proliferative diabetic retinopathy. Despite the many excellent studies of ocular blood flow in diabetes, further work is now needed to firmly establish whether retinal blood flow may be useful as a potential prognostic biomarker and therapeutic target in diabetic retinopathy.

References

1. Anonymous (1993) The effect of intensive treatment of diabetes on the development and progression of long-term complications in insulin-dependent diabetes mellitus. The Diabetes Control and Complications Trial Research Group. N Engl J Med 329:977–986
2. Anonymous (1985) Photocoagulation for diabetic macular edema. Early Treatment Diabetic Retinopathy Study report number 1. Early Treatment Diabetic Retinopathy Study research group. Arch Ophthalmol 103:1796–1806
3. Anonymous (1998) Intensive blood-glucose control with sulphonylureas or insulin compared with conventional treatment and risk of complications in patients with type 2 diabetes (UKPDS 33). UK Prospective Diabetes Study (UKPDS) Group. Lancet 352:837–853
4. Anonymous (1998) Efficacy of atenolol and captopril in reducing risk of macrovascular and microvascular complications in type 2 diabetes: UKPDS 39. UK Prospective Diabetes Study Group. BMJ 317:713–720
5. Anonymous (1998) Tight blood pressure control and risk of macrovascular and microvascular complications in type 2 diabetes: UKPDS 38. UK Prospective Diabetes Study Group. BMJ 317:703–713
6. Anonymous (1991) Early photocoagulation for diabetic retinopathy. ETDRS report number 9. Early Treatment Diabetic Retinopathy Study Research Group. Ophthalmology 98:766–785
7. Adamis AP, Altaweel M, Bressler NM et al (2006) Changes in retinal neovascularization after pegaptanib (Macugen) therapy in diabetic individuals. Ophthalmology 113:23–28
8. Aiello LP, Avery RL, Arrigg PG et al (1994) Vascular endothelial growth factor in ocular fluid of patients with diabetic retinopathy and other retinal disorders. N Engl J Med 331:1480–1487
9. Aiello LP, Clermont A, Arora V, Davis MD, Sheetz MJ, Bursell SE (2006) Inhibition of PKC beta by oral administration of ruboxistaurin is well tolerated and ameliorates diabetes-induced retinal hemodynamic abnormalities in patients. Invest Ophthalmol Vis Sci 47:86–92
10. Aiello LP, Gardner TW, King GL et al (1998) Diabetic retinopathy. Diabetes Care 21:143–156
11. Aiello LP, Pierce EA, Foley ED et al (1995) Suppression of retinal neovascularization in vivo by inhibition of vascular endothelial growth factor (VEGF) using soluble VEGF-receptor chimeric proteins. Proc Natl Acad Sci USA 92:10457–10461
12. Alder VA, Su EN, Yu DY, Cringle SJ, Yu PK (1997) Diabetic retinopathy: early functional changes. Clin Exp Pharmacol Physiol 24:785–788
13. Amos AF, McCarty DJ, Zimmet P (1997) The rising global burden of diabetes and its complications: estimates and projections to the year 2010. Diabet Med 14(Suppl 5):S1–S85
14. Anderson HR, Stitt AW, Gardiner TA, Archer DB (1995) Diabetic retinopathy: morphometric analysis of basement membrane thickening of capillaries in different retinal layers within arterial and venous environments. Br J Ophthalmol 79:1120–1123
15. Antonetti DA, Barber AJ, Bronson SK et al (2006) Diabetic retinopathy: seeing beyond glucose-induced microvascular disease. Diabetes 55:2401–2411
16. Arend O, Wolf S, Jung F et al (1991) Retinal microcirculation in patients with diabetes mellitus: dynamic and morphological analysis of perifoveal capillary network. Br J Ophthalmol 75:514–518
17. Arend O, Wolf S, Remky A et al (1994) Perifoveal microcirculation with non-insulin-dependent diabetes mellitus. Graefes Arch Clin Exp Ophthalmol 232:225–231
18. Arevalo JF, Fromow-Guerra J, Quiroz-Mercado H et al (2007) Primary intravitreal bevacizumab (Avastin) for diabetic macular edema: results from the Pan-American Collaborative Retina Study Group at 6-month follow-up. Ophthalmology 114:743–750
19. Armulik A, Abramsson A, Betsholtz C (2005) Endothelial/pericyte interactions. Circ Res 97:512–523
20. Ashton N (1951) Retinal micro-aneurysms in the non-diabetic subject. Br J Ophthalmol 35:189–212
21. Ashton N (1953) Arteriolar involvement in diabetic retinopathy. Br J Ophthalmol 37:282–292
22. Avery RL, Pearlman J, Pieramici DJ et al (2006) Intravitreal bevacizumab (Avastin) in the treatment of proliferative diabetic retinopathy. Ophthalmology 113:1695.e1–1695.e15
23. Balabanov R, Dore-Duffy P (1998) Role of the CNS microvascular pericyte in the blood–brain barrier. J Neurosci Res 53:637–644
24. Ballantyne AJ, Loewenstein A (1943) The pathology of diabetic retinopathy. Trans Ophthalmol Soc UK 63:95–113
25. Bek T (2000) Histopathology and pathophysiology of diabetic retinopathy. In: van Bijsterveld PO (ed) Diabetic retinopathy. Martin Dunitz Ltd, London, pp 169–188
26. Benjamin LE, Hemo I, Keshet E (1998) A plasticity window for blood vessel remodelling is defined by pericyte coverage of the preformed endothelial network and is regulated by PDGF-B and VEGF. Development 125:1591–1598
27. Berkowitz BA, Kowluru RA, Frank RN, Kern TS, Hohman TC, Prakash M (1999) Subnormal retinal oxygenation response precedes diabetic-like retinopathy. Invest Ophthalmol Vis Sci 40:2100–2105
28. Bertram B, Wolf S, Arend O, Schulte K, Reim M (1992) Retinal circulation and current blood glucose value in diabetic retinopathy. Klin Monatsbl Augenheilkd 200:654–657
29. Bertram B, Wolf S, Fiehofer S, Schulte K, Arend O, Reim M (1991) Retinal circulation times in diabetes mellitus type 1. Br J Ophthalmol 75:462–465
30. Bienvenu K, Granger DN, Perry MA (1995) Flow dependence of leukocyte-endothelial cell adhesion in

postcapillary venules. In: Granger DN, Schmid-Schonbein H (eds) Physiology and pathophysiology of leuckocyte adhesion. Oxford University Press, New York, pp 278–293
31. Bjarnegard M, Enge M, Norlin J et al (2004) Endothelium-specific ablation of PDGFB leads to pericyte loss and glomerular, cardiac and placental abnormalities. Development 131:1847–1857
32. Brazitikos PD, Pournaras CJ, Munoz JL, Tsacopoulos M (1993) Microinjection of L-lactate in the preretinal vitreous induces segmental vasodilation in the inner retina of miniature pigs. Invest Ophthalmol Vis Sci 34:1744–1752
33. Brenner R, Perez GJ, Bonev AD et al (2000) Vasoregulation by the beta1 subunit of the calcium-activated potassium channel. Nature 407:870–876
34. Bresnick GH, Davis MD, Myers FL, de Venecia G (1977) Clinicopathologic correlations in diabetic retinopathy. II. Clinical and histologic appearances of retinal capillary microaneurysms. Arch Ophthalmol 95:1215–1220
35. Budzynski E, Smith JH, Bryar P, Birol G, Linsenmeier RA (2008) Effects of photocoagulation on intraretinal PO_2 in cat. Invest Ophthalmol Vis Sci 49:380–389
36. Bursell SE, Clermont AC, Kinsley BT, Simonson DC, Aiello LM, Wolpert HA (1996) Retinal blood flow changes in patients with insulin-dependent diabetes mellitus and no diabetic retinopathy. Invest Ophthalmol Vis Sci 37:886–897
37. Bursell SE, Clermont AC, Shiba T, King GL (1992) Evaluating retinal circulation using video fluorescein angiography in control and diabetic rats. Curr Eye Res 11:287–295
38. Bursell SE, Takagi C, Clermont AC et al (1997) Specific retinal diacylglycerol and protein kinase C beta isoform modulation mimics abnormal retinal hemodynamics in diabetic rats. Invest Ophthalmol Vis Sci 38:2711–2720
39. Chakrabarti S, Gan XT, Merry A, Karmazyn M, Sima AA (1998) Augmented retinal endothelin-1, endothelin-3, endothelinA and endothelinB gene expression in chronic diabetes. Curr Eye Res 17:301–307
40. Chakrabarti S, Sima AA (1997) Endothelin-1 and endothelin-3-like immunoreactivity in the eyes of diabetic and non-diabetic BB/W rats. Diabetes Res Clin Pract 37:109–120
41. Chakrabarti S, Sima AA (1987) Pathogenetic heterogeneity in retinal capillary basement membrane thickening in the diabetic BB-rat. Diabetologia 30:966–968
42. Chakravarthy U, Gardiner TA (1999) Endothelium-derived agents in pericyte function/dysfunction. Prog Retin Eye Res 18:511–527
43. Chakravarthy U, Hayes RG, Stitt AW, Douglas A (1997) Endothelin expression in ocular tissues of diabetic and insulin-treated rats. Invest Ophthalmol Vis Sci 38:2144–2151
44. Chen W, Esselman WJ, Jump DB, Busik JV (2005) Anti-inflammatory effect of docosahexaenoic acid on cytokine-induced adhesion molecule expression in human retinal vascular endothelial cells. Invest Ophthalmol Vis Sci 46:4342–4347
45. Chibber R, Ben-Mahmud BM, Chibber S, Kohner EM (2007) Leukocytes in diabetic retinopathy. Curr Diabetes Rev 3:3–14
46. Ciulla TA, Harris A, Latkany P et al (2002) Ocular perfusion abnormalities in diabetes. Acta Ophthalmol Scand 80:468–477
47. Clermont AC, Aiello LP, Mori F, Aiello LM, Bursell SE (1997) Vascular endothelial growth factor and severity of nonproliferative diabetic retinopathy mediate retinal hemodynamics in vivo: a potential role for vascular endothelial growth factor in the progression of nonproliferative diabetic retinopathy. Am J Ophthalmol 124:433–446
48. Clermont AC, Brittis M, Shiba T, McGovern T, King GL, Bursell SE (1994) Normalization of retinal blood flow in diabetic rats with primary intervention using insulin pumps. Invest Ophthalmol Vis Sci 35:981–990
49. Clermont AC, Bursell SE (2007) Retinal blood flow in diabetes. Microcirculation 14:49–61
50. Cogan DG, Kuwabara T (1967) The mural cell in perspective. Arch Ophthalmol 78:133–139
51. Cogan DG, Toussaint D, Kuwabara T (1961) Retinal vascular patterns. IV. Diabetic retinopathy. Arch Ophthalmol 66:366–378
52. Cox DH, Aldrich RW (2000) Role of the beta1 subunit in large-conductance Ca(2+)-activated K(+) channel gating energetics. Mechanisms of enhanced Ca(2+) sensitivity. J Gen Physiol 116:411–432
53. Cunha-Vaz JG, Fonseca JR, de Abreu JR, Lima JJ (1978) Studies on retinal blood flow. II. Diabetic retinopathy. Arch Ophthalmol 96:809–811
54. Cunningham ET Jr, Adamis AP, Altaweel M et al (2005) A phase II randomized double-masked trial of pegaptanib, an anti-vascular endothelial growth factor aptamer, for diabetic macular edema. Ophthalmology 112:1747–1757
55. Curtis TM, Major EH, Trimble ER, Scholfield CN (2003) Diabetes-induced activation of protein kinase C inhibits store-operated Ca^{2+} uptake in rat retinal microvascular smooth muscle. Diabetologia 46:1252–1259
56. Curtis TM, Scholfield CN (2004) The role of lipids and protein kinase Cs in the pathogenesis of diabetic retinopathy. Diabetes Metab Res Rev 20:28–43
57. Darland DC, Massingham LJ, Smith SR, Piek E, Saint-Geniez M, D'Amore PA (2003) Pericyte production of cell-associated VEGF is differentiation-dependent and is associated with endothelial survival. Dev Biol 264:275–288
58. de Gooyer TE, Stevenson KA, Humphries P, Simpson DA, Gardiner TA, Stitt AW (2006) Retinopathy is reduced during experimental diabetes in a mouse model of outer retinal degeneration. Invest Ophthalmol Vis Sci 47:5561–5568
59. Dean FM, Arden GB, Dornhorst A (1997) Partial reversal of protan and tritan colour defects with inhaled oxygen in insulin dependent diabetic subjects. Br J Ophthalmol 81:27–30
60. Delaey C, Boussery K, Van de Voorde J (2000) A retinal-derived relaxing factor mediates the hypoxic vasodilation of retinal arteries. Invest Ophthalmol Vis Sci 41:3555–3560

61. Delaey C, Van de Voorde J (2000) Pressure-induced myogenic responses in isolated bovine retinal arteries. Invest Ophthalmol Vis Sci 41:1871–1875
62. Delaey C, Van de Voorde J (2000) Regulatory mechanisms in the retinal and choroidal circulation. Ophthalmic Res 32:249–256
63. Dimmeler S, Assmus B, Hermann C, Haendeler J, Zeiher AM (1998) Fluid shear stress stimulates phosphorylation of Akt in human endothelial cells: involvement in suppression of apoptosis. Circ Res 83:334–341
64. Ditzel J, Standl E (1975) The problem of tissue oxygenation in diabetes mellitus. I. Its relation to the early functional changes in the microcirculation of diabetic subjects. Acta Med Scand Suppl 578:49–58
65. Dumskyj MJ, Eriksen JE, Dore CJ, Kohner EM (1996) Autoregulation in the human retinal circulation: assessment using isometric exercise, laser Doppler velocimetry, and computer-assisted image analysis. Microvasc Res 51:378–392
66. Ejaz S, Chekarova I, Ejaz A, Sohail A, Lim CW (2008) Importance of pericytes and mechanisms of pericyte loss during diabetes retinopathy. Diabetes Obes Metab 10:53–63
67. Engerman RL (1989) Pathogenesis of diabetic retinopathy. Diabetes 38:1203–1206
68. Erickson KK, Sundstrom JM, Antonetti DA (2007) Vascular permeability in ocular disease and the role of tight junctions. Angiogenesis 10:103–117
69. Ernest JT, Goldstick TK, Engerman RL (1983) Hyperglycemia impairs retinal oxygen autoregulation in normal and diabetic dogs. Invest Ophthalmol Vis Sci 24:985–989
70. Essner E, Lin WL (1988) Immunocytochemical localization of laminin, type IV collagen and fibronectin in rat retinal vessels. Exp Eye Res 47:317–327
71. Fadini GP, Agostini C, Avogaro A (2005) Endothelial progenitor cells and vascular biology in diabetes mellitus: current knowledge and future perspectives. Curr Diabetes Rev 1:41–58
72. Falck A, Laatikainen L (1995) Retinal vasodilation and hyperglycaemia in diabetic children and adolescents. Acta Ophthalmol Scand 73:119–124
73. Feke GT, Buzney SM, Ogasawara H et al (1994) Retinal circulatory abnormalities in type 1 diabetes. Invest Ophthalmol Vis Sci 35:2968–2975
74. Feke GT, Tagawa H, Yoshida A et al (1985) Retinal circulatory changes related to retinopathy progression in insulin-dependent diabetes mellitus. Ophthalmology 92:1517–1522
75. Ferris FL III, Patz A (1984) Macular edema. A complication of diabetic retinopathy. Surv Ophthalmol 28 Suppl:452–461
76. Frank RN (2004) Diabetic retinopathy. N Engl J Med 350:48–58
77. Friedenwald JS (1950) Diabetic retinopathy. Am J Ophthalmol 33:1187–1199
78. Gardiner TA, Anderson HR, Stitt AW (2003) Inhibition of advanced glycation end-products protects against retinal capillary basement membrane expansion during long-term diabetes. J Pathol 201:328–333
79. Gardiner TA, Archer DB, Curtis TM, Stitt AW (2007) Arteriolar involvement in the microvascular lesions of diabetic retinopathy: implications for pathogenesis. Microcirculation 14:25–38
80. Gardiner TA, Gibson DS, de Gooyer TE, de la Cruz VF, McDonald DM, Stitt AW (2005) Inhibition of tumor necrosis factor-alpha improves physiological angiogenesis and reduces pathological neovascularization in ischemic retinopathy. Am J Pathol 166:637–644
81. Gardiner TA, Stitt AW, Anderson HR, Archer DB (1994) Selective loss of vascular smooth muscle cells in the retinal microcirculation of diabetic dogs. Br J Ophthalmol 78:54–60
82. Garhofer G, Zawinka C, Resch H, Menke M, Schmetterer L, Dorner GT (2003) Effect of intravenous administration of sodium-lactate on retinal blood flow in healthy subjects. Invest Ophthalmol Vis Sci 44:3972–3976
83. Garner A (1970) Pathology of diabetic retinopathy. Br Med Bull 26:137–142
84. Gidday JM, Park TS (1993) Adenosine-mediated autoregulation of retinal arteriolar tone in the piglet. Invest Ophthalmol Vis Sci 34:2713–2719
85. Gillow JT, Gibson JM, Dodson PM (1999) Hypertension and diabetic retinopathy – what's the story? Br J Ophthalmol 83:1083–1087
86. Gilmore ED, Hudson C, Nrusimhadevara RK et al (2007) Retinal arteriolar diameter, blood velocity, and blood flow response to an isocapnic hyperoxic provocation in early sight-threatening diabetic retinopathy. Invest Ophthalmol Vis Sci 48: 1744–1750
87. Girach A, Lund-Andersen H (2007) Diabetic macular oedema: a clinical overview. Int J Clin Pract 61:88–97
88. Gottfredsdottir MS, Stefansson E, Jonasson F, Gislason I (1993) Retinal vasoconstriction after laser treatment for diabetic macular edema. Am J Ophthalmol 115:64–67
89. Grunwald JE, Brucker AJ, Grunwald SE, Riva CE (1993) Retinal hemodynamics in proliferative diabetic retinopathy. A laser Doppler velocimetry study. Invest Ophthalmol Vis Sci 34:66–71
90. Grunwald JE, Brucker AJ, Schwartz SS et al (1990) Diabetic glycemic control and retinal blood flow. Diabetes 39:602–607
91. Grunwald JE, Riva CE, Baine J, Brucker AJ (1992) Total retinal volumetric blood flow rate in diabetic patients with poor glycemic control. Invest Ophthalmol Vis Sci 33:356–363
92. Grunwald JE, Riva CE, Brucker AJ, Sinclair SH, Petrig BL (1984) Altered retinal vascular response to 100% oxygen breathing in diabetes mellitus. Ophthalmology 91:1447–1452
93. Grunwald JE, Riva CE, Brucker AJ, Sinclair SH, Petrig BL (1986) Effect of panretinal photocoagulation on retinal blood flow in proliferative diabetic retinopathy. Ophthalmology 93:590–595
94. Hamilton CW, Chandler D, Klintworth GK, Machemer R (1982) A transmission and scanning

electron microscopic study of surgically excised preretinal membrane proliferations in diabetes mellitus. Am J Ophthalmol 94:473–488
95. Harhaj NS, Felinski EA, Wolpert EB, Sundstrom JM, Gardner TW, Antonetti DA (2006) VEGF activation of protein kinase C stimulates occludin phosphorylation and contributes to endothelial permeability. Invest Ophthalmol Vis Sci 47:5106–5115
96. Haritoglou C, Kook D, Neubauer A et al (2006) Intravitreal bevacizumab (Avastin) therapy for persistent diffuse diabetic macular edema. Retina 26:999–1005
97. Harris A, Arend O, Danis RP, Evans D, Wolf S, Martin BJ (1996) Hyperoxia improves contrast sensitivity in early diabetic retinopathy. Br J Ophthalmol 80:209–213
98. Hayflick L (1965) The limited in vitro lifetime of human diploid cell strains. Exp Cell Res 37:614–636
99. Hein TW, Xu W, Kuo L (2006) Dilation of retinal arterioles in response to lactate: role of nitric oxide, guanylyl cyclase, and ATP-sensitive potassium channels. Invest Ophthalmol Vis Sci 47:693–699
100. Hellstedt T, Immonen I (1996) Disappearance and formation rates of microaneurysms in early diabetic retinopathy. Br J Ophthalmol 80:135–139
101. Higashi S, Clermont AC, Dhir V, Bursell SE (1998) Reversibility of retinal flow abnormalities is disease-duration dependent in diabetic rats. Diabetes 47:653–659
102. Hill DW (1977) The regional distribution of retinal circulation. Ann R Coll Surg Engl 59:470–475
103. Ishii H, Jirousek MR, Koya D et al (1996) Amelioration of vascular dysfunctions in diabetic rats by an oral PKC beta inhibitor. Science 272:728–731
104. Jorge R, Costa RA, Calucci D, Cintra LP, Scott IU (2006) Intravitreal bevacizumab (Avastin) for persistent new vessels in diabetic retinopathy (IBEPE study). Retina 26:1006–1013
105. Joussen AM, Murata T, Tsujikawa A, Kirchhof B, Bursell SE, Adamis AP (2001) Leukocyte-mediated endothelial cell injury and death in the diabetic retina. Am J Pathol 158:147–152
106. Joussen AM, Poulaki V, Mitsiades N et al (2003) Suppression of Fas-FasL-induced endothelial cell apoptosis prevents diabetic blood-retinal barrier breakdown in a model of streptozotocin-induced diabetes. FASEB J 17:76–78
107. Kaaja R, Loukovaara S (2007) Progression of retinopathy in type 1 diabetic women during pregnancy. Curr Diabetes Rev 3:85–93
108. Kern TS (2007) Contributions of inflammatory processes to the development of the early stages of diabetic retinopathy. Exp Diabetes Res 2007:95103
109. Kern TS, Engerman RL (1995) Vascular lesions in diabetes are distributed non-uniformly within the retina. Exp Eye Res 60:545–549
110. King H, Aubert RE, Herman WH (1998) Global burden of diabetes, 1995–2025: prevalence, numerical estimates, and projections. Diabetes Care 21:1414–1431
111. Klein R, Klein BE, Moss SE (1989) The Wisconsin epidemiological study of diabetic retinopathy: a review. Diabetes Metab Rev 5:559–570
112. Klein R, Klein BE, Moss SE, Cruickshanks KJ (1995) The Wisconsin Epidemiologic Study of Diabetic Retinopathy. XV. The long-term incidence of macular edema. Ophthalmology 102:7–16
113. Klein R, Klein BE, Moss SE, Davis MD, DeMets DL (1984) The Wisconsin epidemiologic study of diabetic retinopathy. II. Prevalence and risk of diabetic retinopathy when age at diagnosis is less than 30 years. Arch Ophthalmol 102:520–526
114. Klein R, Klein BE, Moss SE, Davis MD, DeMets DL (1984) The Wisconsin epidemiologic study of diabetic retinopathy. III. Prevalence and risk of diabetic retinopathy when age at diagnosis is 30 or more years. Arch Ophthalmol 102:527–532
115. Klein R, Klein BE, Moss SE et al (2003) Retinal vascular abnormalities in persons with type 1 diabetes: the Wisconsin Epidemiologic Study of Diabetic Retinopathy: XVIII. Ophthalmology 110:2118–2125
116. Klein R, Meuer SM, Moss SE, Klein BE (1995) Retinal microaneurysm counts and 10-year progression of diabetic retinopathy. Arch Ophthalmol 113: 1386–1391
117. Knaus HG, Folander K, Garcia-Calvo M et al (1994) Primary sequence and immunological characterization of beta-subunit of high conductance Ca(2+)-activated K+ channel from smooth muscle. J Biol Chem 269:17274–17278
118. Kohner EM, Hamilton AM, Saunders SJ, Sutcliffe BA, Bulpitt CJ (1975) The retinal blood flow in diabetes. Diabetologia 11:27–33
119. Kohner EM, Henkind P (1970) Correlation of fluorescein angiogram and retinal digest in diabetic retinopathy. Am J Ophthalmol 69:403–414
120. Kohner EM, Patel V, Rassam SM (1995) Role of blood flow and impaired autoregulation in the pathogenesis of diabetic retinopathy. Diabetes 44:603–607
121. Koya D, King GL (1998) Protein kinase C activation and the development of diabetic complications. Diabetes 47:859–866
122. Kristinsson JK, Gottfredsdottir MS, Stefansson E (1997) Retinal vessel dilatation and elongation precedes diabetic macular oedema. Br J Ophthalmol 81:274–278
123. Kuwabara T, Cogan DG (1963) Retinal vascular patterns. VI. Mural cells of the retinal capillaries. Arch Ophthalmol 69:492–502
124. Laties AM (1967) Central retinal artery innervation. Absence of adrenergic innervation to the intraocular branches. Arch Ophthalmol 77:405–409
125. Li M, Zhao MQ, Kumar Durairajan SS et al (2008) Protective effect of tetramethylpyrazine and salvianolic acid B on apoptosis of rat cerebral microvascular endothelial cell under high shear stress. Clin Hemorheol Microcirc 38:177–187
126. Lin WL, Essner E (1990) Immunogold localization of basement membrane molecules in rat retinal capillaries. Cell Mol Biol 36:13–21

127. Lindblom P, Gerhardt H, Liebner S et al (2003) Endothelial PDGF-B retention is required for proper investment of pericytes in the microvessel wall. Genes Dev 17:1835–1840
128. Linsenmeier RA, Braun RD, McRipley MA et al (1998) Retinal hypoxia in long-term diabetic cats. Invest Ophthalmol Vis Sci 39:1647–1657
129. Linskens MH, Harley CB, West MD, Campisi J, Hayflick L (1995) Replicative senescence and cell death. Science 267:17
130. Lip PL, Belgore F, Blann AD, Hope-Ross MW, Gibson JM, Lip GY (2000) Plasma VEGF and soluble VEGF receptor FLT-1 in proliferative retinopathy: relationship to endothelial dysfunction and laser treatment. Invest Ophthalmol Vis Sci 41:2115–2119
131. Ljubimov AV, Burgeson RE, Butkowski RJ et al (1996) Basement membrane abnormalities in human eyes with diabetic retinopathy. J Histochem Cytochem 44:1469–1479
132. Lopes de Faria JM, Jalkh AE, Trempe CL, McMeel JW (1999) Diabetic macular edema: risk factors and concomitants. Acta Ophthalmol Scand 77:170–175
133. Lorenzi M, Feke GT, Cagliero E et al (2008) Retinal haemodynamics in individuals with well-controlled type 1 diabetes. Diabetologia 51:361–364
134. Matsushita K, Puro DG (2006) Topographical heterogeneity of K(IR) currents in pericyte-containing microvessels of the rat retina: effect of diabetes. J Physiol 573:483–495
135. McGahon M, Needham M, Scholfield CN, McGeown JG, Curtis TM (2009) Ca^{2+}-activated Cl^- current in retinal arteriolar smooth muscle. Invest Ophthalmol Vis Sci 50:364–371
136. McGahon MK, Dash DP, Arora A et al (2007) Diabetes downregulates large-conductance Ca^{2+}-activated potassium beta 1 channel subunit in retinal arteriolar smooth muscle. Circ Res 100:703–711
137. McGahon MK, Dawicki JM, Arora A et al (2007) Kv1.5 is a major component underlying the A-type potassium current in retinal arteriolar smooth muscle. Am J Physiol Heart Circ Physiol 292: H1001–H1008
138. McGahon MK, Zhang X, Scholfield CN, Curtis TM, McGeown JG (2007) Selective downregulation of the BKbeta1 subunit in diabetic arteriolar myocytes. Channels (Austin) 1:141–143
139. Milkiewicz M, Uchida C, Gee E, Fudalewski T, Haas TL (2008) Shear stress-induced Ets-1 modulates protease inhibitor expression in microvascular endothelial cells. J Cell Physiol 217:502–510
140. Miyamoto K, Khosrof S, Bursell SE et al (1999) Prevention of leukostasis and vascular leakage in streptozotocin-induced diabetic retinopathy via intercellular adhesion molecule-1 inhibition. Proc Natl Acad Sci USA 96:10836–10841
141. Miyamoto K, Ogura Y (1999) Pathogenetic potential of leukocytes in diabetic retinopathy. Semin Ophthalmol 14:233–239
142. Mizutani M, Kern TS, Lorenzi M (1996) Accelerated death of retinal microvascular cells in human and experimental diabetic retinopathy. J Clin Invest 97:2883–2890
143. Mott JD, Khalifah RG, Nagase H, Shield CF III, Hudson JK, Hudson BG (1997) Nonenzymatic glycation of type IV collagen and matrix metalloproteinase susceptibility. Kidney Int 52:1302–1312
144. Nagaoka T, Sakamoto T, Mori F, Sato E, Yoshida A (2002) The effect of nitric oxide on retinal blood flow during hypoxia in cats. Invest Ophthalmol Vis Sci 43:3037–3044
145. Nagaoka T, Yoshida A (2006) Noninvasive evaluation of wall shear stress on retinal microcirculation in humans. Invest Ophthalmol Vis Sci 47: 1113–1119
146. Nguyen QD, Tatlipinar S, Shah SM et al (2006) Vascular endothelial growth factor is a critical stimulus for diabetic macular edema. Am J Ophthalmol 142:961–969
147. Nicoletti R, Venza I, Ceci G, Visalli M, Teti D, Reibaldi A (2003) Vitreous polyamines spermidine, putrescine, and spermine in human proliferative disorders of the retina. Br J Ophthalmol 87:1038–1042
148. Nishijima K, Ng YS, Zhong L et al (2007) Vascular endothelial growth factor-A is a survival factor for retinal neurons and a critical neuroprotectant during the adaptive response to ischemic injury. Am J Pathol 171:53–67
149. Nishikawa T, Giardino I, Edelstein D, Brownlee M (2000) Changes in diabetic retinal matrix protein mRNA levels in a common transgenic mouse strain. Curr Eye Res 21:581–587
150. Pannarale L, Onori P, Ripani M, Gaudio E (1996) Precapillary patterns and perivascular cells in the retinal microvasculature. A scanning electron microscope study. J Anat 188(Pt 3):693–703
151. Park JY, Takahara N, Gabriele A et al (2000) Induction of endothelin-1 expression by glucose: an effect of protein kinase C activation. Diabetes 49:1239–1248
152. Patel V, Rassam S, Newsom R, Wiek J, Kohner E (1992) Retinal blood flow in diabetic retinopathy. BMJ 305:678–683
153. Patel V, Rassam SM, Chen HC, Kohner EM (1994) Oxygen reactivity in diabetes mellitus: effect of hypertension and hyperglycaemia. Clin Sci (Lond) 86:689–695
154. Pecsvarady Z, Fisher TC, Darwin CH et al (1994) Decreased polymorphonuclear leukocyte deformability in NIDDM. Diabetes Care 17:57–63
155. Peppiatt CM, Howarth C, Mobbs P, Attwell D (2006) Bidirectional control of CNS capillary diameter by pericytes. Nature 443:700–704
156. Pournaras CJ, Rungger-Brandle E, Riva CE, Hardarson SH, Stefansson E (2008) Regulation of retinal blood flow in health and disease. Prog Retin Eye Res 27:284–330
157. Puro DG (2007) Physiology and pathobiology of the pericyte-containing retinal microvasculature: new developments. Microcirculation 14:1–10

158. Rassam SM, Patel V, Chen HC, Kohner EM (1996) Regional retinal blood flow and vascular autoregulation. Eye 10(Pt 3):331–337
159. Rassam SM, Patel V, Kohner EM (1995) The effect of experimental hypertension on retinal vascular autoregulation in humans: a mechanism for the progression of diabetic retinopathy. Exp Physiol 80: 53–68
160. Riva CE, Logean E, Falsini B (2005) Visually evoked hemodynamical response and assessment of neurovascular coupling in the optic nerve and retina. Prog Retin Eye Res 24:183–215
161. Robinson GS, Pierce EA, Rook SL, Foley E, Webb R, Smith LE (1996) Oligodeoxynucleotides inhibit retinal neovascularization in a murine model of proliferative retinopathy. Proc Natl Acad Sci USA 93: 4851–4856
162. Roy S, Maiello M, Lorenzi M (1994) Increased expression of basement membrane collagen in human diabetic retinopathy. J Clin Invest 93:438–442
163. Schmetterer L, Wolzt M (1999) Ocular blood flow and associated functional deviations in diabetic retinopathy. Diabetologia 42:387–405
164. Scholfield CN, Curtis TM (2000) Heterogeneity in cytosolic calcium regulation among different microvascular smooth muscle cells of the rat retina. Microvasc Res 59:233–242
165. Scholfield CN, McGeown JG, Curtis TM (2007) Cellular physiology of retinal and choroidal arteriolar smooth muscle cells. Microcirculation 14:11–24
166. Sharma NK, Gardiner TA, Archer DB (1985) A morphologic and autoradiographic study of cell death and regeneration in the retinal microvasculature of normal and diabetic rats. Am J Ophthalmol 100:51–60
167. Shiba T, Inoguchi T, Sportsman JR, Heath WF, Bursell S, King GL (1993) Correlation of diacylglycerol level and protein kinase C activity in rat retina to retinal circulation. Am J Physiol 265:E783–E793
168. Simo R, Hernandez C (2008) Intravitreous anti-VEGF for diabetic retinopathy: hopes and fears for a new therapeutic strategy. Diabetologia 51:1574–1580
169. Sinclair SH, Grunwald JE, Riva CE, Braunstein SN, Nichols CW, Schwartz SS (1982) Retinal vascular autoregulation in diabetes mellitus. Ophthalmology 89:748–750
170. Skovborg F, Nielsen AV, Lauritzen E, Hartkopp O (1969) Diameters of the retinal vessels in diabetic and normal subjects. Diabetes 18:292–298
171. Skovborg F, Nielsen AV, Schlichtkrull J, Ditzel J (1966) Blood-viscosity in diabetic patients. Lancet 1:129–131
172. Spaide RF, Fisher YL (2006) Intravitreal bevacizumab (Avastin) treatment of proliferative diabetic retinopathy complicated by vitreous hemorrhage. Retina 26:275–278
173. Stitt AW, Anderson HR, Gardiner TA, Archer DB (1994) Diabetic retinopathy: quantitative variation in capillary basement membrane thickening in arterial or venous environments. Br J Ophthalmol 78:133–137
174. Stitt AW, Gardiner TA, Archer DB (1995) Histological and ultrastructural investigation of retinal microaneurysm development in diabetic patients. Br J Ophthalmol 79:362–367
175. Stitt AW, Li YM, Gardiner TA, Bucala R, Archer DB, Vlassara H (1997) Advanced glycation end products (AGEs) co-localize with AGE receptors in the retinal vasculature of diabetic and of AGE-infused rats. Am J Pathol 150:523–531
176. Sundberg C, Kowanetz M, Brown LF, Detmar M, Dvorak HF (2002) Stable expression of angiopoietin-1 and other markers by cultured pericytes: phenotypic similarities to a subpopulation of cells in maturing vessels during later stages of angiogenesis in vivo. Lab Invest 82:387–401
177. Takagi C, Bursell SE, Lin YW et al (1996) Regulation of retinal hemodynamics in diabetic rats by increased expression and action of endothelin-1. Invest Ophthalmol Vis Sci 37:2504–2518
178. Takagi C, King GL, Clermont AC, Cummins DR, Takagi H, Bursell SE (1995) Reversal of abnormal retinal hemodynamics in diabetic rats by acarbose, an alpha-glucosidase inhibitor. Curr Eye Res 14: 741–749
179. Tang J, Mohr S, Du YD, Kern TS (2003) Non-uniform distribution of lesions and biochemical abnormalities within the retina of diabetic humans. Curr Eye Res 27:7–13
180. Thoumine O, Nerem RM, Girard PR (1995) Changes in organization and composition of the extracellular matrix underlying cultured endothelial cells exposed to laminar steady shear stress. Lab Invest 73: 565–576
181. Tolentino MJ, McLeod DS, Taomoto M, Otsuji T, Adamis AP, Lutty GA (2002) Pathologic features of vascular endothelial growth factor-induced retinopathy in the nonhuman primate. Am J Ophthalmol 133:373–385
182. Tong PC, Lee KF, So WY et al (2004) White blood cell count is associated with macro- and microvascular complications in Chinese patients with type 2 diabetes. Diabetes Care 27:216–222
183. Tooke JE (1995) Microvascular function in human diabetes. A physiological perspective. Diabetes 44: 721–726
184. Trick GL, Edwards P, Desai U, Berkowitz BA (2006) Early supernormal retinal oxygenation response in patients with diabetes. Invest Ophthalmol Vis Sci 47:1612–1619
185. Uemura A, Ogawa M, Hirashima M et al (2002) Recombinant angiopoietin-1 restores higher-order architecture of growing blood vessels in mice in the absence of mural cells. J Clin Invest 110: 1619–1628
186. Valone JA Jr, McMeel JW, Franks EP (1981) Unilateral proliferative diabetic retinopathy. I. Initial findings. Arch Ophthalmol 99:1357–1361
187. Wagener HP, Story DTD, Wilder RM (1934) Retinitis in diabetes. N Engl J Med 211:1131–1137

188. Wallow IH, Geldner PS (1980) Endothelial fenestrae in proliferative diabetic retinopathy. Invest Ophthalmol Vis Sci 19:1176–1183
189. Wangsa-Wirawan ND, Linsenmeier RA (2003) Retinal oxygen: fundamental and clinical aspects. Arch Ophthalmol 121:547–557
190. Williams JM Sr, de Juan E, Jr MR (1988) Ultrastructural characteristics of new vessels in proliferative diabetic retinopathy. Am J Ophthalmol 105:491–499
191. Wolf S, Arend O, Toonen H, Bertram B, Jung F, Reim M (1991) Retinal capillary blood flow measurement with a scanning laser ophthalmoscope. Preliminary results. Ophthalmology 98:996–1000
192. Wong TY, Klein R, Sharrett AR et al (2002) Retinal arteriolar narrowing and risk of diabetes mellitus in middle-aged persons. JAMA 287:2528–2533
193. Wu DM, Kawamura H, Sakagami K, Kobayashi M, Puro DG (2003) Cholinergic regulation of pericyte-containing retinal microvessels. Am J Physiol Heart Circ Physiol 284:H2083–H2090
194. Xia P, Inoguchi T, Kern TS, Engerman RL, Oates PJ, King GL (1994) Characterization of the mechanism for the chronic activation of diacylglycerol-protein kinase C pathway in diabetes and hypergalactosemia. Diabetes 43:1122–1129
195. Yamanishi S, Katsumura K, Kobayashi T, Puro DG (2006) Extracellular lactate as a dynamic vasoactive signal in the rat retinal microvasculature. Am J Physiol Heart Circ Physiol 290:H925–H934
196. Yanoff M (1966) Diabetic retinopathy. N Engl J Med 274:1344–1349
197. Yokota T, Ma RC, Park JY et al (2003) Role of protein kinase C on the expression of platelet-derived growth factor and endothelin-1 in the retina of diabetic rats and cultured retinal capillary pericytes. Diabetes 52:838–845
198. Yoshida A, Feke GT, Morales-Stoppello J, Collas GD, Goger DG, McMeel JW (1983) Retinal blood flow alterations during progression of diabetic retinopathy. Arch Ophthalmol 101:225–227
199. Zimmet PZ (1999) Diabetes epidemiology as a tool to trigger diabetes research and care. Diabetologia 42:499–518

Age-Related Macular Degeneration: Hemodynamic Changes

16

Constantin J. Pournaras, Efstratios Mendrinos, and Jean-Antoine C. Pournaras

Core Messages

- The choriocapillaris layer/Bruch's membrane/retinal pigment epithelium complex represents a single functional unit; the dysfunction of CC/RPE complex induces the age-related and age macular degeneration (AMD)-related changes.
- Choroidal vascular density is reduced, and significant vasoconstriction of the CC may occur.
- In eyes with geographic atrophy, the RPE degenerates first while CC loss is secondary to RPE degeneration. In eyes with exudative AMD, degeneration of the CC layer occurs while RPE is still functional. Resulting ischemia induces production of vasoproliferative factors from the RPE.
- Choroidal neovascularization (CNV) would result from a misbalance between angiogenic factors, including VEGF, and antiangiogenic factors, such as PEDF, endostatin, and thrombospondin-1. These latter are significantly reduced in AMD.
- An association exists between systemic vascular diseases and AMD.
- Choroidal blood flow decreases according to age; it is reduced in patients with AMD according to the degree of severity of AMD; the decreases in flow precedes the formation of CNV, strongly suggesting that these changes may have a role in the development of CNV.
- Choroidal blood flow is dysregulated in patients with exudative AMD.
- The choroidal watershed zone (WZ) courses through the fovea more often in patients suffering from AMD than in age-matched controls, particularly in the presence of CNV. Choroidal neovascularization usually arises within these WZ.
- The choroidal microcirculatory abnormalities observed in eyes with AMD support the role of choroidal ischemia in the pathophysiology of AMD.

16.1 Introduction

Although metabolic changes of the retinal pigment epithelium (RPE) appear to be primarily due to aging, the RPE is also affected by vascular changes in the choroid. Vascular changes in the

C.J. Pournaras, FARVO (✉) • E. Mendrinos, M.D.
Vitreo-retinal Unit, Department of Ophthalmology, Geneva University Hospitals, 22 Rue Alcide Jentzer, 1211, GE 14 Geneva, Switzerland
e-mail: constantin.pournaras@hcuge.ch; efstratios.mendrinos@hcuge.ch

J-.A.C. Pournaras, M.D.
Vitreo-retinal Surgery Unit, Department of Ophthalmology, Jules Gonin Eye Hospital, University of Lausanne, 15 Avenue de France, Lausanne 1004, Switzerland
e-mail: jean-antoine.pournaras@fa2.ch

L. Schmetterer, J.W. Kiel (eds.), *Ocular Blood Flow*,
DOI 10.1007/978-3-540-69469-4_16, © Springer-Verlag Berlin Heidelberg 2012

choroid have potentially deleterious effects on the RPE and, in addition to metabolic changes of the RPE due to senescence, may induce the early clinical findings in age-related macular degeneration AMD. A healthy RPE is necessary for the preservation of the choriocapillaris (CC), since RPE loss induces CC degeneration [29, 62, 85].

Inversely, delayed filling of the choriocapillaris may correlate with the diffuse thickening of Bruch's membrane [97], which is often observed in eyes with neovascular age-related macular degeneration (AMD) [108], and may lead to an increased incidence of geographic atrophy [99].

Prolonged choroidal perfusion was associated to a delayed ERG implicit time [108] and scotopic threshold elevation in dark adapted static perimetry [22].

In addition to environmental influences (smoking, sunlight exposure, and nutritional factors), the risk factors for AMD are similar to those for cardiovascular diseases, such as hypertension, suggesting a vascular role in the development of AMD. This hypothesis appears to be supported by laser Doppler flowmetry (LDF) data demonstrating a significant decrease of choroidal blood flow with age [45] as well as with the presence of AMD [46]. In addition, evidence of increased vascular resistance in the choroidal vasculature is provided by retrobulbar hemodynamics assessment, which indicates increased pulsatility and decreased end-diastolic blood flow velocity of the short posterior ciliary arteries [39].

Blood flow changes affect the metabolism of the adjacent external retinal layers and, through the release of neovascular mediators, can lead to the development of subretinal neovascular tissue [36]. Histopathological studies reveal that abnormal new blood vessels grow from the CC and proliferate through breaks in Bruch's membrane under the retinal pigment epithelium (RPE) and further proliferate into the subretinal space between the RPE and the neurosensory retina [116].

Assessment of choroidal blood flow (ChBF) allows the evaluation of the potential hemodynamic changes during progressive stages of AMD macular changes.

16.2 Choroidal Blood Flow

Choroidal blood flow is about ten times higher than the flow in the gray matter of the brain and four times that of the kidney [10], with flow estimates ranging from 500 to 2,000 ml/min/100 g [5, 7, 9, 11, 15–18], even though a corresponding difference in metabolic requirements does not exist.

This high rate of blood flow [4] is attributable to the low resistance of the choroidal vascular system, a consequence of the unusually large caliber of the choriocapillary lumen. Eighty-five of the total blood flow to the eye is distributed to the choroid and only 4% to the retina. The remaining flow perfuses the ciliary body (10%) and the iris (1%) [9, 10].

The more obvious function of high choroidal blood flow is the delivery of oxygen and nutrients and the removal of metabolic waste; the high blood flow would optimize the partial pressure and concentration gradients for efficient metabolic exchanges between the choroid and the retina. Consequently, the oxygen extraction from the choroidal blood is very low: the arteriovenous difference is only about 3% [6, 9]. As a result of the low oxygen consumption from the choroid, high tissue oxygen partial pressure (PO_2) values (superior to 80 mmHg) are recorded at the choroid [3, 78, 104]. High choroidal PO_2 values are essential to assure about 65 mmHg PO_2 decrease between the choriocapillaris and the inner segments of photoreceptor [132]. In pigs, about 60% of oxygen and 75% of glucose are delivered to the retina by the choroidal circulation, despite the low oxygen extraction from the choroidal blood [130].

The oxygen content of the retinal venous blood in humans [63] and pigs [130] is about 38% compared to the arterial blood.

In man and other primates, the choroidal metabolic support of the retina is supplemented by the retinal vessels, present only in the superficial layers, and retinal blood flow is relatively small, that is, 25–50 ml/min/100 g [5, 7, 9, 12, 18].

16.3 Systemic Vascular Factors and AMD

A great number of risk factors for the development of AMD are identical to those of cardiovascular pathologies. Systemic hypertension [66, 72, 135]; history of coronary, carotid, and peripheral vascular disease [136]; serum cholesterol [71]; dietary fat intake [82, 124]; body mass index [119]; and smoking [89, 123] are considered to be relevant risk factors. Among these potential risk factors, epidemiological studies have shown that only smoking has consistently been associated with AMD [123].

As a consequence of these associations, patients of 49–73 years old presenting early clinical manifestations of AMD could be associated to higher risk of cerebrovascular strokes after 10 years follow-up [140]. Others added that AMD could be linked to higher risk of cardiovascular mortality [25].

There have been reports linking statin use to a lower risk of AMD [51, 83, 84, 139].

Several authors suggested that hypertension could increase the potential risk for the development of AMD based on the effects on choroidal circulation [2, 92]. Observational [66, 128] and prospective studies [72, 135] pointed out the association between hypertension and AMD risk. In the "Beaver Dam Eye Study," the incidence of pigment abnormalities and the risk of AMD manifestation at 10 years were increased in the presence of high systolic blood pressure and pulse pressure at the initial examination [72]. A twofold to threefold risk of neovascular AMD in 10 years was associated with controlled or uncontrolled hypertension under medication at initial examination. A threefold risk of incident late-stage AMD over the next 5 years was observed in people whose systolic blood pressure had increased more than 5 mmHg from baseline to the 5-year follow-up examination, compared with people presenting stable systolic blood pressure values during the same period. The Australian "Blue Mountains Study" demonstrated that focal arteriolar narrowing was associated to the incidence of several AMD signs [138], underlying the association with systemic hypertension.

As studies demonstrated an overlap of risk factors for AMD and cardiovascular disease, suggestion arises that a common disease mechanism may be operative in both AMD and cardiovascular disease and atherosclerosis, resulting in the deposition of lipid in the sclera and in Bruch's membrane. The increasingly rigid sclera would act to encapsulate the ocular vasculature in a more incompressible compartment, leading to a greater degree of systolic-diastolic variation in the blood velocity during the cardiac cycle [35–37]. This working hypothesis gives us the correlation between the alterations observed and systemic factors. Further investigations are needed to know whether AMD-related hemodynamic abnormalities are due to increased sclera rigidity, increased systemic vascular rigidity, or both.

16.4 Retinal Vascular Changes and Retinal Blood Flow in AMD

Retinal vascular changes such as focal arteriolar narrowing, arteriovenous (AV) nicking, and generalized retinal arteriolar and venular narrowing were recently associated with cerebrovascular and cardiovascular outcomes such as stroke [90] and coronary heart disease [137] and may be markers of microvascular damage from disease processes including prolonged hypertension and chronic inflammation [141]. Less is known regarding the associations of focal and generalized retinal arteriolar narrowing, AV nicking, and retinopathy with AMD.

A limited number of population-based cohort studies have investigated the association of retinal vessel diameter and retinal vascular structural changes with the incidence of early (indistinct soft or reticular drusen or combined distinct soft drusen and retinal pigmentary abnormalities) and late (geographic atrophy or neovascularization) AMD.

In the Beaver Dam Eye Study, retinal vascular characteristics appeared to be weakly related and inconsistently associated with AMD [73]. Arteriole-to-venule ratio was only associated with the incidence of soft indistinct drusen and the incidence of RPE depigmentation. Focal retinal arteriolar narrowing was not associated with the incidence of AMD while AV nicking was associated with the incidence of early AMD. The Beaver Dam Eye Study also showed that retinal arteriolar diameter was not related to the 10-year incident early or late AMD and narrower arteriolar diameter was associated only with incident RPE depigmentation [73].

Similarly, the Rotterdam Study found no association between incident AMD and retinal vessel (artery or vein) diameter [67]. This latter finding is consistent with data from the Atherosclerosis Risk in Communities Study [70] and the Blue Mountains Eye Study [77, 138].

In the Blue Mountains Eye Study, focal arteriolar narrowing and AV nicking at baseline were weakly associated and of borderline significance with the 5-year incidence of both early and late AMD [138]. In the Atherosclerosis Risk in Communities Study, focal retinal arteriolar narrowing was associated with RPE depigmentation, although it was not associated with other AMD changes [70].

In the 10-year incidence of early and late AMD and following adjustment for age, gender, smoking, and mean arterial blood pressure, the Blue Mountains Eye Study found that the presence of focal arteriolar narrowing was associated with an increased risk of incident neovascular AMD and geographic atrophy, although only the latter reached statistical significance [77]. Focal arteriolar narrowing was not significantly associated with incident early AMD, whereas the presence of moderate/severe AV nicking at baseline remained significantly associated with incident early and late AMD. No significant association was found between baseline arteriolar or venular caliber and the 10-year incidence of late or early AMD, although a weak association was observed between vessel calibers and pigment abnormalities [77].

Taken together, these population-based data suggest that retinal arteriolar changes are inconsistently related to the incidence of AMD. Arteriovenous nicking seems to be more often associated with AMD than any other retinal vascular sign.

In stratified analyses, this association was found to be present in persons without hypertension, indicating that the AV nicking–AMD link may be independent of hypertensive processes [77]. On the contrary, there seems to be no association between the retinal vessel diameter and incident AMD. The significance and reproducibility of these findings are yet to be determined and deserve further research.

In recent years, a few studies have also investigated retinal blood flow in AMD. Its relation to the disease has not been fully understood. Applying the laser Doppler technique to retinal arteries of eyes with various degrees of AMD, Sato et al. found increasing pulsatility but constant blood flow, with increasing severity of AMD [118]. This can be interpreted as an effect of reduced compliance proximal to the eye and not as increased distal vascular resistance, suggesting a more generalized systemic vascular pathology rather than an intraretinal vascular pathology. In another study, changes in retinal capillary blood flow in AMD were assessed with the Heidelberg retinal flowmeter [107]. These results indicated no change in blood flow in eyes with nonexudative AMD compared with healthy controls. However, reduced capillary blood flow in the disciform stage of late AMD and increased perimacular capillary blood flow were found in the exudative form of late AMD [107]. Such hemodynamic data in late AMD need to be interpreted with caution as it is yet not clear whether abnormalities in retinal capillary blood flow are secondary to an autoregulative reaction or whether they are a primary mechanism in the pathogenesis of AMD. Subsequent studies did not confirm the previous results and found reduced retinal capillary blood flow in eyes with nonexudative AMD [100] and no changes in retinal capillary blood flow in eyes with exudative AMD comparing to normal controls [94]. More data about the changes in retinal blood flow in AMD are warranted.

16.5 Choroidal Hemodynamic Changes in AMD

16.5.1 Choroidal Histopathological Vascular Changes in AMD

16.5.1.1 Choriocapillaris and Bruch's Membrane in Aging and AMD

Macular drusen and focal areas of RPE atrophy, in nonexudative AMD eyes, are associated with varying degrees of loss of the CC. Drusen, a form of deposit on or within Bruch's membrane, is believed to be incompletely digested material from the RPE that cannot traverse Bruch's membrane for removal by the CC. Deposit of the debris on/in Bruch's membrane possibly is a result of CC insufficiency or alternatively may induce CC loss. Hard drusen has been present almost always over intercapillary septa and not over CC lumens, suggesting that this material accumulates on/in Bruch's membrane where transport is probably less efficient [19, 80].

Thickening and formation of deposits on Bruch's membrane have been associated with angiographic filling defects indicating choroidal perfusion abnormalities [97, 126].

16.5.1.2 Anatomical Changes in Choriocapillaris/Bruch's Membrane/RPE in Dry AMD

In geographic atrophy (GA), the photoreceptors and RPE degenerate in a horseshoe-shaped pattern surrounding the fovea; the loss of choroidal vasculature appears to be a secondary event [115].

Quantification of CC number and lumen diameters in cross sections indicates a decrease with age and a further decline both in AMD [105], an increase in capillary density and a decrease in large blood vessel diameter [125], and a narrowing of CC lumen and loss of CC cellularity in AMD [74].

Using the alkaline phosphatase (APase) flat-embedding technique, expressing high constitution of APase activity in choriocapillaris and choroidal veins, on human RPE/Bruch's membrane/CC complex, a two-dimensional analysis of the choroidal vasculature with the observation of the vascular pattern in flat perspective and the choroidal structure in cross sections of areas in which the vasculature has been mapped microscopically was achieved [85].

An important 90% dropout of RPE cells is observed atrophic in areas in the posterior pole with the border of the RPE atrophy poorly defined (Fig. 16.1a, b). The density of choroidal blood vessels in this area was reduced 30–50% compared to the same regions in the control subject; however, no area was completely devoid of choriocapillaris. In cross sections through the area with RPE atrophy, extreme vasoconstriction of surviving CC was apparent (Fig. 16.1c).

In nonsubmacular areas without loss of RPE, the CC had APase activity, wide luminal diameters, and structurally appeared normal (Fig. 16.1d). Clinically undetected choroidal neovascularization was observed in the periphery and the macula and was always associated with surviving RPE cells.

In case of GA with areolar RPE atrophy, a severe 53% reduction in vascular density in the area from disk to the submacular region was observed, associated to an almost complete RPE atrophy. The border of the RPE defect was clearly delineated and coincided closely with the area of decreased choroidal vascular density. Surviving CC in the area of RPE atrophy had significantly smaller luminal diameters than CC in control subjects and in normal areas of the GA eyes. The study of GA subjects demonstrated that the RPE cells atrophied first followed by degeneration of the CC [85]. Therefore, in this form of AMD at least, the loss of choroidal vasculature appears to be a secondary event, suggesting that GA is not of vascular etiology. Interestingly, even in areas with complete RPE atrophy, some CC segments remained viable (APase positive) but severely constricted, contradicting the data found in animal models, suggesting that RPE is essential for survival of CC [75].

The association of surviving RPE cells with CNV in GA specimens suggests that RPE cells may furnish a stimulus for new vessel formation or stabilization [127].

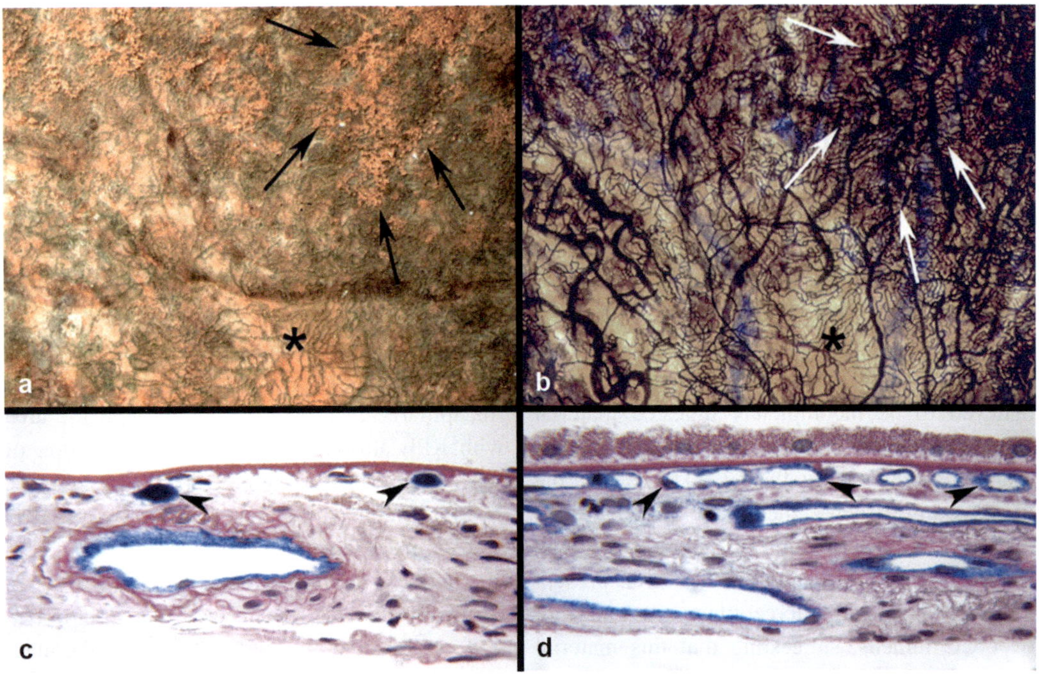

Fig. 16.1 Dry AMD (**a–d**) in APase flat mount choroids (**a**, **b**) and in cross-sections (**c**, **d**). Epi-illuminated (**a**) and transilluminated choroid (**b**) with geographic atrophy showing a region with RPE atrophy and choriocapillaris compromise (*asterisk* in **a** and **b**). A region with surviving RPE and viable choriocapillaris is indicated by the *arrows*. In cross section, the region of RPE atrophy has rarified capillaries that are constricted (*arrowheads* in **c**). In a region with surviving RPE, the choriocapillaris appears normal (*arrowheads* in **d**) (From Lutty and Mcleod [81], with permission of the editor)

16.5.1.3 Anatomical Changes in Choriocapillaris/Bruch's Membrane/RPE in Exudative AMD

In subjects with wet AMD characterized by fluid and hemorrhage beneath the RPE or neurosensory retina due to choroidal neovascularization (CNV), significant attenuation of CC around active CNV and disciform scars were observed. The surviving CC and CNV lumens were always associated with viable RPE cells, and areas without RPE had greatly reduced areas of viable CC as well. In areas completely covered with RPE, there were large areas that lacked APase^{+} CC (Fig. 16.2a, b). The anterior tips of the fan-shaped CNV formation had viable neovascular channels with intense APase activity (Fig. 16.2a) with no choriocapillaris in advance of the CNV (Fig. 16.2c).

It was apparent that RPE was present over areas with extremely attenuated CC, but the RPE was hypertrophic (Fig. 16.2c). In areas with active CNV, the CC was not viable and had extremely hypertrophic RPE covering the CNV (Fig. 16.2d); every case of active CNV was associated with surviving RPE [81].

16.5.2 Choroidal Microcirculation in AMD

16.5.2.1 Angiographic Evaluation of the Choroidal Circulation

In eyes with AMD, fundus angiographic studies have shown evidence of impaired macular choroidal circulation [20, 41, 57, 114, 143]; choroidal perfusion defect was also detected in 42% of the fellow normal eyes in patients with neovascular AMD [108].

The angiographic choroidal filling defects and prolonged filling times in AMD [97, 98] may be related to reduction in capillary density as well as

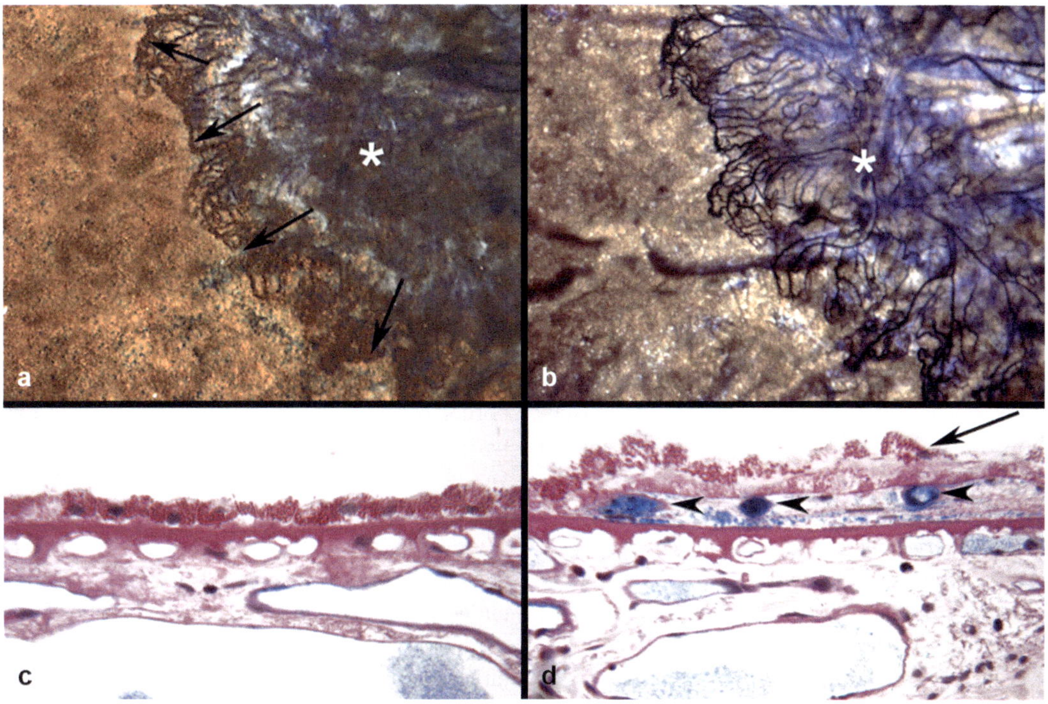

Fig. 16.2 Epi-illuminated (**a**) and transilluminated choroid (**b**) from an eye with a wet form of AMD that is characterized by a choroidal neovascular membrane (*asterisk* in **a** and **b**). The actively growing tips of the neovascularization are intimately associated with viable RPE (*arrows* in **a**). In advance of the active CNV, there are very few viable CC (**b**). Cross section through that choroid in advance of the choroidal neovascular membrane (**c**) demonstrates loss of viable choriocapillaris (no APase activity), but RPE is present. Cross section through the edge of the actively growing neovascularization (*arrowheads* in **d**) shows hypertrophic RPE (*arrow*) (From Lutty and Mcleod [81], with permission of the editor)

narrowing of choriocapillaris lumens [85]. Considering the RPE/Bruch's membrane/CC as a complex, whose function is essential for healthy photoreceptors, any loss in CC function or anatomical deficit could result in debris accumulation and potential RPE hypoxia, dysfunction, and death.

Angiographic reduction in perfusion or filling defects have been related to deposit formation on Bruch's membrane [97, 98, 126] and higher incidence of geographic atrophy [64, 99].

A dye dilution analysis of ICG angiography demonstrated a statistically significant delayed and heterogeneous filling in the choriocapillaris of eyes with nonexudative AMD compared to normal age-matched control eyes [24].

The choroidal arteries and veins, evaluated by ICG video images angiography, indicate that the choroidal arterioles are dilated, fewer, run a straighter course, and possess fewer bifurcations. Their number and the macular fluorescent intensity in the arterial phase of choroidal filling were significantly less in patients with AMD as compared to age-matched normal controls. The mean and maximum caliber of choroidal veins in the macula was increased in AMD eyes. Those findings suggest an implication of poor choroidal perfusion of the macula in the pathogenesis of AMD [93].

16.5.2.2 Choroidal Watershed Zones and Neovascularization

In vivo studies on the choroidal vascular bed have shown that the posterior ciliary arteries and all their branches right down to their terminal arterioles and the choriocapillaris have segmental distribution with no functional anastomoses between them and thus behave as end arteries. The borderline area between the territories of distribution of any two end arteries is defined as watershed zone [55, 59]. During a decrease in the perfusion pressure in the vascular bed of one or more end arteries, the watershed zone, being an area of comparatively poor blood flow supply, is most vulnerable to hypoxia-ischemia [60].

In the brain, development of watershed zone infarcts along the borders of areas of supply by the cerebral arteries is well known [131]. The presence and location of the watershed zone have been implicated in the pathogenesis and the type of visual field defects observed in anterior ischemic optic neuropathy and normal-tension glaucoma.

Following the first observations concerning a relation between CNV and watershed zones [52, 53], an increased incidence has been reported to occur in patients with AMD, especially those with the exudative form, compared to age-matched control subjects [114].

Macular choroidal watershed zones were defined as characteristic vertical, angled, or stellate-shaped zones of early and transient choroidal hypofluorescence, produced by the delayed filling of a portion of the choriocapillaris layer; the subsequent perfusion of the watershed zone from the adjacent choriocapillaris causes its disappearance and the production of an even background fluorescence [41, 58, 59, 114].

In normal subjects, the topographic variation of the watershed zone of the choroid has been shown by fluorescein angiography studies to have typically a vertical pattern [42].

In vivo experimental studies in rhesus monkeys and in man, using fluorescein angiography as the main research technique, modified radically our conception of the choroidal vascular bed. Contrary to the results provided by postmortem cast studies suggesting that the choriocapillaris forms a freely communicating and an uninterrupted vascular bed in the entire choroid, in vivo studies have shown the choroidal vascular bed to be a segmental and an end-arterial system [55, 59, 60], with each choriocapillaris lobule being a functional independent unit with no anastomoses with the adjacent lobules in the living eye [54]. Thus, in vivo studies with fluorescein angiography reveal the actual physiologic circulatory pattern. Similarly, Flower used a technique by which images of only choriocapillaris blood flow are extracted from high-speed ICG angiography in rhesus monkeys and reported that the posterior pole choriocapillaris does not behave as a homogeneous structure. Observed areas of lobular ICG-dye filling were viewed as evidence that choriocapillaris blood flow can be at least functionally segmented [33].

The choroid is therefore an end-arterial tissue and, as such, has watershed zones located at the border between the areas of distribution of any two or more end arteries. The apical parts of the various segments supplied by the short PCAs meet each other in the center of the macula. Multiple watershed zones meet in the macula, and consequently are most vulnerable to ischemic disorders. Thus, the submacular choroid is predisposed to chronic ischemia and neovascularization more than any other part of the posterior choroid.

There is also experimental and clinical evidence that the macular choroid is more vulnerable to ischemia. In experimental animal studies on malignant arterial hypertension, choroidal ischemia was the main feature of the hypertensive choroidopathy, and the choroidal ischemic lesions were most prominent in the macular area, with fundus fluorescein angiography revealing a marked selective, delayed filling of the macular choroid, particularly in its central part [61].

When the perfusion pressure was lowered in the ocular vessels by lowering the systemic blood pressure, fluorescein angiography also revealed delayed filling of the macular watershed zones and of the central choriocapillaris layer [56, 57].

In patients with exudative AMD and choroidal watershed zone(s), three patterns were identified (Table 16.1). The stellate pattern of the watershed zone was the most common one (60%), followed by the vertical (36%), and the angled one (4%). Choroidal neovascularization arose within the watershed zone in 88% of cases [86]. Schematic presentations of the various patterns are illustrated in Fig. 16.3.

The *stellate-shaped watershed zone* seen in 30 (60%) consisted of a vertical part associated with

Table 16.1 Topographic variation of the choroidal watershed zone (WZ) and its relation to choroidal neovascularization (CNV) in exudative age-related macular degeneration

Pattern of the WZ	Number and (%) of patients	Relation between WZ and CNV
Stellate	30 (60%)	CNV arose within the center of the WZ
Vertical/three subgroups	18 (36%)	
Coursing through the optic nerve head with a part extending into the fovea	9/18 (50%)	CNV arose from the part of WZ extending into the fovea
Coursing through the optic nerve head without extending into the fovea	6/18 (33%)	CNV arose within the temporal margin of the WZ
Coursing through the fovea without extending into the optic nerve head	3/18 (16.7%)	CNV arose within the WZ
Angular	2 (4%)	CNV arose within the WZ

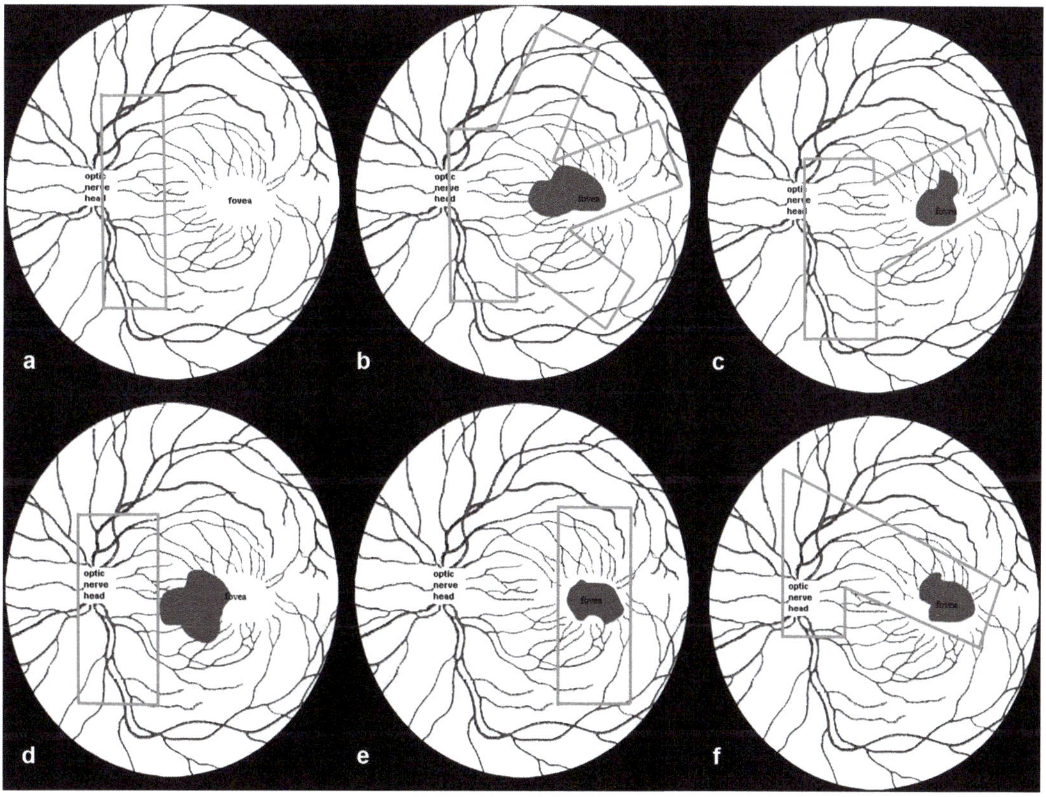

Fig. 16.3 Schematic illustration of the different watershed patterns and their relation to neovascularization in exudative age-related macular degeneration. *Light gray lines* delineate the pattern of the posterior choroidal watershed zone and *dark gray* areas represent neovascularization. (**a**) Vertical pattern coursing through the optic nerve head in a normal subject. (**b**) Stellate pattern with neovascularization arising within the center of the watershed zones. (**c**) Vertical pattern coursing through the optic nerve head and extending into the fovea. Neovascularization arises from the part of the watershed zone extending to the fovea. (**d**) Vertical pattern coursing through the optic nerve head without extension to the fovea. Neovascularization arises from the margin of the watershed zone and extends away from it to the fovea. (**e**) Vertical pattern coursing through the fovea with no extension to the optic nerve head. Neovascularization arises within the watershed zone. (**f**) Angled pattern of the watershed zone coursing through and adjacent to the optic nerve head. Neovascularization arises within the watershed zone

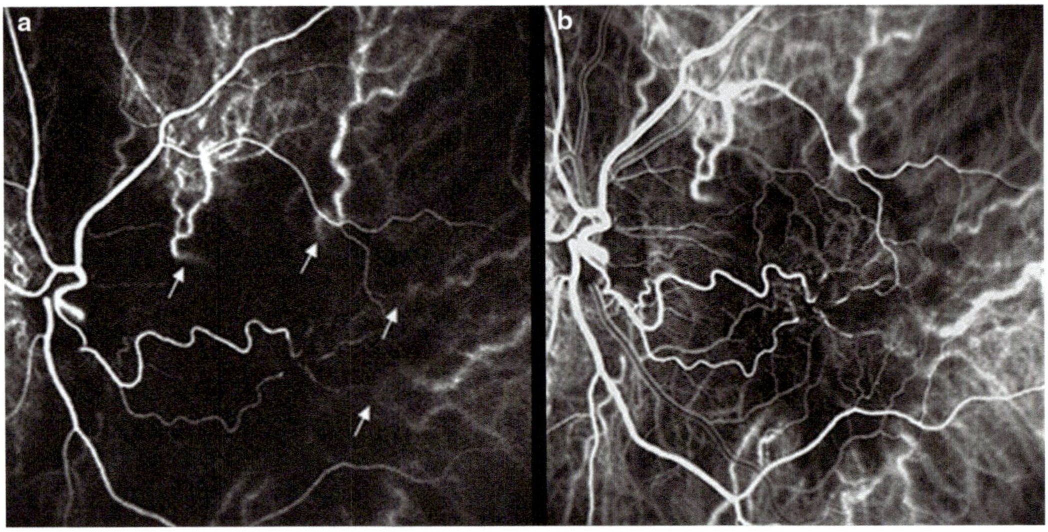

Fig. 16.4 (**a**) Stellate pattern of the watershed zone as shown in indocyanine green angiography. Note the vertical part that corresponds to the watershed zone between the medial and lateral posterior ciliary arteries and multiple small triangular watershed zones between the distal and temporal short posterior ciliary arteries (indicated by *white arrows*). (**b**) Choroidal neovascularization arose within the center of the watershed zones

multiple smaller triangular zones. The vertical part of the stellate pattern was coursing through and/or adjacent to the optic nerve head and corresponded to the watershed zone between the medial and lateral posterior ciliary arteries. The smaller triangular zones corresponded to the watershed zones that exist between the adjacent distal and temporal short posterior ciliary arteries, branches of the lateral posterior ciliary artery [59]. These triangular watershed zones were seen to meet each other in the center of the macula. The choroidal neovascularization arose within the center of these watershed zones in all these cases (Fig. 16.4).

A *vertical orientation of the watershed zone* was observed in 18 (36%) of the 50 patients, and three subgroups could be identified. Nine (50%) of these 18 patients had a vertical zone coursing through the optic nerve head with a part extending into the fovea. In all these cases, the choroidal neovascularization arose from the part extending into the fovea. Six (33.3%) of the 18 patients had a vertical zone coursing through the optic nerve head without extension into the fovea. In all these cases, the choroidal neovascularization arose from the temporal margin of the watershed zone and was seen extending away from it to the fovea. Three (16.7%) of the 18 patients had a vertical zone coursing through the fovea without involvement of the optic nerve head. In all these cases, the choroidal neovascularization arose within the watershed zone.

An angled-shaped watershed zone coursing through the optic nerve head and the fovea was observed in 2 (4%) of 50 patients. In both cases, the choroidal neovascularization arose within the watershed zone.

The choroidal neovascularization arose within the watershed zone in 44 (88%) of the 50 patients, that is, whenever it was coursing through or extending into the fovea. It arose from its margin in 6 (12%) of the 50 patients, corresponding to the subgroup of patients who presented a vertical zone coursing through the optic nerve head without any extension to the fovea. Even when the watershed zone was not concerning the fovea, some portion of it was always noted to involve at least a part of the macular region, that is, either the temporal peripapillary choroid or the maculopapillary bundle.

All types of CNV were observed, and their relatively few numbers did not allow to make any correlation between the different subtypes of neovascular ARMD and a particular pattern of watershed zone.

Giovannini et al. studied the choroidal filling pattern in 145 patients with AMD with or without CNV and looked for a correlation with the CNV. They found a correspondence in 71% of cases between the watershed zones and the site of the CNV. In 59% of cases with AMD, there was delayed choroidal filling compared to 23% of cases in the control population; 52% of CNVs were seen to grow inside or at the margins of zones of delayed choroidal filling [41].

Ross et al. found a higher correspondence between CNV and watershed zones. Thirty-three (91.7%) of their 36 patients with exudative AMD and watershed zones showed the CNV extending directly out and away from the watershed zone or its border [114]. In the study by Ross et al., the watershed zone was seen to course through or adjacent to the fovea in 33 (80.5%) of their 41 patients (with or without CNV). In our study, this was the case in 44 (88%) of the 50 patients. Our higher percentages may be due to the fact that we only considered patients with exudative AMD contrary to the previous studies [41, 114].

There is an interindividual variation in the pattern and the location of the choroidal watershed zone; however, it seems that in normal subjects, it has typically a vertical orientation and involves in most cases the optic nerve head with or without the near choroid. It is less often located between the optic nerve head and the fovea and has not been reported to extend to the fovea in normal subjects. Giuffre evaluated its topographic variation in 357 normal subjects and found that a vertical pattern was always present and was coursing through the optic nerve head with or without the near choroid in about half of these cases (46.8%) and involved the temporal half of the optic nerve head and the near choroid in the other half (45.4%). Very rarely, the watershed zone involved the nasal half of the optic nerve head (3.1%) or the area between the optic nerve head and the fovea (4.8%) [42]. The three control patients demonstrating a watershed zone included in the study by Ross et al. exhibited purely vertical zones, which were coursing directly through (two patients) or adjacent to the optic nerve head (one patient) [114].

In the light of experimental and clinical data, it is logical to consider that the submacular choroid is particularly susceptible to ischemic disorders by virtue of the numerous watershed zones meeting there. Since the center of the watershed zone is the one that receives blood at lower pressure, it follows that a decrease in the perfusion pressure of the choroid causes a decrease in blood flow that is most severe in the center of the watershed zone. In the event of such a reduction in the choroidal perfusion pressure in association with the altered vascular regulation seen in patients with AMD [103], and/or with the AMD-related rarefaction and dysfunction of the choriocapillaris [98], the resulting hypoxia-ischemia could disturb the physiologic balance between angiogenic and antiangiogenic factors predisposing to choroidal neovascularization [76, 142] arising within or from the margin of the macular watershed zone(s).

16.5.2.3 Laser Doppler Flowmetry Evaluation

Laser Doppler flowmetry (LDF) data demonstrated a significant decrease of choroidal blood flow (ChBF) with age [45] as well as with the presence of AMD [46].

In order to explore the choroidal circulatory changes in patients with AMD, Grunwald et al., using laser Doppler flowmetry, showed that there is a systematic decrease in choroidal circulatory parameters with an increase in the severity of AMD features associated with risk for the development of CNV, suggesting a role for ischemia in the development of CNV [47].

In a longitudinal study in order to investigate whether foveolar choroidal blood flow changes are associated with the development of choroidal neovascularization (CNV) in AMD, relative foveolar choroidal blood velocity (ChBVel), volume (ChBVol), and flow (ChBFlow) were assessed in 135 patients with AMD, at baseline (visual acuity of 20/40 or better and no CNV at the time of enrolment) and then annually with laser Doppler flowmetry.

Comparison of foveolar choroidal circulatory measurements at baseline and their change before the development of CNV was made between eyes that had CNV and those that did not.

CNV developed in 28 eyes during the study. Baseline average foveolar ChBVol and ChBFlow in these eyes were 24% and 20% lower than that observed in the 165 eyes in which CNV did not develop. In the eyes with CNV, foveolar ChBVol and ChBFlow decreased by 9.6% and 11.5% before the formation of CNV, whereas in the eyes that did not, they increased by 6.7% and 2.8%, respectively. Eyes with lower baseline foveolar ChBFlow were more likely to show visual loss of three or more lines than were eyes with a higher baseline ChBFlow. The development of CNV and visual loss are associated with lower choroidal circulatory parameters at baseline. This study shows that the decreases in flow preceded the formation of CNV, strongly suggesting that these changes may have a role in the development of CNV [87].

16.5.2.4 Modifications of the Subfoveal Choroidal Blood Flow Regulation

Laser Doppler flowmetry (LDF) data demonstrated a significant decrease of choroidal blood flow (ChBF) with age and the severity of AMD; however, whether this LDF data indeed represent a decrease of choroidal blood flow or are due to changes in the laser light scattering properties of the sampled tissue is still a matter of debate [110, 125]. This scattering probably changes with age and with the presence of typical anatomical changes associated to AMD.

An alternate approach to compare choroidal hemodynamics between normal and AMD eyes is to assess the capability of subfoveal choroidal blood flow to respond to a physiological stimulus, such as an increase of the ocular perfusion pressure (PPm). In healthy volunteers, subfoveal choroidal flow is maintained largely constant in spite of increases in the PPm up to 67% above the baseline value, when PPm is raised by means of isometric exercises [111]. This approach provides data uninfluenced by potential changes of the tissue light scattering assuming, legitimately, that isometric exercise itself does not modify this scattering.

The evaluation of the regulation of choroidal blood flow in the subfoveal region in response to an acute increase of the systemic blood pressure, in eyes with either senile changes of the choroidal vasculature or subfoveal neovascularization, indicates that acute elevation of the ocular perfusion pressure by 20–23% induced by isometric exercise does not induce a significant change of subfoveal choroidal blood flow in elderly subjects with early macular changes; in contrast, in eyes with neovascular AMD, the ChBF parameter significantly increased at the end of squatting [103] (Fig. 16.5).

Thus, regulation of subfoveal choroidal blood flow during moderate increase of PPm induced by isometric exercise seems to be preserved also in aged individuals. In contrast to healthy young volunteers, even with an average increase by 23% in PPm, the LDF ChBF in eyes of patients presenting subfoveal neovascularization increased significantly during squatting, as revealed by the significant correlation between PPm and ChBF changes and the statistically significant increase of ChBF in these eyes (Fig. 16.6).

The increase of ChBF in the subretinal neovascular tissue during isometric exercise results from an insufficient increase in vascular resistance either within this immature neovascular bed and/or within the underlying choroidal vasculature. Therefore, an overperfusion of the neovascular tissue should occur during increases of the perfusion pressure.

The vascular resistance depends mainly upon the contractile state of the smooth muscle of the arterioles irrigating the neovascular tissue and, presumably, of the pericytes of the capillary network [48]. Changes of the vascular resistance during isometric exercise can occur along either the ciliary arteries or the choroidal terminal arterioles or the capillaries within the neovascular tissue since, during the isometric exercise, blood pressure in the ophthalmic artery rises in parallel with that in the brachial artery [112].

The choroidal vascular resistance adaptation is achieved through a mechanism involving either the sympathetic nervous system [5, 8, 32] or the release of vasoconstricting substances by the vascular endothelium [49, 102]. The sympathetic nervous system affects the contraction of the

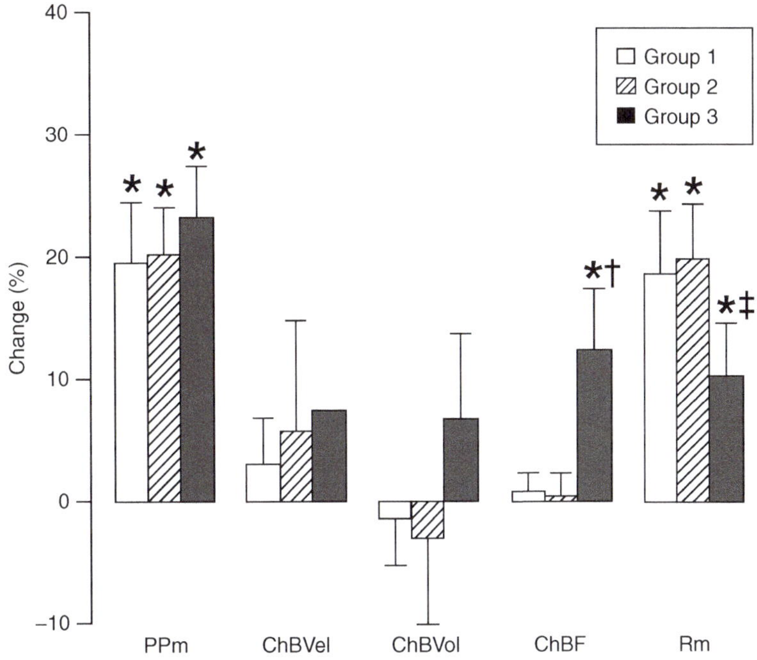

Fig. 16.5 Mean percentage changes of LDF parameters (ChBVel, ChBVol, ChBF), PPm, and resistance (Rm) at the end of squatting, for young, aged normal individuals and AMD groups (respectively group 1, 2, and 3) (*paired *t* test; $P \leq 0.05$). Squatting induced a significant increase in mean PPm that was similar in all groups. At the end of squatting, the LDF parameters in groups 1 and 2 were not significantly different from those at baseline. In contrast, in group 3, the ChBF parameter significantly increased. Changes in ChBF and Rm were significantly different between groups 2 and 3 (†ChBF, $P < 0.0001$ and $P < 0.002$) (In Pournaras et al. [103], reprint with the authorization of the editor)

smooth muscle of choroidal arterioles, and the vascular endothelium affects either the smooth muscle of the arterioles, or the contractile state of pericytes surrounding the capillaries [50].

Loss of neurons and adrenergic nerve terminations in various human systems, including the choroidal circulation, is observed with aging [95]. Furthermore, aging induces morphological changes of the choroidal arterioles vascular wall leading to its stiffening [35, 36, 39] or alterations of the vascular endothelium of these vessels. All these morphological changes could have a detrimental effect on the regulatory capability of the choroidal arterioles.

Abnormal regulation was observed in the subfoveal area including the neovascular tissue for increases of the PPm similar to those reached in young or aged normals. As the LDF signal in the subfoveal and parafoveal areas is dominated by the choroidal circulation [101, 109] with little contribution from the retinal capillary bed, choroidal blood flow dysregulation is confirmed in the patients with neovascular AMD.

A choroidal blood flow dysregulation in neovascular AMD patients could suggest a failure of adaptation of the choroidal vascular resistance, in contrast to what occurs in the aged normal subjects. However, subfoveal neovascularization is expected to scatter the laser light due to the fact that it usually expresses an irregular multilayer interconnecting pattern of capillaries associated with larger irrigating arterioles, draining venules, and fibroblastic tissue [44]. This scatter may be the dominant component of the LDF signal, which would preclude conclusions about choroidal dysregulation.

Within the subfoveal abnormal neovascular bed, alteration of the autonomic system innervating the arterioles irrigating the choroidal neovascular network has not been demonstrated, in contrast to the existence of immature features of the capillaries. The new vessels in the choroid, as

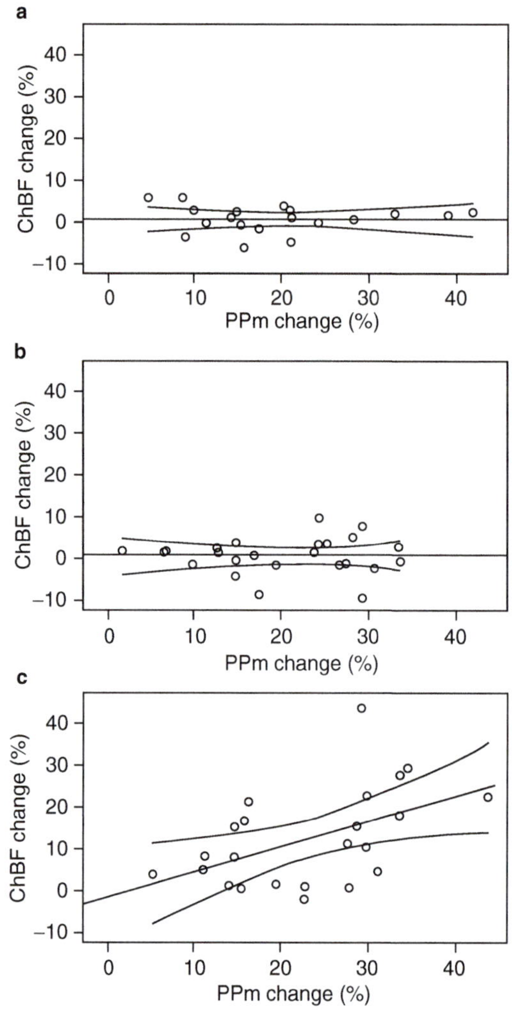

Fig. 16.6 Percentage changes in ChBF as a function of PPm for the three groups (linear fits and 95% confidence intervals). Similar PPm increases for all groups were obtained. For the groups of (**a**) young and (**b**) elderly healthy persons, no correlation was found between these two parameters. (**c**) For the AMD group, linear correlation provided significant values (In Pournaras et al. [103], reprint with authorization of the editor)

those in the retina [88], express immature fenestrated endothelial cells, inadequate number of pericytes, diminished pericyte coverage of the endothelial cells [116], and inadequate sites of contacts between endothelial and pericytes cells membranes [68]. Pericytes express contracting smooth muscle a-actin. Based on the behavior of pericytes in culture, under the effect of contracting or relaxing factors, pericytes can contract or relax to modify capillary diameter [27, 48]. In addition, electronic light microscopy of freshly isolated rat retinas demonstrates capillary relaxation in response to carbonic anhydrase blockers [106, 121]. These findings suggest that pericytes act as capillary diameter regulators within the retinal microvasculature. Assuming that pericytes of the subfoveal new vessels have a similar behavior within the neovascular network, altered contraction capability of capillary pericytes could affect the vascular resistance response of newly formed capillaries of group III eyes to increases of the PPm, resulting in an overperfusion in the neovascular tissue during isometric exercise. Altered vascular resistance within the neovascular tissue in eyes with neovascular AMD has already been revealed under normal conditions of perfusion pressure when fundus pulsation amplitudes measured from healthy retinal areas adjacent to CNV are compared with those obtained from areas associated with CNV [120].

In contrast to normal and aged subjects, a similar elevation of PPm induces blood flow increases within the neovascular tissue in eyes with AMD-related subfoveal neovascularization. This effect is due, most probably, to the inability of the new vessels to increase their flow resistance during isometric exercise. This increase is expected to induce an overperfusion of the neovascular tissue, which may lead to an increase of exudation and bleeding, a potentially threatening condition for the macular function.

16.6 Retrobulbar Hemodynamic Modifications

Several authors have evaluated the retrobulbar circulation in patients with AMD by the means of color Doppler imaging (CDI). Friedman et al. found reduced flow velocities and increased pulsatilities indices in the central retinal artery and short posterior ciliary arteries in a mixed group of patients with exudative and nonexudative AMD, as compared with healthy, age-matched control subjects, suggesting increased resistance of the choroidal vasculature [39]. Related to these results, they proposed a hemodynamic model of the pathogenesis of AMD and suggested that this disorder was caused by a progressive decrease in the compliance of the sclera and the choroidal vessels, leading to an increase in the resistance of the choroid to flow of blood [35–37].

The CDI findings on retrobulbar vessels were confirmed and extended upon by other authors in later studies. Peak systolic (PSV) and end-diastolic velocities (EDV) in the retrobulbar vessels of AMD patients were significantly lower with the resistivity index being significantly higher in these patients when compared to controls [65]. Increased pulsatility and resistivity indices in the posterior ciliary arteries have also been reported in patients with late AMD (CNV, pigment epithelium detachment, disciform scar) [30].

Choroidal perfusion abnormalities seem to be more important in eyes with exudative AMD than in their fellow eyes with nonexudative AMD. In a prospective study of 26 patients with asymmetric AMD, pulsatility and resistivity indices of short posterior ciliary arteries were significantly higher in eyes with CNV compared with their fellow eyes with nonexudative AMD; systolic and diastolic velocities were also lower in eyes with CNV for all vessels, except for the PSV of the nasal posterior ciliary artery [134].

In normal fellow eyes of patients with unilateral exudative AMD, EDV was significantly decreased in the posterior ciliary artery, while both eyes tended to have increased pulsatility indices in the same artery as compared to controls [30]. Furthermore, there was no significant difference in any of the circulatory parameters between normal fellow and exudative AMD eyes. The increased pulsatility in the posterior ciliary artery of both fellow and affected eyes and the lack of significant differences between the two eyes of patients with unilateral AMD suggest that the normal fellow eye has deficient choroidal blood perfusion.

Retrobulbar blood flow abnormalities have been found also in eyes with nonexudative (including geographic atrophy) AMD [23]. Subjects with nonexudative AMD showed reduced blood flow in the nasal and temporal posterior ciliary arteries which supply the choroid. Specifically, subjects with nonexudative AMD showed a consistent trend of lower PSV and EDV in the short posterior ciliary arteries in these patients compared with age-matched control subjects [23]. There was a significant decrease in the EDV of the nasal posterior ciliary arteries in the AMD group, while the resistivity index was not significantly altered in the short posterior ciliary arteries. The EDV was lower, whereas the resistive index was higher in the central retinal artery in the AMD, findings that were highly statistically significant. The combination of reduced flow velocities at a constant resistivity index, seen in nonexudative AMD, is consistent with reduced flow in these ciliary vessels, suggesting that choroidal perfusion is abnormal in this form of AMD. The changes in the central retinal artery, consistent with the findings by Friedman et al. [39], also suggest that there may be a more generalized perfusion abnormality beyond the choroid in patients with AMD or that the central retinal artery exhibits a secondary autoregulatory response to a primary change elsewhere, perhaps decreased retinal metabolic requirements in AMD.

16.7 Modifications of Vascular Blood Flow and Alterations Cascade

In the healthy or in AMD human eye, the choroidal blood flow evaluation and regulation experimentations allow a better understanding of the correlation between the pathological conditions and the choroidal circulation changes.

The results obtained by laser Doppler flowmeter indicate significant reduction of the choroidal blood flow correlated with age and with the presence of AMD disease, compared to control subjects in the same range of age.

Although RPE metabolic modifications due to senescence processes are considered as the principal cause of the early clinical observations of AMD, choroidal vascular modifications seem to have secondary effects affecting the RPE/photoreceptors interactions.

Resistance increase at the level of choroidal circulation consecutive to generalized vascular sclerosis [35, 39] and/or scleral rigidity increase [38], associated to blood flow decrease, modify the osmotic gradient through RPE, leading to metabolic waste accumulation and drusen formation.

Choroidal circulation alterations may affect the metabolic exchanges through RPE and Bruch's membrane, important to maintain visual function. Electrical activity of RPE cells being very sensitive to systemic hypoxia [79], choroidal circulation alterations can affect oxygen and metabolic substrates diffusion through RPE and Bruch's membrane, leading to functional alterations essential for vision due to dysfunction of the metabolic interactions between RPE cells and photoreceptors.

Blood flow alteration at the level of retrofoveal choroidal circulation contributes to the process of neovascularization characteristic of AMD. The fact that patients with high risk of neovascularization present the most severe decrease of choroidal blood flow underlines the role of choroidal ischemia in the development of neovessels in the evolution of AMD [47]. Pathophysiological mechanisms for the development of subretinal neovascular tissue implicate release of neovascular mediators, potentially related to metabolic changes at the level of the external layers of the retina secondary to the decrease of choroidal blood flow.

Anatomical alterations linked to AMD (thickening of Bruch's membrane, drusen formation) increase diffusion distance between choriocapillaris and photoreceptors internal segments. These anatomical alterations associated to the reduction of choroidal blood flow should enhance tissue hypoxia at the area of the photoreceptors inner segments, essential stimulus of vasoproliferative factors which are upregulated by hypoxia [26, 34].

Neovascularization in the eye is also considered to result from an imbalance between stimulatory and inhibitory angiogenic factors [26, 34, 40]. Vascular endothelial growth factor (VEGF) is a likely candidate for the angiogenic stimulus for CNV as VEGF is produced also by RPE [1], and so, it has provided a therapeutic target for CNV [43, 113].

The balance between VEGF and pigment epithelial growth factor (PEDF) in aging and AMD had been suggested that could shift the angiogenic potential in retina [40, 69]. PEDF was purified from the conditioned media of human retinal pigment epithelial cells and was found to be a potent inhibitor of angiogenesis [28] and a neurotrophic factor [129]. PEDF administered intravitreally inhibits aberrant blood vessel growth in VEGF-induced neovascularization [31].

In eyes with AMD, PEDF levels were found significantly reduced in RPE cells, Bruch's membrane, and choroidal stroma in contrast to VEGF immunoreactivity which was not significantly reduced in the RPE/Bruch's membrane/CC complex [14].

Several of the antiangiogenic agents as endostatin, which is the proteolytically cleaved carboxyl terminus globular domain of collagen XVIII (coll XVIII) [117], or thrombospondin-1 (TSP-1), produced by RPE in culture [21, 91], detected in vitreous and aqueous humor [122], are potentially involved in AMD.

Endostatin inhibits endothelial cell proliferation and migration in vitro and potently inhibits angiogenesis in vivo [96], inhibiting the mitogen-activated protein kinase (MAPK) activation in endothelial cells [117]. Endostatin is significantly reduced in CC, Bruch's membrane, and RPE basal lamina in AMD compared with aged control choroids [13]. TSP-1 in Bruch's membrane declines with age, and it is almost absent in AMD, being significantly reduced in Bruch's membrane, CC, and walls of large choroidal blood vessels [133]. The decline in the endogenous antiangiogenic substances potentially favors the development of CNV.

The contribution of the vascular disturbances in the pathogenesis of AMD is actually established. However their precise role is not determined yet; associated to multiple factors affecting the metabolism of the RPE and/or genetic factors, they lead to the appearance of the various phenotypes of AMD.

Support

This work was supported by FNSRS no 3200-061685.00 (CJP)

References

1. Adamis AP, Shima DT, Yeo KT et al (1993) Synthesis and secretion of vascular permeability factor/vascular endothelial growth factor by human retinal pigment epithelial cells. Biochem Biophys Res Commun 193:631–638
2. Age-Related Eye Disease Study Research Group (2000) Risk factors associated with age-related macular degeneration. A case–control study in the age-related eye disease study. Age-Related Eye Disease Study Report Number 3. Ophthalmology 107:2224–2232
3. Alder VA, Cringle SJ, Constable IJ (1983) The retinal oxygen profile in cats. Invest Ophthalmol Vis Sci 24:30–36
4. Alm A (1972) Effects of norepinephrine, angiotensin, dihydroergotamine, papaverine, isoproterenol, histamine, nicotinic acid, and xanthinol nicotinate on retinal oxygen tension in cats. Acta Ophthalmol 50: 707–719
5. Alm A (1977) The effect of sympathetic stimulation on blood flow through the uvea, retina and optic nerve in monkeys (Macaca irus). Exp Eye Res 25:19–24
6. Alm A, Bill A (1970) Blood flow and oxygen extraction in the cat uvea at normal and high intraocular pressures. Acta Physiol Scand 80:19–28
7. Alm A, Bill A (1972) The oxygen supply to the retina. II. Effects of high intraocular pressure and of increased arterial carbon dioxide tension on uveal and retinal blood flow in cats. A study with radioactively labelled microspheres including flow determinations in brain and some other tissues. Acta Physiol Scand 84:306–319
8. Alm A, Bill A (1973) The effect of stimulation of the cervical sympathetic chain on retinal oxygen tension and on uveal, retinal and cerebral blood flow in cats. Acta Physiol Scand 88:84–94
9. Alm A, Bill A (1973) Ocular and optic nerve blood flow at normal and increased intraocular pressures in monkeys (Macaca irus): a study with radioactively labelled microspheres including flow determinations in brain and some other tissues. Exp Eye Res 15: 15–29
10. Alm A, Bill A (1987) Ocular circulation. In: Moses RA, Hart WM (eds) Adler's physiology of the eye. Mosby, St. Louis, pp 183–203
11. Alm A, Stjernschantz J, Bill A (1976) Effects of oculomotor nerve stimulation on ocular blood flow in rabbits after sympathetic denervation. Exp Eye Res 23:609–613
12. Alm A, Tornquist P, Stjernschantz J (1977) Radioactively labelled microspheres in regional ocular blood flow determinations. Bibl Anat 16:24–29
13. Bhutto IA, Kim SY, McLeod DS et al (2004) Localization of collagen XVIII and the endostatin portion of collagen XVIII in aged human control eyes and eyes with age-related macular degeneration. Invest Ophthalmol Vis Sci 45:1544–1552
14. Bhutto IA, McLeod DS, Hasegawa T et al (2006) Pigment epithelium-derived factor (PEDF) and vascular endothelial growth factor (VEGF) in aged human choroid and eyes with age-related macular degeneration. Exp Eye Res 82:99–110
15. Bill A (1962) Calorimetric procedures for the study of the blood flow through the ciliary region and the choroid in rabbits. Acta Ophthalmol (Copenh) 40: 131–148
16. Bill A (1962) Intraocular pressure and blood flow through the uvea. Arch Ophthalmol 67:336–348
17. Bill A, Linder M, Linder J (1977) The protective role of ocular sympathetic vasomotor nerves in acute arterial hypertension. Bibl Anat 16:30–35
18. Bill A, Sperber G, Ujiie K (1983) Physiology of the choroidal vascular bed. Int Ophthalmol 6:101–107
19. Bird AC (1992) Bruch's membrane change with age. Br J Ophthalmol 76:166–168
20. Boker T, Fang T, Steinmetz R (1993) Refractive error and choroidal perfusion characteristics in patients with choroidal neovascularization and age-related macular degeneration. Ger J Ophthalmol 2:10–13
21. Carron JA, Hiscott P, Hagan S et al (2000) Cultured human retinal pigment epithelial cells differentially express thrombospondin-1, -2, -3, and –4. Int J Biochem Cell Biol 32:1137–1142
22. Chen JC, Fitzke FW, Pauleikhoff D et al (1992) Functional loss in age-related Bruch's membrane change with choroidal perfusion defect. Invest Ophthalmol Vis Sci 33:334–340
23. Ciulla TA, Harris A, Chung HS et al (1999) Color Doppler imaging discloses reduced ocular blood flow velocities in nonexudative age-related macular degeneration. Am J Ophthalmol 128:75–80
24. Ciulla TA, Harris A, Kagemann L et al (2002) Choroidal perfusion perturbations in non-neovascular age related macular degeneration. Br J Ophthalmol 86:209–213
25. Clemons TE, Kurinij N, Sperduto RD (2004) Associations of mortality with ocular disorders and an intervention of high-dose antioxidants and zinc in the Age-Related Eye Disease Study: AREDS Report No. 13. Arch Ophthalmol 122:716–726
26. D'Amore PA (1994) Mechanisms of retinal and choroidal neovascularization. Invest Ophthalmol Vis Sci 35:3974–3979

27. Das A, Frank RN, Weber ML et al (1988) ATP causes retinal pericytes to contract in vitro. Exp Eye Res 46:349–362
28. Dawson DW, Volpert OV, Gillis P et al (1999) Pigment epithelium-derived factor: a potent inhibitor of angiogenesis. Science 285:245–248
29. Del Priore LV, Kaplan HJ, Hornbeck R et al (1996) Retinal pigment epithelial debridement as a model for the pathogenesis and treatment of macular degeneration. Am J Ophthalmol 122:629–643
30. Dimitrova G, Tamaki Y, Kato S (2002) Retrobulbar circulation in patients with age-related maculopathy. Eye 16:580–586
31. Duh EJ, Yang HS, Suzuma I et al (2002) Pigment epithelium-derived factor suppresses ischemia-induced retinal neovascularization and VEGF-induced migration and growth. Invest Ophthalmol Vis Sci 43:821–829
32. Ernest JT (1977) The effect of systolic hypertension on rhesus monkey eyes after ocular sympathectomy. Am J Ophthalmol 84:341–344
33. Flower RW (1993) Extraction of choriocapillaris hemodynamic data from ICG fluorescence angiograms. Invest Ophthalmol Vis Sci 34:2720–2729
34. Folkman J, Klagsbrun M (1987) Angiogenic factors. Science 235:442–447
35. Friedman E (1997) A hemodynamic model of the pathogenesis of age-related macular degeneration. Am J Ophthalmol 124:677–682
36. Friedman E (2000) The role of the atherosclerotic process in the pathogenesis of age-related macular degeneration. Am J Ophthalmol 130:658–663
37. Friedman E (2004) Update of the vascular model of AMD. Br J Ophthalmol 88:161–163
38. Friedman E, Ivry M, Ebert E et al (1989) Increased scleral rigidity and age-related macular degeneration. Ophthalmology 96:104–108
39. Friedman E, Krupsky S, Lane AM et al (1995) Ocular blood flow velocity in age-related macular degeneration. Ophthalmology 102:640–646
40. Gao G, Li Y, Zhang D et al (2001) Unbalanced expression of VEGF and PEDF in ischemia-induced retinal neovascularization. FEBS Lett 489:270–276
41. Giovannini A, Mariotti C, Ripa E et al (1994) Choroidal filling in age-related macular degeneration: indocyanine green angiographic findings. Ophthalmologica 208:185–191
42. Giuffre G (1989) Main posterior watershed zone of the choroid. Variations of its position in normal subjects. Doc Ophthalmol 72:175–180
43. Gragoudas ES, Adamis AP, Cunningham ET Jr et al (2004) Pegaptanib for neovascular age-related macular degeneration. N Engl J Med 351:2805–2816
44. Grossniklaus HE, Green WR (2004) Choroidal neovascularization. Am J Ophthalmol 137:496–503
45. Grunwald JE, Hariprasad SM, Dupont J (1998) Effect of aging on foveolar choroidal circulation. Arch Ophthalmol 116:150–154
46. Grunwald JE, Hariprasad SM, Dupont J et al (1998) Foveolar choroidal blood flow in age-related macular degeneration. Invest Ophthalmol Vis Sci 39:385–390
47. Grunwald JE, Metelitsina TI, Dupont JC et al (2005) Reduced foveolar choroidal blood flow in eyes with increasing AMD severity. Invest Ophthalmol Vis Sci 46:1033–1038
48. Haefliger IO, Zschauer A et al (1994) Relaxation of retinal pericyte contractile tone through the nitric oxide-cyclic guanosine monophosphate pathway. Invest Ophthalmol Vis Sci 35:991–997
49. Haefliger IO, Meyer P, Flammer J et al (1994) The vascular endothelium as a regulator of the ocular circulation: a new concept in ophthalmology? Surv Ophthalmol 39:123–132
50. Haefliger IO, Zschauer A, Anderson DR (1994) Relaxation of retinal pericyte contractile tone through the nitric oxide-cyclic guanosine monophosphate pathway. Invest Ophthalmol Vis Sci 35:991–997
51. Hall NF, Gale CR, Syddall H et al (2001) Risk of macular degeneration in users of statins: cross sectional study. BMJ 323:375–376
52. Hayashi K, de Laey JJ (1985) Indocyanine green angiography of choroidal neovascular membranes. Ophthalmologica 190:30–39
53. Hayashi K, de Laey JJ (1985) Indocyanine green angiography of submacular choroidal vessels in the human eye. Ophthalmologica 190:20–29
54. Hayreh SS (1974) The choriocapillaris. Graefes Arch Clin Exp Ophthalmol 192:165–179
55. Hayreh SS (1975) Segmental nature of the choroidal vasculature. Br J Ophthalmol 59:631–648
56. Hayreh SS (1981) Controversies on submacular choroidal circulation. Ophthalmologica 183:11–19
57. Hayreh SS (1983) Macular lesions secondary to choroidal vascular disorders. Int Ophthalmol 6: 161–170
58. Hayreh SS (1983) Physiological anatomy of the choroidal vascular bed. Int Ophthalmol 6:85–93
59. Hayreh SS (1990) In vivo choroidal circulation and its watershed zones. Eye 4(Pt 2):273–289
60. Hayreh SS (2004) Posterior ciliary artery circulation in health and disease: the Weisenfeld lecture. Invest Ophthalmol Vis Sci 45:749–757, 748
61. Hayreh SS, Servais GE, Virdi PS (1986) Fundus lesions in malignant hypertension. VI. Hypertensive choroidopathy. Ophthalmology 93:1383–1400
62. Henkind P, Gartner S (1983) The relationship between retinal pigment epithelium and the choriocapillaris. Trans Ophthalmol Soc U K 103(Pt 4):444–447
63. Hickam JB, Frayser R, Ross JC (1963) A study of retinal venous blood oxygen saturation in human subjects by photographic means. Circulation 27: 375–385
64. Holz FG, Wolfensberger TJ, Piguet B et al (1994) Bilateral macular drusen in age-related macular degeneration. Prognosis and risk factors. Ophthalmology 101:1522–1528
65. Hosal BM, Karakoc G, Gursel E et al (1998) Color Doppler imaging of the retrobulbar circulation in age-related macular degeneration. Eur J Ophthalmol 8:234–238
66. Hyman L, Schachat AP, He Q et al (2000) Hypertension, cardiovascular disease, and age-related macular degen-

eration. Age-Related Macular Degeneration Risk Factors Study Group. Arch Ophthalmol 118:351–358
67. Ikram MK, van Leeuwen R, Vingerling JR et al (2005) Retinal vessel diameters and the risk of incident age-related macular disease: the Rotterdam Study. Ophthalmology 112:548–552
68. Ishibashi T, Inomata H, Sakamoto T et al (1995) Pericytes of newly formed vessels in experimental subretinal neovascularization. Arch Ophthalmol 113:227–231
69. Kim SY, Mocanu C, McLeod DS et al (2003) Expression of pigment epithelium-derived factor (PEDF) and vascular endothelial growth factor (VEGF) in sickle cell retina and choroid. Exp Eye Res 77:433–445
70. Klein R, Clegg L, Cooper LS et al (1999) Prevalence of age-related maculopathy in the Atherosclerosis Risk in Communities Study. Arch Ophthalmol 117:1203–1210
71. Klein R, Klein BE, Franke T (1993) The relationship of cardiovascular disease and its risk factors to age-related maculopathy. The Beaver Dam Eye Study. Ophthalmology 100:406–414
72. Klein R, Klein BE, Tomany SC et al (2003) The association of cardiovascular disease with the long-term incidence of age-related maculopathy: the Beaver Dam Eye Study. Ophthalmology 110:1273–1280
73. Klein R, Klein BE, Tomany SC et al (2004) The relation of retinal microvascular characteristics to age-related eye disease: the Beaver Dam eye study. Am J Ophthalmol 137:435–444
74. Kornzweig AL (1977) Changes in the choriocapillaris associated with senile macular degeneration. Ann Ophthalmol 9(753–756):759–762
75. Korte GE, Reppucci V, Henkind P (1984) RPE destruction causes choriocapillary atrophy. Invest Ophthalmol Vis Sci 25:1135–1145
76. Kvanta A (2006) Ocular angiogenesis: the role of growth factors. Acta Ophthalmol Scand 84:282–288
77. Liew G, Kaushik S, Rochtchina E et al (2006) Retinal vessel signs and 10-year incident age-related maculopathy: the Blue Mountains Eye Study. Ophthalmology 113:1481–1487
78. Linsenmeier RA (1986) Effects of light and darkness on oxygen distribution and consumption in the cat retina. J Gen Physiol 88:521–542
79. Linsenmeier RA, Steinberg RH (1984) Effects of hypoxia on potassium homeostasis and pigment epithelial cells in the cat retina. J Gen Physiol 84:945–970
80. Lutty G, Grunwald J, Majji AB et al (1999) Changes in choriocapillaris and retinal pigment epithelium in age-related macular degeneration. Mol Vis 5:35
81. Lutty GA, McLeod DS (2008) Dégénérescence des choriocapillaires liée à la sénescence et dans la dégénérescence maculaire liée à l'âge. In: Pournaras CJ (ed) Pathologies vasculaires oculaires, rapport de la Societé Francaise de l'Ophtalmologie 2008. Elsevier, Paris, pp 711–716
82. Mares-Perlman JA, Brady WE, Klein R et al (1995) Dietary fat and age-related maculopathy. Arch Ophthalmol 113:743–748
83. McGwin G Jr, Owsley C, Curcio CA et al (2003) The association between statin use and age related maculopathy. Br J Ophthalmol 87:1121–1125
84. McGwin G Jr, Xie A, Owsley C (2005) The use of cholesterol-lowering medications and age-related macular degeneration. Ophthalmology 112:488–494
85. McLeod DS, Taomoto M, Otsuji T et al (2002) Quantifying changes in RPE and choroidal vasculature in eyes with age-related macular degeneration. Invest Ophthalmol Vis Sci 43:1986–1993
86. Mendrinos E, Pournaras CJ (2009) Topographic variation of the choroidal watershed zone and its relation to neovascularisation in age-related macular degeneration. Acta Ophthalmol Scand 87(3):290–6
87. Metelitsina TI, Grunwald JE, DuPont JC et al (2008) Foveolar choroidal circulation and choroidal neovascularization in age-related macular degeneration. Invest Ophthalmol Vis Sci 49:358–363
88. Miller H, Miller B, Zonis S, Nir I (1984) Diabetic neovascularization: permeability and ultrastructure. Invest Ophthalmol Vis Sci 25:1338–1342
89. Mitchell P, Wang JJ, Smith W et al (2002) Smoking and the 5-year incidence of age-related maculopathy: the Blue Mountains Eye Study. Arch Ophthalmol 120:1357–1363
90. Mitchell P, Wang JJ, Wong TY et al (2005) Retinal microvascular signs and risk of stroke and stroke mortality. Neurology 65:1005–1009
91. Miyajima-Uchida H, Hayashi H, Beppu R et al (2000) Production and accumulation of thrombospondin-1 in human retinal pigment epithelial cells. Invest Ophthalmol Vis Sci 41:561–567
92. Miyazaki M, Nakamura H, Kubo M et al (2003) Risk factors for age related maculopathy in a Japanese population: the Hisayama study. Br J Ophthalmol 87:469–472
93. Mori K, Gehlbach PL, Ito YN et al (2005) Decreased arterial dye-filling and venous dilation in the macular choroid associated with age-related macular degeneration. Retina 25:430–437
94. Nowak I, Pilas-Pomykalska M, Czajkowski J et al (2007) Retinal capillary blood flow in exudative age-related macular degeneration. Klin Oczna 109:135–137
95. Nuzzi R, Finazzo C, Grignolo FM (1996) Changes in adrenergic innervation of the choroid during aging. J Fr Ophtalmol 19:89–96
96. O'Reilly MS, Boehm T, Shing Y et al (1997) Endostatin: an endogenous inhibitor of angiogenesis and tumor growth. Cell 88:277–285
97. Pauleikhoff D, Chen JC, Chisholm IH et al (1990) Choroidal perfusion abnormality with age-related Bruch's membrane change. Am J Ophthalmol 109:211–217
98. Pauleikhoff D, Spital G, Radermacher M et al (1999) A fluorescein and indocyanine green angiographic study of choriocapillaris in age-related macular disease. Arch Ophthalmol 117:1353–1358
99. Piguet B, Palmvang IB, Chisholm IH et al (1992) Evolution of age-related macular degeneration with choroidal perfusion abnormality. Am J Ophthalmol 113:657–663

100. Pilas-Pomykalska M, Nowak I, Czajkowski J et al (2005) Retinal capillary blood flow in nonexudative age-related macular degeneration. Klin Oczna 107:60–62
101. Polska E, Luksch A, Ehrlich P et al (2002) Measurements in the peripheral retina using LDF and laser interferometry are mainly influenced by the choroidal circulation. Curr Eye Res 24: 318–323
102. Pournaras CJ (1996) Autoregulation of ocular blood flow. In: Kaiser HJ, Flammer J, Hendrickson Ph (eds) Ocular blood flow. New insights into the pathogenesis of ocular diseases. Karger, Basel, pp 40–50
103. Pournaras CJ, Logean E, Riva CE et al (2006) Regulation of subfoveal choroidal blood flow in age-related macular degeneration. Invest Ophthalmol Vis Sci 47:1581–1586
104. Pournaras CJ, Riva CE, Tsacopoulos M et al (1989) Diffusion of O_2 in the retina of anesthetized miniature pigs in normoxia and hyperoxia. Exp Eye Res 49:347–360
105. Ramrattan RS, van der Schaft TL, Mooy CM et al (1994) Morphometric analysis of Bruch's membrane, the choriocapillaris, and the choroid in aging. Invest Ophthalmol Vis Sci 35:2857–2864
106. Reber F, Gersch U, Funk RW (2003) Blockers of carbonic anhydrase can cause increase of retinal capillary diameter, decrease of extracellular and increase of intracellular pH in rat retinal organ culture. Graefes Arch Clin Exp Ophthalmol 241:140–148
107. Remsch H, Spraul CW, Lang GK et al (2000) Changes of retinal capillary blood flow in age-related maculopathy. Graefes Arch Clin Exp Ophthalmol 238:960–964
108. Remulla JF, Gaudio AR, Miller S et al (1995) Foveal electroretinograms and choroidal perfusion characteristics in fellow eyes of patients with unilateral neovascular age-related macular degeneration. Br J Ophthalmol 79:558–561
109. Riva CE, Cranstoun SD, Grunwald JE et al (1994) Choroidal blood flow in the foveal region of the human ocular fundus. Invest Ophthalmol Vis Sci 35:4273–4281
110. Riva CE, Petrig BL (2003) Laser Doppler techniques in ophthalmology – principles and applications. In: Fankhauser F, Kwasniewska S (eds) Lasers in ophthalmology – basic, diagnostic and surgical aspects. Kugler Publications, The Hague, pp 51–59
111. Riva CE, Titze P, Hero M, Movaffaghy A, Petrig BL (1997) Choroidal blood flow during isometric exercises. Invest Ophthalmol Vis Sci 38:2338–2343
112. Robinson F, Riva CE, Grunwald JE, Petrig BL, Sinclair SH (1986) Retinal blood flow autoregulation in response to an acute increase in blood pressure. Invest Ophthalmol Vis Sci 27:722–726
113. Rosenfeld PJ, Brown DM, Heier JS et al (2006) Ranibizumab for neovascular age-related macular degeneration. N Engl J Med 355:1419–1431
114. Ross RD, Barofsky JM, Cohen G et al (1998) Presumed macular choroidal watershed vascular filling, choroidal neovascularization, and systemic vascular disease in patients with age-related macular degeneration. Am J Ophthalmol 125:71–80
115. Sarks JP, Sarks SH, Killingsworth MC (1988) Evolution of geographic atrophy of the retinal pigment epithelium. Eye 2(Pt 5):552–577
116. Sarks JP, Sarks SH, Killingsworth MC (1997) Morphology of early choroidal neovascularisation in age-related macular degeneration: correlation with activity. Eye 11(Pt 4):515–522
117. Sasaki T, Fukai N, Mann K et al (1998) Structure, function and tissue forms of the C-terminal globular domain of collagen XVIII containing the angiogenesis inhibitor endostatin. EMBO J 17:4249–4256
118. Sato E, Feke GT, Menke MN et al (2006) Retinal haemodynamics in patients with age-related macular degeneration. Eye 20:697–702
119. Schaumberg DA, Christen WG, Hankinson SE et al (2001) Body mass index and the incidence of visually significant age-related maculopathy in men. Arch Ophthalmol 119:1259–1265
120. Schmetterer L, Kruger A, Findl O et al (1998) Topical fundus pulsation measurements in age-related macular degeneration. Graefes Arch Clin Exp Ophthalmol 236:160–163
121. Schonfelder U, Hofer A, Paul M et al (1998) In situ observation of living pericytes in rat retinal capillaries. Microvasc Res 56:22–29
122. Sheibani N, Sorenson CM, Cornelius LA et al (2000) Thrombospondin-1, a natural inhibitor of angiogenesis, is present in vitreous and aqueous humor and is modulated by hyperglycemia. Biochem Biophys Res Commun 267:257–261
123. Smith W, Assink J, Klein R et al (2001) Risk factors for age-related macular degeneration: pooled findings from three continents. Ophthalmology 108: 697–704
124. Smith W, Mitchell P, Leeder SR (2000) Dietary fat and fish intake and age-related maculopathy. Arch Ophthalmol 118:401–404
125. Spraul CW, Lang GE, Grossniklaus HE (1996) Morphometric analysis of the choroid, Bruch's membrane, and retinal pigment epithelium in eyes with age-related macular degeneration. Invest Ophthalmol Vis Sci 37:2724–2735
126. Staurenghi G, Bottoni F, Lonati C et al (1992) Drusen and 'choroidal filling defects': a cross-sectional survey. Ophthalmologica 205:178–186
127. Sunness JS, Gonzalez-Baron J, Bressler NM et al (1999) The development of choroidal neovascularization in eyes with the geographic atrophy form of age-related macular degeneration. Ophthalmology 106:910–919
128. The Eye Disease Case–control Study Group (1992) Risk factors for neovascular age-related macular degeneration. Arch Ophthalmol 110:1701–1708
129. Tombran-Tink J, Chader GG, Johnson LV (1991) PEDF: a pigment epithelium-derived factor with potent neuronal differentiative activity. Exp Eye Res 53:411–414

130. Tornquist P, Alm A (1979) Retinal and choroidal contribution to retinal metabolism in vivo. A study in pigs. Acta Physiol Scand 106:351–357
131. Torvik A (1984) The pathogenesis of watershed infarcts in the brain. Stroke 15:221–223
132. Tsacopoulos M (1979) Physiopathology of the uveal circulation. J Fr Ophtalmol 2:135–142
133. Uno K, Bhutto IA, McLeod DS et al (2006) Impaired expression of thrombospondin-1 in eyes with age related macular degeneration. Br J Ophthalmol 90: 48–54
134. Uretmen O, Akkin C, Erakgun T et al (2003) Color Doppler imaging of choroidal circulation in patients with asymmetric age-related macular degeneration. Ophthalmologica 217:137–142
135. van Leeuwen R, Ikram MK, Vingerling JR et al (2003) Blood pressure, atherosclerosis, and the incidence of age-related maculopathy: the Rotterdam Study. Invest Ophthalmol Vis Sci 44:3771–3777
136. Vingerling JR, Dielemans I, Bots ML et al (1995) Age-related macular degeneration is associated with atherosclerosis. The Rotterdam Study. Am J Epidemiol 142:404–409
137. Wang JJ, Liew G, Wong TY et al (2006) Retinal vascular calibre and the risk of coronary heart disease-related death. Heart 92:1583–1587
138. Wang JJ, Mitchell P, Rochtchina E et al (2004) Retinal vessel wall signs and the 5 year incidence of age related maculopathy: the Blue Mountains Eye Study. Br J Ophthalmol 88:104–109
139. Wilson HL, Schwartz DM, Bhatt HR et al (2004) Statin and aspirin therapy are associated with decreased rates of choroidal neovascularization among patients with age-related macular degeneration. Am J Ophthalmol 137:615–624
140. Wong TY, Klein R, Sun C et al (2006) Age-related macular degeneration and risk for stroke. Ann Intern Med 145:98–106
141. Wong TY, Mitchell P (2004) Hypertensive retinopathy. N Engl J Med 351:2310–2317
142. Zhang SX, Ma JX (2007) Ocular neovascularization: implication of endogenous angiogenic inhibitors and potential therapy. Prog Retin Eye Res 26:1–37
143. Zhao J, Frambach DA, Lee PP et al (1995) Delayed macular choriocapillary circulation in age-related macular degeneration. Int Ophthalmol 19:1–12

The Role of Ocular Blood Flow Abnormalities in the Pathogenesis of Glaucoma

17

Balwantray C. Chauhan

Core Messages

- Glaucoma is a group of optic neuropathies and a leading cause of blindness worldwide. Open-angle glaucoma is the most common form of the neuropathy clinically characterised by pathological optic disc cupping and visual field change.
- Intraocular pressure (IOP) is uniformly recognised as the major known causative risk factor for the development and progression of the neuropathy. Treatment to lower IOP reduces the incidence and progression of glaucoma. Nonetheless, a significant proportion of patients continue to progress in spite of successful IOP reduction, suggesting that in at least some patients, other factors are responsible for the disease. One of the earliest identified additional risk factors is optic nerve head ischaemia.
- Epidemiological studies provide compelling evidence that ocular (particularly diastolic) perfusion pressure is related to glaucoma. A putative result of faulty autoregulation to low perfusion pressure in order to maintain adequate blood flow is an ischaemic component to glaucomatous damage.
- Reduced blood flow in glaucoma has been shown with a wide range of measurement techniques; however, most of these studies cannot determine whether reduced blood flow is the cause or result of the neuropathy.
- A variety of ocular and systemic drugs have an effect on blood flow; however, there is a significant paucity of evidence to prove that increasing optic nerve head blood flow has a favourable impact on disease course.
- Experimental models are providing some insight on the effect of purely ischaemic damage on the optic nerve; however, most evidence suggests that the neuropathy may require the mediation of intraocular pressure.
- Carefully executed research studies and trials are required first to ascertain that patients with reduced ocular blood flow are either more likely to develop glaucoma or progression of existing damage, after controlling the influence of IOP. Only after a strong evidence base can therapeutic trials begin in earnest.

B.C. Chauhan, Ph.D., FARVO
Department of Ophthalmology and Visual Sciences, Dalhousie University, 1276 South Park Street, VG Site, Room 2035, 2W Victoria, Halifax, NS, B3H 2Y9, Canada
e-mail: bal@dal.ca

L. Schmetterer, J.W. Kiel (eds.), *Ocular Blood Flow*,
DOI 10.1007/978-3-540-69469-4_17, © Springer-Verlag Berlin Heidelberg 2012

17.1 Introduction

Glaucoma is a group of optic neuropathies characterised by damage to the optic nerve head and visual field. It is the second leading cause of blindness worldwide with a projected estimate of 60.5 and 79.6 million persons to be affected worldwide by 2010 and 2020, respectively [152]. More significantly, 8.4 and 11.2 million persons will be bilaterally blind from the glaucomas by 2010 and 2020, respectively [152]. There is little doubt that glaucoma represents a major worldwide healthcare problem with significant challenges in the identification of new cases as well as in the management of existing patients with the disease.

Open-angle glaucoma is the commonest form of glaucoma [47, 94], comprising an average of approximately 75% of all glaucomas, though there is considerable variation in this proportion worldwide [152]. It is a chronic age-related neuropathy whose pathogenesis has traditionally been linked to the action of intraocular pressure (IOP) on the optic nerve head [7, 151, 172]. Neurosensory damage occurs to retinal ganglion cells (RGCs) [150] which relay visual information to the brain, in addition to higher order neurons in the visual pathway [26, 73, 188, 197]. There is considerable evidence that RGCs die through programmed cell death or apoptosis [61, 154].

Elevated IOP is the most significant known risk factor for open-angle glaucoma with an exponential increase in disease prevalence and incidence or progression with increasing IOP [67, 115, 117, 170]. Despite the obvious importance of IOP, epidemiological studies show that up to 95% of individuals with statistically elevated IOP do not have open-angle glaucoma [11, 93, 170, 173]. Evidence suggests that the large majority of these individuals will not develop the disease [37, 67, 106]. On the other hand, up to 50% of persons with glaucoma have screening IOP within normal limits [12, 93, 171]. Randomised clinical trials have consistently shown that IOP lowering results in reduced incidence [103] and progression of existing glaucoma [2, 85]; however, patients can continue to deteriorate in spite of successful IOP lowering [2, 85, 103]. Hence, other ocular or systemic factors may either reduce the threshold for IOP-induced damage, for glaucoma to occur at statistically normal IOPs, or act independently to cause glaucoma. Of these factors, optic nerve head blood flow irregularities, through a mechanism of compromised autoregulation [6, 48], vasospasm [53, 63, 74] or microvascular disease [162], is the candidate most widely suspected. The precise mechanism of glaucomatous damage is unknown, but it is likely that changes in the cellular and extracellular environment around the RGC and its axon dictate its fate, irrespective of the primary trigger.

17.2 Potential Mechanisms of Ischaemic Damage in Glaucoma

It is widely thought that the primary insult in glaucoma occurs at the level of the optic nerve head [7, 23, 151], specifically at or around the lamina cribrosa. The lamina is composed of perforated sheets, containing collagen and laminin [88, 175] and is around 250-μm thick in humans [153]. The laminar sheets bridge the scleral canal through which pass the unmyelinated RGC axons and the central retinal artery and vein. The laminar insertion point into the sclera and the material properties of the peripapillary sclera [42, 168] are thought to be highly relevant in explaining the biomechanical changes in the nerve head [22], which may also potentially affect all vessels in the vicinity of the lamina. Significant connective tissue changes occur before or at the onset of RGC loss [22] involving intracellular ion and electrolyte imbalance, and interruption of trophic support [99, 145].

The blood supply to the optic nerve head is detailed in Chap. 1. Briefly, however, the optic nerve blood supply is segmental with the nerve fibre layer supplied exclusively by the central retinal artery; the prelaminar and laminar nerve by the posterior ciliary arteries via the circle of Zinn-Haller; and the retrolaminar nerve by the central retinal artery and short posterior ciliary arteries via the pial arteries [140]. Because of this segmental nature and the considerable differences between the retinal and choroidal circulations, any pathology arising from altered blood flow to the optic nerve head would in turn be expected to be segmental. For example, alterations in retinal blood flow measured by many clinical devices likely do not accurately reflect blood flow alterations in the deeper optic nerve and have limited ability to elucidate any ischaemic pathology in glaucoma.

There are several potential mechanisms of reduced or impaired blood flow to the optic nerve which may be responsible for contributing to the pathophysiology of glaucoma. The evidence base for each of these mechanisms varies considerably. It is possible that these do not act independently of each other, and more importantly they may act in combination with IOP to cause the phenomena of glaucomatous optic neuropathy and visual field damage.

17.2.1 Decreased Ocular Perfusion Pressure

Perfusion to the optic nerve head depends on the difference between ophthalmic artery pressure and IOP. Ocular perfusion pressure (OPP) can be expressed as either the systolic, diastolic or mean perfusion pressure. OPP increases when blood pressure is high and/or IOP is low and decreases when blood pressure is low and/or IOP is elevated. Blood flow to the optic nerve head depends on OPP as well as local resistance to flow (e.g. when elevated IOP compresses veins to increase resistance). OPP changes with changes in blood pressure and IOP, as well as physiologically throughout the 24-h cycle. Under normal conditions, blood flow is maintained at more or less constant levels in the face of changing OPP via a process called myogenic autoregulation [148] where the local resistance is increased primarily by vasoconstriction when the OPP is high and reduced by vasodilation when OPP is low. Blood flow may also be maintained according to the functional demand of the optic nerve via a process known as metabolic autoregulation [148] where the threshold for a minimum concentration of vital metabolites dictates vascular tone.

Ischaemic injury to the optic nerve can potentially occur when blood flow cannot be maintained to sustain optic nerve function. Impaired autoregulation has frequently been postulated as the main reason for ischaemic injury; however, it could also potentially occur when the OPP is too low due to very low blood pressure and/or high IOP in spite of normal autoregulation.

An added complication of a temporarily decreased perfusion, and potentially hypoxia, is reperfusion damage which occurs after blood flow is restored with an increase OPP. Reperfusion injury has been noted in several organ systems, including the eye [143], and may be accompanied with oxidative damage, release of reactive oxygen species [70], specific vasoactive peptides such as endothelin-1 (ET-1) [186] and matrix metalloproteinases (MMPs) [122] responsible for modulating the extracellular matrix. Since remodelling of the optic nerve head extracellular matrix is a hallmark of glaucomatous neuropathy, activation and inhibition of MMPs may be highly relevant.

17.2.2 Autoregulatory Disturbances

In healthy individuals, autoregulation ensures that within a range of OPP, optic nerve head blood flow is maintained at a more or less constant value. At very low OPP, the relationship between OPP and blood flow becomes passive with a proportional decrease in blood flow with decrease in OPP because arteries are maximally dilated. Patients with so-called vascular dysregulation syndrome [50, 53] are thought to have one or more abnormalities that preclude a normal response to maintain nutrient supply to tissues and organs. Hence, blood flow is reduced at higher-than-normal OPP. Among these are vascular endothelial cell damage, atherosclerosis and vasospasm.

Vasospasm is the inability of vessels to dilate fully or in a sustained manner. While vasospasm is likely a multifactorial syndrome, it is associated with systemic conditions such as angina, Raynaud's disease and migraine. Vasospasms can also be triggered by cold, nicotine, stress and potentially local or circulating vasoconstrictors such as ET-1. Irrespective of the cause of vasospasm, it is likely an important factor in the inability of vessels to respond to decreasing OPP and/or high IOP potentially causing ischaemic damage.

17.2.3 Mechanical Compression or Collapse of Vessels

Blood vessels in the optic nerve head are subject to extraluminal pressure due to IOP. When the extraluminal pressure equals or exceeds the intraluminal pressure, the vessel will compress and eventually collapse. This phenomenon is more likely in venules and veins because of the lower venous pressure. Compression or collapse of veins will cause increased resistance to blood flow upstream with potential consequences in patients with impaired autoregulation in whom there may be an inability of arterioles to dilate to cope with the increased resistance.

Laminar vessels traverse parallel and perpendicular to the plane of the lamina. A commonly held view is that direct compression of the laminar beams with glaucomatous optic neuropathy leads to direct physical damage from distortion or kinking of vessels. Increasing evidence of peripapillary scleral involvement in glaucoma [42, 168] suggests that scleral stress and strain may also generate added extraluminal pressure causing potential damage to vessels in and around the circle of Zinn-Haller.

17.2.4 Atherosclerosis

Atherosclerosis is an inflammatory arterial disease which decreases the lumen of the arteries due to accumulation of plaques [125]. It is not possible to measure atherosclerosis in optic nerve vessels non-invasively, and therefore proxy measures such as carotid artery ultrasound [18] or general markers for inflammation, such as serum C-reactive protein or fibrinogen, can be used [111, 189].

According to Poiseuille's law, blood flow is inversely proportional to the length of the vessel and viscosity and directly proportional to the fourth power of the vessel radius [49]. In other words, with vessel length and viscosity constant, a 10% reduction in lumen leads to a 34% reduction in flow, while a 50% reduction in lumen leads to a 94% reduction in flow. Reduction in vessel lumen therefore has a dramatic effect on blood flow with potential for ischaemic damage. In the presence of vasospasm, this situation is exacerbated further.

17.2.5 Vascular Endothelial Factors

The vascular endothelium is responsible for the production of several vasoactive substances. The most important relaxing endogenous factor is nitric oxide (NO) [129]; however, adenosine [46] and prostacyclin [130] also cause vasodilation. Among the contracting factors are ET-1 [195], thromboxane A_2 [130] and prostaglandin H_2 [130].

Endothelial dysfunction can lead to the modulation of any of the endothelium-derived factors compromising blood flow. It has been demonstrated that reperfusion injury can cause endothelial damage via production of reactive oxygen species and inhibit activation of NO [104, 159]. Systemic disease such as hypertension and diabetes can also compromise endothelial-derived vasodilation [40, 55, 160]. Finally, formation of atherosclerotic plaque and lipid deposits interfere with endothelial-derived vasodilation [166, 167]. These systemic disorders are associated with an increase in ET-1 production [120].

17.2.6 Barriers to Nutrient Delivery

Age-related changes to the optic nerve head extracellular matrix include thickening of astrocyte basement membrane [87, 89]. In glaucoma, the additional changes include deposition of new extracellular matrix materials including glial tissue and collagens [21, 86, 90, 131, 132]. In normal optic nerve heads, oxygen and other nutrients transfer from laminar capillaries to astrocytes and the nerve fibre bundles occur via diffusion through the vascular endothelium, pericytes, the extracellular matrix and astrocyte basement membrane. With age- and glaucoma-related changes, this diffusion may become compromised even if blood flow is unaltered.

17.2.7 Circulating Vasoconstrictors

As mentioned above, vascular tone is maintained by locally produced vasoactive compounds such as NO and ET-1. Low concentrations of ET-1 in plasma are found [101, 137, 176]; however, because of the tight blood-retinal barrier, it is theoretically unlikely for circulating ET-1 to have an impact on smooth muscle cells to cause vasoconstriction. Normally, intraluminal levels of ET-1 required to cause vasoconstriction are several log units higher than extraluminal levels because of the direct exposure to ET-1 to smooth muscle cells [196]. Nonetheless, the situation is quite different in diseases where the blood-retinal barrier may be compromised. Similarly, vasoactive substances may leak from the choroidal circulation where the endothelial junctions are not as tight. In certain conditions, it is possible that even erythrocytes may leak and could be a possible cause of the characteristic disc haemorrhages which are almost exclusively associated with glaucoma [4].

Under some conditions, such as cold and stress, concentrations of vasoconstrictors such as ET-1 may increase [101, 137]. In turn, ET-1 may leak from vessels with multiple effects including astrocyte activation [149, 185] in addition to vasoconstriction [31, 196].

17.3 Evidence Base Supporting the Importance of Ischaemia in Glaucoma

As discussed above, there are several mechanisms whereby reduced blood flow and nutrient supply potentially lead to optic nerve damage in glaucoma. Considerable data have been amassed over the last few decades, both from human and animal research on the issue of association, causality and whether treatment to ameliorate blood flow leads to favourable outcomes in glaucoma.

17.3.1 Association and Causality

17.3.1.1 Reduction in Optic Nerve Head Blood Flow

A considerable body of evidence exists to support the notion that blood flow in the optic nerve head, choroid, retina and indeed outside the eye is reduced in glaucoma [51, 52]. While practically every technique used for measuring blood flow shows alterations in glaucoma, it should be noted that the glaucoma populations were different across studies. For example, many studies only reported findings in glaucoma patients with statistically normal IOP based on the assumption that glaucoma due to high IOP is due to IOP whereas ischaemic factors may be responsible for damage at normal IOP.

Fluorescein angiography first was used to show filling defects and delayed filling in the choroid, optic nerve and retina [91, 164], though it is unclear how filling defects and angiographic transit times relate to blood flow because of parameters such as vessel diameter and dye transit to the eye. Pulsatile ocular blood flow is derived from changes in IOP measured during the cardiac cycle [169]. These measurements are converted to change in pulse volume due primarily to change in choroidal volume from systole to diastole. In spite of the fact that the optic nerve head component of choroidal blood flow is very small, several reports show a reduction in pulsatile ocular blood flow in glaucoma [98, 105, 181], indicating more global alterations in blood flow.

Doppler-based techniques have been used widely to assess blood flow in glaucoma. Laser Doppler velocimetry has been used to show reduction in retinal blood flow velocity [75], while laser Doppler flowmetry was used to show reduction in optic nerve head blood flow in glaucoma patients and suspects [72, 147]. Scanning laser Doppler flowmetry measurements from several research laboratories report reduction in retinal and optic nerve head blood flow in glaucoma [76, 127, 138]. Finally, colour Doppler imaging of blood velocities in the retrobulbar vessels including the ophthalmic artery, short posterior ciliary arteries and central retinal vein has been used to show reductions in glaucoma patients [24, 59, 81, 139].

Blood flow alterations may be more pronounced in eyes that have faster glaucomatous progression [163, 194]. Spatial correlations between area of reduced blood flow and visual field damage have also been published [10]. Finally, there is some evidence that blood flow may precede development of glaucoma in studies of patients with unilateral glaucoma showing blood flow alterations in the perimetrically unaffected eye [54, 136].

17.3.1.2 Blood Pressure, Intraocular Pressure and Perfusion Pressure

Many systemic conditions such as hypertension are age-related; hence, it is important to elucidate whether glaucoma and systemic hypertension are co-morbidities. A considerable number of epidemiological studies in glaucoma across diverse populations have been undertaken, yet the relationship between blood pressure and IOP, and blood pressure and glaucoma is complex [38], with possible interactions with racial, genetic and environmental factors.

There is a positive relationship between IOP and both systolic and diastolic blood pressure in European-derived populations [16, 41, 108], a mixed US population [177], Caribbean Blacks [190] and Chinese-derived populations [56, 192]. While statistically significant, every 10-mmHg increase in systolic or diastolic blood pressure accounts for an increase in IOP of less than 0.5 mmHg. Evidence from longitudinal studies [107, 191] also shows that a higher baseline systolic or diastolic blood pressure explains a small (less than 0.5 mmHg) but statistically significant increase in IOP.

Paradoxically, the studies on the relationship between systemic hypertension and glaucoma have yielded opposing results. Epidemiological findings from Italian [16], Dutch [41] and Australian [128] populations show a positive relationship between hypertension and glaucoma, while studies in the United States [177] and Caribbean [114] populations failed to confirm these findings. Furthermore, population-based longitudinal data do not show a relationship between hypertension and incident glaucoma. In fact, the evidence points to the contrary, that is, low systolic blood pressure may be associated with incident glaucoma [118, 119] and progression of existing glaucoma [117], at least in glaucoma with lower IOP.

Perhaps the most equivocal finding relating derivatives of blood pressure to glaucoma is the strong association between diastolic ocular perfusion pressure and the disease (Fig. 17.1). The Baltimore Eye Survey first reported the relationship between the prevalence of glaucoma and diastolic perfusion pressure [177]. There was no effect of diastolic perfusion pressure in the prevalence of glaucoma until the values dropped to below 45 mmHg. The odds of having glaucoma increased by a factor of over 6 in patients with

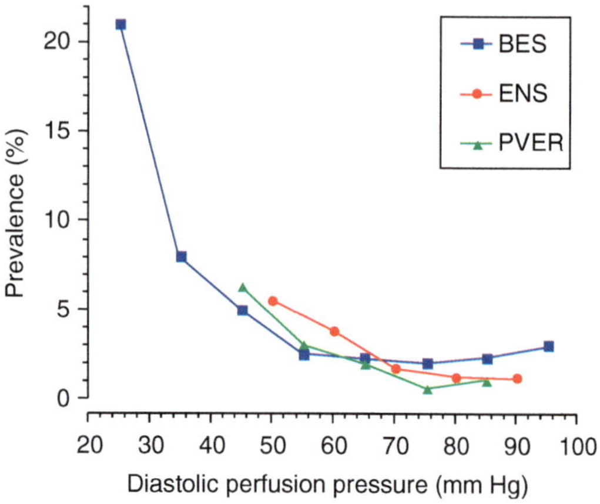

Fig. 17.1 Increasing prevalence of open-angle glaucoma with lower diastolic perfusion pressure (Data are derived from three population-based studies, the Baltimore Eye Survey (BES) [177], the Egna-Neumarkt Study (ENS) [16] and Proyecto VER (PVER) [155])

diastolic perfusion pressure of <30 mmHg compared to those with values >30 mmHg. These findings provided indirect support for the hypothesis that patients with low perfusion pressure may be unable to autoregulate blood supply to the optic nerve at low perfusion pressure; however, it is unlikely that low perfusion pressure alone accounts for all cases of glaucoma as the number of individuals with such low perfusion pressure was small [177].

Other epidemiological studies have subsequently confirmed the relationship between low diastolic perfusion pressure and glaucoma [16, 114, 155]. These findings however are not supported uniformly. The Blue Mountains Eye Study found only a marginally significant relationship between systolic perfusion pressure and glaucoma, though no significant relationship existed between diastolic or mean perfusion pressure and glaucoma [128]. In the Rotterdam Study, the prevalence of glaucoma with lower IOP was *reduced* in patients with diastolic perfusion pressure <50 mmHg while the prevalence of glaucoma with higher IOP was increased in patients with diastolic perfusion pressure <50 mmHg [96].

Finally, two longitudinal studies also confirm the importance of perfusion pressure in glaucoma. The Barbados Eye Study showed that

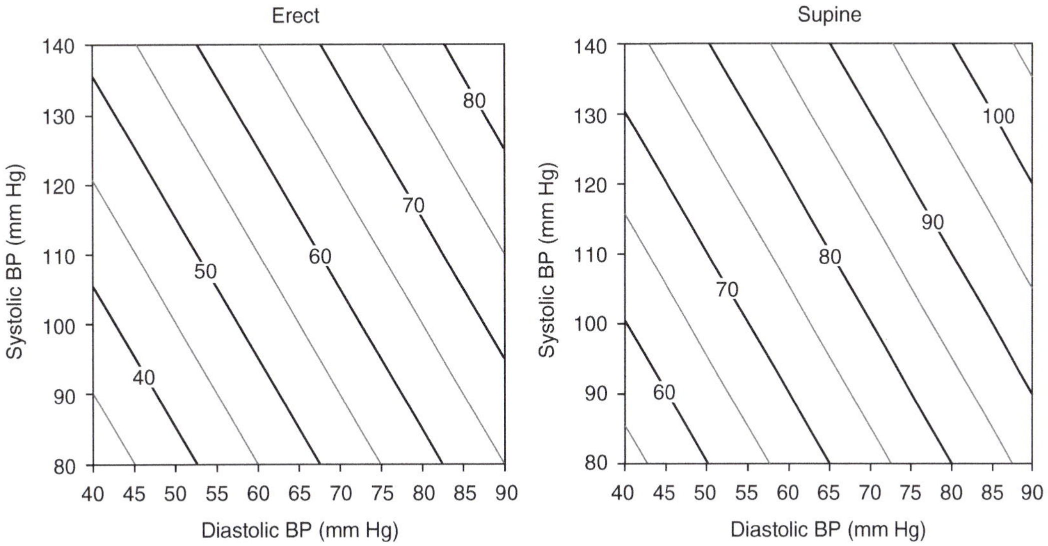

Fig. 17.2 Mean arterial pressure (MAP) at the level of the eye as a function of diastolic (SBP) and systolic (SBP) brachial blood pressure in erect and supine positions. *Diagonal lines* show iso-MAP values. These values are derived with the assumptions that: (1) MAP = DBP + 1/3 (SBP − DBP); (2) the vertical height difference between the heart and eye = 30 cm and (3) 1 $cmH_2O = 0.72$ mmHg. These data show that for a given IOP and blood pressure, ocular perfusion pressure is significantly higher in the supine position than in the erect position

subjects developing glaucoma 9 years after initial assessment had lower baseline systolic, diastolic and mean perfusion pressure [118]. In the Early Manifest Glaucoma Trial, lower systolic perfusion pressure was associated with progression of existing glaucoma [117]. The associations in these longitudinal studies are notable as they were carried out in racially distinct populations.

17.3.1.3 Nocturnal Hypotension

Systemic blood pressure lowers or dips physiologically at night. There is considerable evidence that in at least some glaucoma patients, the level of dipping is exaggerated compared to non-glaucoma subjects, with the potential of hypoperfusion of the optic nerve head contributing to glaucomatous optic neuropathy [68, 83]. This situation may be exacerbated in those patients taking systemic hypotensive drugs. Subsequent research demonstrated nocturnal dipping was associated with progressive glaucomatous damage [69] and that patients with non-progressive glaucoma had nocturnal retrobulbar blood flow measurements that did not differ from healthy subjects [82]. It has also been suggested that non-dipping is also associated with glaucoma progression [39, 178]; hence, these findings are paradoxical in regard to perfusion pressure. Recent research showed that fluctuations in mean ocular perfusion pressure were associated with nocturnal dipping and that the level of fluctuation was related to the level of visual field damage at diagnosis [29].

While ocular blood flow parameters are related to blood pressure and perfusion pressure in glaucoma patients, but not healthy subjects [57], it is unlikely that potential ischaemia of the optic nerve head can occur from a nocturnal reduction in blood pressure alone. For a given blood pressure, the ocular perfusion pressure in the supine position is actually *higher* than in the erect position because the height difference between the heart and eye is eliminated [14]. Figure 17.2 shows mean blood pressure for a range of systolic and diastolic blood pressures in the supine and erect positions. Assuming a systolic blood pressure of 120 mmHg, the mean blood pressure at the level of the eye is approximately equal when the supine diastolic pressure is around 60 mmHg and the erect diastolic pressure is around 90 mmHg. Assuming a diastolic blood pressure of 80 mmHg, the mean blood pressure at the level of the eye is approximately equal when the supine blood pressure is around 80 mmHg and the erect blood pressure is around 140 mmHg. These figures show that erect blood pressures have to be relatively high and supine blood pressures relatively low to obtain the same mean arterial pressure. The approximately 20 mmHg higher mean blood pressure at the level of the eye in the supine compared to erect position is much higher than the increase in nocturnal IOP to have an impact on ocular perfusion pressure.

Hence, while the evidence suggests that potential nocturnal ischaemia may be a contributing factor in glaucoma, it cannot be explained on the basis of reduced blood pressure at night and decreased ocular perfusion pressure. It is possible that blood flow may be reduced at night because of factors such as increased resistance.

17.3.1.4 Vasospasm

There is considerable evidence that vasospasm (or vascular dysregulation) and its surrogate measures such as migraine are associated with glaucoma. The possible association between vasospasm and glaucoma was first published almost 25 years ago [146] when it was reported that glaucoma patients with lower IOP had a higher prevalence of migraine compared to healthy subjects and those with higher IOP with and without manifest glaucoma. Several subsequent studies have confirmed these findings with patient-reported symptoms [36] or with indirect measurements of ocular vasospasm, such as finger blood flow [45, 162] or nailfold capillaromicroscopy [64, 65], which assess peripheral vasospasm.

Two population-based studies on self-reported migraine and glaucoma led to opposing conclusions – one finding an association between migraine and glaucoma [184], while the other did not [109].

Longitudinal studies have also addressed whether the incidence or progression of existing glaucoma is exacerbated by migraine or vasospasm. The Ocular Hypertension Treatment Study (OHTS) did not find an association between the existence of self-reported migraine and the development of glaucoma [67]. The Collaborative Normal Tension Glaucoma Study (CNTGS) showed that among glaucoma patients randomised to no treatment, patients with self-reported migraine were 2.5 times as likely to progress compared to those without migraine [43]. Building on previous research suggesting that vasospastic patients have a more IOP-dependent disease and therefore presumably more responsive to IOP reduction [162], the CNTGS showed that among those randomised to treatment, migraine patients responded more favourably than non-migraine patients [8]. Findings of the CNTGS were not confirmed by the Early Manifest Glaucoma Trial, another trial which also randomised patients to treatment and no treatment [117].

The Canadian Glaucoma Study (CGS) was designed to specifically test whether patients with objectively measured peripheral vasospasm at baseline had a more favourable outcome under a uniform IOP treatment protocol during prospective follow-up [1]. The CGS failed to statistically confirm this hypothesis, though non-statistically significant trends were reported consistently suggesting that vasospasm may be important in glaucoma; however, in some populations, its effect may be quite small [28].

17.3.1.5 Endothelin and Other Circulating Peptides

Plasma levels of ET-1 in glaucoma patients have been reported, with varying results. Some studies show a higher basal concentration of plasma ET-1 [25, 174], but the majority of subsequent studies have failed to confirm this finding [92, 101, 110, 137, 176]. Under different physiological conditions, however, such as a posture change [101] or cold provocation [137], at least some glaucoma patients show an increase in plasma ET-1 concentration compared to control subjects. The mechanisms whereby a systemic increase in ET-1 concentration contributes to glaucomatous optic neuropathy remain to be elucidated.

Other circulating peptides, such as angiotensin-1, serotonin and markers of nitric oxide (e.g. cyclic guanosine monophosphate), have been investigated in plasma and aqueous of patients with glaucoma [60]. Several of these studies show alterations in aqueous concentrations in patients with glaucoma with likely effects on IOP, though the impact of altered plasma levels is not clear.

A possible consequence of vasoconstrictors leaking from blood vessels is the observation of focal narrowing of retinal vessels in the vicinity of the optic disc margin [156, 158]. It has been proposed that vasoconstrictors diffusing from the choroidal circulation are responsible for this observation which has been related to worsening glaucoma in a spatially corresponding location [112]. While there is no evidence to date, similar phenomena in the posterior ciliary circulation may adversely affect optic nerve head blood supply and contribute to glaucomatous neuropathy.

17.3.2 Effects of Treatment

There is an exhaustive list of published studies on the effect of systemic drugs on ocular blood flow [35]. These vary from acute single dose studies to the more uncommon longitudinal studies reporting on the relatively long-term effect of drugs that increase ocular blood flow on visual function. The list of class of compounds used is large and includes calcium channel blockers, angiotensin-converting enzyme inhibitors and nitric oxide synthase inhibitors. Similarly, the effect of topical medication on ocular blood flow and possible preservation of visual function have been performed. Reviews on this topic have been published elsewhere [35, 113], and evidence from only those studies on glaucoma patients in a longitudinal study design will be covered.

17.3.2.1 Calcium Channel Blockers

Calcium channel blockers (CCB) are the most widely investigated systemic drugs for the non-conventional treatment of glaucoma. Among their actions are blockade of calcium entry in vascular smooth muscle cells leading to vasodilation. Numerous CCBs have been investigated in glaucoma, primarily the dihydropyridines, including nifedipine and nimodipine.

The effects of CCB on optic nerve head blood flow are equivocal. Notwithstanding different classes of CCB and blood flow measurement techniques, some studies show an increase [179, 180, 193], while others show no change [15, 66, 79, 157]. Several studies have demonstrated a more favourable outcome in measures of visual function, primarily visual field [62, 97, 134, 161] but also contrast sensitivity [17]. Other studies have failed to confirm this finding [79, 157]. In two studies, neither blood flow parameters nor visual function changed with CCB [79, 157], while in another, a change in visual function was not associated with change in blood flow [15]. Perhaps the most common thread among these studies is the general lack of improvement in blood flow [66, 79, 157], visual function [79, 157] or optic disc change [123] with the peripherally acting CCB, nifedipine.

Recent evidence from the Rotterdam Study suggests that subjects using CCB were at a significantly higher risk of developing glaucoma [133].

17.3.2.2 Topical Adrenergic Antagonists

Topically instilled adrenergic agents are a main class of compounds to reduce IOP in glaucoma patients and suspects. Most of the research on the effects of this class of drug on ocular blood flow comes from studies with non-selective and selective beta-blockers. Similar to the use of systemic drugs, there is considerable conflicting evidence on whether topical adrenergic agents alter ocular blood flow and whether this is related to changes in visual function. Some evidence exists for potentially protective effects of betaxolol [34, 44, 100, 126], but there are a number of non-supporting reports [102, 187]. Further, only one study reported improvements in visual function associated with changes in blood flow; however, no control group was used [183].

17.3.2.3 Topical Carbonic Anhydrase Inhibitors

There is considerable evidence that topical treatment with carbonic anhydrase inhibitors (CAI) increases blood flow measured in a variety of ocular tissues [9, 58, 77, 124]; however, unsurprisingly these results are not universally supported [13]. There is some evidence that changes in blood flow induced by topical CAI may lead to more favourable visual function outcomes [78, 124].

17.3.2.4 Prostaglandin Analogues

Several classes of prostaglandin analogues are used increasingly in glaucoma therapy. While no long-term controlled studies have been published on visual function, the majority of studies indicate that this class of drug leads to little change in ocular blood flow [5, 80, 135, 198].

17.4 Experimental Models of Ischaemia Relating to Glaucoma

Most models of glaucoma involve elevation of IOP. While the potential effects on ocular blood flow have been debated [144], there exist models of ocular ischaemia without alteration in IOP that have been used to study neuronal loss as it may relate to glaucoma. These models may be divided broadly into acute and chronic.

17.4.1 Acute Ischaemia

Models of acute ischaemia involve elevation of IOP in excess of the central retinal artery pressure to produce transient cessation of blood flow and a subsequent reperfusion injury [20, 165]. Typically, a cannula is introduced into the anterior chamber of rodents and IOP is controlled via a pressure head or syringe pump. Ligation of either the optic nerve [182] or central retinal artery [121] has also been proposed.

Time-related (period of ischaemia and animal survival) loss of RGC is observed with these models [165]; however, unlike in clinical glaucoma, the damage is not generally limited to RGCs. It is also difficult to segregate the effects of potential mechanical damage from the very high levels of IOP or optic nerve ligation from purely ischaemic ones. Loss of neurons in other layers of the retina occurs [95] as well as functional loss of the outer retina [71]. Furthermore, no optic disc cupping which is a hallmark of glaucoma has been reported. For these reasons, while these models may help in studying RGC loss, because of their lack of specificity, they are thought to have little relevance to glaucoma.

17.4.2 Chronic Ischaemia

To counter the confounding effects of acute IOP elevation and the acute nature of the models discussed above, a model of chronic ET-1-induced optic nerve ischaemia was described [142]. Using an osmotic minipump, a small volume of ET-1 is delivered to the retrobulbar optic nerve in monkey [141], rabbit [142] or rat [27]. This model leads to reduction in optic nerve head blood flow [27, 32, 141, 142], RCC and axonal loss [27, 33] and in earlier reports to optic disc cupping in rabbits [142], though these findings have not been confirmed in monkeys[30] or in rats [27].

Because ET-1 leads to optic nerve astrocyte proliferation [149, 185] and changes in expression of MMPs and their tissue inhibitors [84], it is possible that other mechanisms in addition to ischaemia may contribute to ET-1-induced optic neuropathy. Recently, RGC loss was reported at ET-1 concentrations sub-threshold for causing optic nerve head ischaemia [185] adding further evidence that ET-1 has multiple effects that lead to RGC loss.

17.5 Summary

17.5.1 Diversity of Evidence

The role of ischaemia in glaucoma has been debated for decades, and it is likely the debate will continue for the upcoming years. The diversity of evidence in the area is not surprising given that:

(a) Open-angle glaucoma likely has multiple phenotypes in which ocular blood flow may have varying degrees of importance. These phenotypes likely interact with genetic, racial and environmental factors.
(b) Blood flow to the optic nerve is complex in terms of both supply and inter-individual and intra-individual variation; therefore, these factors are important when considering the systematic effects of blood flow changes.
(c) Measurement techniques vary. It is unlikely that two techniques provide the same level of information. For example, some are more sensitive to retinal and superficial optic nerve

head blood flow, while others measure more global parameters.

(d) Research studies have used differing definitions for inclusion criteria. Patients with more advanced damage may show different findings to those who are suspects or have minimal damage.

(e) Effects of ischaemia may be difficult or impossible to isolate from other factors such as IOP. It is more than likely that the phenomenon of glaucomatous optic neuropathy requires multiple factors with ischaemia perhaps being one of many. It has been argued that the phenomenon of optic disc cupping requires the action of IOP, however low the actual IOP level may be. Ischaemia may predispose the nerve to damage at a lower IOP, though this does not mean that ischaemia is the only factor in glaucoma with statistically normal or low IOP.

17.5.2 Evidence Base Compared to Intraocular Pressure

It is generally recognised that IOP is the most well-known risk factor for the development and progression of glaucoma. This evidence comes from years of clinical observation and recently from a number of clinical trials [2, 3, 28, 85, 103] which while very different in design (population, treatment method, diagnostic testing, etc.) concluded that IOP is a causative risk factor for glaucoma and whose treatment leads to favourable outcomes compared to non-treatment. All studies have recognised that in spite of successful IOP reduction, some patients continue to deteriorate; hence, other factors must be at play.

In contrast, little evidence exists to support the causative role of optic nerve head ischaemia in the pathophysiology of glaucoma. Often studies have included poor or no control groups and there is an almost complete absence of randomised controlled trials.

The epidemiologist Bradford-Hill postulated on nine criteria required to establish causality from chance association [19]. These levels of evidence have been modified here:

Table 17.1 Evidence base for the role of intraocular pressure and ischaemia in glaucoma

Level of evidence	Intraocular pressure	Ischaemia
Association	Satisfied	Satisfied
Temporality	Satisfied	Not satisfied
Dose-response	Satisfied	Not satisfied
Reversibility	Partially satisfied	Not satisfied

1. *Association*: Evidence that 'cause' occurs more frequently with 'effect' than chance alone.
2. *Temporality*: Evidence that 'cause' precedes 'effect'.
3. *Dose-response*: Evidence that increased degrees of 'cause' leads to increased levels of 'effect'.
4. *Reversibility*: Evidence that reducing the degree of 'cause' leads to a reduction in the frequency or severity of 'effect'.

The levels of evidence ('satisfied', 'partially satisfied' or 'not satisfied') for the role of IOP and ischaemia are summarised in Table 17.1. based on the evidence to date.

17.5.3 Requirements to Strengthen Evidence Base

Unlike establishing the causative role of IOP in the pathogenesis of open-angle glaucoma, no published clinical trials have been carried out to establish the causal role for ischaemia. With respect to temporality, besides clinical reports in patients with unilateral glaucoma, there are no published studies showing definitively that reduced blood flow leads to a higher incidence or progression of existing glaucoma. The OHTS [67] and EMGT [116] clearly showed a dose-response relationship between the level of IOP exposure and glaucoma. No such similar evidence exists for the role of ischaemia.

Multi-centred clinical studies with uniform definitions and measurement techniques are required to establish temporality and dose-response. Finally, if studies on blood flow and glaucoma provide compelling evidence that ischaemia is causative, then studies designed to ameliorate blood flow to prove a beneficial outcome in glaucoma are the ultimate objectives.

Acknowledgment Supported by Grant MOP57851 from the Canadian Institutes of Health Research.

References

1. Canadian Glaucoma Study Group (2006) Canadian Glaucoma Study: 1. Study design, baseline characteristics, and preliminary analyses. Can J Ophthalmol 41:566–575
2. Collaborative Normal-Tension Glaucoma Study Group (1998) Comparison of glaucomatous progression between untreated patients with normal-tension glaucoma and patients with therapeutically reduced intraocular pressures. Am J Ophthalmol 126: 487–497
3. The AGIS Investigators (2000) The Advanced Glaucoma Intervention Study (AGIS): 7. The relationship between control of intraocular pressure and visual field deterioration. Am J Ophthalmol 130:429–440
4. Airaksinen PJ, Tuulonen A, Werner EB (1996) Clinical evaluation of the optic disc and retinal nerve fiber layer. In: Ritch R, Shields MB, Krupin T (eds) The glaucomas. Mosby, St. Louis, pp 617–657
5. Akarsu C, Yilmaz S, Taner P, Ergin A (2004) Effect of bimatoprost on ocular circulation in patients with open-angle glaucoma or ocular hypertension. Graefes Arch Clin Exp Ophthalmol 242:814–818
6. Anderson DR (1983) The mechanisms of damage of the optic nerve. In: Kriegelstein GK, Leydhecker W (eds) Glaucoma update II. Springer, New York, pp 89–93
7. Anderson DR (1983) What happens to the optic disc and retina in glaucoma? Ophthalmology 90:766–770
8. Anderson DR, Drance SM, Schulzer M (2003) Factors that predict the benefit of lowering intraocular pressure in normal tension glaucoma. Am J Ophthalmol 136:820–829
9. Arend O, Harris A, Wolter P, Remky A (2003) Evaluation of retinal haemodynamics and retinal function after application of dorzolamide, timolol and latanoprost in newly diagnosed open-angle glaucoma patients. Acta Ophthalmol Scand 81:474–479
10. Arend O, Remky A, Cantor LB, Harris A (2000) Altitudinal visual field asymmetry is coupled with altered retinal circulation in patients with normal pressure glaucoma. Br J Ophthalmol 84:1008–1012
11. Bankes JLK, Perkins ES, Tsolakis S, Wright JE (1968) Bedford glaucoma survey. Br Med J 1:791–796
12. Bengtsson B (1981) Aspects of the epidemiology of chronic glaucoma. Acta Ophthalmol Suppl 146:1–48
13. Bergstrand IC, Heijl A, Harris A (2002) Dorzolamide and ocular blood flow in previously untreated glaucoma patients: a controlled double-masked study. Acta Ophthalmol Scand 80:176–182
14. Bill A (1984) Circulation in the eye. In: Renkin EM, Michel CC (eds) Handbook of physiology, vol IV, The cardiovascular system. American Physiological Society, Bethesda, pp 1001–1034
15. Boehm AG, Breidenbach KA, Pillunat LE et al (2003) Visual function and perfusion of the optic nerve head after application of centrally acting calcium-channel blockers. Graefes Arch Clin Exp Ophthalmol 241:34–38
16. Bonomi L, Marchini G, Marraffa M et al (2000) Vascular risk factors for primary open angle glaucoma: the Egna-Neumarkt Study. Ophthalmology 107:1287–1293
17. Bose S, Piltz JR, Breton ME (1995) Nimodipine, a centrally active calcium antagonist, exerts a beneficial effect on contrast sensitivity in patients with normal-tension glaucoma and in control subjects. Ophthalmology 102:1236–1241
18. Bots ML, Hofman A, De Jong PT, Grobbee DE (1996) Common carotid intima-media thickness as an indicator of atherosclerosis at other sites of the carotid artery. The Rotterdam Study. Ann Epidemiol 6:147–153
19. Bradford-Hill A (1965) The environment and disease. Association or causation. Proc R Soc Med 58:295–300
20. Buchi ER, Suivaizdis I, Fu J (1991) Pressure-induced retinal ischemia in rats: an experimental model for quantitative study. Ophthalmologica 203:138–147
21. Burgoyne CF, Downs JC (2008) Premise and prediction-how optic nerve head biomechanics underlies the susceptibility and clinical behavior of the aged optic nerve head. J Glaucoma 17:318–328
22. Burgoyne CF, Downs JC, Bellezza AJ, Hart RT (2004) Three-dimensional reconstruction of normal and early glaucoma monkey optic nerve head connective tissues. Invest Ophthalmol Vis Sci 45:4388–4399
23. Burgoyne CF, Downs JC, Bellezza AJ, Suh JK, Hart RT (2005) The optic nerve head as a biomechanical structure: a new paradigm for understanding the role of IOP-related stress and strain in the pathophysiology of glaucomatous optic nerve head damage. Prog Retin Eye Res 24:39–73
24. Butt Z, O'Brien C, McKillop G, Aspinall P, Allan P (1997) Color Doppler imaging in untreated high- and normal-pressure open-angle glaucoma. Invest Ophthalmol Vis Sci 38:690–696
25. Cellini M, Possati GL, Profazio V et al (1997) Color Doppler imaging and plasma levels of endothelin-1 in low-tension glaucoma. Acta Ophthalmol Scand Suppl 224:11–13
26. Chaturvedi N, Hedley-Whyte ET, Dreyer EB (1993) Lateral geniculate nucleus in glaucoma. Am J Ophthalmol 116:182–188
27. Chauhan BC, LeVatte TL, Jollimore CA et al (2004) Model of endothelin-1-induced chronic optic neuropathy in rat. Invest Ophthalmol Vis Sci 45:144–152
28. Chauhan BC, Mikelberg FS, Balaszi AG et al (2008) Canadian Glaucoma Study: 2. risk factors for the progression of open-angle glaucoma. Arch Ophthalmol 126:1030–1036
29. Choi J, Jeong J, Cho HS, Kook MS (2006) Effect of nocturnal blood pressure reduction on circadian fluctuation of mean ocular perfusion pressure: a risk factor for normal tension glaucoma. Invest Ophthalmol Vis Sci 47:831–836
30. Cioffi GA (2005) Ischemic model of optic nerve injury. Trans Am Ophthalmol Soc 103:592–613

31. Cioffi GA, Orgul S, Onda E, Bacon DR, Van Buskirk EM (1995) An in vivo model of chronic optic nerve ischemia: the dose-dependent effects of endothelin-1 on the optic nerve microvasculature. Curr Eye Res 14:1147–1153
32. Cioffi GA, Sullivan P (1999) The effect of chronic ischemia on the primate optic nerve. Eur J Ophthalmol 9(Suppl 1):S34–S36
33. Cioffi GA, Wang L, Fortune B et al (2004) Chronic ischemia induces regional axonal damage in experimental primate optic neuropathy. Arch Ophthalmol 122:1517–1525
34. Collignon-Brach J (1994) Longterm effect of topical beta-blockers on intraocular pressure and visual field sensitivity in ocular hypertension and chronic open-angle glaucoma. Surv Ophthalmol 38(Suppl): S149–S155
35. Costa VP, Harris A, Stefansson E et al (2003) The effects of antiglaucoma and systemic medications on ocular blood flow. Prog Retin Eye Res 22:769–805
36. Cursiefen C, Wisse M, Cursiefen S et al (2000) Migraine and tension headache in high-pressure and normal-pressure glaucoma. Am J Ophthalmol 129:102–104
37. David R, Livingston D, Luntz MH (1978) Ocular hypertension: a comparative follow-up of black and white patients. Br J Ophthalmol 62:676–678
38. Deokule S, Weinreb RN (2008) Relationships among systemic blood pressure, intraocular pressure, and open-angle glaucoma. Can J Ophthalmol 43:302–307
39. Detry M, Boschi A, Ellinghaus G, De Plaen JF (1996) Simultaneous 24-hour monitoring of intraocular pressure and arterial blood pressure in patients with progressive and non-progressive primary open-angle glaucoma. Eur J Ophthalmol 6:273–278
40. Diederich D, Yang ZH, Buhler FR, Luscher TF (1990) Impaired endothelium-dependent relaxations in hypertensive resistance arteries involve cyclooxygenase pathway. Am J Physiol 258:H445–H451
41. Dielemans I, Vingerling JR, Algra D et al (1995) Primary open-angle glaucoma, intraocular pressure, and systemic blood pressure in the general elderly population. The Rotterdam Study. Ophthalmology 102:54–60
42. Downs JC, Suh JK, Thomas KA et al (2005) Viscoelastic material properties of the peripapillary sclera in normal and early-glaucoma monkey eyes. Invest Ophthalmol Vis Sci 46:540–546
43. Drance S, Anderson DR, Schulzer M (2001) Risk factors for progression of visual field abnormalities in normal-tension glaucoma. Am J Ophthalmol 131: 699–708
44. Drance SM (1998) A comparison of the effects of betaxolol, timolol, and pilocarpine on visual function in patients with open-angle glaucoma. J Glaucoma 7:247–252
45. Drance SM, Douglas GR, Wijsman K, Schulzer M, Britton RJ (1988) Response of blood flow to warm and cold in normal and low-tension glaucoma patients. Am J Ophthalmol 105:35–39
46. Dunwiddie TV, Masino SA (2001) The role and regulation of adenosine in the central nervous system. Annu Rev Neurosci 24:31–55
47. Epstein DL (1997) Primary open-angle glaucoma. In: Epstein DL (ed) Chandler and Grant's glaucoma. Lea and Febinger, Philadelphia, pp 183–231
48. Ernest JT (1975) Pathogenesis of glaucomatous optic nerve disease. Trans Am Ophthalmol Soc 73: 366–388
49. Feke GT, Tagawa H, Deupree DM et al (1989) Blood flow in the normal human retina. Invest Ophthalmol Vis Sci 30:58–65
50. Flammer J, Haefliger IO, Orgul S, Resink T (1999) Vascular dysregulation: a principal risk factor for glaucomatous damage? J Glaucoma 8:212–219
51. Flammer J, Orgul S (1998) Optic nerve blood-flow abnormalities in glaucoma. Prog Retin Eye Res 17: 267–289
52. Flammer J, Orgul S, Costa VP et al (2002) The impact of ocular blood flow in glaucoma. Prog Retin Eye Res 21:359–393
53. Flammer J, Pache M, Resink T (2001) Vasospasm, its role in the pathogenesis of diseases with particular reference to the eye. Prog Retin Eye Res 20:319–349
54. Fontana L, Poinoosawmy D, Bunce CV, O'Brien C, Hitchings RA (1998) Pulsatile ocular blood flow investigation in asymmetric normal tension glaucoma and normal subjects. Br J Ophthalmol 82:731–736
55. Fortes ZB, Garcia Leme J, Scivoletto R (1983) Vascular reactivity in diabetes mellitus: role of the endothelial cell. Br J Pharmacol 79:771–781
56. Foster PJ, Machin D, Wong TY et al (2003) Determinants of intraocular pressure and its association with glaucomatous optic neuropathy in Chinese Singaporeans: the Tanjong Pagar Study. Invest Ophthalmol Vis Sci 44:3885–3891
57. Fuchsjager-Mayrl G, Wally B, Georgopoulos M et al (2004) Ocular blood flow and systemic blood pressure in patients with primary open-angle glaucoma and ocular hypertension. Invest Ophthalmol Vis Sci 45: 834–839
58. Fuchsjager-Mayrl G, Wally B, Rainer G et al (2005) Effect of dorzolamide and timolol on ocular blood flow in patients with primary open angle glaucoma and ocular hypertension. Br J Ophthalmol 89: 1293–1297
59. Galassi F, Nuzzaci G, Sodi A et al (1994) Possible correlations of ocular blood flow parameters with intraocular pressure and visual-field alterations in glaucoma: a study by means of color Doppler imaging. Ophthalmologica 208:304–308
60. Galassi F, Renieri G, Sodi A et al (2004) Nitric oxide proxies and ocular perfusion pressure in primary open angle glaucoma. Br J Ophthalmol 88:757–760
61. Garcia-Valenzuela E, Shareef S, Walsh J, Sharma SC (1995) Programmed cell death of retinal ganglion cells during experimental glaucoma. Exp Eye Res 61:33–44
62. Gaspar AZ, Flammer J, Hendrickson P (1994) Influence of nifedipine on the visual fields of patients with optic-nerve-head diseases. Eur J Ophthalmol 4:24–28
63. Gasser P (1989) Ocular vasospasm: a risk factor in the pathogenesis of low-tension glaucoma. Int Ophthalmol 13:281–290

64. Gasser P, Flammer J (1991) Blood-cell velocity in the nailfold capillaries of patients with normal-tension and high-tension glaucoma. Am J Ophthalmol 111: 585–588
65. Gasser P, Flammer J, Guthauser U, Mahler F (1990) Do vasospasms provoke ocular diseases? Angiology 41:213–220
66. Geyer O, Neudorfer M, Kessler A et al (1996) Effect of oral nifedipine on ocular blood flow in patients with low tension glaucoma. Br J Ophthalmol 80:1060–1062
67. Gordon MO, Beiser JA, Brandt JD et al (2002) The Ocular Hypertension Treatment Study: baseline factors that predict the onset of primary open-angle glaucoma. Arch Ophthalmol 120:714–720; discussion 829–730
68. Graham SL, Drance SM (1999) Nocturnal hypotension: role in glaucoma progression. Surv Ophthalmol 43(Suppl 1):S10–S16
69. Graham SL, Drance SM, Wijsman K, Douglas GR, Mikelberg FS (1995) Ambulatory blood pressure monitoring in glaucoma. The nocturnal dip. Ophthalmology 102:61–69
70. Grieshaber MC, Mozaffarieh M, Flammer J (2007) What is the link between vascular dysregulation and glaucoma? Surv Ophthalmol 52(Suppl 2):S144–S154
71. Grozdanic SD, Sakaguchi DS, Kwon YH, Kardon RH, Sonea IM (2003) Functional characterization of retina and optic nerve after acute ocular ischemia in rats. Invest Ophthalmol Vis Sci 44:2597–2605
72. Grunwald JE, Piltz J, Hariprasad SM, DuPont J (1998) Optic nerve and choroidal circulation in glaucoma. Invest Ophthalmol Vis Sci 39:2329–2336
73. Gupta N, Ang LC, Noel de Tilly L, Bidaisee L, Yucel YH (2006) Human glaucoma and neural degeneration in intracranial optic nerve, lateral geniculate nucleus, and visual cortex. Br J Ophthalmol 90:674–678
74. Hafez AS, Bizzarro R, Descovich D, Lesk MR (2005) Correlation between finger blood flow and changes in optic nerve head blood flow following therapeutic intraocular pressure reduction. J Glaucoma 14: 448–454
75. Hamard P, Hamard H, Dufaux J, Quesnot S (1994) Optic nerve head blood flow using a laser Doppler velocimeter and haemorheology in primary open angle glaucoma and normal pressure glaucoma. Br J Ophthalmol 78:449–453
76. Harju M, Vesti E (2001) Blood flow of the optic nerve head and peripapillary retina in exfoliation syndrome with unilateral glaucoma or ocular hypertension. Graefes Arch Clin Exp Ophthalmol 239:271–277
77. Harris A, Arend O, Chung HS et al (2000) A comparative study of betaxolol and dorzolamide effect on ocular circulation in normal-tension glaucoma patients. Ophthalmology 107:430–434
78. Harris A, Arend O, Kagemann L et al (1999) Dorzolamide, visual function and ocular hemodynamics in normal-tension glaucoma. J Ocul Pharmacol Ther 15:189–197
79. Harris A, Evans DW, Cantor LB, Martin B (1997) Hemodynamic and visual function effects of oral nifedipine in patients with normal-tension glaucoma. Am J Ophthalmol 124:296–302
80. Harris A, Garzozi HJ, McCranor L et al (2009) The effect of latanoprost on ocular blood flow. Int Ophthalmol 29:19–26
81. Harris A, Sergott RC, Spaeth GL et al (1994) Color Doppler analysis of ocular vessel blood velocity in normal-tension glaucoma. Am J Ophthalmol 118: 642–649
82. Harris A, Spaeth G, Wilson R et al (1997) Nocturnal ophthalmic arterial hemodynamics in primary open-angle glaucoma. J Glaucoma 6:170–174
83. Hayreh SS, Zimmerman MB, Podhajsky P, Alward WL (1994) Nocturnal arterial hypotension and its role in optic nerve head and ocular ischemic disorders. Am J Ophthalmol 117:603–624
84. He S, Prasanna G, Yorio T (2007) Endothelin-1-mediated signaling in the expression of matrix metalloproteinases and tissue inhibitors of metalloproteinases in astrocytes. Invest Ophthalmol Vis Sci 48: 3737–3745
85. Heijl A, Leske MC, Bengtsson B et al (2002) Reduction of intraocular pressure and glaucoma progression: results from the Early Manifest Glaucoma Trial. Arch Ophthalmol 120:1268–1279
86. Hernandez MR, Andrzejewska WM, Neufeld AH (1990) Changes in the extracellular matrix of the human optic nerve head in primary open-angle glaucoma. Am J Ophthalmol 109:180–188
87. Hernandez MR, Luo XX, Andrzejewska W, Neufeld AH (1989) Age-related changes in the extracellular matrix of the human optic nerve head. Am J Ophthalmol 107:476–484
88. Hernandez MR, Luo XX, Igoe F, Neufeld AH (1987) Extracellular matrix of the human lamina cribrosa. Am J Ophthalmol 104:567–576
89. Hernandez MR, Wang N, Hanley NM, Neufeld AH (1991) Localization of collagen types I and IV mRNAs in human optic nerve head by in situ hybridization. Invest Ophthalmol Vis Sci 32:2169–2177
90. Hernandez MR, Ye H, Roy S (1994) Collagen type IV gene expression in human optic nerve heads with primary open angle glaucoma. Exp Eye Res 59:41–51
91. Hitchings RA, Spaeth GL (1977) Fluorescein angiography in chronic simple and low-tension glaucoma. Br J Ophthalmol 61:126–132
92. Hollo G, Lakatos P, Farkas K (1998) Cold pressor test and plasma endothelin-1 concentration in primary open-angle and capsular glaucoma. J Glaucoma 7:105–110
93. Hollows FC, Graham PA (1966) Intra-ocular pressure, glaucoma, and glaucoma suspects in a defined population. Br J Ophthalmol 50:570–586
94. Hoskins HD, Kass MA (1989) Primary open-angle glaucoma. In: Hoskins HD, Kass MA (eds) Becker-Shaffer's diagnosis and therapy of the glaucomas. The C.V. Mosby Company, St. Louis, pp 277–307
95. Hughes WF (1991) Quantitation of ischemic damage in the rat retina. Exp Eye Res 53:573–582

96. Hulsman CA, Vingerling JR, Hofman A, Witteman JC, de Jong PT (2007) Blood pressure, arterial stiffness, and open-angle glaucoma: the Rotterdam study. Arch Ophthalmol 125:805–812
97. Ishida K, Yamamoto T, Kitazawa Y (1998) Clinical factors associated with progression of normal-tension glaucoma. J Glaucoma 7:372–377
98. James CB, Smith SE (1991) Pulsatile ocular blood flow in patients with low tension glaucoma. Br J Ophthalmol 75:466–470
99. Johnson EC, Deppmeier LM, Wentzien SK, Hsu I, Morrison JC (2000) Chronology of optic nerve head and retinal responses to elevated intraocular pressure. Invest Ophthalmol Vis Sci 41:431–442
100. Kaiser HJ, Flammer J, Stumpfig D, Hendrickson P (1994) Longterm visual field follow-up of glaucoma patients treated with beta-blockers. Surv Ophthalmol 38 Suppl:S156–S159; discussion S160
101. Kaiser HJ, Flammer J, Wenk M, Luscher T (1995) Endothelin-1 plasma levels in normal-tension glaucoma: abnormal response to postural changes. Graefes Arch Clin Exp Ophthalmol 233:484–488
102. Kamal D, Garway-Heath D, Ruben S et al (2003) Results of the betaxolol versus placebo treatment trial in ocular hypertension. Graefes Arch Clin Exp Ophthalmol 241:196–203
103. Kass MA, Heuer DK, Higginbotham EJ et al (2002) The Ocular Hypertension Treatment Study: a randomized trial determines that topical ocular hypotensive medication delays or prevents the onset of primary open-angle glaucoma. Arch Ophthalmol 120:701–713; discussion 829–730
104. Katusic ZS, Vanhoutte PM (1989) Superoxide anion is an endothelium-derived contracting factor. Am J Physiol 257:H33–H37
105. Kerr J, Nelson P, O'Brien C (1998) A comparison of ocular blood flow in untreated primary open-angle glaucoma and ocular hypertension. Am J Ophthalmol 126:42–51
106. Kitazawa Y, Horie T, Aoki S, Suzuki M, Nishioka K (1977) Untreated ocular hypertension. A long-term prospective study. Arch Ophthalmol 95:1180–1184
107. Klein BE, Klein R, Knudtson MD (2005) Intraocular pressure and systemic blood pressure: longitudinal perspective: the Beaver Dam Eye Study. Br J Ophthalmol 89:284–287
108. Klein BE, Klein R, Linton KL (1992) Intraocular pressure in an American community. The Beaver Dam Eye Study. Invest Ophthalmol Vis Sci 33: 2224–2228
109. Klein BE, Klein R, Meuer SM, Goetz LA (1993) Migraine headache and its association with open-angle glaucoma: the Beaver Dam Eye Study. Invest Ophthalmol Vis Sci 34:3024–3027
110. Kunimatsu S, Mayama C, Tomidokoro A, Araie M (2006) Plasma endothelin-1 level in Japanese normal tension glaucoma patients. Curr Eye Res 31: 727–731
111. Labarrere CA, Zaloga GP (2004) C-reactive protein: from innocent bystander to pivotal mediator of atherosclerosis. Am J Med 117:499–507
112. Lam A, Bunya V, Piltz-Seymour JR (2002) Visual field loss in patients with glaucoma who have asymmetric peripapillary focal arteriolar narrowing. Arch Ophthalmol 120:1494–1497
113. Lesk MR, Wajszilber M, Deschenes MC (2008) The effects of systemic medications on ocular blood flow. Can J Ophthalmol 43:351–355
114. Leske MC, Connell AM, Wu SY, Hyman LG, Schachat AP (1995) Risk factors for open-angle glaucoma. The Barbados Eye Study. Arch Ophthalmol 113:918–924
115. Leske MC, Connell AM, Wu SY et al (2001) Incidence of open-angle glaucoma: the Barbados Eye Studies. The Barbados Eye Studies Group. Arch Ophthalmol 119:89–95
116. Leske MC, Heijl A, Hussein M et al (2003) Factors for glaucoma progression and the effect of treatment: the early manifest glaucoma trial. Arch Ophthalmol 121:48–56
117. Leske MC, Heijl A, Hyman L et al (2007) Predictors of long-term progression in the early manifest glaucoma trial. Ophthalmology 114:1965–1972
118. Leske MC, Wu SY, Hennis A, Honkanen R, Nemesure B (2008) Risk factors for incident open-angle glaucoma: the Barbados Eye Studies. Ophthalmology 115:85–93
119. Leske MC, Wu SY, Nemesure B, Hennis A (2002) Incident open-angle glaucoma and blood pressure. Arch Ophthalmol 120:954–959
120. Levin ER (1996) Endothelins as cardiovascular peptides. Am J Nephrol 16:246–251
121. Levy NS, Adams CK (1975) Slow axonal protein transport and visual function following retinal and optic nerve ischemia. Invest Ophthalmol 14:91–97
122. Lu L, Gunja-Smith Z, Woessner JF et al (2000) Matrix metalloproteinases and collagen ultrastructure in moderate myocardial ischemia and reperfusion in vivo. Am J Physiol Heart Circ Physiol 279:H601–H609
123. Lumme P, Tuulonen A, Airaksinen PJ, Alanko HI (1991) Neuroretinal rim area in low tension glaucoma: effect of nifedipine and acetazolamide compared to no treatment. Acta Ophthalmol (Copenh) 69:293–298
124. Martinez A, Sanchez M (2008) Effects of dorzolamide 2% added to timolol maleate 0.5% on intraocular pressure, retrobulbar blood flow, and the progression of visual field damage in patients with primary open-angle glaucoma: a single-center, 4-year, open-label study. Clin Ther 30:1120–1134
125. Massy ZA, Keane WF (1996) Pathogenesis of atherosclerosis. Semin Nephrol 16:12–20
126. Messmer C, Flammer J, Stumpfig D (1991) Influence of betaxolol and timolol on the visual fields of patients with glaucoma. Am J Ophthalmol 112:678–681
127. Michelson G, Langhans MJ, Groh MJ (1996) Perfusion of the juxtapapillary retina and the neuroretinal rim area in primary open angle glaucoma. J Glaucoma 5:91–98

128. Mitchell P, Lee AJ, Rochtchina E, Wang JJ (2004) Open-angle glaucoma and systemic hypertension: the blue mountains eye study. J Glaucoma 13: 319–326
129. Moncada S, Palmer RM, Higgs EA (1991) Nitric oxide: physiology, pathophysiology, and pharmacology. Pharmacol Rev 43:109–142
130. Moncada S, Vane JR (1978) Pharmacology and endogenous roles of prostaglandin endoperoxides, thromboxane A2, and prostacyclin. Pharmacol Rev 30:293–331
131. Morrison JC, Dorman-Pease ME, Dunkelberger GR, Quigley HA (1990) Optic nerve head extracellular matrix in primary optic atrophy and experimental glaucoma. Arch Ophthalmol 108:1020–1024
132. Morrison JC, L'Hernault NL, Jerdan JA, Quigley HA (1989) Ultrastructural location of extracellular matrix components in the optic nerve head. Arch Ophthalmol 107:123–129
133. Muskens RP, de Voogd S, Wolfs RC et al (2007) Systemic antihypertensive medication and incident open-angle glaucoma. Ophthalmology 114: 2221–2226
134. Netland PA, Chaturvedi N, Dreyer EB (1993) Calcium channel blockers in the management of low-tension and open-angle glaucoma. Am J Ophthalmol 115:608–613
135. Nicolela MT, Buckley AR, Walman BE, Drance SM (1996) A comparative study of the effects of timolol and latanoprost on blood flow velocity of the retrobulbar vessels. Am J Ophthalmol 122:784–789
136. Nicolela MT, Drance SM, Rankin SJ, Buckley AR, Walman BE (1996) Color Doppler imaging in patients with asymmetric glaucoma and unilateral visual field loss. Am J Ophthalmol 121:502–510
137. Nicolela MT, Ferrier SN, Morrison CA et al (2003) Effects of cold-induced vasospasm in glaucoma: the role of endothelin-1. Invest Ophthalmol Vis Sci 44:2565–2572
138. Nicolela MT, Hnik P, Drance SM (1996) Scanning laser Doppler flowmeter study of retinal and optic disk blood flow in glaucomatous patients. Am J Ophthalmol 122:775–783
139. Nicolela MT, Walman BE, Buckley AR, Drance SM (1996) Various glaucomatous optic nerve appearances. A color Doppler imaging study of retrobulbar circulation. Ophthalmology 103:1670–1679
140. Onda E, Cioffi GA, Bacon DR, Van Buskirk EM (1995) Microvasculature of the human optic nerve. Am J Ophthalmol 120:92–102
141. Orgul S, Cioffi GA, Bacon DR, Van Buskirk EM (1996) An endothelin-1-induced model of chronic optic nerve ischemia in rhesus monkeys. J Glaucoma 5:135–138
142. Orgul S, Cioffi GA, Wilson DJ, Bacon DR, Van Buskirk EM (1996) An endothelin-1 induced model of optic nerve ischemia in the rabbit. Invest Ophthalmol Vis Sci 37:1860–1869
143. Osborne NN, Casson RJ, Wood JP et al (2004) Retinal ischemia: mechanisms of damage and potential therapeutic strategies. Prog Retin Eye Res 23: 91–147
144. Pang IH, Clark AF (2007) Rodent models for glaucoma retinopathy and optic neuropathy. J Glaucoma 16:483–505
145. Pease ME, McKinnon SJ, Quigley HA, Kerrigan-Baumrind LA, Zack DJ (2000) Obstructed axonal transport of BDNF and its receptor TrkB in experimental glaucoma. Invest Ophthalmol Vis Sci 41: 764–774
146. Phelps CD, Corbett JJ (1985) Migraine and low-tension glaucoma. A case-control study. Invest Ophthalmol Vis Sci 26:1105–1108
147. Piltz-seymour JR, Grunwald JE, Hariprasad SM, Dupont J (2001) Optic nerve blood flow is diminished in eyes of primary open-angle glaucoma suspects. Am J Ophthalmol 132:63–69
148. Pournaras CJ (1996) Autoregulation of ocular blood flow. In: Kaiser HJ, Flammer J, Hendrickson P (eds) Ocular blood flow. Karger, Basel, pp 40–50
149. Prasanna G, Krishnamoorthy R, Clark AF, Wordinger RJ, Yorio T (2002) Human optic nerve head astrocytes as a target for endothelin-1. Invest Ophthalmol Vis Sci 43:2704–2713
150. Quigley HA (1999) Neuronal death in glaucoma. Prog Retin Eye Res 18:39–57
151. Quigley HA, Addicks EM, Green WR, Maumenee AE (1981) Optic nerve damage in human glaucoma. II. The site of injury and susceptibility to damage. Arch Ophthalmol 99:635–649
152. Quigley HA, Broman AT (2006) The number of people with glaucoma worldwide in 2010 and 2020. Br J Ophthalmol 90:262–267
153. Quigley HA, Hohman RM, Addicks EM, Massof RW, Green WR (1983) Morphologic changes in the lamina cribrosa correlated with neural loss in open-angle glaucoma. Am J Ophthalmol 95: 673–691
154. Quigley HA, Nickells RW, Kerrigan LA et al (1995) Retinal ganglion cell death in experimental glaucoma and after axotomy occurs by apoptosis. Invest Ophthalmol Vis Sci 36:774–786
155. Quigley HA, West SK, Rodriguez J et al (2001) The prevalence of glaucoma in a population-based study of Hispanic subjects: Proyecto VER. Arch Ophthalmol 119:1819–1826
156. Rader J, Feuer WJ, Anderson DR (1994) Peripapillary vasoconstriction in the glaucomas and the anterior ischemic optic neuropathies. Am J Ophthalmol 117: 72–80
157. Rainer G, Kiss B, Dallinger S et al (2001) A double masked placebo controlled study on the effect of nifedipine on optic nerve blood flow and visual field function in patients with open angle glaucoma. Br J Clin Pharmacol 52:210–212
158. Rankin SJ, Drance SM (1996) Peripapillary focal retinal arteriolar narrowing in open angle glaucoma. J Glaucoma 5:22–28
159. Rubanyi GM, Vanhoutte PM (1986) Superoxide anions and hyperoxia inactivate endothelium-derived relaxing factor. Am J Physiol 250:H822–H827

160. Saenz de Tejada I, Goldstein I, Azadzoi K, Krane RJ, Cohen RA (1989) Impaired neurogenic and endothelium-mediated relaxation of penile smooth muscle from diabetic men with impotence. N Engl J Med 320:1025–1030
161. Sawada A, Kitazawa Y, Yamamoto T, Okabe I, Ichien K (1996) Prevention of visual field defect progression with brovincamine in eyes with normal-tension glaucoma. Ophthalmology 103:283–288
162. Schulzer M, Drance SM, Carter CJ et al (1990) Biostatistical evidence for two distinct chronic open angle glaucoma populations. Br J Ophthalmol 74: 196–200
163. Schumann J, Orgul S, Gugleta K, Dubler B, Flammer J (2000) Interocular difference in progression of glaucoma correlates with interocular differences in retrobulbar circulation. Am J Ophthalmol 129: 728–733
164. Schwartz B, Rieser JC, Fishbein SL (1977) Fluorescein angiographic defects of the optic disc in glaucoma. Arch Ophthalmol 95:1961–1974
165. Selles-Navarro I, Villegas-Perez MP, Salvador-Silva M, Ruiz-Gomez JM, Vidal-Sanz M (1996) Retinal ganglion cell death after different transient periods of pressure-induced ischemia and survival intervals. A quantitative in vivo study. Invest Ophthalmol Vis Sci 37:2002–2014
166. Shimokawa H, Aarhus LL, Vanhoutte PM (1987) Porcine coronary arteries with regenerated endothelium have a reduced endothelium-dependent responsiveness to aggregating platelets and serotonin. Circ Res 61:256–270
167. Shimokawa H, Flavahan NA, Vanhoutte PM (1989) Natural course of the impairment of endothelium-dependent relaxations after balloon endothelium removal in porcine coronary arteries. Possible dysfunction of a pertussis toxin-sensitive G protein. Circ Res 65:740–753
168. Sigal IA, Flanagan JG, Ethier CR (2005) Factors influencing optic nerve head biomechanics. Invest Ophthalmol Vis Sci 46:4189–4199
169. Silver DM, Farrell RA, Langham ME, O'Brien V, Schilder P (1989) Estimation of pulsatile ocular blood flow from intraocular pressure. Acta Ophthalmol Suppl 191:25–29
170. Sommer A (1989) Intraocular pressure and glaucoma. Am J Ophthalmol 107:186–188
171. Sommer A, Tielsch JM, Katz J et al (1991) Relationship between intraocular pressure and primary open angle glaucoma among white and black Americans. The Baltimore Eye Survey. Arch Ophthalmol 109:1090–1095
172. Sossi N, Anderson DR (1983) Blockage of axonal transport in optic nerve induced by elevation of intraocular pressure. Effect of arterial hypertension induced by angiotensin I. Arch Ophthalmol 101: 94–97
173. Stromberg U (1962) Ocular hypertension. Acta Ophthalmol Suppl 69:1–75
174. Sugiyama T, Moriya S, Oku H, Azuma I (1995) Association of endothelin-1 with normal tension glaucoma: clinical and fundamental studies. Surv Ophthalmol 39(Suppl 1):S49–S56
175. Tengroth B, Rehnberg M, Amitzboll T (1985) A comparative analysis of the collagen type and distribution in the trabecular meshwork, sclera, lamina cribrosa and the optic nerve in the human eye. Acta Ophthalmol Suppl 173:91–93
176. Tezel G, Kass MA, Kolker AE, Becker B, Wax MB (1997) Plasma and aqueous humor endothelin levels in primary open-angle glaucoma. J Glaucoma 6:83–89
177. Tielsch JM, Katz J, Sommer A, Quigley HA, Javitt JC (1995) Hypertension, perfusion pressure, and primary open-angle glaucoma. A population-based assessment. Arch Ophthalmol 113:216–221
178. Tokunaga T, Kashiwagi K, Tsumura T, Taguchi K, Tsukahara S (2004) Association between nocturnal blood pressure reduction and progression of visual field defect in patients with primary open-angle glaucoma or normal-tension glaucoma. Jpn J Ophthalmol 48:380–385
179. Tomita G, Niwa Y, Shinohara H et al (1999) Changes in optic nerve head blood flow and retrobular hemodynamics following calcium-channel blocker treatment of normal-tension glaucoma. Int Ophthalmol 23:3–10
180. Tomita K, Araie M, Tamaki Y, Nagahara M, Sugiyama T (1999) Effects of nilvadipine, a calcium antagonist, on rabbit ocular circulation and optic nerve head circulation in NTG subjects. Invest Ophthalmol Vis Sci 40:1144–1151
181. Trew DR, Smith SE (1991) Postural studies in pulsatile ocular blood flow: II. Chronic open angle glaucoma. Br J Ophthalmol 75:71–75
182. Tsang D, Yew DT, Lam ST (1985) Acute responses of rat retina after optic nerve ligation: a biochemical and histochemical study. Brain Res 336:289–295
183. Turacli ME, Ozden RG, Gurses MA (1998) The effect of betaxolol on ocular blood flow and visual fields in patients with normotension glaucoma. Eur J Ophthalmol 8:62–66
184. Wang JJ, Mitchell P, Smith W (1997) Is there an association between migraine headache and open-angle glaucoma? Findings from the Blue Mountains Eye Study. Ophthalmology 104:1714–1719
185. Wang X, LeVatte TL, Archibald ML, Chauhan BC (2009) Increase in endothelin B receptor expression in optic nerve astrocytes in endothelin-1 induced chronic experimental optic neuropathy. Exp Eye Res 88:378–385
186. Warner TD, Klemm P (1996) What turns on the endothelins? Inflamm Res 45:51–53
187. Watson PG, Barnett MF, Parker V, Haybittle J (2001) A 7 year prospective comparative study of three topical beta blockers in the management of primary open angle glaucoma. Br J Ophthalmol 85:962–968
188. Weber AJ, Chen H, Hubbard WC, Kaufman PL (2000) Experimental glaucoma and cell size, density, and number in the primate lateral geniculate nucleus. Invest Ophthalmol Vis Sci 41:1370–1379

189. Woodburn KR, Lowe GD (1997) Fibrinogen, fibrin turnover, endothelial products and vascular surgery. Br J Surg 84:1059–1064
190. Wu SY, Leske MC (1997) Associations with intraocular pressure in the Barbados Eye Study. Arch Ophthalmol 115:1572–1576
191. Wu SY, Nemesure B, Hennis A, Leske MC (2006) Nine-year changes in intraocular pressure: the Barbados Eye Studies. Arch Ophthalmol 124: 1631–1636
192. Xu L, Wang H, Wang Y, Jonas JB (2007) Intraocular pressure correlated with arterial blood pressure: the Beijing eye study. Am J Ophthalmol 144:461–462
193. Yamamoto T, Niwa Y, Kawakami H, Kitazawa Y (1998) The effect of nilvadipine, a calcium-channel blocker, on the hemodynamics of retrobulbar vessels in normal-tension glaucoma. J Glaucoma 7:301–305
194. Yamazaki Y, Drance SM (1997) The relationship between progression of visual field defects and retrobulbar circulation in patients with glaucoma. Am J Ophthalmol 124:287–295
195. Yanagisawa M, Kurihara H, Kimura S et al (1988) A novel potent vasoconstrictor peptide produced by vascular endothelial cells. Nature 332:411–415
196. Yu DY, Su EN, Cringle SJ, Yu PK (2003) Isolated preparations of ocular vasculature and their applications in ophthalmic research. Prog Retin Eye Res 22:135–169
197. Yucel YH, Zhang Q, Gupta N, Kaufman PL, Weinreb RN (2000) Loss of neurons in magnocellular and parvocellular layers of the lateral geniculate nucleus in glaucoma. Arch Ophthalmol 118:378–384
198. Zeitz O, Matthiessen ET, Reuss J et al (2005) Effects of glaucoma drugs on ocular hemodynamics in normal tension glaucoma: a randomized trial comparing bimatoprost and latanoprost with dorzolamide [ISRCTN18873428]. BMC Ophthalmol 5:6

Ocular Blood Flow in Disease: Other Diseases

18

Makoto Araie and Junko Kami

Core Messages

- Central retinal artery (CRA) occlusion is often associated with intraluminal calcific emboli in the retro-laminar portion of the central retinal artery (CRA).
- Central retinal vein (CRV) occlusion is associated with decrease in CRA and CRV blood velocity and increase in vascular resistance in the CRA.
- Ocular ischemic syndrome is most often caused by severe stenosis of internal carotid artery (ICA) and reversal of ophthalmic artery (OA) blood flow is a feature of ICA occlusion.
- Uveitis such as Behcet's disease or sarcoidosis is reportedly associated with decreased blood velocity in the CRA and posterior ciliary arteries (PCA).
- Non-arteritic anterior ischemic optic neuropathy (NAION) is associated with decreased blood velocity in the CRA, PCA or OA. Disturbance of retrobulbar circulation is also reported in patients with optic neuritis.
- In essential hypertension, blood velocity in the CRA or PCA was decreased, implying increased peripheral vascular resistance, while choroidal blood flow is less affected.
- Graves Ophthalmopathy is typically associated with increased blood velocity in OA, CRA or CRV and decreased blood velocity in the superior ophthalmic vein and possibly choroidal blood flow.

Color Doppler imaging (CDI) has been most widely used to estimate retrobulbar hemodynamics in various ocular disorders and provided new insight into the pathogenesis of the various diseases or conditions concerned [1]. Discussed below are a few clinical entities other than glaucoma, diabetes, and age-related macular degeneration in which ocular blood flow has been studied with a view to creating a better understanding of their underlying pathogenesis.

18.1 Retinal Diseases

Central retinal artery occlusion (CRAO) is considered to be an acute stroke of the eye that results in profound visual loss. Spontaneous recovery rates are poor. Most CRAOs are caused by thromboembolism in the central retinal artery (CRA). Sergott et al., using orbital CDI performed in patients with CRAO, identified the presence of intraluminal calcific emboli in the CRA in 42% of patients. Eighty

M. Araie, M.D., Ph.D., FARVO(✉) • J. Kami, M.D., Ph.D.
Department of Ophthalmology, University of Tokyo School of Medicine, 7-3-1 Hongo, Bunkyo-Ku, Tokyo, 113-8655, Japan
e-mail: araie-tky@umin.net; kami-tky@umin.net

L. Schmetterer, J.W. Kiel (eds.), *Ocular Blood Flow*,
DOI 10.1007/978-3-540-69469-4_18, © Springer-Verlag Berlin Heidelberg 2012

percent of these plaques were located in the retrolaminar portion of the CRA [2]. Although thrombolytic therapy for acute management of CRAO has shown promise in nonrandomized studies with regard to improving visual outcomes, the finding of Sergott et al. may be useful for a clinician to justify not administering systemic anticoagulation in patients with intra-arterial calcific embolus, as opposed to those with an echographically visible platelet-fibrin embolus.

Retinal vein occlusion is one of the most frequent ocular vascular diseases. Ischemic retinal vein occlusion leads to severe impairment of vision. Keyser et al., as well as other authors, reported decreases in peak systolic velocities (PSV) and increases in vascular resistance in CRA of patients with central retinal vein occlusion (CRVO), as compared with unaffected eyes [3, 4]. These findings reflect an impedance to outflow in the retinal circulation. There were no hemodynamic abnormalities observed in the ophthalmic arteries (OA). In the prospective monitoring of both ischemic and nonischemic retinal vein occlusion patients for a year (performed by Arséne et al.), CDI showed persistent impairment of central retinal vein (CRV) velocity whereas CRA velocity was only briefly modified [5]. The rapid normalization of arterial flow was mostly related to the development of collateral channels rather than recanalization of the CRV. Therefore, CRVO appears to be caused by alterations in venous flow, and the severity of the disease may be correlated with the degree of occlusion causing the initial deterioration in arterial flow [4, 6]. Williamson et al. found significant reduction in the minimum velocity of CRV in patients who developed iris neovascularization as compared with patients who did not develop this complication [7].

The ocular ischemic syndrome is the clinical manifestation of chronic ocular hypoperfusion and is most often caused by severe stenosis of the internal carotid artery (ICA) [8, 9]. Reversal of flow in the OA is a well-recognized feature of ICA occlusion [10–12], as a result of development of collateral circulations from branches of the external carotid artery to the circle of Willis [13]. OA flow reversal occurs when there is inadequate crossflow in the circle of Willis from the contralateral ICA, which produces a vascular steal phenomenon accounting for the observed clinical manifestations of iris neovascularization, relative ocular hypotony, and retinal hemorrhages. OA flow reversal can readily be confirmed by CDI. Hu et al. reported that flow velocities in the OA and CRA decrease as the degree of carotid stenosis increases [14–16]. According to Lee et al., CRA flow may not be detected in patients with complete occlusion of the ipsilateral ICA [17].

18.2 Uveitis

The most prominent feature of Behçet's disease is systemic and retinal vasculitis with endothelial dysfunction, the mechanism of which includes arterial and vascular thrombosis [18, 19]. Atilla et al. performed a hemodynamic comparison using CDI in four groups of patients: patients with Behçet's disease (with or without ocular involvement), patients with different etiological uveitis, and a healthy control group [20, 21]. The PSV and end-diastolic velocity (EDV) in the CRA and posterior ciliary artery (PCA) were significantly reduced in patients with Behçet's disease and uveitis with different etiologies when compared with healthy volunteers. The change was absent with respect to the OA. The reduction in blood flow values in patients with Behçet's disease is more evident in those with ocular involvement than in those without ocular involvement [22–26]. This might be the result of the inflammatory process in the occlusive vasculitis, which is frequently observed in the retinal vessels of patients with Behçet's disease. The vasculitic involvement of the PCA is more specific to Behçet's disease and occurs later in the course of the disease [20].

Using CDI, Yamada et al. reported that blood flow velocities in OA in patients with ocular sarcoidosis, characterized by obliterative arteriolitis [27], were significantly lower than those in healthy controls. The flow velocities in orbital vessels may vary with the clinical findings of the patient and the severity of the vascular involvement [28].

18.3 Optic Nerve Disorders

Nonarteritic anterior ischemic optic neuropathy (NAION) is an infarction of the optic nerve head in which blood supply from the short posterior ciliary arteries (SPCA) is compromised at the level of the lamina cribosa, leading to sudden vision loss. Flaharty et al. performed a hemodynamic study of patients with NAION and observed decreased PSV in the CRA, SPCA, and OA [29, 30]. There was an increase in the PSV of all vessels after optic nerve sheath decompression in the patients, although currently, there are no proven treatments for NAION [31].

Modrezejewska et al. observed ocular circulation abnormalities in multiple sclerosis patients with retrobulbar optic neuritis by means of CDI [32]. Systolic, diastolic, and mean velocities of blood flow were diminished in the CRA and SPCA. However, according to Karaali et al., blood flow velocity can increase in OA during the initial period of acute inflammation of the optic nerve [33]. The increase was not observed in other arteries. Akarsu et al. found similar results in the CRA using CDI performed several months after optic neuritis [34]. Reduction in blood flow in arteries occur both in the eyes previously affected by optic neuritis and in contralateral, unaffected eyes. This may indicate that circulation disturbances occur both locally and systemically in patients with optic neuritis.

18.4 Systemic Diseases

Using high-resolution duplex scanner imaging (DSI), Steigerwalt et al. found that PSV and EDV were significantly reduced in the CRA and PCA in patients with essential hypertension. The reduced diastolic component of both CRA and PCA implies increased peripheral vascular resistance. This indicates that there is peripheral vasospasm in hypertensive patients leading to the decreased orbital and ocular blood flow [35]. Niknam et al. demonstrated that systemic hypertension does not have a large effect on the choroidal circulation using laser Doppler flowmetry [36]. Choroidal blood flow is maintained despite an increase in perfusion pressure, suggesting that in humans, the choroidal vasculature may autoregulate [37, 38].

In Graves' ophthalmopathy (GO), venous obstruction is thought to lead to proptosis, periorbital swelling, and chemosis in patients with extraocular muscle changes. Using CDI, Alp et al. found increased velocity in OA, CRA, and CRV and decreased velocity in superior ophthalmic vein (SOV) in patients with GO. The increased blood flow velocities probably resulted from either the cardiovascular effect of hyperthyroidism or from orbital inflammation [39–41]. Tsai et al. observed reduced pulsatile ocular blood flow (POBF) in GO patients [42]. The pulsatile component of ocular blood flow is primarily choroidal [43]. The elevated intraorbital pressure associated with severe autoimmune congestive inflammatory orbitopathy or contraction of the extraocular muscles against the fibrotic adhesions may result in elevated venous pressure and increased resistance of choroidal vessels.

References

1. Erickson SJ, Hendrix LE, Massaro BM, Harris GJ, Lewandowski MF, Foley WD, Lawson TL (1989) Color Doppler flow imaging of the normal and abnormal orbit. Radiology 173(2):511–516
2. Sergott RC, Flaharty PM, Lieb WE Jr, Ho AC, Kay MD, Mittra RA, Savino PJ, Bosley TM (1992) Color Doppler imaging identifies four syndromes of the retrobulbar circulation in patients with amaurosis fugax and central retinal artery occlusions. Trans Am Ophthalmol Soc 90:383–398
3. Keyser BJ, Flaharty PM, Sergott RC, Brown GC, Lieb WE, Annesley WH Jr (1994) Color Doppler imaging of arterial blood flow in central retinal vein occlusion. Ophthalmology 101(8):1357–1361
4. Baxter GM, Williamson TH (1993) Color Doppler flow imaging in central retinal vein occlusion: a new diagnostic technique? Radiology 187(3):847–850
5. Arséne S, Giraudeau B, Le Lez ML, Pisella PJ, Pourcelot L, Tranquart F (2002) Follow up by colour Doppler imaging of 102 patients with retinal vein occlusion over 1 year. Br J Ophthalmol 86(11):1243–1247
6. Baxter GM, Williamson TH (1996) The value of serial Doppler imaging in central retinal vein occlusion: correlation with visual recovery. Clin Radiol 51(6):411–414
7. Williamson TH, Baxter GM (1994) Central retinal vein occlusion, an investigation by color Doppler imaging. Blood velocity characteristics and prediction of iris neovascularization. Ophthalmology 101(8):1362–1372
8. Brown GC, Magargal LE (1988) The ocular ischemic syndrome. Clinical, fluorescein angiographic and carotid angiographic features. Int Ophthalmol 11(4):239–251
9. Costa VP, Kuzniec S, Molnar LJ, Cerri GG, Puech-Leão P, Carvalho CA (1997) Clinical findings and hemodynamic changes associated with severe occlusive carotid artery disease. Ophthalmology 104(12): 1994–2002
10. PITTS FW (1962) Variations of collateral circulation in internal carotid occlusion. Comparison of clinical and x-ray findings. Neurology 12:467–471

11. Hodek-Demarin V, Müller HR (1979) Reversed ophthalmic artery flow in internal carotid artery occlusion. A re-appraisal based on ultrasonic Doppler investigations. Stroke 10(4):461–463
12. Tatemichi TK, Chamorro A, Petty GW, Khandji A, Oropeza LA, Duterte DI, Mohr JP (1990) Hemodynamic role of ophthalmic artery collateral in internal carotid artery occlusion. Neurology 40(3 Pt 1):461–464
13. Chen CS, Miller NR (2007) Ocular ischemic syndrome: review of clinical presentations, etiology, investigation, and management. Compr Ophthalmol Update 8(1):17–28
14. Hu HH, Sheng WY, Yen MY, Lai ST, Teng MM (1993) Color Doppler imaging of orbital arteries for detection of carotid occlusive disease. Stroke 24(8):1196–1203
15. Ho AC, Lieb WE, Flaharty PM, Sergott RC, Brown GC, Bosley TM, Savino PJ (1992) Color Doppler imaging of the ocular ischemic syndrome. Ophthalmology 99(9):1453–1462
16. Mawn LA, Hedges TR 3rd, Rand W, Heggerick PA (1997) Orbital color Doppler imaging in carotid occlusive disease. Arch Ophthalmol 115(4):492–496
17. Lee HM, Fu ER (1997) Orbital colour Doppler imaging in chronic ocular ischaemic syndrome. Aust N Z J Ophthalmol 25(2):157–163
18. Lie JT (1992) Vascular involvement in Behçet's disease: arterial and venous and vessels of all sizes. J Rheumatol 19(3):341–343
19. Schmitz-Huebner U, Knop J (1984) Evidence for an endothelial cell dysfunction in association with Behçet's disease. Thromb Res 34(4):277–285
20. Atilla H, Zilelioğlu G, Ozdemir H, Atilla S, Isik S (1997) Color Doppler imaging in uveitis. Eur J Ophthalmol 7(1):92–100
21. Ozdemir H, Atilla H, Atilla S, Işik S, Zilelioğlu G (1995) Diagnosis of ocular involvement in Behçet's disease: value of spectral and color Doppler sonography. AJR Am J Roentgenol 164(5):1223–1227
22. Oner A, Akal A, Erdogan N, Dogan H, Oner M (2006) Color Doppler imaging of ocular hemodynamic changes in Behçet disease and uveitis patients with different etiologies. Curr Eye Res 31(6):519–523
23. Uçakhan O, Salih M, Altan S, Ozdemir O (1997) Color Doppler ultrasound in ocular Behçet's disease. Eur J Ophthalmol 7(3):256–261
24. Celebi S, Akfirat M, Celebi H, Alagöz G (2000) Color Doppler ultrasonography in ocular Behçet's disease. Acta Ophthalmol Scand 78(1):30–33
25. Duranoğlu Y, Apaydin C, Karaali K, Yücel I, Apaydin A (2001) Color Doppler imaging of the orbital vessels in Behçet's disease. Ophthalmologica 215(1):8–15
26. Isik C, Yagci B, Yildirim C, Yaylali V, Tatlipinar S, Ozden S (2007) Orbital color Doppler imaging in Behçet's disease with or without ocular involvement. Int Ophthalmol 27(1):37–42
27. James DG (1986) Ocular sarcoidosis. Ann N Y Acad Sci 465:551–563
28. Yamada R, Ueno S, Yamada S (2002) Assessment of blood flow in orbital arteries in ocular sarcoidosis. Curr Eye Res 24(3):219–223
29. Flaharty PM, Sergott RC, Lieb W, Bosley TM, Savino PJ (1993) Optic nerve sheath decompression may improve blood flow in anterior ischemic optic neuropathy. Ophthalmology 100(3):297–302
30. Collignon-Robe NJ, Feke GT, Rizzo JF 3rd (2004) Optic nerve head circulation in nonarteritic anterior ischemic optic neuropathy and optic neuritis. Ophthalmology 111(9):1663–1672
31. Arnold AC, Levin LA (2002) Treatment of ischemic optic neuropathy. Semin Ophthalmol 17(1):39–46
32. Modrzejewska M, Karczewicz D, Wilk G (2007) Assessment of blood flow velocity in eyeball arteries in multiple sclerosis patients with past retrobulbar optic neuritis in color Doppler ultrasonography. Klin Oczna 109(4–6):183–186
33. Karaali K, Senol U, Aydin H, Cevikol C, Apaydin A, Lüleci E (2003) Optic neuritis: evaluation with orbital Doppler sonography. Radiology 226(2):355–358
34. Akarsu C, Tan FU, Kendi T (2004) Color Doppler imaging in optic neuritis with multiple sclerosis. Graefes Arch Clin Exp Ophthalmol 242(12):990–994
35. Steigerwalt RD Jr, Belcaro GV, Laurora G, Cesarone MR, De Sanctis MT, Incandela L (1998) Ocular and orbital blood flow in patients with essential hypertension treated with trandolapril. Retina 18(6):539–545
36. Niknam RM, Schocket LS, Metelitsina T, DuPont JC, Grunwald JE (2004) Effect of hypertension on foveolar choroidal haemodynamics. Br J Ophthalmol 88(10):1263–1265
37. Riva CE, Titze P, Hero M, Petrig BL (1997) Effect of acute decreases of perfusion pressure on choroidal blood flow in humans. Invest Ophthalmol Vis Sci 38(9):1752–1760
38. Robinson F, Riva CE, Grunwald JE, Petrig BL, Sinclair SH (1986) Retinal blood flow autoregulation in response to an acute increase in blood pressure. Invest Ophthalmol Vis Sci 27(5):722–726
39. Alp MN, Ozgen A, Can I, Cakar P, Gunalp I (2000) Colour Doppler imaging of the orbital vasculature in Graves' disease with computed tomographic correlation. Br J Ophthalmol 84(9):1027–1030
40. Kurioka Y, Inaba M, Kawagishi T, Emoto M, Kumeda Y, Inoue Y, Morii H, Nishizawa Y (2001) Increased retinal blood flow in patients with Graves' disease: influence of thyroid function and ophthalmopathy. Eur J Endocrinol 144(2):99–107
41. Benning H, Lieb W, Kahaly G, Grehn F (1994) Color duplex ultrasound findings in patients with endocrine orbitopathy. Ophthalmologe 91(1):20–25
42. Tsai CC, Kau HC, Kao SC, Lin MW, Hsu WM, Liu JH, Wei YH (2005) Pulsatile ocular blood flow in patients with Graves' ophthalmopathy. Eye (Lond) 19(2):159–162
43. Langham ME, Farrell RA, O'Brien V, Silver DM, Schilder P (1989) Non-invasive measurements of pulsatile blood flow in the human eye. In: Lambrou GN, Greve EL (eds) Ocular blood flow in glaucoma. Kugler and Ghedini, Amsterdam, pp 93–99

Systemic Diseases: Cardiovascular Disease and Ocular Manifestation

19

Ghazaleh Gouya and Michael Wolzt

Core Messages

- There is insufficient evidence to recommend a routine ophthalmoscopic consultation for all patients with hypertension. If the initial clinical findings are equivocal (e.g., there is borderline or inconsistent hypertension with no other evidence of target-organ damage), an ophthalmoscopic consultation may be useful to supplement risk assessment and treatment decisions. Hypertensive retinopathy, neuropathy, and choroidopathy represent target-organ damage in persons with systemic arterial hypertension. Hypertensive retinopathy in particular is a recognized cardiovascular risk stratification factor. Prompt recognition and accurate diagnosis of hypertensive retinopathy, neuropathy, and choroidopathy have important implications for both the ocular and general health of the individual.

19.1 Atherosclerosis

The lesions of atherosclerosis occur principally in large- and medium-sized elastic and muscular arteries and can lead to ischemia of the heart, brain, or extremities, resulting in chronic ischemia and/or acute infarction. Cardiovascular disease affects not only the major arteries in the systemic circulation—retinal vascular disease can be a manifestation of generalized atherosclerosis or a result of embolic disease originating from the cardiovascular system. Atherosclerosis, hypertension, and diabetes are all thought to be possible risk factors for nonarteritic ischemic optic neuropathy.

Patients with cerebrovascular disease can also develop eye problems. These may be directly related, such as a cortical stroke affecting the visual pathway, or amaurosis fugax, indicating compromise of the anterior cerebral circulation, or problems may arise indirectly, such as corneal exposure secondary to a seventh nerve palsy affecting the orbicularis oculi muscle (which aids lid closure).

19.1.1 Pathogenesis of Atherosclerosis

Because high plasma concentrations of cholesterol, in particular those of low-density lipoprotein (LDL) cholesterol, are one of the principal risk factors for atherosclerosis, the process of atherogenesis has been considered by many to

G. Gouya, M.D. (✉) • M. Wolzt, Ph.D.
Department of Clinical Pharmacology,
Medical University of Vienna,
Weahringer Guertel 18-20, Vienna, A-1090, Austria
e-mail: ghazaleh.gouya@meduniwien.ac.at;
michael.wolzt@meduniwien.ac.at

L. Schmetterer, J.W. Kiel (eds.), *Ocular Blood Flow*,
DOI 10.1007/978-3-540-69469-4_19, © Springer-Verlag Berlin Heidelberg 2012

consist largely of the accumulation of lipids within the artery wall. In fact, the lesions of atherosclerosis represent a series of highly specific cellular and molecular responses that can best be described, in aggregate, as an inflammatory disease [32].

Atherosclerotic lesions are asymmetric focal thickenings of the innermost layer of the artery, the intima. They consist of cells, connective tissue elements, lipids, and debris. Blood-borne inflammatory and immune cells constitute an important part of an atheroma, the remainder being vascular endothelial and smooth muscle cells. Most of these cells in the fatty streak are macrophages, together with some T cells. Fatty streaks are prevalent in young people, never cause symptoms, and may progress to atheromata or eventually disappear.

Possible causes of endothelial dysfunction leading to atherosclerosis include elevated and modified LDL; free radicals caused by cigarette smoking, hypertension, and diabetes mellitus; genetic alterations; elevated plasma homocysteine concentrations; infectious microorganisms such as herpesviruses or *Chlamydia pneumoniae*; and combinations of these or other factors.

19.1.2 Internal Carotid Artery Disease (ICA)

Internal carotid artery supplies blood to the organ of vision; therefore, pathologies of those arteries caused by atherosclerosis may have direct influence on its functioning. Most common syndromes are temporary, which is known as amaurosis fugax, or constant reduction of visual acuity [9]. In fundus examination, central retinal artery occlusion and branch retinal artery occlusion are the most common diagnosis; retinal vein occlusion, anterior ischemic optic neuropathy, and ocular ischemic syndrome are less frequent. An ophthalmologist might be the very first specialist whom the patient with internal carotid artery disease will visit. Proper recognition and diagnosis of disease may protect patient against serious life-threatening complications like stroke.

A history of transient monocular blindness, and visualization of cholesterol (Hollenhorst) plaques at funduscopic examination are indications for diagnostic testing to identify carotid artery occlusive disease. Vascular events that occur in the brain also occur in the eye, giving rise to different ophthalmologic manifestations that range from amaurosis fugax to complete blindness due to central retinal artery occlusion.

Amaurosis fugax, Hollenhorst plaques, and venous stasis retinopathy have the highest predictive value among ocular conditions for carotid artery atherosclerosis, and warrant referral for carotid imaging [23].

19.1.3 Effects on the Ocular Circulation

Although the association of atherosclerosis with ocular vascular disease is well established, the consequences on the ocular vasculature and ocular blood flow regulation are less established. In a study in rhesus monkeys, experimental hyperlipoproteinemia was associated with lipemia retinalis, but not with any retinal vascular changes [43]. In a study in rabbits fed with a high-cholesterol diet, no changes in retinal vasculature were seen, although severe atherosclerotic manifestations were visible in cerebral and carotid arteries [36]. In a primate model, however, an atherogenic diet induced perivascular lesions of the central retinal artery and vein as well as partial occlusion of the vessels because of proliferation of subendothelial cells [26].

By far the largest studies on this topic have been performed by Hayreh [14]. In more than 80 monkeys, no atherosclerotic lesions in the retinal arterioles were seen ophthalmoscopically, although the monkeys were fed with an atherogenic for as long as 15 years and had markedly increased cholesterol plasma levels and severe atherosclerotic alterations in other vascular beds. Patchy atherosclerosis was, however, observed in the intraorbital parts of the ophthalmic artery, central retinal artery, and posterior ciliary arteries.

In the recent years, some but not all epidemiological studies have observed an association between measures of atherosclerosis and a reduced retinal arteriovenous ratio [37]. In the Atherosclerosis Risk In Communities Study, it was observed that smaller retinal arteriovenous ratio is associated with carotid artery plaque and carotid arterial stiffness, but not with carotid intima-media thickness [21]. In the Rotterdam Study, a lower retinal arteriovenous ratio was related to higher carotid intima-media thickness and increased carotid plaque score [18]. In the Cardiovascular Health Study, however, retinal arteriovenous ratio was independent of any direct measure of atherosclerosis [41]. A relation between dyslipidemia and retinal arteriovenous ratio is evident from numerous epidemiological studies [37]. Larger retinal venular caliber appears to be associated with higher triglyceride plasma levels, lower HDL cholesterol plasma levels, and possibly elevated LDL cholesterol. An association has consistently been observed between markers of systemic inflammation and wider retinal vascular venous diameter.

Functionally, the choroidal vasculature shows an abnormal vasoconstrictor response to systemic nitric oxide synthase inhibition indicating for endothelial dysfunction [2]. This sign of ocular endothelial dysfunction can, however, be restored by 28-day administration of simvastatin. Another ocular sign of endothelial dysfunction at the level of the ocular vasculature may be a reduced retinal response to diffuse luminance flicker [8]. Such reduced flicker responses can be seen in patients with diabetes [10], systemic hypertension, and/or hypercholesterolemia and are associated to some degree to endothelial dysfunction as assessed via flow-mediated vasodilatation in the forearm and elevated cholesterol plasma levels [29].

19.1.3.1 Retinal Artery Occlusion

Retinal artery occlusion (RAO) is the ocular equivalent of cerebrovascular disease [31]. Occlusion of a retinal vessel leads to ischemia of retinal tissue supplied by that arterial territory, resulting in acute retinal edema and death of retinal ganglion cells. By definition, a central RAO causes an infarction of the whole central retinal artery and loss of central vision in the affected eye. Branch RAO is an infarction of one of the branches that leaves the optic nerve.

Clinical Characteristics

In most cases, central RAO is due to emboli or thrombi, with the former being more common. It appears, however, that central RAO can also occur due to a severe reduction in ocular perfusion pressure, vasculitis, or ocular compression [15]. Visual acuity is generally poor with central RAO, though it may be completely normal if an existing cilioretinal artery spares the macular region. Visual acuity is reduced infrequently in branch RAO. Reduction in visual field is commensurate with the degree of retinal infarction. This is generally much more severe with central RAO than with branch RAO. A cherry-red spot in the macula is seen frequently in central RAO. Occurrence of irreversible retinal damage occurs approximately 100 min after central RAO [17].

In central RAO, the artery is completely occluded, although in the fluorescein angiogram residual retinal circulation can often be observed. This is either due to collateral circulation via cilioretinal capillary anastomoses within the optic nerve head or due to collateral circulation via pial and intraneural anastomoses of the central retinal artery [15]. It needs, however, also to be mentioned that the fluorescein angiogram can even be normal when the diagnosis of central RAO is made. This can, for instance, occur when the embolus gets dislodged. In addition, it has been hypothesized that central RAO can transiently occur during nocturnal blood pressure dips in patients with ocular ischemia or stenosis of the carotid or ophthalmic artery or due to release of vasoconstrictory serotonin from the atherosclerotic plaques in the carotid arteries [15].

Diagnosis

Artery occlusion may occur on base of plaque rupture of an atherosclerotic lesion or on base of embolic events. To assess the carotid vasculature, magnetic resonance angiography (MRA) or computed tomography angiography (CTA) is preferred because the intracranial ICA can be visualized. RAO has been reported in the presence of intracranial carotid stenosis. Vascular imaging also reveals any unusual abnormalities of the carotid arteries, including dissection or fibromuscular dysplasia. Cardiac sources of embolism in RAO are rare, and the very low yield of finding abnormalities on echocardiography may not warrant workup. Hypercoagulability workups should be considered in patients aged

50 years or much younger. However, there seems little scientific rationale for the use of fibrinolytic therapy in the treatment of central RAO because of a lack of effect on visual acuity [15].

Mortality/Morbidity

The main morbidity from RAO is vision loss [1]. Other factors must be considered since this disease carries the same risk factors as stroke in the brain. These patients usually have an increased incidence of coexistent carotid disease, and a few patients have cardiac disease.

19.1.3.2 Retinal Vein Occlusion (RVO)

Retinal vein occlusion is the second most common retinal vascular occlusive disease after diabetic retinopathy, affecting not only the elderly but also young patients [25]. Retinal vein occlusions may affect the venous blood supply of the entire retina (central RVO), half of the retina (hemicentral RVO), or a quadrant drained by one of the branches of the central retinal vein (branch RVO).

Clinical Characteristics

Retinal vein thrombosis causes increased venous pressure and may lead to retinal capillary decompensation with macular edema and subsequent visual loss [3].

Pathogenesis

The exact pathogenesis of RVO is unclear. Open-angle glaucoma is the most frequent local alteration predisposing to RVO as it compromises venous outflow by increased intraocular pressure. Degenerative changes of the vessel wall, abnormal perivascular changes, and abnormal hematologic factors constitute as the primary mechanisms of vessel occlusion.

RVO is defined as a focal occlusion of a retinal vein at an arteriovenous crossing site. In almost all cases, the RVO occurs at crossing sites where the artery is passing anteriorly (superficially) to the vein. The resultant thickening of the artery appears to cause compression of adjacent veins, a process that may be aggravated because the two vessels are confined within a common adventitial sheath. With increased compression, venous blood flow velocity at the crossing site gradually increases until local shear stress causes endothelial cell loss, thrombus formation, and vein occlusion. Regarding thrombophilic risk factors and RVO, there is only evidence for an association with hyperhomocysteinemia and anticardiolipin antibodies, factors that are known as risk factors for venous thrombosis as well as for arterial vascular disease. The minor effect of FV Leiden mutation and the prothrombin gene mutation (risk factors for venous thrombosis only) suggests that atherosclerosis might be an important factor in the development of RVO.

Diagnosis

Based on this analysis, it seems that the risk profile for arterial disease of the patient with RVO is more important than thrombophilic risk factors. An explanation for this might be the anatomy of the retinal arteries and veins. In investigating a new patient with RVO, one should screen for hypertension, lipid abnormalities, and diabetes mellitus since retinal venous obstruction may be the first presentation of significant cardiovascular morbidity.

19.1.3.3 Ischemic Optic Neuropathy

Ischemic optic neuropathy (ION) includes a variety of disorders that produce ischemia to the optic nerve [12]. By definition, ION is termed anterior if disk edema is present acutely. This clinical finding suggests infarction of the portion of the optic nerve closest to the globe. ION also may be posterior, lying several centimeters behind the globe. In this instance, swelling of the optic nerve head is not visible, and the practitioner will only witness a normal fundus. Specific arteriolar pathology, especially giant cell arteritis, also may be associated with ION. Giant cell arteritis must be ruled out in suspected cases.

Clinical Characteristics

Anterior ION (AION) is due to ischemia of the anterior part of the optic nerve head and typically develops as abrupt, painless monocular vision loss, though a few patients do experience some discomfort. As opposed to optic neuritis in which color vision typically is reduced drastically, the color vision deficit in AION mirrors the degree of decrease in visual acuity The optic nerve head acutely appears edematous, which confirms the anterior nature of this disorder. Hemorrhage on the disk is commonly present.

Patients with AION have certain characteristics. Primarily, this is a disease of Caucasians. Essentially no sex predilection is reported, and average age of onset is in the mid-1960s. These patients generally have an increased incidence of hypertension, hyperlipidemia, and diabetes. Smoking may predispose patients to an earlier onset of the disease. Most investigators believe that ION has no association with carotid stenosis, but this has not been evaluated systematically in a large number of patients. Giant cell arteritis must be ruled out in suspected cases.

The exact mechanism of infarction of the optic nerve head is unknown, but it is believed to be secondary to hypotension and, possibly, watershed infarction between the central retinal artery and the posterior ciliary arteries. Likewise, AION with severe visual loss may occur following blood loss and hypotension, as in gastrointestinal hemorrhage, or during major cardiac or spine surgeries. AION rarely occurs in the setting of hypotension or blood loss alone, and the combination of the two is probably important in the pathogenesis. Simultaneous bilateral AION or posterior ION is not uncommon under these circumstances.

Mortality/Morbidity

The morbidity of this condition specifically concerns vision loss. AION has little association with carotid artery disease; therefore, an attack of AION does not denote a risk for impending stroke, as does amaurosis fugax. ION is a disease of older individuals and most commonly occurs in patients older than 40 years. The average age of onset is the mid-1960s. Smoking and other risk factors may decrease the age of onset in some patients [14].

19.1.3.4 Asymptomatic Retinal Emboli

Background

The most common emboli include (1) cholesterol, (2) fibrin/platelet, and (3) calcific. Their location, size, color, and shape can help distinguish them. Not all types of emboli are associated strongly with occlusive signs in the retina.

Pathophysiology

Cholesterol emboli were first described by Hollenhorst, who correctly surmised the association of these bright refractile particles to carotid disease [4]. Although found more commonly with ICA disease, these emboli also are found in the setting of atherosclerosis of the aortic arch and innominate arteries.

A common question from a referring physician is "What is the risk to my patient who has asymptomatic retinal emboli?" The presence of these emboli, even when asymptomatic, is not entirely benign. The US Physicians Health Study observed an association between emboli and hypertension, smoking, and heterogeneous carotid plaques. After an average follow-up of 3.4 years, these patients had a tenfold increase in stroke, independent of other risk factors of stroke.

Key Message

Patients with cerebrovascular disease can also develop eye problems. These may be directly related, such as a cortical stroke affecting the visual pathway, or amaurosis fugax, indicating compromise of the anterior cerebral circulation, or problems may arise indirectly, such as corneal exposure secondary to a seventh nerve palsy affecting the orbicularis oculi muscle (which aids lid closure).

19.2 Vasculitis

Vasculitis is a group of diseases featuring inflammation of the wall of blood vessels including veins, arteries, and capillaries due to leukocyte migration and resultant damage. While most vasculitides are rare, they generally affect several organ systems involving the ocular vasculature and can cause severe disability [19]. Retinal vasculitis represents a group of disorders characterized by retinal vascular inflammation associated with intraocular inflammation but rarely may develop as an isolated ocular disease. The diagnosis is based on characteristic ophthalmoscopic and fluorescein angiographic features [11].

Common clinical manifestations include retinal vascular sheathing as well as vitreitis and cystoid macular edema. Staining and leakage of affected vessels is typically demonstrated by fluorescein angiography. These clinical and angiographic findings are the result of inflammation within or surrounding the retinal vessels. Retinal vasculitis most commonly occurs in association with a systemic or ocular disease but rarely may develop as an isolated ocular disease. Many patients have been previously diagnosed with an associated disease, but some patients may develop retinal vasculitis as the initial manifestation of an underlying disorder [40]. Thus, it is essential that a thorough history, review of systems, and physical examination are performed in patients with retinal vasculitis. Special forms of vasculitis with reported ocular involvement are specially reported below:

19.2.1 Takayasu's Arteritis (Aortic Arch Syndrome)

The first case of Takayasu's arteritis was described in 1908 by Dr. Mikito Takayasu at the Annual Meeting of the Japan Ophthalmology Society. Dr. Takayasu described a peculiar "wreath-like" appearance of blood vessels in the back of the eye (retina) [24]. In the Western world, atherosclerosis is a more frequent cause of obstruction of the aortic arch vessels than is Takayasu's arteritis.

19.2.1.1 Pathophysiology

Takayasu's arteritis is a chronic vasculitis mainly involving the aorta and its main branches, such as the brachiocephalic, carotid, subclavian, vertebral, and renal arteries, as well as the coronary and pulmonary arteries. Takayasu's arteritis is characterized histologically as "panarteritis" involving all layers of the arterial wall, including intimal fibrous thickening and/or typical atheromatous lesions, destruction of medial smooth muscles and elastic layers, cellular infiltration and collagenous fibrosis in the media, and thickened adventitia with cellular infiltration around vasa vasorum [28].

19.2.1.2 Clinical Characteristics

It induces clinically varied ischemic symptoms due to stenotic lesions or thrombus formation, including blindness, cataract and/or retinal hemorrhage, pulselessness, aortic regurgitation, and/or congestive heart failure due to dilatation of the ascending aorta. More acute progression causes destruction of the media of the arterial wall, leading to the formation of aneurysms and/or dissecting aneurysm or rupture of the involved arteries.

19.2.1.3 Epidemiology

Epidemiologically, it is found mostly in female patients and is more prevalent in Asian and Latin American countries. Autoimmune mechanisms induced by viruses have been suspected in the pathogenesis as HLA analysis demonstrated a statistically significant frequency of B*5201 (B52) B*3902 haplotype in these patients. Although it has been reported worldwide, it shows a predilection for young Asian women. Females with this disease outnumber males by 8:1, and the age of onset is typically between 15 and 30 years.

19.2.1.4 Classification

Four types of late-phase Takayasu's arteritis are described on the basis of the sites of involvement as follows:

Type I—Classic pulseless type that involves the brachiocephalic trunk, carotid arteries, and subclavian arteries

Type II—Combination of types I and III

Type III—Atypical coarctation type that involves the thoracic and abdominal aortas distal to the arch and its major branches

Type IV—Dilated type that involves extensive dilatation of the length of the aorta and its major branches

19.2.2 Behcet's Disease

Although BD, a systemic occlusive vasculopathy, was originally described as a triad of recurrent oral and genital ulcerations along with hypopyon–iritis, the scope of the disease has widened over the years to encompass a spectrum of multiorgan involvement with variable manifestations [33].

19.2.2.1 Clinical Characteristics

Besides CNS involvement, ocular involvement of complete BD, frequently termed ocular BD, is the most serious manifestation due to its association to a poor visual outcome most often despite immunosuppressive treatment [7]. Eye involvement affects 60–80% of BD patients, may cause either anterior uveitis (inflammation in the front of the eye) or posterior uveitis (inflammation in the back of the eye), and sometimes causes both at the same time. Anterior uveitis results in pain, blurry vision, light sensitivity, tearing, or redness of the eye. Posterior uveitis may be more dangerous and vision threatening because it often causes fewer symptoms while damaging a crucial part of the eye—the retina.

19.2.2.2 Pathogenesis

Behcet's disease is one of the few forms of vasculitis in which there is a known genetic predisposition. The pathogenesis of BD remains unclear, but research of the last decades has shown a complex role of genetic factors (HLA-B51) predisposing to inflammation with involvement of the innate immune system (neutrophils, NK cells), perpetuated by the adaptive immune response, most importantly T cells, against infectious and/or auto-antigens.

Eye disease in BD patients is mostly a recurrent nongranulomatous uveitis with necrotizing obliterative retinal vasculitis, which may be found either in the anterior or the posterior segment, or both.

19.2.2.3 Diagnosis

Diagnosis of BD is mostly based on several sets of diagnostic criteria. Today the most widely used are the criteria of the International Study Group for Behcet's Disease from 1990. The criteria put forth by the group include recurrent oral ulceration (at least three occasions in a year). In addition, a patient must also meet two of the following four criteria for Behcet's disease: recurrent genital ulcerations, eye lesions (uveitis or retinal vasculitis), skin lesions (erythema nodosum, lesions, acne), and/or positive "pathergy test."

The pathergy test is a simple test in which the forearm is pricked with a small sterile needle. Occurrence of a small red bump or pustule at the site of needle insertion constitutes a positive test. Although a positive pathergy test is helpful in the diagnosis of Behcet's, only a minority of Behcet's patients demonstrate the pathergy phenomenon (i.e., have positive tests).

Despite aggressive immunosuppressive treatment, the visual prognosis of ocular BD was generally poor to date. Recently, novel biologic drugs, including interferon-*a* and tumor necrosis factor (TNF)-*a* antagonists, have been introduced in the treatment of ocular BD with very promising results and seem for the first time to improve the prognosis of the disease.

19.2.2.4 Epidemiology

In many parts of the world, ocular BD still is a major cause of blindness. Patients with BD, a rare disorder with varying prevalence, are distributed endemically along the ancient Silk Route from Mediterranean countries to the Middle East and the Eastern Asian countries, but they are rarely encountered in Northern Europe, Northern Asia, most of continental Africa, Australia, or North America [38, 44].

19.2.3 Thromboangiitis Obliterans

Buerger's disease (also known as *thromboangiitis obliterans*) is an acute inflammation and characterized by occlusive thrombotic segmental and often multiple inflammatory lesions of medium-sized and small arteries and superficial veins primarily of the limbs. A cerebral form of von Winiwarter–Buerger's disease according to the classification by Lindenberg and Spatz from 1939 presents with thrombus formation and proliferative change of the intimal layer of the pial arteries of the brain and results in widespread cortical necroses. Direct involvement of the ocular vasculature is rarely described in the literature but may be involved secondarily by involvement of the carotid or medial cerebral artery as well as by the involvement of the pial arteries [22].

19.2.3.1 Diagnosis and Clinical Characteristics

Thromboangiitis obliterans is strongly associated with the use of tobacco products, primarily from smoking, and age younger than 45. Definitive diagnosis of thromboangiitis obliterans is often difficult as it relies heavily on exclusion of the conditions. It is associated with younger age and presence of distal extremity ischemia (indicated by claudication, pain at rest, ischemic ulcers or gangrene, exclusion of a proximal source of emboli by echocardiography).

19.2.3.2 Treatment

Complete cessation of smoking remains the cornerstone of therapy, but blockade with antagonists of cannabinoid or endothelin receptors and the use of gene- or cell-based therapy to induce therapeutic angiogenesis have opened up new possibilities for treatment. Use of vascular growth factor and stem cell injections has been showing promise in clinical studies [27].

19.2.4 Temporal Arteritis

Temporal arteritis was first described in 1890 by Hutchinson and was more clearly defined in 1932 by Horton et al. as a granulomatous arteritis of the temporal vessels. The presence of giant cells in the granulomatous infiltration gave rise to the term "giant cell arteritis" (GCA). Visual loss was first described in 1938 by Jennings, and in 1941, Gilmour pointed out that the disease is more generalized and that other medium-sized and large-sized vessels are also affected. The most common vessels affected are the superficial temporal artery, the ophthalmic artery, the posterior ciliary arteries, and the vertebral arteries; occasionally, the central retinal artery may also be involved [30].

Today, it is recognized that GCA may produce a broad range of symptoms related to the affected artery, including headache, jaw claudication, visual loss, and diplopia; constitutional symptoms such as malaise, anorexia, myalgia, and arthralgia are also common. Involvement of the visual system in GCA is considered an ophthalmologic emergency, and it is essential for those physicians involved in the care of patients with GCA to be familiar with the distinctive signs and symptoms, diagnostic testing, and therapeutic guidelines.

19.2.4.1 Epidemiology

GCA occurs most commonly in the elderly, with the incidence increasing with age; there are several studies that suggest a genetic susceptibility in GCA. Familial clustering and monozygotic twin concordance have been reported along with human leukocyte antigen DRB1*04, DRB1*01, and DRW6 associations.

19.2.4.2 Pathogenesis

The inciting event of GCA has been widely researched yet is still not well understood. The series of events leading to GCA has been proposed to begin with activated T cells in the vasa vasorum upregulating in response to an unknown stimulus, and activating macrophages which then form the granulomatous immune reaction and cause ultimate disruption and destruction within the blood vessel wall by expressing metalloproteases and reactive oxygen intermediates. After the initial immune response, the vessel undergoes a healing response to injury, which includes intimal thickening, myofibroblast proliferation, and extracellular matrix deposition, all of which contribute to vascular stenosis and occlusion.

19.2.4.3 Ocular Manifestations

The most common ocular manifestation of GCA is arteritic anterior ischemic optic neuropathy (AAION), which accounts for approximately 80% of visual loss attributed to the disease. AAION typically presents with severe visual loss developing over hours to days in one or both eyes. Simultaneous bilateral ischemic optic neuropathy is more commonly arteritic than nonarteritic. Involvement of the second eye occurs within 14 days in approximately one-third of untreated cases. Ischemia of the retrobulbar portion of the optic nerve, or posterior ischemic optic neuropathy (PION), can also occur in GCA. Upon initial presentation, PION is characterized by acute visual loss without apparent optic nerve abnormalities or edema. Over time, the optic nerve will develop visible atrophy.

19.2.5 Wegener's Granulomatosis

Wegener's granulomatosis is a form of vasculitis that affects the lungs, kidneys, and other organs. Due to its end-organ damage, it can be a serious disease that requires long-term immune suppression. It is named after Dr. Friedrich Wegener, who described the disease in 1936 [34].

19.2.5.1 Pathogenesis

Wegener's granulomatosis is part of a larger group of vasculitic syndromes, all of which feature the presence of an abnormal type of circulating antibody termed ANCAs (antineutrophil cytoplasmic antibodies) and affect small- and medium-sized blood vessels. On histopathological examination, a biopsy will show *leukocytoclastic vasculitis* with necrotic changes and granulomatous inflammation (clumps of typically arranged white blood cells) on microscopy. These granulomas are the main reason for the appellation of "Wegener's granulomatosis," although it is not an essential feature. Unfortunately, many biopsies can be nonspecific, and 50% provide too little information for the diagnosis of Wegener's.

19.2.5.2 Ocular Manifestation

Fifty percent to 60% have ophthalmologic manifestations, which can be a presenting feature in a minority of patients. Orbital disease is the most common manifestation and may result in proptosis, restrictive ophthalmopathy, chronic orbital pain, and, in chronic cases, orbital retraction syndrome and intractable socket pain. Wegener's may also cause injury to the optic nerve, ophthalmoplegia, conjunctivitis, keratitis, scleritis, episcleritis, dacryocystitis, nasolacrimal duct obstruction, dacryoadenitis, uveitis, and retinal vasculitis. Initial treatment is generally with corticosteroids and oral cyclophosphamide. Once remission has been achieved, azathioprine and steroids can be used to maintain remission.

19.2.5.3 Diagnosis

In 1990, the American College of Rheumatology accepted the classification criteria for Wegener's. These criteria were not intended for diagnosis but for inclusion in randomized controlled trials. Two or more positive criteria have a sensitivity of 88.2% and a specificity of 92.0% of describing Wegener's http://emedicine.medscape.com/article/332622-overview.

Nasal or oral inflammation:

- Painful or painless oral ulcers *or* purulent or bloody nasal discharge

Lungs: abnormal chest x-ray with:

- Nodules, infiltrates, *or* cavities

Kidneys: urinary sediment with:

- Microhematuria *or* red cell casts

Biopsy:

- Granulomatous inflammation within the arterial wall *or* in the perivascular area

19.2.6 Kawasaki Disease

Kawasaki disease is a poorly understood self-limited vasculitis that affects many organs. Usually in children (age<4), it affects large, medium, and small vessels, prominently the coronary arteries, and is associated with mucocutaneous lymph node syndrome [5].

19.2.6.1 Clinical Characteristics

Kawasaki disease often begins with a high and persistent fever that is not very responsive to normal doses of paracetamol (acetaminophen) or ibuprofen. The fever may persist steadily for up to 2 weeks and is normally accompanied by irritability. Affected children develop red eyes, red mucous membranes in the mouth, red cracked lips, a "strawberry tongue," iritis, keratic precipitates (detectable by an ophthalmologist but usually too small to be seen by the unaided eye), and swollen lymph nodes. Skin rashes occur early in the disease, and peeling of the skin in the genital area, hands, and feet (especially around the nails and on the palms and soles) may occur in later phases. Some of these symptoms may come and go during the course of the illness.

19.2.6.2 Diagnosis

Classically, 5 days of fever [5] plus at least four out of five criteria:

- Bilateral conjunctival injection
- Injected or fissured lips, injected pharynx, or strawberry tongue
- Erythema of palms/soles, edema of hands/feet, periungual desquamation

- Polymorphous rash
- Cervical lymphadenopathy (at least one node >1.5 cm)

19.3 Vascular Malformations

19.3.1 Hereditary Telangiectasia (Rendu–Osler–Weber Syndrome)

HHT is an autosomal dominant genetic disorder that leads to vascular malformations and is characterized by telangiectasia (small vascular malformations) on the skin and mucosal linings, epistaxis (nosebleeds), and arteriovenous malformations (AVMs) in various internal organs. Skin and mucosa telangiectasias are most remarkable on the tongue, hands/fingers, nose, lips, mouth/throat, and conjunctiva. There is no specific treatment for the condition.

19.3.1.1 Diagnosis

There are four diagnostic criteria [35]. If three or four are met, a patient has *definite HHT*, while two give a possible diagnosis:

- Spontaneous recurrent epistaxis
- Multiple telangiectasias on typical locations (see above)
- Proven visceral AVM (lung, liver, brain, spine)
- First-degree family member with HHT

19.3.1.2 Pathophysiology

The mechanism underlying the formation of vascular malformations is not completely understood, but signaling of transforming growth factor-β1 is most likely to be involved. Possibly, connective tissue is required to support and guide proliferating blood vessels during angiogenesis, and defects in TGF-β signaling adversely affect connective tissue and matrix production.

19.4 Systemic Hypertension and Treatment

Hypertension is sustained elevation of resting systolic BP (≥140 mmHg), diastolic BP (≥90 mmHg), or both. Hypertension with no known cause (primary; formerly, essential hypertension) is most common. Hypertension with an identified cause (secondary hypertension) is usually due to a renal disorder. Usually, no symptoms develop unless hypertension is severe or long-standing. Diagnosis is by sphygmomanometry. Tests may be done to determine cause, assess damage, and identify other cardiovascular risk factors. Treatment involves lifestyle changes and drugs, including diuretics, β-blockers, ACE inhibitors, angiotensin II receptor blockers, and Ca channel blockers [6].

BP increases with age. About two-thirds of people >65 have hypertension, and people with a normal BP at age 55 have a 90% lifetime risk of developing hypertension. Because hypertension becomes so common with age, the age-related increase in BP may seem innocuous, but higher BP increases morbidity and mortality risk.

19.4.1 Etiology

Hypertension may be primary (85–95% of cases) or secondary.

19.4.1.1 Primary Hypertension

Hemodynamics and physiologic components (e.g., plasma volume, activity of the renin–angiotensin system) vary, indicating that primary hypertension is unlikely to have a single cause. Even if one factor is initially responsible, multiple factors are probably involved in sustaining elevated BP (the mosaic theory). In afferent systemic arterioles, malfunction of ion pumps on sarcolemmal membranes of smooth muscle cells may lead to chronically increased vascular tone. Heredity is a predisposing factor, but the exact mechanism is unclear. Environmental factors (e.g., dietary Na, obesity, stress) seem to affect only genetically susceptible people.

19.4.1.2 Secondary Hypertension

Causes include renal parenchymal disease (e.g., chronic glomerulonephritis or pyelonephritis, polycystic renal disease, connective tissue disorders, obstructive uropathy), renovascular disease, pheochromocytoma, Cushing's syndrome, primary aldosteronism, congenital adrenal hyperplasia, hyperthyroidism, myxedema, and coarctation of the aorta. Excessive alcohol intake and use of oral contraceptives are common causes of curable hypertension. Use of sympathomimetics, NSAIDs, corticosteroids, cocaine, or licorice commonly contributes to hypertension.

19.4.2 Pathophysiology

Because BP equals cardiac output (CO) × total peripheral vascular resistance (TPR), pathogenic mechanisms must involve increased CO, increased TPR, or both.

In most patients, CO is normal or slightly increased, and TPR is increased. This pattern is typical of primary hypertension and hypertension due to pheochromocytoma, primary aldosteronism, renovascular disease, and renal parenchymal disease.

In other patients, CO is increased (possibly because of venoconstriction in large veins), and TPR is inappropriately normal for the level of CO. Later in the disorder, TPR increases and CO returns to normal, probably because of autoregulation. Some disorders that increase CO (thyrotoxicosis, arteriovenous fistula, aortic regurgitation), particularly when stroke volume is increased, produce isolated systolic hypertension. Some elderly patients have isolated systolic hypertension with normal or low CO, probably due to inelasticity of the aorta and its major branches. Patients with high, fixed diastolic pressures often have decreased CO.

Plasma volume tends to decrease as BP increases; rarely, plasma volume remains normal or increases. Plasma volume tends to be high in hypertension due to primary aldosteronism or renal parenchymal disease and may be quite low in hypertension due to pheochromocytoma. Renal blood flow gradually decreases as diastolic BP increases and arteriolar sclerosis begins. GFR remains normal until late in the disorder; as a result, the filtration fraction is increased. Coronary, cerebral, and muscle blood flow is maintained unless severe atherosclerosis coexists in these vascular beds.

19.4.3 Pathology and Complications

No pathologic changes occur early in hypertension. Severe or prolonged hypertension damages target organs (primarily the cardiovascular system, brain, kidneys, and eye), increasing risk of coronary artery disease (CAD), MI, stroke (particularly hemorrhagic), and renal failure. The mechanism involves development of generalized arteriolosclerosis and acceleration of atherogenesis. Because of increased afterload, the left ventricle gradually hypertrophies, causing diastolic dysfunction. The ventricle eventually dilates, causing dilated cardiomyopathy and heart failure (HF) due to systolic dysfunction. Thoracic aortic dissection is typically a consequence of hypertension; almost all patients with abdominal aortic aneurysms have hypertension.

19.4.4 Symptoms and Signs

Hypertension is usually asymptomatic until complications develop in target organs. Dizziness, flushed facies, headache, fatigue, epistaxis, and nervousness are not caused by uncomplicated hypertension. Severe hypertension can cause severe cardiovascular, neurologic, renal, and retinal symptoms (e.g., symptomatic coronary atherosclerosis, HF, hypertensive encephalopathy, renal failure).

Retinal changes may include arteriolar narrowing, hemorrhages, exudates, and, with encephalopathy, papilledema. Changes are classified (according to the Keith, Wagener, and Barker classification) into four groups with increasingly worse prognosis: constriction of arterioles only (grade 1), constriction and sclerosis of arterioles (grade 2), hemorrhages and exudates in addition to vascular changes (grade 3), and papilledema (grade 4).

19.4.5 Diagnosis of Hypertension

- Multiple measurements of BP to confirm, single office measurements might not be reliable.

Investigations of target-organ damage and evaluation of risk profile:

- Urinalysis and urinary albumin/creatinine ratio: if abnormal, consider renal ultrasonography.
- Blood tests: fasting lipids, creatinine, K.
- If creatinine is increased, renal ultrasonography.
- If K is decreased, evaluate for aldosteronism.
- ECG: if left ventricular hypertrophy, consider echocardiography.
- Sometimes thyroid-stimulating hormone measurement.
- If BP elevation is sudden and labile or severe, evaluate for pheochromocytoma.

Hypertension is diagnosed and classified by sphygmomanometry. History, physical examination, and other tests help identify etiology and determine whether target organs are damaged. BP must be measured at least twice—first, with the patient supine or seated, then after the patient has been standing for ≥2 min, on three separate days. The average of these measurements is used for diagnosis. BP is classified as normal, prehypertension, or stage 1 (mild) or stage 2 hypertension (see Fig. 19.1).

BP is measured in both arms; if BP in one arm is much higher, the higher value is used. BP is also measured in a thigh (with a much larger cuff) to rule out coarctation of the aorta, particularly in patients with diminished or delayed femoral pulses; with coarctation, BP is significantly lower in the legs. If BP is in the low hypertensive range or is markedly labile, more BP measurements are desirable. BP measurements may be sporadically high before hypertension becomes sustained; this phenomenon probably accounts for "white coat hypertension" in which BP is elevated when measured in the physician's office but normal when measured at home or by ambulatory BP monitoring. (See the Agency for Healthcare Research and Quality's Evidence Report/Technology Assessment summary on the Utility of Blood Pressure Monitoring Outside of the Clinic Setting.) However, extreme BP elevation alternating with normal readings is unusual and possibly suggests pheochromocytoma or unacknowledged drug use.

JNC 7 Classification of blood pressure in adults

Classification	BP (mmHg)
Normal	<120/80
Prehypertension	120–139/80–89
Stage 1	140–159 (systolic) or 90–99 (diastolic)
Stage 2	≥160 (systolic) or ≥100 (diastolic)

Fig. 19.1 JNC seven classification of blood pressure in adults

19.4.5.1 History

The history includes the known duration of hypertension and previously recorded levels; any history or symptoms of CAD, HF, or other relevant coexisting disorders (e.g., stroke, renal dysfunction, peripheral arterial disease, dyslipidemia, diabetes, gout); and a family history of any of these disorders. Social history includes exercise levels and use of tobacco, alcohol, and stimulant drugs (prescribed and illicit). A dietary history focuses on intake of salt and stimulants (e.g., tea, coffee, caffeine-containing sodas, energy drinks).

19.4.5.2 Physical Examination

The physical examination includes measurement of height, weight, and waist circumference; funduscopic examination for retinopathy; auscultation for bruits in the neck and abdomen; and a full cardiac, respiratory, and neurologic examination. The abdomen is palpated for kidney enlargement and abdominal masses. Peripheral arterial pulses are evaluated; diminished or delayed femoral pulses suggest aortic coarctation, particularly in patients <30.

19.4.5.3 Testing

The more severe the hypertension and the younger the patient, the more extensive is the evaluation. Generally, when hypertension is newly diagnosed, routine testing to detect target-organ damage and cardiovascular risk factors is done. Tests include urinalysis, spot urine albumin/creatinine ratio, blood tests (creatinine, K, Na, fasting plasma glucose, lipid profile), and ECG. Thyroid-stimulating hormone is often measured. Ambulatory BP monitoring, renal radionuclide imaging, chest x-ray, screening tests for pheochromocytoma, and renin-Na profiling are not routinely necessary. Peripheral plasma renin activity is not helpful in diagnosis or drug selection.

Depending on results of initial tests and examination, other tests may be needed. If urinalysis detects albuminuria (proteinuria), cylindruria, or microhematuria or if serum creatinine is elevated (≥1.4 mg/dL in men; ≥1.2 mg/dL in women), renal ultrasonography to evaluate kidney size may provide useful information. Patients with hypokalemia unrelated to diuretic use are evaluated for primary aldosteronism and high salt intake.

On ECG, a broad, notched P wave indicates atrial hypertrophy and, although nonspecific, may be one of the earliest signs of hypertensive heart disease. Left ventricular hypertrophy, indicated by a sustained apical thrust and abnormal QRS voltage with or without evidence of ischemia, may occur later. If either of these findings is present, echocardiography is often done. In patients with an abnormal lipid profile or symptoms of CAD, tests for other cardiovascular risk factors (e.g., C-reactive protein) may be useful.

If coarctation of the aorta is suspected, chest x-ray, echocardiography, CT, or MRI helps confirm the diagnosis. Patients with labile, significantly elevated BP and symptoms such as headache, palpitations, tachycardia, excessive perspiration, tremor, and pallor are screened for pheochromocytoma. Patients with symptoms suggesting Cushing's syndrome, a connective tissue disorder, eclampsia, acute porphyria, hyperthyroidism, myxedema, acromegaly, or CNS disorders are evaluated.

19.4.6 Prognosis

The higher the BP and the more severe the retinal changes and other evidence of target-organ involvement, the worse is the prognosis. Systolic BP predicts fatal and nonfatal cardiovascular events better than diastolic BP. Without treatment, 1-year cardiovascular event-free survival is <10% in patients with retinal sclerosis, cotton-wool exudates, arteriolar narrowing, and hemorrhage (grade 3 retinopathy) and <5% in patients with the same changes plus papilledema (grade 4 retinopathy). CAD is the most common cause of death among treated hypertensive patients. Ischemic or hemorrhagic stroke is a common consequence of inadequately treated hypertension. However, effective control of hypertension prevents most complications and prolongs life.

19.4.7 General Treatment

- Weight loss and exercise
- Smoking cessation
- Diet: increased fruits and vegetables, decreased salt, limited alcohol
- Drugs if BP is initially high (>160/100) or unresponsive to lifestyle modifications

Primary hypertension has no cure, but some causes of secondary hypertension can be corrected. In all cases, control of BP can significantly limit adverse consequences. Despite the theoretical efficacy of treatment, BP is lowered to the desired level in only one-third of hypertensive patients in the USA.

For all patients, treatment aims to reduce BP to <140/90 mmHg; for those with a kidney disorder or diabetes, the goal is <130/80 mmHg or as near this level as tolerated. Even the elderly and frail elderly can tolerate a diastolic BP as low as 60–65 mmHg well and without an increase in cardiovascular events. Ideally, patients or family members measure BP at home, provided they have been trained to do so, they are closely monitored, and the sphygmomanometer is regularly calibrated.

19.4.7.1 Lifestyle Modifications

Recommendations include regular aerobic physical activity at least 30 min/day most days of the week; weight loss to a body mass index of 18.5–24.9 (see the Cochrane review abstract "Dieting to reduce body weight for controlling hypertension in adults"); smoking cessation; a diet rich in fruits, vegetables, and low-fat dairy products with reduced saturated and total fat content; dietary sodium [Na+] of <2.4 g/day (<6 g NaCl); and alcohol consumption of ≤1 oz/day in men and ≤0.5 oz/day in women. (See the National Heart Lung and Blood Institute's Dietary Approaches to Stop Hypertension [DASH] Eating Plan.) In stage 1 (mild) hypertension with no signs of target-organ damage, lifestyle changes may make drugs unnecessary. Patients with uncomplicated hypertension do not need to restrict their activities as long as BP is controlled. Dietary modifications can also help control diabetes, obesity, and dyslipidemias. Patients with prehypertension are encouraged to follow these lifestyle recommendations.

19.4.7.2 Drugs

If systolic BP remains >140 mmHg or diastolic BP remains >90 mmHg after 6 months of lifestyle modifications, antihypertensive drugs are required. Unless hypertension is severe, drugs are usually started at low doses. Drugs are initiated simultaneously with lifestyle changes for all patients with prehypertension or hypertension plus diabetes, a kidney disorder, target-organ damage, or cardiovascular risk factors and for those with an initial BP of >160/100 mmHg. Signs of hypertensive emergencies require immediate BP reduction with parenteral antihypertensives.

For most hypertensive patients, one drug is given initially depending on the patient's characteristics and coexisting disorders.

If the initial drug is ineffective or has adverse effects, another drug can be substituted. If the initial drug is only partly effective but well tolerated, the dose can be increased or a second drug with a different mechanism added.

If initial systolic BP is >160 mmHg, two drugs are often used. Options include combining a diuretic with a β-blocker, an ACE inhibitor, or an angiotensin II receptor blocker and combining a Ca channel blocker with an ACE inhibitor or an angiotensin II receptor blocker. Appropriate combination and dose are determined; many are available as single tablets, which improve compliance. For severe or refractory hypertension, three or four drugs may be necessary. The choice of drug is less important than the reduction of blood pressure achieved.

19.5 Hypertensive Retinopathy

Systemic hypertension is a common condition associated with significant morbidity and mortality [39]. Hypertension confers cardiovascular risk by causing target-organ damage that includes retinopathy in addition to heart disease, stroke, renal insufficiency, and peripheral vascular disease. The recognition of hypertensive retinopathy is important in cardiovascular risk stratification of hypertensive individuals [42].

The detection of hypertensive retinopathy with the use of an ophthalmoscope has long been regarded as part of the standard evaluation of persons with hypertension. This clinical practice is supported by both previous and current reports of the Joint National Committee on Prevention, Detection, Evaluation, and Treatment of High Blood Pressure (JNC), which list retinopathy as one of several markers of target-organ damage in hypertension. On the basis of the JNC criteria, the presence of retinopathy may be an indication for initiating antihypertensive treatment, even in people with stage 1 hypertension (blood pressure, 140 to 159/90 to 99 mmHg) who have no other evidence of target-organ damage.

19.5.1 Classification

Hypertensive retinopathy was first described by Marcus Gunn in the nineteenth century in a series of patients with hypertension and renal disease. The retinal signs he observed included generalized and focal arteriolar narrowing, arteriovenous nicking, flame-shaped and blot-shaped retinal hemorrhages, cotton-wool spots, and swelling of the optic disk. In 1939, Keith et al. showed that these signs of retinopathy were predictive of death in patients with hypertension [20].

Keith–Wagener–Barker grade features:

I Mild generalized retinal arteriolar narrowing

II Definite focal narrowing and arteriovenous nipping

III The above and retinal hemorrhages, exudates, and cotton-wool spots

IV Severe grade III and papilloedema

Scheie grade features:

0 No changes

1 Barely detectable arterial narrowing

2 Obvious arterial narrowing with focal irregularities plus light reflex changes

3 Grade **2** plus retinal hemorrhages and exudates

4 Grade **3** plus silver wiring and papilledema

Signs of hypertensive retinopathy are common and are correlated with elevated blood pressure. Recent studies show that some of these signs (e.g., retinal hemorrhages, microaneurysms, and cotton-wool spots) predict stroke and death from stroke independently of elevated blood pressure and other risk factors. Patients with these signs of retinopathy may benefit from close monitoring of cerebrovascular risk and intensive measures to reduce that risk.

19.5.2 Pathophysiology

The retinal circulation undergoes a series of pathophysiological changes in response to elevated blood pressure. In the initial vasoconstrictive stage, there is vasospasm and an increase in retinal arteriolar tone owing to local autoregulatory mechanisms. This stage is seen clinically as a generalized narrowing of the retinal arterioles. Persistently elevated blood pressure leads to intimal thickening, hyperplasia of the media wall, and hyaline degeneration in the subsequent sclerotic stage. This stage corresponds to more severe generalized and focal areas of arteriolar narrowing, changes in the arteriolar and venular junctions (i.e., arteriovenous nicking or nipping), and alterations in the arteriolar light reflex (i.e., widening and accentuation of the central light reflex, or "copper wiring").

This is followed by an exudative stage in which there is disruption of the blood–retina barrier, necrosis of the smooth muscles and endothelial cells, exudation of blood and lipids, and retinal ischemia. These changes are manifested in the retina as microaneurysms, hemorrhages, hard exudates, and cotton-wool spots. Swelling of the optic disk may occur at this time and usually indicates severely elevated blood pressure (i.e., malignant hypertension).

Because better methods for the control of blood pressure are now available in the general population, malignant hypertension is rarely seen. In contrast, other retinal vascular complications of hypertension, such as macroaneurysms and branch-vein occlusions, are not uncommon in patients with chronically elevated blood pressure.

19.5.3 Blood Pressure

Numerous studies have confirmed the strong association between the presence of signs of hypertensive retinopathy and elevated blood pressure. Two studies have further evaluated the effect of a history of elevated blood pressure on the occurrence of specific retinal signs. In both studies, generalized retinal arteriolar narrowing and arteriovenous nicking were associated with an elevation in blood pressure that had been documented 6–8 years before the retinal assessment; the studies were controlled for concurrent blood pressure levels. This association suggests that generalized narrowing and arteriovenous nicking are markers of vascular damage from chronic hypertension. In contrast, other signs (focal arteriolar narrowing, retinal hemorrhages, microaneurysms, and cotton-wool spots) were related to current but not previous blood pressure levels and may therefore be more indicative of the severity of recent hypertension.

Furthermore, the observation of signs of retinopathy in people without a known history of hypertension suggests that these signs may be markers of a prehypertensive state. For example, generalized and focal narrowing of the retinal arterioles has been shown to predict the risk of hypertension in normotensive persons. Other factors unrelated to hypertension (e.g., hyperglycemia, inflammation, and endothelial dysfunction) may also be involved in the pathogenesis of retinopathy.

19.5.3.1 The Risk of Stroke

The strongest evidence of the usefulness of an evaluation of hypertensive retinopathy for risk stratification is based on its association with stroke. It is well known that the retinal circulation shares anatomical, physiological, and embryologic features with the cerebral circulation. An autopsy study of patients with stroke showed a close correlation between retinal and cerebral arteriolar findings. Functional alterations in retinal blood flow in patients with lacunar stroke have also been reported.

Population-based studies in Wisconsin and in Japan have shown that the risks of fatal and nonfatal stroke are two to three times as high in persons with signs of retinopathy as they are in persons who do not have these signs—an association that is independent of cardiovascular risk factors.

19.5.3.2 The Risk of Coronary Heart Disease

There are fewer data regarding the association of hypertensive retinopathy and the risk of coronary heart disease. In the National Health Examination Survey, persons with retinal arteriolar narrowing, as detected on ophthalmoscopy, were two to six times as likely to have preexisting coronary heart disease as those without these changes after the analysis was controlled for the presence or absence of hypertension and diabetes and for serum cholesterol levels.

19.5.4 Treatment

Several studies have shown that signs of hypertensive retinopathy regress with the control of blood pressure, although spontaneous resolution of these signs in the presence of high blood pressure has also been reported. It is unclear whether antihypertensive medications that are thought to have direct beneficial effects on the microvascular structure (e.g., angiotensin-converting enzyme inhibitors) would reduce the damage of retinopathy beyond the reduction effected by lowered blood pressure. In a small study of 28 patients with mild hypertension who were randomly assigned to receive treatment with enalapril or hydrochlorothiazide, opacification of the retinal arteriolar wall was significantly reduced after 26 weeks of treatment with enalapril; no other signs of retinopathy were reduced. In contrast, hydrochlorothiazide did not have any effect on the signs of retinopathy. However, to date, there are no data from prospective, controlled trials that demonstrate that the specific reduction of hypertensive retinopathy also reduces the morbidity and mortality associated with cardiovascular disease. It is also unclear whether the targeting of persons with hypertensive retinopathy for established risk-reducing interventions offers additional advantages over the use of strategies without regard to retinal findings.

19.5.4.1 ACE Inhibitors and the Eye

The circulating renin–angiotensin system (RAS) plays an important role in the control of electrolyte homeostasis and blood pressure [13]. The rate-limiting step in the RAS system is the kidney-derived enzyme renin, an aspartyl protease, which cleaves its substrate angiotensinogen, a liver-derived circulating plasma protein, to form the decapeptide angiotensin (ANG) I. ANG I is then converted by the angiotensin-converting enzyme (ACE) to the octapeptide ANG II, which stimulates release of aldosterone and constriction of blood vessels. Vasoconstriction has been demonstrated in feline retinal blood vessels exposed to NG II, and ANG binding sites have been found in bovine and human retinal vessels.

The published data do not distinguish whether activated RAS in the eye is a cause or an effect of the investigated eye diseases. Logically, treatment with enzyme inhibitors, possessing sufficient lipid solubility to cross the blood–ocular barriers, seems indicated in prophylaxis or therapy.

Evidence of the existence of a local RAS in ocular tissues has been given by biochemical, immunohistochemical, and molecular biological studies. ACE mRNA was found in different parts of human, monkey, dog, rabbit, and chicken eyes [13]. The concentrations of ANG I and ANG II in the anterior uvea, retina, and choroid were reported to be higher than in plasma. These results suggest that locally generated ANG II may have important functions, such as regulation of local blood flow and/or vascular remodeling, in the eye. However, most studies indicate that the RAS has only minor effects on ocular blood flow.

References

1. Augsburger JJ, Magargal LE (1980) Visual prognosis following treatment of acute central retinal artery obstruction. Br J Ophthalmol 64:913–917
2. Bayerle-Eder M, Fuchsjäger-Mayrl G, Sieder A, Polska E, Roden M, Stulnig T, Bischof MG, Waldhäusl W, Schmetterer L, Wolzt M (2002) Effect of pravastatin on responsiveness to *N*-monomethyl-l-arginine in patients with hypercholesterolaemia. Atherosclerosis 160(1):177–184
3. Bhagat N, Goldberg MF, Gascon P et al (1999) Central retinal vein occlusion: review of management. Eur J Ophthalmol 9:165–180
4. Bruno A, Russell PW, Jones WL et al (1992) Concomitants of asymptomatic retinal cholesterol emboli. Stroke 23:900–902
5. Burns JC, Glodé MP (2004) Kawasaki syndrome. Lancet 364:533–544
6. Chobanian AV, Bakris GL, Black HR, Cushman WC, Green LA, Izzo JL Jr, Jones DW, Materson BJ, Oparil S, Wright JT Jr, Roccella EJ, National High Blood Pressure Education Program Coordinating Committee (2003) The seventh report of the Joint National Committee on prevention, detection, evaluation, and treatment of high blood pressure. Hypertension 42:1206–1252
7. Deuter CM, Kötter I, Wallace GR, Murray PI, Stübiger N, Zierhut M (2008) Behçet's disease: ocular effects and treatment. Prog Retin Eye Res 27:111–136
8. Dorner GT, Garhofer G, Kiss B, Polska E, Polak K, Riva CE, Schmetterer L (2003) Nitric oxide regulates retinal vascular tone in humans. Am J Physiol Heart Circ Physiol 285:H631–H636
9. Fisher CM (1952) Transient monocular blindness associated with hemiplegia. Arch Ophthalmol 47:167–203
10. Garhöfer G, Zawinka C, Resch H, Kothy P, Schmetterer L, Dorner GT (2004) Reduced response of retinal vessel diameters to flicker stimulation in patients with diabetes. Br J Ophthalmol 88:887–891
11. George RK, Walton RC, Whitcup SM, Nussenblatt RB (1996) Primary retinal vasculitis. Systemic associations and diagnostic evaluation. Ophthalmology 103:384–389
12. Guyer DR, Miller NR, Auer CL, Fine SL (1985) The risk of cerebrovascular and cardiovascular disease in patients with anterior ischemic optic neuropathy. Arch Ophthalmol 103:1136–1142
13. Van Haeringen NJ (1996) The renin-angiotensin system in the human eye. Br J Ophthalmol 80:99–100
14. Hayreh SS (1999) Retinal and optic nerve head ischemic disorders and atherosclerosis: role of serotonin. Prog Retin Eye Res 18:191–221
15. Hayreh SS (2005) Prevalent misconceptions about acute retinal vascular occlusive disorders. Prog Retin Eye Res 24:493–519
16. Hayreh SS, Joos KM, Podhajsky PA, Long CR (1994) Systemic diseases associated with nonarteritic anterior ischemic optic neuropathy. Am J Ophthalmol 118:766–780
17. Hayreh SS, Zimmerman MB, Kimura A, Sanon A (2004) Central retinal artery occlusion. Retinal survival time. Exp Eye Res 78:723–736
18. Ikram MK, de Jong FJ, Vingerling JR, Witteman JC, Hofman A, Breteler MM, de Jong PT (2004) Are retinal arteriolar or venular diameters associated with markers for cardiovascular disorders? The Rotterdam Study. Invest Ophthalmol Vis Sci 45:2129–2134
19. Jennette JC, Falk RJ, Andrassy K, Bacon PA, Churg J, Gross WL, Hagen EC, Hoffman GS, Hunder GG, Kallenberg CG et al (1994) Nomenclature of systemic vasculitides. Proposal of an international consensus conference. Arthritis Rheum 37:187–192
20. Keith NM, Wagener HP, Barker NW (1939) Some different types of essential hypertension: their course and prognosis. Am J Med Sci 197:332–343
21. Klein R, Sharrett AR, Klein BE, Chambless LE, Cooper LS, Hubbard LD, Evans G (2000) Are retinal arteriolar abnormalities related to atherosclerosis? The Atherosclerosis Risk in Communities Study. Arterioscler Thromb Vasc Biol 20:1644–1650
22. Lindenberg R, Spatz H (1939) Über die Thromboendarteriitis obliterans der Hirngefasse (Cerebrale Form der v. Winiwarter-Burgerschen Krankheit). Virchows Arch 305:531–577
23. McCullough HK, Reinert CG, Hynan LS, Albiston CL, Inman MH, Boyd PI, Welborn MB 3rd, Clagett GP, Modrall JG (2004) Ocular findings as predictors of carotid artery occlusive disease: is carotid imaging justified? J Vasc Surg 40:279–286
24. Numano F (2002) The story of Takayasu arteritis. Rheumatology (Oxford) 41:103–106
25. O'Mahoney PR, Wong DT, Ray JG (2008) Retinal vein occlusion and traditional risk factors for atherosclerosis. Arch Ophthalmol 126:692–699
26. O'Steen WK, Kraeer SL, St Clair RW (1981) Influence of fatty acids and an oral contraceptive on the eyes of a nonhuman primate. Exp Mol Pathol 34:43–51
27. Olin JW (2000) Thromboangiitis obliterans (Buerger's disease). N Engl J Med 343:864–869
28. Park MC, Lee SW, Park YB, Chung NS, Lee SK (2005) Clinical characteristics and outcomes of Takayasu's arteritis: analysis of 108 patients using standardized criteria for diagnosis, activity assessment, and angiographic classification. Scand J Rheumatol 34:284–292
29. Pemp B, Weigert G, Karl K, Petzl U, Wolzt M, Schmetterer L, Garhofer G (2009) Correlation of flicker-induced and flow-mediated vasodilatation in patients with endothelial dysfunction and healthy volunteers. Diabetes Care 32:1536–1541
30. Pineles SL, Arnold AC (2007) Giant cell arteritis. Int Ophthalmol Clin 47:105–119

31. Recchia FM, Brown GC (2000) Systemic disorders associated with retinal vascular occlusion. Curr Opin Ophthalmol 11:462–467
32. Ross R (1999) Atherosclerosis – an inflammatory disease. N Engl J Med 340:115–126
33. Sakane T, Takeno M, Suzuki N, Inaba G (1999) Behçet's disease. N Engl J Med 341:1284–1291
34. Seo P, Stone JH (2004) The antineutrophil cytoplasmic antibody-associated vasculitides. Am J Med 117: 39–50
35. Shovlin CL, Guttmacher AE, Buscarini E, Faughnan ME, Hyland RH, Westermann CJJ, Kjeldsen AD, Plauchu H (2000) Diagnostic criteria for hereditary hemorrhagic telangiectasia (Rendu-Osler-Weber syndrome). Am J Med Genet 91:66–67
36. Slatter DH, Nelson AW, Stringer JM (1979) Effects of experimental hyperlipoproteinemia on the canine eye. Exp Eye Res 29:437–447
37. Sun C, Wang JJ, Mackey DA, Wong TY (2009) Retinal vascular caliber: systemic, environmental, and genetic associations. Surv Ophthalmol 54:74–95
38. Verity DH, Marr JE, Ohno S, Wallace GR, Stanford MR (1999) Behçet's disease, the Silk Road and HLA-B51: historical and geographical perspectives. Tissue Antigens 54:213–220
39. Walsh JB (1982) Hypertensive retinopathy. Description, classification, and prognosis. Ophthalmology 89:1127–1131
40. Walton RC, Ashmore ED (2003) Retinal vasculitis. Curr Opin Ophthalmol 14:413–419
41. Wong TY, Klein R, Sharrett AR, Manolio TA, Hubbard LD, Marino EK, Kuller L, Burke G, Tracy RP, Polak JF, Gottdiener JS, Siscovick DS (2003) The prevalence and risk factors of retinal microvascular abnormalities in older persons: the Cardiovascular Health Study. Ophthalmology 110:658–666
42. Wong TY, Mitchell P (2004) Hypertensive retinopathy. N Engl J Med 351:2310–2317
43. Yanko L, Michaelson IC, Rosenmann E, Ivri M, Lutsky I (1983) Effects of experimental hyperlipoproteinaemia on the retina and optic nerve in rhesus monkeys. Br J Ophthalmol 67:32–36
44. Zouboulis CC, Vaiopoulos G, Marcomichelakis N, Palimeris G, Markidou I, Thouas B, Kaklamanis P (2003) Onset signs, clinical course, prognosis, treatment and outcome of adult patients with Adamantiades-Behçet's disease in Greece. Clin Exp Rheumatol 21 (4 Suppl 30):S19–S26

Index

L. Schmetterer, J.W. Kiel (eds.), *Ocular Blood Flow*,
DOI 10.1007/978-3-540-69469-4, © Springer-Verlag Berlin Heidelberg 2012

R

S

GPSR Compliance

The European Union's (EU) General Product Safety Regulation (GPSR) is a set of rules that requires consumer products to be safe and our obligations to ensure this.

If you have any concerns about our products, you can contact us on ProductSafety@springernature.com

In case Publisher is established outside the EU, the EU authorized representative is:

Springer Nature Customer Service Center GmbH
Europaplatz 3
69115 Heidelberg, Germany

Batch number: 10705770

Printed by Printforce, the Netherlands